Springer Water

Series Editor

Andrey Kostianoy, Russian Academy of Sciences, P. P. Shirshov Institute of
Oceanology, Moscow, Russia

The book series Springer Water comprises a broad portfolio of multi- and inter-disciplinary scientific books, aiming at researchers, students, and everyone interested in water-related science. The series includes peer-reviewed monographs, edited volumes, textbooks, and conference proceedings. Its volumes combine all kinds of water-related research areas, such as: the movement, distribution and quality of freshwater; water resources; the quality and pollution of water and its influence on health; the water industry including drinking water, wastewater, and desalination services and technologies; water history; as well as water management and the governmental, political, developmental, and ethical aspects of water.

More information about this series at http://www.springer.com/series/13419

Hassan Awaad · Mohamed Abu-hashim ·
Abdelazim Negm

Editors

Mitigating Environmental Stresses for Agricultural Sustainability in Egypt

 Springer

Editors
Hassan Awaad
Faculty of Agriculture
Zagazig University
Zagazig, Egypt

Mohamed Abu-hashim
Faculty of Agriculture
Zagazig University
Zagazig, Egypt

Abdelazim Negm
Faculty of Engineering
Zagazig University
Zagazig, Egypt

ISSN 2364-6934   ISSN 2364-8198 (electronic)
Springer Water
ISBN 978-3-030-64325-6   ISBN 978-3-030-64323-2 (eBook)
https://doi.org/10.1007/978-3-030-64323-2

This Springer imprint is published by the registered company Springer Nature Switzerland AG
The registered company address is: Gewerbestrasse 11, 6330 Cham, Switzerland

Preface

The subject of mitigating environmental stresses for sustainable agriculture is essential not only for Egypt but for almost all countries having a similar situation. The book has been produced via wide collaboration of teamwork of 30 distinguished researchers and scientists from different institutions, academic, and research centers with significant concerns regarding environmental stresses, field crops, drought tolerance, heat stress, pollution tolerance, salt stress, natural resources, biotic stresses, laser, seed technology, and molecular tools mycorrhizae.

This volume consists of 21 chapters in 6 parts. The first part is an introduction and contains one chapter which is written by the editors with inputs from the authors to introduce the book to the audiences.

The second part of the volume deals with the improvement of crop tolerance for abiotic stresses. This part consists of 7 chapters. Chapter 2 is titled "Drought Tolerance in Some Field Crops." The authors address the genetic diversity among genotypes, related traits to drought tolerance, genetic behavior, breeding efforts, and biotechnology in rice, maize, barley, and sunflower. While Chapter 3 is dealing with "Performance and Genetic Diversity in Water Stress Tolerance and Relation to Wheat Productivity under Rural Regions." The author addresses breeding efforts and biotechnology to improve drought tolerance. The attention will also be paid to some agricultural procedures that improve water stress tolerance in relation to wheat productivity. On the other hand, Chapter 4 presents the "Heat Stress Tolerance, Challenges and Solutions." The chapter casts light on some concepts related to heat stress, breeding achievements and presents several practical adapting options to the increased temperatures. Moreover, Chapter 5 is titled "Environmental Pollution Tolerance in Crop Plants." The author addresses the source of pollutants and their damages, how can crop plants tolerate air and heavy metal pollutants, resistance mechanisms of crop plants to environmental pollution through morpho-physiological and biochemical characters, the role of plant breeding and biotechnology, besides procedures for pollution control. Furthermore, Chapter 6 is titled "Rapid Screening Wheat Genotypes for Tolerance to Heavy Metals." This chapter highlights the influence of heavy metals on wheat characteristics, and how to screen the tolerance of genotypes to heavy metals, also help to understand the level of tolerance and sensitivity of wheat genotypes for establishing breeding program to improve tolerance in wheat. While, Chapter 7 is titled "Performance, Adaptability and Stability of Promising Bread Wheat Lines across Different Environments." This chapter highlights the performance of wheat genotypes for economic traits, determines their adaptability and stability, employs joint regression and the AMMI method, and copes with environmental changes under the Mediterranean region of Egypt. However, Chapter 8 is titled "Effect of Salt Stress on Physiological and Biochemical Parameters of African Locust Bean {*Parkia biglobosa* (Jacq.) Benth} Cell Suspension Culture." This chapter presents unique informations on *Parkia biglobosa*, salt stress adaptive mechanism, antioxidant activity and phenolic compounds, and establishment protocol for cell suspension culture. The authors also address on electrophoretic, isozyme, and HPLC analyses and also utilized molecular characterization in *Parkia biglobosa* plant.

Part III consists of 5 chapters under the theme "Recent Approaches for Biotic Stress Tolerance." Chapter 9 presents the "Varietal Differences and Their Relation to Brown Rot Disease Resistance in Potato." The authors address on survey, isolation, and identification the causal organism of potato brown rot disease, *Ralstonia solanacearum* in weeds, molecular biology techniques, varietal differences in resistance to the disease, and effect of plant extracts on *R. solanacearum* growth *in vitro*. Chapter 10 is integrated with Chapter 9 where it is titled "Effect of Soil Type and Crop Rotation on Activity of *Ralstonia solanacearum* the Causal Agent of Brown Rot Disease in Potato under Egyptian Condition." This chapter focuses on the biotic abilities of *R. solanacearum* in relation to dispersion and survival, the effect of soil type, soil pH, and crop rotation on potato brown rot disease, and also be paid into designing effective management against the pathogen.

Moreover, Chapter 11 is titled "Advanced Methods in Controlling Late Blight in Potato." This chapter goes to resistant cultivars, alternative control methods, i.e., plant oils and extracts, use of nanotechnology and biocontrol compared to chemical fungicides to reduce the environmental problems on the plants, animals, and then humans. Meanwhile, Chapter 12 is devoted to "Developing Rust Resistance of Wheat Genotypes under Egyptian Conditions." The current chapter provides detailed information about yield losses percentage caused by wheat rusts and genetic variability, seeking for new sources of resistance, genes conferring resistance, breeding efforts, and biotechnology as a foundation for developing durable rust-resistant wheat cultivars. In the meantime, Chapter 13 presents the "Importance of Faba Bean (*Vicia fabae* L.) Diseases in Egypt." This chapter provides knowledge about the impact of the major plant diseases affecting faba bean yield production, yield loss, some control measures for minimizing the diseases, i.e., resistant cultivars, fungicides treatment, and inducing disease resistance, biological control, plant extracts, and growth regulators besides agricultural practices.

Part IV consists of 4 chapters and all are devoted to discussing the "Advanced Procedures in Improving Crop Productivity." Chapter 14 presents and discusses the "Role of He-Ne LASER in Improving Wheat Grain Yield Potentiality." This chapter highlights the application of laser technology as a sustainable, secure, and clean means in agriculture, advantages, and applications on crop plants, especially wheat in improving biotic and abiotic stresses, yield and quality characteristics. While

Chapter 15 is titled "Seed Technology and Improvement Productivity of Field Crops." This chapter focuses on the new technology of seed production from the perspective of seed quality, seed treatments, seed processing, seed storage, seed propagation, seed inspection, and testing seed quality. However, Chapter 16 is about the "Identification of Salt Tolerance Genotypes among Egyptian and Nigerian Peanut (*Arachis hypogaea* L.) Using Biochemical and Molecular Tools." This chapter has been proposed in order to provide information concerning salt tolerance of Egyptian and Nigerian peanut at morphological, biochemical, and molecular levels and application of polymerase chain reaction to amplify the KAT1 gene, while Chapter 17 is about the "Importance of Mycorrhizae in Crop Productivity." The author addresses on background about the arbuscular mycorrhizal fungi, its importance in plant nutrient and improving crop productivity under water deficit and salinity stress environments.

Part V is devoted to the "Sustainability of Environmental Resources from a Crop Production Perspective" and consists of 3 chapters. Chapter 18 is presented to discuss the "Optimizing Inputs Management for Sustainable Agricultural productivity." In this chapter, information is reviewed about the importance of sustainable agriculture for protecting environment and humans, and potential risks of some agricultural practices such as excessive synthetic fertilizer, toxic chemical pesticides, or herbicides in crop production. Also, the importance of organic and bio-fertilizers in the agricultural system is discussed. The attention will also be paid to the advantages of cropping rotation systems for better health of soil, environment, and human. Chapter 19 is about "Maize Productivity in the New Millennium." This chapter summarizes the historical trend of maize production in the world, and in developing countries (Egypt), the challenges faced in maize agriculture must be met to increase productivity, quality, and increase in resource use efficiency and the protection of environmental quality using traditional, modern, and advanced breeding methods. Finally, Chapter 20 is dealing with "Quinoa and Cassava Crops to Increase Food Security in Egypt." The chapter presents a state-of-the-art review on quinoa and cassava internationally and nationally from perspectives of the effect of salinity, water stress, insects and diseases, uses and introduction of Quinoa and Cassava Crops in Egypt to help researchers in Egypt to expand and improve their work with these two important crops.

The book ends with the conclusion chapter where the editors present an update of the book topics and present the most important conclusions and recommendations from all chapters.

The editors want to acknowledge the partial support of the Science and Technology Development Fund (STDF) of Egypt in the framework of the grant no. 30771 for the project titled "A Novel Standalone Solar-Driven Agriculture Greenhouse - Desalination System: That Grows Its Energy And Irrigation Water" via the Newton-Musharafa funding scheme.

The editors want to express their thanks to all who contributed in one way or another to make this high-quality volume a reality and a unique source of knowledge and latest findings in the field of mitigating environmental stresses for sustainable agriculture. We want to appreciate and thank all the authors for their contributions. Their patience and efforts in writing and revising the different versions of the chapters to satisfy the high-quality standards of Springer make it possible to produce this volume and make it a reality. Acknowledgments must be extended to include all members of the Springer team who have worked long and hard to produce this volume and make it a reality for the researchers, graduate students, and scientists around the world.

The editors appreciate too much the efforts done by Dr. Elsayed Mansour, Assistant Professor, Crop Science Department, Faculty of Agriculture, Zagazig University, in increasing the resolution of the figures included in the chapters of the book.

The volume editors would be happy to receive the comments for all the audiences to improve future editions. Comments, feedback, suggestions for improvement, or new chapters for next editions are welcomed and should be sent directly to the volume editors. The emails of the editors can be found inside the books at the footnote of their chapters.

Zagazig, Egypt   Abdelazim Negm
April 2019   Mohamed Abu-hashim
                                                                                                  Hassan Awaad

Contents

Part VI   Conclusions

About the Editors

Dr. Hassan Awaad is the Professor of Crop Breeding in Faculty of Agriculture, Zagazig University. Dr. Awaad got his Ph.D. in 1992 in "Breeding Studies on Some Wheat Crosses," Assistant Professor on 24/2/1998 and Professor on 25/3/2003, General Specialization: Crops, Current Specialization: Crop breeding. He published more than 48 research and articles, published 6 books in the field of physiology, crop breeding, and biotechnology in Arabic language, and produced new promising lines in wheat crop. He is Member of the Permanent Commission for the Promotion of professors and assistant professors, Member of the Egyptian Society of Crop Science, Egyptian Society of Plant Breeding, Egyptian Society of Applied Sciences and National Association of Science and Technology. He is Member and Principal Alternate Researcher and Consultant for several research projects. He supervised 18 Master's and Doctorate's degrees, 14 of which were awarded—arbitration of more than one hundred researches in local and international journals.

Dr. Mohamed Abu-hashim is the Contact person and the Associate partner of EXCEED—SWINDON—Middle East North Africa program "International Network on Sustainable Water and Environmental Management" in developing countries. Dr. Abu-hashim is the manager of Technology Transfer Office (TTO) and the Director of Project Management Unit, F. Ag., Zagazig University. Dr. Abu-hashim received his Ph.D. in the field of Water Resource Management from the Faculty of Civil Engineering and Environmental Science, TU-Braunschweig, Germany. His main subject is agricultural engineering, hydrology modeling, geoecology, and subsidiary subjects: water resource management, environmental risk assessment, and soil sciences. Currently, Dr. Abu-hashim works in several national and international projects; Director and the coordinator of the international afforestation project in MENA region "Development of salt tolerant agricultural practices and afforestation for bioremediation and CO_2 sequestration in the Middle East Region." Dr. Abu-hashim is the Egyptian coordinator of the ERANETMED project funded by EU with the title "Decentralized treatment wetlands for sustainable water management in rural and remote areas of semi-arid regions" in cooperation with several countries. Dr. Abu-hashim is the Principal Investigator of Zagazig University for TEMPUS Project entitled: Establish a new joint master degree in biotechnology applied to agri-science, environment, and pharmacology. 543865-TEMPUS-1-2013-1-EG-TEMPUS-JPCR. Dr. Abu-hashim published several papers related using remote sensing and GIS technique in fields of climate changes and water resource management in Egypt.

Prof. Dr. Abdelazim Negm is a Professor of Water Resources in the Faculty of Engineering, Zagazig University, Egypt, since 2004. He got his Ph.D. in 1992 from Zagazig University (ZU) and his M.Sc. from Ain Shams University in the year 1990. He was the Vice Dean of the Faculty of Engineering of ZU from 12/2008 to 12/2011 and was the Head of the Environmental Engineering Department at Egypt-Japan University of Science and Technology (E-JUST), Egypt 10/2013–9/2015. He has participated in several international projects since 2006 until now. He published more than 300 papers in national and international journals and conferences. Some of his papers were awarded the prize of the best papers. He is the editor of several volumes at the Springer Handbook of Environmental Chemistry series (HEC series). Two volumes were published in May 2017 and one of them was within the top ten in water science for the year 2017. Also, ten volumes were published during the years 2018/2019. He is a reviewer for more than 22 scientific journals and associate editor of few journals. He is a member of the scientific committee of several international conferences.

Part I
Introduction

Chapter 1
Introduction to "Mitigating Environmental Stresses for Agricultural Sustainability in Egypt"

Hassan Auda Awaad, Abdelazim M. Negm, and Mohamed Abu-hashim

Abstract This chapter provides a brief overview of the book, the purpose and the scope of the book. It discusses four categories of coherent topics. The first how to improve crop tolerance to abiotic stresses such as drought, heat stress, salinity and environmental pollution tolerance in crop plants. The second is about recent approaches for biotic stress tolerance. It contains varietal differences and their relation to brown rot disease resistance and effect of soil type and crop rotation on the disease in potato, advanced methods in controlling late blight in potato and importance of faba bean diseases, developing rust resistance in wheat beside the importance of faba bean diseases in Egypt. While the third is devoted to present advanced procedures in improving crop productivity include the role of He–Ne LASER, seed technology, identification of salt tolerance in peanut using molecular tools and importance of mycorrhizae in improving crop productivity. The last theme of the book is about sustainability of environmental resources from a crop production perspective and include sustainable use and optimizing inputs management for natural resources by crop rotations and bio-fertilizers in the agro-climatic zones of Egypt. The chapters under this theme concentrate on different field crops, maize productivity in the New Millennium as well as highlights on Quinoa and Cassava as promising crops to increase food security in Egypt.

Keywords Crop tolerance · Abiotic stresses · Drought · Heat stress · Salt stress · Environmental pollution · Sustainability · Management · Natural resources · Agro-climatic · Egypt · Maize productivity · Quinoa and Cassava · Food security · Biotic stresses · Varietal differences · Disease resistance · Crop rotation · Potato ·

H. A. Awaad (✉)
Crop Science Department, Faculty of Agriculture, Zagazig University, Zagazig 44511, Egypt

A. M. Negm
Water and Water Structures Engineering Department, Faculty of Engineering, Zagazig University, Zagazig 44519, Egypt
e-mail: amnegm@zu.edu.eg

M. Abu-hashim
Soil Science Department, Faculty of Agriculture, Zagazig University, Zagazig 44511, Egypt

© Springer Nature Switzerland AG 2021
H. Awaad et al. (eds.), *Mitigating Environmental Stresses for Agricultural Sustainability in Egypt*, Springer Water,
https://doi.org/10.1007/978-3-030-64323-2_1

3

Faba bean · Rust resistance · Wheat · He–Ne LASER · Seed technology ·
Biochemical tools · Peanut · Mycorrhizae

1.1 Background

Egypt faces many challenges in the current period, namely, abiotic pressures. i.e. drought, heat stress, salt stress, and environmental pollution. Also, biotic stresses represent diseases injury the strategic crops such as wheat, maize, faba bean and potato. So, it is of importance to mitigate environmental stresses through developing tolerant cultivars with applied appropriate agronomic practices and follow recent approaches as tools in improving crop productivity under stress conditions.

Where population growth has become a threat to all aspects of economic growth in Egypt in recent years. With the estimated population of 99.38 million in the year 2018, Egypt ranks 15th in the world (http://www.fao.org/countryprofiles/index/en/? iso3=EGY). A study by the National Planning Institute predicted that the population would reach more than 114 million by 2030 and that by 2050 it would reach 125.9 million. The increase in the Egyptian population is a problem in the light of the imbalance between the population on the one hand and available resources and economic development rates on the other hand. The reports of the Food and Agriculture Organization of the United Nations (FAO), indicated that Egypt ould could not achieve self-sufficiency of food because of water scarcity, severe weather events, pests and agricultural problems. Some experts point out that drought and harsh weather conditions in food-gap-affected areas in Africa in recent decades may be due to climate change.

According to UN water deficit statistics, Egypt is currently facing this problem sharply. The average per capita water is estimated at 660 cubic meters per year. By 2030 this average will be reduced to just 500 cubic meters, making Egypt among the countries facing "absolute deficit" in water.

There are several solutions proposed to address the problem of food shortages, and these solutions attention to the development of new varieties tolerant to stress conditions and resistance to diseases and the adoption of appropriate agricultural practices in addition to modern techniques.

Thus, agricultural development plays a key role in generating the income needed to ensure food security; between half and two-thirds of the world's poor live in rural areas, where agriculture is the dominant sector.

However, agriculture and food experts stress that priority must be given to improving agricultural land productivity. In this regard, over the past 15 years, researchers at Agriculture Research Center have produced more than 300 new agricultural varieties, and hybrids tolerate to drought, high temperatures, and soil salinity levels. These new agricultural varieties succeeded in increasing the productivity of agricultural crops.

1.2   Purpose of the Book

This book concentrated on the earth and environmental resources and how to exploit and deal with them for sustainable agriculture under the Egyptian conditions. Topics such as crop science, plant breeding and biotechnology for tolerance to environmental abiotic stresses are treated. Also, drought, heat, salt, pollutants, as well as biotic stresses such as disease resistance, ways to increase productivity, improving the quality of field crops and reducing the food gap are included. The main reasons behind this approach results from the potential of horizontal expansion in the newly reclaimed lands in Egypt depend on the development of crop varieties high tolerant to stress environmental conditions, and mitigate the impact of climate changes on their productivity. Besides that, the use of modern technologies is an essential tool in the scientific research system to improve crops production through the use of Laser, seed technology, mycorrhiza, and biotechnology to increase the yield of genotypes in sustainable farming systems. Therefore, this volume discusses the ways to increase the productivity of field crops under different environmental circumstances.

The purpose of this book is to broaden the knowledge of the nature of the environmental pressures to which crop plants are exposed in the light of climate change. Also, this scientific work aims to analyse the possible scenarios of the impact of environmental stress on crop plants and the strategies to be followed to avoid the negative effects on the productivity and quality of agricultural crops. Besides, the possibility of entering new crops to the agricultural arena in the desert lands, as well as interest in improving the productivity of cultivated crops in order to narrow the food gap and ensuring food requirements. Where, Egypt is one of the arid regions with a very low rainfall rate of 133 mm per year in the coastal areas, half in the delta and a quarter in central Egypt and is almost absent in Upper Egypt. Crop plants suffer from the effects of environmental stresses such as drought, high temperature, salinity and environmental pollutants, as well as the biotic pressures of plant diseases, which negatively affect crop production. Such environmental stresses affect crop outcome in many regions of the world, especially arid, semi-arid and hot regions such as the Arab region. The areas of horizontal expansion in the Egyptian deserts are one of these areas. Thus, horizontal expansion in the desert outside the cultivated area in the Nile Delta and valley is necessary. The new lands are located mainly on both the east and west sides of the Delta and valley. It scattered over various areas in the country where it covers 1.05 million ha (http://www.fao.org/docrep/008/y5863e/y5863e06. htm). These lands are viewed as an opportunity for increasing agricultural production and ensuring food security in the country. The performance of crop germplasm suffers from abiotic and biotic stresses and its effects on various plant traits. Crop genotypes varied in their response to stress conditions and showed various degrees of tolerance to stresses (Mensah et al. 2006; Awaad 2009; Doaa et al. 2015; Ali and Abdul-Hamid 2017 and Abdel-Motagally and Manal El-Zohri 2018). So, the food security problem is one of the most important challenges facing the achievement of the objectives of the policies and sustainable development in Egypt, which can be overcome through an integrated system in which different sectors of the state cooperate.

1.3   Themes of the Book and Contribution of the Chapters

In addition to the introduction (this chapter) and the conclusions (the last chapter), the volume consists of 4 themes. The first is titled "Improve Crop Tolerance for Abiotic Stresses" which is covered in 7 chapters. The second theme is written in 5 chapters and is titled "Recent approaches for biotic stress tolerance." While the third theme is titled "Advanced procedures in improving crop productivity" is covered in 4 chapters and the last theme is covered in 3 chapters under the title "Sustainability of environmental resources from a crop production perspective."

In the subsections, the main technical elements of the chapters under each theme are presented.

1.3.1   Improve Crop Tolerance for Abiotic Stresses

Chapter 2 highlights the drought tolerance in some important field crops, including rice, barley, maize and sunflower from the following aspects, (a) performance and genetic diversity, (b) related traits to drought tolerance, (c) genetic behavior beside breeding efforts and (d) biotechnology. Previous results concluded that there is a strong relationship between highest mean performances for morpho-physiological and biochemical characters and grain yield/plant under drought conditions, referring to the importance of such traits in improving drought tolerance (Moussa and Abdel-Aziz 2008; Farid et al. 2016). Several statistical parameters were utilized as genetic correlations and path coefficient analysis to identify the most selection criteria associated with water stress tolerance. Researchers have focused on the importance of employing Molecular markers RAPD and ISSR to detect the genetic diversity between various genotypes and to speed up the cycle of the breeding program (Abdel-Ghany 2012).

In integration with Chaps. 2, and 3 is presented to review drought tolerance in a strategically important crop in the world and Egypt i.e., wheat. The chapter discusses how to increase wheat productivity either through vertical expansion by increasing the productivity of the unit area or by horizontal expansion by increasing the area cultivated by drought tolerant cultivars in the new lands. This chapter addresses the critical stages of wheat growth that are affected by water stress (Salter and Goode 1967; Gupta 1997), and grain yield reductions percentage due to water stress. Many researchers explained the plant reaction to water stress through escape, avoidance, and tolerance mechanisms, where mechanisms of water stress resistance differ in different plant species (Foulkes et al. 2007). They also mention the importance of measuring and assessing drought tolerance through some indices such as stress sensitivity index as a useful measurement of comparing yield performance of genotypes between stressful and non-stressful environments (Fischer and Maurer 1978). Interrelationships and relative importance between grain yield and relevant physiological and component characters under both normal and drought conditions have been computed by many investigators (Abd El-Mohsen Dina 2015; Saleh 2011; Ata et al. 2014). This chapter also crippled the possibility of using molecular markers in breeding programs to differentiate between different wheat genotypes (Al-Naggar et al. 2015), also many efforts of transgenic wheat lines have been incorporated in the national wheat-breeding program of Agriculture Research Center, Egypt for further field testing and seed multiplication (Wally 2016). Lastly, the author suggested the possibility of mitigating the effects of water stress hazard through specific agricultural practices with tolerant varieties.

Chapter 4 is proposed to highlight an important factor of environmental stress affecting crop productivity, which is high temperature. This chapter brings to the reader's attention of some concepts related to heat stress, direct and indirect impact of heat stress on the Egyptian Agriculture Sector. Also, this chapter exposure to how crop plants can deal with high temperature through heat resistance mechanisms such as avoidance and thermotolerance. And addresses the capabilities of crop plants that help them to adapt with heat stress whether they are phenological, morpho-physiological or biochemical characteristics of molecular structure (Singh and Ahmad 2003; Awaad 2009; Farooq et al. 2009; Kumari et al. 2012; El Basyoni et al. 2017) are the key for heat stress tolerance. The authors surveyed many genes responsible for the production of heat shock proteins are closely associated with heat stress resistance in several crops. This chapter also concentrates on breeding achievements and biotechnology from aspects of molecular markers, gene transfer, and tissue culture technology. Finally, the authors explained the role of agricultural processes in reducing the impact of heat stress on different crop plants.

From the perspective of assessing the impact of pollutants and their effect on the quality of the environment, crop productivity and how to deal with sources of pollution (Awaad et al. 2013; Pal 2016) suggest Chap. 5. Chapter 5 focuses on the various aspects of the effect of environmental pollutants on crop plants, the role of plant breeding and biotechnology as well as procedures to mitigate the impact of the pollutants. Where, recently, the decrease in wheat yield due to heavy metals pollution is more than 20% besides the effect on quality (EL-Gharbawy 2015). The Codex Alimentarius Commission of the Food and Agriculture Organization of the United Nations (FAO)/World Health Organization (WHO) determined maximum allowable levels of cadmium and inorganic arsenic in crop plants. Testing and evaluation of the water quality of 24 sites between Aswan and Cairo along the Nile, the water quality variations were mainly related to inorganic nutrients and heavy metals (Abdel-Satar et al. 2017). Review on the main research findings on crop adaptation mechanisms revealed the importance of selection, hybridization with molecular genetics using DNA-markers method in improving tolerance of crop plants to environmental pollutants. There are many suggestions to reduce the impact of contaminants on crop plants and the application of laws and compliance with the provisions of international conventions.

Chapter 6 was suggested to integrate with Chap. 5 and comprises detailed information about the effect of heavy metals and their mixture on germination and growth characters of wheat genotypes. This study will help to understand the level of tolerance and sensitivity of wheat genotypes under heavy metals stress conditions for establishing a breeding program for heavy metals tolerance in wheat. The use of tolerance measurements such as seedling phytotoxicity (Chou and Lin 1976) and seedling tolerance index (Iqbal and Rahmati 1992) helps to recognize tolerate or sensitive genotypes. They suggested the importance of recommending the cultivation of tolerant cultivars or could be exploited in a breeding program for high-yielding and tolerance for heavy metals.

On the light of climate change in the world, especially in the Mediterranean region, the aims of Chap. 7 are to identify adaptability and stability of wheat genotypes for earliness, quality and grain yield under diverse environmental combinations. Among the various parameters joint regression (Eberhart and Russell 1966) and additive main effects and multiplicative interaction, AMMI (Gauch 1992) have been reported to measure the adaptability and stability of new cultivars. The author showed that highly significant differences among environments reflect the wide differences in climatic conditions prevailing during the growing seasons. Exploit Stress Tolerance index (TOL) provides a measure of yield stability based on yield loss under stress as compared to non-stressed condition (Rosielle and Hamblin 1981), it identifies tolerant genotypes to environmental stress and confirmed with those previous results by Ali and Abdul-Hamid (2017). Both methodologies of Eberhart and Russell (1966) and AMMI (Gauch 1992) are consistent in describing the stability of Line 1 and Line 5 for grain yield. These genotypes could be useful in wheat breeding programs for improving stability.

Chapter 8 aims to highlight the African locust bean, as known to grow in a diversity of agro-ecological zones ranging from tropical rain forests to arid zones (Millogo-Kone et al. 2008). This chapter stressed the effects of salinity stress on some physiological and biochemical parameters of *Parkia biglobosa* cell suspension culture, at the application of NaCl concentration levels. Also, the establishment of an application protocol for cell suspension culture production from callus under salt stress. They applied protein electrophoresis technique in order to determine the molecular weight of protein, sodium dodecyl sulphate polyacrylamide gel electrophoresis (SDS-PAGE) under denaturing conditions (Laemmli 1970), also isozymes electrophoresis as suggested by Larkindale and Huang (2004). The researcher recommended by using the HPLC technique to separate the phenolic constituents that responsible for *P. biglobosa* to tolerate salinity such as gallic, caffeic, vanillic, ferulic, p-coumaric and salicylic acids.

1.3.2 Recent Approaches to Biotic Stress Tolerance

Chapter 9 comes from the point of view that the bacterial wilt problem is a major constraint for vegetable growers, especially potato and tomato farmers in many regions of the world and Egypt. So this manuscript was carried out in an attempt to study the transmission of *Ralstonia solanacearum*, the causal agent of potato brown rot, through field weeds, survey, isolation, identification and varietal differences in their relation to brown rot disease resistance in potato under Egyptian condition. The problem lies in the greater number of weed varieties as hosts for *R. solanacearum*; it causes disease at least 200 different plant species. For these reasons, isolates of *R. solanacearum* from weeds and other plant species must be identified using morphological, physiological and biochemical tests. Also, advanced Immunofluorescence Antibody Stains (IFAs), Polymerase Chain Reaction (PCR) and Real-time PCR (Taq-Man) techniques could be applied (Hamad 2016). Researchers have been able to divided potato varieties according to tolerance levels and varied from more sensitive to resistant to the pathogen. They also determined some plant extracts from *Corchorus olitorius*, *Solanum nigrum*, *Ricinus communis* and *Portulaca oleracea* against *R. solanacearum*.

Chapter 10 focuses on the biotic abilities of *R. solanacearum* in relation to dispersion, survival, and effects which might be important in designing effective management against the pathogen. This study included the importance of soil type, soil pH and crop rotation, as the most important factors affecting on the causal agent of potato brown rot disease *R. solanacearum*. Application of suitable agronomic practices which led to decreasing pH value might be useful for controlling the causal agent. Crop rotation is considered an important agricultural procedure to reduce the number of weed varieties as hosts for *R. solanacearum*. There is increasing evidence of hosts which under certain condition act as latent or symptomless carriers of infection. Hamad, (2008) have reached that some rotations were more effective in decline the population density.

The purpose of Chap. 11 is to introduce recent information about the late blight disease in potatoes (*Solanum tuberosum* L.) which is arguably one of the most infamous diseases in agriculture. This chapter also focuses on assessment the economic importance of the disease. Also, identification of the fungus using the most vital methods differentiates between the new isolates of the pathogen by numerous techniques, a- the traditional method b- using DNA markers and c- bioinformatics and then control fungus by the best practices. Meanwhile, some chemicals are toxic and dangerous for both the environment and human health. Therefore, the research studies indicate that cultivation of resistant varieties and the application of another methods such as treatment with plant oils, natural extracts as well as the use of nanotechnology and biological control as an alternative to chemical fungicides will reduce environmental problems on plants and animals and then humans.

Based on the destructive effect of rust diseases on the productivity and quality of wheat, the current Chap. 12 is proposed to integrate the system of biological stresses in the light of modern trends in plant breeding. This chapter is divided into three main categories to explain the Egyptian case of wheat rusts, *i.e.*, yellow rust, leaf rust, and stem rust. This chapter provides detailed information about genetic variability, seeking for new sources of resistance, genetic system, and genes conferring resistance, breeding efforts and biotechnology along with yield losses caused by the rust diseases. They emphasized that breeding for resistance is a continuous procedure, and plant breeders need to increase new effective resources in breeding materials (Draz et al. 2015). The authors have suggested that the previous information's will serve as a foundation for developing durable rust-resistant wheat cultivars.

Chapter 13 provides some knowledge about the impact of the major plant diseases as a factor affecting faba bean yield production. This section limps on the most important diseases attacked faba bean plants, i.e. fungal diseases (chocolate spot, rust, root rot, and wilt) and viral diseases (Faba Bean Necrotic Yellow Virus, Bean Yellow Mosaic Virus, and Bean Leaf Roll Virus). These pathogens considered as a main constrains affected growth of the plant and contributed significantly to causing great yield loss both in quantity and quality, herby led to importing considerable amounts of faba bean seeds to fulfill the requirements and demands of the consumers (El-Metwally et al. 2013). Reference studies have suggested the possibility of recommending some measures to reduce the pathogenic effects involve breeding for disease resistance, fungicides treatments, plant extracts, agricultural practices, and others.

1.3.3   Advanced Procedures in Improving Crop Productivity

Chapter 14 focus on the helium-neon laser and its advantages and applications on crop plants, especially wheat in improving yield and quality characteristics. This manuscript is exposed to focuses on the mode of action of biostimulation, advantages, and disadvantages of helium-neon laser and objectives of using the helium-neon laser in agriculture. Also, applications of helium-neon laser in improving crop plants, quality, tolerance to environmental stresses and increasing water use efficiency. Where Chen and Han (2014) and Abaza Ghada et al. (2017) suggest applied Helium-neon laser to improve morphological parameters and the yield of wheat crop.

Chapter 15 highlights the application of the new technology of seed production of field crops. Also, exposed to seed as a basis for development, seed processing, seed treatments, the role of seed storage in agricultural development. High seed quality with genetical purity, high vigor, free from weed seeds and infested or borne diseases needs should be used for planting seed crops. It focuses on the importance of perfect harvesting, seed processing, and suitable storage facilitates for saving and protect the seed for next seed sowing (Ali 2017). It is important to perform seed multiplying of improved varieties to keep the new varieties from mixture or losses their pure genetical. Finally, this chapter emphasizes the need seed inspection and testing must be occurred before sowing in order to obtain high seed quality.

The objectives of Chapter 16 are to screen peanut genotypes from Egypt and Nigeria for salt stress in terms of morphological, biochemical and molecular genetic level. Also, to design specific primers for identifying and sequencing the salt tolerance gene in peanut genotypes. They also emphasized to compare between tolerant and susceptible genotype for salinity tolerance based on SNPs level. Negrao et al. [2017] discuss how to quantify the impact of salinity on different traits, such as relative growth rate, water relations, transpiration, ionic relations, photosynthesis, senescence, yield and yield components. The results of this chapter recommend that

Ismailia1 and Samnut 22 varieties could be involved in peanut salt tolerance breeding programs

Chapter 17 deal with the importance of mycorrhizae in crop productivity from the perspectives of various sides i.e. effects of organic and inorganic fertilizers on the activity of arbuscular mycorrhizal fungi in the rhizosphere, role of arbuscular mycorrhizal fungi on phosphorus and nitrogen uptake by host plants, role of arbuscular mycorrhizal fungi on plants grown under drought and salinity stress. Also, the chapter deliberates the role of arbuscular mycorrhizal fungi on the quality and productivity of crop plants. Reacharch results found that the manipulation of arbuscular mycorrhizal fungi in sustainable agricultural systems will be of marvelous vital for soil fertility and crop yield under severe edapho-climatic conditions (Lal 2009).

1.3.4 Sustainability of Environmental Resources from a Crop Production Perspective

Chapter 18 focuses on how to optimizing inputs management for sustainable agricultural development. In the current chapter, the authors discuss the importance of sustainable agriculture for protecting environment and humans from potential risks of some agricultural practices such as using excessive synthetic fertilizer as well as using toxic chemical pesticides or herbicides in crop productivity. This chapter also exposed to the importance of organic (biochar) and bio-fertilizers and their significance in the agricultural system. The attention will also be paid into the advantages of cropping rotation systems for better health of soil, environment, and human. The research studies suggest that use of modern agricultural technology such as productivity of new seeds, fertilizers, modern irrigation systems and suitable management strategies has saved such serious expectations (Abou El Hassan et al. 2014).

Chapter 19 deals with climate change and environmental degradation that threaten cereal production and global food security. In Egypt, the expected changes, according to the climate change scenarios will cause harmful effect in crop production. So this chapter focuses on the strategic crop maize: consumption and gap, distribution, ecology and growth requirements (air temperature, soil temperature, soil moisture, precipitation, and water requirements). Also concentrate on maize production under Egyptian conditions and challenges to face maize production such as abiotic stresses (drought, heat, poor soil fertility, low soil nitrogen tolerance and soil acidity/aluminum toxicity tolerance) and biotic stresses (insects and diseases). This besides milestones in maize breeding advances and Biotech Interventions. Studies suggest that crop productivity could be increased on both old lands and in the newly reclaimed areas using improved varieties, optimum cultivation practices, high seed-quality and the efficient use of land and water inputs. In addition, with better extension, the gaps between yields obtained by farmers and those obtained by researchers could be narrowed or even closed (FAO 2005).

Finally, Chap. 20 displays on quinoa and cassava crops to increase food security in Egypt. There is a large gap between production and consumption of wheat estimated by 55%, which negatively affects the production of bread in Egypt. This work is exposed to previous research by the Crop Intensification Research Department; Agricultural Research Center in Egypt which indicated that 40% of quinoa flour could be mixed with wheat flour in bread making. Furthermore, their research also indicated that 30% of cassava flour could be mixed with wheat flour (Shams 2011). Thus, the objective of this chapter was to review the studies that was done on quinoa and cassava internationally and nationally to help researchers in Egypt to expand and improve their work with these two important crops from the following aspects, salinity, water stress, insects and diseases, uses internationally as well as introduction of quinoa and cassava in Egypt

The last part consists of one chapter which is the conclusions chapter. It presents an update, the most important conclusions and a set of key recommendations.

Acknowledgment Hassan Awaad, Mohamed Abu-hashim and Abdelazim Negm acknowledge the partial support of the Science and Technology Development Fund (STDF) of Egypt in the framework of the grant no. 30771 for the project titled "A Novel Standalone Solar-Driven Agriculture Greenhouse - Desalination System: That Grows Its Energy And Irrigation Water" via the Newton-Musharafa funding scheme.

References

Abaza Ghada MShM, Gomaa MA, Awaad HA, Atia ZMA (2017) Performance and breeding parameters for yield and its attributes in M2 generation of three bread wheat cultivars as influenced by Gamma and LASER ray. Zagazig J Agric Res 44(6B):2431–2444

Abd El-Mohsen DA (2015) Yield stability of some wheat genotypes under normal and water stress conditions. M. Sc. Thesis, Agronomy Department, Faculity of Agriculture, Zagazig University, Egypt

Abdel-Ghany M (2012) Genetic studies on sunflower using biotechnology. PhD Thesis, Department of Genetics, Faculty of Agriculture, Cairo University, Egypt

Abdel-Motagally FMF, El-Zohri Manal (2018) Improvement of wheat yield grown under drought stress by boron foliar application at different growth stages. J Saudi Soc Agric Sci 17(2):178–185

Abdel-Satar Amaal M, Ali Mohamed H, Goher Mohamed E (2017) Indices of water quality and metal pollution of Nile River, Egypt. The Egyptian J of Aquatic Research 43(1):21–29

Abou El Hassan WH, Hafez EM, Ghareib AAA, Ragab MF, Seleiman MF (2014) Impact of nitrogen fertilization and irrigation on N accumulation, growth and yields of Zea mays L. J Food, Agric Environment 12(3&4):217–222

Ali, A-G.A. (2017). The seed and technological of seed processings and storage. Seed Phycol Storage 8:183–258. Faculty of Agriculture, Zagazig University, Egypt. 2017/23761

Ali MMA, Abdul-Hamid MIE (2017) Yield stability of wheat under some drought and sowing dates environments in different irrigation systems. Zagazig J Agric Res 44(3):865–886

Al-Naggar AMM, Al-Azab KF, Sobieh, SES (2015) Morphological and SSR assessment of putative drought tolerant M3 and F3 families of wheat (*Triticum aestivum* L.). Br Biotechnol J 6(4):174–190

Ata A, Yousaf B, Khan AS, Subhani GhM, Asadullah HM, Yousaf A (2014) Correlation and path coefficient analysis for important plant attributes of spring wheat under normal and drought stress conditions. J Nat Sci Res 4(8):66–73

Awaad HA (2009) Genetics and breeding crops for environmental stress tolerance, I: Drought, heat stress and environmental pollutants. Egyptian Library, Egypt

Awaad HA, Morsy AM, Moustafa ESA (2013) Genetic system controlling cadmium stress tolerance and some related characters in bread wheat. Zagazig J of Agric Res 40(4):647–660

Chen HZ, Han R (2014) He-Ne laser treatment improves the photosynthetic efficiency of wheat exposed to enhanced UV-B radiation. Laser Phys 24:10–17

Chou C, Lin, H (1976) Autointoxication mechanism of *Oryza sativa* I. Phytotoxic effects of decomposing rice residues in soil. J Chem Ecol 2(3):353–367

Doaa RM El-Naggar, Soliman SSA (2015) Evaluation of some mutant lines in three Egyptian bread wheat cultivars for resistance to biotic stress caused by wheat rusts. Egypt J Appl Sci 30(8):254–269

Draz IS, Abou-Elseoud MS, Kamara AM, Alaa-Eldein OA, El-Bebany AF (2015) Screening of wheat genotypes for leaf rust resistance along with grain yield. Ann Agric Sci 60(1):29–39

Eberhart SA, Russell WA (1966) Stability parameters for comparing varieties. Crop Sci 6:36–40

El Basyoni IM, Saadalla S Baenziger, Bockelman H, Morsy S (2017) Cell membrane stability and association mapping for drought and heat tolerance in a worldwide wheat collection. Stainability 9:1–16

EL-Gharbawy SS (2015) Wheat breeding for tolerance to heavy metals pollution. M. Sc. Thesis, Agronomy Department, Faculty of Agriculture, Zagazig University, Egypt

EL-Metwally IM, El-Shahawy TA, Ahmed MA (2013) Effect of sowing dates and some broomrape control treatments on faba bean growth and yield. J Appl Sci 9(1):197–204

FAO (2005) Fertilizer use by crop in Egypt. First version, published by FAO, Rome

Farid MA, Abou Shousha AA, Negm MEA, Shehata SM (2016) Genetical and molecular studies on salinity and drought tolerance in rice (*Oryza sativa* L). J Agric Res 42(2):1–23

Farooq M, Wahid A, Kobayashi N, Fujita D, Basra SMA (2009) Plant drought stress: effects, mechanisms and management. Agron Sustain Dev 29:185–212. https://doi.org/10.1051/agro:200 8021

Fischer RA, Maurer R (1978) Drought resistance in spring wheat cultivars, 1. Grain yield responses. Aust J Agric Res 26(4):897–912

Foulkes MJ, Sylvester-Bradley R, Weightman R, Snape JW (2007) Identifying physiological traits associated with improved drought resistance in winter wheat. Field Crop Res 103(1):11–24

Gauch HG (1992) Statistical analysis of regional trials: AMMI analysis of factorial designs. Elsevier, Amsterdam, The Netherlands, p 278

Gupta, U. S. (1997). Crop improvement, vol 2. Stress Tolerance. Science Publishers, Enfield, NH, USA

Hamad YI (2008) Studies on the transmission of potato brown rot causal organism through weeds in the Egyptian fields. M.Sc. Thesis, Plant Pathology Department, Faculty Agriculture, Zagazig University, Egypt

Hamad YI (2016) Pathological studies on Potato Brown Rot under Egyptian conditions. Ph D Plant Pathology, Agriculture Botany Department Faculty Agriculture, Suez Canal University, Egypt

Iqbal M, Rahmati K (1992) Tolerance of *Albizia lebbeck* to Cu and Fe application. Ekológia, ČSFR 11(4):427–430

Kumari M, Pudake RN, Singh VP, Joshi A (2012) Association of stay green trait with canopy temperature depression and yield traits under terminal heat stress in wheat (*Triticum aestivum* L.). Euphytica. http://dx.doi.org./10.1007/s10681-012-0780-3

Laemmli UK (1970) Cleavage of structural protein during the assembly of the head of Bacteriophage T4. Nature 227:680–685

Lal R (2009) Soil degradation as a reason for inadequate human nutrition. Food Secur 1:45–57

Larkindale J, Huang B (2004) Thermo tolerance and antioxidant systems in *Agrostis stolonifera*: involvement of salicylic acid, abscisic acid, calcium, hydrogen peroxide and ethylene. J Plant Physiol 161:405–413

Mensah JK, Akomeah PA, Ikhajiagbe B, Ekpekurede EO (2006) Effects of salinity on germination, growth and yield of five groundnut genotypes. Afr J Biotechnol 5(20):1973–1979

Millogo-Kone H, Guissou IP, Nacoulma O, Traore AS (2008) Comparative study of leaf and stem bark extracts of *Parkia biglobosa* against enterobacteria. Afr J Trad CAM 5:238–243

Moussa HR, Abdel-Aziz SM (2008) Comparative response of drought tolerant and drought sensitive maize genotypes to water stress. Aust J Crop Sci 1(1):31–36

Negrao S, Schmockel SM, Tester M (2017) Evaluating physiological responses of plants to salinity stress. Ann Bot 119:1–11

Pal P (2016) Detection of environmental contaminants by RAPD method. Int J Curr Microbiol App Sci 5(8):553–557

Rillig MC, Mummey D (2006) Mycorrhizas and soil structure. New Phytol 171:41–53

Rosielle AA, Hamblin J (1981) Theoretical aspects of selection for yield in stress and non-stress environment. Crop Sci 21:943–946

Saleh SH (2011) Performance, correlation and path coefficient analysis for grain yield and its related traits in diallel crosses of bread wheat under normal irrigation and drought conditions. World J Agric Sci 7(3):p270

Salter PJ, Goode JE (1967) Crop responses to water at different stages of growth. Research Review No. 2. Commonwealth Agricultural Bureaux

Shams A (2011) Combat degradation in rain fed areas by introducing new drought tolerant crops in Egypt. Int J Water Resour Arid Environ 1(5):318–325

Singh NB, Ahmed, Z (2003) Seedling vigour as an index for assessing terminal heat tolerance in wheat under irrigated late sown condition. In: 2nd International Conference on Plant Physiolog, Jan 8–12, 2003, New Delhi, India, p 173

Wally A (2016) Agricultural biotechnology annual. USDA, foreign agricultural services, gain report. https://gain.fas.usda.gov/Recent%20GAIN%20Publications/Agricultural%20Biotechnology%20Annual_Cairo_Egypt_11-17-2016.pdf

Part II
Improve Crop Tolerance for Abiotic Stresses

Chapter 2
Drought Tolerance in Some Field Crops: State of the Art Review

Mohammed M. Abd- El-Hamed Ali, Elsayed Mansour, and Hassan Auda Awaad

Abstract Drought stress has a significant negative impact on plant growth, yield and quality of crop plants, particularly under current climate change. Therefore, improving drought tolerance in field crops is essential to increase sustains productivity particularly in arid and semi-arid regions in the Mediterranean basin. There are various selection criteria associated with drought tolerance could be exploited in selecting drought-tolerant genotypes. Combining morpho-physiological and biochemical traits can provide a more completed model of gene-to-phenotype relationships and genotype-by-environment interactions. Integration of recent advances breeding methods as quantitative trait loci, marker-assisted selection and genetic engineering technique with classic plant breeding helps significantly in developing drought-tolerant genotypes accurately and rapidly. This chapter also addresses to genetic diversity among genotypes, related traits to drought tolerance, genetic behavior, breeding efforts and biotechnology in rice, maize, barley and sunflower.

Keywords Drought · Field crops · Breeding methods · Biotechnology · Marker-assisted selection

2.1 Introduction

Global population is increasing while water resources for crop production are decreasing. Limited irrigation water is one of the major stresses that reducing crop production and quality in agricultural systems (Arshadi et al. 2018; Hasanuzzaman et al. 2018; Mansour et al. 2020). Moreover, the importance of drought has become more serious with increasing climatic change and global warming. The increase of air temperature and the decrease of rainfall caused heat stress and drought "in many areas, especially in arid and semi-arid regions" (http://www.tandfonline.com/doi/abs/10.1080/07900629208722539). Egypt suffers from severe water deficit in recent years, facing water shortage amounted about 7 billion cubic meters annually and

M. M. A. Ali · E. Mansour (✉) · H. A. Awaad
Crop Science Department, Faculty of Agriculture, Zagazig University, 44511, Zagazig, Egypt

© Springer Nature Switzerland AG 2021
H. Awaad et al. (eds.), *Mitigating Environmental Stresses for Agricultural Sustainability in Egypt*, Springer Water,
https://doi.org/10.1007/978-3-030-64323-2_2

this may increase in the near future due to the effect of Ethiopian Renaissance Dam (Osman et al. 2016). Drought tolerance is defined as the ability of the genotypes to produce acceptable yield under limited water supply better than the other genotypes (Desoky et al. 2020). Whereas, drought sensitivity is the reduction in yield of the genotypes under drought stress (Gavuzzi et al. 1997; Turner et al. 2001). Drought stress tolerance development is difficult due to the phenomenon of well-built interactions between genotypes and the environmental conditions. Therefore, based on yield loss under water stress conditions with compare to normal conditions, various drought measurements were determined that have been used for identification of drought-tolerant genotypes (Mitra 2001; Ali 2016). While other experiments yet have chosen a mid-point and think in selection under both favorable with combined stress conditions (Mardeh et al. 2006; Abdel-Ghany 2012). Marker-assisted selection approach serves the purpose of efficient transfer of the desirable gene into the elite of crop plants which infers sustainable saving in time compared to conventional backcross breeding. Marker-assisted selection such as RAPD, ISSR and real-time quantitative polymerase chain reaction (RT-qPCR) were used in breeding programs to assess the genetic relationships among genotypes and to improve biotic and abiotic tolerant traits (Nguyen et al. 1997; Premnath et al. 2016).

2.2   Rice

2.2.1   Economic Importance

Rice crop feeds approximately half of the world's population (Uga et al. 2013). The area allotted to rice crop in Egypt was 1.4 million feddan (feddan = 4200 m^2), and total production was 5.7 million tons of grains with an average yield of 4.0 tons fed^{-1} (Faostat 2018). The drought tolerance mechanism is complex, influenced by variants in plant phenology and controlled by numerous quantitative trait loci. Drought is a major constraint to the productivity of rice in worldwide. Drought stress is a major environmental limit to rice growth, production and yield in many regions of the 'world (Luo et al. 2009). Drought stress is a problem in nearly 45% of agricultural areas and the major worldwide restraint to productivity, making it an important and vital research space in scientific reports (Heinemann et al. 2015; Todaka et al. 2015).

Water deficit is an important constraint affecting rice physiological processes that are involved in its growth and development, as well as agricultural productivity. Given the importance of the rice crop economically, seems imperative to study the extent to which different genotypes are tolerated under water-stress conditions. The study of morphological, physiological and biochemical characteristics is important in this regard, in line with the objectives of breeding. Therefore, drought-resistant rice cultivars are an economically important target for crop production and food security.

2.2.2 Mean Performance and Genetic Diversity

Various field trials were conducted under normal and drought conditions to evaluate yield response of different genotypes to drought stress in addition to identify related traits to drought tolerance. Abd-Allah et al. (2010) assessed thirty-three local and exotic rice genotypes including eighteen Egyptian, six Italian and nine Chinese rice genotypes were evaluated under normal and drought conditions at the experimental farm of Rice Research and Training Center. The results revealed highly significant differences among the genotypes for all evaluated traits. Several promising genotypes were identified as drought tolerant under different growth stages, *i.e.* seedling, early and late vegetative, panicle initiation and heading stages. These genotypes displayed associated traits with drought tolerance as early maturity, intermediate tillering ability and plant height, root depth, root thickness, root volume, dry root: shoot ratio, plasticity in leaf rolling and unrolling. Additionally, these genotypes exhibited high water use efficiency and water application efficiency. These genotypes were Giza 178, Giza 182, GZ5121, GZ 6296-12-1-2-1-1, GZ 8310-7-3-2-1, GZ 8367-11-8-3-2, GZ 8372-5-3-2-1, GZ 8375-2-1-2-1, GZ 8450-19-6-5-3, GZ 8452-7-6-5-2, GZ 1368-S-4, Augusto and SIS R215.

El-khoby et al. (2014) crossed six rice genotypes with different water stress tolerance at the experimental farm of Rice Research and Training Center, Sakha, Kafr Elsheikh, Egypt. Three crosses viz; cross I [Tsuyuake (tolerant) × Sakha 103 (sensitive)], cross II [Zenith (tolerant) × Sakha 104 (moderate)] and cross III {BL1 (moderate) × Sakha 106 (sensitive)} were produced. Results showed that high variances between the six parents for all root characters, grain yield and its associated traits under water stress circumstances. Parent Tsuyuake produces the highest estimates for most traits, whereas, the lowest values were noted for Sakha 106 rice cultivar. The mean values of F_1 were higher than the highest parent for root volume, root/shoot ratio, in cross I, root length, root fresh weight, plant height and grain yield plant^{-1} in crosses I and II and number of panicles plant^{-1} in the three crosses.

To evaluate drought effect on rice growth and productivity, as well as the influence of silicon to enhance rice tolerance to drought stress. Ibrahim et al. (2018) performed an experiment in clay soil. They assessed five levels of soil moisture content (SMC) (70, 80, 90, 100, and 120% of the soil saturation point) and five rates of Silicon (Si) (0, 2.1, 4.2, 6.3, and 8.4 mg Si/10 plants). They detected significant reduction in plant height, rice straw, root yield, and grain yield by 32, 52, 36, and 27%, in the same order, with reducing SMC from 120 to 70%. On the other hand, Si significantly increased the aforementioned traits by 38, 97, 49, and 106%, respectively. The combination of SMC of 120% and 8.4 mg Si rate presented the maximum plant growth traits, while SMC of 70% and 0 mg Si rate recorded minimum plant growth traits. Furthermore, grain yield and aboveground biomass could be maintained under reducing water supply, if Si were supplied.

At the Rice Research and Training Center (RRTC), Sakha, Kafr Elsheikh, Egypt, during 2014 and 2015 growing seasons, Bassuony and Anis (2016) evaluated some

root characters of five rice genotypes for drought and well-watered conditions. Genotypes were differed significantly by sowing conditions. Growth rate (g day^{-1}) varied
considerably amongst the considered genotypes. Genotype Nerica 4 gave a much
greater growth rate compared to other genotypes. Drought stress leads to a decreased
total root length for all genotypes. Root volume decreased differentially with a range
from 24.53 to 19.17 under the drought situation. Numbers of total roots differed
significantly under the two sowing situations. Under well-watered conditions, the
number of roots increased compared to drought situation. Under well-watered root,
the dry weight was higher compared to drought. Also, root to shoot ratio decreased
under drought stress.

Variation in root morphology, some physiological traits, yield and its components
and genetic diversity among some Egyptian rice varieties and some upland rice accessions under normal and drought conditions were studied by Sedeek et al. (2010).
Four upland rice accessions originated from Cote de Ivoire (IRAT112, IRAT170,
WAB450-I-B-P-105-HB and Yun Len62) compared with three Egyptian rice varieties
(Giza177, Sakha101 and GZ745613-6-5-3) were evaluated. Significant variation was
observed in most of the investigated root traits such as root length, root volume, root
thickness and root dry weight among the upland rice accessions and Egyptian rice
varieties. Also, there are significant variation in some physiological traits which were
related to fitness and productivity, including yield and its components under normal
and drought conditions. This indicated that the upland rice accessions are tolerant to
drought via its have good root system compared with the Egyptian rice varieties which
were sensitive to drought. Most of the traits showed significant and highly significant
correlation. Furthermore, among sixteen rice genotypes evaluated under drought and
normal conditions, Gaballa (2018a) found that the effect of years was highly significant for leaf rolling and the number of tillers plant^{-1}. Genotype GZ 5310-20-3-3 had
the desirable value of sterility percentage under normal conditions, while Giza 107
gave the best value under drought conditions. The heaviest mean value of 100-grain
weight was recorded by E. Yasmine cv. Moreover, the highest grain yield plant^{-1} was
produced by Giza 103, Giza 105 and Giza 107 under normal conditions, while Giza
107 and GZ 8372-5-3-2-1 gave the highest value under drought conditions. Also,
Gaballa (2018b) tried four treatments, *i.e.* flash irrigation every 12 days, flash irrigation every 12 days + application of kinetin (1gL^{-1}), flash irrigation every 12 days +
application of kinetin (2gL^{-1}) and normal irrigation. Registered highly significant
differences among treatments, genotypes and their interaction in most traits. The most
desirable mean values towards root length, root volume, number of roots plant^{-1},
days to heading, number of tillers plant^{-1} and number of panicles plant^{-1} resulted
from Giza 179. Sakha 105 gave the best values of root: shoot ratio and chlorophyll
content. The greatest values of leaf rolling, flag leaf area, plant height, harvest index
and sterility percentage were recorded by GZ 8714-7-1-1-2. The superior values for
plant height and panicle length were found by GZ 9781-3-2-2-6 as well as for highest
root thickness, 100-grain weight and grain yield plant^{-1} by GZ 9792-13-1-1-2.

2.2.3 Related Traits to Drought Tolerance

The drought tolerant genotype is that has higher grain yield compared to the others under the same water deficit conditions. The improvement of breeding for drought tolerance in rice is slow due to the complication of the drought environment (Fukai and Cooper 1995). Several drought-tolerance mechanisms and characters were identified in rice *i.e.* drought escape through appropriate phenology, root traits, avoidance of specific dehydration and drought recovery. The physiological characters are considered very important for growth environmental conditions. The root system plays an essential role in water uptake and extraction from the deep soil layers. Therefore, root system traits *i.e.* deep root, high intensity root length in depth are valuable in this context. Ekanayake et al. (1985) presented that selection of root types must be successful in early segregating generations if the selection is proficient on such traits. Furthermore, relations between root traits and other agronomic traits as plant height, tiller number, and shoot weight were positive and significant. Root traits significantly associated with field drought tolerance as well as with leaf water potential that strengthening the role of root traits in maintaining high leaf water potential under drought stress. Abd-Allah et al. (2010) reported that certain evaluated agronomic traits as number of tillers plant^{-1}, number of panicles plant^{-1}, 100-grain weight and panicle weight indicated significant genotypic correlation with grain yield. Likewise, number of filled grains panicle^{-1} displayed highest direct contribution (0.63), and similarly it showed highest indirect contribution (0.867) followed by 100-grain weight (0.850) to grain yield. Path coefficient analysis demonstrated that number of panicles plant^{-1}, 100-grain weight, number of filled grains panicle^{-1}, panicle weight should be improved to increase grain yield under both normal and drought conditions. El-khoby et al. (2014) reported that grain yield plant^{-1} was highly significant and positively correlated with root length, root number plant^{-1}, root volume, panicle length, number of panicles plant^{-1} and 100-grain weight. On the other hand, the grain yield plant^{-1} was highly significant and negatively correlated with plant height, with similar results were reported by Muthuramu et al. (2010) and Hosseini et al. (2012). Under drought conditions, Gaballa (2018b) documented positive and highly significant correlation among rice grain yield plant^{-1} and other traits as root length, root volume, number of roots plant^{-1}, root thickness, root: shoot ratio, chlorophyll content, relative water content, flag leaf area, plant height, number of tillers plant^{-1}, panicle length, 100-grain weight, and number of panicles plant^{-1}. Furthermore, negative correlation has been detected with leaf rolling and sterility percentage.

Various stress-responsive genes, encoding antioxidant enzymes, oxidoreductases, kinases, and detoxification proteins, were prompted by drought and contributed to adjust the balance of reactive oxygen species ROS metabolism in the drought-tolerant cultivars (Degenkolbe et al. 2009; Yang et al. 2010). Besides, many genes were expressed under water stress lead to alternate physiological and phenotypic reactions amid genotypes (Lenka et al. 2011). Under water-limiting environments, the genotypes that reserve high leaf water potential usually present best growth. Nevertheless, it is not known whether genotypic variation in leaf water potential is uniquely caused by root traits or other reasons. Osmotic adjustment is promising, because it resists the effects of a rapid shortage of leaf water potential. There is a genotypic difference in expression of green leaf retention as physiological character related to prolonged droughts (Fukai and Cooper 1995). Additionally, Farid et al. (2016) reported that rice crosses Sakha 102 × Giza 178, Sakha 102 × A22, Giza 178 × WAB 54-125 and Sakha 105 × WAB 54-125 exhibited the highest performances of leaf chlorophyll content, proline content, number of panicles plant^{-1} and grain yield plant^{-1} under drought conditions, indicating to their importance in improving drought tolerance in rice.

Drought always causes oxidative injury to cells as a result of excessive generation of reactive oxygen species ROS in crop plants (Selote et al. 2004; Moumeni et al. 2011). The drought-tolerant rice genotypes showed lower H_2O_2 level and enhanced antioxidative enzyme activity compared to drought-sensitive ones (Rabello et al. 2008). Likewise, Zhang et al. (2012) proposed that the genotype Yangdao-6 seedlings showed drought tolerance associated with greater antioxidant enzyme activity and less ROS accumulation rather than Yangdao-2 genotype.

2.2.4 Genetic Behavior

Tolerance to water stress in rice is a complex trait (Fukai and Cooper 1995). Ekanayake et al. (1985) reported that the ability of rice genotypes to tolerate drought stress is related with root system traits. The inheritance of root traits of rice (*Oryza sativa* L.) was assessed using the parents, F_1, F_2 and F_3 populations of cross IR20 × MGL-2. Polygenic inheritance was detected for the root traits. The F1 plants had root systems described by thick, deep roots with a higher lateral and vertical distribution rather than the low parent (IR20), with positive and significant heterosis. The additive and dominant genetic effects were equally contributed to the expression of the traits. Progeny parent regression of F_3 and F_2 and the narrow sense heritability estimations in F_3 were high for root thickness (0.61 and 0.80), root dry weight (0.56 and 0.92) and root length density (0.44 and 0.77) as documented by Ekanayake et al. (1985). Consequently, selection of root types, based on individual plant performance, would be successful in early segregating generations. Besides, El-khoby et al. (2014) evaluated three crosses viz; cross I {Tsuyuake (tolerant) × Sakha 103 (sensitive)}, cross II {Zenith (tolerant) × Sakha 104 (moderate)} and cross III {BL1 (moderate) Sakha 106 (sensitive)}. They recorded highly significant positive heterosis and heterobeltiosis for some root traits and grain yield and its relevant traits especially in crosses I and II. The additive genetic component was higher than the dominance one for root volume and grain yield plant^{-1} in cross II, root fresh weight and days to 50% heading in crosses II and III, root:shoot ratio in crosses I and III, as well as, plant height and panicle length in the three crosses. Heritability in broad sense ranged from low to intermediate and high in the three crosses. However, it was low in the narrow sense. Furthermore, they recorded low genetic advance (3.99) for root number plant^{-1} to high (30.15) for root length in the cross III. Tsuyuake and Zenith might be useful genotypes in breeding program for water stress condition. Moreover, Bassuony and Anis (2016) assess some root traits of some rice genotypes for drought and well-watered conditions. The magnitude of heterosis manifested over mid-parent, and better parents exist. Highly significant estimates of heterosis as a deviation from mid-parent and better parent were revealed in all crosses for growth rate (g/day) and root length (cm) under normal and drought situations (Bassuony and Anis 2016). Zhang et al. (2012) identified numerous differentially expressed genes between seedlings of genotypes YD2 and YD6. For examples, *CuZnSOD*, *CAT-A*, *GSR-1*, *CYP71A1*, *CYP72A1*, *CYP72A5*, and *CYP86A1* were notably induced by drought in YD6, but mildly or not altered in YD2.

2.2.5   Breeding Efforts and Biotechnology

The genetic improvement of drought tolerance is addressed through the conventional approach by selecting for yield and its stability over locations and years. Du to of low heritability of yield under stress and inherent variation in the field, such selection programs are expensive and slow in getting progress. Nguyen et al. (1997) presented that the ability of root systems to provide for evapotranspiration demand from deep soil moisture and capacity for osmotic adjustment are considered main drought tolerance traits in rice. Selection for these traits still requires extensive investments facilities and is subjected to challenges of repeatability due to environmental inconsistency. Molecular markers associated with root traits and osmotic adjustment are being identified, that lead to marker-assisted selection. Furthermore, transgenic tolerant rice to water stress and osmotic stress have been documented. Moreover, various researches on genetic engineering of osmoprotectants, such as proline and glycine betaine, into the rice plant for drought tolerance improvement are in progress. Basi (2008) noticed expression of a putative tyrosine kinase gene (Pup1 candidate gene #43) under P and drought in NIL 14-4 but not in NIL 14-6 or Nipponbare. Tolerance specific gene #18 and gene #43 is absent from the Nipponbare genome but are present in drought-tolerant wild rice species from Asia and Africa, e.g. *O. longistaminata*, *O. barthii*, and *O. rufipogon*, proposing that these genes might be ancient and were lost in modern, stress-sensitive genotypes. Madabula et al. (2016) assessed the phenotypic response of 6 Brazilian rice varieties and 2 different crosses between them under drought conditions. Four genes related to auxin response and root modifications (*OsGNOM1 CRL4*, *OsIAA1*, *OsCAND1* and *OsRAA1*) were detected. Through real-time quantitative polymerase chain reaction (RT-q PCR), displayed that all genotypes lengthened its roots in response to drought, especially the derived 2 hybrids. The expression of these genes is modified in response to drought stress, and *OsRAA1* has a very special behavior, constituting a target for the future. Under Egyptian conditions, several drought-tolerant varieties have been developed through the National Rice Research Program characterized by a short growth period and water-saving *i.e.* Giza 179, Giza 182, Sakha 102, Sakha 103 and Sakha 107.

2.3 Maize

2.3.1 Economic Importance

Maize (*Zea mays* L.) is a major crop in the United States and worldwide. Drought is one of the most vital problems facing crop cultivation in most areas of the world. It is also the most important environmental limitations and pressures affecting the growth and productivity of maize plants. The production and stability of the crop are strongly affected by drought stress conditions. So, improved drought tolerance in maize is a top priority for corn breeding programs. In Egypt, the average productivity of maize was 8.3 tons ha^{-1} with a total production of about 8.6 million tons of grains (Faostat 2018). The performance of crops are a highly complex phenomenon under water stress condition and negative affected (Reynolds et al. 2006). Effects of water stress on maize depend on the growth stage of prevalence, duration of prevalence, and intensity of stress. Consequently, research on irrigation and water management has focused on crop productivity responses to water provide (Chen et al. 2010; Köksal 2011). It is interesting to mention that when drought stress starts to effect on the plant at the reproductive stage, the plant lessens the demand for carbon by decreasing the size of the sink. Therefore, reduction in leaf size, stem extension and root proliferation, the flower may drop pollen may die and ovule may abort (Blum 1996; Farooq et al. 2009). The yield reduction under drought stress was higher at the reproductive stage than at the vegetative and grain filling stages (Fatemi et al. 2006). Whereas, Khodarahmpour and Hamidi (2012) found that the production of maize grain yield reduced 60% due to stress condition at grain filling stages. Furthermore, Ali (2016) registered reduction % due to water stress valued (32.83%) for grain yield (ard./fed.), (15.60%) for plant height, (14.99%) for ear length, (13.99%) for ear leaf area, (12.17%) for 100-kernel weight, (10.63%) for number of kernels row^{-1}, (6.98%) for ear diameter, (6.64%) for leaf relative water content and (4.22%) for number of rows ear^{-1}.

2.3.2 Mean Performance and Genetic Diversity

Terminal drought stress is one of the most important environmental stress factors which can cause a significant reduction in maize production. Barutcular et al. (2016) conducted two field experiments with some maize hybrids in two cropping seasons, 2014 and 2015 under two moisture levels (normal irrigation and water deficit-stress) at the grain filling stage. Results revealed that, yield and major yield traits of hybrids harmfully affected due to terminal drought stress, and causing a reduction in yield with comparing normal irrigation situation. Water stress significantly affected on maize hybrids and there was high variation among hybrids, which could be befitting for screening the genotypes. Hybrids 71May69, Aaccel and Calgary were revealed less reduction in grain yield under terminal drought stress. Genotypes with high stress sensitivity index and tolerance index were expressed as highly sensitive to drought and only appropriate for irrigated conditions.

Mean values of the maize parental lines and their respective twenty crosses for earliness and grain weight plant^{-1} have been assessed under stress (D) and normal (N) conditions by Khaled et al. (2013). The results revealed that the mean performance of days to 50% tasseling for 9 parental lines varied from extreme earliness of line A3 (53.3 days in drought and 55.7 days in favorable environments) to lateness of line C12 (80.3 days in drought and 84 days in favorable environments). The cross (A3 × B3) had the best mean values for earliness (52.7 days and 54.3 days) under drought and normal stress conditions, respectively. As for grain weight plant^{-1}, the range of mean performances varied from 24.0 to 37.5 g for B8 and B5 lines, respectively under normal conditions. However, the range was narrower extending from 15.9 g for line B10 to 22.1 g for line B5 under drought stress. The crosses (C16 × C12) and (B10 × B3) had the highest grain yield under normal and drought stress conditions, respectively. The results of drought sensitivity index (DSI) displayed values of 0.38, 0.39, 0.74, and 0.99 for the parental lines B8, A3, C12 and C16, respectively, indicating relative drought resistance. At the same time, Maciej et al. (2013) revealed that the DSI values fluctuated from 0.381 to 0.650. While, B10 and C1 were the most sensitive maize parental lines. Regarding the F_1 crosses, 9 out of the 20 F1 crosses showed relative drought resistance.

During the three growing seasons 2011, 2012 and 2013 at Experimental Farm, Fac. of Agric., Zagazig University, Ali (2016) evaluated half diallel crosses among eight yellow maize inbred lines i.e. Z12 (P1), Z15 (P2), Z167 (P3), Z147 (P4), Z40 (P5), Z56 (P6), Z58 (P7) and Z103 (P8) under well-watered and water stress environments. Results presented in Table 2.1 showed that the following crosses had the most desirable sensitivity index to drought tolerance i.e. SC. 168, (P2 x P5), (P5 x P7) and (P5 x P8) for days to 50% silking; (P3 x P7), (P2x P6), (P4 x P6) and TWC.352 for anthesis to silking interval; (P7 x P8) for plant height; (P1 x P5), (P1 x P6) and (P1 x P8) for ear leaf area; (P2 x P3) and (P2 x P8) for leaf water content; (P2 x P5), (P1 x P7), (P1 x P6) and (P1 x P5) for ear diameter; (P5 x P6), (P2 x P7) and (P3 x P7) for ear length; (P3 x P6), (P3 x P7), TWC.352 and (P7 x P8) for number rows/ear; (P4 x P5), (P2 x P7) and (P7 x P8) for number of kernels/row; (P5 x P8), (P4 x P5) and TWC.352 for 100-kernels weight and (P2 x P7), (P3 x P7), (P5 x P6), (P1 x P6) and (P7 x P8) for grain yield.

In comparison among drought-sensitive and drought-tolerant maize genotypes in respect to morph physiological and biochemical characters components, Tůmová et al. (2018) exhibited that genotype CE704 had significantly higher values of number of visible leaves and plant height compared to maize genotype 2023 under drought conditions. While genotype 2023 subjected to drought attained strong symptoms of senescence for the older leaves, rather than genotype CE704, which developed normally even in stress situation. Drought significantly reduced the shoot dry mass and total leaf area in genotype 2023 nevertheless not in genotype CE704, resultant in genotypic variances under drought conditions. Also, the root biomass in stressed plants of genotype CE704 was a greater rather than genotype 2023. Drought stress was decreased leaf osmotic potential values in both genotypes compared to control.

Table 2.1 The mean performance of 28 F$_1$ maize crosses and two check varieties for drought sensitivity index (DSI) for all studied traits and drought tolerance index (DI) and stress tolerance index (STI) for grain yield only (Ali 2016)

Crosses	Days to 50% silking	Anthesis-silking interval (ASI)	Plant height (cm)	Ear leaf area (cm²)	Leaf water content%	Ear diameter (cm)	Ear length (cm)	No. rows ear⁻¹	No. kernels /row	100—kernel weight(g)	Grain yield (ard. fed.⁻¹)		
											DSI	STI	DI
P1 X P2	2.11	2.47	0.75	0.42	0.86	0.59	0.72	2.05	1.74	1.46	1.06	0.58	0.60
P1 X P3	1.89	1.74	1.22	0.53	1.39	0.87	0.93	1.31	1.08	1.16	1.24	0.61	0.53
P1 X P4	1.54	0.81	0.98	0.87	1.09	0.91	1.23	0.89	1.56	0.86	1.08	0.77	0.67
P1 X P5	0.67	1.78	0.72	0.08	0.70	0.51	0.50	1.33	1.58	1.66	0.76	0.75	0.84
P1 X P6	1.28	−1.68	0.88	0.23	0.70	0.50	1.64	0.36	1.05	0.85	0.96	0.69	0.70
P1 X P7	0.68	4.74	0.85	1.37	0.86	0.47	1.89	0.49	1.05	1.96	1.03	0.70	0.67
P1 X P8	1.43	1.18	1.11	0.31	0.87	0.77	1.04	1.32	1.28	1.41	1.16	0.57	0.55
P2 X P3	0.86	4.31	1.01	0.71	0.52	−0.24	1.42	0.83	0.65	1.20	1.04	0.97	0.79
P2 X P4	1.02	1.35	1.09	1.33	1.74	−0.31	1.27	0.48	0.76	1.83	1.02	0.83	0.73
P2 X P5	0.38	−1.59	1.40	0.37	1.18	0.40	1.03	1.76	0.98	0.68	0.99	0.63	0.66
P2 X P6	1.72	0.14	1.16	0.94	0.95	0.85	0.57	0.55	0.77	1.10	1.04	0.46	0.54
P2 X P7	1.02	1.00	1.26	2.02	0.94	0.77	0.28	0.55	0.39	0.88	0.74	0.45	0.66
P2 X P8	1.24	0.82	0.82	1.76	0.52	0.68	1.17	0.93	0.83	0.96	0.93	0.61	0.68
P3 X P4	1.67	2.34	1.00	1.38	1.21	1.66	1.12	1.91	0.68	1.14	1.07	0.66	0.63
P3 X P5	0.64	0.95	1.20	0.92	1.05	1.11	1.56	−0.17	1.22	0.47	1.20	0.65	0.57
P3 X P6	0.57	3.47	0.93	0.67	1.35	2.43	1.36	0.11	1.46	0.72	1.01	0.84	0.75
P3 X P7	1.62	0.11	1.11	0.69	0.92	1.33	0.51	0.20	0.69	0.91	0.67	0.73	0.87
P3 X P8	1.24	1.63	0.96	0.89	0.64	1.14	1.02	1.25	0.68	0.41	0.95	0.57	0.64
P4 X P5	0.88	0.96	0.93	1.09	0.91	1.81	0.94	−0.17	0.17	0.31	0.91	0.65	0.71

(continued)

Table 2.1 (continued)

Crosses	Days to 50% silking	Anthesis-silking interval(ASI)	Plant height (cm)	Ear leaf area (cm²)	Leaf water content%	Ear diameter (cm)	Ear length (cm)	No. rows ear⁻¹	No. kernels /row	100—kernel weight(g)	Grain yield (ard. fed.⁻¹)		
											DSI	STI	DI
P4 X P6	1.56	0.26	1.12	1.67	1.45	1.41	0.88	1.03	1.02	0.87	1.31	0.52	0.46
P4 X P7	0.57	3.32	0.77	2.20	0.77	0.63	1.21	1.49	0.79	1.23	1.00	0.81	0.74
P4 X P8	0.62	1.32	0.79	1.20	0.96	1.22	0.99	1.24	1.35	1.26	0.92	1.04	0.89
P5 X P6	0.86	−0.33	0.78	0.84	0.75	0.90	0.07	0.33	0.67	0.89	0.81	0.51	0.67
P5 X P7	0.32	1.12	1.09	0.69	0.65	1.89	0.73	1.35	0.51	0.71	0.84	0.56	0.69
P5 X P8	0.31	0.99	1.27	0.36	0.87	1.85	0.93	1.62	1.20	0.25	1.03	0.83	0.73
P6 X P7	0.94	0.78	1.17	0.32	0.62	1.14	0.99	1.04	0.98	0.11	1.03	0.47	0.55
P6 X P8	0.92	0.99	0.90	1.31	1.02	0.54	0.74	1.56	0.54	1.25	0.97	0.83	0.76
P7 X P8	1.14	1.11	0.65	1.08	1.45	0.79	0.53	0.31	0.39	0.91	0.92	0.47	0.60
Checks													
SC. 168	0.27	2.02	0.93	1.28	1.52	0.76	0.68	1.77	1.58	1.39	1.03	0.91	0.76
TWC. 352	0.55	0.28	0.98	1.03	1.33	1.38	1.05	0.16	1.37	0.40	0.97	0.71	0.71
LSD 0.05	0.88	2.53	0.37	1.09	0.66	1.31	0.86	ns	0.76	0.90	0.31	0.10	0.15
LSD 0.01	1.17	3.37	0.49	1.46	0.88	1.75	1.15	ns	1.01	1.20	0.41	0.13	0.20

Ardab maize = 140 kg grains, Feddan = 4200 m²

Transpiration rate was reduced in genotype CE704 than in genotype 2023 after 14 days without watering. Also, both stomatal conductance and net photosynthetic rate were reduced by drought, and the changes in the mean values of these traits were less evident in genotype CE704 than in genotype 2023. This was revealed in the presence of significant differences among both maize genotypes in the stomatal conductance under drought conditions. The drought-stressed plants of genotype CE704 had more carotenoids and chlorophyll in the leaves compared to genotype 2023. Also, drought stress reduced the efficiency of the primary photosynthetic procedures. There were obvious variances between the drought-stressed plants and control, mainly for the factors describing the performance index PI_{ABS} and electron transport within the photosystem (PS) II reaction centre, such as ψ_{E0}, φ_{P0} and φ_{E0}. These differences were more obvious in the genotype 2023. The drought stressed plants had higher values for cell membrane injury than in the control. This increase was higher in genotype 2023 than in genotype CE704. The genotype 2023 also exhibited slightly higher peroxidation of membrane lipids based on the malondialdehyde content (MDA) compared to genotype CE704. Regarding to proline content, genotype CE704 was also characterized by higher percent in the leaves compared to genotype 2023, and drought stress induced a further elevation of this osmoprotectant content in genotype CE704 (Tůmová et al. 2018).

2.3.3 Related Traits to Drought Tolerance

Maize is more sensitive than most other cereal crops to drought stresses at flowering stage, where yield losses can be severe through reductions or barrenness in kernels per ear (Bolanos and Edmeades 1996; Al-Naggar et al. 2000). Moussa and Abdel-Aziz (2008) tested maize genotypes Trihybrid 321 (drought sensitive) and Giza 2 (drought-tolerant) for water stress condition produced by irrigating the pots with polyethylene glycol (PEG) solutions of 0.0, -5, -10 and -20 bars. The two genotypes responded differently under water stress and control conditions. The tolerant genotype Giza 2 displayed lower accumulation of malondialdehyde (MDA) and H_2O_2 content related to increasing activities of catalase, peroxidase compounds, and superoxide dismutase, under water stress situations. The superoxide dismutase activity as antioxidant increased constantly with increasing drought in both the genotypes, but the percent of antioxidant was higher in Giza-2 drought-tolerant. The lower membrane injury and higher water retention capacity have been detected in Giza-2. Further, Trihybrid 321 displayed resistance to water stress via the above adjustments. Water stress was resulted by the accumulation of free proline and gylycinebetain in both cultivars. The level of enhancement in both omsolytes was lower in Trihybrid 321 than Giza 2. Therefore, the accumulation of gylycinebetaine and free proline in the maize leaves genotypes can be used as the possible indicator for water stress tolerance. Giza 2 had higher relative water content under both water stress and non-stress conditions. Thus, the high relative water content is helping the tolerant maize genotype to achieve physio-biochemical processes more efficiently under drought environments than sensitive genotype. While, the contents of endogenous brassi-nosteroids (BRs) were considered in two genotypes differed in their water stress sensitivity. The attendance of 28-norbrassinolide in rather high quantities (1–2 pg. mg^{-1} fresh mass) is reported for the first time in the leaves of monocot plants. The drought-resistant genotype was considered by a significantly higher content of total endogenous BRs (28-norbrassinolide and particularly typhasterol) compared with the drought-sensitive genotype which exhibited higher levels of 28-norcastasterone. The differences observed between CE704 and 2023 genotypes in content of total endogenous BRs are probably related with their different levels of drought sensi-tivity, which was confirmed at different levels of plant biochemistry, physiology and morphology (Tůmová et al. 2018).

Tolerant maize genotypes were characterized by having shorter anthesis to the silking interval (ASI) (Bolaños and Edmeades 1993; Ali 2016), greater number of kernels ear^{-1} and more ears $plant^{-1}$ (Ribaut et al. 1997). Barutcular et al. (2016) reported a positive relationship between stress indices, grain yield, geometric mean productivity (GMP), drought resistance index (DRI), harmonic mean (HM), stress tolerance index (STI), mean production (MP), and Yield index (YI). Hereby could be used as the best selection indices for identifying the tolerant maize genotypes under terminal drought stress.

Under drought situations, Ziyomo and Bernardo (2013), recorded strongest genetic correlation between anthesis to silking interval (ASI) and grain yield (−0.77). Grain yield also under drought stress was strongly associated with leaf chlorophyll content, plant height, and leaf senescence. The genetic correlation was 0.61 between grain yield in the drought and control experiments, indicating that inbred lines with superior testcross performance in control condition also tend to perform well under drought environment.

Ahmed (2013) added that correlation coefficients between the studied variables showed that only the ear height and number of kernels row^{-1} were negatively corre-lated with grain yield and the highest correlations were observed between grain yield and grain weight under drought condition. While under control conditions, number of kernels/m^2 was highly associated with grain yield, thus, the hybrids with larger number of kernels should be selected to increase grain yield under irrigated condi-tion. Whereas, number of kernels row^{-1} and grain weight ear^{-1} could be used as an important traits for improvement grain yield productivity under drought stress at the grain growth stage.

To identify secondary characters for selection of high grain yield under high plant density joint with water stress at flowering, Al-Naggar et al. (2016) chose six different maize inbred lines in tolerance to high density and drought stress at flowering (three lines tolerant *i.e.* Sk-5, L-20 and L-53, and three lines sensitive *i.e.* Sd-7, L-18 and L-28) for diallel crosses. During two seasons, parents and hybrids were evaluated under two different environments; well-watered and low density of 47,600 plants ha-1 and drought stress with high density of 95,200 plants ha-1. They recorded strong favorable and significant genetic correlations between stress tolerance index or grain yield $plant^{-1}$ with all yield components for inbred lines and crosses and days to anthesis, ear height, plant height, leaf angle and barren stalks for crosses. Therefore, low days to anthesis, leaf angle, ear height, and high rows ear^{-1}, 100-kernel weight, kernels row^{-1}, kernels plant^{-1}, ears plant^{-1}, might be considered secondary characters to drought and high-density tolerance. The optimum selection environment for grain yield plant^{-1} is the drought and high-density tolerance environment for crosses and well-watered and low density environment for inbred lines. Moreover, Ali (2016) recorded positive and significant genotypic and phenotypic correlations between grain yield and each of leaf relative water content (0.488** and 0.307**), ear leaf area (0.443** and 0.355**), ear length (0.783** and 0.647**), ear diameter (0.691** and 0.546**), number of rows ear^{-1} (0.291* and 0.237), number of kernels row^{-1} (0.486** and 0.451**), 100-kernels weight (0.659** and 0.543**) and drought sensitivity index (0.484** and 0.388**, respectively), but had negative correlations with days to 50% silking (-0.034 and 0.004) and anthesis to silking interval (-0.572** and -0.491**, respectively) (Table 2.2). Path coefficient analysis showed that ear length exhibited the largest direct effect on grain yield (0.340) followed by drought sensitivity index (0.251), leaf relative water content (0.231), ear leaf area (0.182), number of kernels row^{-1} (0.171), ear diameter (0.135) and number of rows ear^{-1} (0.104) as shown in Table 2.3.

2.3.4 Genetic Behavior

Additive gene action and partial dominance were revealed for harvest index and plant height under normal and stress conditions, however the over-dominance was recorded for 100-grain weight and kernels ear row^{-1} (Hussain 2009). Furthermore, Khaled et al. (2016) showed that parents versus crosses, as an indication of average heterosis over crosses were highly significant under the two environments for the studied traits. The two main effects of "females" and "males" were highly significant under drought and normal conditions for all traits, indicating the prevailing of additive gene action. Mean square due to the "males × females" interaction was also highly significant in both conditions, reflecting the importance of dominance variance in the genetics of these characters. Estimates of heterosis showed that flowering of 9 and 10 out of 20 crosses were significant flowered than their mid-parents with negative heterosis values varied from (-4.07% to -20.56%) and (-2.94% to -19.29%) under normal and drought environments, respectively. As for grain weight, estimates of heterosis were positive and highly significant for all crosses under both conditions. Heterotic values ranged from 28.45% to 208.36% for crosses (C16 × B5) and (C16 × C12), respectively under normal conditions. Whereas, the heterotic estimates were increased and varied from 78.45% to 286.03% for crosses (C15 × B3) and (B10 × B3), respectively under drought conditions. Commonly, the superiority of some crosses over their mid parents reflects the important role of non-additive genetic variance in the inheritance of these traits.

General combining ability effects (GCA) was estimated by Ali (2016) as shown in Table 2.4, as combined over two environments normal and drought stress. Positive GCA effects were desirable for all studied traits, except for the silking date, anthesis

Table 2.2 Genotypic (rg) and phenotypic (rph) correlation coefficients as calculated from the combined analysis of variance of various metric traits in yellow maize genotypes across two environments (Ali 2016)

		Anthesis-silking interval (ASI)	Plant height (cm)	Ear leaf area (cm²)	Leaf water content %	Ear diameter (cm)	Ear length (cm)	No. rows ear⁻¹	No. kernels row⁻¹	100 kernel weight (g)	DSI for grain yield	Grain yield (ard. fed.⁻¹)
Days to 50% silking	rg	0.342**	-0.257	0.228	0.325*	-0.177	-0.158	-0.337**	0.083	-0.102	-0.009	-0.034
	rph	0.236	-0.120	0.089	0.222	-0.126	-0.071	-0.147	0.052	-0.062	-0.011	0.004
ASI	rg	1.000	0.023	-0.159	-0.122	-0.330*	-0.226	-0.213	-0.350**	-0.383**	0.165	-0.572**
	rph	1.000	0.025	-0.139	-0.053	-0.304*	-0.212	-0.193	-0.333*	-0.333*	0.079	-0.491**
Plant height	rg		1.000	0.555**	-0.276*	0.425**	0.628**	-0.209	0.078	0.398**	0.621**	0.166
	rph		1.000	0.482**	-0.137	0.356**	0.510**	-0.144	0.069	0.335**	0.207	0.157
Ear leaf area	rg			1.000	-0.125	0.212	0.546**	-0.216	0.482**	0.634**	0.031	0.443**
	rph			1.000	-0.099	0.149	0.433**	-0.160	0.416**	0.542**	-0.011	0.355**
Leaf water content %	rg				1.000	0.343**	0.050	0.176	-0.208	-0.026	0.558**	0.488**
	rph				1.000	0.097	0.045	0.107	-0.125	0.008	0.119	0.307*
Ear diameter	rg					1.000	0.719*	0.519**	0.197	0.517**	0.809**	0.691**
	rph					1.000	0.536**	0.431**	0.127	0.374**	0.324*	0.546**
Ear length	rg						1.000	-0.011	0.453**	0.729**	0.829**	0.783**
	rph						1.000	-0.008	0.412**	0.617**	0.345**	0.647**
No. Rows/ear	rg							1.000	-0.035	-0.023	0.246	0.291*
	rph							1.000	-0.069	0.032	0.079	0.237
No. kernels/Row	rg								1.000	0.428**	-0.252	0.486**
	rph								1.000	0.376**	-0.042	0.451**

(continued)

Table 2.2 (continued)

		Anthesis-silking interval (ASI)	Plant height (cm)	Ear leaf area (cm²)	Leaf water content %	Ear diameter (cm)	Ear length (cm)	No. rows ear⁻¹	No. kernels row⁻¹	100 kernel weight (g)	DSI for grain yield	Grain yield (ard. fed.⁻¹)
100 kernel	rg									1.000	0.439**	0.659**
	rph									1.000	0.201	0.543**
DSI for grain yield	rg										1.000	0.484**
	rph										1.000	0.388**

*, ** Significant at $P = 0.05$ and $P = 0.01$, respectively Ardab maize = 140 kg grains, Feddan = 4200 m²

Table 2.3 Direct (Diagonal) and indirect effects of some agronomic traits on grain yield of yellow maize genotypes across two environments relative to phenotypic correlation (rph) (Ali 2016)

Characters	Days to 50% silking	Anthesis-silking interval (ASI)	Plant height (cm)	Ear leaf area (cm^2)	Leaf water content %	Ear diameter (cm)	Ear length (cm)	No. rows ear^{-1}	No. kernels row^{-1}	100 kernel weight (g)	DSI for grain yield	Correlation with grain yield
Days to 50% silking	0.034	−0.065	0.020	0.016	0.051	−0.017	−0.024	−0.015	0.009	−0.002	−0.003	0.004
Anthesis-silking interval (ASI)	0.008	−0.277	−0.004	−0.025	−0.012	−0.041	−0.072	−0.020	−0.057	−0.010	0.020	−0.491
Plant height	−0.004	−0.007	−0.169	0.088	−0.032	0.048	0.173	−0.015	0.012	0.010	0.052	0.157
Ear leaf area	0.003	0.039	−0.081	0.182	−0.023	0.020	0.147	−0.017	0.071	0.016	−0.003	0.355
Leaf water content %	0.007	0.015	0.023	−0.018	0.231	0.013	0.015	0.011	−0.021	0.000	0.030	0.307
Ear diameter	−0.004	0.084	−0.060	0.027	0.022	0.135	0.182	0.045	0.022	0.011	0.082	0.546
Ear length	−0.002	0.059	−0.086	0.079	0.010	0.072	0.340	−0.001	0.070	0.019	0.087	0.647
No. Rows/ear	−0.005	0.054	0.024	−0.029	0.025	0.058	−0.003	0.104	−0.012	0.001	0.020	0.237
No. kernels/row	0.002	0.092	−0.012	0.076	−0.029	0.017	0.140	−0.007	0.171	0.011	−0.011	0.451
100 kernel	−0.002	0.092	−0.056	0.099	0.002	0.050	0.210	0.003	0.064	0.030	0.051	0.543
DSI for grain yield	0.000	−0.022	−0.035	−0.002	0.028	0.044	0.117	0.008	−0.007	0.006	0.251	0.388
Residual =	0.495											

Table 2.4 General combining ability (GCA) effects for grain yield and other agronomic traits combined over two environments (Ali 2016)

Inbred lines	Days to 50% silking	Anthesis-silking interval (ASI)	Plant height (cm)	Ear leaf area (cm²)	Leaf water content%	Ear diameter (cm)	Ear length (cm)	No. rows ear⁻¹	No. kernels row⁻¹	100 kernel weight(g)	Grain yield (ard. fed.⁻¹)
P1 (Z12)	-0.29	-0.23^*	-3.50	-25.41^{**}	0.17	0.16^*	-0.48^*	1.36^{**}	-0.09	-0.61	0.15
P2 (Z15)	0.90^*	-0.44^{**}	-0.45	-17.24^{**}	0.05	0.08	-0.12	-0.48	-0.41	-0.15	-0.45
P3 (Z167)	-1.02	0.12	3.12	-2.28	0.67	0.21^*	0.60^*	2.16^{**}	-1.06^*	-0.12	0.89^*
P4 (Z147)	0.26	-0.11	23.63^*	47.82^{**}	-0.20	0.17^*	1.87^{**}	-0.22	1.15^*	1.98^{**}	1.43^*
P5 (Z40)	1.11^*	0.42^{**}	-2.58	34.45^*	0.19	-0.07	-0.80^*	-0.88^*	0.39	-0.17	-0.27
P6 (Z56)	0.20	0.40^{**}	-1.98	-7.60	0.06	-0.26^*	-0.28	-1.66^{**}	-1.99^{**}	-0.64	-0.92^*
P7 (Z58)	-1.57^*	-0.25^*	1.49	-11.73	-1.22^*	-0.16^*	-0.58^*	-0.69^*	0.11	-0.48	-1.35^*
P8 (Z103)	0.42	0.09	-19.73^{**}	-18.01^*	0.28	-0.12^*	-0.22	0.41	1.91^*	0.20	0.53
LSD 0.05 (gi)	0.85	0.15	3.76	15.05	0.72	0.11	0.39	0.50	0.74	0.66	0.61
LSD 0.01 (gi)	2.23	0.39	9.85	39.44	1.88	0.30	1.02	1.31	1.94	1.72	1.59
LSD 0.05 (gi-gi)	1.29	0.23	5.69	22.76	1.09	0.17	0.59	0.75	1.12	0.99	0.92
LSD 0.01 (gi-gi)	3.38	0.59	14.90	59.63	2.85	0.45	1.55	1.98	2.94	2.60	2.40

*, ** Significant at $P = 0.05$ and $P = 0.01$, respectively Ardab maize $= 140$ kg grains, Feddan $= 4200$ m²

to the silking interval and plant height which exhibited negative values indicate a tendency towards earliness and shortness. Therefore, short plant height might be more resistant to stalk breakage, lodging and increasing plant density. The results indicate that, the parental line P7 (Z58) possessed negative and significant GCA effects for days to 50% silking, P1 (Z12), P2 (Z15) and P7 (Z103) for anthesis to silking interval, in the desired direction of earliness as well as P8 (Z103) for short plants. With respect to the ear leaf area, P4 (Z147) and P5 (Z40) had positive and significant GCA effects. None of the parents recorded positive and significant GCA effects for leaf water content. Positive and significant GCA effects were observed in P1 (0.16), P3 (0.21) and P4 (0.17) for ear diameter; P3 (0.6) and P4 (1.87) for ear length; P1 (1.36) and P3 (2.16) for number of rows ear^{-1}; P4 (1.15) and P8 (1.91) for number kernels row^{-1}; P4 (1.98) for 100-kernels as well as P3 (0.89) and P4 (1.43) toward higher-yielding ability. This result indicated that the two inbred lines P3 (Z167) and P4 (Z147) could be considered as good combiners for improving hybrids with yielding ability as well as inbred lines P7 (Z58) and P8 (Z103) for earliness and short plants, respectively.

Added that the ratio of GCA/SCA variances was more than unity for days to 50% silking, plant height and number of rows ear^{-1}, indicating the major role of additive gene effects in controlling the genetic mechanism of these characters over water environments. In contrast, the ratio of variance GCA to variance SCA was blown one for anthesis to the silking interval, ear leaf area, leaf relative water content, ear diameter, ear length, number of rows ear^{-1}, number of kernels row^{-1}, 100-kernel weight and grain yield. This emphasized that non-additive gene action was the prevailed type in controlling these traits. Narrow sense heritability estimates were high (> 50%) for days to 50% silking, plant height, ear diameter and number of rows ear^{-1}, moderate for ear leaf area (41.68%) and ear length (45.55%), and low (< 30%) for anthesis to silking interval, leaf relative water content, number of kernels row^{-1}, 100-kernel weight and grain yield over two environments. Moreover, Al-Naggar et al. (2016) recorded high narrow sense heritability (h^2n) for days to anthesis, leaf angle, ear height, rows ear^{-1}, kernels plant^{-1}, kernels row^{-1}, 100-kernel weight and ears plant^{-1} under both drought stress and high density of 95,200 plants ha^{-1} and well-watered and low density of 47,600 plants ha^{-1} environments.

2.3.5 Breeding Efforts and Biotechnology

The genetic diversity has been considered more efficiently linking polymorphism from the DNA, morphological and biochemical labels. Genetic variability in maize genotypes has been measured by morphological characters (Louette and Smale 2000; Beyene et al. 2005). DNA polymorphism analyses are great tools for studying and describing germplasm resources (Powell et al. 1996). Random amplified polymorphic DNA (RAPD) markers have been used in describing genetic diversity among maize genotypes (Moeller and Schaal 1999; Beyene et al. 2005). The level of association between DNA marker-based genetic similarity and agronomic description might differ between different crop species. In maize, a close correlation was established (Messmer et al. 1993). Hence, it is necessary to determine within each species whether agronomic characterization and DNA marker-based genetic similarity gave similar information about the genetic divergence among existing germplasm. In this respect, Okumus (2007) used 160 primers were to screening maize accessions and 14 primers were found to be valuable for RAPD assay and were exploited to amplify genomic DNA of 17 maize accessions. A total of 62 fragments were produced by seven primers with an average of 8.86 fragments per primer. The bands created in the RAPD experiments were strong and weak. The range of number amplification bands by each primer varied from 5 (OPW-08) to 13 (OPAT-08). The range of these fragments is in 99 (OPW-08) to 943 bp (OPAT08). Two maize inbred lines Sd-63 and Gm-18 were selected as drought-tolerant and drought sensitive, respectively to obtain the F_1 generation and then selfed to produce the F_2 generation. The two inbred lines and their F_1 and bulks of the two extreme F_2 plant groups (the most sensitive F_2 group and the most tolerant F_2 group) were verified against six RAPD primers. This analysis revealed that, four RAPD primers (A01, A05, A06 and B08) out of the six developed molecular markers for drought tolerance in maize. These RAPD markers can be expressed as reliable molecular markers related to drought tolerance in maize (Ahmed 2013). Bawa et al. (2015) revealed that the parent populations TAIS03, GUMA03-OB, IDW-C3-SYN-F2 and KOBN03-OB, and F1 hybrid populations IWD x GUMA03, IWD x TAIS03, IWD x KOBN03, TAIS03 x IWD, DT x TAIS03, DT x SISF03, DT x KOBN03, TAIS03 x DT, TAIS03 x SISF03, SISF03 x GUMA03, SISF03 x KOBN03, SISF03 x TAIS03, GUMA03 x DT,GUMA03 x KOBN03 and GUMA03 x IWD displayed bands across 9 drought-linked SSR markers. Thus, they contain the quantitative traits responsible for drought tolerance. Moreover, Li et al. (2016) identified the QTLs that control drought tolerance-related traits under well-watered conditions and water-stressed conditions using joint linkage analysis of the CN-NAM population. They identified 8–23 QTLs for the seven drought-related traits under well-watered, explained 23.7–66.3% of the total phenotypic variation. However, they recognized 8–20 QTLs under water-stressed conditions, explained 20.2–55.4% of the total phenotypic variation (Ahmed 2013). The genome-wide association analysis recognized 365 single nucleotide polymorphisms (SNPs) related to drought-related traits, and these SNPs were found in 354 candidate genes. Fifty-two of these genes exhibited significant variation expression in the inbred line B73 under the well-watered and water-stressed conditions.

In Egypt, Agriculture Research Center has succeeded in producing high yielding hybrid cultivars of white maize *i.e.* single crosses 10, 128, 129, 130, 131 and 132, and three-way crosses 310, 314, 321, 324 and 329, and the yellow maize crosses, such as single crosses 162, 166, 167, 168, 173, 176, 178 and 180 and three-way crosses of 352, 353 and 368 (Anonymous 2019).

2.4 Barley

2.4.1 Economic Importance

Barley is one of the oldest cultivated crops in the world with a high adaptive capacity. It displays high tolerance to adverse environmental conditions compared to other cereal crops (Gürel et al. 2016; Yang et al. 2017). Consequently, it is commonly grown in marginal areas and lands suffering water shortage and salinity. It is one of the most important crops with uses ranging from food and feed production representing the fourth most abundant cereal acreage and production after wheat, maize and rice (Faostat 2018). It is principally used for animal and poultry feeding, in addition to malt and some uses in the pharmaceutical industry. It contains 3–7% β-glucan, which is a very important dietary fiber that has health benefits (Oscarsson et al. 1996). For that reason, recently there is an increasing interest for human consumption due to its nutritional and healthy values especially hull-less barley as an ideal type for achieving this goal (Biel and Jacyno 2013).

Global population is increasing while water resources for crop production are decreasing. Limited irrigation water is one of the major stresses that reduces crop production and quality in agricultural systems (Arshadi et al. 2018; Hasanuzzaman et al. 2018). Moreover, the importance of drought has become more serious with increasing climatic change and global warming. The increase of air temperature and the decrease of rainfall caused heat stress and drought in many areas, especially in arid and semi-arid regions. Egypt suffers from severe water deficit in recent years, facing water shortage amounted about 7 billion cubic meters annually, and this may increase in the near future due to the effect of Ethiopian Renaissance Dam (Osman et al. 2016).

The drought has different impacts over barley growing stages. During the reproductive development stage, drought is a key factor affecting number of spikes/m^2. During spikelet initiation, drought leads to reduce grain set and grain number spikelet^{-1}, while during grain filling period it leads to reduce individual grain weight. Moreover, drought at the beginning of grain filling has negative effects on grain weight and grain yield more than during the late grain filling period (Samarah et al. 2009).

Drought tolerance is defined as the ability of the genotypes to produce acceptable yield under limited water supply better than the other genotypes. Whereas, drought sensitivity is the reduction in yield of the genotypes under drought stress (Turner et al. 2001; Desoky et al. 2020).

2.4.2 Mean Performance and Genetic Diversity

Barley germplasm provides a very fruitful source of genes and rich sources of genetic variation for improving drought tolerance. The genotypes exhibit different ability to produce acceptable yield under water deficit conditions. Accordingly, it is essential to screen the genetic potentiality of these genotypes under different water regimes. Additionally, evaluation the performance of barley genotypes under stress as well as favorable conditions is important at the beginning of breeding programs to identify suitable genotypes for environments, which helps in improving crop productivity (Mansour et al. 2017).

Various trials were performed under normal and water deficit conditions to display the genotypic variation and identify drought-tolerant and sensitive genotypes. In this context, Mansour et al. (2017) evaluated seventeen barley genotypes under different irrigation levels using drip irrigation system under sandy soil conditions. The treatments included four irrigation levels (severely-low 1200 m^3 ha^{-1}, low 2400 m^3 ha^{-1}, medium 3600 m^3 ha^{-1}, and high 4800 m^3 ha^{-1}). Plants exposed to water stress showed a significant decrement in plant height and yield attributes in comparison with well-watered plants. The drought-tolerant genotypes managed to produce more yield with higher water use efficiency (WUE) compared to drought-sensitive genotypes. Furthermore, the genotypes were classified into three groups using clustering procedure based on grain yield and four drought tolerance indices (mean productivity (MP), geometric mean productivity (GMP), stress tolerance index (STI) and yield index (YI)) (Fig. 2.1). The first group (a) included six genotypes which presented high tolerance indicesm and they considered as drought-tolerant genotypes (Giza-123, Giza-124, Giza-126, Giza-133, Giza-127 and Giza-128). The second group (b) contained

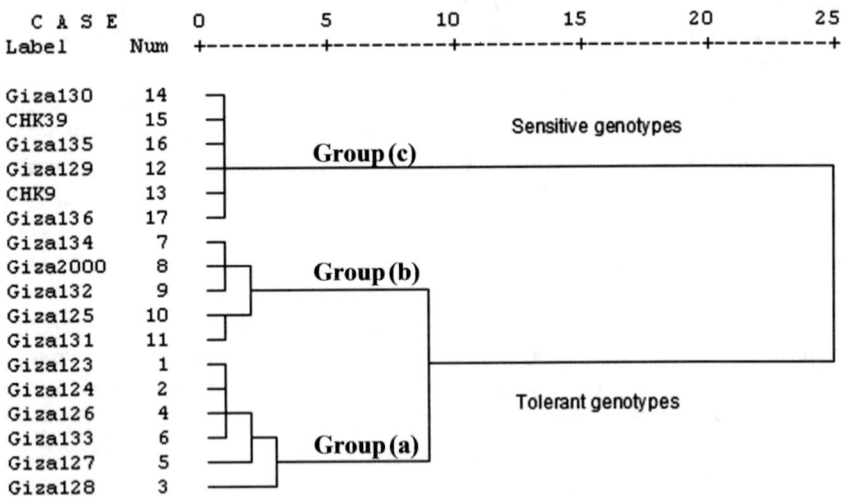

Fig. 2.1 Dendrogram of the phenotypic distances among seventeen barley genotypes under water stress and non-water stress conditions based on grain yield and the drought-tolerant indices, with cutting dendrogram obtained from Ward method in distance five. Group (a) represents drought tolerant genotypes, group (b) represents moderate drought-tolerant genotypes, and group (c) represents drought-sensitive genotypes (Mansour et al. 2017)

five genotypes which intermediated group in the drought-tolerance indices and were considered as moderate drought-tolerant genotypes (Giza-134, Giza-2000, Giza-132, Giza-125 and Giza-131). And finally the third group (c) presented six genotypes which displayed lowest tolerance indices and were considered as drought-sensitive genotypes (Giza-130, CHK-39, Giza-135, Giza-129, CHK-9, and Giza-136).

Furthermore, the results showed that drought-tolerant genotypes (group a) produced much greater grain yield and aboveground biomass with less irrigation water amounts compared with drought-sensitive genotypes (group c). Greater yield in drought-tolerant genotypes could be explained by greater yield attributes for those genotypes under water stress conditions (Fig. 2.2).

Additionally, the performance of fifteen barley genotypes under normal and under limiting water was studied by EL-Shawy et al. (2017). The evaluated genotypes included twelve genotypes from ICARDA, Egyptian landrace and two local varieties (Rihane-3 and Giza-126). Under limited water, the genotypes were irrigated only at sowing using approximately 1200 m^3/ha, (water stressed environment), while under normal conditions, the genotypes were irrigated at sowing (1200 m^3 ha^{-1}), followed by 1800 and 1850 m^3 ha^{-1}, 45 and 75 days after sowing, respectively. Result revealed sufficient genetic variability among the genotypes and importance of selection under stress conditions. The mean squares of irrigation regimes explained most of the variations for all the traits in both growing seasons, indicating the importance of evaluating genotypes under different irrigation regimes. Water deficit highly affected yield and its components in both growing seasons. Moreover, it was observed that among the studied genotypes, Line-2 had early heading and maturity which indicates that this genotype could be used as a source of earliness in breeding program. Besides, the genotypes; Line-7 and Line-11 exhibited the highest yield potential under water deficit in both seasons. This indicates that these genotypes could be considered drought-tolerant genotypes and could be exploited in the barley breeding programs for developing high yielding and drought-tolerant varieties.

Likewise, the response of fifteen barley genotypes under normal and water stress conditions was assessed by Mariey and Khedr (2017). The normal condition was irrigated twice after sowing; i.e. 45 days after sowing (at the tillering stage) and 75 days after sowing (at booting stage). While the water stress condition was given just sowing irrigation. They found that Giza-2000, Giza-126 and Giza-131 had the highest performance for grain yield and its components under normal and water stress conditions. Therefore these genotypes could be considered drought-tolerant and they could be recommended for using in breeding program for realizing high-yielding genotypes under normal and water stress conditions.

Moreover, El-Hashash and Agwa (2018) tested sixteen barley genotypes including fourteen exotic genotypes from ICARDA and two checks cultivars (Giza-123 and Giza-2000) under different stress conditions. The used genotypes were evaluated under severe drought stress, moderate stress and non-stressed conditions. The severe stress experiment was irrigated only once at sowing. While, the non-stress experiment was irrigated three timesm i.e., at sowing, after 30 days from sowing (at the tillering stage) and after 75 days from sowing (at booting stage). On the other hand, the moderate stress experiment was irrigated twice at sowing and after 30 days from

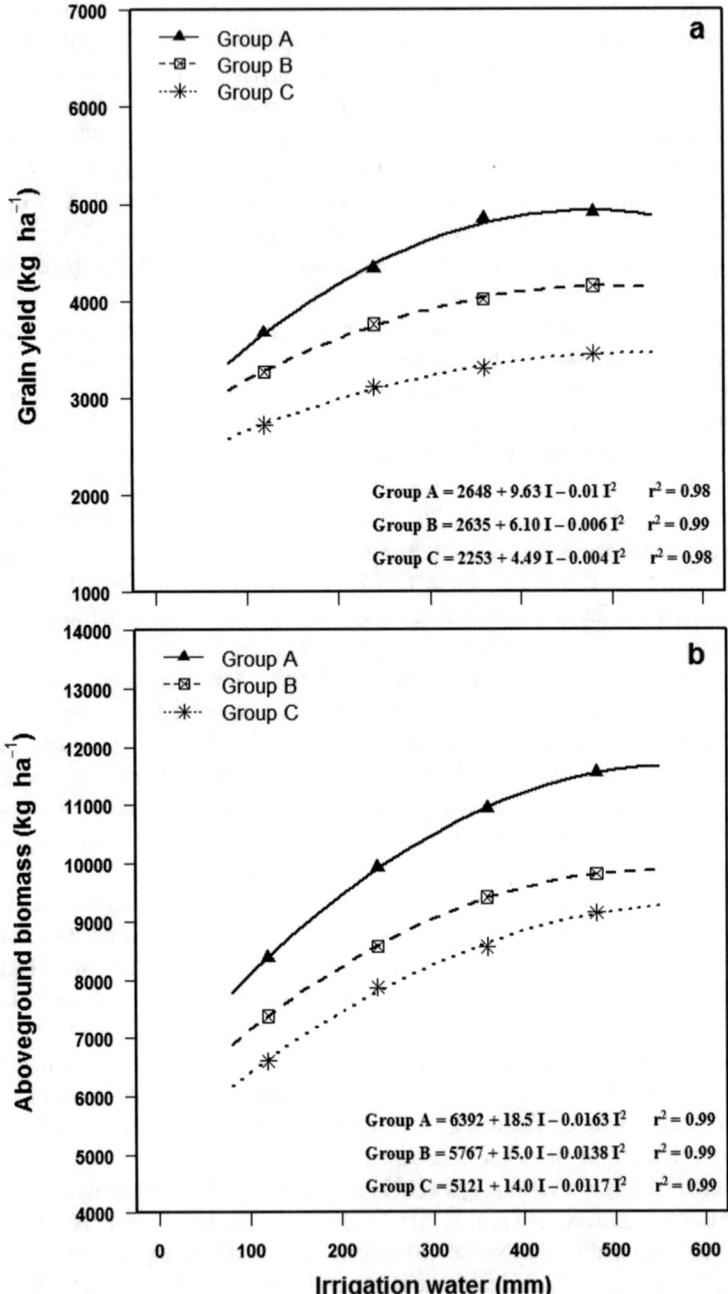

Fig. 2.2 Grain yield (a) and aboveground biomass (b) of the three groups of barley genotypes in response to irrigation water (Mansour et al. 2017)

sowing. The results revealed that all studied traits exhibited high performance under non-stress, followed by mild stress and then severe stress for most investigated genotypes. Most evaluated genotypes were better than the checks cultivars under the three irrigation treatments. Furthermore, it was observed that the genotypes; Line-1, Line-4, Line-6 and Line-10 recorded the highest values for grain yield and its components under the three irrigation treatments. The findings suggest that these genotypes are reliable promising candidates for developing barley varieties under stress conditions in Egypt.

2.4.3 Related Traits to Drought Tolerance

Breeding for drought-tolerant genotypes based on traits associated with drought tolerance facilitates the selecting process. Some responses have been observed in plants induced by drought stress. The responses have been investigated and associated with morphological and physiological traits (Ludlow and Muchow 1990).

The most important morphological traits in a relationship with drought tolerance in barley are deep and prolific root system, leaf rolling, erect leaves and peduncle length. The deep and prolific root system is a very important adaptive trait for maximizing water uptake and improve productivity under drought stress (Ludlow and Muchow 1990). Additionally, leaf rolling could be considered a drought avoidance mechanism and it is very useful to increase drought tolerance. As, it reduces leaf area and transpiration under water deficit conditions (Clarke 1986; Kadioglu et al. 2012). Erect leaves also allow minimum surface area exposed to the sun which reducing transpiration loss (Sinha and Patil 1986; Reynolds et al. 1999). Similarly, peduncle length could be a useful trait associated with drought tolerance. It was found a strong positive correlation between peduncle length and grain yield under drought stress (Acevedo and Ceccarelli 1989; Kaya et al. 2002).

On the other hand, the physiological traits as leaf cuticular wax, relative water content, proline accumulation and osmotic adjustment are associated with drought tolerance. Various studies showed that leaf cuticular wax could protect the plants against drought where it increases the reflection of solar radiation and reduces water loss (Wójcicka 2015). Relative water content also is a useful indicator to assess the water status of a plant where it associates with physiological activities of plants (Mansour et al. 2020). It is a sensitive variable and quickly responds to environmental conditions, therefore, it could be considered a reliable trait to screen the genotypes for drought tolerance (Rachmilevitch et al. 2016). Moreover, proline accumulation is one of the responses of plants to drought stress (Hanson and Hitz 1982). In drought-tolerant genotypes proline is more synthesized than the sensitive ones under the drought stress (Chandrasekar et al. 2000). Furthermore, the osmotic adjustment is a tolerance mechanism that plants use to cope with drought stress. It refers to a reduction in water potential due to the net accumulation of organic and inorganic solutes as a response to water deficit (González et al. 2008; Nayyar and Walia 2004). This allows the turgor potential to be kept higher and helps limit the effects of water stress on the opening of stomata, photosynthesis and growth (González et al. 2008; Mansour et al. 2020). Besides, carbon isotope discrimination (CID) also correlated positively with grain yield and water use efficiency under drought stress (Farquhar and Richards 1984; Bort et al. 1998; Teulat et al. 2002). Consequently, it could be a useful indicator under water limiting condition.

2.4.4 Genetic Behavior

Drought tolerance represents a complex quantitative trait (polygenic) is governed by multiple loci with multiple components associated with plant water status affecting plant productivity (Al Abdallat et al. 2014; Mir et al. 2012). Accordingly, identification of genetic behavior of drought tolerance and associated traits is important. However, the literature on exploration the genetic control of the related traits to drought tolerance and its mechanism in barley is meager (Madhukar et al. 2018).

Nakhjavan et al. (2009) studied the gene action of some of quantitative traits of barley (plant height, spike length, grain yield in plant, weight of 1000 grain, harvest index, days to heading and days to physiological maturity) under two conditions, normal irrigation and terminal drought. The parents with F_1, F_2, BC_1, and BC_2 generations produced from crosses were evaluated. Results indicated that additive, dominance and epistasis effects presented importance in inheritance of all evaluated traits under normal irrigation and drought conditions. Broad-sense heritability ranged between 0.34 to 0.63 and narrow sense heritability between 0.25 to 0.53 under normal condition. But under terminal drought condition broad and narrow sense heritability varied between 0.48 to 0.77 and 0.29 to 0.62, respectively.

Furthermore, Moustafa (2014) evaluated the type of gene action of grain yield and its components using the six populations (P_1, P_2, F_1, F_2, BC_1 and BC_2) of three barley crosses under two levels of field capacity (50 and 70%). The scaling test provided evidence for the presence of non-allelic interaction for studied traits in all crosses under study. The additive (d), dominance (h) and digenic (additive × additive (i), additive × dominance (j) and dominance × dominance (I)) gene action were significant and involved in the inheritance of most traits under both levels of field capacity.

Likewise, Raikwar (2015) performed generation mean analysis using three crosses to study the nature and magnitude of gene effects for yield and its component traits in barley. The results of this study showed the importance of both additive and non-additive types of gene interaction for all the traits studied. Additionally, the additive × additive (i) type gene interaction and duplicate epistasis also detected suggesting the possibilities of obtaining transgressive segregants in later generations. It was found that the additive × additive (i) epistatic effect was more important and higher than the dominance × dominance (l) effect in the inheritance of a number of effective tillers plant^{-1} and spike length. On the other hand, dominance × dominance epistatic effect was more important in inheritance of grain weight spike^{-1}, 1000-grain weight and grain yield plant^{-1}. Therefore, selection for these traits could be fruitful if delayed to minimize dominance and epistasis effects in the advanced generations.

Besides, Mansour (2017) estimated the type of gene action controlling grain yield and related traits of five barley crosses. These crosses used six-population model to determine the intra and inter-allelic gene interactions controlling the inheritance of related traits in barley. It was found that the dominance × dominance component was greater in magnitude than other components in most traits, indicating that these traits were greatly affected by dominance and its non-allelic interactions.

Similarly, Madhukar et al. (2018) assessed the generation mean analysis for yield and drought related-traits of four barley crosses under rainfed and irrigation conditions. The results suggested that the additive gene effect (d) was more import than the dominance (h) one for 100-grain weight, proline content and plant height. Accordingly, selection for these traits would be effective in the early generations. On the other hand, the dominance (h) gene effect was more important than additive (d) in the inheritance of days to maturity and chlorophyll content. Furthermore, the epistasis was observed for chlorophyll content, stomatal conductivity and grain yield. Also, the dominance × dominance (l) interaction was larger than the additive × additive (i) and additive × dominance (j) effects. Their finding revealed that grain number spike^{-1}, grain weight/spike, and grain yield along with stomatal conductance were predominantly influenced by dominance (h) and dominance × dominance (l) gene action. Therefore, the selection of these traits could be difficult in the early generations and is better to be delayed to advanced generations.

Additionally, Madakemohekar et al. (2018) produced six generations (P_1, P_2, F_1, F_2, BC_1 and BC_2) from four different crosses combination in barley to perform generation mean analysis under irrigated and rainfed conditions. The joint scaling test revealed to presence of non-allelic interactions in all four crosses for grain yield and related traits. However, presence of the epistasis effect varied with crosses as well as traits. The generation mean analysis showed that both additive and dominant types of gene effects were important for most studied traits. In general, dominance gene effects (h) were higher than additive gene effects (d) under both irrigated and rainfed conditions. The gene effects indicate that additive component was predominant over dominant for the majority of the traits under rainfed condition. While among the epistatic interaction the major role played by additive × additive type of epistasis which was followed by additive × dominance and dominance × dominance in the four crosses. Which displays that transgressive segregants obtained from these crosses may perform well in the next generations and variety can be developed by selection from these breeding materials.

2.4.5 Breeding and Biotechnology Efforts

Although, classic plant breeding methods have and still present achievement in developing drought-tolerant genotypes, it is a slow process. Therefore, it needs to be integrated with the recent advances breeding methods which improve the process of developing drought-tolerant genotypes (Colmer et al. 2006; Cattivelli et al. 2008; Ashraf 2010). Under Egyptian conditions, several drought-tolerant varieties have been developed of six-row barley, *i.e.* Giza 124, Giza 125, Giza 126, Giza 2000, Giza 132, Giza 133, Giza 134, Giza 137, Giza 138, Giza 129, Giza 130 and Giza 131.

Devolving new barley varieties characterized with drought tolerance depends on generating new allele combinations and subsequent assessing under drought stress then selecting the desirable phenotypes during selfing generations (Nevo 1992). The reliable selection during early generation is difficult due to the heterozygous. Therefore, performing reliable selection requires reaching to acceptable level of homozygosity. On the other hand, the biotechnology tools present rapid achieving homozygosity in barley by producing doubled haploid lines either from immature pollen grains by anther or microspore culture, or through interspecific crosses between barley and *H. bulbosum* with subsequent chromosome elimination (Pickering 1992).

Additionally, mutations induced by radiation or chemical treatments, have also been used for creating genetic variation which could lead to increase crop productivity under drought stress (Parry et al. 2009). Besides, the biotechnology facilitates identification genomic locations of genes controlling traits related to drought tolerance (Lanceras et al. 2004). Detecting stable quantitative trait loci (QTLs) associated with drought tolerance is necessary for utilizing marker-assisted selection (MAS) in breeding. Association of QTLs related to drought tolerance with molecular markers or candidate genes helps breeders to detect the presence or absence of a given QTLs in selected plant material (Collins et al. 2008). QTLs in barley associated with drought tolerance in Mediterranean basin as (Teulat et al. 2001, 2002, 2003; Diab et al. 2004; Wójcik-Jagła et al. 2018).

Marker-assisted selection (MAS) approach is an important method for developing drought-tolerant genotypes not only because of its less time consuming but also labor and cost effective (Kiriga et al. 2016). Furthermore, utilization of MAS allows QTL pyramiding and introducing multiple QTLs related to different drought phenotypic traits into one plant material to improve drought and yield associated traits (Kosova et al. 2014).

The understanding of regulatory networks controlling the drought stress response was enhanced by recent biotechnology tools. This has led to practical approaches for engineering drought tolerance in plants and providing drought tolerant genotypes (Ritala et al. 1994; Umezawa et al. 2006). Genomic analysis has provided gene discovery and allowed genetic engineering using different functional related to drought tolerance in plants (Umezawa et al. 2006). Accordingly, the candidate genes could be transferred to other plant materials during short period compared to classical breeding methods.

2.5 Sunflower

2.5.1 Economic Importance

Sunflower (*Helianthus annuus* L.) has become an important oil crop in the world. The total area reached around 24.8 million hectares worldwide with average productivity of 1.66 ton ha^{-1}. gave total production of 41.3 million metric tons. In Egypt, the total area was about 10,000 hectares with an average production of 2.5 ton ha^{-1} gave total production 25,000 tons (Faostat 2018). The oil has found widespread receipt as a great quality, edible oil, rich in the unsaturated fatty acids, oleic and linoleic acids, vitamin E and contains about 25% proteins.

Egypt suffers from a severe shortage in the production of edible vegetable oils from many problems, leading to a decrease in the local production of oil crops, which led to the failure to meet the needs of domestic consumption (Hassan and Sahfique 2010). Sarvari et al. (2017) reveal that drought stress is a serious adverse factor that limits sunflower growth and productivity. Sunflower is classified as a low to medium drought sensitive crop. The amount and distribution of water have a significant effect on achene and oil yield of sunflower (Krizmanic et al. 2003; Reddy et al. 2003; Iqbal et al. 2005). Conversely, oil quality of sunflower has not been significantly affected by drought condition (Petcu et al. 2001; Pekcan et al. 2015).

2.5.2 Mean Performance and Genetic Diversity

Varietal variances were described for most growth and yield characters (Abou Khadrah et al. 2002; Sharief et al. 2003). Abdel-Wahab et al. (2005) found that among three sunflower hybrids i.e. Euroflour, XF4731 and Vidoc, hybrid XF4731 surpassed the other hybrids for seed yield and oil yield. Sunflower hybrids G-101 and 64-A-93 were the best and most tolerant to drought.. Plant height and dry matter stress tolerance indices for hybrids reduced with increasing drought stress. Whereas an increase in root length stress index was found in the tested sunflower hybrids Dry matter stress index were found to be a reliable indicator of drought tolerance in sunflower (Ahmad et al. 2009). Yasein (2010) verified two sunflower varieties viz. Sakha 53 and Giza 102 under sandy soil conditions and showed that Sakha 53 exceeded Giza 102 in plant height, head diameter, number of seeds/head, seed weight/head, seed yield and seed oil content. Sunflower gave a great genetic variation for osmotic adjustment in response to water shortage (Hussain et al. 2014). Moreover, Pekcan et al. (2015) showed that the differences in sunflower yield varied according to the seasons due to severe heat and drought in the period of growth affected seed yield by 30–35% reduction. Some inbred lines displayed positive responses and some inbred lines attained lost 60% of the control in seed yield. Drought tolerance of male inbred lines in conflict of stress conditions ranged from 50–100% in 1000-seed weight and 70–100% in oil yield. However, oil content of inbred lines was not affected by drought stress; conversely, most of their oil content was increased in a stress environment.

Furthermore, Salem et al. (2013) evaluated 12 sunflower genotypes at three levels of water supply i.e. adequate ($7140 \text{ m}^3 \text{ ha}^{-1}$), moderate ($4760 \text{ m}^3 \text{ ha}^{-1}$) and severe ($2380 \text{ m}^3 \text{ ha}^{-1}$) water regimes. The results demonstrated highly significant differences between the tested genotypes for leaf chlorophyll content, transpiration rate and leaf water content. Combined analyses evidence that L350 had the highest leaf chlorophyll content followed by L460, while the lowest value was recorded by L38. Sunflower genotypes L38 and L20 displayed the highest transpiration rate and L350 was the lowest. Also, the highest leaf water content was recorded by the genotype L11, furthermore L20 and L990 attained higher leaf water content than the remaining genotypes. There was a significant interaction between genotype and the water supply levels for both transpiration rate and the leaf water content. Both transpiration rate and leaf water content decreased significantly for all the genotypes with a reduction in water supply. When the plants received adequate water ($7140 \text{ m}^3 \text{ ha}^{-1}$), the highest transpiration rate was recorded for genotype L11, followed by L38 and L20, whereas L350 attained the lowest transpiration rate. In this treatment, L20 had the highest leaf water content, and Giza 102 and Sakha 53 were the lowest. In the case of moderate drought level, L20 had the highest transpiration rate followed by L38 and L990, while L350 had the lowest value. The highest leaf water content was recorded for L11, though a number of other genotypes had comparable values. The lowest content was found for L38. Under severe drought ($2380 \text{ m}^3 \text{ ha}^{-1}$), L8 had the highest transpiration rate followed by L990 and L11, while Sakha 53 and L460 had the lowest rates. The highest leaf water content was recorded for genotype L11, followed by L350, which was on par with several other genotypes. With an adequate water regime, L990 produced the highest seed yield, while L11 was the lowest. In the case of both moderate and severe drought, L38 had the highest seed yield, while L350 and L11 were the lowest ones. The interaction between genotypes and water regimes had a significant effect on the seed and oil yields. When the plants were provided with an adequate water regime, L990 had the highest oil yield, on par with L8, L460 and Giza 102, while L11 had the lowest. In the case of moderate or severe drought, L38 produced the highest oil yield, followed by L20, L460 and Giza 102 on par at both levels, together with L235 and L8 in the severe water regime. Genotypes L350 and L11 gave the lowest oil yield at both moderate and severe water supply levels.

2.5.3 Related Traits to Drought Tolerance

Acclimation of plants to water stress is due to different traits and the occurrence of adaptive changes in morphological, physiological and biochemical properties such as changes in plant construction, growth, osmotic adjustment and antioxidant defenses

(Duan et al. 2007). Evolution in plant protection systems occurs through the production of antioxidants including enzymatic and non-enzymatic protection to reduce oxidation from ROS (Davar et al. 2013). Scavenging of Reactive oxygen species ROS and decreasing their damaging effects might associate with drought tolerance of crops. The elimination of ROS requires the activity of many enzymes such as APX1, CAT2, SOD3, POX4, MDHAR5, DHAR6, GR7 as well as non-enzymatic i.e. ascorbate, phenolic compounds, carotenoids, glutathione, glycine betaine, proline, sugar, polyamines (Gill and Tuteja 2010; Karuppanapandian et al. 2011).

Sarvari et al. (2017) assessed physio-biochemical changes and antioxidant enzymes activities of six sunflower lines under normal and irrigation at 60 and 40% of the field capacity. They recorded significant differences among sunflower lines for physio-biochemical and enzymes activity under drought stress condition. Genotypes C104 and RHA266 were best tolerant to drought stress. The resistant genotypes had a better osmotic adjustment to maintain their water potential under drought and normal situations. The highest relative water content was correlated with genotypes LR55 and RHA266 at 60% and the lines C104 and RHA266 at 40% drought stress. Genotype LR25 showed the highest rise in chlorophyll under 60% and the genotype RHA266 under 40% drought stress. The highest increase in carotenoid content was registered in the genotype C100 at 60% and the genotype RHA266 at 40% drought stress. The genotype C100 had the highest carotenoid content in comparison with the others under both drought stress conditions. The highest proline content was recorded in genotypes C104 and LR25 at 60% and 40% drought stress, respectively. The lowest values of malon dialdehyde belonged to the genotypes LR4 and RHA266 under 60% drought, however genotype LR55 exhibited lowest values at 40% drought stress. Genotype C100 gave the highest catalase activity at 60% and in genotypes C104 and RHA266 at 40% drought stress. Genotypes C104 and RHA266 had well water status and osmotic adjustment and tolerate to drought stress. The highest glutathione reductase activity was detected in genotypes C100 and RHA266 at 60% and genotype C104 at 40% drought stress. Both genotypes C100 and LR4 gave the highest ascorbate peroxidase activity at 60 and 40% drought stress, respectively. The highest guaiacol peroxidase activity belonged to genotypes LR25 and LR55 under 60 and 40% of drought stress, respectively. Santhosh (2014) showed that sunflower genotypes DRSF-113 and EC-602063 recorded better values for morphological, physiological and yield characters over other genotypes under both moisture stress and control situations. Genotypes DRSF113 and EC-602063 showed better tolerance under stress conditions. The different plant features helps to maintain plant water status or enable the plant to tolerate stress like root traits, leaf area, leaf water potential, photosynthesis, growth rates and rapid recovery after stress, water use efficiency, and biochemical traits such as chlorophyll and membrane stability, proline, osmoregulation and protecting enzymes.

Results of correlation revealed the importance of transpiration rate, plant height, 100-achene weight and leaf water content as selection criteria for improvement of sunflower yield under normal condition. While, under water-stressed state, head weight, head diameter, number of achene and chlorophyll content showed positive and significant correlation with seed yield plant^{-1}. Head diameter and number of achene under both conditions and chlorophyll content under water-stressed condition have a positive direct effect on seed yield plant^{-1}. Also results based on path coefficient analysis recorded by Awaad et al. (2016) showed that maximum direct effect on achene yield plant^{-1} was accounted for transpiration rate and plant height with values of 12.941% and 12.219%. Whereas, moderate direct effects were recorded by both 100-achene weight and leaf water content with values of 7.128 and 7.779%. Moreover, the other remaining three characters, i.e., leaf chlorophyll content, achene oil content and head diameter had less contributions as exhibited 1.917, 0.438 and 0.081%, respectively. The highest indirect effects on achene yield plant^{-1} variation were observed for transpiration rate via plant height followed by transpiration rate via 100-achene weight, leaf water content via 100-achene weight, plant height via 100-achen and leaf chlorophyll content via plant height with values of 8.442, 5.530, 4.579, 3.181 and 2.858, respectively (Table 2.5). According to the total contribution of the studied characters on achene yield/plant variation, it could be arranged as follows, transpiration rate (22.778%), plant height (20.413%), 100-achene weight (13.939%), leaf water content (12.796%), leaf chlorophyll content (4.997%), achene oil content (2.571%) and then head diameter (0.871%). Generally, it could be concluded that the studied characters accounted for 78.365% of the achene yield/plant variation. However, the residual effect was 21.635%.

2.5.4   Genetic Behavior

The study of genetic behavior is important before starting the implementation of the breeding program to choose suitable breeding method to improve the economic characteristics in sunflower. Ortis et al. (2005) indicated the predominant role of the additive gene action in controlling plant height,1000—achene weight and seed oil content of sunflower. Whereas, Mijic et al. (2008) showed that general combining ability variance was larger than specific combining ability for achene yield, oil content and oil yield. Moreover, general combining ability variance was greater than specific combining ability variance for yield, head diameter and oil content (Machikowa et al. 2011). 1000-achene weight, achene number head^{-1} and oil yield were inherited under both additive and dominant effects. Plant height and oil content were governed by additive effects, however over dominant effect was noticed for achene yield (Ghaffari and Mirzapour 2011).

Salem and Ali (2012) aimed to select sunflower parents with good general combining ability (GCA) and crosses with best specific combining ability (SCA) effects under different levels of water supply i.e. adequate (7140 m^3 ha^{-1}), moderate (4760 m^3 ha^{-1}) and severe (2380 m^3 ha^{-1}). Seven sunflower inbred lines were crossed in a 7 × 7 half diallel cross fashion. The resultant 21 hybrids were evaluated along with their parents to yield contributing characters and oil content. Mean squares of genotype x environment interactions were highly significant for plant height, transpiration rate, yield and oil content, suggesting the differential response of sunflower genotypes to water stress environments. The variance due to general combining

Table 2.5 Direct and indirect effect of various metric traits of sunflower genotypes on achene yield plant^{-1} across three environments (Awaad et al. 2016)

S.O.V.	CD	RI %	Total contribution on achene yield plant^{-1}
Leaf water content % (X_1)	0.07779	7.779	12.796
Leaf chlorophyll content (X_2)	0.01917	1.917	4.997
Transpiration rate (X_3)	0.12941	12.941	22.778
Plant height (X_4)	0.12219	12.219	20.413
Head diameter (X_5)	0.00081	0.081	0.871
100-achene weight (X_6)	0.07128	7.128	13.939
Achene oil content % (X_7)	0.00438	0.438	2.571
X_1 x X_2	0.01746	1.746	
X_1 x X_3	0.02367	2.367	
X_1 x X_4	0.00537	0.537	
X_1 x X_5	0.00056	0.056	
X_1 x X_6	0.04579	4.579	
X_1 x X_7	0.00748	0.748	
X_2 x X_3	0.01270	1.270	
X_2 x X_4	0.02858	2.858	
X_2 x X_5	0.00101	0.101	
X_2 x X_6	0.00525	0.525	
X_2 x X_7	0.00132	0.132	
X_3 x X_4	0.08422	8.422	
X_3 x X_5	0.00615	0.615	
X_3 x X_6	0.05530	5.530	
X_3 x X_7	0.01470	1.470	
X_4 x X_5	0.00678	0.678	
X_4 x X_6	0.03181	3.181	
X_4 x X_7	0.00713	0.713	
X_5 x X_6	0.00074	0.074	
X_5 x X_7	0.00055	0.055	
X_6 x X_7	0.00207	0.207	
R^2	0.78365	78.365	78.365
Residual	0.21635	21.635	21.635
Total	1.00000	100	100.000

ability was greater than that of specific combining ability for the studied characters, indicating the preponderance of additive gene action. The general combining ability effects of the parents revealed that the inbred lines L350, L460, L990 and L770 proved to be good combiners for achene yield, while the parents L38, L990 and L235 were found to be promising general combiners for oil % content. Inbred lines L38, L11 and L235 were good candidates for drought tolerance. L350 was a good combiner for seed yield, while L38 proved to be good combiner for oil content and drought tolerance. On the basis of SCA, the cross L38 x L350 was identified as promising seed yield and oil content. On the other hand, Awaad et al. (2016) assessed heterosis as a percentage of mid-parents for leaf water content, leaf chlorophyll content and transpiration rate of sunflower. Cross combinations P_1 x P_2, P_1 x P_6, P_4 x P_5 and P_4 x P_6 gave significant positive heterotic effects for leaf water content at adequate water supply. Meanwhile, no significant positive heterosis was found among all crosses at both moderate and severe stress treatments. Also, significant positive heterosis was found for leaf chlorophyll content in the F_1 crosses P_1 x P_2 and P_4 x P_6 at adequate water supply; P_1 x P_3, P_2 x P_3, P_2 x P_7 and P_3 x P_7 at moderate and P_2 x P_3 and P_3 x P_4 at severe drought stress as presented in Table 2.6. On the other hand, desirable significantly negative heterosis has been registered for transpiration rate by only one cross (P_5 x P_6) at adequate water supply; P_1 x P_7, P_2 x P_3, P_2 x P_4, P_2 x P_5, P_2 x P_7, P_3 x P_4, P_3 x P_5, P_3 x P_7, P_5 x P_6 and P_5 x P_7 at moderate and all crosses at severe stress (Salem and Ali 2012). These results were desired for the breeder where the genotypes, which maintain high levels of leaf water content with high ability to reduce transpiration rate might be considered as tolerant to water stress. It is interesting to note that the sunflower cross ($P_1 \times P_6$) produced the maximum value of heterosis for achene yield plant^{-1} under adequate water supply (22.77%) and moderate (22.58%) level with DSI value less than 1.0 and had significantly negative heterosis for plant height (-10.52%) and the transpiration rate (-44.93%) at severe stress. Therefore it could be considered the promising one and classified as tolerant to water stress.

2.5.5 Breeding Efforts and Biotechnology

Drought can be managed by modifying the plant morphology or incorporating required attributes that help the plants to mitigate the effect of drought stress effectively (Yordanov et al. 2000). Thus, genetic modification is the most successful and cheapest strategy to deal with drought. Because modifications in plant morphology and physiology are heritable, then transferred into breeding materials will be a permanent source of drought tolerance. There are several breeding strategies for drought stress tolerance in crop plants like induction of earliness for drought escape, modification of some plant traits related to drought tolerance and the introduction of traits associated with high yield. Strategy for breeding drought stress depends upon the intensity, frequency and timing of drought occurrence. Also, mutations affecting seed oil fatty acid composition are very useful for getting a novel oil composition (Lacombe et al. 2000). The presence of mutant genes to increase levels of individual

Table 2.6 Heterosis over mid-parent (M.P.) for physiological characters of sunflower crosses of half-diallel analysis in three environments (Awaad et al. 2016)

Character	Leaf water content (%)			Leaf chlorophyll content (SPAD)			Transpiration rate (mg H_2O/cm²/h)		
Water supply Genotype	Adequate water supply	Moderate stress	Severe stress	Adequate water supply	Moderate stress	Severe stress	Adequate water supply	Moderate stress	Severe stress
P1 x P2	5.51**	−3.67	−8.99**	4.97*	6.43	3.97	10.00**	−7.69	−41.25**
P1 x P3	0.51	−2.50	−9.82**	0.00	7.81*	1.60	3.75	−7.26	−45.07**
P1 x P4	2.35	−3.27	−6.24**	−0.10	0.83	1.56	10.98**	−1.47	−23.53**
P1 x P5	1.74	−5.73	−9.63**	3.06	6.72	−1.48	8.16*	1.33	−42.11**
P1 x P6	3.66*	−5.08	−5.64*	0.57	4.42	−0.28	8.33*	3.13	−44.93**
P1 x P7	−1.86	−4.53	−7.77**	−2.47	7.34	−4.96	8.24*	−20.19**	−55.00**
P2 x P3	0.84	0.65	−12.91**	−0.03	11.80**	8.12*	2.17	−13.57**	−53.57**
P2 x P4	1.79	−0.06	−15.09**	1.11	0.62	3.49	2.27	−15.49**	−40.74**
P2 x P5	0.19	2.21	−11.69**	−0.56	5.41	4.29	−0.98	−11.54**	−55.81**
P2 x P6	2.07	0.23	−10.66**	2.10	4.83	1.49	3.81	−6.47	−50.00**
P2 x P7	−1.05	−2.58	−12.92**	−2.54	8.84*	−2.48	2.15	−25.83**	−70.00**
P3 x P4	2.05	−5.41	−15.14**	3.85	4.09	8.23*	2.38	−11.43**	−44.00**
P3 x P5	−0.73	1.68	−16.97**	−0.69	7.49	1.07	0.00	−9.21*	−65.79**
P3 x P6	0.01	0.29	−18.51**	1.42	−0.73	2.55	−1.04	−2.67	−73.53**
P3 x P7	−2.13	−6.81*	−18.10**	−2.81	12.58**	−1.11	2.25	−10.00**	−73.44**
P4 x P5	3.24*	0.13	−15.68**	−4.66	−1.08	−2.67	0.00	−3.95	−21.65**
P4 x P6	3.54*	−4.24	−10.95**	6.57*	−3.44	2.21	−3.33	−0.61	−22.22**
P4 x P7	2.52	−3.81	−10.20**	−8.74*	−5.46	−1.37	2.35	−8.87	−28.75**
P5 x P6	−0.49	−4.60	−25.21**	−3.54	2.42	1.73	−10.00**	−19.88**	−105.71**
P5 x P7	1.69	−3.97	−18.51**	0.07	4.21	−3.10	−1.01	−17.36**	−114.06**

(continued)

Table 2.6 (continued)

Character	Leaf water content (%)			Leaf chlorophyll content (SPAD)			Transpiration rate (mg H_2O/cm^2/h)		
Water supply Genotype	Adequate water supply	Moderate stress	Severe stress	Adequate water supply	Moderate stress	Severe stress	Adequate water supply	Moderate stress	Severe stress
P6 x P7	0.75	−6.35	−15.87**	−0.34	2.16	−2.58	3.92	−7.14	−121.67**

*,**, Significant at $P = 0.05$ and $P = 0.01$, respectively

fatty acids and for different forms and levels of tocopherol leads to improved oil quality and composition of sunflower hybrids (Skoric et al. 2008).

Consequently, the genetic improvement of plant cultivars for water stress tolerance and the ability of a crop plant to produce maximum yield over a wide range of stress and non-stress conditions has been a principal objective of breeding programs for a long time (Chachar et al. 2016; Moustafa et al. 1996). Some tolerant genes were determined in wild types and transferred into cultural ones (Kaya 2014). On the other hand, especially in controlled conditions, sunflower genotypes might be screened to produce tolerant lines to drought stress with simultaneous selection (Geetha et al. 2012; Soorninia et al. 2012). Under Egyptian conditions, drought-tolerant varieties have been developed through breeding programs such as Sakha 53 and Giza 102.

Marker-assisted selection approach serves the purpose of efficient transfer of the desirable gene into elite sunflower inbreds which infers sustainable saving in time compared to conventional backcross breeding. Markers/QTLs would be useful in the marker-assisted selection in breeding program to improve oil quality (Premnath et al. 2016). Molecular such as RAPD and ISSR and the biochemical like protein markers were used to assess the genetic relationships among genotypes. In this respect, based on genetic similarity using the Dice coefficient, Abdel-Ghany (2012) studied morphological traits, oil content and fatty acids of six sunflower genotypes. Genotypes showed insignificant differences among them except for plant height and 1000-seed weights. Bani Suef, Aswan and Beharia genotypes exhibited a high proportion of polyunsaturated fatty acids. Molecular markers results revealed that the genetic similarity ranged from 9% to 77% and 24.1% to 96.6% for the RAPD and ISSR, respectively. Cluster analysis based on similarity matrices of ISSR and combined data separated Giza 102 from all the other genotypes, while the other genotypes fall in a second cluster. RAPD produced 29 unique markers for five of the six genotypes. However, ISSR technology has revealed the ability to discriminate 22 unique markers in the four sunflower genotypes from the six genotypes under study. The biochemical analysis showed slight differences among the six genotypes. Furthermore, biotechnological techniques based on tissue culture were utilized to define the best genotypes and the best explant suit different genotypes. Different regeneration media with different concentrations of BA and NAA showed that the shoot tips of developing the highest rate of response and El-wadi El-gidid genotypes showed the best ones for shoot formation in different concentration of hormones rather than the other genotypes.

Ali et al. (2018) estimated a number of morphological characters, along with relative water content in 100 inbred sunflower lines under natural and water stress during two successive years. At the molecular level, 30 simple sequence repeat (SSR) primer pairs and 14 inter-retrotransposon amplified polymorphism (IRAP) also 14 retrotransposon-microsatellite amplified polymorphism (REMAP) primer combinations were used for DNA fingerprinting of the genotypes. Most of the studied characters exhibited lower values under water stress than natural states. Maximum and minimum decreases were observed in yield and oil percentage. Broad-sense heritability for the examined characters ranged from 0.20–0.73 and 0.10–0.34 under natural and water stress states, respectively. In the verified samples, 8.97% of the

435 possible locus pairs of the SSRs represented significant linkage disequilibrium levels. The SSR analysis identified 22 and 21 markers for the studied characters under natural and water stress states, respectively. The corresponding values were 50 and 37 using retrotransposon-based molecular markers. Some detected markers were common among the characters under water stress and natural states.

2.6 Conclusions

Based on the above discussions, we can state the following conclusions:

1. Drought stress has significant negative effect on plant growth, yield components, grain quality and yield of crop plants, particularly under current climate change.
2. Drought-tolerant genotypes produce considerably greater yield compared with drought-sensitive genotypes especially under water deficit conditions. Therefore, utilizing drought-tolerant genotypes is essential to increase productivity particularly in arid and semi-arid regions in the Mediterranean.
3. There are various morphological and physiological traits associated with drought tolerance could be exploited in selecting drought-tolerant genotypes; as deep and prolific root system, leaf rolling, erect leaves, peduncle length, leaf cuticular wax, relative water content, proline accumulation, osmotic adjustment and antioxidants.
4. Integration of recent advances breeding methods (as quantitative trait loci, Marker-assisted selection and genetic engineering technique with classic plant breeding help significantly in developing drought-tolerant genotypes accurately and rapidly.

2.7 Recommendations

The chapter highlights the following recommendation:

1. Combining morpho-physiological and biochemical traits can provide a more complete model of gene-to-phenotype relationships and genotype-by-environment interactions to tolerate drought stress.
2. Integration of recent improvement procedures as DNA-marker assisted selection and gene manipulation with classic plant breeding aids in improving drought-tolerant genotypes accurately and rapidly.

Acknowledgements Hassan Awaad and Elsayed Mansour acknowledge the partial support of the Science and Technology Development Fund (STDF) of Egypt in the framework of the grant no. 30771 for the project titled "A Novel Standalone Solar-Driven Agriculture Greenhouse - Desalination System: That Grows Its Energy And Irrigation Water" Via The Newton-Musharafa Funding Scheme.

References

Abd-Allah A, Ammar M, Badawi A (2010) Screening rice genotypes for drought resistance in Egypt. J Plant Breed Crop Sci 2(7):205–215

Abdel-Ghany M (2012) Genetic studies on sunflower using biotechnology. PhD Thesis, Department of Genetics, Faculty of Agriculture, Cairo University, Egypt

Abdel-Wahab A, Rhoden E, Bonsi C, Elashry M, Megahed SE, Baumy T, El-Said M (2005) Productivity of some sunflower hybrids grown on newly reclaimed sandy soils, as affected by irrigation regime and fertilization. Helia 28(42):167–178

Abou Khadrah S, Mohamed A, Gerges N, Diab Z (2002) Response of four sunflower hybrids to low nitrogen fertilizer levels and phosphorine biofertilizer. J Agric Res, Tanta Univ 28(1):105–118

Acevedo E, Ceccarelli S (1989) Role of the physiologist-breeder in a breeding program for drought resistance conditions. In: Baker FWG (ed) Drought resistance in cereals, CAB International pp 117–139

Ahmad S, Ahmad R, Ashraf MY, Ashraf M, Waraich EA (2009) Sunflower (*Helianthus annuus* L.) response to drought stress at germination and seedling growth stages. Pak J Bot 41(2):647–654

Ahmed M (2013) Development of some molecular markers for drought tolerance in maize (*Zea mays* L.). Asian J Crop Sci 5(3):312–318

Al Abdallat A, Ayad J, Elenein JA, Al-Ajlouni Z, Harwood W (2014) Overexpression of the transcription factor HvSNAC1 improves drought tolerance in barley (*Hordeum vulgare* L.). Mol Breed 33(2):401–414

Ali MMA (2016) Estimation of some breeding parameters for improvement grain yield in yellow maize under water stress. Mansoura J Plant Prod 7(12):1509–1521

Ali SG, Darvishzadeh R, Ebrahimi A, Bihamta MR (2018) Identification of SSR and retrotransposon-based molecular markers linked to morphological characters in oily sunflower (*Helianthus annuus* L.) under natural and water-limited states. J Genet 97(1):189–203

Al-Naggar A, Atta M, Ahmed M, Younis A (2016) Direct vs indirect selection for maize (*Zea mays* L.) tolerance to high plant density combined with water stress at flowering. J Appl Life Sci Int 7(4):2394–1103

Al-Naggar A, El-Ganayni A, El-Sherbeiny H, El-Sayed M (2000) Direct and indirect selection under some drought stress environments in corn (*Zea mays* L.). J Agric Sci, Mansoura Univ 25(1):699–712

Anonymous (2019) Recommendations techniques in maize cultivation. ARC Giza, Egypt

Arshadi A, Karami E, Sartip A, Zare M (2018) Application of secondary traits in barley for identification of drought tolerant genotypes in multienvironment trials. Aust J Crop Sci 12(1):157–67

Ashraf M (2010) Inducing drought tolerance in plants: Recent advances. Biotechnol Advances 28(1):169–183

Awaad H, Salem A, Ali MMA, Kamal K (2016) Expression of heterosis, gene action and relationship among morpho-physiological and yield characters in sunflower under different levels of water supply. J Plant Prod, Mansoura Univ 7(12):1523–1534

Barutcular C, Sabagh AE, Konuskan O, Saneoka H, Yoldash KM (2016) Evaluation of maize hybrids to terminal drought stress tolerance by defining drought indices. J Exp Biol Agric Sci 4(6):610–616

Basi S (2008) Molecular and histological study of root tissues in relation to phosphorous uptake efficiency and drought tolerance in rice. Ph. D Thesis, University of Bonn, Germany

Bassuony NN, Anis GB (2016) Evaluation of drought tolerance and heterosis of some root characters in rice (*Oryza sativa* L.). J Agric Res, Kafr El-Shaikh Univ 42(1):82–97

Bawa A, Isaac AK, Abdulai MS (2015) SSR markers as tools for screening genotypes of maize (*Zea mays* L.) for tolerance to drought and *Striga hermonthica* (Del.) Benth in the Northern Guinea Savanna Zone of Ghana. Res Plant Biol 5(5):17–30

Beyene Y, Botha AM, Myburg AA (2005) A comparative study of molecular and morphological methods of describing genetic relationships in traditional Ethiopian highland maize. Afr J Biotechnol 4(7):586–595

Biel W, Jacyno E (2013) Chemical composition and nutritive value of spring hulled barley varieties. Bulg J Agric Sci 19(4):721–727

Blum A (1996) Crop responses to drought and the interpretation of adaptation. Plant Growth Regul 20(2):135–148

Bolaños J, Edmeades G (1993) Eight cycles of selection for drought tolerance in lowland tropical maize. I. Responses in grain yield, biomass, and radiation utilization. Field Crop Res 31(3–4):233–252

Bolanos J, Edmeades G (1996) The importance of the anthesis-silking interval in breeding for drought tolerance in tropical maize. Field Crop Res 48(1):65–80

Bort J, Araus J, Hazzam H, Grando S, Ceccarelli S (1998) Relationships between early vigour, grain yield, leaf structure and stable isotope composition in field grown barley. Plant Physiol Biochem 36(12):889–897

Cattivelli L, Rizza F, Badeck F, Mazzucotelli E, Mastrangelo AM, Francia E, Mare C, Tondelli A, Stanca AM (2008) Drought tolerance improvement in crop plants: an integrated view from breeding to genomics. Field Crop Res 105(1–2):1–14

Chachar MH, Chachar NA, Chachar Q, Mujtaba SM, Chachar S, Chachar Z (2016) Physiological characterization of six wheat genotypes for drought tolerance. Int J Res-Granthaalayah 4:184–196

Chandrasekar V, Sairam RK, Srivastava G (2000) Physiological and biochemical responses of hexaploid and tetraploid wheat to drought stress. J Agron Crop Sci 185(4):219–227

Chen P, Haboudane D, Tremblay N, Wang J, Vigneault P, Li B (2010) New spectral indicator assessing the efficiency of crop nitrogen treatment in corn and wheat. Remote Sens Environ 114(9):1987–1997

Clarke JM (1986) Effect of leaf rolling on leaf water loss in *Triticum spp*. Can J Plant Sci 66(4):885–891

Collins NC, Tardieu F, Tuberosa R (2008) Quantitative trait loci and crop performance under abiotic stress: where do we stand? Plant Physiol 147(2):469–86

Colmer TD, Flowers TJ, Munns R (2006) Use of wild relatives to improve salt tolerance in wheat. J Exp Bot 57(5):1059–1078

Davar R, Darvishzadeh R, Majd A (2013) Changes in antioxidant systems in sunflower partial resistant and susceptible lines as affected by Sclerotinia sclerotiorum. Biologia 68(5):821–829

Degenkolbe T, Do PT, Zuther E, Repsilber D, Walther D, Hincha DK, Köhl KI (2009) Expression profiling of rice cultivars differing in their tolerance to long-term drought stress. Plant Mol Biol 69(2):133–153

Desoky EMD, Mansour E, Yasin MAT, El-Sobky EEA, Rady MM (2020) Improvement of drought tolerance in five different cultivars of Vicia faba with foliar application of ascorbic acid or silicon. Spanish J Agri Res 18(2):e0802

Diab AA, Teulat-Merah B, This D, Ozturk NZ, Benscher D, Sorrells ME (2004) Identification of drought-inducible genes and differentially expressed sequence tags in barley. Theor Appl Genet 109(7):1417–1425

Duan B, Lu Y, Yang Y, Li C, Korpelainen H, Berninger F (2007) Interactions between water deficit, ABA, and provenances in Picea asperata. J Exp Bot 58(11):3025–3036

Ekanayake I, O'toole J, Garrity D, Masajo T (1985) Inheritance of root characters and their relations to drought resistance in rice. Crop Science 25(6):927–933

El-Hashash E, Agwa A (2018) Genetic parameters and stress tolerance index for quantitative traits in barley under different drought stress severities. Asian J Crop Sci 1(1):1–16

El-khoby WMH, El-lattef ASMA, Mikhael BB (2014) Inheritance of some rice root characters and productivity under water stress conditions. Egypt J Agric Res 92(2):529–549

El-Shawy E, El-Sabagh A, Mansour M, Barutcular C (2017) A comparative study for drought tolerance and yield stability in different genotypes of barley (*Hordeum vulgare* L.). J Exp Biol Agric Sci 5(2):151–162

Faostat (2018) Food and Agriculture Organization of the United Nations Statistical Database

Farid MA, Abou Shousha AA, Negm MEA, Shehata SM (2016) Genetical and molecular studies on salinity and drought tolerance in rice (*Oryza sativa* L). J Agric Res, Kafr El-Shaikh Univ 42(2):1–23

Farooq M, Wahid A, Kobayashi N, Fujita D, Basra S (2009) Plant drought stress: Effects, mechanisms and management. Agron Sustain Dev 29(1):185–212

Farquhar G, Richards R (1984) Isotopic composition of plant carbon correlates with water-use efficiency of wheat genotypes. Funct Plant Biol 11(6):539–552

Fatemi R, Kahrarian B, Ghanbari A, Valizadeh M (2006) The evaluation of different irrigation regims and water requirement on yield and yield components of corn. J Agric Sci 12:133–141

Fukai S, Cooper M (1995) Development of drought-resistant cultivars using physiomorphological traits in rice. Field Crop Res 40(2):67–86

Fukai S, Cooper M (1996) Stress physiology in relation to breeding for drought resistance: a case study of rice in physiology of stress tolerance in rice. In: Proceeding of International Conference on Stress Physiology of Rice, Rarendra Deva University of Agriculture and Technology, 28 February to 5 March Lucknow, India pp 123–149

Gaballa M (2018a) Agronomic traits and drought tolerance indices for some rice genotypes under drought stress and non-stress conditions. The Seventh Field Crops Conference, Egypt, 18–19 December, Abstract No. 11

Gaballa M (2018b) Kinetn application on some rice genotypes performance under water deficit. The Seventh Field Crops Conference, Egypt, 18–19 December, Abstract No. 23

Gavuzzi P, Rizza F, Palumbo M, Campanile R, Ricciardi G, Borghi B (1997) Evaluation of field and laboratory predictors of drought and heat tolerance in winter cereals. Can J Plant Sci 77(4):523–531

Geetha A, Sivasankar A, Prayaga L, Suresh J, Saidaiah P (2012) Screening of sunflower genotypes for drought tolerance under laboratory conditions using PEG. Sabrao Journal of Breeding and Genetics 44(1):28–41

Ghaffari MFI, Mirzapour M (2011) Combining ability and gene action for agronomic traits and oil content in sunflower (*Helianthus annuns* L.) using F1 hybrids. Crop Breed J 1(1):75–87

Gill SS, Tuteja N (2010) Reactive oxygen species and antioxidant machinery in abiotic stress tolerance in crop plants. Plant Physiol Biochem 48(12):909–930

González A, Martin I, Ayerbe L (2008) Yield and osmotic adjustment capacity of barley under terminal water-stress conditions. J Agron Crop Sci 194(2):81–91

Grzesiak M, Waligórski P, Janowiak F, Marcińska I, Hura K, Szczyrek P, Głąb T (2013) The relations between drought susceptibility index based on grain yield (DSI GY) and key physiological seedling traits in maize and triticale genotypes. Acta Physiol Plant 35(2):549–565

Gürel F, Öztürk ZN, Uçarl C Rosellini D (2016) Barley genes as tools to confer abiotic stress tolerance in crops. Frontiers in Plant Science 7(1137):1–6

Hanson A, Hitz W (1982) Metabolic responses of mesophytes to plant water deficits. Annu Rev Plant Physiol 33:163–203

Hasanuzzaman M, Shabala L, Brodribb TJ, Zhou M, Shabala S (2018) Understanding physiological and morphological traits contributing to drought tolerance in barley. J Agron Crop Sci 1–12

Hassan M, Sahfique F (2010) Current situation of edible vegetable oils and some propositions to curb the oil gap in Egypt. Nat Sci 8(12):1–7

Heinemann AB, Barrios-Perez C, Ramirez-Villegas J, Arango-Londoño D, Bonilla-Findji O, Medeiros JC, Jarvis A (2015) Variation and impact of drought-stress patterns across upland rice target population of environments in Brazil. J Exp Bot 66(12):3625–3638

Hosseini S, Sarvestani Z, Pirdashti H (2012) Responses of some rice genotypes to drought stress. Int J Agric: Res Rev 2(4):475–482

Hussain I (2009) Genetics of drought tolerance in maize (*Zea mays* L.). Ph D Thesis, Department of Plant Breeding and Genetics, University of Agriculture, Faisalabad, Pakistan

Hussain S, Iqbal J, Ibrahim M, Atta S, Ahmed T, Saleem MF (2014) Exogenous application of abscisic acid may improve the growth and yield of sunflower hybrids under drought. Pak J Agric Sci 51(1):49–58

Ibrahim MA, Merwad A, Elnaka EA (2018) Rice (*Oryza Sativa* L.) tolerance to drought can be improved by silicon application. Commun Soil Sci Plant Anal 49(8):945–957

Ilarslan R, Kaya Z, Tolun A et al (2001) Genetic variability among Turkish pop, flint and dent corn (*Zea mays* L. spp. Mays) races: Enzyme polymorphism. Euphytica 122(1):171–179

Iqbal N, Ashraf M, Ashraf M, Azam F (2005) Effect of exogenous application of glycinebetaine on capitulum size and achene number of sunflower under water stress. Int J Biol Biotechnol 2(3):765–771

Kadioglu A, Terzi R, Saruhan N, Saglam A (2012) Current advances in the investigation of leaf rolling caused by biotic and abiotic stress factors. Plant Sci 182:42–48

Karuppanapandian T, Moon JC, Kim C, Manoharan K, Kim W (2011) Reactive oxygen species in plants: their generation, signal transduction, and scavenging mechanisms. Aust J Crop Sci 5(6):709–725

Kaya Y (2014) Sunflower, in Alien Gene transfer in crop plants, vol. 2, Springer pp 281–315

Kaya Y, Topal A, Gonulal E, Arisoy RZ (2002) Factor analysis of yield traits in genotypes of durum wheat (*Triticum durum*). Indian J Agric Sci 72(5):301–303

Khaled A, El-Sherbeny G, Elsayed H (2013) Genetic relationships among some maize (*Zea mays* L.) genotypes on the basis of gene action and RAPD markers under drought stress. Egypt J Genet Cytol 42(1):73–88

Khaled AMI (2008) Combining ability and types of gene action in yellow maize (*Zea mays* L.). PhD. Thesis, Agronomy Department, Faculty of Agriculture, Assiut University, Egypt

Khaled A, El-Sherbeny G, Elsayed H (2016) Genetic relationships among some maize (*Zea mays* L.) genotypes on the basis of gene action and RAPD markers under drought stress. Egypt J Genet Cytol 42(1):73–88

Khodarahmpour Z, Hamidi J (2012) Study of yield and yield components of corn (*Zea mays* L.) inbred lines to drought stress. Afr J Biotechnol 11(13):3099–3105

Kiriga WJ, Yu Q, Bill R (2016) Breeding and genetic engineering of drought-resistant crops. Int J Agric Crop Sci 9(1):7

Köksal ES (2011) Hyperspectral reflectance data processing through cluster and principal component analysis for estimating irrigation and yield related indicators. Agric Water Manag 98(8):1317–1328

Kosova K, Vitamvas P, Urban MO, Kholova J, Prášil IT (2014) Breeding for enhanced drought resistance in barley and wheat–drought-associated traits, genetic resources and their potential utilization in breeding programmes. Czech J Genet Plant Breed 50(4):247–261

Krizmanic M, Liovic I, Mijic A, Bilandzic M, Krizmanic G (2003) Genetic potential of os sunflower hybrids in different agroecological conditions. Sjemenarstvo 20(5):237–245

Lacombe S, Guillot H, Kaan F, Millet C, Bervillé A (2000) Genetic and molecular characterization of the high oleic content of sunflower oil in Pervenets. In: Proceedings of the 15th International Sunflower Conference, Toulouse, France, pp 12–15

Lanceras JC, Pantuwan G, Jongdee B, Toojinda T (2004) Quantitative trait loci associated with drought tolerance at reproductive stage in rice. Plant Physiol 135(1):384–399

Lenka SK, Katiyar A, Chinnusamy V, Bansal KC (2011) Comparative analysis of drought-responsive transcriptome in Indica rice genotypes with contrasting drought tolerance. Plant Biotechnol J 9(3):315–327

Li C, Sun B, Li Y, Liu C, Wu X, Zhang D, Shi Y, Song Y, Buckler ES, Zhang Z (2016) Numerous genetic loci identified for drought tolerance in the maize nested association mapping populations. BMC Genom 17(894):1–11

Louette D, Smale M (2000) Farmers' seed selection practices and traditional maize varieties in Cuzalapa, Mexico. Euphytica 113(1):25–41

Ludlow MM, Muchow RC (1990) A critical evaluation of traits for improving crop yields in water-limited environments. In: Advances in agronomy, Brady NC, Editor pp 107–153

Luo R, Wei H, Ye L, Wang K, Chen F, Luo L, Liu L, Li Y, Crabbe MJC, Jin L (2009) Photosynthetic metabolism of C3 plants shows highly cooperative regulation under changing environments: a systems biological analysis. Proc Natl Acad Sci 106(3):847–852

Machikowa T, Seating C, Funpeng K (2011) General and specific ability for quantitative characters in sunflower. J Agric Sci, Mansoura Univ 3(1):91–95

Madabula FP, Santos RSD, Machado N, Pegoraro C, Kruger MM, Maia LCD, Sousa ROD Oliveira ACD (2016) Rice genotypes for drought tolerance: Morphological and transcriptional evaluation of auxin-related genes. Bragantia 75(4):428–434

Madakemohekar A, Prasad L, Prasad R (2018) Generation mean analysis in barley (*Hordeum vulgare* L.) under drought stress condition. Plant Arch 18(1):917–922

Madhukar K, Prasad L, Lal J et al. (2018) Generation mean analysis for yield and drought related traits in barley (*Hordeum vulgare* L.). Int J Pure Appl Biosci 6(1):1399–1408

Mansour E, Abdul-Hamid MI, Yasin MT, Qabil N, Attia A (2017) Identifying drought-tolerant genotypes of barley and their responses to various irrigation levels in a Mediterranean environment. Agric Water Manag 194:58–67

Mansour E, Moustafa ESA, Desoky EM, Ali MMA, Yasin MAT, Attia A, Alsuhaibani N, Tahir MU, El-Hendawy S (2020) Multidimensional evaluation for detecting salt tolerance of bread wheat genotypes under actual saline field growing conditions. Plants 9(10):1324

Mansour M (2017) Genetic analysis of earliness and yield component traits in five barley crosses. J Sustain Agric Sci 43(3):165–173

Mardeh A, Ahmadi A, Poustini K, Mohammadi V (2006) Evaluation of drought resistance indices under various environmental conditions. Field Crop Res 98(2–3):222–229

Mariey SA, Khedr RA (2017) Evaluation of some Egyptian barley cultivars under water stress conditions using drought tolerance indices and multivariate analysis. J Sustain Agric Sci 43(2):105–114

Messmer MM, Melchinger AE, Herrmann RG, Boppenmaier J (1993) Relationships among early European maize inbreds: II. Comparison of pedigree and RFLP data. Crop Sci 33(5):944–950

Mijić A, Kozumplik V, Kovačević J, Liović I, Krizmanić M, Duvnjak T, Marić S, Horvat D, Šimić G, Gunjača J (2008) Combining abilities and gene effects on sunflower grain yield, oil content and oil yield. Period Biol 110(3):277–284

Mir RR, Zaman-Allah M, Sreenivasulu N, Trethowan R, Varshney RK (2012) Integrated genomics, physiology and breeding approaches for improving drought tolerance in crops. Theor Appl Genet 125(4):625–645

Mitra J (2001) Genetics and genetic improvement of drought resistance in crop plants. Curr Sci 80:758–763

Moeller D, Schaal B (1999) Genetic relationships among Native American maize accessions of the Great Plains assessed by RAPDs. Theor Appl Genet 99(6):1061–1067

Moumeni A, Satoh K, Kondoh H, Asano T, Hosaka A, Venuprasad R, Serraj R, Kumar A, Leung H, Kikuchi S (2011) Comparative analysis of root transcriptome profiles of two pairs of drought-tolerant and susceptible rice near-isogenic lines under different drought stress. BMC Plant Biol 11(174):1–17

Moussa HR, Abdel-Aziz SM (2008) Comparative response of drought tolerant and drought sensitive maize genotypes to water stress. Aust J Crop Sci 1(1):31–36

Moustafa ESA (2014) Genetic analysis of grain yield and its components in barley under drought stress condition. Egypt J Plant Breed 18(2):211–223

Moustafa M, Boersma L, Kronstad W (1996) Response of four spring wheat cultivars to drought stress. Crop Sci 36(4):982–986

Muthuramu S, Jebaraj S, Gnanasekaran M (2010) Association analysis for drought tolerance in rice (*Oryza sativa* L.). Res J Agric Sci 1(4):426–429

Nakhjavan S, Bihamta M, Darvish F, Sorkhi B, Zahravi M (2009) Mode of some of barley quantitative inheritance traits in normal irrigation and terminal drought stress conditions using generation mean analysis. New Find Agric 3(2):203–222

Nayyar H, Walia DP (2004) Genotypic variation in wheat in response to water stress and abscisic acid-induced accumulation of osmolytes in developing grains. J Agron Crop Sci 190(1):39–45

Nevo E (1992) Origin, evolution, population genetics and resources for breeding of wild barley, *Hordeum spontaneum* in the Fertile Crescent. In: R. Shewry (ed.) Barley: genetics, biochemistry, molecular biology and biotechnology. CAB International, Wallindford, Oxon, UK. pp 19–43

Nguyen HT, Babu RC, Blum A (1997) Breeding for drought resistance in rice: physiology and molecular genetics considerations. Crop Sci 37(5):1426–1434

Okumus A (2007) Genetic variation and relationship between Turkish flint maize landraces by RAPD markers. Am J Agric Biol Sci 2:49–53

Ortis L, Nestares G, Frutos E et al (2005) Combining ability analysis for agronomic traits in sunflower (*helianthus annuus* l.). Helia 28:125–134

Oscarsson M, Andersson R, Salomonsson AC, Åman P (1996) Chemical composition of barley samples focusing on dietary fibre components. J CeR Sci 24(2):161–170

Osman R, Ferrari E, McDonald S (2016) Water scarcity and irrigation efficiency in Egypt. Water Econ Policy 2(04):1–28

Parry MA, Madgwick PJ, Bayon C, Tearall K, Hernandez-Lopez A, Baudo M, Rakszegi M, Hamada W, Al-Yassin A, Ouabbou H (2009) Mutation discovery for crop improvement. J Exp Bot 60(10):2817–2825

Pekcan V, Evci G, Yilmaz MI, Nalcaiyi ASB, Erdal SC, Cicek N, Ekmekci Y, Kaya Y (2015) Drought effects on yield traits of some sunflower inbred lines. Poljoprivreda i Sumarstvo 61(4):101–107

Petcu E, Arsintescu A, Stanciu D (2001) The effect of drought stress on fatty acid composition in some Romanian sunflower hybrids. RomIan Agric Res 15:39–42

Pickering R (1992) Haploid production: approaches and use in plant breeding. Genet, Mol Biol Biotechnol 113:511–539

Powell W, Morgante M, Andre C, Hanafey M, Vogel J, Tingey S, Rafalski A (1996) The comparison of RFLP, RAPD, AFLP and SSR (microsatellite) markers for germplasm analysis. Mol Breed 2(3):225–238

Premnath A, Narayana M, Ramakrishnan C, Kuppusamy S, Chockalingam V (2016) Mapping quantitative trait loci controlling oil content, oleic acid and linoleic acid content in sunflower (*Helianthus annuus* L.). Mol Breed 36(106):1–7

Rabello AR, Guimarães CM, Rangel PH, da Silva FR, Seixas D, de Souza E, Brasileiro AC, Spehar CR, Ferreira ME, Mehta Â (2008) Identification of drought-responsive genes in roots of upland rice (*Oryza sativa* L). BMC Genom 9(485):1–13

Rachmilevitch S, DaCosta M, Huang B (2016) Physiological and biochemical indicators for stress tolerance, in Plant-environment interactions. CRC Press, Boca Raton, FL, pp 334–368

Raikwar R (2015) Generation mean analysis of grain yield and its related traits in barley (*Hordeum vulgare* L.). Electron J Plant Breed 6(1):37–42

Reddy GKM, Dangi KS, Kumar SS, Reddy AV (2003) Effect of moisture stress on seed yield and quality in sunflower. Helianthus Annu J Oilseeds Res 20(2):282–283

Reynolds M, Rajaram S, Sayre K (1999) Physiological and genetic changes of irrigated wheat in the post–green revolution period and approaches for meeting projected global demand. Crop Sci 39(6):1611–1621

Reynolds MP, Rebetzke G, Pellegrinesci A, Trethowan R (2006) Drought adaptation in wheat. Drought adaptation in cereals. In: J. Ribaut (ed.) Haworth food & agricultural products press, CRC Press, New York, pp 402–436

Ribaut J, Jiang C, Gonzalez-de-Leon D, Edmeades G, Hoisington D (1997) Identification of quantitative trait loci under drought conditions in tropical maize. 2. Yield components and marker-assisted selection strategies. Theor Appl Genet 94(6–7):887–896

Ritala A, Aspegren K, Kurtén U, Salmenkallio-Marttila M, Mannonen L, Hannus R, Kauppinen V, Teeri TH, Enari T (1994) Fertile transgenic barley by particle bombardment of immature embryos. Plant Mol Biol 24(2):317–325

Salem A, Ali M (2012) Combining ability for sunflower yield contributing characters and oil content over different water supply environments. J Am Sci 8(9):227–233

Salem A, Omar A, Ali M (2013) Various responses of sunflower genotypes to water stress on newly reclaimed sandy soil. Acta Agron Hung 61(1):55–69

Samarah N, Alqudah A, Amayreh J, McAndrews G (2009) The effect of late-terminal drought stress on yield components of four barley cultivars. J Agron Crop Sci 195(6):427–441

Santhosh B (2014) Studies on drought tolerance in sunflower genotypes. PhD thesis, Department of Crop Physiology, Acharya NG Ranga Agricultural University, Rajendranagar, Hyderabad

Sarvari M, Darvishzadeh R, Najafzadeh R, Maleki H (2017) Physio-biochemical and Enzymatic Responses of Sunflower to Drought Stress. J Plant Physiol & Breed 7(1):105–119

Sedeek S, Kazutoshi O, Abdelkhalik O (2010) Genotypic variation of some Egyptian and upland rice genotypes in some physio-morphological traits and microsatellite DNA under drought condition. Mansoura Univ J Agric Chem Biotechnol 1(3):141–155

Selote DS, Bharti S, Khanna-Chopra R (2004) Drought acclimation reduces O_2—accumulation and lipid peroxidation in wheat seedlings. Biochem Biophys Res Commun 314(3):724–729

Sharief A, El-Kalla S, Sultan M, El-Bossaty N (2003) Response of some short duration cultivars of soybean and sunflower to intensive cropping. Sci J King Faisal Univ 4:95–104

Sinha N, Patil B (1986) Screening of barley varieties for drought resistance. Plant Breed 97(1):13–19

Skoric D, Jocic S, Sakac Z, Lecic N (2008) Genetic possibilities for altering sunflower oil quality to obtain novel oils. Can J Physiol Pharmacol 86(4):215–221

Soorninia F, Toorchi M, Norouzi M, Shakiba MR (2012) Evaluation of sunflower inbred lines under drought stress. Univers J Environ Res Technol 2(1):70–76

Teulat B, Merah O, Sirault X, Borries C, Waugh R, This D (2002) QTLs for grain carbon isotope discrimination in field-grown barley. Theor Appl Genet 106(1):118–126

Teulat B, Merah O, Souyris I, This D (2001) QTLs for agronomic traits from a Mediterranean barley progeny grown in several environments. Theor Appl Genet 103(5):774–787

Teulat B, Zoumarou-Wallis N, Rotter B, Ben Salem M, Bahri H, This D (2003) QTL for relative water content in field-grown barley and their stability across Mediterranean environments. Theor Appl Genet 108(1):181–188

Todaka D, Shinozaki K, Yamaguchi-Shinozaki K (2015) Recent advances in the dissection of drought-stress regulatory networks and strategies for development of drought-tolerant transgenic rice plants. Front Plant Sci 6(84):1–20

Tůmová L, Tarkowská D, Řehořová K, Marková H, Kočová M, Rothová O, Čečetka P, Holá D (2018) Drought-tolerant and drought-sensitive genotypes of maize (Zea mays L.) differ in contents of endogenous brassinosteroids and their drought-induced changes. PloS one 13(5):e0197870

Turner NC, Wright GC, Siddique K (2001) Adaptation of grain legumes (pulses) to water-limited environments. Adv Agronom 71:193–231

Uga Y, Sugimoto K, Ogawa S, Rane J, Ishitani M, Hara N, Kitomi Y, Inukai Y, Ono K, Kanno N (2013) Control of root system architecture by DEEPER ROOTING 1 increases rice yield under drought conditions. Nat Genet 45(9):1097–1102

Umezawa T, Fujita M, Fujita Y, Yamaguchi-Shinozaki K, Shinozaki K (2006) Engineering drought tolerance in plants: discovering and tailoring genes to unlock the future. Curr Opin Biotechnol 17(2):113–122

Wójcicka A (2015) Surface waxes as a plant defense barrier towards grain aphid. Acta Biol CracNsia S Bot 57(1):95–103

Wójcik-Jagła M, Fiust A, Kościelniak J, Rapacz M (2018) Association mapping of drought tolerance-related traits in barley to complement a traditional biparental QTL mapping study. Theor Appl Genet 131(1):167–181

Yang S, Vanderbeld B, Wan J, Huang Y (2010) Narrowing down the targets: towards successful genetic engineering of drought-tolerant crops. Mol Plant 3(3):469–490

Yang Y, Wang Q, Chen Q, Yin X, Qian M, Sun X, Yang Y (2017) Genome-wide survey indicates diverse physiological roles of the barley (Hordeum vulgare L.) calcium-dependent protein kinase genes. Sci Rep 7(1):5306

Yasein M (2010) Some agronomic factors affecting productivity and quality of sunflower crop (*Helianthus annuus* L.). Ph.D. Thesis, Agronomy department, Faculty of Agriculture, Zagazig University, Egypt

Yordanov I, Velikova V, Tsonev T (2000) Plant responses to drought, acclimation, and stress tolerance. Photosynthetica 38(2):171–186

Zhang H, Pan X, Li Y, Wan L, Li X, Huang R (2012) Comparison of differentially expressed genes involved in drought response between two elite rice varieties. Mol Plant 5(6):1403–1405

Ziyomo C, Bernardo R (2013) Drought tolerance in maize: Indirect selection through secondary traits versus genomewide selection. Crop Sci 53(4):1269–1275

Chapter 3
Performance and Genetic Diversity in Water Stress Tolerance and Relation to Wheat Productivity Under Rural Regions

Hassan Auda Awaad

Abstract Wheat is considered as the most strategic crop overall the world and Egypt. Under Egyptian conditions there is a gap between wheat production and consumption. Wheat grows in approximately 1.26 million hectare giving total production about 8.86 million tonnes, while the actual consumption reaches up to about 16 million tons. With the trend of cultivating the new lands, which are suffering from water stress. Therefore, the current chapter came to provide information about fundamentals of wheat crop to tolerate water stress, plant characteristics, genetic behavior, genetic diversity and source of tolerance to water stress. This chapter also addresses to breeding efforts and biotechnology to improve drought tolerance. The attention will also be paid into some agricultural procedures that improve water stress tolerance in relation to wheat productivity

Keywords Wheat · Water stress · Plant characteristics · Genetic behavior · Genetic diversity · Breeding efforts · Biotechnology · Agricultural procedures · Productivity

3.1 Introduction

Water is one of the most essential factors determining the cultivation and production of wheat and food security. Low rainfall and changing its patterns from season to another have led to drought around the world (Lobell et al. 2011). An analysis of published research results from 1980 to 2015 showed a 21% reduction in wheat yield (*Triticum aestivum* L.) as a result drought on a global scale (Daryanto et al. 2016).

Osman et al. (2016) stated that Egypt suffers from a severe water shortage in recent years, facing water shortage about 7 billion cubic meters annually. This shortage will increase in the future due to the effect of the Ethiopian Renaissance dam.

Egypt is one of the arid regions with a very low rainfall rate of 133 mm year $^{-1}$ in coastal areas, halves in the delta and a quarter in central Egypt and is almost

H. A. Awaad (✉)
Crop Science Department, Faculty of Agriculture, Zagazig University, 44511, Zagazig, Egypt

© Springer Nature Switzerland AG 2021
H. Awaad et al. (eds.), *Mitigating Environmental Stresses for Agricultural Sustainability in Egypt*, Springer Water,
https://doi.org/10.1007/978-3-030-64323-2_3

absent in Upper Egypt. An alternative solution to the potential for cultivating drought-affected land is to cultivate varieties that use water more efficiently to reduce water consumption of the crop, and capable of growing, developing and producing accepted yield levels under water shortages, this is beside proper agricultural management. This goal is one of the important strategies for plant breeders and crop scientists to meet the growing human needs of food and clothing, and to solve the problem of food shortages suffered by humans in several parts of the world and in Egypt (Awaad 2009).

The performance of wheat germplasm suffers from drought stress and its effects on various plant traits. Bread and durum wheat genotypes varied in their response to water stress conditions, and showed various degrees of tolerance to stress. Also, stability analysis revealed a differential response of wheat varieties to various drought stress environments (Ali and Abdul-Hamid 2017; Abdel-Motagally and Manal El-Zohri 2018).

Statistics indicate that the area of wheat amounted to approximately 3 million feddan and total production was 8.68 million tons of grains with an average yield 2.89 tonnes/feddan (Anonymous 2017). On the other hand, domestic consumption is reached about 16 million tons, pointing to a gap between production and consumption. Therefore, it seems important to increase productivity either through vertical expansion by increasing the productivity of the unit area or by horizontal expansion by increasing the area cultivated in the new lands. Hence, the total wheat production might be improved by developing high yielding varieties tolerant to water stress and simultaneously applying good agricultural managements to ensure high production at different locations.

3.2 Concept of Water Stress and Its Impact on Wheat Plants

Water stress is defined as the exposure of crop plants to conditions of water shortage to the extent that the quantity of water absorbed by the roots is less than the lost by transpiration. Where the water potential of the plant and cells turgor is low to the extent that affect the functioning of biological and physiological processes, resulting in a low rate of dry matter production and partitioning in the form of an economic crop. In general, water stress is due to deficiency of soil moisture or the inability of the plant to obtain its water needs.

3.2.1 Impact of Water Stress on Wheat Plants

The adverse effects of water stress on wheat plants could be summarized as follows:

1. Negative impact on the growth and development of plants through the impact on plant physiology from germination to maturity.

2. Poor germination and impaired seedling establishment, and the effect on the plant population at harvest.
3. Reduce plant growth mainly by influencing cellular division, enlargement, and differentiation.
4. Damage to chloroplast and loss of chlorophyll protein and affect the activity of photosynthesis due to increased respiratory rate than the rate of photosynthesis.
5. Shortage the number and area of green leaves and rolling and burning the leaves, and thus reduce photosynthetic efficiency.
6. Ears blight and the lack of size of floral structures and lack of grain filling.
7. Early aging of leaves and stems and low grain size.
8. Reduced activity of enzymes controlled in the pathways of synthesis of carbohydrates, protein and fat.
9. Severe damage to the proteins, disturbs their synthesis, inhibit major enzymes and damage cell membranes
10. Accumulation of some osmotic substances such as proline, glycine, betaine, benitol, and some sugars such as fructose.
11. Increase the plant content of abscisic acid and ethylene.
12. Decreased turgor pressure and slow level of photosynthesis under water stress circumstances principally limit the leaf expansion
13. Photosynthetic disturbances and lack of ADP and ATP.
14. Inhibiting the activity of ribosomal DNA, messenger mRNA and severe damage to the proteins, disturb their synthesis and inactivate major enzymes.
15. Occurrence of genetic changes and mutations in the genetic makeup of the plant with the intensity of moisture stress.
16. Lack of the rates of photosynthesis and transition of nutrients and thus a significant decrease in Fresh and dry weights of yield and its components.

3.3 Importance of Coordination Between Plant Breeders and Specialists in Other Related Fields of Science

The success of plant breeders in producing new genotypes depends on full knowledge of the crop and the existing coordination with specialists in the related field sciences that contribute to the acquisition of an ideal genotype. The cooperation of plant breeders with genetics and genetic engineering professionals is important for the identification of genetic characteristics and understanding of the nature of inheritance and identification of genes in the traits associated with drought tolerance and methods of transferring traits. Where there is a large stock of genetic information in crop populations that can be adapted through traditional breeding programs or biotechnological techniques to become more suited to dry environments.

Noteworthy cooperative efforts between physiologists and breeders have been devoted to understanding and dealing with such complex of morpho-physiological traits. The cooperation between plant breeders and plant physiologists is useful in providing information about physiological and biochemical processes, enzymatic changes and plant work mechanics under stress conditions. Also, ways of estimating the different plant traits associated with drought tolerance. Also, techniques of evaluation and identifying drought tolerant genotypes.

The cooperation between plant breeders and specialists in the field of diseases and insects is imperative in artificial inoculations and estimates reaction of wheat genotypes and isolation of resistant genotypes to pests. The cooperation with specialists in the field of soil science is necessary to provide information on the soil properties, wilting point, field capacity and other soil indicators.

Cooperation between plant breeders and technologists is important in quality assessments, improving the content of the teleological components and the biological value of wheat products

The role of plant breeders come in using screening tools in selecting the most appropriate genotypes as parents in breeding programs carrying phenological, morphological, physiological or biochemical characteristics associated with drought tolerance. Also, exploit biotechnology techniques such as gene markers and gene transfer technology as recent approach in the field of breeding to water stress tolerance.

Also, significant cooperative between plant breeders and agronomists have been excited to increase the productivity of cultivars under drought conditions, and attention to agricultural practices, and choose the most tolerant cultivars with appropriate fertilization programs and supplementary irrigation to maintain the numbers of survival plants and improve growth and productivity under these extreme conditions.

3.4 Critical Stages of Water Stress Impact on Wheat

The crop plant undergoes critical periods during its life cycle. The effect of water stress on wheat productivity depends on the growth stage in which the stress occurs. Critical periods are defined as stages where the plant suffers from a lack of soil moisture, the yield decreases significantly statistically. The critical stages of growth that are affected by water stress are the three stages (tillering, anthesis, and grain filling) (Salter and Goode 1967; Gupta 1997). Under Egyptian condition, Hamada et al (2008) found that grain yield reductions comprised 35.29 and 33.62% when water stress was formed at heading and grain filling, respectively. At Enshas, El-Sharkia Governorate, Egypt, Atta and Swelam (2012) showed that exposing wheat plants of Giza 168 cultivar to drought stress decreased its yield and its attributes Table 3.1. The highest grain yield reduction was occurred by skipping two irrigations at tillering (I2), followed by heading (I3), milk-ripe (I4) and then dough-ripe stages (I5) by 33.0, 25.0, 20.0 and 12.0%, respectively as compared to well irrigated plants. Skipping irrigation at tillering stage had maximum value of yield response factor to water deficit (Ky = 2.75), while the minimum value 0.60 was scored by skipping irrigation at dough-ripe stage. Yield response factor to water deficit (Ky) is influenced by the losses in wheat yield and the adjustment required in water supply. Therefore, exposing wheat plants to drought stress at tillering stage induced a great yield loss

Table 3.1 Effect of water stress at different stage of plant age on crop water use efficiency WUEc, relative grain yield decrease, relative evapotranspiration deficit and yield response factor (Ky) as average of two growing seasons on Giza 168 wheat cultivar (Atta and Swelam 2012)

Trait Growth stage	Grain yield (Kg/fed.*)	Actual Evapotranspiration (cm)	WUEc	Relative grain yield decrease	Relative Evapotranspiration deficit	Yield response factor (Ky)
I_1	2.100	42.15	49.88	–	–	–
I_2	1.410	37.09	38.20	0.33	0.12	2.75
I_3	1.567	35.79	43.78	0.25	0.15	1.67
I_4	1.685	35.29	47.75	0.20	0.16	1.25
I_5	1.848	33.84	54.61	0.12	0.20	0.60

I_1: Full irrigation (without skipping any irrigation)
I_2: The same as I_1 except skipping two irrigations during tillering stage
I_3: The same as I_1 except skipping two irrigations during heading stage
I_4: The same as I_1 except skipping two irrigations during milk-ripe stage
I_5: The same as I_1 except skipping two irrigations during dough-ripe stage
*Feddan $= 4200$ m^2
(Atta and Swelam 2012)

and produced higher value of (Ky) than any other irrigation regimes which indicated low ability of wheat plants to tolerate water stress. Ky values decreased gradually with plant age when plant were exposed to water stress.

3.5 Adaptation Mechanisms for Water Stress Conditions

Many researchers explained the plant reaction to water stress through escape, avoidance, and tolerance mechanisms. The different mechaisms of water stress resistance differ in various plant species (Foulkes et al. 2007) and illustration in Fig. 3.1 (Awaad 2009).

Water stress escape: Wherein a plant secures its phonological development to escape drought through critical stages of its development, through the rapid of germination and growth of seedlings and sensitivity to the photoperiod and early maturity in order to escape the effects of drought stress especially that occurred at the end of the season.

Water stress avoidance: It means the plant ability to maintain the internal balance of water in a way that serves the plant and helps it to grow normally under drought stress conditions. This achieved through the ability to store water and increase the resistance of the plant to losing water for reuse when needed.

Water stress tolerance: Is defined as the capability of genotypes to produce acceptable yield under limited water better than the other genotypes (Ramrez and Kelly 1998). Thus means the ability of the plant to maintain essential functions when plant tissues become under actual pressure due to ambient water deficit and

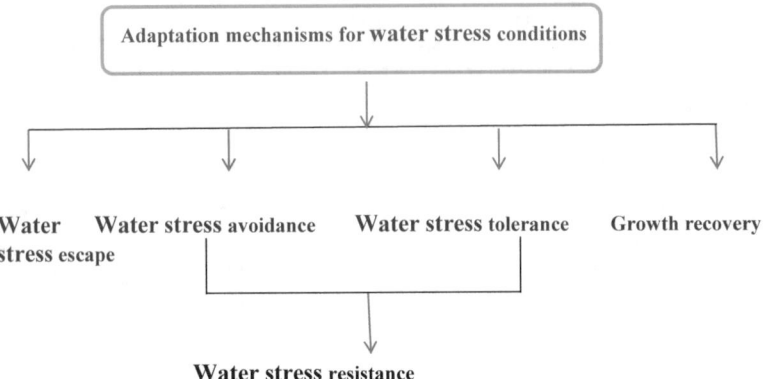

Fig. 3.1 Adaptation mechanisms for water stress conditions (Awaad 2009)

decreasing soil moisture content and the lack of sufficient supply of water. In this case, the plant suffers from internal water stress, but it has the ability to tolerant drought conditions by two main factors: Firstly: Maintain the wall pressure by the osmoregulation and increasing the elasticity of cell walls and resistance to shrinkage. Secondly: Dehydration tolerance by protoplasm tolerance, proline accumulation, and cell wall properties. In grain filling period, high-tolerant drought varieties are characterized by increased accumulation of proline, carbohydrate and high osmotic adjustment. Thus maintaining photosynthetic activity at a higher rate than the less drought-tolerant varieties. From crop improvement opinion, tolerance is considered to be previous and more useful than drought escaping or avoidance.

Water stress resistance: Is the ability of the plant to maintain satisfactory water balance and turgidity when exposed to water stress conditions. Also, plant grows, develop and reproduce naturally and survive without damage and give a good level of yield under soil moisture deficit conditions.

Recovery of growth: It means the ability of the genotype to recovery of active growth after exposure to water deficit conditions, when improved environmental conditions and the availability of water supply. The ability to recover growth is one of the factors that enable the plant to complete its life cycle. Recovery of growth is a key objective in breeding programs to develop and improve the adaptation of varieties to stress environmental conditions.

3.6   Foundations of Wheat Crop Tolerance to Water Stress

3.6.1   Plant Characteristics and Their Genetic Behavior Relevant to Water Stress Tolerance

Wheat genotypes display specific modifications in their growth patterns and physiological process to deal with the sever effects of water stress. There are several criteria for assessing plant tolerance to water stress. These include unique morphological, physiological and biochemical properties that support and help the plant to tolerate water shortage. Plant characteristics related to drought resistance can be divided into the following major groups:

3.6.1.1   Phenological Characteristics

Early maturity leading to escape from the effects of moisture stress. Early ripening helps to maintain crop yield potentiality when soil moisture decreases. Earliness makes the plant more able to avoid drought conditions, although exposed to it. Early maturity is also considered the most important phenological characteristics that may be used for screening and selecting genetic materials in breeding programs under water stress conditions. In this regard, Al-Naggar and Shehab-El-Din (2012) showed the superiority of Line 1 and Line 6 in the plant grain yield by 8.5 and 29.5%, respectively, was accompanied by significantly higher earlier in comparison with Giza 168 for heading by about (6.2 and 10.7 days) and maturity by (2.0, 3.3 days) under water stress conditions. Abd-Elmohsen, Dina (2015) showed that the early maturity wheat cultivars Gemmeiza 11, Sids12, and Shandweel 1 were more superior in grain yield than the other 16 genotypes under water deficit condition. On the other side, El-Rawy and Hassan (2014) recorded strongly positively correlated between grain yield/plant and the most heritable character flowering time in collection of wheat genotypes under water stress condition.

Gene expression and heritability varied from normal to stress. In four parental wheat varieties and their three cross populations exposed to favorable and water stress. At Sakha, Kafer El-Sheikh Governorate, Egypt, Sultan et al. (2010) found that values in earliness characters appeared to be reduced from favorable to stress as a mechanism to avoid water stress. They found also that additive and epistasis gene effects were significant and important in the genetics of days to heading, days to maturity, and grain filling period with major role of epistasis in the inheritance of days to maturity as well as additive in the genetics of grain filling rate under both environments. Heritability in narrow sense reduced from favorable to stress in most cases.

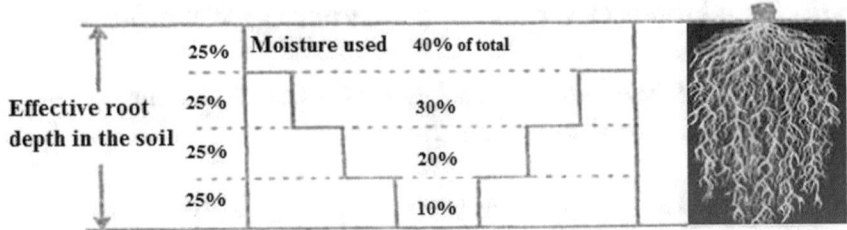

Fig. 3.2 Moisture extraction system of the root region (Modified after Balasubramaniyan and Palaniappan 2001)

3.6.1.2 Morphological Characters

Root System

Plant roots are an important factor in determining adaption of wheat genotypes to drought conditions. It maintaining the water balance within the plant and avoiding the effects of water deficit.

The depth, spread and branching of roots play a major role in the plant's ability to meet water and nutrient requirements and then stabilize crop levels. Field crops vary in the effective root depth. Plants of deep root varieties are characterized by drought avoidance compared to plants with shallow root. Wheat is a moderately-deep rooted crop with a depth of 90 cm. This varies depending on species, varieties, soil properties and agricultural practices.

The region of root hairs is the largest area in which the water is performs through it to the plant wall. It works to increase the absorption rate in different crops. Figure 3.2 shows the effective depth of root extension in some field crops and shows that 40% of the total moisture absorbed by the plant is extracted from the first quarter of the soil profile, 30% from the second, 20% from the third and 10% from the fourth quarter.

The mechanism of adaptation to water stress conditions depends on two factors: Firstly: a dense root mass occupies more space from the soil surface. The secondly: deeper root system able to extract water from the great depths of the soil. The behavior of the root system varies under different environments where the roots grow naturally in the irrigated land, while under drought conditions and due to the depletion of available moisture in the soil, most of the spread and distribution of roots is concentrated in the deeper layers of the soil.

Wheat cultivars with large root system were characterized by drought tolerance, which have a large root depth of more than 110 cm during the flowering period. Its inheritance is governed by many genes in the winter wheats. Both additive and dominance gene action were shown to inherit both number of embryonic roots and crown length with the predominance of additive component in the genetics of crown length. Dominance gene action is the most important in inheriting the number of embryonic roots and the germination % under moisture stress conditions—8 bar (Islam et al. 1999). The genotype Chakwal-86 was characterized by the maximum fresh root weight rather than MT-2/13 under the highest osmotic stress conditions (Zaid et al. 2016). Furthermore, Thapa et al. (2018) conducted three cultivars (TAM 111, TAM 112 and TAM 304) under two water treatments (wet: adequate water and dry: water-limited). Generally, TAM 111 and TAM 112 used more water for cumulative evapotranspiration (ET) and gave more tillers and greater root mass and shoot mass in comparison to TAM 304. In the dry treatment, TAM 112 had 67 and 81% more grain yield than TAM 111 and TAM 304, correspondingly.

Leaf Waxes

A dense of epicuticular wax would be beneficial in plant adaptation to water stress where it decreases plant water loss. The improved drought tolerance is achieved by increasing the reflection of solar radiation and reducing the cuticular permeability for water loss and thus maintaining greater leaf water potential.

The presence of leaf epicuticular wax on the surface of the leaves and stems is a determining factor for water loss and a dry-avoidance characteristic. Wild plant leaves that grow under dry environments have a thick waxy layer compared to plants that grow under normal conditions. This explains the benefits gained from the conserved water for varieties with sensitive stomata for closure. Uddin and Marshall (1988) tested collection of bread wheat (*Triticum aestivum* L.) and durum wheat (*T. turgidum* L. var. durum) varieties for epicuticular wax content and their relationship with spectral reflectance. They detected greatly significant differences (1.51–2.80 mg/dm^2) in the amount of epicuticular wax (EW) between the varieties. Level of EW was significantly enhanced with water stress. Epicuticular wax is correlated with water stress tolerance. EW content under stress and normal conditions were positively associated ($r = 0.85$, $p < 0.01$). So selection for this trait could be practiced in either environment.

Moreover, Al-Bakry (2007) observed significant differences among six wheat genotypes for epicuticular wax content and agronomic traits. Mutant line GWM4 produced the highest quantity of (4.30 mg/dm^2), while the lowest amount was attained by the parent Sids 1. Line GWM5 is characterized by a higher number of spikes/plant and grain yield/plant with low epicuticular wax (3.27), but higher than its parent Sids 1 (2.26). The phenotypic coefficient of variation was 0.31 and the corresponding genotypic one was 0.29. Furthermore, Al-Naggar et al. (2004a) showed that the three bread wheat variants *i e.* a glaucous (waxy) mutant (V1) selected from the irradiated Egyptian cultivar Sids 1, and the two mutants (V2) selected from cross Sids1 x Giza 164 and (V3) selected from cross Sakha 8 × Giza 164 were tolerant to drought stress. Gue et al. (2016) confirmed that leaf cuticular wax is related to drought tolerance. Cuticular wax content was found to be augmented under water-stress circumstances rather than well-irrigated environments. In tolerant wheat near-isogenic lines, cuticular wax content appeared to be more associated with drought-tolerance measurements (tolerance index and stress sensitivity index), has reasonably high water potential and relative water content. Six genes governing wax biosynthesis have been detected and are located on the subsequent wheat chromosomes: *W1* and

IW1 on 2BS, *W2* and *IW2* on 2DS, *W3* on 2BS, and *IW3* on 1BS (Tsunewaki and Ebana 1999; Adamski et al. 2013; Wu et al. 2013; Wang et al. 2014; Zhang et al. 2015).

Awns Characteristics

From the botanical point of view, the awn is a long needle-like structure that forms at the end of the lemma in the florets of wheat plants. It plays a role in shattering resistance and protection from animals, birds and contributes to photosynthesis activity. Under conditions of hot and dry environments, physiological functions are affected, and the long awn is a good visual selection criterion for improving wheat grain yield, as a result of the contribution of awns in photosynthesis process (Blum (1986). The importance of long awns is also associated with its high ratio between carbon exchange and transpiration rate (Blum 1985; Hosseiniet al. 2012), and great reasonable heat transfer accountable for a cooler canopy and improved yield (Weyhrich et al. 1994). While, Foulkes et al. (2007) showed less important role of awns under moderate water stress in the UK than under more severe environments. However, when Taheri et al. (2011) evaluated 17 spring wheat lines with variable responses to water stress, they registered positive and significant association between stress tolerance index and awn length, spike length and plant height. Anonymous (2020) showed that the Egyptian drought-tolerant bread wheat cultivars are characterized by the existence of awn such as Sakha 94, Sakha 95, Misr 1, Misr 2, Shandweel 1, Giza 171, Gemmeiza 7, 9, 10, 11 and 12. Also, drought-tolerant durum wheat cultivars Bani Sweif 5 and Bani Sweif 6 characterized by long awn. Therefore could be grown under stress conditions. Three major dominant inhibitors of awn development (*Hd*, *B1* and *B2*) are recognized in hexaploid wheat (Yoshioka et al. 2017)

Leaf Features

Leaf water condition is highly affected by the morphological characteristics of the leaf such as the size, shape, angle, the degree of its reflection of the radiation and longevity. Drought-resistant genotypes are characterized by small size and low surface area and volume. These characters increases the ability of plant to maintain leaf water potential compared to species adapted to humid environments. Wheat varieties tolerant to drought were found to have less transpiration surface and higher stomata resistance compared to sensitive ones (Golestani Araghi and Assad 1998).

Under rain fed as stress condition, Darwish and El-Hosary, (2003) found that wheat cultivar Basribey 95 was characterized by the lowest flag leaf area (27.2 cm^2), but gave high level of yield (447.0 g/m^2). Whereas Cumshuriyat 75 genotype was the highest in flag leaf area (44.9 cm^2) but attained moderately low grain yields (398.0 g/m^2). Saleh (2011) recorded a positive and significant correlation between grain yield and flag leaf area with relative contribution (11.90%) under normal. Whereas, under stress, the association was positive and significant with reduced relative importance to (6.24%). Qabil (2017) showed that the parental wheat cultivar Misr 1, Gemmeiza 9 and Gemmeiza 11 tolerant or moderate tolerant to drought exhibited flag leaf area values 35.57, 33.88 and 33.79 cm^2 under normal condition (four irrigations were done at tillering, jointing, flowering and grain filling stages, in total 2000 m^3/fed.), but valued 20.44, 22.49 and 23.41 under drought stress (irrigation was prevented after tillering stage up to maturity in total once irrigation after sowing with 600 m^3/fed.

The phenomenon of leaf rolling in cereal grasses is a unique response to drought stress and leads to a decrease in the transpiration rate by 50 to 70% (Gusta and Chen 1987). The reduced leaf angle besides rolling the leaf blade under water stress reduces efficiently the radiation load on plants (Boyer 1996).

Stay green leaves and late aging are unique characters as keys to tolerate the terminal drought. It is considered as a function to describe the tolerance of strains in the post-flowering phase. This visual observation is a real guide to the response of genotypes to drought tolerance. These strains have the advantage of extending the activity of photosynthesis during or after grain filling period or both and the accumulation of more assimilates. Stay green of the wheat leaf was expressed an important determinant of yield under water stress and normal conditions. It behaved as quantitative trait where many genes on the chromosomes 2D and 2B governing the behavior of the character (Verma et al. 2004), and the additive gene action played an important role in the inheritance of stay green leaves (Simon 1999).

3.6.1.3 Physiological Characteristics

To better recognize the physiological base of drought tolerance in wheat cultivars, it seems important to argument the following traits in relative to drought tolerance

Rapidity of Germination and Seedling Development

Seed germination represents the first stage of crop plant growth to counteract the adverse environment. The ability of seeds to germinate and developed under drought conditions is associated with the genetic makeup potentials that adapt to the surrounding conditions. The results of the studies differed in respect to the association between the ability to germinate under drought conditions and the tolerance of plants to these conditions in the advanced stages of growth. It is likely that this correlation is not steady. As wheat shows a high level of tolerance until the germination is complete, but its seedling are highly sensitive to moisture stress after emergence from the soil (Blum et al. 1980).

In another studies, no correlation was found between the ability of wheat seeds to germinate under stress conditions and seedling growth rate. On the other hand, there was a significant correlation between seedling ability to grow in drought conditions and the ability of adult plants to tolerate these stress conditions. Stability of seedling and its ability to live is associated with the merstimic activity and the green leaves area (Clarke and Toenly-Smith 1984).

Crop varieties vary in speed and proportion of seed germination under water stress conditions or when exposed to osmoticums under laboratory conditions. Significant variances were detected between the parents and their crosses of bread wheat in germination rate under water stress conditions (-8 bar) with the importance of non-additive gene action in controlling the inheritance of this character (Islam et al. 1999). While, Zaid et al. (2016) noticed the decrease in the germination rate in wheat strains, or delayed due to water stress.

Mickky and Aldesuquy (2017) showed that at early seedling stage, polyethylene glycol-induced drought had negative impact on seedling indicators *i.e.,* plumule and radicle length, number of adventitious roots and seedling biomass and water content. Severe drought has significant effect on germination measurements and seedling membrane features, was more pronounced than moderate stress. Based on Stress Impact Index, Egyptian cultivar Sids 13 appeared to be the most drought-tolerant, while Shandweel 1 was most sensitive one at juvenile growth stage.

Features and Resistance of Stomatal

Leaves stomatal of the plant play an vital role in influencing and controlling many of physiological processes, storage and utilization of energy, photosynthesis and respiration. It also acts as a mechanism of protection by reducing the water lost by transpiration during water stress periods.

The ability of stomata to regulate the water lost in the crop is subject under genetic control. This is done either by reducing the amount of water lost by transpiration or by increasing water use efficiency through the partial closure of stomata during periods of moisture stress.

Crop species responded to water deficiency with decreases in stomatal conductance before the critical water potential brought about. Some species exhibited wider safety margins for evading stress and these differences could be associated with the specific drought strategy showed by water saver or water spender (Levitt 1972). Wheat genotypes vary in the amount of water stress required to close the stomata and reached (-31 bar) in wheat at the grain filling stage (Turner 1979).

Darwish and El-Hosary and (2003) recorded differences between Egyptian and Turkish wheat varieties in the number of stomata on the leaf surface under stress conditions. Genotype Menemen 88 had the highest number of stomata, while Basribey 95 recorded the lowest number. Morsy (2003) recorded genetic differences in the number of stomata between wheat varieties and strains under water stress conditions. Whereas, Akram et al. (2014) recorded positive and significant association between stomatal conductivity and grain yield under water stress.

It is interest to note that, the behavior of the high content of abscisic acid associated with the activity of stomatal conductivity is inherited as a simple trait with high heritability (Austin et al. 1982). Both additive and dominance gene action with a great role of additive controlling the inheritance of stomatal resistance, leaf temperature and transpiration rate with moderate heritability estimates (30-50%) in narrow sense and high (< 50%) in broad sense (El-Borhamy (2004). Similarly, El- Rawy and Hassan (2014) recorded high broad-sense heritability for stomata frequency (59.0%), and stomata length (54.0%). Grain yield/plant was strongly positively associated with stomata length and stress tolerance index, while negatively associated with stomata frequency and drought sensitivity index. Consequently, highly heritable traits forcefully associated with grain yield under stress conditions mainly stomata frequency and length can be used as reliable guides for choosing high-yielding genotypes more tolerant to drought stress.

Relative Water Content

The relative water content of the leaf is an important physiological characteristic of the plant response to drought conditions and an important criterion in the breeding programs. Increasing the leaf relative water content is an indication of the adaptation of wheat genotypes to water stress and a determinant of drought resistance rather than escape mechanism. The relative water content of stem apex in the cereals contributes to the water immunity of the apical meristems. It maintains relatively high water content, although the nominal osmotic pressure of the cells decreases to - 50 bar. Increasing the relative water content of plant leaves also increases the cell's ability to retain its cell wall potential. Wheat genotypes varied in this respect, for example, at - 20 osmotic potential, the Durum wheats retained 90% of its water content, while the relative water content of the bread wheat decreased at a lower stress level (-16 bar). At -25 osmotic potential, the durum wheats were maintained by 83% of its water content, while the water content was reduced in bread wheat to 64% (Morgan 1977).

The increase in the leaf relative water content contributes to improve grain yield in tolerant wheat varieties. In comparison between tolerant TAM W -101 and Sturdy drought sensitive wheat varieties, the TAMW-101 wheat cultivar was characterized by higher relative water content accompanied by higher grain yield compared with the Sturdy one (Schonfeld et al. 1988). Akram et al. (2014) recorded a positive and highly significant relationship between leaf relative water content and grain yield under water deficit conditions. Furthermore, Saleh (2011) recorded positive and significant association between grain yield and relative water content under normal with relative contribution of (12.46%), while under stress it reduced to be (5.37%) in wheat grain yield variation.

The analysis of the six populations on five wheat crosses between six parents of bread wheat varied in their tolerance to drought, Salem et al. (2003) showed that the simple genetic model was appropriate for the interpretation the inheritance of leaf relative water content in three crosses. Whereas the complex genetic model was valid in the explaining the inheritance of transpiration rate, osmotic pressure, proline content and chlorophyll in most cases.

Leaf relative water content as physiological character was found to be associated with drought stress tolerance in wheat, Morsy (2014) revealed that over dominance gene action was the prevailed type controlling the inheritance of leaf relative water content in most crosses with moderately low narrow sense heritability under water stress. In continuous, Egyptian wheat cultivar Sakha 8 (tolerant) and exotic Pishtaz (sensitive) as parents were involved in six populations under drought stress conditions. Golparvar (2012) showed significant difference between generations with over-dominance for both relative water content and grain filling rate. Narrow sense heritability was low (27.20%) for relative water content and moderate (34.30%) for grain filling rate. Indirect selection via grain filling rate is healthier than relative water content as indirect selection measures to improve wheat grain yield under drought stress.

Osmotic Potential

Osmotic potential is defined as the ability of water molecules to transport from a hypotonic solution (more water, less solutes) to a hypertonic solution (less water, more solutes) across a semi permeable membrane. Plant cells can have great turgor pressures through their great concentrations of organic solutes. These solutes attract water into the cells by osmosis process, where water flowing through semipermeable membranes that prevent the passage of solutes and allow the passage of water.

It has been observed that high osmotic adjustment genotypes extract more water from the soil compared to lower genotypes in this ability. Crop species and varieties within species vary in their ability to osmotic adjustment. Genotypes with higher osmotic adjustment characterized by its ability to maintain its water content under water stress conditions of the cells for a longer period than those that deficiency the ability. The main role of the effect of osmotic control on the crop is illustrated by the following phenomena:

1. Tension of root tissues and penetration of soil and increase the ability to extract water.
2. Improving water use efficiency.
3. Improving the longevity of the green surface.
4. Maintain the transition rate of metabolic products to the grain.
5. Improved yield and increased harvest index.

It is interest to note that the yield of high-osmotic wheat varieties increased by 60% compared to the lower varieties of this capacity under conditions of decreasing water supply (Morgan et al. 1986 and Blum and Pnuel 1990). Moreover, Moustafa et al. (1995) exposed two Egyptian (Giza 165 and Gemmeiza 1) and two exotic (Klassic, and SPHE3) bread wheat varieties from different geographic origins to water stress imposed by withholding irrigation at various growth stages. The tried treatments were: no stress; 10 d stress at tillering; and 10 d stress at heading. Water stress caused great differences in yield and yield components. Under 10 d stress at heading, Giza 165, Gemmeiza 1, and SPHE3 had yield reductions of 44%, 43%, and

18%, respectively, whereas Klassic had a yield increase of 4%. The average slopes of the turgor vs leaf water potential plots were 0.56 for Giza 165 and Gemmeiza 1 vs 0.43 for Klassic and SPHE3 at heading stage, with significant variance between the two groups. Stress was more obvious when applied at heading.

Early wheat cultivars recorded the highest values of osmotic control and drought tolerance compared to late ones (Salem and Kamel 1996 and Morsy 2003). Darwish and El-Hosary (2003) showed that genotype Basribey 95 was higher in osmotic pressure and grain yield, but Cumshuriyat 75 displayed lower osmotic pressure with lower grain yield among the other eight wheat varieties under rain fed condition.

Zhang et al. (2009) found that water shortage accelerated the remobilization of stem in Westonia wheat cultivar but not in Kauz. The profile of stem water soluble carbohydrate accumulation and loss was negatively associated with them RNA concentration of 1-FEH. The expression of 1-FEH w3 may offer a better indicator when associated with osmotic potential and green leaf retention.

Leaf Chlorophyll Content

Chlorophyll is green substance that traps light energy from the sun, and then uses it to incroporate carbon dioxide and water into sugars in photosynthesis process (https://www.sciencedaily.com/terms/chlorophyll.htm). Chlorophyll content is strongly affected by drought conditions as a result of its devastation and the entry into the leaf aging and death.

Chlorophyll content and chlorophyll fluorescence are associated with photosynthesis, and have been used to ascertain drought tolerant wheat genotypes (Sayed 2003; Kumar et al. 2012). Preserving higher chlorophyll content for a longer period of plant life is one of the methodologies for improving grain yield under water stress conditions (Guo et al. 2008).

Chlorophyll stability index and the high content of photosynthesis pigments are evidence of the plant's ability to tolerate moisture stress. Akram et al. (2014) recorded positive and highly correlation between chlorophyll content and grain yield under water shortage conditions. Eftekhari et al. (2016) showed that leaf chlorophyll content was controlled by both additive and dominance but the role of dominance effect is more important under water stress condition. Leaf chlorophyll content appeared to be more heritable character (Salem et al. 2003). Whereas, Qabil (2017) found that narrow sense heritability for flag leaf chlorophyll content was found to be reduced from normal (25.06%) to drought stress environment (14.38%). This could be due to the effect of moisture stress on gene expression

Photosynthesis Rate

Photosynthesis process is one of the most vital physiological mechanisms for adaptation to water stress conditions. Green plants in photosynthesis process convert light energy into chemical energy. Light energy is captured and used to convert water, carbon dioxide, and minerals into energy-rich organic compounds and oxygen is released.

Water use efficiency, stomatal conductance and transpiration rate as physiological characteristics, help the plant to maintain photosynthetic activity and complete the metabolism synthesis of proteins, fatty compounds, nucleic acids and other organic matter when the plant is exposed to water stress. The scientists differed about the relationship between the behavior of the stomatal and photosynthesis during stress. The rate of photosynthesis may continue without change or the rate of CO_2 metabolism may be affected under stress conditions. Drought-tolerant cultivars maintain sensible photosynthetic leaf area under water deficit rather than drought-avoidant cultivars (Baker 1989). Therefore, photosynthesis is a real criterion for selection to water deficit tolerance during early generations in breeding wheat programs.

In this regard, the drought-tolerant Tam w-101 genotype was distinguished by great photosynthetic efficiency and the low influence of CO_2 metabolism processes under water stress from the Sturdy sensitive one. Genetic differences in the photosynthesis efficiency were attributed to an increase in the number of closed stomatal in the sensitive genotypes compared to the drought-tolerant ones. On the contrary, a significant reduction was detected in the CO_2 exchange rate and the ratio of net photosynthesis to water loss in other wheat cultivars under water stress situations (El-Hafid et al. 1998). Ding et al. (2018) showed that the distribution of the stomata contributes to maintaining the wheat varieties with high photosynthesis rates, even under drought conditions. Chlorophyll content and net photosynthetic rate were significantly less influenced by water stress in wheat genotype Pubing 143 than Zhengyin 1. Under water stress, Thapa et al. (2018) found that flag leaf of genotype TAM 112 at mid-grain filling period had lesser stomatal conductance, CO_2 concentration, transpiration rate and net photosynthetic rate, however higher photosynthetic water use efficiency compared to TAM 111 and TAM 304. Therefore, TAM 112 was more capable than TAM 111 and TAM 304 in the development of physiological mechanisms to cope with water stress. Photosynthesis, water efficiency, relative water content and loss of water from the leaves were inherited according to the simple genetic model with high ($< 50\%$) narrow sense heritability, this is attributed to the control of a few major genes (Malik and Wright 1995).

Transpiration Rate

Transpiration is the process of losing water and its evaporation from aerial parts of crop plants like leaves, stems and flowers. Water is an essential factor for plant. The plant uses a small amount of water that is consumed by the roots and consumed for growth and metabolism, while the remaining 97-99.5% is lose by transpiration and guttation (https://en.wikipedia.org/wiki/Transpiration). Transpiration cools the plant canopy and adjusts the osmotic pressure of cells. It also allows mass flow of mineral nutrients and water from the root to the shoot system. Transduction is a vital physiological characteristic associated with plant tolerance to water stress conditions.

The low rate of transpiration causes the plant to maintain its water content under stress conditions, while it is associated with low photosynthesis.

It is known that increased soil moisture stress is associated with a significant decrease in the rate of transpiration in the tolerant varieties, while water availability is accompanied by an increase in the transpiration rate. Drought-tolerant wheat varieties were characterized by low transpiration rates and increased cell protoplasm tolerance under drought circumstances. Changhai et al. (2010) observed that transpiration efficacies of wheat varieties Shi4185 and Kenong199 declined with decreasing soil water content. Transpiration rate of Kenong199 reduced significantly, whereas the rate in Jinmai47 and Luohan2 were increased. This proposes that a strong non-stomatal limitation of photosynthesis occurred in variety Shi4185, especially to Kenong199 rather than Jinmai47 and Luohan2 which were relatively less influenced by non-stomatal limitation. Akram et al. (2014) recorded positive and high correlation between the rate of transpiration and grain yield under water stress conditions. Morsy (2014) revealed that transpiration rate is controlled by dominance gene action in most crosses with moderately low narrow sense heritability under water stress.

3.6.1.4   Biochemical Characteristics

Proline Content

Proline plays an significant role in osmoregulation and serves as a source of energy, carbon as well as nitrogen in addition to protecting various plant enzymes from heat and water deficit (Paleg et al. 1981). Proline and other amino acids accumulate in plant tissues and act as osmotic defense counter to the damage of water stress. It is therefore a selection indicator in crop breeding programs to drought tolerance. Drought-tolerant plants are characterized by high content of proline. It is noteworthy that under normal conditions and irrigation water, a permanent oxidation of the produced proline in the plant as it is converted into glutamic acid and other components, while under stress conditions the proline oxidation rate decreases and accumulates, therefore proline considered an indicator of stress.

The soluble sugar content and proline content in the leaves increased significantly under drought stress in wheat genotypes (Changhai et al. 2010). Hameed et al. (2011) stated that wheat varieties varied in the accumulation rate of proline at different stages under conditions of low soil moisture. They also stressed the role of proline in tolerating wheat genotypes to drought stress.

Salem et al. (2003) and Morsy (2014) showed proline content as biochemical content was related to water stress tolerance. Both water shortage and heat stress upregulate the expression of Heat shock proteins HSPs in wheat (Pandey et al. 2015). Additive gene action played an important role in controlling the inheritance of proline content with moderate to high narrow sense heritability under water stress. Whereas, Iftekhar Eftekhari et al. et al. (2016) revealed that proline content was controlled by dominance gene effects with moderate (35%) narrow sense heritability.

Abscisic Acid

Abscisic acid is defined by its role in the mechanical closure and opening of the stomata. Increased acid content in plant leaves was observed under water stress conditions. The basis for this depends on the fact that the activity of the stomata is governed by a complex physiology, the most important is the abscesic acid (Hartung and Slovik 1991). Field crops i.e. wheat, rice, sorghum, millet, etc. are different in their ability to produce abscisic acid. The abscisic acid is produced in green plastids and remains in the leaves of plants that do not suffer from stress. When the plant exposed to stress, the plastid membranes become more permeable for the abscisic acid and allow it to move to the epidermis cells, including the guard cells where it then works to close the stomatal. Another strategy for abscisic acid is also produced in the root system and then passes through the sapwood to the leaves and leads to the closure of the stomata (Davies et al. 1994).

Accumulation of abscisic acid is inherited as a simple mendalian basic and that there is an opportunity to improve the tolerance of the stress conditions by selecting the absicic acidic content (Clarke and Toenly-Smith 1984). Hameed et al., (2011) showed the importance of abscisic, proline, tryptophan and glycine betaine in wheat tolerant to drought stress conditions.

Antioxidants

Antioxidants play a vital role in regulating the work of different enzyme genes. It adds importance to the effectiveness of drought tolerance mechanisms. Antioxidants preserves the functional properties of cells, especially those related to photosynthesis based on the production of antioxidants such as ascorbic acid, glutathione, peroxidase, catalase, superoxide dismutase and others. The production of antioxidants is an important role in the face of the very low water content of the leaves, which is caused by drought intensification that cannot be avoided by drought-prevention mechanisms. Hameed et al. (2011) indicated the importance of superoxide dismutase in tolerating drought stress conditions of wheat genotypes. Mickky and Aldesuquy (2017) showed that moderate drought stress can activate peroxidase, polyphenol oxidase and ascorbic peroxidase of wheat genotypes. While, severe stress was found to be inhibited the enzyme activity. Enzyme activity of catalase, superoxide dismutase and glutathione reductase was contrariwise retarded by water stress either at moderate or severe level. Dominant gene effects played the importance role in the expression of Catalase enzyme, ascorbate peroxidase and guaiacol peroxidase enzyme with low to moderate narrow sense heritability under drought stress conditions as detected by Eftekhari et al. (2016).

Cell Membrane Stability

The improved penetrability and extracellular ions leakage were used as criteria of cell membrane stability and as a screening test for stress tolerance. Thus, the recognized membrane-stable genotypes can be exploited as parental genotypes in breeding programs for drought tolerance.

Drought stress affect the biochemical and physiological processes comprising plant cell membrane role. Water stress could destroy cell membrane integrity by stimulating lipid peroxidation with clear rise in membrane leakage and subsequent reduction in its stability. Mickky and Aldesuquy (2017) detected significant disrupt membrane integrity of evaluated wheat genotypes under water deficit. Sids 13 appeared to be the most tolerant cultivar followed by Misr 1, Misr 2, Gemmeiza 9, Gemmeiza 11, Sids 12, Sakha 93, Sakha 94, and Giza 186 and lastly Shandaweel 1 was the maximum sensitivity.

El-Basyoni et al. (2017) indicated significant differences between wheat genotypes regarding cell membrane stability CMS for drought treatment. Relative cell membrane injury fluctuated from 27.0 to 54.5% with a mean of 39.8% under Poly Ethylene Glycol as inducer for osmoticum. Moreover, several SNPs markers were found to be significantly related with CMS. Most of tested SNPs are linked with vital functional genes, that control solute transport through the cell membrane and other biochemical activities associated with water pressure tolerance. They added that tolerant genotypes have proven to be more productive under conditions of field stress. Ciulca et al. (2017) showed that the stability of membrane reflects the capability of cell tissues to hold electrolytes at water stress condition by retentive the membrane structure of the leaf cell integral. This trait has been exploited as selection guide for drought in wheat. Cell membrane stability is controlled by over dominance with the expression of different genes.

3.7   Yield Performance in Relation to Water Stress Tolerance

Yield Performance of wheat under differing environments was measured by stress sensitivity index (Fischer and Maurer 1978) as a useful measurement of comparing yield performance of genotypes between stressful and non-stressful situations. It expresses the separate effects of yield capacity and stress sensitivity on yield. Therefore, lower sensitivity index is expressed as indicator with higher stress tolerance.

"Water shortage is one of themain abiotic stresses, which harmfully affects crop growth and yield" (http://citeseerx.ist.psu.edu/viewdoc/download?doi=10.1.1.323. 1932&rep=rep1>). Wheat genotypes greatly differed in their responses to water stress. Drought caused higher reduction in grain yield and its contributors, i.e. flag leaf area, plant height, spike length, number of spikes/ plant, number of spikelets/spike, number of kernels/ spike and 1000-kernel weight as well as days to heading and relative water content (Saleh 2011).

Abd El-Mohsen, Dina (2015) showed that grain yield/plant has a positive and significant genetic correlation with each of Plant height, 1000- grain weight and number of grains/spike under normal and drought conditions. Otherwise, grain yield/plant exhibited significantly negative genetic correlation with each of days to 50% heading and days to 50% maturity under water stress conditions, suggests that early heading and maturity genotypes could be avoid stress condition (Table 3.2). Based on path coefficient analysis, selection for 1000- grain weight followed by number of grains/spike, number of spikes/m^2 and plant height might be effective for improving grain yield under normal irrigation condition. However, plant height followed by 1000- grain weight and then number of grains/spike could be considered as selection traits for improving grain yield under water stress environment.

To assess reliability of drought tolerance measurements, El-Rawy and Hassan (2014) evaluated fifty genotypes of bread wheat at three environments: normal (clay fertile soil, E1), 100% (E2), and 50% (E3) field water capacity under sandy calcareous soil. They obtained moderate to high heritability estimates in broad-sense for 1000-kernel weight (0.47), spike length (0.38), plant height (0.54), flowering time (0.73), stomata frequency (0.59), and stomata length (0.54). Grain yield/plant was strongly positively associated with grain weight/spike, number of tillers, plant height, days to flowering, stomata length, stress tolerance index DSI, yield stability index, whereas it negatively associated with stomata frequency and DSI in E2 and E3, respectively. Thus, highly heritable traits strongly associated with grain yield under stress conditions may be used as dependable criteria for selecting high yielding and tolerant genotypes. Saleh (2011) tested seven parents of bread wheat and their 21 F1 crosses under two water regimes, i.e. normal irrigation (plants gave 5 irrigations during growth season) and water stress (plants gave 3 irrigations where the 2nd and 4th irrigations were prevented during vegetative and anthesis stages, respectively). Mean performance of wheat genotypes differed significantly in their responses to water regimes in respect to yield contributing traits. Drought caused great decrease in grain yield and its attributes, i.e. flag leaf area, plant height, spike length, number of spikes/plant, number of spikelets/spike, number of kernels/spike and 1000-kernel weight as well as days to heading and relative water content. Wheat genotypes; Cham 6, Line 1, Line 2 and Maryout 5 were superior in the most traits under both water regimes. Moreover, the parents Cham 6, IB 18, Giza 168 and Line 1 were the better drought tolerant based on drought sensitivity index. Positive and significant phenotypic correlation was found between grain yield/ plant and each of flag leaf area, relative water content, number of kernels/ spike, 1000-kernel weight and number of spikes/ plant under both environments. Path coefficient analysis showed that flag leaf area, relative water content under both water regimes followed by number of spikes/ plant under drought stress proved to be the major contributors in grain yield variation. Thus, these traits would be more important for yield improvement under the target treatments.

A comparison among 25 wheat genotypes conducted in drought and normally irrigated conditions were evaluated for various morpho-physiological and yield traits.

Table 3.2 Genotypic correlation coefficient of different metric traits of wheat genotypes under normal and water stress conditions across 10 environments (Abd El-Mohsen 2015)

Grain yield/plant		1000- grain weight		Number of spikes/m²		Number of grains/spike		Plant height		Days to 50% maturity		Traits
D	I	D	I	D	I	D	I	D	I	D	I	
−0.533**	−0.025	0.013	−0.064	0.666**	0.325*	−0.239	−0.163	−0.053	0.175	0.838**	0.798**	Days to 50% heading
−0.598**	−0.097	−0.035	−0.120	0.527**	0.238	−0.240	−0.233	−0.280	−0.220			Days to 50% maturity
0.586**	0.441**	−0.224	0.335*	−0.189	−0.120	0.297*	0.286					Plant height
0.443**	0.420**	0.083	0.224	−0.443**	−0.701**							Number of grains/spike
−0.740**	−0.244	−0.401**	−0.301*									Number of spikes/m²
0.403**	0.652**											1000- grain weight

I: Normal irrigation D: Drought stress *, ** Significant at 0.05 and 0.01 levels of probability, respectively

Ata et al. (2014) showed that highest yield was registered for the genotype V-11164 under normal and V-11168 in drought conditions, whereas most stable genotype was V-11168 for yield in drought conditions. Path analysis revealed that relative water content has positive direct effect on yield. Under both environments, number of grains per spike and pike length might be used for direct selection for the yield. Plant height, 1000-grian weight and relative water content also contributed positively for the grain yield. Canopy temperature can similarly be utilized as unique trait to select the best tolerant genotype under water stress. Yield performance and drought sensitivity index utilized six bread wheat genotypes, four of them were local cultivars i.e., Sakha 95, Giza 168, Sids 1 and Gemmeiza 9, and two are Russian cultivars (Onomly and Dobary) of diverse origin crossed in half diallel parents of F1 crosses. These materials were evaluated under irrigated one (stress) or irrigated three times (normal irrigation). Drought sensitivity index for days to maturity indicated that the best parent was Sids 1 and the cross Onomly x Gemmeiza 9, while for grain yield/plant the best parent was Gemmeiza 9 and the cross Onomly x Sids 1 (Abd El-Aty and El-Borhamy 2007). They added that mean squares for genotypes, general and specific combining abilities for drought sensitivity index were highly significant for days to heading, days to maturity, plant height, number of grains/spike and grain yield/plant. Drought tolerant cultivar Sids 1 gave negative significant general combining ability effect (GCA) for drought sensitivity index for earliness and grain yield/plant. Whereas, Giza 168 gave negative and significant general combining ability effect for drought sensitivity index for plant height and number of grains/spike. This finding indicates that the intrinsic performance of their general combining ability. At this moment, selection to improve previous traits could be practiced either on mean performance or GCA effects basis with similar efficiency. Significant or highly significant negative specific combining ability effects for drought sensitivity index expressed in earliness were registered in crosses Giza 168 × Gemmiza 9, Dobary x Sids 1 and Omanly x Gemmeiza 9 as well as positive and significant for grain yield/plant in cross Sakha 93 × Giza 168. It is of great interest to note that the most crosses which included parents Giza 168 and Dobary were tolerant expressed in most traits, indicating the importance of these parents for improving drought tolerance. Moreover, Gomaa et al. (2014) crossed eight bread wheat parents of diverse origin (Sahel1, Giza-168, Misr-1, Gemmieza-9, Gemmieza-11, Line 1, Line 2, Line 3) in a half-diallel model to produce their 28 F1 hybrids evaluated under normal and water stress conditions in farm of Nubaria Agricultural Research Center, Egypt. Based on drought sensitivity index, parental genotypes Sahel 1, Giza 168, Gemmieza 9 and F1's crosses (Sahel 1 X Gemmeiza 9) and (Gemmieza 9 × Line 1) followed by (Sahel 1 X Misr 1), (Line 1 X Line 3) and (Giza 168 X Gemmeiza 11) were the best tolerant to drought stress. On the contrary, the parental genotypes Misr 1, Gemmieza 11, Line 3 and F1's crosses (Sahel 1 X Gemmeiza 11) and (Misr 1 X Line-1) (Gemmieza 9 x Line 3) were more sensitive ones. Moreover, Both Line 1, Line 2 were moderate tolerant to drought stress. Both additive and dominance gene action were significant and involved in the inheritance of both earliness and grain yield/plant. Performance of parental Egyptian bread wheat cultivar Gemmeiza 11 and F1 crosses (Line 1 × Gemmeiza 11), (Misr 1 × Line 2) and (Gemmeiza 9 × Gemmeiza 11) showed drought sensitivity index (DSI)

Table 3.3 Drought sensitivity index (DSI) of grain yield for some bread wheat genotypes under surface and sprinkler irrigation systems (Modified after Ali and Abdul-Hamid 2017)

Genotypes	Surface irrigation		Sprinkler irrigation
	Season 2013/2014	Season 2015/2016	
Line 7	0.68	0.89	0.81
Line 9	0.75	0.89	0.72
Line 15	0.87	0.56	0.86
Line 16	1.02	1.07	0.94
Giza 168	1.06	1.10	1.23
Line 31	0.86	0.75	0.80
Line 32	1.23	1.23	1.12
Line 33	1.10	1.16	1.24
Sahel 1	1.21	1.06	1.00
Misr 1	0.96	1.23	1.13

as respects to their grain yield/ plant were less than unity. Thus, these genotypes were considered as more tolerant to drought stress. Moreover, parental wheat cultivars Misr 1 and Gemmeiza 9 and F1 crosses (Line 1 × Gemmeiza 9) and (Misr 1 × Gemmeiza 9) had DSI values near one, so these genotypes were considered as moderate tolerant. Otherwise, Line 1 and Line 2 and F1 crosses (Line 1 × Misr 1), (Line 1 × Line 2), (Misr 1 × Gemmeiza 11) and (Line 2 × Gemmeiza 9) had DSI values more than 1.0, hence previous genotypes were sensitive to drought stress. Furthermore, Ali and Abdul-Hamid (2017) estimated drought sensitivity index and revealed that, Line 7, Line 9, Line 15 and Line 31 exhibited DSI < 1 under both surface and sprinkler irrigation. Therefore, they are more tolerant to drought stress. On the other side, the wheat genotypes Line 16, Line 33, Giza 168 and Sahel 1 were moderately tolerant to drought conditions. Whereas, Misr 1 changed in its behavior from one season to another and from irrigation system to another and Line 32 was sensitive one as presented in Table 3.3.

3.8 Genetic Diversity and Sources of Water Stress Tolerant Genotypes

Crop varieties vary in their ability to tolerate drought conditions and water deficit, some are tolerant and others are sensitive. The sensitivity of the same variety varies according to the stage of its growth. Genetic variability between local and exotic varieties, germplasm collections, relatives, and wild species of crop carrying drought-resistant genes are important as genetic resources of resistance genes to commercial cultivars. The following is an explanation of the most important variations and sources of drought resistance in the wheat crop.

Under Egyptian circumstance, wheat varieties exhibited substantial genetic variability in their tolerance to drought conditions. Local bread wheat varieties Sids 1, Sids 12 Sids 13, Sids 14, Sakha 93, Sakha 94, Sakha 95, Gemmeiza 3, Gemmeiza 5, Gemmeiza 7, Gemmeiza 9, Gemmeiza 10, Gemmeiza 11 and Gemmeiza 12, Giza 168, Giza 171 and Shandweel 1 are characterized by tolerance to water stress conditions. Also, local durum or macaroni wheat varieties Sohag 2, Sohag 4, Sohag 5, Beni Swif 1, Beni Swif 4, Beni Swif 5, and Beni Swif 6 are classified as more tolerant to water deficit conditions. On the other hand, the varieties Giza 162 and Sakha 61 are less tolerant (Anonymous 2020).

Newly fifth generation strains characterized by high tolerant to abiotic stresses has been developed by Desert Research Institute of the Arab Republic of Egypt, namely Mariout 3, Siwa Oasis 18 and Siwa Oasis 25 compared to the Mariout 7, Mariout 16, Mariout 20, Mariout 22 and Sakha 8 (Afiah and Darwish 2003). Whereas, in another study, Sakha 8 and the imported strains of Omtel-1, Mrbll were more early and superior in grain yield and leaf content of proline and chlorophyll content (A + B) with values of drought sensitivity index (DSl < 1) compared to the Portuguese genotype Korifla and their hybrids Korifla x Mrbll, Korifla x Sakha 69 (Amar 2003). In continuous, Rashed et al. (2010) found that wheat grain yield of the tolerant parent (Sahel 1) and the F1 plants displayed higher mean values than the sensitive parent line 13, wich recorded a high reduction in productivity under drought condition. Thus, the parent Sahel 1 and their F1 plants might be expressed as more tolerant to drought condition. They identified two negative and four positive RAPD markers can be expressed as reliable markers for water stress tolerance in Egyptian wheat genotypes.

Qabil (2017) identified parental bread wheat cultivar Gemmeiza 11 and F1 crosses (Line 1 × Gemmeiza 11), (Misr 1 × Line 2) and (Gemmeiza 9 × Gemmeiza 11) as more tolerant to drought stress as respects to their grain yield/ plant (DSI > 1(. Besides, parental wheat cultivars Misr 1 and Gemmeiza 9 and F1 crosses (Line 1 × Gemmeiza 9) and (Misr 1 × Gemmeiza 9) had DSI values near unity, thus these genotypes were considered as moderate tolerant to drought stress. On the contrary, Line 1 and Line 2 and F1 crosses (Line 1 × Misr 1), (Line 1 × Line 2), (Misr 1 × Gemmeiza 11) and (Line 2 × Gemmeiza 9) exhibited DSI < 1, therefore these genotypes were sensitive to drought stress.

Molecular markers have been exploited in breeding programs and detected higher percentage of polymorphisms among wheat genotypes (Abdel-Tawab et al. 2003, Awaad et al. 2010 and Al-Naggar et al. 2015).

Eight genotypes of wheat (three parents *i.e.* Sids 12, Sakha 94 and Gemmeiza 9 and five promising derivatives after mutagenesis by Gamma rays, LASER and EMS) were tested under stress environment where, soil texture at 1-25 cm was sandy loam and at 25-50 cm was sandy of Experimental Farm of the Plant Research Department, Nuclear Research Center, Atomic Energy Authority, Inchas, Egypt under. Abaza (2017) revealed high percentage of genetic diversity between wheat genotypes based on similarity matrix. Results in Table 3.4 showed that the highest similarity was observed between the cultivar Sk-94 and Sd-12 with similarity coefficient value of

Table 3.4 The similarity values between the different wheat genotypes (Abaza 2017)

	Sk-94 cont.	Sk-94 400 Gy	Sd-12 cont.	Gm-9 250 Gy	Gm-9 1 h LASER	Gm- 9 cont.	Sd-12 0.4 EMS	Sk-94 2 h LASER
Sk-94 cont.	1.00							
Sk-94 cont.	0.82	1.00						
Sk-94 cont.	0.94	0.88	1.00					
Sk-94 cont.	0.64	0.58	0.70	1.00				
Sk-94 cont.	0.88	0.82	0.94	0.76	1.00			
Sk-94 cont.	0.76	0.82	0.82	0.64	0.88	1.00		
Sk-94 cont.	0.70	0.64	0.76	0.82	0.82	0.82	1.00	
Sk-94 cont.	0.76	0.58	0.70	0.52	0.76	0.64	0.58	1.00

0.94, while the highest genetic diversity coefficient was obtained by the genotypes Sk-94 2 h LASER and Gm-9 250 Gy with value of 0.52.

The dendrogram resulting from the UPGMA cluster analysis illustrated that the evaluated genotypes could be divided into two main clusters (Fig. 3.3). The first cluster contained only one genotype called Sk-94 2 h LASER. The second cluster was divided into two sub-clusters, the first sub-cluster contained two genotypes (Gm-9 250 Gy and Sd-12 0.4 EMS). The second sub cluster contained 5 genotypes (Sd-12 Cont., Sk-94 Cont., Gm-9 cont., Sk-94 400 Gy and Gm-9 1 h LASER).

3.9 Breeding Efforts

The choice of parents with the desired characteristics and crossing between them and the evaluation of morpho-physiological and crop yielding indicators under target environment are considered the best ways to judge the degree of adaptation of breeding materials. The breeder in the breeding programs follows the method of selection, hybridization and mutations.

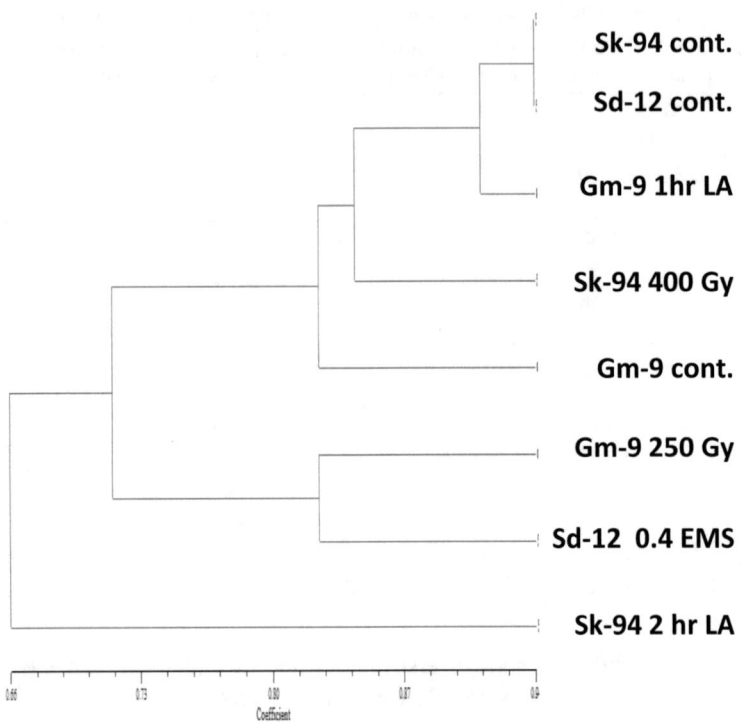

Fig. 3.3 A dendrogram based UPGMA cluster analysis showing the relationships between the different genotypes after mutagen treatments (Abaza 2017)

3.9.1   Breeding Strategies

Several strategies can be followed when the breeder wants to improve cultivars under water stress conditions, they can be listed as follows:

The first strategy:

Introgression of gene or genes of simple inherited desired trait associated with drought tolerance to cultivated adapted varieties by backcrossing technique.

The second strategy:

The development of multiple crosses by hybridization of the local adapted and high yield variety with non-adapted drought tolerant lines, or hybridization of one non-adaptable drought-tolerant parent with several local adapted varieties to produce multiple crosses high yielding ability and drought tolerant.

The third strategy:

The hybridization of the local cultivar, stable in yield with drought-tolerant genotype. Then practiced the simultaneous selection of segregated generations to produce genetic recombinations combining high yield and drought tolerance. This is important strategy in this area (Rajaram et al. 1996).

Under Egyptian conditions, plant breeders produced through selection, and hybridization a collection of bread wheat cultivars that tolerant to different environmental conditions such as Giza 168, Giza 171, Gemmeiza 11, Gemmeiza 12, Sids 1, Sids 6, Sids 12, Sids 13, Sids 14, Sakha 94, Sakha 95, Misr 1, Misr 2 and Misr 3 as well as durum wheat varieties *i.e.* Sohag 3, Beni Suef 1 and Beni Suef 3.

3.9.2 Mutations

The breeder sometimes uses physical or chemical mutagens or a combination of the two to produce new genetic recombination that can be exploited as parents in breeding programs or release of improved new cultivars.

For the past fifty years, the Food and Agriculture Organization of the United Nations (FAO) and the International Atomic Energy Agency (IAEA) have played a decisive role in supportive their Member States in using induced mutations to develop new crop varieties. The fruitful application of gamma rays and further physical and chemical mutagens in crop breeding over the past 90 years has improved crop biodiversity and yield potentialty across the world. Plant biotechnologies are vital to the operative application of mutation breeding procedures, to produce good adapted genotypes to climate change. The use of mutation induction joined with biotechnologies, genomics and molecular marker methods can speed breeding programs, through select desired mutants and rapid multiplication (FAO/IAEA 2018).

Numerous genotypes were produced, which included many economic crops such as wheat, barley, rice, cotton, and others. Under the Egyptian conditions, some mutants have been developed by gamma rays like Mutant 7, mutant 8 and mutant 12 from Giza 164 variety, and Mutant 19 from Sahel 1. The Mutants 7 and 19 were characterized by high water use efficincy and drought tolerance (Sobieh and Ragab 2005). Moreover, Awaad et al. (2018) produced promising M_3 wheat mutants *i.e.*, Sd-12 0.3EMS and Sd-12 0.4EMS, Sk-94 350 Gy, Sk-94 400 Gy, Sk-94 2 h LASER and Gm-9 0.3EMS were high in productivity, flag leaf chlorophyll content and flag leaf area.

3.10 Role of Biotechnology in Improving Water Stress Tolerance

3.10.1 Molecular Markers

Molecular markers and identification of drought tolerant genes is an important step in the breeding program. Marker-assisted selection utilizes genetic markers to identify polymorphism and genetic differentiation among desirable wheat genotypes. The success relies on choosing accurate marker-trait associations. DNA markers are proved to be powerful tools to evaluate the genetic diversity, identification and testing genetic purity in several crops such wheat (Selim et al. 2010, Hussein et al. 2013 and Suprasanna and Jain 2017; Tiegu et al. 2007). Successful applications have been achieved in improving wheat tolerance to drought conditions. In this concern, Al-Naggar et al. (2004bb) utilized RAPD markers to characterize the genetic differences among six Egyptian bread wheat genotypes, three drought variants (V1, V2 and V3) and three Egyptian cultivars (Sakha 8, Giza 164 and Sids 1). The highest genetic similarity (96.0%) was detected between Sids 1 and the waxy mutant (Variant 1) while the lowest (86.4%) was found between Variant 3 and its maternal parent (Sakha 8). The waxy mutant (V1) was identified by one positive single marker amplified by OPB-15 primer at the molecular weight of 580.388 bp. Al-Naggar et al. (2015) selected twelve families included 7 M_3 families; two (SF2 -Sakha-93 and SF3-Giza 168) were selected under water stress condition WS, and five (SF1-Aseel-5, SF4-Giza-168, SF5-Giza-168, SF6- Sahel-1 and SF7-Sahel-1) selected under well water. They obtained three families (SF9, SF10 and SF11) chosen under WS, from the F2 of Sd4 X Mr5, Sk61 XAs5 and Sk61 X Sk93, individually. These genotypes characterized by high yielding, drought tolerant with low yield decrease due to water stress. They utilized 15 SSR primers for PCR amplification of the 12 selections and their parents. The SSR technique showed that the 12 families are genetically different from their 7 parents, with 86.67% polymorphism. The genetic diversity varied from 30% to 88%. Both mutants SF3 and SF4 displayed very high genetic diversity (42 and 40%, respectively) with their common parent (Giza-168). Suggesting that gamma rays were effective in inducing genetic variation of Giza168 towards high grain yield/plant under WS conditions. SSR assay identified seven unique bands i.e. 5 positive and 2 negative for three drought tolerant genotypes SF3, SF4 and Aseel-5. These bands could be utilized as markers linked to drought tolerance in bread wheat improvement programs. Haiba et al. (2016) evaluated six durum wheat cultivars for drought. Drought sensitivity index showed that Bani-swaif 4 was the best drought tolerant variety, while Sohag 3 was the most sensitive one. The tolerant, sensitive plants and their F1 and F2 were used for molecular signs of drought tolerance using RAPD and ISSR techniques. RAPD technique exhibited 4 positive and 5 negative markers while ISSR revealed 6 positive and 6 negative markers. Therefore, these markers might be considered as dependable markers of drought tolerance in durum wheat.

Furthermore, Abaza (2017) tested some bread wheat genotypes using PCR based RAPD analysis using six primers as illustrated in Table 3.5. Primer OBE-04 generated two bands with 100% polymorphism. The size of bands ranged between100-200 bp. Both Primers OBF-06 and OPB-19 produced three fragments which were polymorphic with 100% polymorphism. The size of bands for OBF-06 and OPB-19 ranged between100-600 and 150-550 bp, respectively. Primer OBP-07 gave two bands which varied between 150 and 500 bp. No polymorphic band was detected with zero% polymorphism. Figure 3.4 showed the pattern of amplification product of primer OBP-07. Also, primer OPB-08, gave three fragments ranged between 200-600 bp, but not gave any polymorphic fragments with zero% polymorphism. Whereas, the highest number of bands (4) was generated from primer OBP-09, and the four bands

Table 3.5 RAPD analysis of wheat mutant genotypes using six primers (Abaza 2017)

Primer	Size of bands (bp)	Total number of bands	Number of polymorphic bands	Percentage of polymorphic bands
OPE-04	100–200	2	2	100
OPF-06	100–600	3	3	100
OPB-07	150–500	2	0	0
OPB-08	200–600	3	0	0
OPB-09	100–600	4	4	100
OPB-19	150–550	3	3	100
Total		17	12	71

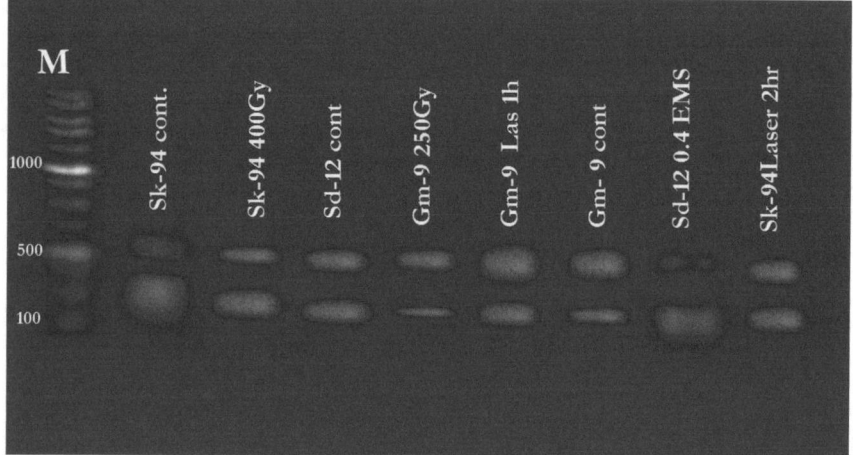

Fig. 3.4 RAPD pattern obtained by primer OPE-07; M (100 Base plus DNA marker *Ladder*) (Abaza 2017)

were polymorphic with 100% polymorphism. The size of bands ranged between 100 and 600 bp.

Accordingly, the six primers produced 17 amplified fragments among them 12 were found polymorphic with 71% polymorphism. The number of polymorphic bands per locus ranged from zero (primer OBP-07 and OBP-08) to 4 (primer OBF-09) with an average number of 2.0 bands per locus. In this respect, Basyouny (2010) demonstrated that primers produced reliable and reproducible banding pattern and that the number, size of amplified DNA fragments and polymorphic bands varied among primers.

3.10.2 Gene Transfer Technology

Using recent techniques and gene transfer technology, it has been possible to produce drought-tolerant plants. The Plant Research Institute in California, USA, has been transfer drought resistance from drought-tolerant wheat relatives to the cultivated wheat genotypes.

Egyptian scientists have also succeeded in producing drought-tolerant genetically modified wheat through transferring a gene from barley. The resulting plants were characterized by increased yield and plant height in field assessment evaluations relative to the local variety. The study recommended the importance of introducing these genotypes into breeding programs to drought tolerance in the Mediterranean region as well as agriculture in desert lands (Bahieldin 2004).

The wheat program focuses on the evaluation of transgenic wheat lines in field trials under drought stress. Scientists in Agricultural Genetic Engineering Research Institute AGERI have released drought-tolerant wheat by transferring the barley HVA1 gene into wheat varieties. In 2009, the National Biosafety Committee (NBC) approved the field trails of transgenic wheat lines. Currently, the novel lines were incorporated in the national wheat-breeding program of Agriculture Research Center, Egypt for supplementary field testing and seed multiplication (Wally 2016).

3.11 Agricultural Procedures to Support Wheat Productivity Under Water Stress Conditions

The effects of water stress hazard can be mitigated through the subsequent agricultural practices, to mitigate the effects of water stress on the critical periods of the wheat growth.

3.11.1 Land Leveling

Laser leveling is a procedure of smoothing the surface of the land ($\pm$ 2 cm) from average elevation by using laser-equipped drag buckets. This practice is well famous for achieving greater levels of accuracy in leveling of the land and offers higher possible for water savings and greater grain yields (Jat et al. 2006). It increases yield, improves homogeneity of crop maturity and diminishes weeds and water required for soil preparation. When laser leveling is applied in different crops and crop patterns, led to water savings of 15-30% (Anonymous 2009). Under Nubarya region as sandy soil stress environment, Egypt, Abdelraouf et al. (2014) tried factors of shortage irrigation (100% Irrigation Requirements "IR", 80% IR, 60% IR and 40% IR) and land leveling practices (conventional and laser). Interaction effect between land leveling and irrigation on irrigation water use efficiency (IWUE) of wheat showed that the maximum values were noticed at adding 100% irrigation requirements x Laser. However, insignificant difference was detected between 100% irrigation requirements x laser > 80% irrigation requirements x laser > and 60% irrigation requirements x laser. Consequently, it can save 40% of irrigation water through adding 60% irrigation requirements with laser land leveling system to irrigate wheat under conditions of sandy soil environments.

3.11.2   Raised Bed Method

Planting method is one of the vital agricultural practices which affect wheat productivity. Raised bed is widely used in developed countries as an advanced system of productivity. Raised bed rises crop yield by 10-20% with the suitable cultivar, saved 30-40% of irrigation water, reduced seed rate, promoted higher nitrogen use efficiency, lessens production cost over the conventional method (BARI 2006). In farmers' fields, Zagazig district, Egypt, sowing wheat on raised bed method of 120 cm width gave the best saving (14.5%) in irrigation water more than the treatment of 75 cm raised bed (Swelam et al. 2015).

3.11.3   Cultivate Tolerant Varieties

Attention must be given to choose the appropriate varieties, tolerant to water stress in areas expected to be cultivated under Egyptian conditions. A range of varieties, whether bread or macaroni wheat, referred to in Sect. 7, can tolerate water stress. It also has satisfactory yield levels, high technological properties and disease resistance.

3.11.4   Sowing Date

Sowing date is one of the most vital agricultural determinants of wheat crop production, due to its effect on sequence of vegetative and reproductive stages. Adjust sowing date in wheat is an important agricultural procedures under sustainable agriculture to maintain yield stability levels under environmental variables and water scarcity.

Substantial reduction in wheat grain yield and its contributors has been registered with delaying sowing date due to terminal drought stress, which reduces grain filling period (Awaad 2001).

To cope with the effects of adverse effects, Hozayn and Abd El-Monem (2010) studied tolerance of wheat cultivar (Sakha-93) to two late sowing (23/12 and 23/1) dates, compared to the optimum date (23/11) as influenced by arginine (0.0, 2.5 and 5.0 mM) in Egypt. Delayed sowing produced manifest decrease in biological and economic yield, by decreasing spike length and weight, spike grain weight, spikes/m^3, and 1000-grain weight. The decrease in economic yield/fed reached to 10.39 and 41.22% when delaying sowing wheat to 23/12 and 23/1, respectively relative to optimum date.

According to the technical recommendations under Egyptian circumstances, optimum performance of growth, grain yield its contributing were obtained when wheat genotypes grown at optimum sowing date i.e. from 15 to 30 November in Marine face and from 10 to 25 November in Upper Egypt. Where the early or delay of those dates led to yield reduction by about 25% (Anonymous 2020).

3.11.5 Plant Density

Plant density plays a significant role in the wheat productivity, where the main stem and its spike are the mainstay of the yield under sandy soil stress conditions, where plants are subjected to moisture stress.

In order to avoid the production of tillers, which often fail to produce spikes for insufficient assimilates represented in meeting the needs under these circumstances. Therefore, the number of productive tillers is of great importance in the crop yield. So, it is recommended to cultivate wheat under sandy areas that suffer from stress conditions at a seeding rate of more than 70 kg/fed. where the wheat grain yield was increased as a result of the large increase in the number of spikes per unit area with the addition of 20 m^3 farm yard manure and 120 kg N/ fed. in splits (Anonymous 2020).

3.11.6 Fertilization

Wheat production depends on the availability of essential nutrients for plant growth. The supply of wheat plant by the necessary elements is one of the main factors affecting the growth and productivity of the crop, especially under poor land conditions. Abdelkhalek et al. (2015) found that yield of Misr 1 wheat cultivar was decreased with increasing irrigation events and nitrogen levels, and reached the highest estimates at three irrigation treatments (3 events) and 90 kg N/fed. Irrigating wheat 4 events through growing seasons and apply of 75 kg N/fed. in the form of ammonia provided the maximum values of yield and its components under North Nile Delta of Egypt. They added that, Misr-1 cultivar was distinguished and produce the highest values of yield attributes and yield response to water factor (K_y) followed by Misr-2, but Sakha-94 exhibited the lowermost values in all measured characters. So, Misr-1 cultivar appeared to be more tolerant to water stress followed by Misr-2 and Sakha-94.

Mohamed, Manal et al. (2016) revealed that foliar spraying wheat cultivar Misr 1 with either potassium K or zinc Zn significantly increased growth characters, yield and yield components as well as nutrient concentration of wheat leaves, straw and grains as compared to control plants. Moreover, the highest observations were obtained by the dual spraying of both zinc and potassium at high level of K (1000 ppm) and Zn (1000 ppm).

Ali (2017) evaluated 16 bread wheat genotypes under El-Khattara region represents sandy soil condition at three nitrogen levels (50, 80 and 120 kg/fed) splitted in six equal doses with sprinkler irrigation system. They applied recommended dose of phosphate and potassium fertilizer at rates of 150 kg/fed (15.5% P2O5) and 50 kg/fed (48% K2O), respectively before sowing for phosphate fertilizer and after 20 days from sowing for potassium fertilizer. The ideal genotypes based on their response to the studied environments were Gemmeiza 7, Sakh 93 and Line 8 for earliness, Line 2, Line 7, Line 8 and Gemmeiza 9 for plant height, Line 4 Line 7, Gemmeiza 7 and Gemmeiza 9 for 1000- grain weight, Line 1, Giza 168, Gemmeiza 9 and Misr 1 for grain yield (ard./fed.) and biological yield (tonnes/fed.). These genotypes were beneficial in wheat breeding programs aiming to improve previous characters under soil fertility stress in newly cultivated sandy soils. In continuous, in sandy soils under the sprinkler irrigation system, Anonymous (2020) refer to wheat fertilization at a rate of 200 kg of super phosphate (15% P2O5)/ fed. before planting and during ground service. Also 50 kg potassium sulfate (48% P2O)/ fed added after a month of cultivation. Also, 120 kg/ N in the form of ammonium nitrate 33.5% nitrogen or ammonium sulfate 20.6% nitrogen splitting into 9-10 equal doses. Boron application significantly increased growth traits under water stress level (50% from the amount of water consumption for wheat) rather than B-untreated plants and get higher yield production and so, better economic profits of wheat (Abdel-Motagally and Manal El-Zohri 2018).

3.11.7 Irrigation System

Irrigation systems play an critical role in improving the efficiency of water use under limited water supply. For example, subsurface drip irrigation system increase water use efficiency as a result of low evaporation of the soil surface, where only the root or partial root areas are irrigated. In contrast, in sprinkler irrigation system, moisturizing the entire field area (Mansour et al. 2014). On the other hand, surface irrigation causes significant loss of water. Under sandy loam soil, Nubaria, Behaira Governorate, Egypt, Mansour and Abd El-Hady (2014) revealed that subsurface drip irrigation at 20 cm has a promote effect on wheat grain and straw yield of wheat cultivar Gemmeiza 9, with increase value of 6.9 and 5.7%, respectively as compared to surface drip system. Whereas the increase percentage was 1.7 and 1.8% comparing subsurface drip irrigation at 10 cm with surface drip irrigation system for both grain and straw yield, respectively. Furthermore, in comparison among 29 bread wheat lines (*Triticum aestivum* L.), 5 durum wheat lines (*Triticum durum* L.) and 4 commercial check varieties under different irrigation systems for drought stress. Under El-Kattara and Ghazalla regions of Egypt, Ali and Abdul-Hamid (2017) showed that wheat genotypes exhibited higher grain yield under drip irrigation than sprinkler and surface flood irrigation systems.

3.11.8   Humic and Amino Acids Treatment

Use of organic fertilizer and humic acid were proposed as alternatives to inorganic fertilization and a rapid source of nitrogen, in addition to their role in improving stress tolerance, enhance soil physical properties and complex metalions and improve plant growth and yield (Khan et al. 2010). Asal et al. (2015) reported that application of humic acid improved root growth and that was directly correlated with enhanced uptake of macro and micronutrients.

To mitigate the adverse effects on wheat yield in Egypt, Hozayn and Abd El-Monem (2010) suggest the role of arginine (0.0, 2.5 and 5.0 mM) in improving the tolerance of cultivar Sakha-93 to two late sowing (23rd of December. and 23rd of January.) dates, in comparison of the optimum date (23rd of November). Foliar application of arginine with 2.5 and 5.0 mM at normal or delayed sowing date displayed significant increase in yield and its components compared to the control. The increases were more obvious at 2.5 mM of arginine which induce increases by 19.23, 20.53 and 25.51% in economic yield at normally, 30 and 60 days delay, respectively. Also, results shows that, 2.5 mM arginine encourage 8.0% increase in grain yield over the plants sowing late at 23rd Dec. and reduce the reduction in grain yield from 41.22 to 26.22% at 23rd of Jan. sowing date. So, arginine could mitigate the adverse effect of climate change by late sowing of wheat and decrease the expected reduction of economic yield under semi-arid regions of irrigated agriculture. To decrease hazards of water stress on wheat cultivar Sakha 94, Hammad Salwa and Ali (2014) revealed that addition of amino acids and yeast extract increased relative water content, photosynthetic pigments, total soluble sugars, total carbohydrates, total free amino acids, enzyme activities, minerals (NPK % and uptakes), yield, its contributing characters and grain quality.

At Siwa Oasis, Matrouh governorate, Egypt, Attia and Shaalan (2015) study the response of wheat cultivar Sakha 94 to humic acid rates (8.3, 10.7 and 13.1 kg/ha) and organic fertilizer rates (35, 47.5 and 60 m^3/ha) at varying Siwa conditions. Combined analyses showed significant differences between locations, organic fertil-izer rates and applied humic acid rates for plant height, number of spikes/m^2, number of grains/spike, biological yield and harvest index. Increasing humic acid rate from 8.3 to 10.7 and from 8.3 to 13.1 resulted in an increase in grain yield by 7.1 and 13.6%, respectively. Furthermore, under newly reclaimed sandy saline soil condi-tions, Egypt, Kandil et al. (2016) showed that foliar spraying of wheat cultivars with mixture of humic and amino acids gave the highest estimates of yield attributes and increased grain and straw yields, protein and carbohydrates contents in grains by 23.29, 7.50, 10.98 and 78.15%, respectively in comparison of the control.

3.12 Conclusions

Water stress affect wheat productivity in several regions of the world, and caused yield reduction by about 25%. In the light of the increase in Egyptian population and extension of agriculture to marginal lands where wheat plants face the effects of water shortage. Hence, approaches should be advanced to manage with the climate change for mitigating the harmful effects of water stress on wheat production. There are several practical options for adapting to water stress. The first option is exploited genetic diversity and sources of drought tolerance in releasing new cultivars. The second option is focused on biotechnology in improving water stress tolerance by using molecular markers and gene transfer technology. The third option is adjusted agricultural techniques with meteorological data, and follows appropriate fertigation programs to avoid the injurious effects of water stress.

3.13 Recommendations

The effects of water stress on wheat plants can be reduced through the following procedures:

- Exploited genetic diversity and sources of drought tolerance in releasing new cultivars
- Cultivate the most tolerate genotypes to water stress.
- Application the appropriate agricultural procedures
- Taking into account the critical periods of wheat life and avoid stress through irrigation.
- Application the appropriate fertilization programs and irrigation systems.

References

Abaza GMShM (2017) Mutation induction in bread wheat using gamma and laser rays. PhD, Agronomy Department, Faculty of Agriculture, Zagazig University, Egypt

Abd El-Aty MSM, El-Borhamy HS (2007) Estimates of combining ability and sensitivity index in wheat diallels crosses under stress and normal irrigation treatments. Egyptian J of Plant Breed 11 (2):651–667, Special Issue, Proceed Fifth Plant Breeding Conference, May 27, 2007 (Giza)

Abd El-Mohsen DA (2015) Yield stability of some wheat genotypes under normal and water stress conditions. M Sc Thesis, Agron Department Faculty of Agriculture, Zagazig University, Egypt

Abdelkhalek AA, Darwesh RKh, El-Mansoury Mona AM (2015) Response of some wheat varieties to irrigation and nitrogen fertilization using ammonia gas in North Nile Delta region. Ann Agric Sci 60(2):245–256

Abdel-Motagally FMF, El-Zohri Manal (2018) Improvement of wheat yield grown under drought stress by boron foliar application at different growth stages. J Saudi Soc Agric Sci 17(2):178–185

Abdelraouf RE, Mohamed MH, Pipars Sabreen Kh, Bakry Bakry A (2014) Impact of laser land leveling on water productivity of wheat under deficit irrigation conditions. Curr Res Agric Sci 1(2):53–64

Abdel-Tawab FM, Fahmy Eman M, Bahieldin A, Mahmoud Asmahan A, Mahfouz HT, Eissa Hala F, Moseilhy O (2003) Marker assisted selection for drought tolerance in Egyptian bread wheat (*Triticum aestivum* L.). Egypt J Genet Cytol 32:34–65

Adamski NM, Bush MS, Simmonds J, Turner AS, Mugford SG, Jones A et al (2013) The inhibitor of wax 1 locus (Iw1) prevents formation of β- and OH-β-diketones in wheat cuticular waxes and maps to a sub-cM interval on chromosome arm 2BS. Plant J 74:989–1002. https://doi.org/10.1111/tpj.12185

Afiah SAN, Darwish IHI (2003) Response of selected F_5 breed wheat lines under abiotic stress conditions. In: Proceedings Third P1 Breed Conference April 26 (2003), (Giza). Egypt J Plant Breed 7(1):181–193

Akram M, Iqbal RM, Jamil M (2014) The response of wheat (*Triticm aestivum* L.) to integrating effects of drought stress and nitrogen management. Bulgarian J Agricultural Science 20(2):275–286

Al-Bakry MRI (2007) Glaucous wheat mutants. I. Agronomic performance and epicuticular wax content. Egypt J Plant Breed 11(1):1–9

Ali MMA (2017) Stability analysis of bread wheat genotypes under different nitrogen levels. J Plant Prod, Mansoura Univ 8(2):261–275

Ali MMA, Abdul-Hamid MEI (2017) Yield stability of wheat under some drought and sowing dates environments in different irrigation systems. Zagazig J Agric Res 44(3):865–886

Al-Naggar AMM, Al-Azab KhF, Sobieh SES, Atta MMM (2015) Morphological and SSR assessment of putative drought tolerant M_3 and F_3 families of wheat (*Triticum aestivum* L.). Br Biotechnol J 6(4):174–190

Al-Naggar AMM, Ragab AEI, Youssef SS, Al-Bakry RIM (2004) New genetic variation in drought tolerance induced via irradiation and hybridization of Egyptian cultivars of bread wheat. Egypt J Plant Breed 8:353–370

Al–Naggar AMM, Youssef SS, Ragab AEI, Al- Bakry MR (2004b) RAPD assessment of new drought tolerant varieties derived via irradiation and hybridization of some Egyptian wheat cultivars. Egypt J Plant Breed 8:255–271

Al-Naggar AMM, Shehab-Eldeen MT (2012) Predicted and actual gain from selection for early maturing and high yielding wheat genotypes under water stress conditions. Egypt J Plant Breed 16(3):73–92

Ammar SElMM (2003) Estimates of genetic variance for yield and its components in wheat under normal and drought conditions. Egypt J Plant Breed 7(2):93–110

Anonymous (2009) Conservation Agriculture: Resource productivity and efficiency. IVth World Congress on Conservation Agriculture, New Delhi, PACA, 1st Floor, NASC Complex, DPS Marg, Pusa, New Delhi—110 012 INDIA, 2009 Available www.conserveagri.org, 2009

Anonymous (2017) Wheat Production and Consumption Economic, Affairs sector Agricultural Research Center, Giza, Egypt

Anonymous (2020) Recommendations Techniques in Wheat Cultivation. Agricultural Research Center, Giza, Egypt

Asal MW, Badr EA, Ibrahim OM, Ghalab EG (2015) Can humic acid replace part of the applied mineral fertilizers? A study on two wheat cultivars grown under calcareous soil conditions. Int J ChemTech Res 8(9):20–26

Ata A, Yousaf B, Khan AS, Subhani GhM, Asadullah HM, Yousaf A (2014) Correlation and path coefficient analysis for important plant attributes of spring wheat under normal and drought stress conditions. J Nat Sci Res 4(8):66–73

Atta YI, Swelam AA (2012) Effect of water deficit on wheat yield and some water relations in sandy soil at east- delta region. Egypt J of Applied Sci 27(12B):570–591

Attia MA, Shaalan AM (2015) Response of wheat 'Triticum aestivum L.' to humic acid and organic fertilizer application under varying Siwa Oasis conditions. J Agric VetY Sci 9(9): 81–86

Austin RB, Hanson, Quarrie SA (1982) Abscisic acid and drought resistance in wheat, millet and rice. In: Drought Resistance in crops with emphasis on rice. Los Banos, Philippines, IRRI, pp 171–180

Awaad HA (2001) The relative importance and inheritance of grain filling rate and period and some related characters to grain yield of bread wheat *(Triticum aestivum L.)* In: Proceedings The Second Pl Breed Conference, October 2 (Assiut Univ), pp 181–198

Awaad HA (2009) Genetics and breeding crops for environmental stress tolerance, I: drought, heat stress and environmental pollutants. Egyptian Library, Egypt

Awaad HA, Attia ZMA, Abdel-lateif KS, Gomaa MA, Abaza Ghada MShM (2018) Genetic improvement assessment of morpho-physiological and yield characters in M_3 mutants of bread wheat. Menoufia J Agric Biotechnol 3:29–45

Awaad HA, Yousef MAH, Moustafa ESA (2010) Identification of genetic variation among bread wheat genotypes for lead tolerance using morpho—physiological and molecular markers. J Am Sci 6(10):1142–1153

Bahieldin A (2004) Egyptian scientists produce drought–tolerant GM wheat. SciDev- Net 14 Oct 2004

Baker FG (1989) Drought resistance in cereals. CAB International UK

Balasubramaniyan P, Palaniappan SP (2001) Principals and practices of agronomy. Agrobios India, New Delhi

BARI (Bangladesh Agricultural Research Institute) (2006) Krushi Projukti Hatobi (Hand book of Agrotechnology) (4th Edn). Bangladesh Agric Res Ins Joydevpur, Gazipur, pp 9–15

Basyouny MAL (2010) Implication of gamma rays and in vitro techniques for improving tolerance of durum wheat to some stress conditions. M Sc Thesis, Department of Agricultural Botany, Faculty of Agriculture, Cairo University, Egypt

Blum A (1985) Photosynthesis and transpiration in leaves and ears of wheat and barley varieties. J Exp Bot 36:432–440

Blum A (1986) The effect of heat stress on wheat leaf and ear photosynthesis. J Exp Bot 37:111–118

Blum A, Pnuel Y (1990) Physiological attributes associated with drought resistance on wheat cultivars in a Mediterranean environment. Aust J Agric Res 41:799–810

Blum A, Sinmena B, Ziv O (1980) An evaluation of seed and seedling drought tolerance tests in wheat. Euphytica 29:727–736

Boyer SJ (1996) Advances in drought tolerance in plants. Adv Agron 56:187–218

Changhai S, Baodi D, Yunzhou Q, Yuxin L, Lei S, Mengyu L, Haipei L (2010) Physiological regulation of high transpiration efficiency in winter wheat under drought conditions. Plant Soil Environ 56(7):340–347

Ciulca S, Mados E, Ciulca A, Sumalan R, Lugojan C (2017) The assessment of cell membrane stability as an indicator of drought tolerance in wheat. In: 17th International Multidisciplinary Scientific GeoConference SGEM 2017. www.sgem.org. SGEM2017 Conference Proceedings, ISBN 978-619-7408-12-6 / ISSN 1314-2704, 29 June–5 July, 2017, 17 (61):1097–1104

Clarke JM, Toenly-Smith TF (1984) Screening and selection techniques for improving drought resistance. In: Vose PB, Blixt SG (eds) Crop breeding a contemporary basis. Pergamon pr NY, pp 137–162

Cueuz L, Erdei L (1996) Improving the drought tolerance of winter wheat in breeding program. In: 5th International Wheat Conference June 10–14, Ankara, Turkey

Darwish IHI, El-Hosary AA (2003) Response of wheat verities to nitrogen under rainfed and irrigation. Egypt J Plant Breed 7(1):241–251

Daryanto S, Wang L, Jacinthe PA (2016) Global synthesis of drought effects on maize and wheat production. PLoS ONE 11(5):e0156362

Davies WJ, Tardieu F, Trejo CL (1994) How do chemical signals work in plants that grow in drying soil? Plant Physiol 104:309–314

Ding H, Liu D, Liu X, Li Y, Kang J, Lv J, Wang G (2018) Photosynthetic and stomatal traits of spike and flag leaf of winter wheat *(Triticum aestivum* L.) under water deficit. Photosynthetica 56 (2):687–697

Eftekhari A, Baghizadeh A, Abdolshahi A, Yaghoobi MM (2016) Analysis of physiological traits and grain yield in bread wheat under drought stress conditions. An Int J 8(2):305–317

El-Basyoni ID, Saadalla M, Baenziger S, Bockelman H, Morsy S (2017) Article cell membrane stability and association mapping for drought and heat tolerance in a worldwide wheat collection. Sustain 9(1606):2–16

El-Borhamy HS (2004) Genetic analysis of some drought and yield related characters in spring wheat varieties (Triticum aestivum L.). J Agric Sci Mansoura Univ 29(7):3719–3729

El-Hafid R, Smith DH, Karrou M, Samir K (1998) Physiological responses of spring durum wheat cultivars to early-season drought in a Mediterranean environment. Ann Bot 81(2):363–370

El-Rawy MA, Hassan MI (2014) Effectiveness of drought tolerance indices to identify tolerant genotypes in bread wheat (Triticum aestivum L.). J Crop Sci Biotechnol 17(4):255–266

FAO/IAEA (2018) International symposium on plant mutation breeding and biotechnology 27–31 August 2018 Vienna, Austria

Fischer RA, Maurer R (1978) Drought resistance in spring wheat cultivars, 1. Grain yield responses. Aust J Agric Res 26 (4):897–912

Foulkes MJ, Sylvester-Bradley R, Weightman R, Snape JW (2007) Identifying physiological traits associated with improved drought resistance in winter wheat. Field Crop Res 103(1):11–24

Golestani Araghi S, Assad MT (1998) Evaluation of four screening techniques for drought resistance and their relationship to yield reduction ratio in wheat. Euphytica 103:293–299

Golparvar AR (2012) Heritability and mode of gene action determination for grain filling rate and relative water content in hexaploid wheat. Genetika 44(1):25–32

Gomaa MA, El-Banna MNM, Gadalla AM, Kandil EE, Ibrahim RHA(2014) Heterosis, combining ability and drought susceptibility index in some crosses of bread wheat (Triticum aestivum L.) under water stress conditions. Middle East J Agric Res 3(2):338–345

Gue J, Xu W, Yu X, Shen H, Li H, Cheng D, Liu A, Liu J, Liu C, Zhao S, Song J (2016) Cuticular wax accumulation is associated with drought tolerance in wheat near-isogenic lines. Front Plant Sci 7:1809

Guo P, Baum M, Varshney RK, Graner A, Grando S, Ceccarelli S (2008) QTL for chlorophyll and chlorophyll fluorescence parameters in barley under post flowering drought. Euphytica 163:203–214

Gupta US (1997) Crop improvement, vol 2 stress tolerance. Science Publishers, Enfield, NH, USA

Gusta LV, Chen TH (1987) The physiology of water and temperature stress. In: Heyne EG (ed) Wheat and wheat improvement, 2nd edn. American Society of Agronomy, Madison, WI, pp 115–150

Haiba AAA, Youssef MAH, Heiba ASS, Ali HBM, Ibrahim AS (2016) Identification of RAPD and ISSR Markers for Drought Stress in Some Egyptian Durum Varieties. Egypt J Environ Res 4:23–39

Hamada AA, Abu-Grab OS, Khalil FA, Kalaf RM, Khalifa MA (2008) Physiological evaluation of twelve wheat genotypes to drought. Egypt J Plant Breed 12:135–156

Hameed A, Bibi N, Akhter J, Iqbal N (2011) Differential changes in antioxidants proteases and lipid peroxidation in flag leaves of wheat genotypes under different levels of water deficit conditions. Plant Physiol Biochem 49:178–185

Hammad Salwa AR, Ali OAM (2014) Physiological and biochemical studies on drought tolerance of wheat plants by application of amino acids and yeast extract. Ann Agric Sci 59(1):133–145

Hartung W, Slovik S (1991) Physiochemical properties of plant growth regulators and plant tissues determine their distribution: Stomatal regulation by abscisic acid in leaves. New Phytol 119:361–382

Hosseini SM, Poustini K, Siddique KHM, Palta JA (2012) Photosynthesis of barley awns does not play a significant role in grain yield under terminal drought. Crop Pasture Sci 63:489–499

Hozayn M, Abd El-Monem AA (2010) Alleviation of the potential impact of climate change on wheat productivity using arginine under irrigated Egyptian agriculture. Economics of drought and drought preparedness in climate change context, Options Mediterranean's, A, No 95

Hussein MHA, Abdel-Hamid AM, Hussein BA, El-Morshedy MA, Nasseef JE (2013) The suitability of RAPD markers in identifying some hexaploid wheat crosses. World Appl Sci J 21(5):732–738

Islam MS, PSL Srivastava, Deshmukh PS (1999) Genetic studies on drought tolerance in wheat. II. Early seedling growth and vigour. Ann Agric Res 20(2):190–194

Jat ML, Parvesh C, Raj G, Sharma SK, Gill MA (2006) Laser Land Leveling: A Precursor Technology for Resource Conservation. Rice-Wheat Consortium Technical Bulletin Series 7. New Delhi, India: Rice-Wheat Consortium for the Indo-Gangetic Plains pp 48

Kandil AA, Sharief AEM, Seadh SE, Altai DSK (2016) Role of humic acid and amino acids in limiting loss of nitrogen fertilizer and increasing productivity of some wheat cultivars grown under newly reclaimed sandy soil. Int J Adv Res Biol Sci 3(4):123–136

Khan RU, Rashid A, Khan MS, Ozturk E (2010) Impact of humic acid and chemical fertilizer application on growth and grain yield of rainfed wheat (*Triticum aestivum* L.). Pakistan J Agric Res 23 (3–4):113–121

Kumar S, Sehgal SK, Kumar U, Prasad PVV, Joshi AK, Gill BS (2012) Genomic characterization of drought tolerance-related traits in spring wheat. Euphytica 186:265–276

Levitt J (1972) Response of plants to environmental stresses, 1st edn. Academic Press, New York

Li P, Chen J, Wu P (2012) Evaluation grain yield and three physiological traits in 30 spring wheat genotypes across three irrigation regimes. Crop Sci 52:110–121

Lobell DB, Schlenker W, Costa-Roberts J (2011) Climate trends and global crop production since 1980. Science 333, 616–620. 10.1126/Science 1204531

Malik TA, Wright D (1995) Genetics of some drought resistant traits in wheat. Pak J Agric Sci 32(4):256–261

Mansour HA, Abd El-Hady M (2014) Performance of irrigation systems under water salinity in wheat production. J Agric VetY Sci 7(7):19–24

Mansour HA, Gaballah MS, Abd El-Hady M, Ebtisam IE (2014) Influence of different localized irrigation systems and treated agricultural wastewater on distribution uniformities, potato growth, tuber yield and water use efficiency. Int J Adv Res 2(2):143–150

Mickky BM, Aldesuquy S (2017) Impact of osmotic stress on seedling growth observations, membrane characteristics and antioxidant defense system of different wheat genotypes. Egypt J Basic Appl Sci 4(1):47–54

Mohamed Manal F, Thalooth AT, Amal GA (2016) Performance of wheat plants in sandy soil as affected by foliar spray of potassium and zinc and their combination. Int J Chem Tech Res 9(7):715–725

Morgan JM (1977) Changes in diffusive conductance and water potential of wheat plants before and after anthesis. Aust J Plant Physiol 4:75–86

Morgan JM, Hare RA, Fletcher RJ (1986) Genetic Variation in osmoregulation in bread and durum wheats and its relationship to grain yield of field environments. Aust J Agric Res 37:61–70

Morsy AM (2003) Performance of grain yield for some wheat genotypes under stress by chemical desiccation. Ph D Agron Department, Faculty of Agriculture, Zagazig University, Egypt

Morsy AM (2014) Gene action and heritability in some cross populations in durum wheat under normal and water stress conditions. Zagazig J Agric Res 41(3):435–450

Moustafa MA, Boersma L, Kronstad WE (1995) Response of four spring wheat cultivars to drought stress. Crop Sci Abstract 36(4):982–986

Osman R, Ferrari F, McDonald SM (2016) Water scarcity and irrigation efficiency in Egypt. Global Water Econ Pol 2 (4):165009, 1–28

Paleg LG, Douglas TJ, van Daal A, Keech DB (1981) Proline, betaine and other organic solutes protect enzymes against heat inactivation. Aust J Plant Physiol 8(1):107–114

Pandey P, Ramegowda V, Senthil-Kumar M (2015) Shared and unique responses of plants to multiple individual stresses and stress combinations: physiological and molecular mechanisms. Front Plant Sci 6:723

Qabil N (2017) Genetic analysis of yield and its attributes in wheat (*Triticum aestivum* L.) under normal irrigation and drought stress conditions. Egypt J Agron 39(3):337–356

Rajaram S, Braun HJ, Van Ginkel M (1996) GIMMYT''' S approach to breed for drought tolerance. Euphytica 92:147–153

Ramirez P, Kelly JD (1998) Traits related to drought resistance in common bean. Euphytica 99:127–136

Rashed MA, Sabry SRS, Atta AH, Mostafa AM (2010) Development of RAPD markers associated with drought tolerance in bread wheat (*Triticum aestivum*). Egypt J Genet Cytol 39:131–142

Saleh SH (2011) Performance, correlation and path coefficient analysis for grain yield and its related traits in diallel crosses of bread wheat under normal irrigation and drought conditions. World J Agric Sci 7(3):270

Salem AH, Eissa MM, Bass AH, Awaad HA, Morsy AM (2003) The genetic system controlling some physiological characters and grain yield in bread wheat *(Triticum aestivum L.)*. Zagazig J Agric Res 30(1):51–70

Salem AH, Kamel NH (1996) Increasing drought tolerance in wheat plants. In: 5th International wheat Cong, June 10–14, Ankara, Turkey p 210

Salter PJ, Goode JE (1967) Crop responses to water at different stages of growth. Res Rew No 2, Commonwealth Agric. Bureaux

Sayed OH (2003) Chlorophyll fluorescence as a tool in cereal crop research. Photosynthetica 41:321–330

Schonfeld MA, Johnson RC, Carver BF, Mornhinweg DW (1988) Water relations in winter wheat as drought resistance indicators. Crop Sci 28:526–530

Selim S, Orabi1 MM, Abdel-Hafez AAM, Hussein SHM (2010) Identification of some local frankia strains based on physiological and molecular variation Pak. J Biotechnol 7(1–2):57–65

Simon MR (1999) Inheritance of flag leaf angle, flag leaf area and flag leaf duration in four wheat crosses. Theor Appl Genet 98:310–314

Sobieh SESS, Ragab AI (2005) Evaluation of induced gamma ray mutations in breed wheat tolerance to drought. Egypt J Appl Sci 20(7):50–64

Sultan MS, Abd El-latif AH, Abdel- Moneam MA, El-Hawary MNA (2010) Genetic parameters for som earliness characters in four crosses of bread wheat under two water regime conditions. Egyptian J of Plant Breed 14(1):117–133

Suprasanna P, Jain SM (2017) Mutant resources and mutagenomics in crop plants. Emir J Food Agric 29(9). https://doi.org/10.9755/ejfa.2017.v29.i9.86

Swelam AA, Manal AH, Osman EAM (2015) Effect of raised bed width and nitrogen fertilizer level on productivity and nutritional status of bread wheat. Egypt J of Applied Sci 30(3):223–234

Taheri S, Saba J, Shekari F, Abdullah TL (2011) Effects of drought stress condition on the yield of spring wheat (*Triticum aestivum*) lines. Afr J Biotechnol 10(80):18339–18348

Thapa S, Reddy SK, Fuentealba MP, Xue Q, Rudd JC, Jessup KE, Devkota RN, Liu S (2018) Physiological responses to water stress and yield of winter wheat cultivars differing in drought tolerance. J of Agron and Crop Sci 204(4):347–358

Tiegu W, Qunce H, Weisen F (2007) Rapd and SSR polymorphism mutant line of transgenic wheat mediated by low energy ion beam. Plantasma Sci Technol 9(5):1

Tsunewaki K, Ebana K (1999) Production of near-isogenic lines of common wheat for glaucousness and genetic basis of this trait clarified by their use. Genes Genet Syst 74:33–41

Turner NC (1979) Drought resistance and adaptation to water deficits in crop plants. In: Mussell H, Stables RC (eds) Stress physiology in crop plants. Wiley Interscience, New York, pp 343–372

Uddin JM, Marshall DR (1988) Variation in epicuticular wax content in wheat. Euphytica 38(1):3–9

Verma V, Foulkes MJ, Worland AJ, Sylvester- Bardley R, Caligari PDS, Snape JD (2004) Mapping quantitative trait loci for flag leaf senescence as a yield determinant in winter wheat under optimal and stressed environments. Euphytica 13:255–263

Wally A (2016) Agricultural Biotechnology Annual. USDA, Foreign Agricultural Services, Gain Report. https://gain.fas.usda.gov/Recent%20GAIN%20Publications/Agricultural%20Biotechnology%20Annual_Cairo_Egypt_11-17-2016.pdf

Wang J, Li W, Wang W (2014) Fine mapping and metabolic and physiological characterization of the glume glaucousness inhibitor locus Iw3 derived from wild wheat. Theor Appl Genet 127:831–841

Weyhrich RA, Carver BF, Smith EL (1994) Effects of awns suppression on grain yield and agronomic traits in hard red winter wheat. Crop Sci 34:965–969

Wu H, Qin J, Han J, Zhao X, Ouyang S, Liang Y et al (2013) Comparative high-resolution mapping of the wax inhibitors Iw1 and Iw2 in hexaploid wheat. PLoS ONE 8:e84691. https://doi.org/10.1371/journal.pone.0084691

Yoshioka M, Julio CMI, Ohno R, Kimura T, Enoki H, Nishimura S, Nasuda S, Takumi S (2017) Three dominant awnless genes in common wheat: fine mapping, interaction and contribution to diversity in awn shape and length. PLoS One 12(4): e0176148

Zaid C, Chachar NA, Chachar QI, Mujtaba SM, Chachar GA, Chachar S (2016) Identification of drought tolerant wheat genotypes under water deficit condition. Int J Res 4(2):2394–3629

Zhang J, Dell B, Conocono E, Waters I, Setter T, Appels R (2009) Water deficits in wheat: fructan exohydrolase (1-FEH) mRNA expression and relationship to soluble carbohydrate concentrations in two varieties. New Phytol 181:843–850

Zhang Z, Wei W, Zhu H, Challa GS, Bi C, Trick HN et al (2015) W3 is a new wax locus that is essential for biosynthesis of β-Diketone, development of glaucousness, and reduction of cuticle permeability in common wheat. PLoS ONE 10:e0140524. https://doi.org/10.1371/journal.pone.0140524

Chapter 4
Heat Stress Tolerance, Challenges and Solutions

Hassan Auda Awaad, Mohamed Abu-hashim, and Abdelazim M Negm

Abstract Heat stress is a serious threat to crop production due to increased ambient temperature in many regions of the world. Most crop plants are sensitive to heat stress that led to progressive decreasing in yields at temperatures above the optimum level. Projected increases in temperature will negatively affect yield production and threaten the food security. Key adaptive mechanisms in crops that grown under the effect of heat stress including; rapid growth and early maturity, vitality of the leaves, chlorophyll content, stomatal conductivity and the transpiration rate, relative water content, antioxidants, heat shock proteins and cell membrane thermostability. These plant traits may contribute to enhance stress tolerance of plants. Identification of heat resistance genes and the development of new cultivars, whether in breeding methods or genetic engineering, is an effective means of reducing the harmful effects of heat stress. So, follow agricultural procedures such as cultivate the tolerant cultivars and adjusted crop irrigation dates with meteorological data and proper fertigation programs are other options in reducing the effects of heat stress waves.

Keywords Heat stress · Adaptive mechanisms · Heat shock proteins · Breeding methods · Genetic engineering · Tolerant cultivars · Agricultural procedures · Production

H. A. Awaad (✉)
Crop Science Department, Faculty of Agriculture, Zagazig University, 44511, Zagazig, Egypt

M. Abu-hashim
Soil Science Department, Faculty of Agriculture, Zagazig University, 44511, Zagazig, Egypt

A. M. Negm
Water and Water Structures Engineering Department, Faculty of Engineering, Zagazig University, 44519, Zagazig, Egypt

© Springer Nature Switzerland AG 2021
H. Awaad et al. (eds.), *Mitigating Environmental Stresses for Agricultural Sustainability in Egypt*, Springer Water,
https://doi.org/10.1007/978-3-030-64323-2_4

4.1 Introduction

A gradual increase in global temperature is observed as a result of human activities and increased levels of atmospheric CO_2 which absorb and capture a large amount of infrared and geothermal radiation emitted from the soil surface. That led to the occurrence of global warming and the Greenhouse effect. Hickey (2017) shows approximately 1% chance that global warming could be at or below 1.5 degrees, which set by the 2016 Paris Agreement. This statistically-based projections which published July 31 in Nature Climate Change show that about a 90% chance that temperatures could be increased in this century by 2.0 to 4.9 °C. In this respect, FAO (2015), Fig. 4.1 represent the worst case scenario projected global warming.

High temperatures affect crop productivity in several regions of the world, especially arid and hot regions such as the Arab region. The areas of horizontal expansion in the Egyptian deserts are one of these areas.

With the increase in Egyptian population and extension of agriculture to marginal lands where the various environmental stresses. Water and heat are major environmental factors affecting crop productivity. Therefore, breeding tolerant varieties to high temperature and follow appropriate agricultural practices are considered as important procedures to cope with heat stress.

The progress achieved in this area is still limited because heat stress tolerance is an multifaceted function that controlled by polygenes (Irmak 2016). The amount of yield obtained under heat stress depends on several major determinants. Crop plants have many morpho-physiological and biochemical adaptations which help them to

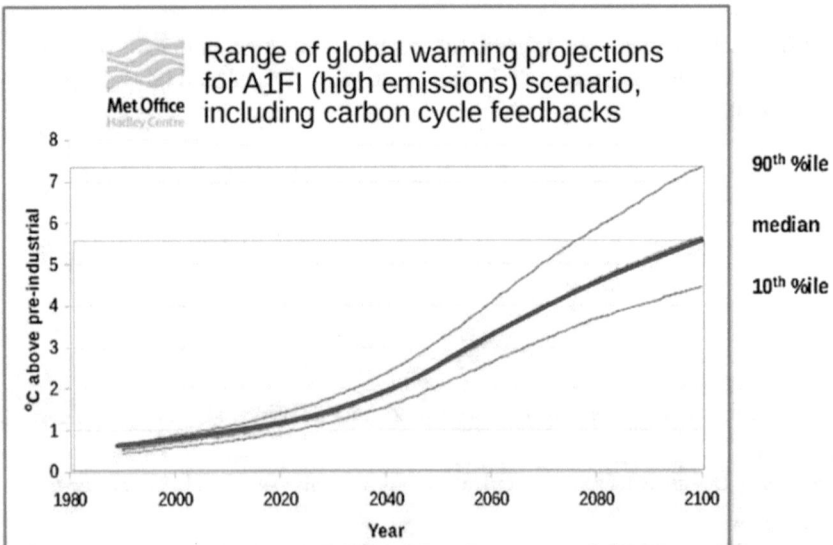

Fig. 4.1 The current world emission scenario that projected the global warming (FAO 2015)

adapt with high temperature. Earliness, cooler canopies, stay green, transpiration rate, photosynthetic rate, water use efficiency and the contribution of stored products in stems, bases and sheaths of leaves to improve grain filling under different heat stress conditions. (Reynolds et al. 1998; Moursi 2003). One of the desirable triat in this approach is maintaining high leaf chlorophyll content as it indicates a low degree of photo inhibition for photosynthetic apparatus at high temperature (Ristic et al. 2007; Talebi 2011). The ability of plant to maintain leaf chlorophyll content under high-temperature stress is correlated with grain yield and its components (Ali et al. 2010).

It is worth mentioning that different crop species and varieties vary in their sensitivity and tolerance to heat stress. Crop plants grown in cold seasons are more sensitive than plants grown in hot seasons. The optimum temperature of each crop varies from one geographical region to another, from one crop to another, and from one cultivar to another.

However, whenever the plant has an appropriate temperature day and night, the rate of photosynthesis increased from the rate of catabolism, then increase dry weight of the plant as shown in the following equation: Plant dry weight = Photosynthesis - Respiration = Net photosynthesis. So this chapter aimed to (i) focus light on impact of heat stress on the Egyptian agriculture sector, (ii) nature of resistance to heat stress, (iii) role of breeding achievements, biotechnology and agronomic practices.

4.2 Concepts Related to Heat Stress

4.2.1 Heat Stress

Heat stress is defined as the rise of air temperature beyond a threshold level for a certain period that sufficient to cause permanent and irreversible damage to the functions, plant growth and development. Heat stress is a complex function of intensity, duration, and the rate of the increase in air temperature. Heat damage is attributable to the absorption of leaves and organs of the plant solar energy rays, which leads to rise their temperature from air temperature.

It should be noted that the high temperature during the day period leads to direct damage and indirect effects on the plant. Direct damage arises with the high temperature of the plant tissue, resulting in severe damage, especially in the structural and functional aspects. Indirect damage is the result of increasing the water requirement of plants due to increased evaporation, leading to increase respiratory rate and the lack of relative water content in the plant.

4.2.2 Vant Hoff Concept (Q10 Coefficient)

Is a base formulated by the Netherlands chemist Van Huff, which links the speed of the chemical reaction to the temperature? It states that the chemical reaction speed is doubled to four times when the temperature rises by 10 degrees. Vant Hoff concept is applied primarily to biochemical reactions. These reactions are assumed to multiply with each 10% increase in reaction temperature. Some crop physiologists suggested applying this concept to the plant dry weight and predicted that the dry weight of the plant would be doubled in geometrical sequence with every 10% increase in the air temperature provided that this increase is within the thermal range from 20 to 30 °C. Considering the value 30 °C is sometimes followed by low dry weight in some field crops due to increased respiratory rate than photosynthesis.

4.2.3 The Concept of Thermal Units

Crop growth and maturity depends upon the rate of heat units which need to reach the target stage. The term "Degree-day" is used to express these units = the number of temperatures above the critical limit of that crop × number of days of growth. It is necessary to determine growth periods, planting dates, ripening and harvesting dates. It can also help farmers to applied agricultural transactions and in determining the crop species that can be grown in a given zone.

This concept assumes that the dry weight of the plant is increased in numerical sequence with each increase in the number of thermal units collected during the crop growth season above a minimum level called zero growth. This differed from one crop to another being 50 °F in maize and 55 °F in sorghum. In the United States of America, the appropriate number of thermal units is 3700 units to obtain a high yield of maize. Brown (2013) showed that environmental temperatures affect the development of biological organisms. In addition, heat unit systems quantify the thermal environment of the living organisms and used in phenology models which could explain growth and development of organism to local weather climate conditions.

4.2.4 Daily Thermoperiodicity Concept

Thermoperiodicity, is defined as the growth and plant flowering responses to alternation of warm and cool periods. Moreover, fluctuations of daily temperature could produce dramatic effects on the growth and flowering of crops. Sunrise and sunset that arise a difference between day and night temperature in a regular daily rhythm, increases the daytime heat with the sunrise and low nighttime temperature at sunset.

The daily thermal rhythm has a significant effect on field crops. Low night temperatures cause a decrease in respiration rate and therefore a decrease in the loss rate of assimilates represented during the day. Also, the high temperature of the day leads to an increase in the rate of photosynthesis in the day within the limits of the temperature of each crop ranging from 26 to 31 °C for cold season crops and 32–37 °C for hot season crops. This means that within each range, the temperature is suitable for photosynthesis.

The planting dates of crops, especially summer crops, are determined in such a way as to ensure an appropriate daily thermal rhythm characterized by daytime increase in temperature compared to a decrease in night temperature. In maize, summer agriculture is characterized by an increase also in daytime temperature and a relatively low night temperature compared to Nile agriculture, which loses this feature and therefore has an abnormal daily temperature.

The example of cotton cultivation is also carried out during February and March compared to late planting in April, which is the case for all summer crops where it is assumed that night temperature should be less than 25°C. If higher, it increases the plant content of ethylene (C_2H_4), the first cause of fall of buds.

4.2.5 Seasonal Thermoperiodicity Concept

The movement of the earth around the sun takes a whole year (365 days). Then there is a difference in temperature depending on the position of the earth from the sun, this results a seasonal rhythm heat. Egypt has four seasons for sustainable agricultural. Considering the seasonal temperature, the temperature changes between the seasons, and the planting dates are determined in such a way as to ensure that the season is suitable for the growth and development of crops.

One of the most famous examples of selecting a suitable planting date that provides optimal seasonal temperature is the selection of wheat crop cultivation in Egypt during the last half of November in the governorates of the face of the sea.

Where wheat needs a seasonal rhythm temperature, which provides a low temperature after germination so that the plants meet the needs of cold, i.e., vernalization requirements. With a longer period of the vernalization, chance of tillering was increased, which increases with reducing temperature from 20 to 10 cm.

Wheat plants after the fertilization phase need to gradually increase the temperature from 20 to 25 °C until the leaves become more increase in elongation and the area. When the ears are headed, wheat plants need to increase the temperature from 25 to 30 °C. This helps in increasing transfer of the assimilates represented to the grains and increases yield. However, any crop plant needs a specific seasonal rhythm and on the basis of which the planting date of the crop is determined so that the plant has all the appropriate thermal requirements for each stage during its life stages.

An example of a crop response to seasonal temperature is the increase in productivity of rice in Egypt, which has been ranked the first in terms of productivity by improving agricultural transactions for high-yield varieties. The most important of these transactions is the selection of suitable planting date so that growth will be carried out during the end of April or the first of May. This provides relatively low temperature at night compared to the day for rice plants during the two months (May and June). When the plants are headed in July, the temperature is high enough to fill the grain in August. Therefore, there is a response from the farmers to early rice cultivation to achieve high yields.

4.3   Heat Stress Impact on the Egyptian Agriculture Sector

The results of the Agricultural Meteorology and Climate Change Research Unit of the the Agricultural Research Center—Egypt showed that the results of long-range prediction using simulations and different climate change scenarios, which resulting rise in temperature of the surface of the earth will negatively affect the productivity of Egyptian agricultural crops. This causing a severe shortage in the productivity of the main crops in Egypt and increasing water consumption. The most important results of studies conducted in this regard are the following (El-Marsafawy Samia 2007):

1. The yield of the wheat crop will be decreased by 9% if the temperature degree increased by 2 °C, and the rate will decrease to about 18% if the temperature is about 3.5 °C. Furthermore, water consumption of this crop would be increased by 2.5% compared to corresponding under current weather conditions.
2. Barley productivity could be declined by 18% (by 2050) and its water consumption will decline by about 2%.
3. The productivity of maize crop could be decreased by 19% with the middle of this century (temperature rises by 3.5 °C) compared to productivity under current conditions, and will increase water consumption by about 8%.
4. The sorghum crop will decline by about 19% and its water consumption will increase by about 8%.
5. The productivity of the rice crop could be declined by 11% and its water consumption will increase by about 16%.
6. The productivity of the soybean crop would be severely influenced under climate change conditions. The average reduction rate at the level of the Republic by mid-century will be about 28% and its water consumption will increase by about 15%.
7. Sunflower crop will decline about 27% and will increase water consumption by about 8%.
8. Tomato crop is very sensitive to high temperature and the productivity would be decreased by 14% if the temperature rises by 1.5 °C, and this crop shortage will reach 50% if the temperature rises 3.5 °C.
9. Sugar production from cane crop will drop by about 25% and its water consumption will increase by 2.5%.

10. Climate change could have a positive effect on cotton yield productivity, and its productivity may increase by 17% when the temperature of the air is rises about 2 °C. The rate of increase in this crop will rise to about 31% at 4 °C. Nevertheless, its water consumption would be increased by 10% compared to its water consumption under current weather conditions.

4.4   Physiological and Biochemical Impacts of Heat Stress

Effects of heat stress on crop plants could be identified as follows:

1. Low germination rate
2. Lack of plant growth due to lack of expansion and development of plant organs and decrease of cell water content and osmotic pressure.
3. Effect on photosynthesis and its various components, Hill reactions and reduce the levels of metabolism compounds during vegetative growth due to low carbon dioxide metabolism.
4. Lack of partitioning the assimilates between the stems, leaves, roots and fruit parts.
5. Severe damage to the reproductive structures and the vitality of the pollen grains and lack of seed setting.
6. Influence on the starch content of cereals due to inhibition of starch enzyme activity.
7. Lack of content of ascorbic acid, pigments and chlorophyll.
8. Deficiency of dissolved protein content and increased levels of free proline.
9. Increase the accumulation of abscisic acid, and thus decrease stomatal conductivity and carbon dioxide metabolism.
10. Deformation of shoot system parts and burning the top of the leaves and flowers.
11. Significant decrease in yield and its components due to incomplete grain filling and the direction of the plant to premature aging.

Therefore, the modification of plant metabolism and biological processes associated with the formation of protoplasm by gene manipulation, both in traditional breeding methods and in biotechnology, is an important procedure for enhancing the performance of the crop varieties with the conditions of heat stress. Where advanced studies indicate the possibility of genetic improvement of the crop product under environmental stress conditions. In practice, the selection of the crop cultivar, the appropriate planting date for the crop growth stages in the region and the combination of fertilizers with available water are considered appropriate agricultural practices to improve plant growth and minimize the negative effect that resulted from the high temperature on crop productivity.

4.5   Critical Periods of Crop Plants to Heat Stress

High temperature effects on crop plants depend on the growth phase in which this stress occurs, depending on the crop spices and the nature of its growth. Critical period knowledge is important in screening programs and evaluating crop germplasm to demonstrate their ability to tolerate high temperatures. The following are the critical stages of heat stress effect on some field crops.

1. Wheat and barley: fertilization and grain filling period.
2. Rice: flowering and grain filling.
3. Maize and sorghum: flowering and grain filling period.
4. Pearl millet: emergence
5. Faba bean: flowering stage.
6. Soybeans: flowering and pods formation.
7. Cotton: flowering and boll formation.
8. Brassica: flowering.
9. Tomato: flowering.

4.6   Nature of Resistance to Heat Stress

The nature of the resistance to heat stress in crop plants is divided according to the the acclimatization mechanism of crop varieties as shown in Fig. 4.2 to the following models

4.6.1   Heat Avoidance

Heat avoidance is one of the main components of thermal stress resistance. Where the temperature of the plant is less than the temperature of the comparison plants, or the atmospheric air by about 10 °C, in which case the leaves retain their vitality.

The ability of the plant to modify the composition of the leaf, whether through volume (medium) or the ability to circumvent and change the position of the edge of the leaf at the high temperatures plays a significant effect to avoide the thermal stress effects. The smooth and green leaves, which reflect a large part of the sun's rays compared to the dark green leaves, contribute to the energy-saving of the plant. The stomatal conductivity and the rate of transpiration and their cooling effect, result in a reduction of 23% of the plant heat, and then adapt the varieties to high temperature conditions.

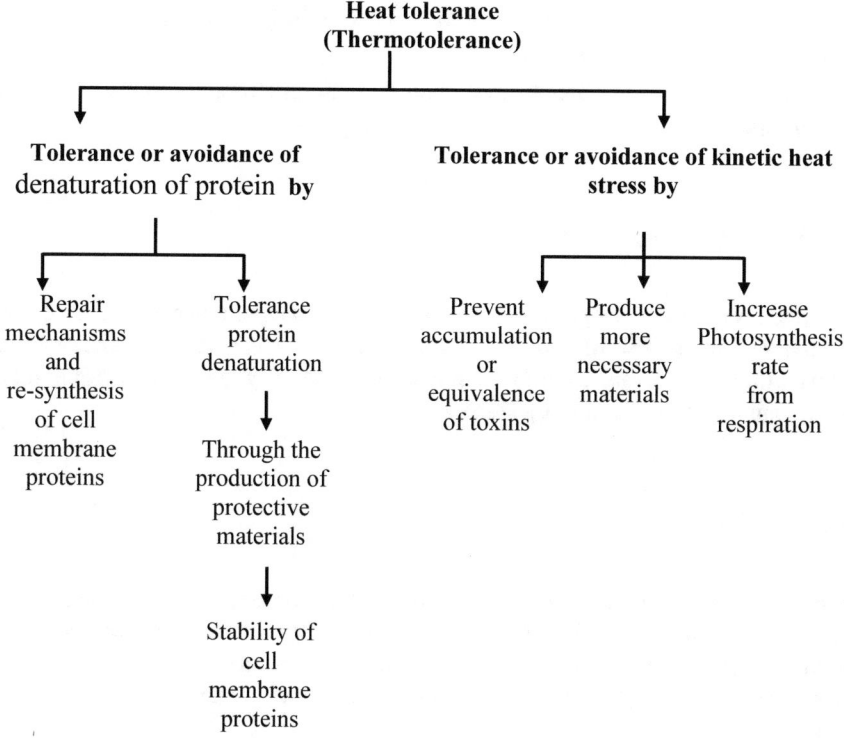

Fig. 4.2 Different models of heat stress tolerance

4.6.2   Heat Tolerance (Thermotolerance)

Heat tolerance is defined as plant organ ability to survive under conditions of heat stress. This is determined genetically by the stability of proteins and nucleic acids or their return to normal state through the repair mechanisms, if they have partial denture. This is in addition to the plant's ability to prevent the accumulation or deactivation of toxins, persistence of photosynthesis and the lack of respiration rate. Heat tolerance depends on the availability of K^+, phosphorus, glucose and various energy sources. As well as the active transfer of calcium Ca^{2+} from the soil to the cells.

Tolerance and avoidance generally require the availability of protective mechanisms by producing substances that inhibit coagulation and increase the alkalinity of the protoplasm, with the strength of protein bonds and the availability of free water.

4.7 Classification of Crop Plants According to Temperature

Crop plants are divided according to their thermal requirements to the following two groups Table 4.1

4.7.1 Cold Season Crops

The plants of these crops germinate and grow at low temperatures. Plants have low thermal requirements if not available do not enter the flowering stage. The optimal temperature ranges from 26 to 31 °C such as wheat, barley, rye, oats, triticale, faba been, lentils, chickpeas and Egyptian clover.

4.7.2 Hot Season Crops

The plants of these crops germinate and grow within a temperature range higher than those of the cold season. The optimum temperature for these crops various between 32 to 37 °C, representing the maximum limit for cold season crops. Examples of this group of crops are sugarcane, maize, sorghum, rice, millet, cotton, soybeans, sesame and tobacco.

Table 4.1 Minimum, optimum and maximum temperatures for cold and hot season crops

Temperature crop	Minimum temperature (°C)	Optimum temperature (°C)	Maximum temperature (°C)
Cold season crops	0–5	26–31	32–37
Hot season crops	15–18	32–37	44–50

From: Anonymous (2016)

4.8 Is It Possible to Mitigate Extreme Heat Stress on Crop Plants? How Crop Plants Can Deal with Heat Stress?

4.8.1 Traits Relevant to Heat Stress Tolerance

4.8.1.1 Plant Traits

Crop plants have much potential that help them to adapt and cope with heat stress. These capabilities include traits whether they are phenological, morpho-physiological or biochemical characteristics of molecular structure. The importance of these traits can be discussed as follows:

Phenological Traits

Rapid Growth and Early Maturity

Early growth is a fast and early criterion for heat tolerance. The characteristics of the earliness and the movement of current assimilates from the source to the sink, grain filling period, grain yield and high quality, crop varieties vary in rapid of germination and seedling vigor. The correlation between the speed of emergence and the growth vigor at seedling stage with the high production of dry matter at maturity was observed in nine wheat genotypes that tolerant to the heat stress compared with the other sensitive ones. Where, Singh and Ahmed (2003) identified seedling vigor and earliness as a field observations and early indications of terminal heat stress tolerance in wheat.

Early maturing provides an escape mechanism under unfavorable conditions and has been identified to be a good approach of wheat breeding for heat stress tolerance (Awaad 2009). In this respect, El- Moselhy Omnya (2015) revealed that the Egyptian wheat cultivars Sakha 93, Gemmeiza 11, and Sids 12 exhibited good level of early heading with 90.86, 91.17 and 91.75 days and accumulated 957.53, 961.22 and 963.06 growing degree days (GDD), respectively, of them Gemmeiza 11 and Sids 12 had high yield potentiality. Whereas, genotype Line 16 was the latest one (97.63 days) with GDD (1034.37) despite that it has high yield potential. Early heading had positive and significant correlation with canopy temperature depression at moderate sowing date (10th December) as well as flag leaf chlorophyll content and number of spikes/m^2 at late sowing (30th December). Early maturity showed positive and significant correlation with canopy temperature depression under late sowing (30th December) one. Early heading was more heritable character, whereas, grain filling period and rate were low narrow sense heritability (Awaad 2001). Under two locations i.e. Sohag and Assiut, heat stress reduced average number of days to heading by 9 days at late sowing date on 26th December rather than favorable (14th November) one as revealed by Hamam and Khalid (2009). Fifty-eight spring wheat genotypes were investigated for heat stress tolerance using three sowing dates by Hamam (2013).

The earliest genotypes were No. 4, 23, 24, 25, 31, 33, 34, 42, 47 and 58 revealed the highest decrease in days to heading under late (middle December) as compared with favorable (first November) sowing date.

Physiological Traits

Vitality of the Leaves and Stay Green Longevity

Leaf attributes and qualities represent the importance in many field crops such as sorghum, maize, wheat, etc. where the plant can sustain photosynthesis with high efficiency and stability of starch. This helps to strengthen the plant during grain filling period. Farooq et al. (2009) mentioned that photosynthesis is one of the main physiological phenomena that affected by the heat stress and the drought periods in plants. The vitality and small size of the leaf played an efficient role in improving pod filling and giving satisfactory yield levels in chickpeas under heat and soil moisture stress conditions (Gupta et al. 2003). However, Rao et al. (2003) observed with the extreme heat stress a significant decrease in relative water content, nutrient uptake rates such as nitrogen, phosphorus and potassium. Also, photosynthetic levels, growth parameters, accumulation of dry matter in the legumes of fodder cowpea and chickpeas were decreased. In wheat, El-Moselhy Omnya (2015) registered significantly positive association between flag leaf area and each of chlorophyll content, number of grains/spike and 1000-grain weight under different levels of heat stress.

Canopy Temperature

The leaves absorb a large amount of sufficient radiant energy to raise its temperature to a deadly extent and this energy would be converted into heat energy. However, the temperature of the leaves does not rise above the temperature of air environment except by a few degrees Celsius, as thermal energy dissipates and loses to the surrounding environment by some natural means such as radiation, conduction and pregnancy. Respiration is one of the phenomena that leading to cooling of the plant canopy.

Reynolds et al. (2001) indicated that CTD could be used to be a good parameter to select genotypes among early generation bulks of heterogeneous populations.

Cooler canopy temperature (CT) is important trait that help the plant to keep the physiological functions under elevated temperature and has been convenient with grain yield under hot irrigated conditions (Reynolds et al. 1994; Kumari et al. 2012).

Genetic resources vary in their degree of tolerance to high temperatures by maintaining the temperature of the plant's canopy when exposed to heat stress. A low temperature of the Canopy characterizes the varieties of thermal stress tolerance. Where, Rane (2003) found that low-temperature wheat varieties were more tolerant to heat stress during the grain filling period.

The temperature of the leaves by 5 °C above ambient atmosphere is equivalent to a 30% reduction in the relative humidity of the surrounding atmosphere. Craufured et al. (1999) mentioned that increasing the temperature of leaf tissue is mainly associated with the radiation load and indirectly to the lack of water absorption, which has a significant impact on water use efficiency, water consumption, leaf quality and Δ values (which measure the difference in $^{13}C/^{12}C$ for air and leaf). They found that Virginia peanuts varieties gave low values of Δ and had a positive effect on water use compared to Spanish ones at 27 °C. However, with high temperature to 34, there was a significant decrease in water use efficiency.

Under late sowing date (30th December D3) as heat stress environment, El-Moselhy Omnya (2015) revealed that most of wheat genotypes tended to increase their canopy temperature depression CTD compared to 20th November (recommended D1) and 10th December (moderate late sowing date D2). At Kafr Al-Hamam location, at D3, Misr 2 showed the highest CTD (7.99) followed by Giza 168 (7.50), Line 16 (7.49), Misr 1 (7.39), Sids 1 (7.27) and Shandweel 1 (7.22). Whereas, at Sids location, Gemmeiza 11 showed the highest CTD (7.62) followed by Giza 168 (7.34), Sids 12 (7.31), Sids 13 (7.24), Line 14 (7.04) and Misr 2 (7.02). Quiet differences in correlation coefficients were noticed from sowing date to another. Under the first sowing date, wheat grain yield was positive and significantly associated with canopy temperature depression and hence for the grain protein content as well as with chlorophyll content under the third sowing date, implying that improving one or more of these traits could result in high grain yield.

Chlorophyll

Chlorophyll fluorescence has extensively been used in assessing genotypes responses to environmental stress. Chlorophyll fluorescence measurements are an indicator of photosynthesis stability, leaf activity and light reactions during periods of heat stress. Therefore, the estimation of chlorophyll radiation as a direct criterion in the screening of the genotypes for high-temperature tolerance. Recurrent selection is a useful procedure to obtain genes responsible for thermal stress tolerance based on chlorophyll fluorescence. The optimal temperature for reproductive growth stage and photosynthesis in wheat is close to 20 cm, but exposure to plants at a temperature higher than 35 °C before physiological maturity limits the wheat productivity in many regions. This is due to the sensitivity of the Thylakoid membrane and the associated reaction by heat stress, which changes the amount of light energy, absorbed from PSII to PS I and changes the pattern of chlorophyll fluorescence. Therefore, chlorophyll fluorescence is an indicator of the safety of the thylakoid membrane and a measure of the relative efficiency of electron transport from PSII to PS I. In this context, Yang et al. (2002) recorded a significantly positive correlation between the relative stability of chlorophyll fluorescence values with wheat grain yield that affected by the post-flowering heat stress. Varieties of wheat varied in chlorophyll fluorescence. The tolerant wheat varieties Ventnor and Debeira showed relative stability and high ability to restore the previous level of chlorophyll fluorescence when the thermal shock treatment at 40 °C for 16 h daily, and 35 °C for 8 h at night for 3 days was followed by optimum heat treatment at 20 °C for 16 h daily, and 15 h for 8 h a night.

Maintaining high leaf chlorophyll is considered a desirable trait as it reveal the low degree of photoinhibition of photosynthetic apparatus at high temperature (Ristic et al. 2007; Talebi 2011). Furthermore, Ali et al. (2010) shown that, under high-temperature stress, the ability of plant to maintain leaf chlorophyll content is associated with grain yield and yield components.

Mondal et al. (2013) reported that chlorophyll content had a significant correlation with 1000-grain weight across locations. However, there were no significant correlations between chlorophyll content and grain yield at individual locations and/or across locations. Meanwhile, El- Moselhy Omnya (2015) registered positive and significant correlation between flag leaf chlorophyll content and each of number of grains/spike and 1000-grain weight under different regime of heat stress and with the depression in canopy temperature under moderate heat stress.

Stomatal Conductance and Transpiration Rate

The size of the stomatal apertures play an important role in controlling the regulation of transpiration rate in plants. In addition, transpiration rate is influenced by the evaporative process at the atmosphere that surrounding the leaf as; temperature, relative humidity, wind and incident sunlight. Moreover, soil temperature and soil water supply can influence stomatal opening, and thus the transpiration rate. So, transpiration considered one of the main parameters that affect the water loss by a plant by the leaves (http://www.knowpia.com/pages/Transpiration). Transpiration serves to evaporatively cool plants. Lu et al. (1997) mentioned that photosynthetic rate, higher stomatal conductance and smaller leaf areas are essential to achieve higher heat tolerance and cotton yield. Subjected crop plants to environmental stress disturbs water relations in plant. These relations are influenced by leaf water potential, transpiration rate, leaf and canopy temperature, and stomatal conductance that affect the stomatal behavior in efficient manner (Farooq et al. 2009).

Nevertheless, if the heat stress is extremely coupled with dry wind on the plant canopies, the reduction in transpiration rate and the magnitude of stomatal closure would be greater for corn and soybean. Where, Irmak (2016) showed that the transpiration rate under water stress conditions and plant stomatal conductance were observed to be lower compared to low fertility conditions. These applications facilitated the process to modify the stomatal function and enhanced the physiological and metabolic processes which support increasing heat stress tolerance and upholding high tissue water potential.

Omara (2010) practiced selection for stomatal frequency under heat stress conditions in five populations of bread wheat under Egyptian condition. He obtained significantly positive response to selection for the frequency of the high stomatal on 1000-grain weight and grain yield under normal conditions and heat stress. That the grain yield and 1000-grain weight were increased with an average of 11.58% and 8.53%, respectively. Based on the molecular basis, Shahinnia et al. (2016) indicated that stomatal traits could be a mechanism efficiently increasing wheat yield at specific loci and could be used in recombinant lines as a proxy to track the target QTL.

Relative Water Content

The water retaining ability and relative water content are considered to be associated with various biochemical and physiological processes, morphological changes, dry matter accumulation and yield. It is therefore one of the important measures to heat stress tolerance in a extreme range of different crop plants at several stages of growth and development, not only to tolerance the high temperature, but also tolerance to drought, salinity and nutrient deficiency. For example, high temperature tolerant chickpeas were characterized by high yields and higher content of relative water in tissues. This is therefore a criterion used for screening large numbers of genotypes during isolating generations and judging the improvement of yield under the severe stress conditions (Deshmukh et al. 2003). Chickpea genotypes RSG 143-1, BGD 72 resistant to heat stress were characterized by the highest productivity and highest level of relative water content (Gupta et al. 2003). Furthermore, Abou Gabal Ashgan and Zaitoun Amera (2015) found that the best maize inbred for relative water content under stressed treatment was DT190 (71.21) and followed by DT62 (62.48). While, the lowest relative water content for the stressed treatment was noticed by INB209 (43.48). Hereby, DT190 and DT62 are classified as relatively tolerant to drought and heat, while INB209 and INB176 are sensitive.

Molecular Base

There are many cellular chemical compounds that affect the heat stress tolerance in vegetative tissues of crop plants during growth stages. They include heat shock proteins, cell membrane stability, antioxidants, stem reserve of carbohydrate mobilization and antioxidants.

Heat Shock Proteins

There is much research on the role and importance of heat shock proteins (HSPs) in resisting plants heat stress since 1991 till now. HSPs significantly important in two points: firstly, HSPs displayed diversity role. Secondly, for the heat tolerance, these proteins may play a vigorous role in acquiring of heat tolerance.

The results of genetic studies have indicated that many genes responsible for production of the heat shock proteins that are closely associated with the heat stress resistance such as *HSP 101, HSP 70,* and *HSP 27* in the *Arabidopsis* (Lee and Schöffl 1996; Hong and Vierling 2000); *HSP 17* in carrot cells (Malik et al. 1999); 104 and 90 kDa polypeptides in rice seedlings (Pareek et al. 1995); *HSP101* in maize (Young et al. 2001); *HSP21* in pea and *HSP22* in soybean chloroplast (Boston et al. 1996) and HMW HSPs and low molecular weight LMW HSPs in wheat (Natu et al. 2003). For example, *HSP 70* is repairing process of various proteins. *HSP-101, HSP-27* proteins also play essential roles in different regions of the cell and the reorganization of the carbon structure of plant protection under stress conditions (Boston et al. 1996; Shawky et al. 2005).

Heat stress effect on heat shock proteins (HSP) synthesis was detected with the seven wheat parents, Badr Asmaa (2017) revealed that the seedling exposed to heat stress produced heat shock proteins in significant amount after exposed to 45 °C. The most stable cultivars were Sakha 93 where it is thermoplastic reactions under heat shock conditions did not change compared with the control treatment. The cultivars Misr 2, Sids 1 and Giza 168 produced the maximum number of heat shock proteins (two bands) 71, 83 KDa, 37, 42 KDa, respectively, otherwise Gemmieza 11 produced the lowest number (one band) 42 KDa. The quantity and number of proteins produced in both cultivars Gemmieza 7 and Gemmieza 10 affected by the thermal shock and the number of bands decreased under stress treatment. Expression of candidate HSP gene was tested using total RNA extract of the two wheat cultivars (Sids 1 and Gemmieza 10) utilizing Real Time PCR. Sids1 exhibited high gene expression with gene copy number 3.61 compared to Gemmieza 10 which had low gene expression with gene copy number 0.05. The observed expression of hsp22 (mitochondrial) might contribute to heat tolerance.

Cell Membrane Thermostability

Cell membrane stability (CMS) is the criteria that used extensively in assessing the tolerance of different genotypes to stress. Although it is a small part of a large number of properties and processes, it interacts to give the plant resistance to stress during different growth stages. Electrolytes level as indicator on perturbation of membrane permeability a significant increased with exposure period increase to high temperature. Increased the cell permeability and leakage of ions out of it was used as essential indicator of screening genotypes and cell membrane stability for stress tolerance. Under Egyptian condition, ElBasyoni et al. (2017) registered significant variations among 2111 spring wheat accessions regarding the CMS. However, several SNPs were observed to be heavily associated with the CMS, associated well with important functional genes, and related to abiotic stress tolerance. Tolerant wheat genotypes under field stress conditions appeared to be more productive. Thus, integrity of the membrane genotypes would be used in wheat breeding programs for heat stress tolerance. In this respect, under heat stress conditions, Omara (2010) practiced selection for cell membrane thermostability in five populations of bread wheat. He detected significant correlation (0.64) between cell membrane thermal stability and the wheat grain yield. Also, positive and significant response to selection for cell membrane thermal stability on wheat grain yield was obtained. Badr Asmaa (2017) revealed that the most tolerant wheat genotypes to heat stress were found to be higher in cell membrane stability accompanied with low ions leakage. Whereas, in maize as summer crop in Egypt, Abou Gabal Ashgan and Zaitoun Amera (2015) detected high degree of genetic variability among maize inbred lines in cell membrane stability. Where, INB 176 (46.67) and DT 196 (45.45) registered under Normal conditions highest values of cell membrane injury, conversely, the lowest injuries 29.69 and 35.00 resulted from DT 62 and DT 235, respectively.

On the other hand, the highest injury of cell membrane 75.63 and 74.07 under heat and drought stress were recorded for inbred lines INB 209 and DT 61, respectively.

While, the lowest values 44.00 and 45.16 were for DT 190 and DT 62, respectively. Therefore, both genotypes can be considered as tolerant to environmental changes. In oil crops, positive and significant association between unsaturated fatty acids contents and the high-temperature tolerant were detected by Iba (2003). In legume crops, negative correlation was recorded between chickpeas seed yield under stress and cell membrane damage at 50% flowering (r = −0527), 20 days after flowering (r = −0.689), 40 days after flowering (r = −0869) (r = −0.483) at pods formation. Genotypes RSG 143-1, BGD 72 resistant to heat stress were characterized by the highest yield and highest level of relative water content with minimal damage to the cell membrane (Gupta et al. 2003).

Stem Reserve of Carbohydrate Mobilization

Decreasing in photosynthesis rate was observed to be reduced in wheat from 11 to 32% and the leaf surface activity from 0 to 40% with a temperature rise from 22 to 32 °C (Al-Khatib and Paulsen 1990). Therefore, pre-anthesis assimilate stocked with cereals stems such as wheat and barley are a source of support grain filling during the terminal heat stress period. Wheat varieties vary in their ability to take advantage of the reserve stored in the stem and guide it to fill grains during stress periods.

A significant correlation was obtained between the amount of stored reserves of carbohydrates in wheat stem and barley varieties tolerate to high temperature and the quantity of yield. Ventnor, Debeira, Newton, Trigo 3 wheat varieties were characterized by a high yield through their ability to move pre-anthesis reserves in stem to support the growth of the grain during the post-anthesis period and were classified as high heat tolerant varieties (Yang et al. 2002).

Chemical desiccation agents sprayed on the wheat varieties with sodium chloride and potassium chloride within 14 days after flowering. This procedure lead to variation among genotypes in their tolerance to the stress and their ability to move the stem reserve mobilization to support grain filling period. Moursi (2003) found that wheat genotypes Bocro 4, Sakh 69 and Shi#4414/Crow "S" were tolerant to stress as they exhibited high carbohydrate reserves, while Gemmeiza 5, Giza 168 and Sahel 1 were more sensitive.

Antioxidants

A recently, discovered group of metabolic products of the secondary plant metabolism called antioxidants plays an essential role in protecting cellular system and tolerant plants to stress conditions. Examples of these substances are peroxides, catalysts, tocopherols i.e. α, β, γ and δ tocopherol, ascorbic acid, folic acid, glutathione and others. The function of antioxidants is protected the cells from the destructive act of free radicals of active oxygen species (ROS) that could be produced under conditions of the environmental stress, which leads to the demolition of pigments, lipids, protein and DNA. Plant antioxidants play a vital role in reducing the destruction process. Rao et al. (2003) showed that the active oxygen species production and the corresponding antioxidant activity would determine the sensitivity or tolerance of legumes genotypes to heat stress. The levels of peroxidase activity were increased with high temperature in cowpea and chickpeas. This is supported by the increased activity of Superoxide dismutase, Catalase, Glutathione reeducates, Ascorbate peroxidase with increased heat stress. These enzymes activity were more pronounced in tolerant wheat varieties compared to sensitive ones (Sairam et al. 2003).

In lupine, El-Enany Magda et al. (2013) tested two *Lupinus* species *albus* (white) and *luteus* (yellow) under different time and temperature regimes (30, 35, 40, 42 °C/30 min and 42 °C for 1, 2, 3 and 4 h, respectively) compared with the constant degree of 25 °C as a control in growth chambers. Electrophoretic patterns of the peroxidase isozyme detected four bands under the control and three bands under different temperatures levels in both lupine species. Polyphenol oxidase isozyme appeared three bands for *L. albus* with different intensity, while the band (No.3) was disappeared under control and different temperatures levels. Furthermore, Siddiqui et al. (2015) tested collection of faba bean genotypes under different levels of temperature. They concluded that the lowest reduction in growth performance was registered in Egyptian genotype 'C5', due to its better antioxidant activities, proline content, total chlorophyll and leaf relative water content. Genotype 'C5' was observed to be the efficient heat stress tolerant compared to the 'Espan' was most sensitive one.

4.8.2 Breeding Achievements

It has been possible through the efforts of plant breeders to develop a many of field crop varieties to tolerate the high temperature with morpho-physiological and biochemical characteristics. The National Wheat Improvement Program in Egypt over the last 25 years has succeeded in increasing wheat productivity from 3.3 tons per hectare in 1981 to 5.1 tons per hectare in 1991 to about 6.9 tons per hectare in 2016, an increase of 109% (Anonymous 2017).

Under Egyptian conditions, plant breeders were able to develop a collection of bread wheat cultivars that able to tolerant the heat stress such as Giza 160, Giza 168, Sids 1, Sids 6, Sids 12, Sids 14, Sakha 94, Sakha 95, and durum wheat varieties i.e. Sohag 3, Beni Suef 1 and Beni Suef 3. Moreover, barley varieties Giza 130, Giza 131 and Giza 2000.

In rice, it was possible to develop and produce early maturity varieties characterized by a short growth period of about 125 days avoiding unfavorable environmental conditions i.e. Giza 177, Giza 179, Giza 182, Sakha 103, Sakha 105 and black rice. These are important achievements at the Egyptian national level. This provides about 30% of the rice water requirements, equivalent to 2.5–3.0 billion cubic meters of irrigation water.

The hybrid vigor phenomenon was also exploited in the production of hybrid maize in commercial production whether they are single or triple tolerant to adverse effects. As well as single sorghum crosses, Mina and Horus, and varieties Sandweel

2 and Shandweel 6 with the coldest canopy and high yield, well adapted under Upper Egypt conditions of rural areas.

Breeding programs have played an important role in improving relative water content in soybean varieties and avoiding low rates of photosynthesis and other physiological activities occurring in older varieties. This has encouraged the cultivation of soybeans under high temperature conditions such as Clark, Crawford and Calland (Atta Allah 2001), Giza 21, Giza 22, Giza 35 and Giza 111 (Anonymous 2020).

The varietal hybridization with pedigree selection has contributed to development the cotton varieties Giza 75, Giza 76, Giza 77, Giza 80, Giza 83 Giza 89, Giza 90, Giza 91 to be grown under less favorable environmental conditions (Anonymous 2019).

In terms of selection and the use of molecular markers, new strains of lentil strains 1, 2, 3 and 4 were found to be superior rather than recommended cultivar Giza 9 in yield characteristics at Assiut and Al-Matinah in Upper Egypt. RAPD-PCR technique was reported in the definition and diagnosis of promising lentil strains (Abo-Elwafa et al. 2003).

4.9 Biotechnology

According to climate analysis, climate conditions changes and the continuous that rise in atmospheric temperature undoubtedly affect the productivity of many cultivated crops. Therefore, attention should be given to discover and use of new genes and alleles that can resist heat stress. Also, improve the efficiency of breeding and selection methods using molecular markers and gene transfer techniques. Therefore, there are new trends in genetic modification technology in crop plants, which can be listed as follows:

1. Improve selection methods and the efficiency of selected genes.
2. Accurate identification of genetic loci for thermal stress resistance.
3. Deepening in the study of the structure and topography of the genome to find out the best loci for gene transfer.
4. Better identification of genetic control loci.
5. Improve the possibility of transferring large fragments of genetic materials.
6. Genotype modeling of cultivars in specific environments.

4.9.1 Molecular Markers

Molecular markers indicate the location of genes and alleles relevant with heat stress resistance. PCR technology has been used to diagnose and isolate the encoded *GPAT* gene for the production of Trienoic fatty acids associated with high-temperature acclimatization and confirmation of its transfer (Verma and Santha 2003). Eighty-three complete gene sequences and 89 incomplete genes for wheat, maize, barley and rice plants were responsible for heat shock proteins production and for identification of a huge number of similar counterpart genes in their synthesis of proteins, especially *HSP70* and *HSP90* (Lee and Schöffl 1996 and Lee and Vierling 2000). Barakat et al. (2004) succeeded in using five primers in distinguishing the genetic differences between bread wheat varieties Sohag 1, Gemmeiza 1, Sakha 69, Giza 167, and the different somaclones resulting from them, which differ in their tolerance to heat stress. RAPD analysis gives 42 amplifications, all of which were polymorphic. The similarity % between wheat varieties and somaclonal variations ranged from 41.3 to 84.8%. In rice, Amer et al (2016) utilized Polymerase chain reaction (PCR) to identify the *DREB2A* gene presence in the transgenic tissues of Sakha104 and Giza178, in improving their heat tolerance. However, in lentil, RAPD-PCR technique have been used to identify and differentiate newly derived Lines by Abo El-wafa et al. (2003) who found the new band 1078 bp in Line 3 and 1355 bp in the selected Line 4. Bands 2805, 965, 560, 1066, 1127, 3636 bp were positive biomarkers for ILL 6004, ILL 5753, ILL 6005, Line 4, Line 3 and Line 2, respectively. High homogeneity values are attributed to the lack of genetic diversity among them for their involvement in origin within the breeding program.

4.9.2   Gene Transfer and Tissue Culture Technology

Genetic engineering and gene transfer techniques are used to alter the gene expression of heat shock proteins (HSPs) and resistance genes in some model plants. New genes of plant and non-plant species were also used and transferred to the target genotypes. Ali et al. (1992) produced heat-resistant wheat cells and plants by treating callus wheat varieties Giza 157, Sakha 69 and Sohag 2 by DNA extracted from *Bacillus* sp at 35 °C and 40 °C. Sohag 2 cultivar give the highest percentage of transgenic plants resistant to heat at 40 °C followed by Giza 157 and Sakha 69. Thus, genetic improvement of heat resistance is accessible fast and efficient equivalent to five times as compared to traditional methods.

Field-testing of the new transgenic cotton hybrids using advanced recombinant DNA technology were prepared and supervised by the Agricultural Genetic Engineering Research Institute (AGERI), in cooperation with the Cotton Research Institute (CRI) of the Agricultural Research Center. The program aims to preserve cotton germplasm quality, increase cotton productivity by resisting biotic factors, and tolerating abiotic environmental conditions (salt, heat, and drought) for the sensitive types when cultivated mainly in the new reclaimed areas (Momtaz 2002).

Tissue culture has been successfully employed for crop tolerance to heat stress and the production of high-temperature resistant strains. The genetic differences in temperature tolerance were estimated at the level of cell cultures and the whole plant levels in three wheat varieties, by subjecting cell suspensions to thermal stress 37 °C for 24 h or 50 °C for 1 h. The analysis of heat shock proteins HSP in genotypes TAM

101 and ND 7532 showed the production of large amount of heat shock proteins of molecular weight 16 and 17 KDa compared with the sensitive ones at 37 °C for 24 h. The formation of heat shock proteins with molecular weight 22 and 23 KDa differed according to species and heat treatments (Wang and Nguyen 1989). Under Egyptian conditions, embryo culture technology helped to identify high tolerant-temperature wheat genotypes. In this regard, Barakat et al. (2004) reported the impact of high temperature on the cells resulting from the cultivation of mature embryos of four varieties of wheat are Sohag 1, Gemmeiza 1, Giza 167, and Sakha 69 subjected to 35 °C, at intervals selected under heat stress treatments i.e. H1 = control (25 $\pm$ 25 °C), H2 = 10 h continuous exposure to 35 °C, H3 = 3 h exposure per day at 35 °C with a total of 75 h, H4 = 3 h exposure per day at 35 °C with a total of 81 h and H5 = 3 h exposure per day to 35 °C with a total of 99 h. The results revealed that, weight of the callus was significantly affected by different genotypes. Significant differences were obtained between the thermal treatments and the genotypes in the callus which produce embryos and the formation of vegetative groups, while interaction between them was insignificant. Both varieties Gemmeiza 1 and Giza 176 give the highest percentage of the embryonic callus (32.49%) and (32.39%), respectively during the various heat treatments, while the lowest yield was (69.5%) and (11.47%), respectively. Gemmeiza 1 was produced the highest (20.01%) formation of vegetative groups during the heat treatments. Whereas, Amer et al. (2016) prepared a well-adopted technique for regeneration and transformation of Egyptian rice cultivars for enhancing heat tolerance. Among various sugar types, maltose with concentration of 4% was deliver the maximum proportion of callus induction. Thus, for shoot regeneration 2 mg/L kinetin plus 0.2 mg/L NAA selected as the best concentration. In addition, Sakha104 cv. displayed the maximum regeneration ability compared to Giza178 which was selected for the transformation experiments. Scutellum calli, from two to three-week-old, were co-cultivated with *Agrobacterium tumefaciens* harboring *DREB2A* gene that driven by the CAMV35S promoter, *GUS* gene, and hygromycin resistance gene. For Egyptian rice, OD600 = 0.8 evidenced to be the best concentration of *Agrobacterium*. That optimum co-cultivation circumstances were 3 days and 25 °C. GUS assay and polymerase chain reaction established the existence of the transgenes.

4.10   Agronomic Practices

In the light of continuing scenarios of industrial human activity scenarios, and carbon dioxide emissions, which increases global warming? This requiring world and sustained global efforts to mitigate heat stress (Peters et al. 2018). The strategy of enhancing productivity is depended on improving the efficiency of photosynthesis process and partition assimilates into the plant organs, combined with agricultural procedures.

4.10.1 Sowing Date

Most field crops in arid and semi-arid areas suffer from the heat stress effects. Where the high temperature is considered as one of the most essential environmental parameters that used for determinants affecting the yield potentiality and quality of the crop varieties. Under Upper Egypt conditions, research results indicate that the heat stress effect due to the delay of sowing date from 15 November to 15 December was more pronounced in its effect on reducing days to maturity and grain weight, while the effect was simple on days to heading, number of spikes/m^2 and plant height (Abdel-Shafi et al. 1999). Therefore, to avoid the effects of temperature rise on critical periods of crop growth and development, must be cultivated the most tolerant genotypes to heat stress at the appropriate time and follow the proper fertilization and irrigation programs.

Fifteen bread wheat genotypes (included two local check cultivars and 13 promising lines) were evaluated under optimum (23rd November) and late (21st December) sowing dates in two seasons by Menshawy (2007). The interactions of genotype x sowing date was significant for grain yield. Both 1000-grain weight and grain yield were more sensitive to sowing dates than number of spikes/m^2 and number of grains/spike. The comparisons indicated that the optimum sowing date resulted in higher 1000- grain weight by 7.3 gm, and grain yield by 3.5 ardab/fed. than the late sowing. Positive and significant relationship was detected between days to heading (days or thermal units) and heat sensitivity index. Therefore, early genotypes are more tolerant to heat stress, Assessing heat tolerance was done using twelve wheat genotypes under numerous environments i.e. two locations (Sohag and Assiut) at two sowing dates 14th November (favorable) and 26th December (heat stress), nitrogen fertilizer levels during three years. Hamam and Khaled (2009) showed that grain yield was influenced significantly by the impact of sowing dates, years, locations, genotypes, and nitrogen fertilizer levels. Grain yield of the different genotypes fluctuated from 10.40 to 15.47 ard/fed. for TRI 2586 and Bocro-4//kauz"S" genotypes, respectively. Under Assiut as heat stress location, mean performance of Johara-19 was improved from 13.76 ard/fed. at the favorable sowing date with 70 kg/fed. N fertilizer to 14.50 ardab/fed. at late sowing date with 100 kg/fed. N fertilizer. Giza 168 maintain productivity from optimum sowing date with 70 kg/fed. N fertilizer (12.04 ard/fed.) to the late sowing date due to increasing the nitrogen fertilization up to 100 kg/fed. N fertilizer produced 12.29 ard/fed.

El-Moselhy Omnya et al. (2015) showed that based on the genotype focused scaling for grain yield (ard./fed.) under diverse twelve environments as illustrated in Fig. 4.3, genotypes Line 27, Misr 1 and Sakha 93 were the most desirable and stable, of them, Misr 1 have high yield potentiality, whereas Line 27 was moderate yield potentiality and Sakha 93 was the lowest one. Whereas, the genotypes Misr 2, Sids 1, Line 14, Line 16, Giza 168 and Sids 12 were unstable and more responsive to environmental changes, from them, Misr 2, Sids 12, Line 14 and Line 16 have high yield potentiality. Sids 1 was moderate yield potentiality and Giza 168 was the lowest one. The best genotypes with respect to E2 and E10 were Misr 1, Sakha 93,

Genotypes	Environments
1. Misr 1	E1. Kafer Al-Hamam (D1) 2010/2011
2. Misr 2	E2. Kafer Al-Hamam (D2) 2010/2011
3. Sakha 93	E3. Kafer Al-Hamam (D3) 2010/2011
4. Giza 168	E4. Sids (D1) 2010/2011
5.Gemmeiza 11	E5. Sids (D2) 2010/2011
6.Shandawel 1	E6. Sids (D3) 2010/2011
7. Sids 1	E7. Kafer Al-Hamam (D1) 2011/2012
8. Sids 12	E8. Kafer Al-Hamam (D2) 2011/2012
9. Sids 13	E9. Kafer Al-Hamam (D3) 2011/2012
10.Line 27	E10. Sids (D1) 2011/2012
11. Line14	E11. Sids (D2) 2011/2012
12. Line 16	E12. Sids (D3) 2011/2012

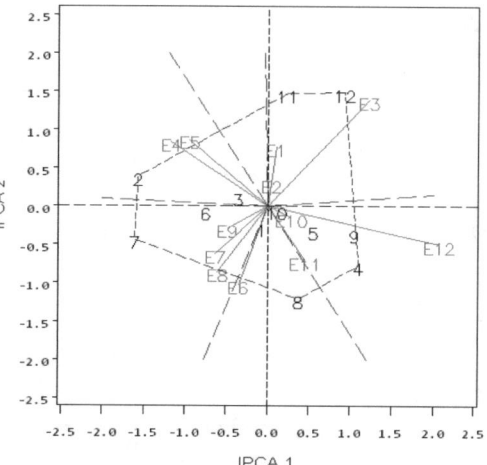

Fig. 4.3 Graphic display of the GE biplot for twelve bread wheat genotypes across twelve environments for grain yield ard./fed*. (ardab wheat = 150 kg., feddan = 4200 m^2) assessed (G1–G12) and the twelve environments considered (E1–E12) in the AMMI model (El-Moselhy, Omnya et al. 2015)

Gemmeiza 11 and Line 27; E9 for Shandweel 1; E1 for Line 14; Line 16 for E3; Giza 168 and Sids 13 for E12; Sids 1 for E6, E7 and E8 and Misr 2 for E4 and E5. The environments E3 and E12 were the discriminative environments.

Heat sensitivity index (HSI) indicated that Egyptian local wheat cultivars Giza 168 and Gemmeiza 11 as well as Line 27 exhibited H.S.I values less than unity under both seasons and locations as well as the combined, revealing that these cultivars were tolerant to the effect of the heat stress as presented in Table 4.2 (El-Moselhy Omnya et al. 2015).

Rice

Rice is sensitive to the stress that caused by high-temperature at almost the stages of its growth. That grain yields decline during the growing season by 10% for each 1 °C increasing in minimum temperature (Peng et al. 2006). Suitable temperature during plant growth ranged from 20 to 37 °C. Rice needs sunlight and lack of light intensity leads to underdevelopment (Anonymous 2020).

Usually rice is sown during the first half of May and should not delay the cultivation of nurseries for that date to avoid unfavorable environmental conditions which lead to a significant decrease in yield, and transplanted 25–30 days after planting to give high and stable yield. That the sowing date of rice plays a vital role in enhancing and improving its growth and increasing its yield. Rice crop sowing date is vital for three major causes; firstly, it ensures that vegetative growth occurs during a period of high levels of solar radiation and suitable temperatures. Secondly, the optimal planting time for each genotype declares that the minimum temperature at night is warmer. Thirdly, sowing on time ensures proper temperatures at the grain filling period, thus achieving better grain quality (Farrell et al. 2003). Early sowing dates (May 10th) had

Table 4.2 Heat sensitivity index of grain yield for 12 bread wheat genotypes (EL Moselhy Omnya 2015)

Genotypes	Season 2010/2011		Season 2011/2012		Overall H.S.I$_{1.3}$
	Kafer Al-Hamam	Sids	Kafer Al-Hamam	Sids	
	H.S.I$_{1.3}$	H.S.I$_{1.3}$	H.S.I$_{1.3}$	H.S.I$_{1.3}$	
Misr 1	0.85	1.17	0.99	0.95	0.99
Misr 2	1.26	1.11	0.96	1.47	1.20
Sakha 93	0.95	1.85	1.26	1.34	1.35
Giza 168	0.95	0.51	0.94	0.97	0.84
Gemmeiza 11	0.63	0.81	0.77	0.73	0.73
Shandweel 1	1.31	0.90	0.84	1.17	1.06
Sids 1 3/ 3/	1.23	0.92	1.00	1.23	1.10
Sids 12	1.39	0.64	1.41	0.84	1.07
Sids 13	0.88	0.52	1.08	0.59	0.77
Line 27	0.88	0.78	0.90	0.98	0.88
Line 14	0.85	1.37	1.18	1.04	1.11
Line 16	0.75	1.30	0.72	0.73	0.87

H.S.I$_{1.3}$: between favorable sowing date "20th November" and late sowing date 30th December "sever heat stress"

reflect the effect on number of panicles/m^2, number of filled grains/ panicle, 1000-grain weight, grain and straw yields/fed. Delaying sowing date sharply decreased leaf area index and dry matter production. In addition, delaying the rice sowing date up to June 15th had significantly decreased the number of days to heading El-Khoby (2004).

Under Sakha Agriculture Research Station, Egypt. Abou-Khadrah et al (2014) investigate the impact of planting method (transplanting and drill) and sowing dates (April 15th, May 1st and May 15th) on growth, yield and its components using different varieties such as of GZ 7112, GZ 9057 and Sakha 105. The results showed that using transplanting method for the variety GZ 9057 sown on April 15th significantly gave higher flag leaf angle, number of tillers/m^2, panicle length, panicle density (%), panicle weight, and number of grains/panicle. Moreover, the results showed that the highest estimates of rice grain yield were 4.31 and 4.65 t/fed. that was recorded by variety GZ 9057 under transplanting method with early sowing date.

Maize

Maize yield decrease usually with increasing the temperatures, that Lobell and Field (2007) revealed that yield decreased by 8.3% with increasing the temperature by 1 °C without any considering any complicating effect related to the water stress.

Higher mean values of early planting date may be due to the optimum environmental conditions that led to increasing the vegetative growth and consequently delaying days to flowering as revealed by Abd-Elaziz (2014) and Al-Falahy (2015). Under two different sowing dates (June 1st and July 1st), 15 maize parents (2 testers

and 13 inbred lines) crossed in the line x tester scheme for yield and its components. Hefny (2010) displayed significant differences for yield contributing traits for
each of the both sowing dates and combined data. Furthermore, delaying in sowing
dates caused a reduction in all yield contributing traits, excluding the ear diameter and rows/ear. The results showed that the sowing date had great influences on
gene expression for flowering traits, ear weight/plant, rows/ear and yield/plant which
reflecting the effects of changing in additive gene action at different planting dates.

   The resultant 45 single maize crosses along with two checks (S.C.10 and S.C.
30k8) were evaluated on two planting dates (Normal, May, 15th and Late, June
15th). Sedhom, Youstina et al. (2016) found that two crosses $P_2 \times P_8$ and $P6 \times P_8$
exhibited desirable heterotic effects in days to 50% silking, leaf chlorophyll content
and grain yield, especially under the late planting date. Sedhom Youstina (2016)
added that the best protein (13.18%) and oil (6.06%) content were registered by the
cross $P_4 \times P_7$ and $P_1 \times P_4$, respectively in the late planting date. Factor analysis
divided the eleven variables into four factors accounted for 74.58% and 74.71%
of the variability in the dependence structure in the first planting date Fig. 4.4 and
Table 4.3 and second planting date Fig. 4.5 and Table 4.4, respectively. The traits of
each factor and their contribution are varied from early to late sowing date. Selection
for the most important yield traits particularly rows/ear, grains/row, 100 grain weight
and shelling % would lead to maximizing total maize grain yield.

   Faba bean

The date of planting of faba bean is one of the factors determining the production of
the crop. Therefore, the latter half of October in Upper Egypt and early November in
Delta and Middle Egypt is the most suitable time for growing faba bean. The early or

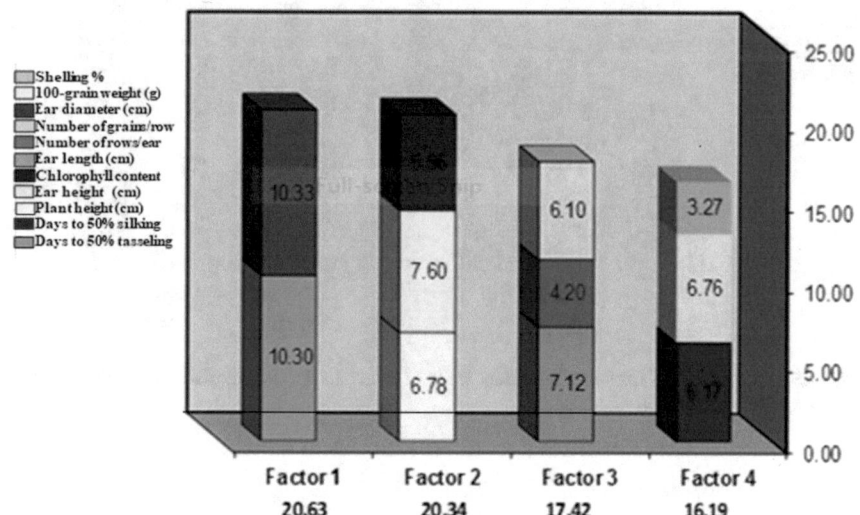

Fig. 4.4 Factor loading for some important of maize in the first planting date (May 15th) (Sedhom
Youstina 2016)

Table 4.3 Summary of factor loading for some important traits of maize in the first planting date (May 15th) (Sedhom Youstina 2016)

Variables	Loading	Percentage of total
Factor 1		20.63
days to 50% tasseling	0.934	49.95
days to 50% silking	0.936	50.05
Factor 2		20.34
Plant height (cm)	0.832	33.33
Ear height (cm)	0.933	37.38
Chlorophyll content	0.731	29.29
Factor 3		17.42
Ear length (cm)	0.884	40.87
Number of rows/ear	0.521	24.09
Number of grains/row	0.758	35.04
Factor 4		16.19
Ear diameter (cm)	0.775	38.08
100-grain weight (g)	0.849	41.72
Shelling percentage	0.411	20.20
Cumulative variance		74.58

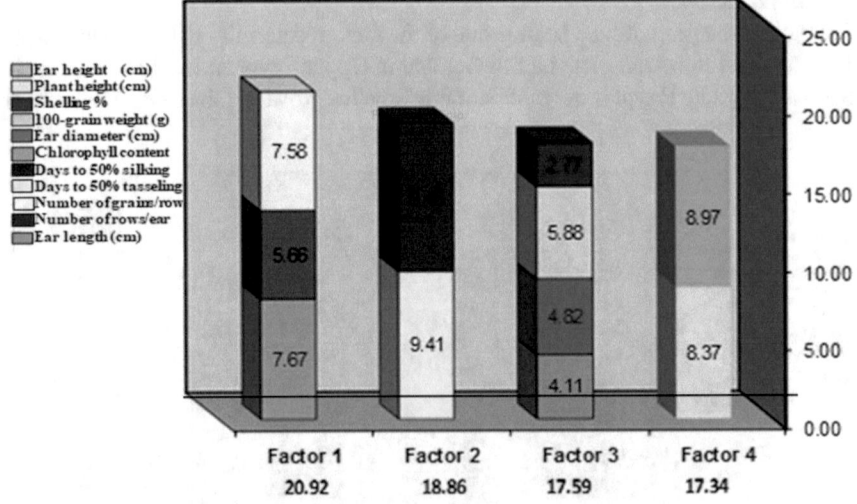

Fig. 4.5 Factor loading for some important traits of maize in the second planting date (June 15th) (Sedhom Youstina 2016)

Table 4.4 Summary of factor loading for some important traits of maize in the second planting date (June 15th) (Sedhom Youstina 2016)

Variables	Loading	Percentage of total
Factor 1		20.92
Ear length (cm)	0.897	36.69
Number of rows/ear	0.662	27.8
Number of grains/row	0.886	36.24
Factor 2		18.86
Days to 50% tasseling	0.931	49.89
Days to 50% silking	0.935	50.11
Factor 3		17.59
Chlorophyll content	0.604	23.39
Ear diameter (cm)	0.708	27.42
100-grain weight (g)	0.863	33.42
Shelling percentage	0.407	15.76
Factor 4		17.34
Plant height (cm)	0.891	48.27
Ear height (cm)	0.955	51.73
Cumulative variance		74.71

delayed dates lead to a 30–40% (Anonymous 2020). Performance of ten genotypes as affected by three sowing dates i.e. October 15, November 1 and November 15 at El-Mattana Research Station, Luxor Governorate, Egypt was investigated by Bakhiet et al. (2015). The completion of flowering was affected from sowing date to another. It was 44.36 day on the latest sowing and 43.33 day on the second sowing one. Highly significant effect of sowing dates was recorded on days to maturity with highest values (138.46 and 131.40 day) for the first sowing date during the first and second seasons, respectively. The highest values for seed yield (2.49 and 2.41 kg/plot) was verified at the medium sowing date during the first and the second seasons, respectively. While, the lowest values (2.10 and 2.19 kg/plot) for the trait was recorded on the latest sowing date in the first and the second season, respectively. This could be interpreted to the effect of unfavorable condition on physiological process. Under El-Fayoum location, Egypt, three irrigation intervals (20, 30 and 40 days) and a suggested three sowing dates (22nd October, 7th and 22nd November) were investigated on Nubaria1, Nubaria 2, Giza 3 improved, Sakha1 and Sakha 2 varieties. Ali Ekram et al. (2017) found that irrigation intervals (20 days) surpassed the others two irrigation intervals in plant height, seed yield and carbohydrate %. While, the early sowing on 22nd October gives the highest values of all seed yield contributors and carbohydrate %. Nubaria1 was classified as the first one in number of branches, 100-seed weight, pods and seed weight/plant, seed and straw yields and carbohydrate %. While, the maximum values of 100-seed weight, pods and seed weight/plant and seed yield (ton/fed.) were obtained by irrigation interval by 30 days with early sowing for Nubaria1 variety.

Soybean

Forecasts of climate change scenarios suggest a decline in soybean productivity to more than 50% as a result of heat stress (Schlenker and Roberts 2009). Pod development and seed filling in soybean are considered the most critical periods to environmental stress. Foroud et al. (1993) mentioned that stressful conditions reduced the yield of soybean. That the soybean plant ages from beginning bloom stage through the maturity of the seed, its ability to be compensated under stressful conditions decreased and yield losses increased.

At different stages of growth, panting date is one of the important factors in determining productivity of soybean cultivar and its tolerates stress under the environmental conditions. Kandil et al. (2013) carried out four separate experiments using six soybeans genotypes; Giza 21, Giza 22, Giza 111, H2L12, H30 and H32 evaluated under different planting dates i.e., 20th April, 5th May, 20th May and 5th June. The results showed that planting on 5th May produced highest values in all yield contributing traits, while the highest values of net assimilation rate and crop growth rate were obtained from planting on 5th June. In addition, the cultivar Giza 21 was consistently produced higher 100-seed weight, seed yield (ton/ha), protein and oil yields (kg/ha) than the remaining genotypes, while H30 line was the lowest one. Nevertheless, delaying planting dates to the 5th June altered the vegetative stage, and significantly pronounced with the line H30 for the net assimilation rate and the crop growth rate. Thus, the obtained results showed that planting in the first May is an efficient strategy to increase soybean yield. This might be due to avoid unsuitable environmental conditions. Onat et al. (2017) revealed the same phenomena that soybean cultivars differ in their sensitivity to high temperatures. That the seed yield was lower (29.5%) in 2016 season compared to 2015 season. This may be due to higher daily maximum air temperatures in 2016 compared to 2015.

Cotton

Cotton is grown during the month of March when the appropriate weather conditions are adopted. The planting date depends mainly on soil temperature. Agriculture should be maintained when soil temperature is about at 15 °C for 10 consecutive days at a depth of 20 cm at 8 am. Agriculture at the appropriate time will lead to reduction of the first fruiting node, increase No. of open bolls with large volume and early maturity, increase yield, quality of grade, and lint yield besides minimize the incidence of pests. In this respect, when Awaad and Nassar (2001) evaluate the stability and adaptability of 11 cotton varieties under 18 different environments which are the combination of two soil textural types of clay soil (Zagazig region) and sandy soil (Khattara Farm) and three planting dates (21th March, 11th April and 2nd May) during three seasons. The local variety Giza 89 was shown as more stability and wide adaptability under different environments for seed cotton yield and length uniformity (%). While the most adapted varieties under less favorable conditions were Giza 80 for lint %, seed oil content, earliness and length uniformity (%); Giza 77 for seed cotton yield, length uniformity (%), 2.5% span length, and fiber strength; Giza 75 for seed cotton yield and boll weight. This indicates the ability of these varieties to tolerate heat stress as induced by delaying planting date under less favorable conditions of Khattara region. In continuous, El- Sayed and El-Menshawy (2006) found that date of planting had a significant effect on growth, yield and its components traits. The early planting of Giza 88 on March 25 led to a significant increase rather than 25 April on the nodal position of the first sympodium, number of days to first open boll, earliness %, boll weight, number of open bolls/plant, seed cotton yield/fed., seed index and fiber fineness, while plant height was reduced.

4.10.2 Fertigation

Soil fertility and water supply are considered the most efficient parameters for plant growth. This leads to improvement of plant status under environmental stress conditions.

Wheat

Maintaining adequate moisture conditions in the soil is an essential strategy to reduce the heat stress impact on crop plants. That at flowering stage, one day of high temperature, lead to 25% reduction in wheat grain yield (Hall 2010). Fertilization and irrigation are the most agricultural procedures that supporting the plant growth under high temperature conditions. Vulnerability of wheat to extreme weather under climate change scenarios (Ibrahim et al. 2012) studied fertigation treatments on wheat. That nitrogen fertilizer such as ammonium nitrate with a rate of 400 kg/ha, potassium sulfate at rate of 100 kg/ha and phosphoric acid (60%) at the rate of 125 kg/ha. The results revealed that, the highest crop yield and water use efficiency were obtained under irrigation supply using 1.2 and 0.8 of Etc, respectively, and fertigation supply in 80% of application time. They recommended irrigating wheat under sandy soil condition with an amount of either 1.0 or 0.8 of ETc with fertigation application in 80% of application time to improve growth and yield, and to mitigate injury on wheat caused by extreme climate change (https://scielo.conicyt.cl/pdf/jsspn/v12n2/art01.pdf).

Faba bean

Heat stress considers one of the main abiotic stresses that affect the yield productivity of faba bean. The high yield genotypes under terminal heat stress used to improve the faba bea heat tolerance and facilitate the possibilities of extending faba bean production in non-traditional regions. Thus, Abdelmula and Abuanja (2007) simulated the impact of heat stress on faba been by sowing it at three dates with interval of 14 days using twenty-two genotypes at several regions; Shambat, Khartoum, Sudan. The genotypes C.52/1/1/1 and C.42, under extreme heat stress, were more stable and revealed the highest yield with moderately tolerant. Efficient correlation was observed between seed yield/plant and each of plant height, number of podded nodes/main stem, number of pods/plant, 100-seed and weight of dry matter/plant. So, could be used as selection criteria for improving faba bean under heat stress conditions. Furthermore, Tayel and Sabreen (2011) studied the effect of skipping two irrigations (using drip irrigation regimes) and phosphorous levels (10, 15 and 20 kg P_2O_5/fed) on two faba bean varieties at different growth stages. Variety Giza Blanca surpassed Giza 461 in all yield characters excluding the plant height.

Under arid and semi-arid regions and deficit irrigation, faba bean could be water stressed at podding stage and achieves comparative growth traits and saves 11.4% of irrigation requirement, relative to the control. Furthermore, Alderfasi and Alghamdi (2010) showed highest plant vigor by using 75% of water holding capacity with the fertilization rates 100 and 200 kg/ha of phosphate and potassium, respectively. However, El- Gizawy and Mehasen (2009) found that using 30 kg P_2O_5/fed mixed with the phosphate dissolving bacteria extremely increased plant height, number of branches/plant, 100-seed weight and seed and straw yields/fed.

Under Northern coast of Egypt, Hegab et al. (2014) studied the response of faba bean cultivar Sakha 3 to five different sowing dates (1st of October, mid of October and 1st of November, mid of November and 1st of December) and three different irrigation levels 0.60, 0.80 and 1.00 of (IR) using drip irrigation. The results revealed that the vegetative growth characters, biological yield and seed yield were decreased as sowing date delayed beyond the 1st of November and with the application rate of 0.60 of (IR). However, the highest seed yield obtained by the 1.00 irrigation treatment. The 0.60 irrigation level gave the maximum water use efficiency on the third sowing date (1st November), followed by the second sowing date (mid. October). However, the lowest water use efficiency was noticed by delaying the sowing date (1st December). The lowest irrigation level (0.60 IR) joint with sowing date on 1st November gave the highest contents of protein and carbohydrate in the seeds (http:// independent.academia.edu/hEGABA).

Soybean

Soil fertility may also affect the degree to which heat and water stress affect crop production. Soil fertility could be enhanced by applying macronutrients and micronutrients under the extreme heat stress. That applications activated physiological processes and altered stomatal function which helped in improving the thermal stress tolerance. In sandy soil as stress environment, Soliman et al. (1995) investigated the effect of inoculation of soybeans with *A. brasilense* and *B. japonicum* either solely or in mixture, with using different nitrogen fertilizer levels. The results showed that the highest values of nodules number and fresh weight were obtained at rate of 20 kg N/ha, while decreased with increasing the nitrogen rate 40 kg N/ha. Nevertheless, with increasing the N fertilizer levels, the N uptake and the dry weight of the above ground parts was increased. Biofertilizer only or at the treatment received organic manure + yeast produced maximum seed oil content (Mekki and Ahmed Amal 2005). Otherwise, organic singly or when it associated with biofertilizer improve protein content.

Cotton

Effect of different fertilizers levels on the cotton yield were described such as nitrogen, potassium, and Mepiquat chloride (MC) to identify their effects on the Egyptian cotton cultivar Giza 86 by Zakaria Sawan et al. (2006). That applying nitrogen at 143 kg/ha combined with using foliar application of potassium at rate of 319 g/ha and MC at 48 + 24 g/ha enhanced the growth and yield. Furthermore, Attia et al. (2008) investigated Egyptian cotton variety Giza 89 and the effect of three first post planting irrigation times with ten fertilization treatments on growth, yield and fiber characters. The times of first post planting irrigation in a separate experiment were 20, 30 and 40 days from planting irrigation time. Adding the first irrigation after 40 days from planting reveled the highest values of growth traits, fiber strength and fiber length uniformity ratio (%). However, application of the first irrigation after 30 days from planting gave the highest estimates of most yield and its components in comparison with after 20 days from planting. Fertilizing cotton plants with organic fertilizer + bio-fertilizer + 66% from recommended NPK provided maximum values of the traits studied.

4.10.3   Intercropping

Intercropping strategy reduce the heat load on components of intercropping types, increase the efficiency of environmental resources, and mitigate the heat stress impact on plants. Gaballah and Ouda Samiha (2008) obtained higher yield under a system of 1:2 soybean/maize intercropping pattern with using irrigation by evaporation pan coefficient equal to 1.2, compared to sole maize or sole soybean planting. While, under 1.0 and 0.8 pan evaporation coefficients (7 and 14% reduction in irrigation water than the control, respectively), the lowest decrease in both soybean and maize yield was registered from 1:2 soybean/maize intercropping pattern with highest water use efficiency.

4.11   Conclusions

High temperatures affect crop productivity in many regions of the world, especially arid, semi-arid and hot regions such as the Arab region. The areas of horizontal expansion in the Egyptian deserts are one of these areas. In the light of the increase in Egyptian population and extension of agriculture to marginal lands where crop plants face various adverse effects. Water and heat are major environmental factors affecting crop productivity. Therefore, strategies should be developed to cope with the climate change for mitigating the negative impacts of extreme heat stress on crop productivity. There are several practical options for adapting to increased temperatures. The first option is exploited DNA-markers technology to diagnose and isolate the gene(s) associated with high-temperature acclimatization and confirmation of its transfer to sensitive cultivars. The second option is growing early maturing and tolerant crop cultivars to heat stress. The third option is adjusted crop irrigation dates with meteorological data, and follow appropriate fertigation programs to avoid the harmful effects of high heat waves.

4.12   Recommendations

The effects of heat stress can be addressed and reduced through the following procedures:

– cultivate the most tolerate genotypes to heat stress at the appropriate time
– Application the appropriate agricultural procedures
– Follow appropriate fertilization and irrigation programs.
– Taking into account the interest of the critical periods of plant life and avoid heat stress through irrigation.
– Link crop irrigation dates with meteorological data to avoid the harmful effects of high heat waves.

Acknowledgements   The authors acknowledge the partial support of the Science and Technology Development Fund (STDF) of Egypt in the framework of the grant no. 30771 for the project titled "A Novel Standalone Solar-Driven Agriculture Greenhouse - Desalination System: That Grows Its Energy And Irrigation Water" via the Newton-Musharafa funding scheme.

References

Abd-Elaziz MA (2014) Diallel analysis for some agronomic attribute of maize under different planting dates. Ph.D thesis, Faculty of Agriculture, Benha University, Egypt
Abdelmula AA, Abuanja IK (2007) Genotypic responses, yield stability, and association between traits among some of Sudanese Faba bean (*Vicia faba* L.) genotypes under heat stress. Tropentag 2007 University of Kassel-Witzenhausen and University of Göttingen, October 9–11, 2007, Conference on International Agricultural Research for Development, p 1–7
Abdel-Shafi AM, Abdel-Ghani AM, Tawfelis MG, Mossaad MG, Moshref MKH (1999) Screening of wheat germplasm for heat tolerance in Upper Egypt. Proceedings of First Pl Breed Conference December 4, (Giza) Egypt J Plant Breed 3:77–87
Abdo-Elwafa A, Taghian AS, El-Aref HM, El-Sayed E, Abd-Elkrim MA (2003) Response to pedigree selection for seed yield and molecular genetic markers for new selected lines of lentil. Assiut J Agric Sci 34(2):1–23
Abou Gabal Ashgan A, Zaitoun Amera F (2015) Seed oil content and fatty acids composition of maize under heat and water stress. Alexandria Sci Exch J 36(3):274–281
Abou-Khadrah SH, Abo-Youssef MI, Emad MH, Amgad AR (2014) Effect of planting methods and sowing dates on yield and yield attributes of rice varieties under DUS. Sci Agric 8(3):133–139
Alderfasi AA, Alghamdi SS (2010) Integrated water supply with nutrient requirements on growth, photosynthesis, productivity, chemical status and seed yield of faba bean. Am Eurasion J Agron 3(1):8–17
Al-Falahy MAH (2015) Estimation combining ability, heterosis and some genetic parameters across four environments using full diallel cross method. Int J Pure Appl Sci Technol 26(1):34–44
Ali AMM, El–Hinnawy MA, El-Kholy HK, Hogran A (1992) Genetically engineered sodium chloride and *Puccinia recondite* tolerant wheat cells and plants. Egypt J Appl Sci 7(8):675–690
Ali Ekram M, EL-Sherif AMA, Mohamed MS (2017) Performance of five Faba bean varieties under different irrigation intervals and sowing dates in newly reclaimed soil. Int J Agron Agric Res 10(4):57–66

Ali MB, Ibrahim AMH, Hays DB, Ristic Z, Jianming F (2010) Wild tetraploid wheat (*Triticum turgidum* L) response to heat stress. J Crop Imp 24:228–243

Al-Khatib K, Paulsen GM (1990) Photosynthesis and productivity during high temperature stress of wheat genotypes from major world regions. Crops Sci 30:1127–1132

Amer A, Eid S, Aly U (2016) Assessment of various factors for high efficiency transformation of Egyptian rice involving DREP2A gene. Int J ChemTech Res 9(12):201–213

Anonymous (2016) Crop Principals. Crop Science Department, Faculty of Agriculture, Zagazig University, Egypt

Anonymous (2017) Wheat production and consumption. Economic Affairs Sector ARC, Giza, Egypt

Anonymous (2019) Recommendation techniques of field crops. ARC, Giza, Egypt

Anonymous (2020) Recommendation techniques of field crops. ARC, Giza, Egypt

Attia AN, Sultan MS, Said EM, Zina AM, Khalifa AE (2008) Effect of the first irrigation time and fertilization treatments on growth, yield, yield components and fiber traits of cotton. J. Agron 7:70–75

Atta Allah SAA (2001) Performance of some soybean cultivars at three N fertilization levels in newly reclaimed sandy soil. J Agric Res of Develop 21(1): 155–173

Awaad HA (2001) The relative importance and inheritance of grain filling rate and period and some related traits to grain yield of bread wheat (*Triticum aestivum* L.). Proceedings of 2nd P1 Breed Conference, Assiut University, October 2nd 2001, P 181–198

Awaad HA (2009) Genetics and breeding crops for environmental stress tolerance, I: Drought, heat stress and environmental pollutants. Egyptian Library, Egypt

Awaad HA, Nassar MAA (2001) Genotype x environment interaction for yield and fiber quality in cotton (*Gossypium barbadense*, L.). J Adv Agric Res 6(2):337–361

Badr Asmaa MSH (2017) Breeding wheat for heat tolerance. MSc thesis, Department of Agronomy, Faculty of Agriculture, Ain Shams University

Bakhiet MA, Rania AREl, Raslan MA, Nagat GA (2015) Genetic variability, heritability and correlation in some faba bean genotypes under different sowing dates. World Appl Sci J 33(8):1315–1324

Barakat MN, Mohamed AE, El-Metalny AM, Omar AA (2004) In vitro selection and identification of wheat somaclones tolerance to high temperature via RAPD markers. Alex J Agric Res 29(2):23–32

Boston RS, Viitanene PV, Vierling EV (1996) Molecular chaorones and protein folding in plants. Plant Mol Biol 32:191–222

Brown PW (2013) Heat Units. College of Agriculture and Life Sciences Cooperative Extension. AZ1602, P 1–6

Craufured PQ, Wheeler TR, Ellis RH, Summerfield RJ, Williams JH (1999) Effect of temperature and water deficit on water-use efficiency, carbon isotope discrimination, and specific leaf area in peanut. Crop Sci 39(1):136–142

Deshmukh PS, Sairam RK, Kushwaha SR, Singh TP (2003) Simple indices for measuring stress tolerance in crop plants. Abst: 2nd International Conference of Plant Physiology, January, 8–12, 2003, New Delhi, India P 148

ElBasyoni I, Saadalla M, Baenziger S, Bockelman H, Morsy S (2017) Cell membrane stability and association mapping for drought and heat tolerance in a worldwide wheat collection. Sustainability 9:1–16

El-Enany Magda AM, Sherin AM, Sara EIE (2013) Gene expression of heat stress on protein and antioxidant enzyme activities of two lupinus species. J Appl Sci Res 9(1):240–247

El-Gizawy NKHB, Mehasen SAS (2009) Response of faba bean to bio. Mineral phosphorus fertilizers and foliar application with Zinc. World Appl Sci J 6(10):1359–1365

El-Khoby WM (2004) Study the effect of some cultural practices on rice crop. Ph.D thesis, Faculty of Agriculture Kafr Elsheikh, Tanta University, Egypt

El-Marsafawy S) 2007) Climate change and its impact on the agriculture sector in Egypt. http://www.radcon.sci.eg/environment2/ArticlsIdeasDetails.aspx?ArticlId=35

El-Moselhy Omnya M (2015) Gene action and stability for some characters related to heat stress in bread wheat. Ph.D thesis, Agronomy Department, Faculty of Agriculture, Zagazig University, Egypt

El-Moselhy Omnya M, Ali AAG, Awaad HA, Sweelam AA (2015) Phenotypic and genotypic stability for grain yield in bread wheat across different environment. Zagazig J Agric Res 42(5):913–926

El-Sayed EA, El-Menshawy M (2006) Influence of indolacetic acid (IAA) application under different planting dates on growth and yield of Giza 88 cotton cultivar. Egypt J Agric Res 84(2):505–519

FAO (2015) Climate change and food security, UN Food & Agricultural Organization FAO. http://www.climatechange-foodsecurity.org/fao.html

Farooq M, Wahid A, Kobayashi N, Fujita D, Basra SMA (2009) Plant drought stress: effects, mechanisms and management. Agron Sustain Dev 29:185–212

Farrell TC, Fox K, Williams RL, Fukai S, Lewin LG (2003) Avoiding low temperature damage in Australia's rice industry with photoperiod sensitive cultivars. Proceedings of the 11th Australian Agronomy Conference. Deakin University, Geelong, VIC, Australia, February 2–6

Foroud N, Mundel HH, Saindon G, Entz T (1993) Effect of level and timing of moisture stress on soybean yield components. Irri Sci 13:149–155

Gaballah MS, Ouda Samiha A (2008) Effect of water stress on the yield of soybean and maize grown under different intercropping patterns. Twelfth International Water Technology Conference, IWTC12 2008, Alexandria, Egypt 1–14

Gupta SC, Sharma N, Singh K, Afria BS, Saini RS (2003) Physiological responses of chickpea genotypes to thermal and moisture stress. Abstract: 2nd International Conference of Plant Physiology, January 8–12, 2003, New Delhi, India, P 234

Hall G (2010) Heat stress can reduce wheat yield by a quarter. Hart Field-Site Group Inc PO Box 939 CLARE SA 5453

Hamam KA (2013) Response of bread wheat genotypes to heat stress. Jordan J Agric Sci 9(4):486–506

Hamam KA, Khaled AGA (2009) Stability of wheat genotypes under different environments and their evaluation under sowing dates and nitrogen fertilizer levels. Aust J Basic Appl Sci 3(1):206–2017

Hefny M (2010) Genetic control of flowering traits, yield and its components in maize (Zea mays L.) at different sowing dates. Asian J Crop Sci 2(4):236–249

Hegab ASA, Fayed MTB, Maha MAH, Abdrabbo MAA (2014) Productivity and irrigation requirements of faba-bean in north delta of Egypt in relation to planting dates. Ann Agric Sci 59(2):185–193

Hickey H (2017) Earth likely to warm more than 2 degrees this century. http://www.washington.edu/news/2017/07/31/earth-likely-to-warm-more-than-2-degrees-this-century/ Search UW News, NIH grants: HD054511, HD070936

Hong SW, Vierling E (2000) Mutants of Arabidopsis thaliana defective in the acquisition of tolerance to high temperature stress. Proc Nath Acad Sci USA 97:4392–4397

Iba k (2003) Trienoic acids and plant tolerance to tempreture. Abstract: 2nd International Conference of Plant Physiology, January 8–12, 2003, New Delhi, India, P 146

Ibrahim MM, Ouda Samiha A, Taha A, El Afandi G, Eid SM (2012) Water management for wheat grown in sandy soil under climate change conditions. J Soil Sci Plant Nutr 12(2):195–210

Irmak S (2016) Impacts of extreme heat stress and increased soil temperature on plant growth and development. Published by IANR Media CropWatch Privacy Policy. https://cropwatch.unl.edu/2016/impacts-extreme-heat-stress-and-increased-soil-temperature-plant-growth-and-development

Kandil AA, Sharief AE, Morsy AR, El-Sayed AIM (2013) Influence of planting date on some genotypes of soybean growth, yield and seed quality. J Biol Sci 13:146–151

Kumari M, Pudake RN, Singh VP, Joshi A (2012) Association of stay green trait with canopy temperature depression and yield traits under terminal heat stress in wheat (*Triticum aestivum* L.). Euphytica 190(1):87–97

Lee JH, Schöffl F (1996) An *Hsp 70* antisense gene affects the expression of *HSP 70/ HSP 70*, the regulation of HSF, and the acquisition of thermotolerance in transgenic *Arabidopsis thaliana*. Mol Gen Genet 252:11–19

Lee JH, Vierling E (2000) A small heat shock protein cooperates with heat stress shock protein 70 systems to reactive a heat-denatured protein. Plant Physiol 122:189–198

Lobell DB, Field CB (2007) Global scale climate–crop yield relationships and the impacts of recent warming. Environ Res Lett 2:014002. https://doi.org/10.1088/1748-9326/2/1/014002

Lu ZM, Chen J, Percy RG, Zeiger E (1997) Photosynthesis rate, stomatal conductance and leaf area in two cotton species (*G. barbadense* and *G. hirsutum*) and their relation to heat resistance and yield. Aust J Plant Physiol 240:693–700

Malik TA, Wright D, Virk DS (1999) Inheritance of net photosynthesis and transpiration efficiency in spring whaet, *Triticum aestivum* L., under drought. Plant Breed 118(1):93–95

Mekki BB, Ahmed AG (2005) Yield and seed quality of soybean (*Glycine max* L.) As affected by organic, biofertilizer and yeast. Appl Res J Agric Biol Sci 1(4):320–324

Menshawy AMM (2007) Evaluation of some early bread wheat genotypes under different sowing dates. Proceedings of Fifth Plant Breeding Conference. May 27, 2007 (Giza), Egyptian J Plant Breed 11(1):25–40 (Special Issue)

Momtaz OA (2002) Current status and prospects of transgenic Egyptian cotton. Agricultural Genetic Engineering Research Institute, Egypt. https://www.icac.org/meetings/plenary/61cairo/ …/tis/momtaz.pdf

Mondal S, Singh RP, Crossa J, Huerta-Espino J, Sharma I, Chatrath R, Singh GP, Sohu VS, Mavi GS, Sukaru VSP, Kalappanavarge IK, Mishra VK, Hussain M, Gautam NR, Uddin J, Barma NCD, Hakim A, Joshi AK (2013) Earliness in wheat: A key to adaptation under terminal and continual high temperature stress in South Asia. Field Crop Res 151:19–26

Moursi AM (2003) Performance of grain yield for some wheat genotypes under stress by chemical desiccation. Ph.D thesis, Agronomy Department, Faculty of Agriculture, Zagazig University, Egypt

Natu OS, Savitha M, Babu S, Udayakumar M (2003) Heat response of wheat cultivars differing in the thermotolerance. Abstract: 2nd International Conference of Plant Physiology, January 8–12, 2003, New Delhi, India, P 246

Omara MK (2010) Selection for cell membrane thermostability and stomatal frequency under drought and heat stress conditions in wheat (*Triticum aestivum* L.). Assiut J Agric Sci 41(2):74–100

Onat B, Bakal H, Leyla G, Halis A (2017) The effects of high temperature at the growing period on yield and yield components of soybean [*Glycine max* (L.) Merr] varieties. Turk J Field Crops 22(2):178–186

Pareek A, Singla-Pareek SL, Grover A (1995) Immunological evidence for accumulation of two high-molecular-weight (104 and 90 kDa) HSPs in response to different stresses in rice and in response to high temperature stress in diverse plant genera. Plant Mol Biol 29(2):293–301

Peng S, Huang J, Sheehy JE, Laza RC, Visperas RM, Zhong X, Centeno GS, Khush GS (2006) Rice yields decline with higher night temperature from global warming, Proc Natl Acad Sci 101(27):9971–9975

Peters GP, Andrew RM, Boden T, Canadell JG, Ciais P, Quéré CL, Marland C, Raupach MR, Wilson C (2018) The challenge to keep global warming below 2 °C. Springer Nature, © 2018 Macmillan Publishers Limited, part of Springer Nature

Rane J (2003) Evaluation of advanced wheat accessions high of temperature tolerance. Abstract: 2nd International Conference of Plant Physiology, January 8–12, 2003, New Delhi, India, P 254

Rao JSP, Rao GTR, Vijayalakshmi K (2003) Effect of combined temperature and moisture stresses on physiological and biochemical parameters of grain legumes. Abstract: 2nd International Conference of Plant Physiology, January 8–12, 2003, New Delhi, India, p 455

Reynolds MP, Balota M, Delgado MIB, Amani J, Fischer RA (1994) Physiological and morphological traits associated with spring wheat yield under hot irrigated conditions. Aust J Plant Physiol
2:717–730

Reynolds MP, Ortiz-Monasterio JI, McNab A (2001) Application of physiology in wheat breeding.
CIMMYT, El Batan, Mexico

Reynolds MP, Singh RP, Ibrahim A, Ageeb OAA, Larque-Saavedra A, Quick JS (1998) Evaluating physiological traits to complement empirical selection for wheat in warm environments.
Euphytica 100:84–95

Ristic Z, Bukovnik U, Vara Prasad PV (2007) Correlation between heat stability of thylakoid
membranes and loss of chlorophyll in winter wheat under heat stress. Crop Sci 47:2067–2073

Sairam RK, Srivastava GC, Agarwal S, Meena PC (2003) Plant antioxidant system: Defense against
multiple abiotic stresses. Abstract: 2nd International Conference of Plant Physiology, January
8–12, 2003, New Delhi, India, P 148

Schlenker W, Roberts MJ (2009) Nonlinear temperature effects indicate severe damages to US
crop yields under climate change. Proc Natl Acad Sci USA 106:15594–15598. https://doi.org/
10.1073/pnas.0906865106

Sedhom Youstina AS (2016) Gene action controlling yield contributing traits using RAPD markers
in maize. MSc thesis, Agronomy Department, Faculty of Agriculture, Zagazig University, Egypt

Sedhom Youstina AS, Ali MMA, Awaad HA, Rabie HA (2016) Heterosis and factor analysis for
some important traits in new maize hybrids. Zagazig J Agric Res 43(3):711–728

Shahinnia F, Roy JL, Laborde B, Sznajder B, Kalambettu P, Mahjourimajd S, Tilbrook J, Fleury
D (2016) Genetic association of stomatal traits and yield in wheat grown in low rainfall
environments. BMC Plant Biology BMC series–open, inclusive and trusted 201616:150

Shawky A, Fayad A, Heakal MY, Dhindsa R, Mansour A (2005) The existence and regulation
of heat—responsive HSP27 and MAPAPK2 proteins in tobacco cells. Zagazig J Agric Res
32(2):491–499

Siddiqui MH, Mutahhar YAl, Al-Qutami MA, Al-Whaibi MH, Anil G, Ali HM, MS Al-Wahibi
(2015) Morphological and physiological characterization of different genotypes of faba bean
under heat stress. Saudi J Biol Sci 22(5):656–663

Singh NB, Ahmed Z (2003) Seedling vigour as an index for assessing terminal heat tolerance
in wheat under irrigated late sown condition. Abstract: 2nd International Conference of Plant
Physiology, January 8–12, 2003, New Delhi, India, P 173

Soliman S, Galal YGM, Elghandour I (1995) bio fertilization of soybean in sandy soils of Egypt
using N-15 tracer technique. Proceedings of the second Arab Conference on the peaceful uses of
atomic energy. Folia Microbiol 40(3):321–326

Talebi R (2011) Evaluation of chlorophyll content and canopy temperature as indicators for drought
tolerance in durum wheat (*Triticum durum* desf.). Aust Basic Appl Sci 5:1457–1462

Tayel MY, Sabreen KhP (2011) Effect of irrigation regimes and phosphorus level on Two *Vica faba*
Varieties: 1- Growth traits. J Appl Sci Res 7(6):1007–1015

Verma M, Santha IM (2003) Cloning of glycerol –3-phosphate acyltransferase gene from *Bassica
juncea*. Abstract: 2nd International Conference of Plant Physiology, January 8–12, 2003, New
Delhi, India P 472

Wang WC, Ntuyen HT (1989) Terminal stress evaluation of suspension cell cultures in winter wheat.
Plant Cell Rep 8:108–111

Yang J, Sears RG, Gill BS, Paulsen GM (2002) Genotypic differences in utilization of assimilate sources during maturation of wheat under chronic heat and heat shock stresses. Euphytica
125:179–188

Young TE, Ling J, Geisler-Lee CJ, Tanguay RL, Caldwell C, Gallie DR (2001) Developmental and
thermal regulation of the maize heat shock protein, HSP101. Plant Physiol 127(3):777–791

Zakaria Sawan M, Mahmoud MH, El-Guibali Amal H (2006) Response of yield, yield components,
and fiber properties of Egyptian Cotton (*Gossypium barbadense* L.) to Nitrogen fertilization and
foliar-applied potassium and mepiquat chloride. J Cotton Sci 10:224–234

Chapter 5
Environmental Pollution Tolerance in Crop Plants

Hassan Auda Awaad

Abstract There are several environmental pollutants which spread in the atmosphere around the crop plants. The most important of these pollutants are ozone, sulfur dioxide, nitrogen dioxide, peroxyacetyl nitrate, hydrogen chloride, ethylene, carbon dioxide and hydrocarbons that are harmful to the plant. It has also increased the negative and harmful effects of heavy metals on plant, animal and human, which cause different types of pollution of air, soil and water such as lead, cadmium, mercury, nickel, cobalt, copper, zinc and others as a result of various non-inhalers human activities. These contaminants reach the plant either by fall of dust on the surfaces of leaves or by the use of chemical fertilizers, pesticides and irrigation water with low-quality. These contaminants cause severe damage in chloroplast and destroy the chlorophyll and reduce photosynthetic activity. Pollutants also affect building of amino acids and proteins and lead to a lack activity of DNA, RNA and gene expression, and that is reflected in the end on lower crop productivity and quality. However, the biological scientists pointed out the possibility of reducing pollution and prevention of risks in different ways through use high-tolerant genotypes, phytoremediation and use of certain micro-organisms that possess specialized systems of enzymes such as mycorrhizal fungi and algal Nostoc and Anabaena, besides the use of biotechnology techniques. God Almighty grants the crop plants with lines of defense, mechanics and characters to adapt and tolerate the impact of environmental pollutants include avoidance, tolerance and resistance. These mechanics are subjected under the genetic control system.

Keywords Environmental pollutants · Heavy metals · Resistance mechanisms · Genetic control · Biotechnology · Phytoremediation

H. A. Awaad (✉)
Crop Science Department, Faculty of Agriculture, Zagazig University, 44511, Zagazig, Egypt

© Springer Nature Switzerland AG 2021
H. Awaad et al. (eds.), *Mitigating Environmental Stresses for Agricultural Sustainability in Egypt*, Springer Water,
https://doi.org/10.1007/978-3-030-64323-2_5

5.1 Introduction

Pollution, a word of general meaning, meaning the appearance of something undesirable in an inappropriate place. With rapid urbanization and the development of modern industry, many soil pollution cases have been monitored with heavy metals and other pollutants in different regions of the world. Assessing the impact of pollutants on crop plants is an important objective in improving the quality of food.

The world's population is estimated to be around 10 billion by 2050 (https://verone twork.wordpress.com/2013/05/19/major-cultural-realms-of-the-world), which will result in severe food shortages and quality under agricultural adjustments. Herby, crops adapted to the adverse environmental factors must be developed to cope with the increase in the world population. FAO (2014) reveal that net worldwide greenhouse gas emissions produced "from agriculture, forestry, and other land usees exceeded 8 billion metric tons of CO_2" https://www.epa.gov/ghgemissions/global-greenhouse-gas-emissions-data). It produces a significant share of global greenhouse gas emissions and destroys habitats. These trends threaten the sustainability of food schemes and undermine the world's ability to meet its food requirements. In another report of FAO (2018) stated that excessive use of nitrogen fertilizers and the spread of environmental pollutants from heavy metals such as arsenic, cadmium, lead and mercury lead to harmful effects on plant metabolism and reduced crop productivity and quality. When these pollutants enter the food chain, they pose risks to food safety, water resources, rural livelihoods and human health.

Assessing the impact of pollutants is important in examining the quality of the environment and how to deal with sources of pollution (Pal 2016). The high concentrations of ozone, sulfur and nitrogen oxides, and the diffusion of heavy element particles from environmental contaminants, affect significantly on the air quality. Pollutants also affect plants, human and animal health (Awaad et al. 2013). In some estimates, the decrease in wheat yield due to heavy metals pollution is more than 20% beside the effect on quality (EL-Gharbawy 2015). Hence it is of the importance of reducing the factors that lead to the environmental pollution and reduce their risks. This chapter focuses on the various aspects of the effect of environmental pollutants on crop plants, crop adaptation mechanisms, role of plant breeding and biotechnology as well as procedures to mitigate the impact of the pollutants.

5.2 Source of Pollutants

5.2.1 Natural Resources

These are not the income of humans, which are those gases resulting from volcanoes, forest fires and storm-induced dust and examples of natural pollutants including:

1. Sulfur dioxide, hydrogen fluoride, and hydrogen chloride.

2. Nitrogen oxides resulting from electrostatic discharge of thunderclouds.
3. Hydrogen sulfide from the extraction of natural gas from the ground, volcanoes or the presence of sulfur bacteria.
4. Ozone gas generated in the air or due to electrical discharge in the clouds.
5. Falling dust left over from comets and meteorites to the surface layers of the atmosphere.
6. Salts that spread in the air by wind and air storms.
7. Plant pollen grains.
8. Fungi, bacteria and various microbes that spread in the air, either from the soil or as a result of rotting animals and dead birds and human waste.
9. Radioactive materials such as those found in some of the crustal rocks or resulting from the ionization of some gases by cosmic rays.

5.2.2 An Unnatural Source

This source is caused by the human activities being is more dangerous than the previous and disrupts the composition of the natural air as well as in the ecological balance. The resulting damage can be reduced, the most important of which are:

1. Use of fuel in industry.
2. Means of land, sea and air transport.
3. Radioactivity.
4. Leakage of ultraviolet radiation, which harms crops such as vegetables, fruit and field crops and threatens agricultural resources.

5.3 Tolerance to Air Pollutants

a. **Ozone**

Ozone is an active form of oxygen, each of which contains 3 atoms (O_3). Ozone is formed when atmospheric air oxygen is exposed to the effect of ultraviolet radiation from the sun. Some of its molecules dissolve into active atoms. Some of these atoms then combine again with oxygen molecules forming ozone. Ozone is formed by natural factors such as lightning.

$$O_2 \longrightarrow \text{Ultraviolet radiation } O + O \quad \text{activate oxygen atoms}$$

$$O + O_2 \rightarrow O_3 \quad \text{Ozone molecule}$$

Ozone is naturally present in the upper atmosphere and protects the earth from the effects of ultraviolet radiation. The concentration of ozone is reached 0.1–0.2 ppm at a height of 20–30 km from the surface of the earth. The ozone layer is the protective

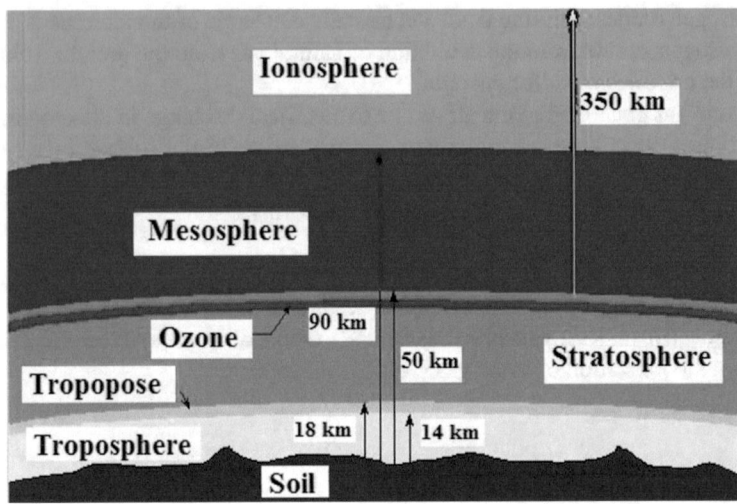

Fig. 5.1 Atmospheric layers (*Source* http://csep10.phys.utk.edu/astr161/lect/earth/atmosphere. html)

shield of life on Earth's surface, absorbing a large amount of ultraviolet radiation passing from the sun to the surface of the earth. Thus ozone protecting all organisms from the dangers of ultraviolet light.

The protective ozone layer prolongs from 8 km (upper troposphere) up through the whole stratosphere. Increased ozone in the troposphere layer due to pollution and the lack of stratosphere leads to an imbalance in the atmosphere which is contributing to global warming and atmosphere. Ozone is not the only one responsible for global warming, but also carbon dioxide, chlorofluorocarbons, nitrogen oxides and methane. The erosion of the ozone layer causes the leakage of ultraviolet radiation to the earth, high temperature of the atmosphere and the earth and the occurrence of harmful effects on human and animal health and the effect on the plant (Fig. 5.1)

Ozone stress has become more acute in the last decade for many plant species. According to indicators related to the effects of ozone on crops, it is considered one of the most important environmental factors that effects on crop production worldwide (Ludwikow and Sadowski 2008).

The reports confirm that the world will continue to suffer for a long time from the effects of ozone layer erosion, even if it stops immediately from the production of chemicals that cause corrosion of this layer. What is eroded from the ozone layer in a year is reconstituted after 100 years because of chemical gases such as freon and chlorofluorocarbons has a large destructive force, as one chlorine atom can destroy 100,000 ozone molecules over its life in the stratosphere (EPA 2017). With the amount of ozone decreases in the stratosphere, more UV rays hit earth, heat it and increase the risk of skin cancer.

Ozone damage and ultraviolet radiation leakage on crop plants:

The erosion of the ozone layer and the leakage of ultraviolet radiation result in the following damages:

- Lack of activity of the enzyme RuBPCase and affect the efficiency of photosynthesis.
- Reduces the strength of chloroplast in lipid metabolism.
- Breaks permeability of the cellular membrane.
- Lack of amino acid building and lack of protein content.
- Reduces glucose and fructose content in plant leaves.
- Reduces the flow of carbohydrates from leaves to fruit parts.
- Ozone stress and carbon dioxide change gene expression and the synthesis of materials responsible for protection.
- Affects the activity of RNA in wheat leaves.
- lack of leaf area, chlorophyll content, net photosynthesis, acceleration of aging, falling leaves, crop growth rate and low yield (Fig. 5.2).
- It affects the growth and development of crop plants by inducing oxidative stress and activating programmed cell death, apparent leaf injuries, low carbon assimilation, poor stomatal conductivity, and reduced water use efficiency (Martin et al. 2000; Joo et al. 2005).

b. **Nitrogen Oxides:**

There are seven nitrogen oxides that can be found in the surrounding air (Jarvis et al. 2010). Nitrous oxide (N_2O) is a greenhouse gas with major anthropogenic sources contributing to its global abundance (~0.3 ppm). However, nitric oxide (NO)

Fig. 5.2 Effect of ozone on potato (*Source* https://extension.umd.edu/learn/air-pollution-effects-vegetables)

and nitrogen dioxide (NO_2) are the two main nitrogen oxides related to combustion sources. Ambient concentrations of the two gases varied greatly based on local sources and sinks, but can surpass a total concentration ($NO + NO_2$) of 500 $\mu g/m^3$ in dense urban regions. Nitrous acid (HONO) is a common pollutant that spreads in the environment, and is generated by the reaction of nitrogen dioxide with water.

The first and second nitrogen dioxide is one of the most polluting gases of the surrounding medium. Nitrogen monoxide is formed from the combustion of coal, oil and gases in the air, as well as from fertilizer plants, welding operations and detonation of explosives. Nitrogen oxide is oxidized at high temperatures to nitrogen dioxide.

Concentrations of nitrogen oxides in urban environments are more than 10–100 times those of rural areas, reflecting the risk of industrial sources when measured by natural resources. The exposure of plant leaves to a concentration of 4–8 ppm of nitrogen dioxide for one hour lead to the necrosis of leaves and affect the efficiency of chlorophyll in photosynthesis process. Increases the damage on leaves when plants are exposed to concentration at 25 ppm for an hour, especially in the growth stage, while leaves of adult plants are more resistant to gas. Agrawal and his collaborators (2003) in India monitored negative effects of nitrogen oxides on growth parameters i.e. specific leaf area, specific leaf weight, leaf duration, net photosynthesis and enzymatic activity, which was ultimately reflected in a lack of wheat grain yield.

To protect human health, nitrogen dioxide is regulated according to the Clean Air Act. The current standard is based on an average of measurements taken over 12 months that should not exceed 0.053 ppm (Daniel 2005).

c. **CO, NO_x and SO_2**

Sulfur dioxide (SO_2) is considered one of the most damaging environmental pollutants. The toxicity of sulfur dioxide depends on the concentration and the duration of exposure (Li and Yi 2012). The phytotoxicity of the air pollutants CO, NOx and SO_2 gases is due to toxic molecular species like those from the combustion of CO_2 to CO, NOx to NO_2, and O_3, sulphite (SO_3^{2-}) and bisulphite HSO_3^-. The detoxification reaction of such compounds also induce generate reactive oxygen species ROS, such as hydroxyl radical ($OH^\bullet$) and hydrogen peroxide. Reactive oxygen species attacks biomolecules and damages nucleic acids, proteins and lipids (Mittler et al. 2004; Foyer and Noctor 2005). Harmful effect of sulfur dioxide was detected on photosynthesis as a result of the stomatal closure, toxicity and destruction of plant cells (Yi et al. 2012) this eventually leads to lower yields (Heyneke et al. 2013).

To identify the effect of carbon monoxide (CO), nitroxide (NO_x) and sulfur dioxide (SO_2) on the production of ROS, photosynthesis and ascorbate–glutathione pathway in strawberry, Muneer et al. (2014) revealed that both oxygen (O_2) and hydrogen peroxide (H_2O_2) contents increased in CO, NO_x and SO_2 treated leaves. With the closure of stomata, photosynthesis activity decreased, accompanied by a decrease in protein, carbohydrates and sucrose as a result of the production of reactive oxygen species under the stress of prolonged exposure to gas. Verma and Chandra (2014)

Light injury   Moderate injury   Heavy injury

PAN-damaged milkweed leaves

Fig. 5.3 Effect of PAN on milkweed leaves (*Source* https://extension.umd.edu/learn/air-pollution-effects-vegetables)

noticed that higher relative water content helps in preserving water balance and offers resistance during osmotic and drought stress. It clarifies the fact that the less resistant plants to SO_2 fumigation tends to lose more water content than resistant varieties.

d. **Pyroxy Acetyl Nitrate:**

Pyroxy Acetyl Nitrate is a powerful oxidizing agent that is generated as ozone from the effect of ultraviolet radiation on nitrogen oxides (NOx) in the existence of oxygen and active hydrocarbons and escalating with automotive exhaust and other incomplete combustion products. Peroxy acetyl nitrate affects plants at low concentrations. Exposure of plants to a concentration of 0.02 ppm of pyroxy acetyl nitrate for 5 h leads to oxidation and leaf toxicity, dehydration and impact on the quantity and quality of the crop productivity.

Dry deposition of (PAN) is identified to have a phytotoxic influence on crop plants under photo-chemical smog situations, however it may also lead to higher production and threaten species richness of vulnerable ecosystems in remote regions. Moravek et al. (2015) found that air masses at the site of Mainz Finthen Airport in Rhineland-Palatinate, Germany were influenced through two different pollution regimes, which led to median diurnal PAN mixing rates fluctuating from 50 to 300 ppt through unpolluted and from 200 to 600 ppt through polluted episodes. Figure 5.3 show damage of PAN on milkweed leaves.

5.4 Tolerance to Heavy Metals Pollutants

The elements of manganese, zinc and copper are enzymatic catalysts, but these metals are toxic and dangerous in certain concentrations. The most dangerous of these minerals in the environment is that they cannot be analyzed by bacteria and other natural processes as well as their provenness, which enables them to propagate far from their source sites or sources. Perhaps the most dangerous is due to the ability of some to bioaccumulate in the tissues and organs of living organisms in the environment water or soil. Some heavy metals have radioactive properties, that is, they act as radioactive isotopes. So these metals will pose double risks to the environment in terms of being toxic and radioactive at the same time, as in radioactive zinc 65 and uranium 235 (Al-Saadi 2008).

The soil is contaminated by heavy metals i.e. lead, mercury and cadmium, which reach to the soil with wastes buried in the soil, with contaminated irrigation water, or as a result of the airborne residues of these metals. These elements are highly toxic at highly concentrated in the tissues of plants and fruits, as it passes through food chain to the human.

Recently, the negative impacts of heavy metals on plants, animals and humans have increased as a result of the spread of many industries such as iron, steel, cement, chemical industries, soap, fertilizers, textile and automotive exhausts (Fig. 5.4). These

Fig. 5.4 Cement factories of the most dangerous air pollutants (*Source* Modified after, https://www.brighthub.com/environment/green-living/articles/43655.aspx)

sources cause of different types of pollution for air, soil and water with many elements such as cadmium, lead, mercury, nickel, cobalt, copper, zinc, silver... and others. Jaishankar et al. (2014) found that metal toxicity depends upon the absorbed dose, the route and duration of exposure, i.e. acute or chronic. This leads to various disturbances and excessive damage due to oxidative stress caused by the formation of free radicals. Several studies have shown the harmful effect of increasing the concentrations of heavy metals in soil solution and their adverse effect on crop production and the negative impact on food safety and marketing quality. Even more serious is the adverse effects on human and animal health through the food chain (Gill 2014). Table 5.1 shows the toxicity limits of environmental pollutants.

Therefore, international bodies and organizations interested with food safety have set limits to these elements that should not be exceeded in order to protect human health. For example, the European Community identified cadmium in small grain crops range 0.1 mg/kg grains as a guideline and 0.2 mg/kg as a maximum level for the marketing process (Council of Europe 1994).

In wheat, studies indicated increased accumulation of cadmium, lead and copper in wheat varieties with phosphate, nitrogen and potassium fertilizers. The response of genotypes varied from season to season, indicating the effect of the environment on plant response to heavy elements (Grant and Bailey 1998; El-Kalla et al. 2004).

Table 5.1 Critical limits of environmental pollutants

Environmental pollutant	Critical limit
*Firstly: Air Pollutants	
O_3	100 μg/m^3 (average allowable value in eight hours)
NO_2	400 μg/m^3 (average allowable value per year)
SO_2	20 μg/m^3 (average allowable value within 24 h)
*Secondly: Heavy metals	
Cadmium	5–30 ppm
Cobalt	15–50 ppm
Chromium	5–30 ppm
Copper	20–100 ppm
Mercury	1–3 ppm
Lead	30–300 ppm
Selenium	5–30 ppm
Zinc	100–400 ppm
Nickel	10–100 ppm

(*Source* https://apps.who.int/iris/.../WHO_SDE_PHE_OEH_06.02_eng.pdf?...**Davis et al. 1978)

In rice, when testing 60 samples collected from four areas in northern Iran, the average concentration of cadmium and lead was 0.41 and 2.23 mg/kg dry weight, respectively, which is higher than the FAO/WHO Manual (Khaniki and Zazoli 2005).

In polished rice, the Codex Alimentarius Commission of the Food and Agriculture Organization of the United Nations (FAO)/World Health Organization (WHO) determined maximum allowable levels of Cadmium and inorganic arsenic in 2008 and 2014 by 0.4 mg/kg and 0.2 mg/kg, respectively (Codex Alimentarius Commission 2014).

In maize, Hussain et al. (2013) found that the concentration of 9–12 mg cadmium chloride/kg dry weight is most influential on physiology and dry matter production in maize

In vegetable crops, the permissible limits are 0.1 for lead, 0.05 for cadmium and 5.0 mg for copper/kg fresh weight according to EU (2001).

Mirecki et al. (2015) study translocation and accumulation of four heavy metals, cadmium (Cd), lead (Pb), copper (Cu) and zinc (Zn) in 10 different plants i.e. corn (*Zea mays* L.), bean (*Phaseolus vulgaris* L.), potato (*Solanum tuberosum* L.), onion (*Allium cepa* L.), pepper (*Capsicum annuum* L.), tomato (*Solanum lycopersicum* L.), lettuce (*Lactuca sativa* L.), Swiss chard (*Beta vulgaris* subsp.vulgaris L.), cabbage (*Brassica oleracea* var.capitata L.), plantain (*Plantago major* L.) in samples from two locations (unpolluted-Leposavić and polluted-Kosovska Mitrovica, Kosovo province). The results stated that transfer factors -TF (heavy metals from soil to plants) are dependent on each other and comparison of the transfer factor for various crop species has significance only if all other conditions are equal. Heavy metals accumulate in plant species in varying densities. Cd and Zn accumulated the most with the transfer factor of 1.0–10, followed by Cu with TF of 0.1–1.0, while Pb had the lowest accumulation with TF usually 0.01–0.1. They detected that TF reduced when the plants were planted in the soil with higher level of heavy metals. When the growth occurs on the same type of soil, the heavy metals accumulation in different species decreases in the following descending order, grains < root vegetables < fruit vegetables < leaf vegetables. They have concluded that plant species (*Plantago major* L.) was detected to be a bioaccumulator. As illustrated in Table 5.2 values attained of Zn, indicated that all studied species had Zn accumulation capacity in their organs. Cultivated plants in this study had a high Zn concentration (8.25–49.5 mg/kg dry weight). The Zn accumulation capacities of *Plantago major* in leaves (1.125 mg/kg) revealed that these plants can be used as possible bioindicators of Zn pollution. Cabbage exhibited significantly higher levels of Pb (3.75 mg/kg) and Cd (0.50 mg/kg) than the other vegetables. The variance in level of heavy metal contamination between cabbage and other species was attributed to their morphophysiological variances in terms of heavy metal content, exclusion, accumulation, foliage deposition and retention efficiency. In plant species (plantain, cabbage and chard) which exhibited high Cd levels were also recorded high Zn and Pb levels.

Intawongse and Dean (2006) grown lettuce, spinach, radish and carrot were on compost that was contaminated before at various concentrations of Cd, Cu, Mn, Pb and Zn and Control (unadulterated compost). They registered significant differences between plant species in their metal uptake, e.g. spinach accumulated a great content

Table 5.2 Heavy metals content in plant from polluted area - Kosovska Mitrovica Concentration mg/kg dry matter (Mirecki et al. 2015)

Concentration mg/kg dry matter			
Crop species	Cd	Pb	Zn	Cu
Corn	0.01 ± 0.00	0.06 ± 0.01	8.25 ± 0.33	2.76 ± 0.22
Bean	0.03 ± 0.00	0.09 ± 0.01	12.97 ± 0.30	3.15 ± 0.27
Potato	0.03 ± 0.00	0.19 ± 0.05	16.15 ± 0.25	4.12 ± 0.43
Onion	0.10 ± 0.00	0.56 ± 0.08	23.12 ± 0.47	5.30 ± 0.41
Tomato	0.12 ± 0.01	0.71 ± 0.19	25.55 ± 0.68	7.21 ± 0.52
Pepper	0.15 ± 0.01	1.80 ± 0.16	35.75 ± 1.45	8.24 ± 0.67
Lettuce	0.28 ± 0.02	2.05 ± 0.23	38.50 ± 1.62	9.23 ± 0.78
Chard	0.25 ± 0.04	2.75 ± 0.25	39.70 ± 2.12	10.88 ± 0.90
Cabbage	0.50 ± 0.06	3.75 ± 0.37	49.25 ± 3.27	15.05 ± 1.05
Plantain	17.70 ± 1.60	265.00 ± 16.50	1125.00 ± 45.50	420.50 ± 21.05

($\pm$, standard deviation)

of Mn and Zn whereas relatively lower concentrations were detected for Cu and Pb in their tissues. Based on the in vitro gastrointestinal study, results show that metal bioavailability different widely from element to another and according to different crop species. The highest extent of metal releasing was registered in lettuce (Mn, 63.7%), radish (Cu, 62.5%), radish (Cd, 54.9%), radish (Mn, 45.8%) and in lettuce (Zn, 45.2%).

5.5 Damage Mechanisms of Heavy Metals and Effects on Crop Plants

Heavy metals enter the plant tissues actively during biochemical processes and are stored as inactive compounds in cells or membranes, affecting the chemical structure of the plant and may show no apparent damage to the plant. The fate of the heavy metals and their role in the plant depend on the following processes:

1. Absorption by plant and transition through it.
2. Enzymatic processes occurring within the plant.
3. Concentration of heavy metals and the various images that exist on them.
4. Limits of deficiency and toxicity of the element.

The negative and harmful effects of heavy metals on crop plants could be summarized as follows:

1. Affects the water balance within the plant organs
2. Imbalance in the vital functions and metabolic processes of various important compounds as a result of the occupation of heavy elements sites of vital compounds.

Fig. 5.5 Symptoms of
cobalt toxicity on sugar beet
(Wallace 1943)

3. Oxidation of the cell membranes lipids and the production of free radicals, which lead to dysfunction of the cell membrane and lack of cell turgid pressure.
4. Interaction with the group of glutathione and phosphate and active groups in energy compounds ATP, ADP necessary for the energy cycle, and inhibition of their activity.
5. Damage the nuclei of the root cell tips and reduce the absorption of nutrients and their transition from roots to the vegetative.
6. Reduce the total chlorophyll content, carotenoids and influence on the activity of photosynthesis.
7. Causing damage in chloroplast cells and the incidence of premature maturity.
8. Reduce of development and low strength of seedlings and metabolism indicators and failure of vegetative growth sometimes.

Figures 5.5 and 5.6 show adverse effects and visual symptoms due to increase the concentration of cobalt and cadmium heavy metals on sugar beet and maize, respectively.

5.6   Water Pollution

Water is an essential component for all living organisms and about him God Almighty said: (And made us of water every living thing) Al-Anbiya / 30. Water covers about 71% of the land, and is about 65% of the human body, 70% of vegetables, about 90% of fruits.

Fig. 5.6 Symptoms of cadmium toxicity on maize (Awaad 2009)

River Nile is the major source of water in Egypt, and suffers from pollution in some regions (Wahaab and Badawy 2004). The river is still able to recover in virtually all the locations, with very little exception, as a result the high dilution ratio. The residues of organo-chlorine insecticides were detected in virtually all locations and were below the limit set by the World Health Organization (WHO).

The high level of pollution caused by the low level of the Nile water becomes the main problem facing Egypt in the future; particularly after implementation the Ethiopia Dam building. In general, water contamination with heavy metals is a serious environmental problem because of its potential toxicity to humans and the environment, and it can bioaccumulate through the food chain (Ezzat et al. 2012; Goher et al. 2014).

When Amaal et al. (2017) tested the water quality for 24 sites between Aswan and Cairo along the Nile River, water quality differences were greatly related to inorganic nutrients and heavy metals Irrigation of plants with low quality wastewater leads to contamination of ground water and increases the proportions of nickel, cobalt, cadmium and lead elements in leaves and green parts of vegetable and field crops. Moreover, when fed to the human suffering from fatigue and nervous stress, headache, nausea, dizziness, vomiting, colic, diarrhea, liver and kidney damage, lungs and bones, high blood pressure and other serious health problems. Mostafa (2001) assessing the validity of the use of different sources of irrigation water, i.e. the Nile water, agricultural drainage water and sewage on the state of heavy elements and productivity of the broad bean plants. Pointed out that the concentration of heavy metals from nickel, cobalt, cadmium and lead differed from water source to another, and the levels were close to the critical limit in the case of sewage. Research studies of Dorgham et al. (2003) indicated that heavy metals like lead, cadmium, cobalt and nickel are found in maize leaves and grains and in parts that are eaten green from vegetable crops such as cucumbers, peppers, eggplants and okra due to irrigation with low-quality drainage water.

5.7 Stress-Sensing in Plants

While air pollutants or heavy metals HMs enter the cytosol, mechanisms of inter-cellular are activated in detoxification (Fig. 5.7). In such a condition, detoxification of heavy metals is done through excluding of these pollutants from root cells (Burzyn´ski et al. 2005), heavy metal transfer and chelation (Rauser 1999). Low-molecular-weight organic acids are involved in HMs detoxification through two approaches: firstly, as plant exudates, organic acids raise extracellular precipitation of HMs through the chelation oxidation-reduction reaction in plant rhizosphere; secondly, through chelation process and sequestrate of HMs pollutants in the vacuole (Rauser 1999; Ryan et al. 2001). This method of detoxification was proved at the excess of Al, Zn, and Ni in nutrient medium.

5.8 Resistance Mechanisms of Crop Pants to Environmental Pollutants

God Almighty grant the crop plants with lines of defense, mechanics and charac-ters to adapt and tolerate the impact of environmental pollutants include avoidance, tolerance and resistance as follows (Awaad 2009):

Avoidance: The ability of the plant to prevent the absorption of excessive amounts of toxic elements.

Tolerance: The appearance of visible symptoms of damage to pollutants with the capability of the plant to tolerate this damage.

Resistance: The ability of the plant to resist the high levels of toxic element, a genetic trait governed by genetic factors.

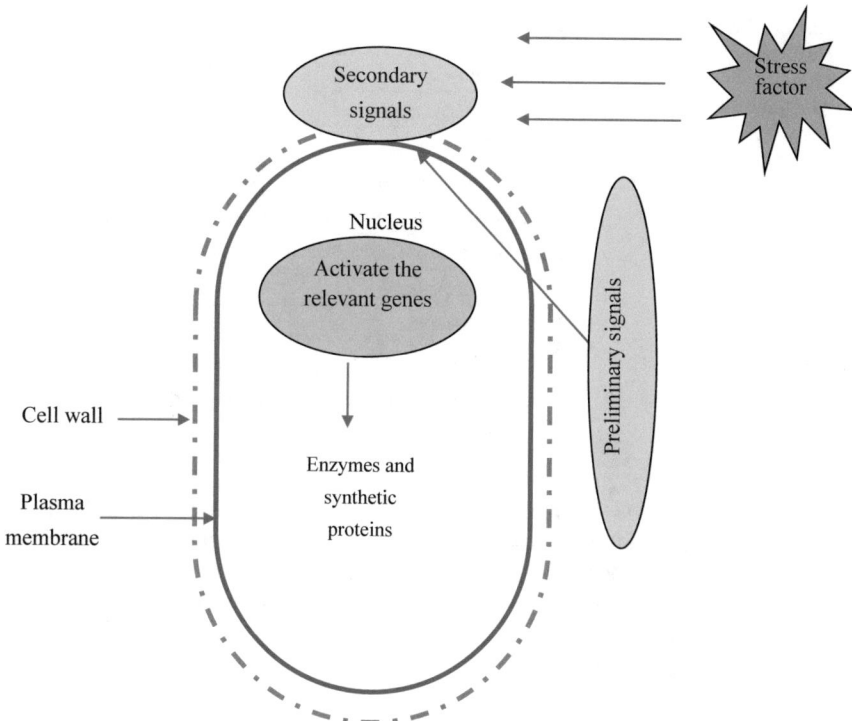

Fig. 5.7 Plant stress sensing begins by detecting stress at the cell surface, then forming secondary signals that move to the nucleus, activating genes related to plant tolerance to stress (From: Physiology and Molecular Biology of Plant During Water Stress, Dr. Jaber Mokhtar Abu Jadallah, Botany Department, Faculty of Science, Damietta)

5.9 Morpho-Physiological and Biochemical Characters Related to Environmental Pollutants Tolerance

5.9.1 Morphological Characters

5.9.1.1 Leaf Area and Angle

Plant leaf is the lung of nature that renews its air and gives it the exorcism of life, namely oxygen. Awaad et al. (2010) showed a significant negative association between flag leaf area and lead stress index (−0.581) as shown in Table 5.3, with a relative contribution of 12.25% in the tolerance level of lead stress in bread wheat. In this regard, El-Gharbawy (2015) observed that the bread wheat genotypes strain 1 and strain 2 which was broader flag leaf area were characterized by high grain yield under the stress of the heavy metals mixture i.e., zinc, lead and cadmium.

Table 5.3 The association between lead sensitivity index and some morpho-physiological and yield characteristics in bread wheat (Awaad et al. 2010)

Character	Normal	Lead stress
Proline content	0.034	−0.609*
leaf chlorophyll content	0.074	−0.743**
Flag leaf area	−0.399	−0.581*
No. of fertile spikelets/spike	0.308	0.347
No. of sterile spikelets/spike	−0.485	−0.587*
No. of productive tillers/plant	−0.474	0.121
No. of grains/spike	0.231	0.167
1000- grain weight	−0.354	−0.291
Grain yield/fed	−0.161	−0.239

* and ** significance at 0.05 and 0.01 levels of probability, respectively

The leaf angle plays an important role in influencing the amount and the distribution of light at different levels within the vegetative canopy to avoid environmental pollutants. The results of the studies showed improved productivity of maize, sorghum and wheat varieties with waxy-reflective leaves under environmental stress conditions (Awaad 2009). The correlation of erect leaves with high yield potential was noted (Fageria et al. 2006).

5.9.2   *Physiological Characters*

5.9.2.1   Chlorophyll Content

The photosynthesis of plants is done by the use of water, light and carbon dioxide in the presence of chlorophyll, resulting in producing carbohydrate which leads to the production of carbohydrates necessary for the growth of the plant and release oxygen. Plants need chlorophyll, which gives plants their green color, is important and essential in photosynthesis process. A significant relationship was found between chlorophyll stability and resistance to ozone damage in legume crops and tobacco (Knudson et al. 1977; Poornima et al. 2003). In bread wheat, Awaad et al. (2010) registered negative and highly significant relationship between the flag leaf chlorophyll content and lead sensitivity index (−0.734**) as in Table 5.3 with a relative contribution of 22.268% in the tolerance level of lead.

5.9.2.2   Stomata Behavior

Changes in micro-morphological parameters such as stomatal guard cell, wall thickness of guard cell, the degree of stomata conductivity, increase in the rate and length of the trichome had been recorded as structural barriers to limit entry of SO_2 as revealed by Verma and Chandra (2014). In this respect, differences between maize, wheat, triticale and mustard in leaf stomatal conductance, cuticular resistances in relation to ozon were registered by Lamaud et al. (2009). Alteration in the efficiency of physiological processes was also detected where, nitrogen use efficiency and water use efficiency were observed to be increased under stress condition caused by Sulphur dioxide fumigation (Choi et al. 2014). Moravek et al. (2015) added the importance of the existence of high stomatal conductance and mesophyllic resistance in limiting the uptake of Peroxy acetyl nitrate.

Stomata of ozone-resistant genotypes were observed partially closed when exposed to ozone, and then recover when the air is free of pollution. However, sensitive genotypes remain open, gas exchange, damage to parenchyma cells, and tissue burning. Stomata behavior is associated with hypersensitivity and partial stomata closer at high concentrations of ozone and sulfur dioxide in wheat (Mckee et al. 1995). The higher accumulation of abscisic acid leads to closes the stomatal in the stress-prone plant. Stomatal conductivity also gets decreased in the case of high SO_2 exposure thus causing inhibition in gas exchange and photosynthesis and respiration processes (Liu et al. 2017). The same behavior has been observed in European lolium (Robinson et al. 1998). The Egyptian beans cultivar Giza 3 is distinguished by the high resistance to ozone damage because it has a good ability to control the closure of the stomata to prevent the entry of gas (Madkour 1998).

5.9.3 Biochemical Characters

5.9.3.1 Cell Membrane Stability

Ozone and heavy metal resistant—genotypes have an antioxidant-protective device that efficiently protects cellular membranes from harmful peroxides caused by physiological deviations.

The antioxidant enzymes activity, proline content and proportions of the main lipid fractions of cell membranes were detected after 7 days ozone treatment under vitro culture of winter wheat cells (Rudolphi-Skórska and Sieprawska 2016). They also added that changes of malondialdehyde (MDA) level, activity of antioxidative enzymes and of composition of membrane lipids demonstrate cell tolerance to ozone stressful conditions. The changes in the structure of phospholipids resulting from ozone stress indicated the existence of tocopherol and gallic acid efficiently reversed these changes. Galactose groups in polar part of galactolipids may be works as a "trap" for ROS, decreasing their access to the interior of the cell membranes. The wheat callus cells use this mechanism as a means of acclimatization. This enhances the combined effect between tocopherol and gallic acid in the presence of ozone sterss.

Cell membrane stability was found to be associated with the tolerance of maize strains to the stress of environmental pollutants and considered as important selection criterion (Hussain et al. 2013). Wheat varieties tolerance to ozone damage was appeared to be associated with cell membrane integrity (Li et al. 2013).

5.9.3.2 Antioxidants

The increase in the rate of accumulation and activity of Reactive Oxygen Species ROS scavenger molecules can act as a signal to induce defense reactions to CO, NO_x and SO_2 gas stress. increasing the level of antioxidant enzymes plays a important role in plant defense due to which strawberry plants can be utilized as a hyper accumulator to maintain environmental pollution, whereas, the defense ability cannot sufficiently mitigate oxidative damage under prolonged exposure of CO, NO_x and SO_2 pressure (Muneer et al. 2014).

The high concentration of ascorbic acid, gives more resistance to the plant, and it is seen that the plants at contaminated site shows more Air Pollution Tolerance Index (APTI) values to SO_2 than that of non-polluted sites (Thara et al. 2015). Ascorbic acid is mainly found in numerous experiments on *Vigna radiata, Solanum esculentum, Zea mays*. Increased Ascorbic acid content has been detected in response to prolonged SO_2 fumigation for 45 Days (Chauhan 2015).

Activity of both catalase and peroxidase enzymes has been detected to be increased significantly in stressed plants has a high ability to scavenge Reactive Oxygen Species (Verma and Chandra 2014) and reduces the peroxidised lipid, and thus decreasing the cytotoxic effect on plant cells (Yoshimura et al. 2004). Moreover, Brahmachari and Kundu (2007) stated that plant defense mechanism deal with stress either by inhibiting gas entry or through detoxifying the excess sulphur and removing the reactive oxygen species. superoxide dismutase, peroxidase, polyphenol oxidase, play an important role in the detoxifying process and a significant amount of toxic sulphur is detoxified through forming S-containing sulphur compounds. This phenomenon is being used for producing SO_2 resistant cultivars via over-expression of cysteine synthase like genes.

In Arabidopsis, difference expression of 2780 genes has been detected in response to sulphur dioxide fumigation. Up-regulation of genes of Cytochrome P450, heat shock proteins and pathogen-related proteins were observed in Arabidopsis shoot. Beside these genes, several antioxidant encoding genes for peroxidases, glutathione peroxidases and superoxide dismutase also get up-regulated (Li and Yi 2012).

Several studies have revealed that some antioxidant enzymes are involved in plant reaction under different abiotic stresses. The invoking of antioxidant enzymes is thought to be a protective response of plants against abiotic stress. Resistance of varieties depends on the internal physiological equilibrium of the concentration of superoxide dismutase and glutathione reductase enzymes, which are observed in the brown mustard resistant genotype Progress to sulfur dioxide pollution, and the resistant genotype Pasas Jai Kisan to cadmium stress (Qadir et al. 2004).

In the atmosphere of industrial centers in Russia, peroxidase activity was identified as an indicator of pollution stress up to 40 km from the source of pollution in crop varieties exposed to sulfur, copper and nickel pollution (Roitto et al. 1999).

The synthesis glutathione enzyme activity and ascorbic acid are increased following the exposure of leaves of wheat and faba bean to heavy metals (Huang et al. 1993; Kumar and Sharma 2003; Rudolphi-Skórska and Sieprawska 2016).

5.9.3.3 Amino and Fatty Acids Content

Lignin, proline and other amino acids played a main defensive role in the oxidative stress induced by sulphur dioxide (Brahmachari and Kundu 2007).

The synthesis of proline and histidine, are increased following the exposure of leaves of wheat, faba bean and soybean to heavy metals and play multi-functional roles in the maintenance of osmotic balance, and protect the stability of the cytoskeleton. In addition to this, proline also helps in scavenging free radicals (Huang et al. 1993; Kumar and Sharma 2003; Awaad et al. 2010, 2013). Results of research studies indicated that the level of catalase activity in cucumber plant was increased under copper stress (Janicka-Russak et al. 2012). Detoxification of SO_2 and recovery of photosynthesis was expressed as another strategy to reduce the SO_2 toxicity is to incorporate sulphur into amino acids like cysteine and methionine (Dittrich et al. 1992). Also, a study on Arabidopsis showed that rise in the content of water soluble non-protein sulphydryl and glucosinolates also act as biological sink for excess Sulphur after exposure to SO_2 (Kooij et al. 1997).

5.9.3.4 Phytohormone

Maintaining crop productivity under adverse environmental pressures is the main challenge facing agricultural workers. Polymines including putrescine, spermidine and spermine as plant natural compounds play an significant role as the low-molecular-weight aliphatic compounds in stress tolerance. Studies indicate the participation of polymers in several physiological processes such as cell growth, development, responsiveness and tolerance to various environmental stresses (Gill and Tuteja 2010).

Polyamines play a fundamental role in stress tolerance by several evidences through stimulating enzyme production, activating transcription levels and stimulating the expression of genes responsible to environmental stress tolerance (Liu et al. 2015). In this regard, polyamines, spermidine, spermine, and their precursor putrescine have been defined as endogenous plant growth regulators or intracellular messengers mediating physiological reactions (Alcázar et al. 2006; Tiburcio et al. 2014).

On the other words, plant hormones like ethylene, jasmonic acid, salicylic acid and abscisic acid appeared to be vital in the regulation of ozone stress tolerance (Tuominen et al. 2004; Gomi et al. 2005; Yaeno et al. 2006; Ludwikow and Sadowski

2008). Plant hormones play a major role in growth, development and plant responses to different stress factors. In Arabidopsis, *FAD7* gene responsible for salicylic acid was found to be expressed as a result of ozone exposure, and subsequently stimulates various defense-related genes (Yaeno et al. 2006).

5.9.3.5   Carbohydrate Content

Soluble sugar content is an important component of plants which acts as a resource of energy. Plants manufacture sugars through photosynthesis and breakdown during respiration. Research results indicate an increase in soluble sugar in air pollution resistant genotypes (Kameli and Losel 1993; Ludlow 1993).

The concentration of soluble sugars is indication of the physiological status of a plant and it regulates the sensitivity of crop genotypes to air pollution (Tripathi and Gautam 2007). Various reports showed that soluble sugars accumulate in different parts of plants in response to diverse environmental pressures, and play a protective role against environmental stresses (Prado et al. 2000; Finkelstein and Gibson 2001; Saxena and Kulshrestha 2016). A great increase in soluable sugar content and phenol was detected in tolerated wheat genotypes after exposure to air pollutants (Agrawal et al. 2003). The genes responsible of sucrose biosynthesis, cell wall metabolism, glycolysis, pentose phosphate pathway, as well as genes regulating enzymes of the Tricarboxylic acid cycle were induced in response to ozone (Gupta et al. 2005; Ludwikow and Sadowski 2008).

5.10   Genetic System and Nature of Gene Action Controlling Inheritance of Environmental Pollutants Tolerance

A study of the nature of inheritance of environmental pollutants is considered useful in the determination of the type of trait is simple or complex and the efficiency of inheritance from the parents to offspring. In this respect, genetic differences are internal contributors to the resistance or sensitivity of individuals to toxins associated with pollutants (Joneidia et al. 2019) Then draw the line of breeding program to produce high-end cultivars. Table 5.4 shows the results of research studies related to the genetic system and nature of gene action controlling inheritance of environmental pollutants tolerance.

Table 5.4 Genetic system and nature of gene action controlling inheritance of environmental pollutants tolerance

Crop	Genetic behaviour	Reference
Wheat:		
Tolerance to ozone damage	Many genes control the signaling pathways associated with ozone stress	Li et al. (2013)
Tolerance to cadmium stress	The additive gene action High heritability (> 50%)	Clarke et al. (1997) Awaad et al. (2013)
Tolerance to lead stress	The additive gene action—High heritability (> 50%)	EL-Gharbawy (2015)
Tolerance to zinc stress tolerance	Dominance gene action—low to medium heritability	EL-Gharbawy (2015)
Rice:		
Resistance to ozone damage	High heritability (> 80%)	Sohn-jaekeun et al. (2002)
Tolerance to cadmium stress	The additive and overdominance gene action	Wu-Shutu et al. (2000)
	Moderate to high heritability	Wu-Shutu et al. (2000)
Corn:		
Tolerance to ozone damage	More than a pair of genes—additive gene action	(Cameron et al. 1970; Cameron 1975)
Tolerance to cadmium, zinc and copper stress	Many genes controlling the signaling pathways associated with cadmium stress tolerance	Nocito et al. (2006)
Cucumber:		
Tolerance to cadmium stress	*Five genes CsHA2, CsHA3, CsHA4, CsHA8, CsHA9 UNDER9* are increasingly expressed under the cadmium stress	Janicka-Russak et al. (2012)
Tomato:		
Tolerance to cadmium, lead and nitrate stress	Overdominance with high hybrid vigour	Kilchevsky et al. (1999)
Tobacco:		
	At least four genes involved in polyamine synthesis responsible to ozone tolerance	van Buuren et al. (2002)

(continued)

Table 5.4 (continued)

Crop	Genetic behaviour	Reference
Arabidopsis:		
	61 genes involved in responsive to cadmium metal stress	Yang et al. (2005)
	23 genes are involved in NO_2 control	Takahashi and Morikawa (2014)
	Differential expression of 2780 genes has been observed in response to sulphur dioxide stress	Li and Yi (2012)
Atriplex halimus:		
	Nine bands showed down-regulation of NADH dehydrogenase and Sedoheptulose-bisphosphatase, six regulated genes representing transcription factors, membrane transporters and ROS detoxification	El-Bakatoushi et al. (2015)

5.11 Role of Plant Breeding and Biotechnology in the Development of Cultivars Tolerant to Environmental Pollutants

5.11.1 Breeding Methods

5.11.1.1 Pedigree and Selection Procedures

Natural or artificial selection plays an important role in improving the tolerance of crop strains and cultivars to environmental pollutants. Several cultivars of vegetables and fruit crops were developed through selection procedures that were more tolerant to nuclear radiation under the conditions of the Chernobyl region of Russia (Kilchevsky et al. 1999). Yang et al. (2000) selected three cultivars of rice tolerant to lead stress i.e. Kumnung, KH-2 J and CH-55.

Under Egyptian conditions, developing a variety of field crop cultivars with morph physiological and biochemical characteristics tolerant to environmental pollutants has been done. Plant breeders developed a variety of cultivars of wheat that can tolerant to pollutants stress such as Giza 168, Giza 171, Sakha 94, Sakha 95 and Sids 1, Sids 12, Sids 13, Sids 14, Misr 1, Misr 2, Misr 3, Shandwill 1, Gemmeiza 10, Gemmeiza 11, Gemmeiza 12, Sohag 3, Beni Suef 1, Beni Suef 3, Beni Suef 4, Beni Suef 5 and Beni Suef 6. Plant breeders were also able to produce many varieties of rice, pulses and others, which vary in the level of tolerance (Anonymous 2018).

5.11.1.2   Hybrid Production

Hybridization has contributed to the development of many crop hybrids, which are characterized by tolerance to environmental pollutants. The breeder under Egyptian conditions succeeded in producing many white, yellow of single or three-way maize crosses. Also, producing many hybrids of sorghum i.e. Shandaweel 1, Shandaweel 2, Shandaweel 6 and hybrid 301, 302, 303, 304, 305 and 306 tolerant to environmental stress factors. Other achievements were obtained by producing Egyptian hybrid rice 1 and hybrid rice 2. Also, release tomato hybrids Brigid, Madeer, California, Heinz 2710 HZ2710, Mena, Master 100, Superdard, Hadeel, Toshka, Shabah and Fahd.

5.11.2   Biotechnology

5.11.2.1   DNA-Markers

DNA markers play an important role in the identification of resistance genes in genetic resources, as assistance for selection, gene transfer and rapid production of new cultivars during a short time.

Various DNA-Markers i.e. RFLP, RAPD, AFLP, SSR ISSR tests have been developed to evaluate the toxicity of environmental contaminants. The novel application of RAPD fingerprinting method for detecting genotoxin induced DNA damage to plant from contaminated sites is well practiced today (Liu et al. 2009).

Erturk et al. (2013) established how DNA polymorphism and genomic template stability (GTS) estimate was significantly affected in Chromium polluted maize seedling by RAPD. DNA damage in the root tip of maize seedling under Cadmium stress became evident by the presence and/or absence of DNA band in the treated samples rather than the control was detected by RAPD technique (Shahrtash et al. 2010). They have claimed that DNA polymorphism detected by RAPD can be expressed as a useful tool to detect environmental contaminants. In their respect, Qurainy and his coworkers (2010) utilized RAPD polymorphism to detect genotoxicity of Cd, Pb and Zn in *Eruca sativa*. Only three decamer primers used out of twenty gave single and polymorphic bands, but other 13 primers generated 5 bands. Multiple metal genotoxicity assessment showed same remarkable result in RAPD profile of *Urtica dioica* (Gjorgieva et al. 2013). DNA damage was distinguished by RAPD in barley seedling treated with Cd (30–120 mg/L) (Liu et al. 2009). Aslam et al. (2014) revealed that DNA alterations identified by RAPD analysis offered a beneficial biomarker assay for the assessment of genotoxic effects of heavy metals in *Capsicum annum*.

A single allele that controls the high content of cadmium in the durum wheat was determined by RAPD technique (Penner et al. 1995) and another allele controlling low cadmium content in durum wheat (Clarke et al. 1997). RAPD-PCR had determined of bands associated with lead stress tolerance among 10 bread wheat genotypes (Fig. 5.8) (Awaad et al. 2010). Also identify of 4 genetic loci responsible

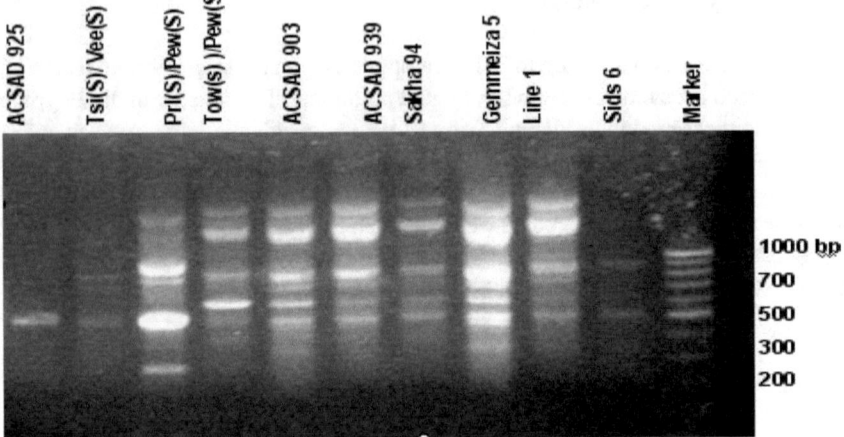

Fig. 5.8 Variation between strains and varieties of bread wheat at the molecular level in lead stress tolerance (Awaad et al. 2010)

for soybean resistance to manganese toxicity (Kassem et al. 2004), in addition to, identify of specific bands associated with zinc and lead stress in hibiscus (Bhaduri and Fuleka 2015).

5.11.2.2   Gene Transfer Technology

Utilized gene transfer technique of biolistic *MerA* gene to produce genetically modified rice plants that were able to extract toxic mercury from the soil and turn it into an ineffective volatile image and clean the environment (Heaton et al. 2003).

– Beside The completion of the production of genetically modified maize that accumulated lower amounts of heavy metals *i.e.* copper, nickel, cadmium, strontium, chromium, zinc and mercury compared to the parents (Rui et al. 2007), and maize strain MTM-2 tolerant to cadmium stress (Hussain et al. 2013).

Hysen et al. (2004) developed of genetically modified mustard plants capable of eliminating selenium toxicity and other canola and tobacco genotypes modified with *MerA* and *MerB* genes capable of eliminating mercury toxicity (Meagher et al. 2000 and De Filippis 2013).

– Development of genetically modified tobacco plants characterized by high tolerance to cadmium, copper, selenium and nickel by increasing the synthesis of the amino acid systeine synthase (Kawashima et al. 2004). While, Brychkova et al. (2007) produced transgenic lines of Arabidopsis with up-regulation *AtSO* gene showed higher resistance against SO_2 as it revealed delayed senescence than the wild lines. Besides *SO* gene, *Msr-A* and *Fds* genes play more positive roles in SO_2 tolerance. Furthermore, development of transgenic Arabidopsis plants with Overexpressing SaCu/Zn SOD has greater Cd uptake ability existed in roots, increased oxidative stress resistance and gave useful information for understanding the importance of SaCu/Zn SOD in response to abiotic stress (Yunxing Zhang 2017). Whereas, Aken (2008) developed transgenic poplars (*Populus*) with overexpressing a mammalian cytochrome P450, a family of enzymes involved in the metabolism of toxic compounds. The transgenic plants showed improved performance with respect to the metabolism of trichloroethylene and the removal of various toxic volatile organic contaminants, including vinyl chloride, carbon tetrachloride, chloroform and benzene.

5.12 Procedures for Pollution Control

5.12.1 Air Pollutants

1. Spreading environmental awareness among the members of society.
2. Use of continuous monitoring systems for emitted pollutants
3. Activation of legislation and legal controls
4. developing strains and cultivars of crops more tolerant to air pollutants
5. Use of biological filters that protect the environment from exhaust effects.
6. Use of gas washing and purifying devices
7. Limiting the impact of volatile organic compounds.
8. Use of adsorption systems such as activate carbon
9. Use of thermal and catalysts
10. Use of cooling condensers
11. Use of steam recovery systems.

5.12.2 Heavy Metals

5.12.2.1 Biological Procedures

In recent years, numerous plant species have been exploited as bioindicators of genotoxicity (Pal 2016). Some of sensitive plants are used as biologic sensors, such as sunflower and rice, as evidence to detect soil contamination with heavy metals. Mechanisms dealing with heavy metals can be listed as follows:

1. Development of strains and cultivars more tolerant to heavy metal stress.
2. Phytoremediation: Some plant species that belong to families, Gramineae, Leguminosea, Crusiferae and Chenopodiaceae are used to clearance soil and water from contaminants of mineral pollutants and to purify the environment

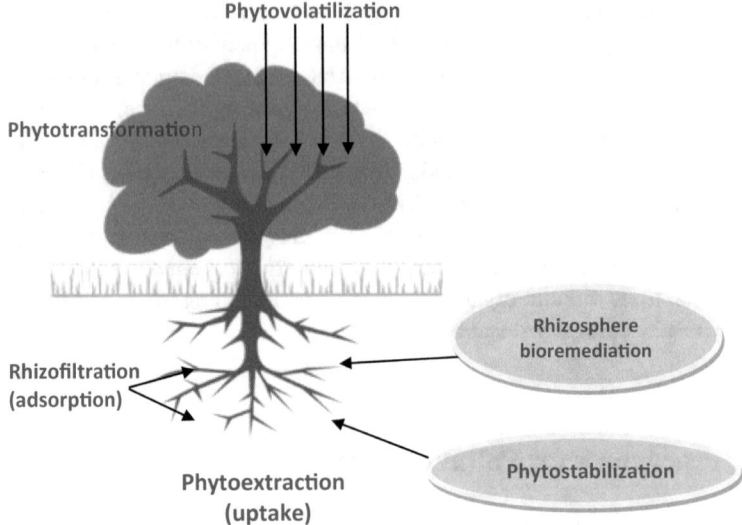

Fig. 5.9 Different ways of plant bioremediation for pollutants (Modified after: https://en.wikipe dia.org/wiki/Phytoremediation)

from pollution. This depends on the plant uptake of pollutants through the roots, the storage of pollutants in the plant, their volatilization by the plant, or their plant representation, or any combination of these mechanisms (Fig. 5.9).

3. Use water lentil or duckweed plant to treat lead, zinc, iron, manganese and copper pollution, and use mustard to treat the contamination of selenium and some canola types such as Darkar in the treatment of cadmium contamination (Qadir et al. 2004).

4. Exploitation of organisms in the environment to create a natural balance that transforms toxic elements such as copper, nickel, mercury, cadmium and others into a non-toxic status.

5. Use of microorganisms such as Mycorrhiza and *Pseudomonas putida* have special systems of enzymes that can perform oxidation and reduction operations and analyze and feed these compounds and reduce zinc, iron and copper contamination (Ramadan 2003).

6. Use of Anabaena and Nostoc algae possess the ability to absorb large amounts of cadmium, lead and nickel (Madkour et al. 2003).

7. Use of bio-fertilization with organisms capable of making beneficial changes in the root rhizosphere in reducing soil, water and air pollution compared to mineral fertilization (El-Kalla et al. 2004).

8. Use of biotechnology in bioremediation by employing some types of bacteria carrying genes responsible for the activity of mercury toxicity destruction enzymes and reducing Hg^{++} to inactive state Hg°.

9. Treatment with algebraic or abscisic acid, limits the movement and absorption of heavy metals such as nickel and cadmium from the roots to the shoot system and thereby minimizing their severity (Rubio et al. 1994).
10. Use of chelating materials.

5.12.2.2 Procedural Means

1. Application of laws and compliance with the provisions of international conventions
2. Develop regulations to reduce greenhouse gas emissions
3. Expanding of the wastewater treatment
4. Reduce the dumping off solid waste and factory waste products.
5. Rationalization in the use of pesticides and chemical fertilizers.

5.13 Conclusions

It is known that environment suffers from several species of environmental pollutants which cause different types of pollution for the air, soil and water. Despite the great importance of reducing the sources of pollutants and their damage on crop plants. Identify mechanisms responsible for crop tolerance for environmental pollutants is of particular importance. Understanding morpho-physiological and biochemical bases helps to improve tolerance of crop plants for environmental pollutants. The genetic system and nature of gene action controlling the inheritance of environmental pollutants tolerance give plant breeders the ability to choose the appropriate breeding program to release new cultivars more tolerant to environmental pollutants. DNA-markers and gene transfer in addition to bioremediation manners help to decrease the harmful effects of environmental pollutants on plants that reach the animals and humans through the food chain.

5.14 Recommendations

In light of the abovementioned, the following recommendations can be made:

– Utilize of environmental-friendly technologies for cleaning up polluted soils.
– Development of crop cultivars tolerant to the impact of environmental pollutants
– Exploit of genetically engineered plants that have enzymatic systems to detoxification of pollutants
– Application of environmental safety measures in industrial facilities
– Spreading environmental awareness among community groups
– Develop regulations to reduce greenhouse gas emissions.

References

Agrawal SB, Singh A, Rathore D (2003) Impact of air pollution on wheat grown in periurban area of Allahabad City. Abstr: 2nd International Cong of Plant Physiology January 8–12, 2003, New Delhi, India, P 443

Aken BV (2008) Transgenic plants for phytoremediation: helping nature to clean up environmental pollution. Trends Biotechnol 26(5):225–227

Alcázar R, Marco F, Cuevas JC, Patron M, Ferrando A, Carrasco P et al (2006) Involvement of polyamines in plant response to abiotic stress. Biotechnol Lett 28:1867–1876

Al-Saadi H (2008) Ecology Science. Dar Al-Yazuri Scientific, p 424

Amaal MA, Mohamed HA, Mohamed EG (2017) Indices of water quality and metal pollution of Nile River, Egypt. Egypt J Aquat Res 43(1):21–29

Anonymous (2018) Recommendations techniques in field crops. ARC, Giza, Egypt

Aslam R, Ansari MYK, Choudhary S, Bhat TM, Jahan N (2014) Genotoxic effects of heavy metal cadmium on growth, biochemical, cyto-physiological parameters and detection of DNA polymorphism by RAPD in *Capsicum annuum* L. an important spice crop in India. Saudi J Biol Sci 21:465–472

Awaad HA (2009) Genetics and breeding crops for environmental stress tolerance, I: drought, heat stress and environmental pollutants. Egyptian Library, Egypt

Awaad HA, Morsy AM, Moustafa ESA (2013) Genetic system controlling cadmium stress tolerance and some related characters in bread wheat. Zagazig J Agric Res 40(4):647–660

Awaad HA, Yousef MAH, Moustafa ESA (2010) Identification of genetic variation among bread wheat genotypes for lead tolerance using morpho—physiological and molecular markers. J Am Sci 6(10):1142–1153

Bhaduri AM, Fulekar MH (2015) Biochemical and RAPD analysis of *Hibiscus rosasinensis* induced by heavy metals. Soil Sedim Contam Int J 24(4):411–422

Brahmachari S, Kundu S (2007) SO2 stress: its effect on plants, plant defense responses and strategies for developing enduring resistance. Int Adv Res J Sci Eng Technol 4(7):303–309

Brychkova G, Xia Z, Yang G, Yesbergenova Z, Zhang Z, Davydov O, Fluhr R, Sagi M (2007) Sulfite oxidase protects plants against sulphur dioxide toxicity. Plant J 50:696–709

Burzyn´ski M, Migocka M, Kłobus G (2005) Cu and Cd transport in cucumber (*Cucumis sativus* L.) root plasma membranes. Plant Sci 168:1609–1614

Cameron JW (1975) Inheritance in sweet corn for resistance to acute ozone injury. J Am Soc Hort 100:577–579

Cameron JW, Johnson H, Taylor OC, Otto HW (1970) Differential susceptibility of sweet corn hybrids to field injury by air pollution. Hort Sci 5:217–219

Chauhan A (2015) Effect of SO2 on ascorbic acid content in crop plants-first line of defence against oxidative stress. Int J Innovative Res Dev 4(11):8–13

Choi D, Toda H, Kim Y (2014) Effect of sulfur dioxide (SO2) on growth and physiological activity in Alnus sieboldiana at Miyakejima Island in Japan. Ecol Res 29:103–110

Clarke JM, Leisle D, Kopytko GL (1997) Inheritance of cadmium concentration in five durum wheat crosses. Crop Sci 37:1722–1726

Codex Alimentarius Commission (2014) CODEX STAN 193-1995. General standard for contaminants and toxins in food and feed. http://foodnara.go.kr/codex/download.do?addPath=/hubfiles/codex/data2&fileName=STAN%20193-1995.pdf&fileNameOri=STAN%20193-1995.pdf

Council of Europe (1994) Lead, cadmium and mercury in food: assessment of dietary intakes and summary of heavy metal limits of foodstuffs. Counci of Europe Press Strasbourg, France

Daniel SG (2005) Human responses to nitrogen dioxide. EPA Grant Number: R828112C043 United States Environmental Protection Agency

Davis RD, Beckett PHT, Wollan E (1978) Critical levels of twenty potentially toxic elements in young spring barley. Plant Soil 49:395

De Filippis LF (2013) Crop improvement through tissue culture. Improvement, pp 289–346

Dittrich APM, Pfanz H, Heber U (1992) Oxidation and reduction of SO2 by chloroplasts and formation of sulfite addition compound. Plant Physiol 98:738–744

Dorgham EA, Osman A, El-Sisi SEM, Abdel Latif EA (2003) Risk assessment of waste water irrigation on the agricultural Lands. Egypt J Appl Sci 18(8):342–353

El-Bakatoushi R, Alframawy AM, Tammam A, Youssef D, El-Sadek L (2015) Molecular and physiological mechanisms of heavy metal tolerance in *Atriplex halimus*. J Int Phytoreme 17(9):789–800

EL-Gharbawy SS (2015) Wheat breeding for tolerance to heavy metals pollution. MSc Thesis, Agronomy Department, Faculty of Agriculture, Zagazig University, Egypt

El-Kalla SE, Sharief AE, Abdalla AM, Leilah AA, El-Awamy SA (2004) Improvement of wheat productivity by using some agricultural practices for reducing environmental pollution. Zagazig J Agric Res 31(3):813–827

EPA (2017) Ozone layer protection basic, ozone layer science. United States Environmental Protection Agency, EPA Web Archive, the January 19, 2017 Web Snapshot. FAO Media Relations, Rome, Italy

Erturk FA, Agar G, Nardemir G, Arslan E, Bozari S (2013) Genetic changes in maize seedlings induced by chromium pollution using RAPD analysis. Proceedings of ICOEST, 18–21

European Commission (2001) Commission Regulation (EC) No 466/2001 of 8 March 2001 setting maximum levels for certain contaminants in foodsuffs. Official J Eu Communities L77:1–13

Ezzat SM, Mahdy HM, Abo-State MA, Abd El Shakour EH, El-Bahnasawy MA (2012) Water quality assessment of River Nile at Rosetta branch: impact of drains discharge. MEJSR 12(4):413–423

Fageria NK, Baligar VC, Clark RB (2006) Physiology of crop production. The Haworth Press, New York

FAO (2014) Agriculture, forestry and other land use emissions by sources and removals by sinks (89 pp 3.5 M, About PDF) Exit Climate, Energy and Tenure Division, FAO

FAO (2018) Soil pollution comes under scrutiny: global soil partnership annual meeting focuses on "black soils" and data-sharing initiatives

Finkelstein RR, Gibson SI (2001) ABA and sugar interactions regulating development: cross-talk or voices in a crowd. Curr Opin Plant Biol 5:26–32

Foyer CH, Noctor G (2005) Oxidant and antioxidant signaling in plants: a reevaluation of the concept of oxidative stress in a physiological context. Plant, Cell Environ 28:1056–1071

Gill M (2014) Heavy metal stress in plants: a review. Int J Adv Res 2(6):1043–1055

Gill SS, Tuteja N (2010) Polyamines and abiotic stress tolerance in plants. Plant Signal Behav. 5(1):26–33

Gjorgieva D, Tatjana KP, Tatjana R, Katerina B (2013) Influence of heavy metal stress on antioxidant status and DNA damage in Urtica dioica. Biomed Res Int, Article ID 276417, 6 pages. http://dx. doi.org/10.1155/2013/276417

Goher ME, Hassan AM, Abdel-Moniem IA, Fahmy E-SS (2014) Evaluation of surface water quality and heavy metal indices of Ismailia Canal, Nile River, Egypt. Egypt J Aquat Res 40(2014):225–233

Gomi K, Ogawa D, Katou S, Kamada H, Nakajima N, Saji H et al (2005) A mitogen-activated protein kinase NtMPK4 activated by SIPKK is required for jasmonic acid signaling and involved in ozone tolerance via stomatal movement in tobacco. Plant Cell Physiol 46:1902–1914

Grant CA, Bailey LD (1998) Nitrogen phosphorus and zinc management effects on grain yield and cadmium concentration in two cultivars of durum wheat. Can J Plant Sci/Rev Can Phytotech 78(1):63–70

Gupta P, Duplessis S, White H, Karnosky DF, Martin F, Podila GK (2005) Gene expression patterns of trembling aspen trees following long-term exposure to interacting elevated CO_2 and tropospheric O_3. New Phytol 167:129–141

Heaton ACP, Rugh CL, Kim T, Wang NJ, Meagher RB (2003) Toward detoxifying mercury—polluted aquatic sediments with rice genetically engineered for mercury resistance. Environ Toxicol Chem 22(12):2940–2947

Heyneke E, Strauss AJ, Van Heerden PDR, Strasser RJ, Krüger GHJ (2013) SO2-drought interaction on crop yield, photosynthesis and symbiotic nitrogen fixation in Soybean (Glycine Max). Photosynthesis research for food, fuel and the future, 612–615

Huang L, Murrary F, Yang X (1993) Response of nitrogen metabolism parameters to sublethal SO2 polluion in wheat [*Triticum aestivaum* cv. Wilgoyne (Ciano/Gallo)] under mild NaCl stress. Environ Exp Bot 33(4):479–493

Hussain I, Shamim A, Muhammad AA, Rizwan R, Ejaz HS, Muhammad I (2013) Response of maize seedlings to cadmium application after different time intervals. ISRN Agronomy Article ID 169610, 9 pages. http://dx.doi.org/10.1155/2013/169610

Hysenet T, Terry N, Pilon-Smith EAH (2004) Exploring the selenium phytoremediation, potential of transgenic Indian mustard overexpressing ATP sulfurylase of cystathionine gamma synthase. Int J Phytorem 6(2):111–118

Intawongse M, Dean JR (2006) Uptake of heavy metals by vegetable plants grown on contaminated soil and their bioavailability in the human gastrointestinal tract. J Food Addit Contam 23(1):36–48

Jaishankar M, Tseten T, Anbalagan N, Mathew BB, Beeregowda KN (2014) Toxicity, mechanism and health effects of some heavy metals. Interdisc Toxicol 7(2):60–72

Janicka-Russak M, Katarzyna K, Marek B (2012) Different effect of cadmium and copper on H$^+$-ATPase activity in plasma membrane vesicles from *Cucumis sativus* roots. J Exp Bot 26:1–19

Jarvis DJ, Adamkiewicz G, Heroux M, Rapp R, Kelly FJ (2010) WHO guidelines for indoor air quality: selected pollutants. Geneva: Copyright © 2010, World Health Organization. http://www.euro.who.int/pubrequest

Joneidia Z, Mortazavia Y, Memarib F, Roointanc A, Chahardoulid B, Rostamid S (2019) The impact of genetic variation on metabolism of heavy metals: Genetic predisposition? Biomed Pharmacother 113:1–12

Joo JH, Wang S, Chen JG, Jones AM, Fedoroff NV (2005) Different signalling and cell death roles of heterotrimeric G protein alpha and beta subunits in the Arabidopsis oxidative stress response to ozone. Plant Cell 17:957–970

Kameli A, Losel DM (1993) Carbohydrate and water stress in wheat plants under water stress. New Phytol 125:609–614

Kassem MA, Meksem K, Kang CK, Njiti VN, Kilo V, Wood AJ, Lightfoot DA (2004) Loci underlying resistance to manganese toxicity mapped in a soybean recombinant inbred line population of 'Essex' x 'Forrest. Plant Soil 260:197–204

Kawashima AU, Noji CG, Nakamura M, Ogra Y, Suzuki KT, Saito K (2004) Heavy metal tolerance of transgenic tobacco plants over—expressing cysteine synthase. Biotech Lett 26(2):153–157

Khaniki GRJ, Zazoli MA (2005) Cadmium and lead contents in rice (*Oryza sativa*) in the North of Iran. Int J Agri Biol 7:1026–1029

Kilchevsky A, Khotylyova L, Peshich V, Kogotko L, School A, Gavrilov A, Kruk A, Scarascia GT, Porceddu E, Pagnotta MA (1999) Breeding of vegetable with minimum pollutant accumulation. Genetics and breeding for crop quality and resistance. Proceedings of the XV Eucarpia Congress, Viterbo, Italy, September 20–25, 1998. 1999, 313–322

Knudson LL, Tibbitts TW, Edwards GE (1977) Measurement of ozone injury by determination of leaf chlorophyll concentration. Plant Physiol 60:606–608

Kooij TAW, Kok JD, Haneklaus S, Schnug E (1997) Uptake and metabolism of sulphur dioxide by Arabidopsis thalianana. New Phytol 135:101–107

Kumar S, Sharma SS (2003) Stomatal movements as affected by heavy metal stress in vicia faba. Abstr.:2nd International Conference of plant physiology, January 8–12, New Delhi, India p 618

Lamaud E, Loubet B, Irvine M, Stella P, Personne E, Cel-lier P (2009) Partitioning of ozone deposition over a developed maize crop between stomata and non-stomatal uptakes, using Eddy-covariance flux measurements and modelling. Agr Forest Meteorol 149:1385–1396

Li CH, Wang TZ, Li Y, Zheng YH, Jiang GM (2013) Flixweed is more competitive than winter wheat under ozone pollution: evidences from membrane lipid peroxidation, antioxidant enzymes and biomass. PLoS ONE 8(3):e60109

Li L, Yi H (2012) Effect of sulfur dioxide on ROS production, gene expression and antioxidant enzyme activity in Arabidopsis plants. Plant Physiol Biochem 58:46–53

Liu JH, Wang W, Wu H, Gong X, Moriguchi T (2015) Polyamines function in stress tolerance: from synthesis to regulation. Front Plant Sci 6:827

Liu W, Yang YS, Li PJ, Zhou QX, Xie LJ, Han YP (2009) Risk assessment of cadmium contaminated soil on plant DNA damage using RAPD and physiological indices. J Hazard Mater 161:878–883

Liu Y, Li Y, Li L, Zhu Y, Liu J, Li G, Lin H (2017) Attenuation of sulfur dioxide damage to wheat seedlings by coexposure to nitric xide. Bull Environ Contam Toxicol. https://doi.org/10.1007/s00 128-017-2103-9

Ludlow FK (1993) Carbohydrate metabolism in drought stress leaves of pigeonpea (*Cajanus cajana*). J Exp Bot 44:1351–1359

Ludwikow A, Sadowski J (2008) Gene networks in plant ozone stress response and tolerance. J Integr Plant Biol 50(10):1256–1267

Madkour SA (1998) Defense mechanisms against ozone injury in bean (*Phaseolus vulgaris* L.). Mansoura Univ 23(7):3203–3225

Madkour MA, Ibrahim SE, El-Kassas HI, Salem SY (2003) Pyto and bioremediation of soil and water polluted with heavy metals. Egypt J Appl Sci 18 (3B):694–712

Martin MJ, Farage PK, Humphries SW, Long SP (2000) Can the stomatal changes caused by acute ozone exposure be predicted by changes occurring in the mesophyll? A simplification for models of vegetation response to the global increase in tropospheric elevated ozone episodes. Aust J Plant Physiol 27:211–219

McKee IF, Farage PK, Long SP (1995) The interactive effects of elevated CO2 and O3 concentration on photosynthesis of spring wheat. Photosynth Res 45:111–119

Meagher RB, Rugh CL, Kandasamy MK, Gragson G, Wang NJ (2000) Engineered phytoremediation of mercury pollution in soil and water using bacterial genes. In: Terry N, Banuelos G (eds) Phytoremediation of contaminated soil and water. Lewis Publishers, Boca Raton, pp 203–221

Mirecki M, Agič R, Šunić L, Milenković L, Ilić ZS (2015) Transfer factor as indicator of heavy metals content in plants. Fresen Environ Bull 24(11c):4212–4219

Mittler R, Vanderauwera S, Gollery M, Van-Breusegem F (2004) Reactive oxygen gene network of plants. Trends Plant Sci 9:490–498

Moravek A, Stella P, Foken T, Trebs I (2015) Influence of local air pollution on the deposition of peroxyacetyl nitrate to a nutrient-poor natural grassland ecosystem. Atmos Chem Phys 15:899–911

Mostafa MM (2001) Nutrition and productivity of board bean plants as affected by quality and source of irrigation water. Zagazig J Agric Res 28(3):517–532

Muneer S, Kim TH, Choi BC, Lee BS, Lee JH (2014) Effect of CO, NO_x and SO_2 on ROS production, photosynthesis and ascorbate–glutathione pathway to induce *Fragaria × annasa* as a hyperaccumulator. Redox Biol 2:91–98

Nocito FF, Clarissa L, Barbara C, Pierre F, Jean-Claude D, Gian AS (2006) Heavy metal stress and sulfate uptake in maize roots. Plant Physiol 141(3):1138–1148

Pal P (2016) Detection of environmental contaminants by RAPD method. Int J Curr Microbiol App Sci 5(8):553–557

Penner GA, Clarke J, Bezte LJ, Leiste D (1995) Identification of RAPD markers linked to a gene governing cadmium uptake in durum wheat. Genome 38:543–547

Poornima R, Padmanabha PV, Shankar AG, Udayakumar M (2003) Role of ferritin and glycine betaine in oxidative stress tolerance: A transgenic approach Abst: International Conference of Plant Physiology, January 8–12, 2003, New Delhi, India, p 264

Prado FE, Boero C, Gallarodo M, Gonzalez JA (2000) Effect of NaCl on germination, growth and soluble sugar content in Chenopodium quinoa willd seeds. Bot Bull Acad Sin 41:27–34

Qadir S, Qureshi MI, Jabvd S, Abdin MZ (2004) Phenotypic variation in phytoremediation potential of Brassica juncea cultivars exposed to Cd stress. Plant Sci 167(5):1171–1181

Qurainy F, Alameri AA, Khan S (2010) RAPD profile for the assessment of genotoxicity on a medicinal plant: Eruca sativa. J Med Plants Res 4(7):579–586

Ramadan AM (2003) Studies on bioremediation soils and plants: 4- Micronutrients distribution in safflower plant organs grown in soils under different treatments. Egypt J Appl Sci 18(1):386–396

Rauser WE (1999) Structure and function of metal chelators produced by plants. Cell Biochem Biophys 31:19–48

Robinson MF, James H, Mansfield TA (1998) Disturbances in stomatal behaviour caused by air pollutants. J Exp Bot 49:461–469

Roitto M, Ahonen-Jonnarth U, Lamppu J, Huttunen U (1999) Apoplastic and total peroxidase activites in scots pine needles at subarctic polluted sites. Eur J Forest Pathol 29(6):399–410

Rubio MI, Escring I, Martinze-Cortina C, Lopez-Benet FJ, Sanz A (1994) Cadmium and Nickel accumulation in rice plants: effects on mineral nutrition and possible interactions of abscisic and gibberellic acids. Plant Growth Regul 20:151–157

Rudolphi-Skórska E, Sieprawska A (2016) Adaptation of wheat cells to short-term ozone stress: the impact of α-tocopherol and gallic acid on natural and model membranes. Acta Physiol Plant 38(85):2–11

Rui YK, Guo J, Huang KL, Jin YH, Luo YB (2007) Application of ICP-MS to the detection of heavy metals in transgenic corn. PubMed 27(4):796–798

Ryan PR, Delhaize E, Jones DL (2001) Function and mechanism of organic anion exudation from plant roots. Annu Rev Plant Physiol Plant Mol Biol 52:527–560

Saxena P, Kulshrestha U (2016) Biochemical effects of air pollutants on plants. In: Saxena P, Kulshrestha UC (eds) Plant responses to air pollution, pp 59–70

Shahrtash M, Zaden SM, Mohabatkar H (2010) Cd induced genotoxicity detected by the random amplification of polymorphism DNA in the maize seedling root. J Cell Mol Res 2(1):42–48

Sohn JK, Lee JJ, Kwon YS, Kim KK (2002) Varietal differences and inheritance of resistance to ozone stress in rice (Oryza sativa L). SABRAO J Breed Genet 34(2):65–71

Takahashi M, Morikawa H (2014) Nitrogen dioxide is a positive regulator of plant growth. New Phytol 201:1304

Thara SB, Kumar NKH, Jagannath S (2015) Micro-morphological and biochemical response of Muntingia calabura L. and Ixora coccinia L. to air pollution. Res Plant Biol 5(4):11–17

Tiburcio AF, Altabella T, Bitrián M, Alcázar R (2014) The roles of polyamines during the lifespan of plants: from development to stress. Planta 240:1–18

Tripathi AK, Gautam M (2007) Biochemical parameters of plants as indicators of air pollution. J Environ Biol 28:127–132

Tuominen H, Overmyer K, Keinänen M, Kollist H, Kangasjärvi J (2004) Mutual antagonism of ethylene and jasmonic acid regulates ozone-induced spreading cell death in Arabidopsis. Plant J 39:59–69

Van Buuren ML, Guidi L, Fornale S, Ghetti F, Franceschetti M, Soldatini GF et al (2002) Ozone-response mechanisms in tobacco: implications of polyamine metabolism. New Phytol 156:389–398

Verma V, Chandra N (2014) Biochemical and ultrastructural changes in Sida cordifolia L. and Catharanthus roseus L. to auto pollution. Int Sch Res Not 2014(2): 11

Wahaab RA, Badawy MI (2004) Water quality assessment of the River Nile system: an overview. Biomed Environ Sci 17:87–100

Wallace T (1943) The diagnosis of mineral deficiencies in plants by visual symptoms. color pictures of mineral deficiencies in plants—1943. http://customers.hbci.com/~wenonah/min-def/sugrbeet.htm

Wu–Shutu YC, Kuo B, Thseng F, Wu S, Yu C, Kuo B, Thsenge F (2000) Diallel analysis of cadmium tolerance in seedling rice. SABRAO J Breed and Genet 32(2):57–61

Yaeno T, Saito B, Katsuki T, Iba K (2006) Ozone-induced expression of the Arabidopsis FAD7 gene requires salicylic acid, but not NPR1 and SID2. Plant Cell Physiol 47:355–362

Yang YY, Jung JY, Song WY, Suh HS, Youngsook L (2000) Identification of rice genotypes with high tolerance or sensitivity to lead and characterization of the mechanism of tolerance. Plant Physiol 124(3):1019–1026

Yang Z, Tian L, Latoszek-Green M, Brown D, Wu K (2005) Arabidopsis ERF4 is a transcriptional repressor capable of modulating ethylene and abscisic acid responses. Plant Mol Biol 58:85–596

Yi H, Yin J, Liu X, Jing X, Fan S, Zhang H (2012) Sulfur dioxide induced programmed cell death in Vicia guard cells. Ecotoxicol Environ Saf 78:281–286

Yoshimura K, Miyao K, Gaber A, Takeda T, Kanaboshi H, Miyasaka H, Shigeoka S (2004) Enhancement of stress tolerance in transgenic tobacco plants overexpressing Chlamydomonas glutathione peroxidase in chloroplasts or cytosol. Plant J 37(1):21–33

Zhang Y (2017) Overexpressing the Sedum alfredii Cu/Zn Superoxide Dismutase Increased Resistance to Oxidative Stress in Transgenic Arabidopsis. Front Plant Sci 8:1–13

Chapter 6
Rapid Screening Wheat Genotypes for Tolerance to Heavy Metals

S. S. Elgharbawy, M. I. E. Abdelhamid, E. Mansour, and A. H. Salem

Abstract The phytotoxicity of heavy metals released through anthropogenic activities conclusively reduces crops growth and productivity. The current study was carried out to assess the effect of some heavy metals (zinc [Zn], lead [Pb] and cadmium [Cd]) and their mixture on growth of eight wheat genotypes in early stage under laboratory conditions. The results showed that the heavy metals significantly reduced germination percentage, root and shoot growth. Zn had the least negative effect, while mixture of the three elements had the most drastic effect followed by Cd and then Pb. Wheat genotypes varied in their response to heavy metal stress. Moreover, it could be concluded that wheat genotype Gemmeiza-11 had the maximum tolerance against Zn, Pb, Cd and their mixture followed by Misr-1 and Sids-12, while, Sids-13 and Gemmeiza-9 displayed the lowest tolerance. The tolerant genotypes in early growth stages could be used commercially under heavy metals stress as well as could be utilized as donors for developing promise cultivars destined for agricultural production under heavy metals stress.

Keywords Heavy metals · Phytotoxicity · Wheat genotypes · Seedling vigor

6.1 Introduction

Pollution is a serious problem for human, animal and agriculture in worldwide (Al-Othman et al. 2016). Some of the pollutant substances called heavy metals exist in the environment through different sources. Their principal sources include emissions from the rapidly expanding industrial areas, land application of fertilizers, animal manures, sewage sludge, pesticides and using wastewater in irrigation due to water shortage in arid and semi-arid regions (Ding et al. 2013; Khan et al. 2017). The common heavy metals found at contaminated sites are zinc (Zn), lead (Pb), cadmium (Cd), chromium (Cr), arsenic (As), copper (Cu), mercury (Hg), and nickel (Ni), (Wuana and Okieimen 2011). Some of these metals are essential micronutrients

S. S. Elgharbawy · M. I. E. Abdelhamid · E. Mansour (✉) · A. H. Salem
Agronomy Department, Faculty of Agriculture, Zagazig University, 44519, Zagazig, Egypt

© Springer Nature Switzerland AG 2021   175
H. Awaad et al. (eds.), *Mitigating Environmental Stresses for Agricultural Sustainability in Egypt*, Springer Water,
https://doi.org/10.1007/978-3-030-64323-2_6

for certain physiological processes in plant as enzyme metabolism and photosynthesis like Zn, Fe, Ni and Cu (Zeng et al. 2015; Ali et al. 2018). Although, at high levels they become toxic pollutants and cause growth inhibition and have a notable adverse effects on crop productivity (Maleva et al. 2009). Since, their phytotoxicity inhibits transpiration, photosynthesis, chlorophyll synthesis and disturbing carbohydrate metabolism. Furthermore, they lead to reduction of germination percentage, inhibition in root growth and damage of lipids, proteins and DNA. On the other hand, Hg, Cd and Pb are nonessential and highly cytotoxic for plant (Lequeux et al. 2010).

Wheat germplasm provides very fruitful source of genes and rich sources of genetic variation for improving tolerance to heavy metal which can provide a solution for the current environmental pollution (Ali et al. 2018; Alybayeva et al. 2014; Rabnawaz et al. 2017). The genotypes display different ability to produce acceptable yield under heavy metal stress. Subsequently, it is needed to screen the genetic potentiality of the genotypes under heavy metal stress to evaluate their effect on plant growth and productivity to identify tolerant genotypes (Alybayeva et al. 2016). Under these circumstances, tolerant genotypes to heavy metal could be recommend for using in breeding program for high-yielding. The negative effects of these elements on plant are increased by raising their concentration in the air and soil ecosystem or aqueous medium. There are several studies emphasizing the negative influence of heavy metals Zn, Pb and Cd on seed germination, plant growth, function and grain yield (Abraham et al. 2013; Bücker-Neto et al. 2017).

Wheat belongs to Poaceae family and it is a major cereal crop in many parts of the world (Mansour et al. 2020). Keeping in view the importance of wheat and the excessive use of Zn, Pb and Cd in various industries, the present study has been designed to study the effect of heavy these metals and their mixture on germination and seedling growth of eight wheat genotypes. This study will help to understand the level of tolerance and sensitivity of wheat genotypes under heavy metals stress conditions for establishing a breeding program for heavy metals tolerance in wheat.

6.2   Materials and Methods

The present study was performed at the seed laboratory, Agronomy Department, Faculty of Agriculture, Zagazig University, Egypt. Seeds of eight wheat genotypes presented in Table 6.1 were provided by wheat section, Agricultural Research Center, Cairo, Egypt. The health and robust seeds of each genotype were surface sterilized with Sodium hypochlorite (5%) for ten minutes to prevent any fungal contaminate followed by washings with distilled water. Ten seeds were placed in Petri dishes (90 mm diameter) on filter paper (What man No. 42). Metal treatment of Zn, Pb, Cd and their mixture were prepared using zinc sulphate, lead acetate and cadmium carbonate with concentrations of 500 mg/L for Zn and 250 mg/L for each of Pb and Cd. These toxic doses were previously used by Gough et al. (1979) and Chen et al. (2003). In addition, the distilled water was used as control. The experimental design was completely randomized with five replicates. The Petri dishes were kept under

Table 6.1 Name, pedigree, year of release and origin of the 16 spring wheat genotypes used in this study

Genotype	Pedigree	Origin	Year of release
Misr-1	OASIS/KAUZ//4*BCN/3/2*PASTOP CMss00Y01881T-050 M-030Y-030 M-030WGY-33 M-0Y-0S.	CIMMYT	2010
Sids-12	BUC//7C/ALD/5/MAYA74/ON//1160-147/3/BB GLL/4/HAT"S"/6/MAYA/VUL/CMH 74A.630/4*SX.SD7096-4SD-1SD-1SD-0SD	Egypt	2008
Sids-13	ALMAZ 19 = KAUZ "S"//TSI/SNB"S" ICSBW 1-0375- 4AP-2AP-030AP-0APS- 3AP- 0APS- 050AP- 0AP- 0SD	Egypt	2010
Gemmeiza-9	Ald"S"/Huac"S"//CMH74A.630/5Xcgm4583-5GM-1GM-0GM	Egypt	1999
Giza-168	MIL/BUC//SeriCM93046-8 M-0Y-0 M-2Y-0B	Egypt	1999
Gemmeiza-11	Bow"s"/Kvz"s"//7c/seri82/3/Giza168/Sakha61CGM7892-2GM-1GM-2GM-1GM-0GM	Egypt	2010
Line-1	SAKHA 93/SIDS 6 CGZ(16)GM-2GM-OMG	Egypt	–
Line-2	GIZA 168/SIDS 7 CGZ(7)4GM-2GM-OMG	Egypt	–

laboratory condition for 12 days. After that, the number of germinated seeds was counted and germination percentage was calculated and maximum root and shoot length were estimated. Also, fresh weight of seedling roots and shoots was measured. Besides, dry weight of seedling roots and shoots biomass was determined using oven at 80 °C for 24 h.

Seedling phytotoxicity (SP) and seedling tolerance index (STI) were estimated using the following formulas: $\mathbf{SP} = \frac{\text{Seedling length of control} - \text{Seedling length of treatment}}{\text{Seedling length of treatment}} \times 100$ according to Chou and Lin (1976).

$\mathbf{STI} = \frac{\text{Mean seedling length in metal solution}}{\text{Mean seedling length in distilled water}} \times 100$ according to Iqbal and Rahmati (1992). Where seedling length is the sum of root and shoot length.

The seed germination and seedling growth data were statistically analyzed using SAS software (version 9.3, SAS Institute, Cary, NC, USA). In addition, the least significant differences were estimated at the 5% significance level.

6.3   Results

6.3.1   Influence of Heavy Metals on Germination

Essentially, heavy metals reduce seed germination, plant growth and grain yield due to their effect on photosynthetic activity, chlorophyll synthesis and antioxidant enzymes (Murzaeva 2004; Rabnawaz et al. 2017; Di Toppi and Gabbrielli 1999). The results of present study indicated that the metals Zn, Pb, Cd and their mixture had significant drastic effect on seed germination (Fig. 6.1). Mixture of the three metals had the most decreasing effects on germination percentage (59%) followed by Cd (51%), Pb (41%) and then Zn (30%). The reduction in germination percentage of wheat genotypes may be attributed to the interference of metals ions (Shaikh et al. 2013), disturbance of osmotic balance (Ashraf 2004; Mansour et al. 2020), quantitative and qualitative changes of proteins (Qureshi et al. 2007), impaired basic steps in the metabolism of carbohydrates and amino acids (Rahoui et al. 2010), changes in the

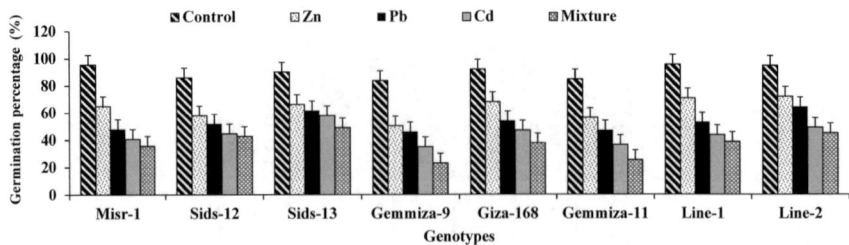

Fig. 6.1 Impact of heavy metals on germination percentage. The bars on the top of the columns represent the LSD (P < 0·05)

activity of autooxidating enzymes, peroxidating lipids and increasing active oxygen in plants (Taulavuori et al. 2005; Zhang et al. 2007).

The different heavy metals have different sites of action within the plant and the toxic response differs between heavy metal (Gill 2014.). The results suggested that Zn had the least decreasing effects on germination percentage of wheat seeds. In this context, Drab et al. (2011) compared the effects of Cd, Cu, Pb and Zn on the germination of seeds of the higher plants and found that Zn had the least decreasing effect which supported the results obtained in the current study.

The inhibitory effects of heavy metals on germination percentage varied among wheat genotypes. Sids-13 performed better under heavy metal stress followed by Line-2, while Gemmeiza-9 had poor performance and lowest germination percentage. Which pointed out that Gemmeiza-9 may be less tolerant to heavy metals stress. Similar observations in wheat genotypes had been observed by Ahmad et al. (2012).

6.3.2   Influence of Heavy Metals on Root Characteristics

The impacts of heavy metals and their mixture on root length, root fresh weight and root dry weight are shown in Fig. 6.2A, B, and C. It is obvious that the heavy metals significantly suppressed the growth of root. The mixture had the most negative effects on root length (69%), root fresh weight (55%) and root dry weight (66%) followed by Cd (53, 38 and 45%, the traits in the same order), Pb (35, 24 and 30%), and Zn (19, 13 and 19%) as compared to the control.

Some elements as Zn are essential micronutrients in low levels for physiological processes in plant, while Cd and Pb are highly toxic contaminants that caused reduction in both new cell formation and cell elongation in the extension region of roots (Liu et al. 2005). In the present study Cd had the most adverse effect on the root growth compared to Pb and Zn. In this respect, Farooqi et al. (2009) found that Cd was highly toxic to root growth compared to Pb. On the other hand, Shaikh et al. (2013) showed that Zn was the least harmful to root growth of wheat.

The negative effects of heavy metals on root growth varied among wheat genotypes. The genotypes; Gemmeiza-11, Giza-168 and Sids-13 performed better under heavy metal stress, while Sids-12 and Misr-1 had poor performance compared to the other genotypes and had the lowest root growth. Similar findings in wheat had been recorded by Ahmad et al. (2012).

6.3.3   Influence of Heavy Metals on Shoot Characteristics

The investigated heavy metals either separately or their mixture had reduced shoot length, shoot fresh weight and shoot dry weight of the evaluated wheat genotypes (Fig. 6.3A, B, and C). Mixture of the three metals had the most harmful effect on

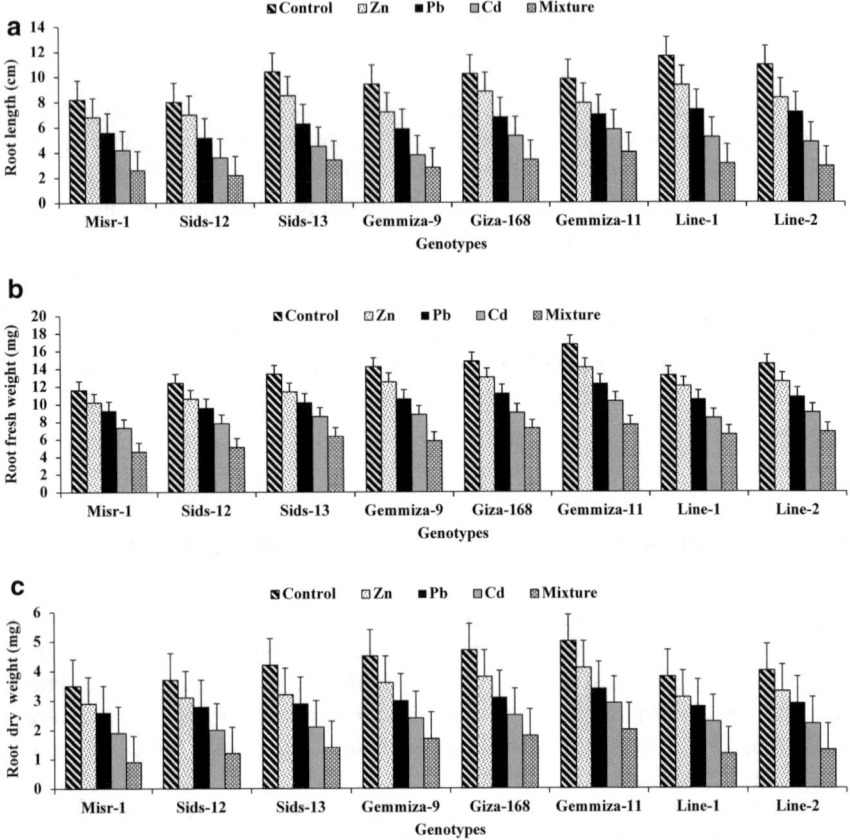

Fig. 6.2 Impact of heavy metals on root length (A), root fresh weight (B) and root dry weight (C) for the eight wheat genotypes. The bars on the top of the columns represent the LSD (P < 0·05)

shoot length (64%), shoot fresh weight (42%)and shoot dry weight (37%) followed by Cd (51, 23 and 25%, the traits in the same order), Pb (31, 17 and 16%) and Zn (15, 10 and 8%) as compared to the control. The obtained results suggested similar effects as root characteristics.

The effects of used metals and their mixture varied among the wheat genotypes on shoot characteristics. It is evident that wheat genotype Gemmeiza-11 performed better under heavy metal stress, followed by Gemmeiza-9, Giza-168 and Line-1 compared to the other wheat genotypes. On the other hand, the genotypes Misr-1, Sids-12 and Sids-13 exhibited the lowest shoot growth under heavy metal stress compared to the other wheat genotypes. The genotypic differential response to heavy metal stress was also reported by Awaad et al. (2010), Ahmad et al. (2012) and Bhatti et al. (2013).

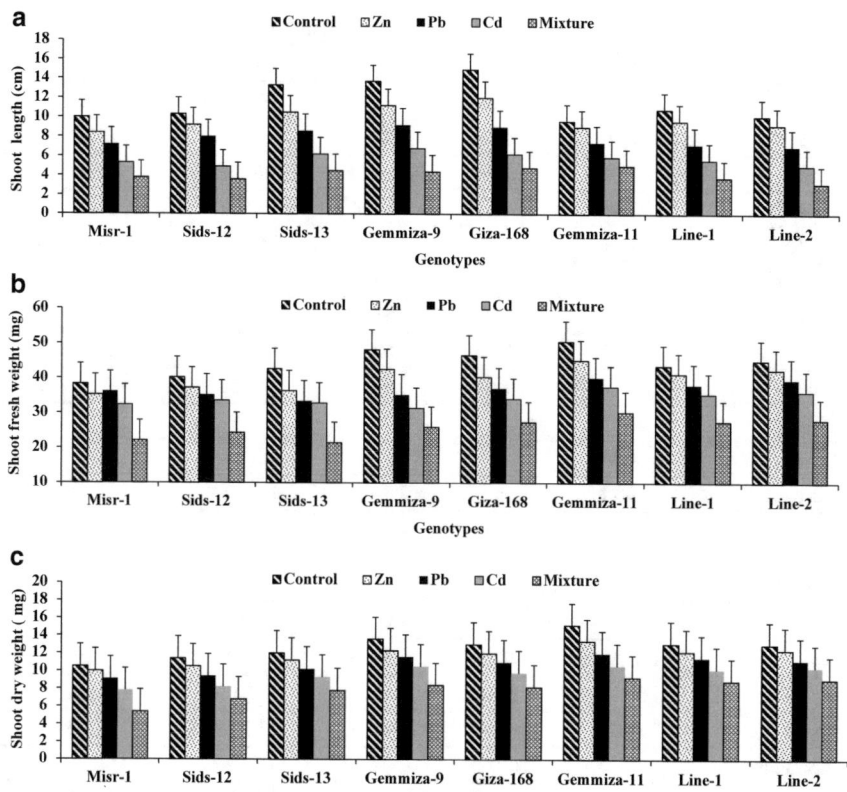

Fig. 6.3 Impact of heavy metals on shoot length (A), shoot fresh weight (B) and shoot dry weight (C) for the eight wheat genotypes. The bars on the top of the columns represent the LSD (P < 0·05)

6.3.4 Phytotoxicity Index

Phytotoxicity index of the used elements either individually or in combination on wheat seedling was calculated and the results are shown in Table 6.2. It can be seen that the mixture had the maximum phytotoxicity index (66.3%) as the mean of all studied wheat genotypes, followed by Cd (51.1%), Pb (32.4%) and then Zn (16.3%). Accordingly, the toxicity of selected heavy metals to young seedling can be arranged in the rank order of inhibition as mixture > Cd > Pb > Zn. Cadmium is easily taken up by plants because, it is quite mobile element in water and soil ecosystems. Cd has an adverse reputation for being highly toxic and threatening to plant growth. On the other hand, Zn could be considered slightly toxic (Shaikh et al. 2013).

Phytotoxicity of the used metals to wheat seedling varied among wheat genotypes. Gemmeiza-11 exhibited the least seedling phytotoxicity index under heavy metal stress. Whereas, Sids-13 and Gemmeiza-9 attained the maximum values of seedling

Table 6.2 Phytotoxicity index for the eight wheat genotypes under heavy metals stress

Genotypes	Zn	Pb	Cd	Mixture	Mean
Misr 1	16.57	29.54	47.83	64.73	**39.67**
Sids 12	11.16	28.39	53.14	68.48	**40.29**
Sids 13	20.36	37.34	54.83	66.92	**44.86**
Gemmeiza 9	20.49	34.45	53.95	68.73	**44.41**
Giza 168	17.24	36.87	54.21	67.35	**43.92**
Gemmeiza 11	12.91	25.67	39.85	53.72	**33.04**
Line 1	15.46	34.74	51.82	69.35	**42.84**
Line 2	16.32	32.25	53.27	70.92	**43.19**
Mean	**16.31**	**32.41**	**51.11**	**66.28**	
LSD_H	**4.03**				
LSD_G	**6.35**				
$LSD_{H \times G}$	**8.72**				

Table 6.3 Tolerance index for the eight wheat genotypes under heavy metals stress

Genotypes	Zn	Pb	Cd	Mixture	Mean
Misr-1	83.50	70.50	52.20	35.30	**60.38**
Sids-12	88.90	72.00	46.90	31.60	**59.85**
Sids-13	79.70	62.70	45.20	33.10	**55.18**
Gemmeiza-9	79.60	65.60	46.10	31.30	**55.65**
Giza-168	82.80	63.20	45.80	32.70	**56.13**
Gemmeiza-11	87.10	74.40	60.20	46.30	**67.00**
Line-1	84.60	65.30	48.20	30.70	**57.20**
Line-2	83.70	67.80	46.80	29.10	**56.85**
Mean	**83.74**	**67.69**	**48.93**	**33.76**	
LSD_H	**5.02**				
LSD_G	**7.43**				
$LSD_{H \times G}$	**9.45**				

phytotoxicity index. These results are in accordance with the findings of Ahmad et al. (2012) and Bhatti et al. (2013).

6.3.5 Tolerance Index

The tolerance index of tested wheat genotypes under three metals and their mixture is showed in Table 6.3. Among the tested wheat genotypes, Gemmeiza-11 had the maximum tolerance index against Zn (87.1%), Pb (74.4%), Cd (60.2%) and mixture of the three heavy metals (46.3%). Otherwise, Gemmeiza-9 exhibited the lowest values (79.6, 65.6, 46.1 and 31.3% in the same order). The genotypic differential responses to heavy metals could be due to the ability of some genotypes to withstand against heavy metals stress, while the others could not implement that. Similar finding was also reported by Gang et al. (2013).

The drastic effect of the mixture on wheat genotypes may be due to the high dose of the three elements together. In this respect, Farooqi et al. (2009) reported that the effect of heavy metals on wheat genotypes depend on the amount of toxic substance taken up from a given environment. Low tolerance of wheat genotypes was recorded due to Cd stress followed by Pb. Similar finding was reported by Munzuroglu and Geckil (2002) who arranged the inhibitory effects of heavy metals in the following sequence: Cd > Pb > Zn in wheat which is in line with the results of present study.

6.4   Conclusions

1. The heavy metals; zinc (Zn), lead (Pb), cadmium (Cd) and their mixture had significant drastic effect on seed germination, root and shoot growth.
2. The mixture of the three elements has the most diminishing effects followed by cadmium (Cd) and then lead (Pb), while zinc (Zn) has the lowest negative effect.
3. Wheat genotypes differ in their response to heavy metal stress. Gemmeiza-11 and Giza-168 exhibited the maximum tolerance against the three elements and their mixture, while Misr-1 and Sids-13 displayed the lowest tolerance.

6.5   Recommendation

1. The tolerant genotypes, as Gemmeiza-11 and Giza-168, which exhibited maximum tolerance against heavy metals could be recommended to be culti-vated commercially under heavy metals stress. While, it should avoid cultivation of the sensitive cultivars under heavy metals stress as Misr-1 and Sids-13.
2. The tolerant genotypes to heavy metal stress could be exploited as donors in wheat breeding program for developing promise cultivars destined for agricultural production under heavy metals stress.

References

Abraham K, Sridevi R, Suresh B, Damodharam T (2013) Effect of heavy metals (Cd, Pb, Cu) on seed germination of *Arachis hypogeae* L. Asian J Plant Sci Res 3(1):10–12
Ahmad I, Akhtar MJ, Zahir ZA, Jamil A (2012) Effect of cadmium on seed germination and seedling growth of four wheat (*Triticum aestivum* L.) cultivars. Pak J Bot 44(5):1569–1574

Ali Z, Mujeeb-Kazi A, Quraishi UM, Malik RN (2018) Deciphering adverse effects of heavy metals on diverse wheat germplasm on irrigation with urban wastewater of mixed municipal-industrial origin. Environ Sci Pollut 25(19):18462–18475

Al-Othman ZA, Ali R, Al-Othman AM, Ali J, Habila MA (2016) Assessment of toxic metals in wheat crops grown on selected soils, irrigated by different water sources. Arab J Chem 9:S1555–S1562

Alybayeva R, Kenzhebayeva S, Atabayeva S (2014) Resistance of winter wheat genotypes to heavy metals. IERI Procedia. 8:41–45

Alybayeva R, Kruzhaeva V, Alenova A, Salmenova I, Asylbekova A, Sadyrbaeva A (2016) The genetic potential of wheat resistance to heavy metals. Bioeng Biosci 4(3):34–41

Ashraf M (2004) Some important physiological selection criteria for salt tolerance in plants. Flora-Morphol Distrib Funct Ecol Plants 199(5):361–376

Awaad H, Youssef M, Moustafa E (2010) Identification of genetic variation among bread wheat genotypes for lead tolerance using morpho-physiological and molecular markers. J Am Sci 6(10):1142–1153

Bhatti K, Anwar S, Nawaz K, Hussain K, Siddiqi E, Sharif R, Talat A, Khalid A (2013) Effect of heavy metal lead (Pb) stress of different concentration on wheat (*Triticum aestivum* L.). Middle East J Sci Res 14(2):148–154

Bücker-Neto L, Paiva ALS, Machado RD, Arenhart RA, Margis-Pinheiro M (2017) Interactions between plant hormones and heavy metals responses. Genet Mol Biol 40(1):373–386

Chen S-h, Zhou Q-x, Sun T-h, Li P-j (2003) Rapid ecotoxicological assessment of heavy metal combined polluted soil using canonical analysis. J Environ Sci 15(6):854–858

Chou C, Lin H (1976) Autointoxication mechanism of *Oryza sativa* I. phytotoxic effects of decomposing rice residues in soil. J Chem Ecol 2(3):353–367

Di Toppi LS, Gabbrielli R (1999) Response to cadmium in higher plants. Environ Exp Bot 41(2):105–130

Ding C, Zhang T, Wang X, Zhou F, Yang Y, Yin Y (2013) Effects of soil type and genotype on lead concentration in rootstalk vegetables and the selection of cultivars for food safety. J Environ Manage 122:8–14

Drab M, Greinert A, Kostecki J, Grzechnik M (2011) Seed germination of selected plants under the influence of heavy metals. Civil Environ Eng Rep, pp 47–57

Farooqi Z, Iqbal MZ, Kabir M, Shafiq M (2009) Toxic effects of lead and cadmium on germination and seedling growth of *Albizia lebbeck* (L.) Benth. Pak J Bot 41(1):27–33

Gang A, Vyas A, Vyas H (2013) Toxic effect of heavy metals on germination and seedling growth of wheat. J Environ Res Dev 8(2):206–213

Gill M (2014) Heavy metal stress in plants: a review. Int J Adv Res 2(6):1043–1055

Gough LP (1979) Element concentrations toxic to plants, animals, and man. Geol Surv Bull 1466:1–80

Iqbal M, Rahmati K (1992) Tolerance of *Albizia lebbeck* to Cu and Fe application. Ekológia ČSFR 11(4):427–430

Khan ZI, Ahmad K, Rehman S, Siddique S, Bashir H, Zafar A, Sohail M, Ali SA, Cazzato E, De Mastro G (2017) Health risk assessment of heavy metals in wheat using different water qualities: implication for human health. Environ Sci Pollut Res 24(1):947–955

Lequeux H, Hermans C, Lutts S, Verbruggen N (2010) Response to copper excess in Arabidopsis thaliana: impact on the root system architecture, hormone distribution, lignin accumulation and mineral profile. Plant Physiol Biochem 48(8):673–682

Liu W, Zhao J, Ouyang Z, Söderlund L, Liu G (2005) Impacts of sewage irrigation on heavy metal distribution and contamination in Beijing, China. Environ Int 31(6):805–812

Maleva MG, Nekrasova GF, Malec P, Prasad M, Strzałka K (2009) Ecophysiological tolerance of Elodea canadensis to nickel exposure. Chemosphere 77(3):392–398

Mansour E, Moustafa ESA, Desoky ESM, Ali MMA, Yasin MAT, Attia A, Alsuhaibani N, Tahir MU, El-Hendawy S, (2020) Multidimensional Evaluation for Detecting Salt Tolerance of Bread Wheat Genotypes Under Actual Saline Field Growing Conditions. Plants 9(10):13–24.

Munzuroglu O, Geckil H (2002) Effects of metals on seed germination, root elongation, and coleoptile and hypocotyl growth in *Triticum aestivum* and *Cucumis sativus*. Arch Environ Contam Toxicol 43(2):203–213

Murzaeva S (2004) Effect of heavy metals on wheat seedlings: activation of antioxidant enzymes. Appl Biochem Microbiol 40(1):98–103

Qureshi MI, Qadir S, Zolla L (2007) Proteomics-based dissection of stress-responsive pathways in plants. J Plant Physiol 164(10):1239–1260

Rabnawaz A, Akram Z, Khan K, Ahmad Q (2017) Performance of wheat genotypes under cadmium contamination of soil. J Plant Sci Curr Res 1(003):1–4

Rahoui S, Chaoui A, El Ferjani E (2010) Membrane damage and solute leakage from germinating pea seed under cadmium stress. J Hazard Mater 178(1–3):1128–1131

Shaikh IR, Shaikh PR, Shaikh RA, Shaikh AA (2013) Phytotoxic effects of heavy metals (Cr, Cd, Mn and Zn) on wheat (*Triticum aestivum* L.) seed germination and seedlings growth in black cotton soil of Nanded, India. Res J Chem Sci 3(6):14–23

Taulavuori K, Prasad M, Taulavuori E, Laine K (2005) Metal stress consequences on frost hardiness of plants at northern high latitudes: A review and hypothesis. Environ Pollut 135(2):209–220

Wuana RA, Okieimen FE (2011) Heavy metals in contaminated soils: A review of sources, chemistry, risks and best available strategies for remediation. ISRN Ecol 402647:1–20

Zeng X, Wang Z, Wang J, Guo J, Chen X, Zhuang J (2015) Health risk assessment of heavy metals via dietary intake of wheat grown in Tianjin sewage irrigation area. Ecotoxicol 24(10):2115–2124

Zhang F-Q, Wang Y-S, Lou Z-P, Dong J-D (2007) Effect of heavy metal stress on antioxidative enzymes and lipid peroxidation in leaves and roots of two mangrove plant seedlings (*Kandelia candel* and *Bruguiera gymnorrhiza*). Chemosphere 67(1):44–50

Chapter 7
Performance, Adaptability and Stability of Promising Bread Wheat Lines Across Different Environments

Hassan Auda Awaad

Abstract Adaptability and stability are important indicators in light of climate change in the world, especially in the Mediterranean region. Therefore, eight bread wheat promising lines and three check cultivars were evaluated for days to 50% heading, grain protein content and grain yield (ard./fed.*) under twelve divers environments which were the combinations of three planting densities (350, 400 and 450 seeds/m^2) × two locations (Ghazaleh and El-khatara) × two seasons (2014/2015 and 2015/2016). Stability analysis of variance revealed highly significant G × E "linear" for days to 50% heading, grain protein content and grain yield (ard./fed.). Phenotypic stability parameters indicated that wheat genotypes Line 5 and Line 7 and Misr 1 were highly adapted to favorable environments for days to 50% heading; Line 2, Line 4 and Line 5 for grain protein content and Line 3, Line 8 and Giza 168 for grain yield. On the contrary, wheat cultivars Sakha 94 and Giza 168 were highly adapted to stress environments for days to 50% heading; Line 7, Giza 168 and Misr 1 for grain protein content and Line 2, Line 4, Line 6, Sakha 94 and Misr 1 for grain yield. Furthermore, wheat genotypes could be grown under wide range of environments were Line 1, Line 2, Line 3, Line 4, Line 6, Line 8 for days to 50% heading, Line 1, Line 3, Line 6, Line 8 and Sakha 94 for grain protein content and Line 1, Line 5 and Line 7 for grain yield. The most desired and stable genotypes were Line 1, Line 4 and Line 8 for earliness, Line 8 for grain protein content and Line 1 and Line 5 for grain yield. The most desirable and stable genotypes were Line 1, Line 2 and Line 4 and Line 8 for earliness; Line 8 for grain protein content and Line 1 and Line 5 for grain yield. The AMMI analysis of variance showed that 43.06, 6.52 and 23.33% of the total sum of squares were attributable to environmental, genotypic and GEI effects for grain yield/fed., respectively. According to GE biplot and ASV, Line 2, Line 3, Line 5 and Line 4 were more stable for days to 50% heading; Line 1, Line 4, Line 8 and Line 2 for grain protein content as well as Line 4, Line 1, Line 2, Line 5 and Misr 1 for grain yield. Tolerance index (TOL) showed that the most tolerant wheat genotypes to environmental stress were Sakha 94 followed by Line 6, Misr 1, Line 2 and Line 4. Both models of Eberhart and Russell (Crop Science 6:36–40, 1966) and AMMI (Gauch in Elsevier, Amsterdam, Netherlands, 1992) are consistent

H. A. Awaad (✉)
Crop Science Department, Faculty of Agriculture, Zagazig University, 44511, Zagazig, Egypt

© Springer Nature Switzerland AG 2021
H. Awaad et al. (eds.), *Mitigating Environmental Stresses for Agricultural Sustainability in Egypt*, Springer Water,
https://doi.org/10.1007/978-3-030-64323-2_7

in describing the stability of Line 2 and Line 4 for days to 50% heading; Line 8 for grain protein content as well as Line 1 and Line 5 for grain yield. These genotypes could be useful in wheat breeding programs for improving stability.

Keyword Bread wheat genotype × environment interaction · Performance · Adaptability · Phenotypic · Stability · AMMI model

7.1 Introduction

In light of climate change and its negative impacts on the agricultural sector. It is important to develop and release new genotypes and cultivars of wheat and determine their adaptability and stability to cope with environmental changes. Where, worldwide warming is the most important environmental challenges to humanity's future. A fluctuation in productivity of field crops in response to the environmental changes contributes to the increase of potential risk in the agricultural sector. Conceivable climate change effect could have crucial effect on agricultural production and subsequent socio-economic impacts. Potential effects of climate change on crop productivity have received immense attention during the last decades (Tao et al. 2008). Therefore, understanding the association between climate and crop production is fundamental to determine possible impacts of future climate and development of adaptation measures.

The results of long-range prediction using simulations and different climate change scenarios indicated that climate change and the resulting rise in temperature of the earth surface would negatively affect the productivity of Egyptian agricultural crops, causing a severe shortage in the productivity of most of the main food crops in Egypt in addition to increasing water consumption. El-Marsafawy (2007) revealed that yield of the wheat crop would be decreased by about 9% if the temperature rises 2 °C, and the rate will decrease to about 18% if the temperature is about 3.5 °C. The water consumption of wheat crop will increase by about 2.5% compared to corresponding under current weather conditions. Hickey (2017) displays a mere 1% chance that warming can be at or below 1.5 °C, the target set by Paris Agreement in 2016. The recent statistically-based projections, published July 31 in Nature Climate Change, show a 90% chance that temperatures will rise this century by 2.0–4.9 °C (https://phys.org/news/2017-07-earth-degrees-century.html).

Wheat is the most important cereal crop in Egypt as a major source of nourishment. Increasing production per unit area and grain protein content with a good level of earliness are very important from the point of view of adaptability and stability. This is essential to avoid terminal drought, and heat stresses and achieves global food security (Lamaoui et al. 2018). Georgopoulou et al. (2017) indicate that climate change may affect the agriculture sector dependent on their agricultural activity and the region, whereas adaptation can alleviate adversative effects of climate change. Therefore, the task of breeder is to evaluate genotypes grown at different environments to enable selection of these genotypes, which are suitable for a wider range of environments or a specific environment.

Wheat plants are one of dense growing crops which have a high compensatory capability, hence plant density is very imperative agronomic practices that regulate wheat yield and are mainly influenced by both the genotype and planting density. Isabelle et al. (2015) tried seeding rate (250, 300, 350, 400 and 450 seeds/m^2) of some wheat cultivars. The seeding rate generally affected yield components but did not affect yields. Increasing seeding rate slightly increased the grain protein content of wheat cultivars. Whereas Abd El-Hady et al. (2018) performed 300, 400 and 500 seeds/m^2 on grain yield of wheat, recorded significant differences between wheat genotypes Giza 171, Sids 12 and Gemmeiza 11 in grain yield. Dense planting 500 seeds/m^2 produce the highest grain yield/fed. as compared to the other two planting densities.

To identify stable genotypes for divergent environments, utilization of G × E interaction is of fundamental importance. G × E interactions are of important in the evaluation and development of wheat cultivars. G × E interaction increases with more differences between the cultivars in diverse environments or from changes in relative ranking of the genotypes (Allard and Bradshaw 1964). Selection of various wheat genotypes under stress environments is the main tasks of plant breeders for utilizing the genetic variation to produce stress tolerant wheat cultivars (Khan and Mohammed 2016).

Different statistical procedures have been reported to find out the adaptability and stability of new cultivars such as joint regression (Eberhart and Russell 1966) and additive main effects and multiplicative interaction, AMMI (Gauch 1992). Eberhart and Russell (1966) suggested that the regression coefficient (b_i) and deviation from the regression coefficient (S^2d) might predict stable genotype. The genotypes are grouped based on the size of their regression coefficients, less than, equal to or higher than one and based on the size of the variance of the regression deviations (equal to or different from zero). Those genotypes with regression coefficients higher than one would be more adapted to favorable environmental conditions, while genotypes with regression coefficients less than one would be adapted to unfavorable environments, and those with regression coefficients equal to one will be average adapted to all environments. Thus, a genotype with unit regression coefficient ($b_i = 1$) and deviation insignificantly different from zero ($S^2d = 0$) is said to be the most stable genotype. Many investigators have assessed the phenotypic stability of yield performance in wheat genotypes, Aly and Awaad (2002), Abd El-Shafi et al. (2014), El-Moselhy et al. (2015), Ali (2017), Ali and Abdul-Hamid (2017) and Siddhi et al. (2018). Furthermore, Hamam et al. (2015) computed stability and genetic diversity of thirty-six wheat genotypes under eight environments. Combined analyses of variance showed that days to heading and grain yield were significantly affected by years, locations, sowing dates and genotypes. Stability analysis revealed highly significant variances among genotypes for both traits. The partitioning of the genotype × environment interaction, as revealed by Env.+ (G × Env.), Env. (Linear) were highly significant for days to heading and grain yield. G × E (linear) was highly significant for both traits. Genotypes No. 5, 6, 14, 19, 20, 22, 24 and 32 exhibited stability for grain yield and are useful in the breeding program in developing new wheat genotypes with tolerance to heat stress conditions. The Additive Main effect and Multiplicative Interaction (AMMI) method suggested by Gauch (1992) was a significant advance in the analysis and interpretation of G × E interaction. The AMMI stability value ASV measure was proposed by Purchase (1997) and Purchase et al. (2000). The AMMI stability value is the distance from zero in a two-dimensional scattergram of IPCA 1 score against IPCA 2. A genotype with least ASV is the most stable. In this respect, many investigators used AMMI method for evaluating yield stability, of them Najafian et al. (2010), Farshadfar et al. (2011), Hagos and Abay (2013), Mohamed et al. (2013), Ali (2017) and Nasab et al. (2019) applied AMMI analysis to the yield data on various environments and they registered specific adaptation for several genotypes to specific environments.

Moreover, Elbasyoni (2018) applied AMMI model and found that analysis of variance showed significant effect of the environments, genotypes, and genotype × environment interaction for number of days to flowering and Grain yield (tons/ha). The variance of the environments attributed to 63.2%, while the variance attributed to genotypes was 14.6% and that for genotype × environment interaction was 22.2%. Genotype-environment interaction was highly significant, indicating differential response of genotypes to environments for a number of days to flowering and Grain yield (tons/ha). Cultivar "Sids12" was stable and exceeded other verified genotypes across environments. However, Gemmeiza 9 was more stable and surpassed other genotypes across environments under the recommended sowing date. Also, cultivar "Gemmeiza 12" was the ideal for the late sown condition. The main purposes of this study were to evaluate yield potentiality and to characterize the adaptability and stability of eight promising bread wheat promising lines and three check cultivars using the joint regression analysis and the AMMI method under the conditions of the Mediterranean region of Egypt.

7.2　Materials and Methods

7.2.1　Plant Materials and Experimental Layout

Eight bread wheat (*Triticum aestivum* L.) promising lines produced by the author, Agronomy Dept., Fac. of Agric. Zagazig Univ., Egypt through pedigree method of breeding program initiated from 1999/2000, were evaluated with three check cultivars under twelve environments. The pedigree and origin of the promising lines and check cultivars are presented in Table 7.1. The twelve environments were the combinations of three planting densities, i.e. (350 seeds/m^2, 400 seeds/m^2 and 450 seeds/m^2), two locations (Agriculture Research Stations, Fac. Agric., Zagazig Univ. at Ghazaleh and El-Khattara) and two seasons (2014/2015 and 2015/2016) as given in Table 7.2. Soil properties of the experimental sites Ghazaleh and Elkhatara are presented in Table 7.3. Also, monthly total precipitation (mm) and an average of the minimum

Table 7.1 Pedigree and origin of the eleven bread wheat genotypes used in this study

No.	Pedigree
Line 1	Gemmeiza9/Pata10//ALD "S" Cr1Zag–Zag190–Zag18–Zag20–Zag12–Zag15–Zag4–0Zag
Line 2	Gemmeiza9/Pata10//ALD "S" Cr1Zag–Zag180–Zag70–Zag20–Zag14–Zag32–Zag30–0Zag
Line 3	Gemmeiza 9/Pata10//ALD "S" Cr1Zag–Zag200–Zag36–Zag54–Zag30–Zag50–Zag27–0Zag
Line 4	Gemmeiza9/Pata10//ALD "S" Cr1Zag–Zag85–Zag80–Zag25–Zag14–Zag19–Zag6-0Zag
Line 5	Gemmeiza9/Pata10//ALD "S" Cr1Zag–Zag70–Zag30–Zag24–Zag15–Zag12–Zag6-0Zag
Line 6	Gemmeiza9/Pata10//ALD "S" Cr2Zag–Zag65–Zag55–Zag22–Zag18–Zag9–Zag12–0Zag
Line 7	Sakha69/Sahel1Cr2Zag–Zag78–Zag14–Zag10–Zag19–Zag3–Zag6-0Zag
Line 8	Sakha69/Sahel1Cr2Zag–Zag100–Zag35–Zag22–Zag12–Zag9–Zag4–0Zag
Sakha 94	Opata/Rayon//KauZ.CMBW90Y3180-0T0PM-3Y-010M-010Y-10M-15Y-0Y-0AP-0S.
Giza 168	MIL/BUC//seri.CM93046-8M-0Y-0M-2Y-OB
Misr 1	OASIS/KAUZ//4*BCN/3/2*PASTOP CMss00Y01881T-050M-030Y-030M-030WGY-33M-0Y-0S.

Table 7.2 Description of the field trails during the two seasons 2014/2015 and 2015/2016

Environments	Code	Latitude	Longitude
E1(Ghazaleh 2014/2015 350 seeds/m^2)	E1	30.6° N	31.6° E
E2(Ghazaleh 2014/2015 400 seeds/m^2)	E2	30.6° N	31.6° E
E3(Ghazaleh 2014/2015 450 seeds/m^2)	E3	30.6° N	31.6° E
E4(Ghazaleh 2015/2016 350 seeds/m^2)	E4	30.6° N	31.6° E
E5(Ghazaleh 2015/2016 400 seeds/m^2)	E5	30.6° N	31.6° E
E6(Ghazaleh 2015/2016 450 seeds/m^2)	E6	30.6° N	31.6° E
E7(El-khattara 2014/2015 350 seeds/m^2)	E7	30.6° N	32.3° E
E8(El-khattara 2014/2015 400 seeds/m^2)	E8	30.6° N	32.3° E
E9(El-khattara 2014/2015 450 seeds/m^2)	E9	30.6° N	32.3° E
E10(El-khattara 2015/2016 350 seeds/m^2)	E10	30.6° N	32.3° E
E11(El-khattara 2015/2016 400 seeds/m^2)	E11	30.6° N	32.3° E
E12(El-khattara 2015/2016 450 seeds/m^2)	E12	30.6° N	32.3° E

Table 7.3 Soil properties of the experimental sites Ghazaleh and Elkhatara

Soil properties	Ghazaleh	El-Khattara
Soil particles distribution		
Sand (%)	20.61	94.18
Silt (%)	31.82	4.35
Clay (%)	47.57	1.47
Soil texture	Clay	Sandy
Calcium carbonate ($CaCO_3$, g kg^{-1})	6.14	6.80
Organic matter (g kg^{-1})	10.34	6.30
pH	8.02	8.07
Electrical conductivity EC (dsm^{-1})	1.94	0.64
Soluble cations and anions (mmolc $^{L-1}$)*		
Calcium (Ca^{++})	5.22	1.67
Magnesium (Mg^{++})	4.37	0.95
Sodium (Na^+)	4.52	2.43
Potassium (K^+)	5.39	1.37
Bicarbonate (HCO_3^-)	6.08	2.17
Chlorine (Cl^-)	6.58	2.68
Sulphate ($SO_4^=$)	6.84	1.54
Available nutrient (mg kg^{-1} soil)		
Nitrogen (N)	57.32	30.52
Phosphorus (P)	8.15	5.49
Potassium (K)	149.3	79.34

Source Central Laboratory of Faculty of Agriculture, Zagazig University, Zagazig Egypt

and maximum temperatures during the growing seasons for the experimental sites Ghazaleh (GH) and El-Khattara (KH) are given in Table 7.4. The aims of this study were to detect the performance of new promising lines, the interaction between bread wheat genotypes and environments as well as estimate the stability parameters of the studied genotypes for days to 50% heading, grain protein content and grain yield (ard./fed.). Protein content was estimated using the micro-kjeldahl method to estimate the total nitrogen in the grain and multiplied by 5.75 to get the percentage of protein according to AOAC (1980). The experimental design at each environment was a randomized complete block design with three replications. Each plot consisted of 6 rows, 3 m long and 15 cm apart. Seeds were hand sown in drills. Sowing dates were on 24 and 25th of November in the 1st and 2nd seasons, respectively. Data were recorded on days to 50% heading, grain protein content and grain yield (ard./fed.). All other cultural practices were applied as recommended in each location.

Table 7.4 Monthly total precipitation (mm) and an average of the minimum and maximum temperatures during the growing seasons for the sites of the experimental, Ghazaleh (GH) and El-Khattara (KH)

Growing seasons and locations

Meteorological variable	Months	2014/2015		2015/2016	
		GH	KH	GH	KH
Precipitation (mm/month)	November	12.0	9.0	9.3	10.7
	December	11.0	9.1	12.2	13.2
	January	12.7	12.8	11.3	12.3
	February	13.2	13.4	15.5	16.5
	March	8.6	7.6	10.7	8.6
	April	1.6	1.0	0.2	0.0
	May	–	–	–	–
Minimum temperature (°C)	November	13.3	14.4	14.4	15.0
	December	10.3	10.8	9.6	9.8
	January	7.1	7.9	6.8	7.7
	February	7.7	8.7	9.2	10.0
	March	10.7	11.3	11.3	12.8
	April	11.7	12.2	13.2	14.3
	May	16.8	17.4	17.4	18.2
Maximum temperature (°C)	November	25.4	26.2	26.0	27.0
	December	22.7	22.6	20.6	19.7
	January	18.9	18.4	18.8	18.2
	February	20.3	19.5	24.0	23.8
	March	25.0	26.2	26.2	27.0
	April	28.5	29.6	33.1	34.6
	May	34.1	34.6	34.3	35.0

7.2.2 Statistical Analysis

Combined analyses of variance over environments and differences among means were tested using L.S.D. test at the 0.05 level according to Steel et al. (1997). Stability parameters were calculated according to the two models of Eberhart and Russell (1966) and additive main effects and multiplicative interaction method (AMMI) as proposed by Gauch (1992).

7.2.2.1 Tolerance Index (TOL)

Rosielle and Hamblin (1981) defined stress tolerance (TOL) as the difference in yield between stress and non-stress environments.

TOL $= Ys - Yp$. Where, Ys and Yp are the average grain yield of a genotype under stress (El-Khattara) and normal conditions (Ghazaleh), respectively.

7.3   Results and Discussion

7.3.1   Analysis of Variance

The combined analyses of variance for days to 50% heading, grain protein content and grain yield (ard./fed.) (Table 7.5) showed highly significant differences among environments for the forgoing traits, suggesting that the environments under study were different. Moreover, a highly significant effect due to years (Y) was obtained for days to 50% heading and grain protein content. This result reflects the wide

Table 7.5 The combined analyses of variance over two years, two locations, three planting densities and eleven wheat genotypes for studied traits

S.O.V	d.f	Days to 50% heading		Grain protein content %		Grain yield (ard./fed.[a])	
		SS	MS	SS	MS	SS	MS
Environments (E)	11	930.78	84.62**	88.30	8.03**	3458.42	314.40**
Reps/Env.	24	22.46	0.94**	7.02	0.29	399.99	16.67**
Years (Y)	1	7.17	7.17**	0.68	0.68**	8.60	8.60
Y × L	1	7.66	7.66**	1.41	1.41**	2.55	2.55
Loc. (L)	1	780.45	780.45**	54.57	54.57**	3086.51	3086.51**
Planting density (D)	2	64.38	32.19**	8.25	4.13**	253.21	126.61**
Y × D	2	8.01	4.00**	0.66	0.33	36.16	18.08
L × D	2	50.04	25.02**	21.44	10.72**	69.55	34.78**
Y × L × D	2	13.08	6.54**	1.29	0.65**	1.83	0.92
Genotypes (G)	10	776.60	77.66**	23.27	2.33**	523.78	52.38**
G × E	110	219.33	1.99**	43.11	0.39**	1874.16	17.04**
G × Y	10	9.04	0.90	3.33	0.33**	132.48	13.25**
G × L	10	41.07	4.11**	8.90	0.89**	444.84	44.48**
G × D	20	61.95	3.10**	13.64	0.68**	358.68	17.93**
G × Y × L	10	18.97	1.90**	1.20	0.12	142.31	14.23*
G × Y × D	20	24.56	1.23	5.04	0.25*	197.24	9.86
G × L × D	20	45.76	2.29**	5.63	0.28*	317.98	15.90**
G × Y × L × D	20	17.97	0.90	5.37	0.27**	280.63	14.03**
Error	240	195.09	0.81	32.86	0.14	1775.58	7.40

*, **Significant at 0.05 and 0.01 levels of probability, respectively. [a]Ardab = 150 kg, Feddan = 4200 m^2

differences in climatic conditions prevailing during the growing seasons. The main effect of locations (L) and planting densities (D) was highly significant for all studied traits. The studied genotypes (G) had highly significant differences for all traits, revealing that these genotypes were genetically different for genes controlling these traits.

Highly significant G × E items were detected for the studied traits. They provide evidence that the studied bread wheat genotypes are differed in their response to the environmental conditions. It is essential to determine the degree of stability for each genotype.

The first order interaction of years × locations (Y × L) differed significantly for days to 50% heading and grain protein content, indicating the different influences of climatic conditions on locations. Also, significant interactions between locations × planting densities (L × D) were found for all traits, whereas only years × planting densities (Y × D) had significant interaction for days to 50% heading. The genotype × locations interaction component (G × L) and (G × D) were accounted for the most of total G × E interaction for studied traits, indicating that locations and planting densities had the major effect on the relative genotypic potential of these traits. Moreover, the combined analyses of variance showed significant interactions between genotypes and years (G × Y) for grain protein content and grain yield (ard./fed.).

The second order (Y × L × D) interaction had highly significant for days to 50% heading and grain protein content; (G × Y × L) for days to 50% heading and grain yield (ard./fed.), (G × Y × D) for grain protein content and (G × L × D) for days to 50% heading, grain protein content and grain yield (ard./fed.). Moreover, the third order (G × Y × L × D) interaction was significant for grain protein content and grain yield (ard./fed.). These results reflected the importance of environmental factors of each year, location and planting density on the performance of wheat genotype regarding the previous traits. Similar results were attained by Tammam and Abd El Rady (2010), Tawfelis et al. (2011), Ali (2017) and Ali and Abdul-Hamid (2017).

7.3.2 Mean Performance of Wheat Genotypes Under Different Environments

Mean performance of days to heading for the studied eleven genotypes are shown in Table 7.6. It is interesting to mention that number of days to 50% heading was reduced from D1, D2 to D3 under Ghazaleh and El-Khattara regions in the first and second seasons.

Bread wheat genotypes Line 1, Line 4, Giza 168, Sakha 94 and Misr 1 were the earliest, among them Line 1, Line 4 and Giza 168 exhibited a good level of grain yield overall environments compared with the other genotypes. The most earliness wheat genotypes were Line 1 with the value of 79.5 days in the 1st season and 77.33 days

Table 7.6 Mean performance for eleven wheat genotypes under different environments for days to 50% heading

Genotype	Ghazaleh						El-Khattara						Combined
	2014/2015			2015/2016			2014/2015			2015/2016			
	D1	D2	D3	D1	D2	D3	D1	D2	D3	D1	D2	D3	
Line 1	82.53	80.17	79.50	82.33	80.21	78.87	79.00	77.70	77.47	78.07	77.67	77.33	79.24
Line 2	83.53	82.67	82.00	83.00	82.40	81.73	80.77	79.83	78.77	79.70	81.37	79.67	81.29
Line 3	86.07	85.03	83.40	85.33	83.40	83.70	81.77	81.03	80.50	80.67	83.00	81.33	82.94
Line 4	83.07	82.10	81.07	83.10	81.17	80.43	80.07	79.80	78.53	78.67	80.73	78.67	80.62
Line 5	85.30	84.67	83.70	84.00	84.37	83.87	81.40	81.23	80.47	78.33	81.67	81.40	82.53
Line 6	84.47	82.33	82.03	83.67	80.40	80.33	79.07	82.17	78.70	79.00	80.73	79.07	81.00
Line 7	87.73	88.67	85.43	88.60	86.00	84.90	82.93	83.17	81.60	83.68	84.67	82.90	85.00
Line 8	83.80	83.67	83.10	83.67	82.77	82.83	80.77	79.40	80.67	80.17	81.00	80.70	81.88
Sakha 94	81.27	82.37	80.50	80.83	81.33	82.13	78.83	80.33	79.00	78.66	80.67	79.27	80.43
Giza 168	80.27	80.80	79.67	79.33	80.00	80.43	77.67	79.00	78.03	77.40	78.00	78.17	79.23
Misr 1	83.40	82.97	82.60	83.87	82.70	81.93	80.40	79.13	78.73	78.80	78.83	77.27	80.89
Mean	83.77	83.22	82.09	83.43	82.25	81.98	80.24	80.25	79.31	79.56	80.76	79.53	81.26
L.S.D. 0.05	1.59	1.15	1.18	2.06	1.90	1.28	1.66	1.38	1.11	1.80	1.55	1.43	0.42

D1, D2 and D3 refer to 350 seeds/m^2, 400 seeds/m^2 and 450 seeds/m^2

in D3 in the 2nd one; Giza 168 with values of 79.33 and 77.40 days in D1 in the 2nd season, under Ghazaleh and El-Khattara regions, respectively.

Under El-Khattara region as sandy soil stress environment, Line 4 was the early valued of 78.53 days in D3 in the 1st season and the 2nd season also in D1, and D3 valued 78.67 day. In this respect, Hameed et al. (2003) found that increasing plant density affected significantly number of days to heading.

Mean performance of grain protein content for the studied twelve genotypes (Table 7.7) differed from that environment to another and from genotype to another. At Ghazaleh location, grain protein content was significantly higher in D2 as compared to D1 and D3. Otherwise under El-Khattara location, grain protein content was higher in D3 as compared to D1 and D2, during the two seasons, as heat shock proteins are produced at high rates under heat stress and are believed to have a protective role to high temperature stress. Since, there was a decline in the overall grain yield under El-Khattara region. Therefore, it is obvious that the combined effect of sandy soil condition, low rainfall, higher temperatures during grain filling period have been the main cause of higher protein content in the stress environments as a result of dilution effect (Tables 7.3 and 7.4).

Under Ghazaleh location, maximum grain protein content was recorded in Line 6 11.97% in D2 in the 1st season, Line 7 12.10% in D3 in the 2nd season, Misr 1 12.03 and 11.97% in D1in both seasons, respectively. Whereas, under El-Khattara location, maximum grain protein content (>12%) was recorded in Line 4, Giza 168 and Misr 1 in D3 in the 1st and 2nd season as well as Line 7 and Line 8 in D1, D2 and D3 in the 2nd one. This may be due to the dilution effect as a result of environmental stress. In this respect, Isabelle et al. (2015) tried seeding rate (250, 300, 350, 400 and 450 seeds/m^2) of some wheat cultivars. The seeding rate generally affected yield components but did not affect yields. Increasing seeding rate slightly increased grain protein content of wheat cultivars.

Based on general mean, mean performance of grain yield (ard./fed.) (Table 7.8) for the studied eleven genotypes showed slight increase under Ghazaleh location from D1 26.08 to D2 26.47 (ard./fed.) but decrease in D3 valued 23.76 (ard./fed.) in the first season. Whereas, it exhibited maximum average value 26.86 (D2) in the 2nd season rather than D3 which recorded 24.42 (ard./fed.). Also, under Ghazaleh location, L1, L8 and L7 gave maximum grain yield valued 30.53 and 29.81; 30.77 and 30.71 as well as 28.84 and 28.20 (ard./fed.) in D1 and D2 in the 1st season as well as 29.03 and 28.27; 30.97 and 29.30 as well as 27.14 and 31.53 (ard./fed.), in the same respective order in the 2nd season. Line 3 valued 31.82, and Line 5 valued 29.10 (ard./fed.) in D2 in the 2nd season, respectively. A higher estimate of grain yield under Ghazaleh location was attributed to suitable environments with soil fertility, adequate water and other inputs. Therefore, it is obvious in Table 7.3 that rich Ghazaleh region to organic matter, 10.34 g kg^{-1}, macro and micro-nutrients with suitable temperatures (Table 7.4) are factors that led to increasing grain yield.

Under El-Khattara location, general mean was decreased and valued 21.01, 19.82 and 18.78 as well as 20.64, 20.33 and 19.38 from D1, D2 and D3, in both seasons, respectively. Line 6 in D2 and Line 7 in D1 produced higher grain yield with values of 25.77 and 25.40 ard/fed in the 2nd season, respectively. Furthermore, Line 1, Line

Table 7.7 Mean performance for eleven wheat genotypes under different environments for grain protein content

| Genotype | Ghazaleh | | | | | | El-Khattara | | | | | | Combined |
| | 2014/2015 | | | 2015/2016 | | | 2014/2015 | | | 2015/2016 | | | |
	D1	D2	D3	D1	D2	D3	D1	D2	D3	D1	D2	D3	
Line 1	10.63	10.70	10.43	10.37	11.03	10.63	11.47	11.67	12.10	11.40	11.47	11.87	11.15
Line 2	10.57	10.80	10.27	10.73	10.93	10.20	10.83	10.70	12.17	11.17	11.67	11.73	10.98
Line 3	10.67	10.97	10.73	11.03	10.97	10.13	11.50	11.90	11.97	11.53	12.03	11.80	11.27
Line 4	10.40	10.67	10.83	10.60	10.93	10.80	11.00	11.67	12.80	11.03	11.57	12.73	11.25
Line 5	10.53	10.80	10.50	11.10	10.87	10.40	11.43	11.93	12.10	11.53	11.57	12.43	11.27
Line 6	10.97	11.97	10.57	11.03	11.60	10.37	12.10	11.97	12.40	11.50	11.60	11.50	11.46
Line 7	11.17	11.90	11.88	11.27	11.83	12.10	11.87	11.93	12.80	12.27	12.40	12.87	12.02
Line 8	11.37	11.57	10.87	11.13	12.03	11.83	11.70	11.60	12.70	12.67	12.17	12.83	11.87
Sakha 94	11.07	11.07	10.83	10.33	11.87	10.87	10.93	11.50	12.80	11.27	10.63	11.67	11.24
Giza 168	11.60	11.23	11.97	10.73	11.70	11.50	11.50	11.83	12.73	10.87	11.83	12.53	11.67
Misr 1	12.03	11.63	11.10	11.97	11.63	10.60	11.47	12.00	12.20	12.27	11.27	12.23	11.70
Mean	11.00	11.21	10.91	10.94	11.40	10.86	11.48	11.79	12.43	11.59	11.66	12.19	11.44
L.S.D. $_{0.05}$	0.546	0.554	0.570	0.651	0.704	0.478	0.701	0.543	0.517	0.566	0.693	0.913	0.171

D1, D2 and D3 refer to 350 seeds/m^2, 400 seeds/m^2 and 450 seeds/m^2

Table 7.8 Mean performance for eleven wheat genotypes under different environments for grain yield (ard./fed.)

Genotype	Ghazaleh						El-Khattara						Combined
	2014/2015			2015/2016			2014/2015			2015/2016			
	D1	D2	D3	D1	D2	D3	D1	D2	D3	D1	D2	D3	
Line 1	30.53	29.81	25.43	29.03	28.27	22.83	22.27	21.37	21.13	19.77	22.17	20.23	24.40
Line 2	23.31	20.55	22.07	22.82	24.70	25.27	20.20	19.28	19.77	17.87	17.97	18.57	21.03
Line 3	25.83	27.38	28.13	26.87	31.82	25.88	19.12	19.07	19.17	18.20	19.81	17.87	23.26
Line 4	26.30	27.93	25.70	26.23	26.47	25.87	19.77	21.93	20.54	22.96	20.80	22.67	23.93
Line 5	25.76	23.98	21.17	28.53	29.10	25.00	22.65	18.56	19.10	22.87	18.37	20.47	22.96
Line 6	24.70	25.98	21.07	25.75	23.19	23.73	22.40	22.18	17.40	17.77	25.77	18.03	22.33
Line 7	28.84	28.20	21.08	27.14	31.53	22.43	22.10	18.17	17.24	25.40	20.47	20.27	23.57
Line 8	30.77	30.71	23.38	30.97	29.30	26.60	18.84	18.03	16.93	21.63	19.53	18.63	23.77
Sakha 94	22.64	22.53	18.46	21.18	22.10	22.87	22.47	18.68	18.07	22.15	19.33	17.99	20.71
Giza 168	27.10	28.23	29.90	28.97	25.00	26.30	21.07	21.30	17.51	16.50	19.77	18.30	23.33
Misr 1	21.07	25.90	23.94	27.23	23.97	21.84	20.23	19.57	19.78	21.92	19.63	20.16	22.10
Mean	26.08	26.47	23.76	26.79	26.86	24.42	21.01	19.83	18.78	20.64	20.33	19.38	22.85
L.S.D. 0.05	4.930	6.794	5.602	4.697	3.661	4.138	4.806	4.065	3.073	5.589	2.766	3.894	1.257

D1, D2 and D3 refer to 350 seeds/m², 400 seeds/m² and 450 seeds/m² Ardab = 150 kg, Feddan = 4200 m²

5, Line 6, Line 7 and Sakha 94 in D1; Line 6 in D2 in 1st season and Line 4, Line 5, Sakha 94 in D1 and Line 1 in D2 as well as Line 4 in D3 in the 2nd season, all produced grain yield about 22 ard/fed.

According to number of days to 50% heading (Table 7.6) and those obtained from wheat grain yield (Table 7.8), wheat genotypes Line 1, Line 4, Line 8 and Giza 168 were early coupled with higher grain yield, with good level of protein content for only Line 8 and Giza 168 (Table 7.7), among them Line 1 was moderate tolerant and Line 4 was tolerant to stress based on TOL values (Table 7.9). Whereas, Line 7 attained a good level of yield averaged 23.57 ard/fed., but it was late 85.0 day in early and moderate tolerant. Therefore, in regions of the terminal stress, breeder seeks for genotypes of shorter grain filling duration to escape or at least to minimize the determine effect of heat stress on grain yield like Line 1, Line 4, Line 8 and Giza 168. Also, wheat grain yield was higher in Loc. 1 (Ghazaleh) compared with Loc. 2 (El-Khattara). This might be due to the favorable conditions of precipitation, temperatures and soil properties in Ghazaleh rather than El-Khattara (Tables 7.3 and 4). Mekonnen (2017) tried four levels of seeding rates (100, 125, 150, and 175 kg ha^{-1}), and showed that agronomic traits like days to heading, days to physiological maturity and grain yield of wheat were significantly influenced by seeding rate.

Under Ghazaleh region, Line 3 and Line 7 acted well in D2 during 1st and 2nd seasons but behaved opposite trend El-Khattara region. This refers to better dry matter accumulation under moderate density (450 seeds/m^2) where competition was decreased interplants and intra spikelets. Similar finding was recorded by Ghorbani and Basiri (2013) and Isabelle et al. (2015). The fluctuation in the performance of grain yield from genotype to another or from environment to environment could be due to G × E interaction effect. Where set of genes controlling grain yield could be changed from environment to another. Similar findings were registered by Tawfelis (2006), Menshawy (2007), Mohammadi (2012), Hamam et al. (2015) and Elbasyoni (2018).

Table 7.9 Tolerance index (TOL) for eleven wheat genotypes for grain yield (ard./fed.)

Genotypes	Ghazaleh	Khatarra	Tolerance index (TOL)
Line 1	27.65	21.16	6.49
Line 2	23.12	18.79	4.33
Line 3	27.56	18.87	8.69
Line 4	26.37	21.45	4.92
Line 5	25.59	20.33	5.26
Line 6	24.07	20.59	3.47
Line 7	26.54	20.60	5.94
Line 8	28.62	18.93	9.69
Sakha 94	21.63	19.78	1.85
Giza 168	27.58	19.00	8.58
Misr 1	23.99	20.21	3.78

Ardab = 150 kg, Feddan = 4200 m^2

7.3.3   Tolerance Index (TOL)

Tolerance index (TOL) offers a measure of yield stability according to yield loss under stress as compared to non-stressed condition (Rosielle and Hamblin 1981). It is interesting to mention that wheat genotypes could be classified into three categories based on TOL values (Table 7.9), the first category was tolerant to stress condition and include Sakha 94, L6, Misr 1, L2 and L4, these genotypes exhibited lower TOL values. The second category was moderate tolerant and include L5, L7 and L1, whereas, the third category was sensitive and include L8, L3 and Giza 168. Similar results were obtained by Ali and Abdul-Hamid (2017).

7.3.4   Phenotypic Stability Analysis

Mean square of joint regression analysis of variance for days to 50% heading, grain protein content and grain yield of the eleven bread wheat genotypes under twelve environments (Table 7.10) revealed highly significant differences among genotypes (G), environments (E) and the G × E interaction for all traits. This indicated the presence of genetic and environmental variation regarding wheat genotypes for the studied traits. Environment + Genotype × Environment (E + G × E) had highly significant effects for all traits. The G × E interaction was further partitioned into linear and non-linear (pooled deviation) components. Mean squares due to environment (linear) were highly significant for all traits, indicating that differences existed

Table 7.10 Joint regression analysis of variance over twelve environments for eleven wheat genotypes for studied traits

S.O.V	df	Days to 50% heading		Protein content %		Grain yield (ard./fed.)	
		SS	MS	SS	MS	SS	MS
Model	131	642.23	4.90**	51.56	0.39**	1952.12	14.90**
Genotype (G)	10	258.87	25.89**	7.76	0.78**	174.59	17.46**
Environment (E)	11	310.26	28.21**	29.43	2.68**	1152.81	104.80**
G × E	110	73.11	0.66**	14.37	0.13**	624.72	5.68**
E + G × E	121	383.37	3.17**	43.80	0.36**	1777.53	14.69**
Environment (linear)	1	310.26	310.26**	29.43	29.43**	1152.81	1152.81**
G × E (linear)	10	16.17	1.62**	1.47	0.15**	132.65	13.26**
Pooled deviation	110	56.94	0.52*	12.91	0.12**	492.07	4.47**
Pooled error	240	65.03	0.27	10.95	0.05	591.86	2.47

*, **Significant at 0.05 and 0.01 levels of probability, respectively Ardab = 150 kg, Feddan = 4200 m^2

between environments and revealed predictable component shared G × E interaction with unpredictable. The linear interaction (G × E linear) was highly significant when tested against pooled deviation for all traits, revealing genetic variances among genotypes for their regression on the environmental index, so it could be proceeded in the stability analysis (Eberhart and Russell 1966) for these traits. Similar results obtained from used phenotypic stability of yield performance in wheat genotypes, Aly and Awaad (2002), Hamam and Khaled 2009 and Siddhi et al. (2018). Furthermore, Hamam et al. (2015) computed stability and genetic diversity of thirty-six wheat genotypes under eight environments. Combined analyses of variance showed that days to heading and grain yield were significantly affected by years, locations, sowing dates and genotypes. Stability analysis revealed highly significant differences among genotypes for both traits. The partitioning of the genotype × environment interaction, as revealed by Env.+ (G × Env.), Env.(Linear) were highly significant for days to heading and grain yield. G × E (linear) was highly significant for both traits. Genotypes No. 5, 6, 14, 19, 20, 22, 24 and 32 exhibited stability for grain yield and are valuable in the breeding program in developing new wheat genotypes with tolerance to heat stress conditions.

The non-linear responses as estimated by pooled deviations from regressions were significant, demonstrating that differences in linear response between genotypes across environments did account for all the G × E interaction effects. Thus, the fluctuation in the performance of genotypes grown in different environments was fully predictable. Highly significant effects for G × E interaction for various wheat traits were previously reported by Abd El-Shafi et al. (2014), El-Ameen (2012), El-Moselhy et al. (2015), Ali (2017) and Ali and Abdul-Hamid (2017).

7.3.4.1 Phenotypic Stability Parameters

Days to 50% Heading

Eberhart and Russell (1966) suggested that an ideal genotype is the one which has the highest yield across a wide range of environments, a regression coefficient (b_i) value of 1.0 and deviation mean squares of zero. Therefore, a genotype with unit regression coefficient ($b_i = 1$) and deviation not significantly different from zero ($S_{di}^2 = 0$) is said to be the most stable one. The estimates of phenotypic stability parameters according to Eberhart and Russell (1966) for eleven bread wheat genotypes grown under twelve environments for mentioned traits are given in Table 7.11.

Regression coefficient (b_i) of bread wheat genotypes ranged from 0.56 (Giza 168) to 1.32 (Misr 1) for days to 50% heading (Table 7.11), indicating a lot of variation of linear response of genotypes to environmental changes. However, (b_i) values were not deviated significantly from unity in wheat genotypes Line 1, Line 2, Line 3, Line 4, Line 6, Line 8 for this trait, indicating that these genotypes could be grown under a wide range of environments. Two wheat cultivars Sakha 94 and Giza 168 exhibited $b_i < 1$ and significant. On contrast, the wheat genotypes Line 5 and Line 7 and Misr 1 had $b_i > 1$ and significant. According to Breese (1969) genotypes

Table 7.11 Genotype means over 12 environments and stability parameters of the 11 wheat genotypes for various traits

Genotypes	Days to 50% heading					Grain protein content %					Grain yield (ard./fed.)				
	Mean ($\bar{X}$)	b_i	S^2_{di}	ASV	Rank	Mean ($\bar{X}$)	b_i	S^2_{di}	ASV	Rank	Mean ($\bar{X}$)	b_i	S^2_{di}	ASV	Rank
Line 1	79.24	1.00	0.48	1.07	7	11.15	1.14	0.04	0.34	1	24.40	1.06	3.17	0.66	2
Line 2	81.29	0.92	0.16	0.13	1	10.98	1.20*	0.04	0.71	4	21.03	0.66**	2.66	0.94	3
Line 3	82.94	1.16	0.18	0.39	2	11.27	1.03	0.12*	1.09	9	23.26	1.42**	5.31*	2.02	7
Line 4	80.62	0.96	0.21	0.43	4	11.25	1.34**	0.07	0.67	2	23.93	0.82*	1.56	0.32	1
Line 5	82.53	1.21*	0.64*	0.42	3	11.27	1.23*	0.09	0.87	6	22.96	1.01	3.94	1.15	4
Line 6	81.00	1.03	1.10**	1.44	9	11.46	0.98	0.17**	0.76	5	22.33	0.67**	6.31**	1.58	6
Line 7	85.00	1.27*	0.63*	0.97	6	12.02	0.75*	0.10*	0.89	7	23.57	1.11	8.06**	2.60	10
Line 8	81.88	0.91	0.37	0.69	5	11.87	0.97	0.09	0.68	3	23.77	1.67**	1.84	2.09	8
Sakha 94	80.43	0.65*	0.54*	1.59	11	11.24	1.07	0.17**	1.15	11	20.71	0.63*	6.00*	2.41	9
Giza 168	79.23	0.56*	0.40	1.47	10	11.67	0.69**	0.19**	1.14	10	23.33	1.31*	7.19**	2.73	11
Misr 1	80.89	1.32*	0.73**	1.36	8	11.70	0.62**	0.19**	1.08	8	22.10	0.64*	3.14	1.26	5
Mean	81.26					11.44					22.85				
L.S.D	0.42					0.17					1.26				

*, **Significant at 0.05 and 0.01 levels of probability, respectively

Ardab = 150 kg, Feddan = 4200 m²

with a regression coefficient more than unity would be adapted to more favorable environments. While those with a coefficient less than one would relatively be better adapted to less favorable conditions.

The deviations from regression (S_{di}^2) ranged from 0.16 (Line 2) to 1.10 (Line 6). The stable genotypes with lowest and insignificant S_{di}^2 values were Line 1, Line 2, L 3, Line 4, Line 8 and Giza 168. On the contrary, the unstable genotypes with the highest and significant S_{di}^2 values were Line 5, Line 6, Line 7, Sakha 94 and Misr 1.

Based on a simultaneous consideration of the three stability parameters ($\bar{X}$, b_i and S_{di}^2), the most desired and stable genotype would be Line 1 with an earliest mean $\bar{X} = 79.24$, $b_i = 1.00$ and $S_{di}^2 = 0.48$, Line 2 with $\bar{X} = 81.29$, $b_i = 0.92$ and $S_{di}^2 = 0.16$ and Line 4 with $\bar{X} = 80.62$, $b_i = 0.96$ and $S_{di}^2 = 0.21$ and Line 8 with $\bar{X} = 81.88$, $b_i = 0.91$ and $S_{di}^2 = 0.37$. These genotypes could be useful in wheat breeding programs for improving earliness.

Grain Protein Content

Phenotypic stability parameters revealed that regression coefficient (b_i) for grain protein content of eleven bread wheat genotypes ranged from 0.62 (Misr 1) to 1.34 (Line 4), indicating the genetic variability among wheat genotypes in their regression response for grain protein content (Table 7.11). The (b_i) values deviated significantly from unity (bi > 1) in Line 2, Line 4 and Line 5 and less than unity ($b_i < 1$) in Line 7, Giza 168 and Misr 1. On the other side, wheat genotypes Line 1, Line 3, Line 6, Line 8 and Sakha 94 had (b_i) values did not deviate significantly from unity. Therefore, these bread wheat genotypes were well adapted under a wide range of environments in respect to grain protein content.

The deviations from regression (S_{di}^2) for grain protein content ranged from 0.04 (Line 1 and Line 2) to 0.19 (Giza 168 and Misr 1). The stable wheat genotypes with lowest (S_{di}^2) values and not significantly different from zero were Line 1, Line 2, Line 4, Line 5 and Line 8. In contrast, the remaining genotypes were unstable as they exhibited the highest and significant (S_{di}^2) values.

Based on simultaneous consideration of the three stability parameters ($\bar{X}$, b_i and S_{di}^2), the most stable genotype would be Line 8 with mean $\bar{X} = 11.87$, $b_i = 0.97$ and $S_{di}^2 = 0.09$.

Grain Yield (ard./fed.)

Phenotypic stability indicated that regression coefficient (b_i) for grain yield of eleven bread wheat genotypes ranged from 0.64 (Misr 1) to 1.67 (Line 8), indicating the genetic variability among bread wheat genotypes in their regression response for grain yield (Table 7.11). The (b_i) values were deviated significantly from unity (bi > 1) in Line 3, Line 8 and Giza 168, showing greater sensitivity to environmental changes and were relatively suitable in favorable environments with soil fertility, adequate water and other inputs. Meanwhile, the (b_i) values deviated significantly and less than unity ($b_i < 1$) in Line 2, Line 4, Line 6, Sakha 94 and Misr 1. Thus they were adapted to less favorable environments. On the other side, wheat genotypes Line 1, Line 5 and Line 7 had the (b_i) values did not deviate significantly from unity. Therefore, these wheat genotypes could be grown under a wide range of environments.

The deviations from regression (S_{di}^2) for grain yield varied from 1.56 to 8.06 for Line 4 and Line 7, respectively. The stable wheat genotypes with lowest S_{di}^2 values and not significant were Line 1, Line 2, Line 4, Line 5, Line 8 and Misr 1. Conversely, the unstable bread wheat genotypes with the highest and significant S_{di}^2 values were Line 3 (5.31*), Line 6 (6.31**), Line 7 (8.06), Sakha 94 (6.00*) and Giza 168 (7.19**).

The desirable and stable wheat genotypes according to the three stability parameters ($\bar{X}$, b_i and S_{di}^2) for grain yield were Line 1 with a mean yield $\bar{X} = 24.40$, bi $= 1.06$ and the $S_{di}^2 = 3.17$ and Line 5 ($\bar{X} = 22.96$, bi $= 1.01$ and $S_{di}^2 = 3.94$). These previous genotypes gave mean estimates above grand mean, and their regression coefficients (b_i) did not differ significantly from unity, with minimum deviation mean squares (S_{di}^2).

7.3.5   Additive Main Effects and Multiplicative Interaction Method (AMMI)

The model of additive main effects and multiplicative interaction (AMMI) combines the analysis of variance for the genotype and environment main effects and the principal components analysis of the genotypes-environments interaction. AMMI use the standard analysis of variance method, after the AMMI model is separated the additive variance from the multiplicative variance, then applies PCA to the interaction (residual) part to give a new set of coordinate axes that more efficiently explain the interaction patterns (Shafii et al. 1992). A genotype is described as stable if its first and second correspondence analyses scores are near zero (Lopez 1990; Kang 2002).

7.3.5.1   Days to 50% Heading

The analyses of variance showed that environments (E), wheat genotypes (G) and the G × E interaction mean squares were highly significant for days to 50% heading (Table 7.12). The IPCA scores of bread wheat genotypes in the AMMI and SREG analyses were significant for IPCA1 and IPCA2. Variance components (%) due to the sum of squares varied from 36.22% for genotypes, 43.41% for environments and 10.23% for GEI. IPCA 1 score explained 32.61% and IPCA 2 had 26.73% of the total GEI for AMMI models. Also, the IPCA 1 score explained 81.90% and IPCA 2 had 7.11% of the total GEI for SREG models.

A bread wheat genotype with the smaller AMMI stability value (ASV) is considered more stable. According to the ASV ranking in Table 7.12 and either Fig. 7.1,

Table 7.12 AMMI analysis of variance over twelve environments for various traits

Source of variation	df	Days to 50% heading			Grain protein content %			Grain yield (ard./fed.*)		
		Sum of square	Mean of square	Percent	Sum of square	Mean of square	Percent	Sum of square	Mean of square	Percent
Environment (E)	11	930.78	84.62**	43.41	88.30	8.03**	45.38	3458.42	314.40**	43.06
Reps/Env.	24	22.46	0.94		7.02	0.29**		399.99	16.67**	
Genotype (G)	10	776.60	77.66**	36.22	23.27	2.33**	11.96	523.78	52.38**	6.52
G × E	110	219.33	1.99**	10.23	43.11	0.39**	22.16	1874.16	17.04**	23.33
IPCA1	20	71.53	3.58**	32.61	17.05	0.85**	39.55	624.26	31.21**	33.31
IPCA2	18	58.63	3.26**	26.73	10.10	0.56**	23.45	497.83	27.66**	26.56
IPCA3	16	39.80	2.49**	18.14	5.83	0.36**	13.51	247.59	15.47**	13.21
IPCA4	14	19.25	1.38**	8.78	4.75	0.34**	11.02	155.14	11.08	8.28
IPCA5	12	15.69	1.31	7.15	2.62	0.22	6.08	110.65	9.22	5.90
IPCA6	10	8.54	0.85	3.90	1.47	0.15	3.42	91.49	9.15	4.88
IPCA7	8	2.75	0.34	1.25	0.71	0.09	1.64	77.70	9.71	4.15
IPCA8	6	2.51	0.42	1.15	0.39	0.07	0.91	54.42	9.07	2.90
IPCA9	4	0.45	0.11	0.21	0.17	0.04	0.39	13.39	3.35	0.71
IPCA10	2	0.20	0.10	0.09	0.02	0.01	0.06	1.67	0.83	0.09
Pooled Error	240	195.09	0.81		32.86	0.14		1775.58	7.40	
Total	395	2144.26			194.57			8031.93		

*Ardab = 150 kg, Feddan = 4200 m²; **Significant at 0.01 level of probability

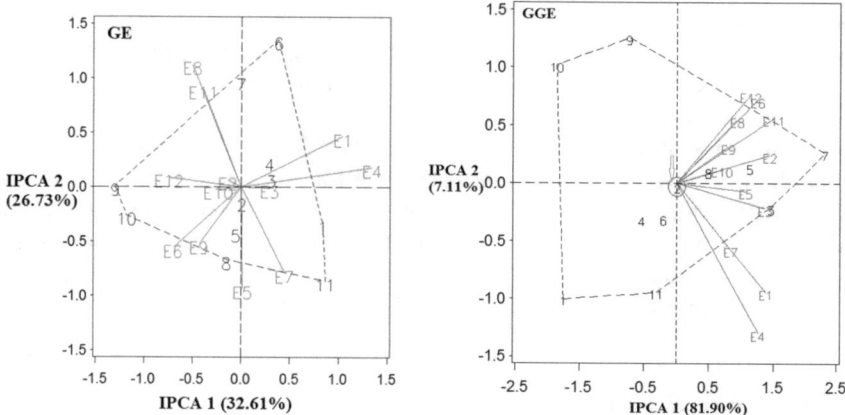

Fig. 7.1 Graphics display of the GE and GGE biplots for eleven wheat genotypes (assessed G1–G11) and 12 environments (assessed E1–E12) in the AMMI and SREG models, respectively for days to 50% heading

Line 2, Line 3, Line 5 and Line 4 were more stable (0.13, 0.39, 0.42 and 0.43, respectively), while Line 8, Line 7, Line 1, Misr 1, Line 6, Giza 168 and Sakha 94 were unstable.

Figure 7.1 show the graphical display of the GEI biplot for eleven bread wheat genotypes (assessed Line 1 to Misr 1) and 12 environments (assessed E1-E12) in the AMMI and SREG models for days to 50% heading.

The bread wheat genotypes and environments that were located far away from the origin were more responsive. Environments E1, E4, E5 and E8 were the most differentiating environments, while environments E3, E2, E10 and E12 were less reactive. Furthermore, the vertex wheat genotypes Line 6, Sakha 94, Giza 168 and Misr 1 were located far away from the origin, were more responsive to environmental changes. Therfore, these genotypes could be considered as specifically adapted, as they have the longest distance from the origin in their direction and genotypes of wheat with long vectors were identified as either the best or the poorest performers in the environment. Based on the genotype-focused scaling, Line 2, Line 3, Line 5 and Line 4 were desirable. They located near the origin and less responsive than the corner wheat genotypes.

Concerning GGE biplot for the SREG model, Fig. 7.1 show a graphical display of the GGE biplot for 11 wheat genotypes for days to 50% heading assessed (G1–G11) and the 12 environments considered (E_1–E_{12}) in the SREG model. An ideal wheat genotype should have the lowest mean performance for days to 50% heading and be absolutely stable (i.e., perform the best under all environments). Line 2 was the ideal wheat cultivar, it had the lowest vector length of the lower wheat genotype and with zero GEI, as represented by the arrow pointing to it (Fig. 7.1).

The angle between the vectors of the two environments is associated with the correlation coefficient among them (https://www.scionresearch.com/__data/assets/

pdf_file/0007/5596/NZJFS_38_12008_Di). The environments E_{12}, E_6, E_8, E_{11}, E_9, E_{10} and E_2 were positively associated because all angles among them were lesser than 90°, while the environments E_4, E_1 and E_7 had negatively correlated with E_{12}, E_6, E_8, E_{11} and E_9 (see Fig. 7.1).

The ideal test environment was E10. It had small IPCA1 scores and small (absolute) IPCA2 scores (more representative of the overall environments). The favorable environments were E_2, E_{10}, E_5 and E_3, while the unfavorable ones were E_1 and E_4. Ali (2017) found that variance components % varied from 21.34% for genotypes, 67.36% for environments and 6.07% for GEI. Genotypes Sakha 93, Gemmeiza 7, Line 8 and Gemmeiza 10 were more stable for days to 50% heading,

7.3.5.2   Grain Protein Content

AMMI analysis showed that environments (E), genotypes (G) and the G × E interaction mean squares were highly significant for grain protein content (Table 7.12). The IPCA scores of wheat genotypes in the AMMI and SREG models were significant for IPCA1 and IPCA2. Variance components (%) due to the sum of squares varied from 11.96% for bread wheat genotypes, 45.48% for environments and 22.16% for GEI. IPCA 1 score explained 39.55% and IPCA 2 had 23.45% of the total GEI for the AMMI model. Moreover, For the SREG model, IPCA 1 score exhibited 48.07%, and IPCA 2 had 21.86% of the total GGEI.

A wheat genotype with the least ASV is expressed the most stable, so as given in Table 7.12 and illustrated in Fig. 7.2, bread wheat genotypes Line 1, Line 4, Line 8 and Line 2 were the most desired and stable genotypes valued 0.34, 0.67, 0.68 and 0.71, respectively. Otherwise, bread wheat genotypes Line 6, Line 5, Line 7, Misr

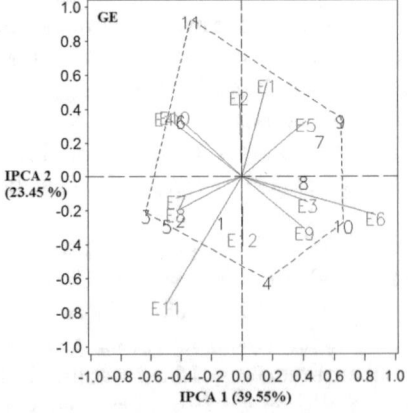 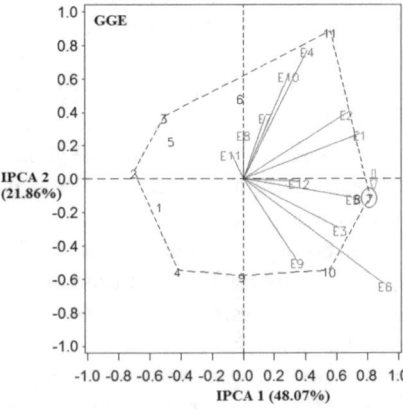

Fig. 7.2 Graphics display of the GE and GGE biplots for eleven wheat genotypes (assessed G1–G11) and 12 environments (assessed E1–E12) in the AMMI and SREG models, respectively for grain protein content

1, Line 3, Giza 168 and Sakha 94 were unstable for grain protein content and more responsive to the environmental changes.

GEI biplot graph for the AMMI showed that Environments E_6, E_{10}, E_4 and E_{11} were the most differentiating environments for grain protein content. On the other side, environments E_3, E_9, E_2, E_5, E_1, and E_{12} were less responsive for this trait. Furthermore, the vertex wheat genotypes Line 3, Sakha 94, Giza 168 and Misr 1 were sited far away from the origin, which were more responsive to environmental changes and are considered as specially adapted wheat genotypes. Based on the genotype-focused scaling, the bread wheat genotypes Line 1, Line 2, Line 7 and Line 8 were the desirable and stable genotypes.

GGEI biplot graph for the SREG model showed that Line 7 was ideal wheat genotype for grain protein content, it had the largest vector length of the higher genotype and with lowest GEI, as symbolized by arrow pointing to it in Fig. 7.2. Wheat genotype Line 8 was more desirable genotype, as it was found closer to the ideal genotype. The environments E_{11}, E_8, E_7, E_{10}, E_4, E_2 and E_1 were positively correlated because all angles among them were lesser than 90°, while the environments E_9, E_6 and E_3 had negatively correlated with E_{11}, E_8 and E_7, the angle among them was higher than 90°. The ideal test environment was E_{12}. It had large IPCA1 scores and small IPCA2 scores. The favorable environments were E_5 and E9, but the unfavorable ones were E_6 and E_4. In this respect, Taghouti et al. (2010) tested twelve wheat cultivars under five locations demonstrating a series of environments in three growing seasons. They showed significant effects of genotype, environment and $G \times E$ for protein content. For protein content, where the environmental effect was larger than that of genotype and $G \times E$ effect, several environmental trials are required to define the protein content of a genotype.

7.3.5.3 Grain Yield (ard./fed.)

AMMI analysis of variance showed that environments (E), wheat genotypes (G) and the $G \times E$ interaction mean squares were highly significant for grain yield (ard./fed.) (Table 7.12). The IPCA scores of a wheat genotype in the AMMI and SREG analyses were significant for IPCA1 and IPCA2. Variance components (%) attributed to the sum of squares varied from 6.52% for genotypes, 43.06% for environments and 23.33% for GEI. IPCA 1 score had 33.31%, and IPCA 2 had 26.56% of the total GEI for AMMI models. For the SREG model, IPCA 1 score exhibited 39.63%, and IPCA 2 had 22.80% of the total GGEI.

In respect to ASV for grain yield as given in Table 7.12 and illustrated in Fig. 7.3 the bread wheat genotypes Line 4, Line 1, Line 2, Line 5 and Misr 1 were the most desired and stable genotypes valued 0.32, 0.66, 0.94, 1.15 and 1.26, respectively. Otherwise, the other bread wheat genotypes Line 6, Line 3, Line 8, Sakha 94, Line 7 and Giza 168 were unstable for this trait and more responsive to the environmental changes.

GE biplot graph for the AMMI model illustrated that environments E_5, E_{10}, E_7 and E_3 were the most differentiating environments for grain yield, they were located far

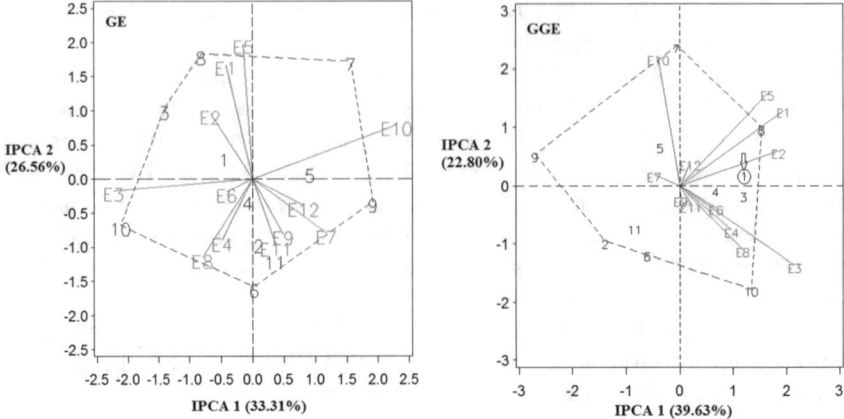

Fig. 7.3 Graphics display of the GE and GGE biplots for eleven wheat genotypes (assessed G1–G11) and 12 environments (assessed E1–E12) in the AMMI and SREG models, respectively grain yield (ardab/fed.)

away from the origin (Fig. 7.3). Whereas, environments E_6, E_{12}, E_9, E_4 and E_2 were less responsive for grain yield. Furthermore, the vertex genotypes Line 7, Line 9, Line 6, Line 10, Line 3 and Line 8 were placed far away from the origin, which were more responsive to environmental changes and are classified as specially adapted genotypes. Based on the genotype-focused scaling, the bread wheat genotypes Line 4, Line 1, Line 2, Line 5 and Line 11 were the desirable and stable, these wheat genotypes were located near the origin, they were less responsive than the corner genotypes.

GGE biplot graph for the SREG model as illustrated in Fig. 7.3, Line 1 was ideal wheat genotype for grain yield, it had the highest vector length of the high yielding genotypes and with zero GE, as represented by the dot with an arrow pointing to it in Fig. 7.3. A wheat genotype is more desirable if it is sited closer to the ideal wheat genotype. Thus Line 3 and Line 4 were desirable genotypes. The environments E_7 with E_{10}, E_{12} with (E_5 and E_1) and E_6 with E_8 were positively correlated. Whereas, the environments E_7 and E_{10} had negatively correlated with E_6, E_4, E_8, E_3 and E_{11}, the angle among them was higher than $90°$. The ideal test environment was E_2, it had large IPCA1 scores and small IPCA2 scores. The favorable environments were E_1 and E5, but the unfavorable ones were E_7, E_8 and E_9 for grain yield. It is interesting to mention that both models of Eberhart and Russell (1966) and AMMI (Gauch 1992) are consistent in describing stability of Line 2 and Line 4 for days to 50% heading; Line 8 for grain protein content as well as Line 1 and Line 5 for grain yield (ard./fed.).

Many investigators applied the AMMI model to the yield data on various environments for evaluating yield stability, of them Najafian et al. (2010), Farshadfar et al. (2011), Hagos and Abay (2013), Mohamed et al. (2013) and Ali (2017) and they registered specific adaptation for several genotypes to specific environments.

Moreover, Elbasyoni (2018) applied AMMI model and found that analysis of variance indicated a significant effect of the environments, genotypes, and genotype × environment interaction for a number of days to flowering and grain yield (tons/ha). Added that the variance of the environments was 63.2%, whereas the variance due to genotypes was 14.6% and that for genotype × environment interaction was 22.2%.

7.4 Conclusions

Highly significant G × E "linear" was registered for days to 50% heading, grain protein content and grain yield. Phenotypic stability parameters showed that wheat genotypes Sakha 94 and Giza 168 were highly adapted to stress environments for days to 50% heading; Line 7 and Misr 1 for grain protein content and Line 2, Line 4, Line 6, Sakha 94 and Misr 1 for grain yield. The most desired and stable genotypes were Line 1, Line 4 and L8 for earliness, Line 8 for grain protein content and Line 1 and Line 5 for grain yield. According to GE biplot and ASV, the most stable genotypes were Line 2, Line 3, Line 5 and Line 4 for days to 50% heading; Line 1, Line 4, Line 8 and Line 2 for grain protein content as well as Line 4, Line 1, Line 2, Line 5 and Misr 1 for grain yield. Also, the most tolerant wheat genotypes to environmental stress were Sakha 94 followed by Line 6, Misr 1, Line 2 and Line 4, while the other wheat genotypes exhibited various degrees of sensitivity.

7.5 Recommendations

In the light of climate change and its negative impacts on the agricultural sector, it is important to develop and release new genotypes and cultivars of wheat and determine their adaptability and stability to cope with environmental changes. Also, understanding the association between climate and wheat crop yield is important to identify potential impacts of future climate and to develop adaptation measures. Therefore, multiple experimental trials are necessary in order to determine the promising and adapted cultivars.

Acknowledgements Hassan Awaad acknowledges the partial support of the Science and Technology Development Fund (STDF) of Egypt in the framework of the grant no. 30771 for the project titled "A Novel Standalone Solar-Driven Agriculture Greenhouse - Desalination System: That Grows Its Energy And Irrigation Water" via the Newton-Musharafa funding scheme.

References

Abd El-Shafi MA, Gheilth EMS, Abd El-Mohsen AA, Suleiman HS (2014) Stability analysis and correlations among different stability parameters for grain yield in bread what. Sci Agric 2(3):135–140

Abd El-Hady S, Ali AAG, Omar AEA, El-Sobky EEA (2018) Influence of sowing date, varietal differences and planting density on productivity of wheat crop (*Triticum aestivum* L.). Zagazig J Agri Res 45(1):1–21

Ali MMA (2017) Stability analysis of bread wheat genotypes under different nitrogen fertilizer levels. J Plant Production, Mansoura Univ 8(2):261–275

Ali MMA, Abdul-Hamid MEI (2017) Yield stability of wheat under some drought and sowing dates environments in different irrigation systems. Zagazig J Agri Res 44(3):865–886

Allard RW, Bradshaw AD (1964) Implication of genotype × environmental interaction in applied plant breeding. Crop Sci 4:503–508

Aly AA, Awaad HA (2002) Partitioning of genotype × environment interaction and stability for grain yield and protein content in bread wheat. Zagazig J Agric Res 29 (3):999–1015

AOAC (1980) Association of Official Agriculture Chemist. Official Method of Analysis, 13th ed, Washington, DC, USA

Breese EL (1969) The measurement and significance of genotypes environment interaction in grasses. Heredity 24:27–44

Eberhart SA, Russell WA (1966) Stability parameters for comparing varieties. Crop Sci 6:36–40

El-Ameen T (2012) Stability analysis of selected wheat genotypes under different environments in Upper Egypt. Afr J Agric Res 7(34):4838–4844

Elbasyoni IS (2018) Performance and stability of commercial wheat cultivars under terminal heat stress. Agronomy 8(4):37. https://doi.org/10.3390/agronomy8040037

El-Marsafawy S) 2007) Climate change and its impact on the agriculture sector in Egypt. http://www.radcon.sci.eg/environment2/ArticlsIdeasDetails.aspx?ArticlId=35

El-Moselhy Omnya M, Ali AAG, Awaad HA, Sweelam AA (2015) Phenotypic and genotypic stability for grain yield in bread wheat across different environments. Zagazig J Agric Res 42(5):913–926

Farshadfar E, Mahmodi N, Yaghotipoor A (2011) AMMI stability value and simultaneous estimation of stability and yield stability in bread wheat (*Triticum aestivum* L.). AJCS 5(13):1837–1844

Gauch HG (1992) Statistical analysis of regional trials: AMMI analysis of factorial designs. Elsevier, Amsterdam, Netherlands, p 278

Georgopoulou E, Mirasgedis S, Sarafidis Y, Vitaliotou M, Lalas DP, Theloudis I, Giannoulaki KD, Dimopoulos D, Zavras V (2017) Climate change impacts and adaptation options for the Greek agriculture in 2021–2050: a monetary assessment. Clim Risk Manage 16:164–182

Ghorbani MH, Basiri M (2013) Plant density effect on growth and seed yield of wheat of saline soils and rainfed condition. Electron J Crop Prod 6(2):57–72

Hagos HG, Abay Fetien (2013) AMMI and GGE biplot analysis of bread wheat genotypes in the Northern part of Ethiopia. J Plant Breed Genet 1(2013):12–18

Hamam KA, Khaled AGA (2009) Stability of wheat genotypes under different environments and their evaluation under sowing dates and nitrogen fertilizer levels. Aus J Basic Appl Sci 3(1):206–217

Hamam KA, Khaled AGA, Zakaria MM (2015) Genetic stability and diversity in yield components of some wheat genotypes through seasons and heat stress under different locations. J Plant Production, Mansoura Univ 6(3):349–370

Hameed E, Wajid AS, Shad AA, Bakht J, Muhammad T (2003) Effect of different planting dates, seed rate and nitrogen levels on wheat. Asian J Plant Sci 2(6):467–474

Hickey H (2017) Earth likely to warm more than 2 degrees this century. http://www.washington.edu/news/2017/07/31/earth-likely-to-warm-more-than-2-degrees-this-century/Search UW News, NIH grants: HD054511, HD070936

Isabelle D, Vanasse A, Pageau D, Dion Y (2015) Seeding rate and cultivar effects on yield, yield components and grain quality of spring spelt in eastern Canada. Can J Plant Sci 95(5):841–849

Kang MS (2002) Genotype-environment interaction: progress and prospects. In: Kang MS (ed) Quantitative genetics, genomics and plant breeding. CABI Publishing, Wallingford, Oxon, UK, pp 221–243

Khan FU, Mohammed F (2016) Application of stress selection indices for assessment of nitrogen tolerance in wheat (*Triticum aestivum* L). J Anim Plant Sci 26(1):201–210

Lamaoui M, Jemo M, Datla R, Bekkaoui F (2018) Heat and drought stresses in crops and approaches for their mitigation. Front Chem 6:26

Lopez J (1990) Estudio de la base geneetica del contenido en taninos condensadaos en la semilla de las habes (*Vicia faba* L.). Doctoral dissertation, University of cardoba, Spain

Mekonnen A (2017) Effects of seeding rate and row spacing on yield and yield components of bread wheat (*Triticum aestivum* L.) in Gozamin District, East Gojam Zone, Ethiopia. J Biol Agric Healthc 7(4):19–37

Menshawy AMM (2007). Evaluation of some early bread wheat genotypes under different sowing dates. Proceeding of the Fifth Plant Breeding Conference, May 27, 2007 (Giza), pp 25–40

Mohamed NEM, Said AA, Amein KA (2013) Additive main effects and multiplicative interaction (AMMI) and GGE biplot analysis of genotype x environment interaction for grain yield in bread wheat (*Triticum aestivum* L.). Afr J Agric Res 8(42):5197–5203

Mohammadi M (2012) Effect of kernel weight and source-limitation on heat grain yield under heat stress. Afr J Biotech 11(12):2931–2937

Najafian G, Kaffashi AK, Jafer-Nezahd A (2010) Analysis of grain yield stability in hexaploid wheat genotypes grown in temperate regions of Iran using additive main effects and multiplicative interaction. J Agr Sci Tech 12:213–222

Nasab SS, Nejad GM, Nakhoda B (2019) Yield stability in bread wheat germplasm across drought stress and non-stress conditions. Agronomy J 111:175–181

Purchase JL (1997) Parametric analysis to describe Genotype × Environment interaction and yield stability in winter wheat. Ph. D. Thesis, Department of Agronomy, Faculty of the Free State, Bloemfontein, South Africa

Purchase JL, Hatting H, Van Deventer CS (2000) Genotype × environment interaction of winter wheat (*Triticum aestivum* L.) in South Africa: II. Stability analysis of yield performance. South Afr J Plant Soil 17:101–107

Rosielle AA, Hamblin J (1981) Theoretical aspects of selection for yield in stress and non-stress environments. Crop Sci 21:943–946

Shafii B, Mahler KA, Price WJ, Auld DL (1992) Genotype × Environment interaction effects on winter rapeseed yield and oil content. Crop Sci 32:922–927

Siddhi S, Patel JB, Patel N (2018) Stability analysis in bread wheat (*Triticum aestivum* L.) J Pharmacognosy and Phytochem 7(4):290–297

Steel RGD, Torrie JE, Dickey DH (1997) Principles and producers of statistics: a biometrical approach, 3rd ed, Mc Graw Hill, New York

Taghouti M, Gaboun F, Nsarellah N, Rhrib R, El-Haila M, Kamar M, F Abbad-Andaloussi F, Udupa SM (2010) Genotype × environment interaction for quality traits in durum wheat cultivars adapted to different environments. Afr J Biotechnol 9(21): 3054–3062

Tammam AM, Abd El Rady AG (2010) Genetical studies on some morpho-physiological traits in some bread wheat crosses under heat stress conditions. Egypt J Agric Res 89(2):589–604

Tao F, Yokozawa M, Liu J, Zhang Z (2008) Climate-crop yield relationship at provincial scales in China and the impact of recent climate trends. Clim Res 38:83–94

Tawfelis MB (2006) Stability parameters of some bread wheat genotypes (*Triticum aestivum*) in new and old lands under upper Egypt conditions. Egypt J Plant Breed 10(1):223–246

Tawfelis MB, Khieralla KA, El Morshidy MA, Feltaous YM (2011) Genetic diversity for heat tolerance in some bread wheat genotypes under Upper Egypt conditions. Egypt J Agric Res 89(4):1463–1480

Chapter 8
Effect of Salt Stress on Physiological and Biochemical Parameters of African Locust Bean {*Parkia biglobosa* (Jacq.) Benth.} Cell Suspension Culture

Mohamed S. Abbas, Hattem M. El-Shabrawi, Mai A. Selim, and Amira Sh. Soliman

Abstract Salinity is one of the major and increasing abiotic stress. African Locust Bean is a multipurpose tree having a good potentiality and suitability to distribute under Egyptian conditions. This chapter was conducted to investigate the effects of NaCl stress on some physiological and biochemical parameters of *Parkia biglobosa* cell suspension culture, at the application of NaCl concentration levels of (0, 100, 200, 300, 400, 500, 600 and 700 mM) during 18 days. Results showed that the highest percentage of viability and concentration of viable cells/ml observed at treatments 500 mM at 12th day with significant differences than the remaining treatments. NaCl treated cell suspension culture increased Na^+ accumulation and declined accumulation of K^+, Ca^{2+}, P^{3+} and N^{3+}, while Na/K ratio increased steadily as a function of external NaCl. Proline content of treated cell suspension at 200 mM reached the highest rate at 18th day with significant difference than the remaining treatments then gradually decreased up to concentration 700 mM recorded the lowest proline content. The variation in protein patterns band induced in all different cultures, especially in callus cells and 600 mM NaCl recorded the most variations than other cultures. NaCl stress enhanced the activity of peroxidase (POX) and glutathione peroxidase (GPX) isozymes, these two enzymes have 2 forms and they have the maximum enzyme activity at 500 mM NaCl concentration. Total phenols content in different *P. biglobosa* extraction from the mother plant, callus and salt-stressed suspension cells (0, 500 and 600 mM NaCl) and it,s supernatant were assessed. By using HPLC could separate the phenolic constituents that responsible for *P. biglobosa* to tolerate salinity

M. S. Abbas (✉) · A. Sh. Soliman
Natural Resources Department, Faculty of African Postgraduate Studies, Cairo University, 12613 Cairo, Egypt

H. M. El-Shabrawi
Plant Biotechnology Department, National Research Centre, 12622 Cairo, Egypt

M. A. Selim
Industrial Technical Institute, Matarya Technology College, Ministry of Higher Education, 11714 Cairo, Egypt

© Springer Nature Switzerland AG 2021
H. Awaad et al. (eds.), *Mitigating Environmental Stresses for Agricultural Sustainability in Egypt*, Springer Water,
https://doi.org/10.1007/978-3-030-64323-2_8

215

such as gallic, caffeic, vanillic, ferulic, p-coumaric and salicylic acids, it is used as a marker for salt tolerance.

Keywords *Parkia biglobosa* · NaCl stress · Cell viability · Ion accumulation · Proline · Isozymes · Total phenols · HPLC

8.1 Introduction

Parkia biglobosa (Jacq.) Benth. commonly named the African locust bean, it is perennial deciduous tree belongs to the Fabaceae family. It is known to grow in a diversity of agro-ecological zones ranging from arid zones to tropical rain forests and therefore found scattered around of West Africa (Millogo-Kone et al. 2008).

Salinity are known to affect many physiological and metabolic processes leading to growth reduction. High concentration of NaCl causes ion imbalance and osmotic stress in numerous plants (Maggio et al. 2000). The high levels of antioxidant in plants were reported to have greater resistance to oxidative damage (El-kahoui et al. 2005).

Plant cells synthesize chemical compounds which act as a protective and repair system to control metabolism and to defend themselves against the damaging effects of the environmental conditions (Singh et al. 2015). It was observed that there is a relationship between salinity and drought in plants for induction of proline content (Ramanjulu and Sudhakar 2001). Contrary, some scientists revealed that there was no increase in proline content under salinity conditions (Koca et al. 2007; Kumar et al. 2003).

Plants are equipped with an enzymatic and non-enzymatic antioxidant system to protect cell membranes and organelles from ROS damaging effects. It has been reported that a significant correlation exists between the activities of antioxidative enzymes and the salt tolerance of plants (Nader et al. 2005).

The isozymes could be used as a biochemical marker to study the tolerance of the plant to stress (Zhang et al. 2013). Sen and Alikamanoglu (2011) showed that peroxidase enzyme (POX) plays a significant role as anti-oxidant response against salt stress in plants. Naik and Devaraj (2016) reported that an increase the activity of glutathione peroxidase (GPX) levels subjected to salt stress.

Phenolic acids Accumulation is a symptom of adverse environmental conditions and the production of different classes of phenolic acids produced via the phenyl-propanoid system is dependent on the nature of stress exposed to plants (Weisskopf et al. 2006).

Phenolic compounds are a class of antioxidant agents which act as free radical terminators (Pourmorad et al. 2006). They are considered as a major group of compounds that contribute to the antioxidant activities of plant materials, because of their scavenging ability on free radicals due to their hydroxyl group (Djeridane et al. 2006).

Plant cell culture techniques provide an alternative source for the production of secondary metabolites and as a tool for the elucidation of secondary metabolite biosynthesis (Daud and Keng 2006). Using in vitro techniques, the rapid growth of callus and cell suspension cultures, from which secondary metabolites are to be extracted, can be obtained (Verpoorte et al. 2002). Nutritional elements, environmental conditions, and hormone regimes are the major factors affected the production of secondary metabolites is (Roberts 1988).

In vitro culture technique is useful tool to study the salt stress response of suspension cells in controlled and uniform conditions. Thus, avoiding complications arising from physiological and structural variability of the whole plant. The plant cell culture studies allow isolation and selection of salt tolerant lines to elucidate the mechanism of tolerance operating at the cellular level (Niknam et al. 2006) and the possibility of developing salt tolerant (Kumar et al. 2008).

Cell and tissue cultures offer monitoring plant responses to salinity at biochemical and physiological levels (Yang et al. 2010). However, it appears that little information is available regarding the effect of salinity on the growth and productivity of multipurpose *P. biglobosa* tree. Therefore, in the present study, *P. biglobosa* cell suspension cultures have been used for the first time to investigate the salt stress adaptive mechanism, correlate with antioxidant activity and phenolic compounds under salt stress. Also, the establishment of an application protocol for cell suspension culture production from callus under salt stress, accumulation of ions and proline, evaluation of some variations in different cultures using molecular characterization and determination of phenolic compounds in different cell cultures of *P. biglobosa* plant.

8.2   Materials and Methods

8.2.1   Plant Material

The seeds were extracted from matured pods of *P. biglobosa* tree grown in Nigeria which collected by Nasir Hassan Wagini, Department of Biology, Faculty of Natural and Applied Sciences, Umaru Musa Yar'adua University, Katsina, Nigeria. Seeds were scarified by concentrated H_2SO_4 (98%) for 5 min then surface sterilized with 60% commercial Clorox solution (contains 5.25% sodium hypochlorite) for 25 min. The seeds were then germinated on MS medium (Murashige and Skoog 1962) supplemented with 3% sucrose and 0.2% gerlite and maintained at 25 ± 1 °C for 16/8 h day photoperiod for two weeks.

For callus induction, stem explants (0.5–1 cm long were excised from two-weeks-old in vitro seedlings as mentioned above) of *P. biglobosa* were transferred to MS medium supplemented with 1 mg/l 2,4-D and 50 mg/l citric acid. The cultures were incubated in the growth chamber at 25 ± 1 °C and exposed to 16/8 h day for 8 weeks.

8.2.2   Establishment of Cell Suspension Culture Under Salt Stress Conditions

According of Abbas et al. (2018) for initiation of suspension culture, 0.5 g of friable callus derived from stem explant were transferred to Erlenmeyer flasks (250 ml), each containing 50 ml liquid MS medium supplemented with 1.0 mg/l 2, 4-D, 3% (w/v) sucrose and different levels of NaCl salt (0, 100, 200, 300, 400, 500, 600 and 700 mM). pH of the medium was adjusted to 5.8 before autoclaving the medium at 120 °C for 20 min. The flasks were agitated at 120 rpm on a gyratory shaker and incubated at 25 ± 1 °C 16/8 h day photoperiod (Padgett and Leonard 1994; Fang et al. 2005) and cultures were maintained for 18 days. Cell numbers counting by using hemocytometer and data were collected as follows (viable cells, non-viable cells, the total number of cells, the percentage of viability and the concentration of viable cells/ml).

8.2.3   Effect of Salt Stress on Ions Uptake of Parkia biglobosa Cell Suspension Culture

Sodium, potassium, calcium, nitrogen and phosphorus content were determined in the oven-dried material. Hundred mg of the dry cells were extracted in 50 ml of deionized water with continuous shaking and estimated on microprocessor-based ion analyzer (Elico, India) using ion specific electrode sodium, potassium, calcium, nitrogen and phosphorus (Na^+, K^+, Ca^{2+}, N^{3+}, P^{3+}) Na/K ratio was calculated as the method of Faithfull (2002).

8.2.4   Effect of Salt Stress on Proline Content of Parkia biglobosa Cell Suspension Culture

Free proline content was determined colorimetrically in fresh samples according to Bates et al. (1973). 500 mg of fresh cell suspension culture treated with different concentration of NaCl (0, 100, 200, 300, 400, 500, 600 and 700 mM) was homogenized in 10 ml of 3% sulpho-salicylic acid solution in an ice bath. The homogenate was centrifuged at 10.000 rpm for 15 min to remove debris at 4 °C. Then, 2 ml of the supernatant was mixed with 2 ml of acid-ninhydrin and 2 ml of glacial acetic acid in a test tube and heated on boiling water bath at 100 °C for 1 h, then cooled in ice bath to stop the reaction.

   The contents of the test tube were shaken vigorously with 4 ml of toluene for 20 s. So, the chromophore-containing organic phase was separated from the hydrated phase, and the absorbance of the resulting organic phase was measured at 520 nm by using spectrophotometer UV-240. The concentration of proline in all samples was estimated by referring to a standard curve prepared using graded concentrations of L-proline and data were expressed as μmoles g^{-1} fresh weight.

8.2.5   Electrophoretic Analysis: (Biochemical Markers)

8.2.5.1   Protein Electrophoresis

In order to determine the molecular weight of protein, sodium dodecyl sulphate polyacrylamide gel electrophoresis (SDS-PAGE) under denaturing conditions were used as described by Laemmli (1970), using 12.5% acrylamide concentration. 500 mg of different plant samples (callus and salt-treated suspension cells) were ground in liquid nitrogen. The fine grounded powder was transferred to Eppendorf tubes and homogenized 500 μl of cold phosphate extraction buffer (100 mM, pH 7.0) and left in the refrigerator overnight, then centrifuged for 10 min at 13.000 rpm under 4 °C.

The supernatant containing total protein fraction transferred to new Eppendorf and stored at -20 °C. 50 μL of the supernatant was mixed with 50 μL sample buffer (0.0625 M Tris-HCl pH 6.8, 2% SDS, 10% glycerol, 5% 2-mercaptoethanol) then boiled at 100 °C for 3 min and centrifuged. 20 μL of bromophenol (loading dye) added to each tube of the sample before loading the samples.

Samples and protein marker were loaded in gels which were put in electrophoresis vertical unit that covered to the DC current. The gel was run at 150 V until the bromophenol blue dye reached the lower end of the gel. The gel were removed from the electrophoresis unit and staining in Commassie brilliant blue R-250 [Commassie 0.25% + methanol 40% + acetic acid 10%] and destained with destaining solution [methanol 40% + acetic acid 10% + distilled water 50%]. The molecular weight of polypeptide bands was calculated from a calibration curve of molecular weight marker (MW) standard of Pharmacia (Uppsala, Sweden) that used to estimate the molecular weight of the sample ranging from 12 to 170 KD.

8.2.5.2   Isozymes Electrophoresis

Enzymes Extraction: The extraction for the enzymes was done as suggested by Larkindale and Huang (2004). Where 500 mg of different plant samples (callus cells and salt-treated suspension cells) were harvested and ground to a fine powder in liquid nitrogen. Then the ground powder was homogenized in 1.5 ml of cold phosphate buffer (100 mM, pH 7.0) containing 1% polyvinyl pyrrolidone (PVP) and 1 mM EDTA and then centrifuge at 4 °C for 15 min at 10000 $\times$ g. The supernatant was separated and stored on ice until the assay of enzyme activity.

Gel Assay for Analysis of ROS Scavenging Enzymes: Changes in proteins having isozymic activity of the ROS scavenging enzymes were studied using PAGE under non-reduced, non-denatured conditions at 4 °C according to the method suggested by Laemmli (1970). Specific conditions were maintained for keeping native protein intact. PAGE was carried out for peroxidase and glutathione peroxidase.

Peroxidases: (1)-10% acrylamide-bisacrylamide gel was prepared, and 50 μg proteins were loaded. (2)-Staining: Gel was incubated for 10 min in a solution composed of 10 mM guaiacol, 50 mM potassium phosphate buffer of pH 7.0 and 10 mM H_2O_2 solution. Orange coloured bands showing the activity of pox were photographed immediately.

Glutathione peroxidase: (1)-Same extract as for superoxide dismutase was used. 10% acrylamide-bisacrylamide gel was prepared, and 50 μg proteins were loaded. (2)-Staining: Staining was done according to the methods suggested by Kho et al. (2004).The gel was submerged in a 50 mM Tris-HCl buffer (pH 7.9) containing 13 mM GSH and 2 mM H_2O_2 for 15 min, followed by incubation in a solution containing 1.2 mM 3-(4,5-dimethylthiazol-2-yl)-2,5-diphenyltetrazolium bromide (MTT) and 1.6 mM phenazine methosulfate (PMS). After this, achromatic bands on a purple background showing the activity of GPX were photographed immediately.

8.2.6 Chemical Analysis

8.2.6.1 Determination of Total Phenolic Contents

The total phenolic content of the extracts were estimated colorimetric (UV/VIS. Spectrophotometer, Jenway, England) by the Folin-Ciocalteu method, according to Singleton and Ross (1965), using Gallic acid as a standard. Plant extracts were prepared according to a standard protocol, 2 g of plant material was transferred to a dark bottle and mixed with 40 ml ethanol 80% for 24 h at room temperature, sonicate for 10 min, volume adjusted to 50 ml by ethanol 80% then infusions were filtered through Whatman No. 1 filter paper.

Supernatants were evaporated to dryness under vacuum at 40 °C using rotary evaporator. The obtained extracts were kept in sterile sample tubes and stored in a refrigerator at 4 °C, ethanolic solution of the extract in the concentration of 1 mg/ml was used in the analysis. The reaction mixture was prepared by mixing 0.5 ml of a methanolic solution of extract, 2.5 ml of 10% Folin-Ciocalteu's reagent dissolved in water and 2.5 ml 7.5% $NaHCO_3$.

Blank was prepared, containing 0.5 ml ethanol, 2.5 ml 10% Folin-Ciocalteu's reagent dissolved in water and 2.5 ml of 7.5% of $NaHCO_3$. The samples were thereafter incubated in a thermostat at 45 °C for 45 min. The absorbance was determined using spectrophotometer at λ max $= 765$ nm. The samples were prepared in triplicate for each analysis, and the mean value of absorbance was obtained.

The same procedure was repeated for the standard solution of gallic acid, and the calibration line was construed. Based on the measured absorbance, the concentration of phenolics was read (ug/ml) from the calibration line; then the content of phenolics in extracts was expressed in terms of gallic acid equivalent (ug of GA/g of extract).

8.2.6.2   Identification and Quantification of Phenolic Constituents

Phenolic constituent's contents in the different extracts were determined by HPLC according to Schneider (2014). Seven—microliter samples were analyzed by using Agilent1260 infinity HPLC Series (Agilent, USA), equipped with Quaternary pump, a Zorbax Eclipse plus C18 column 100 mm × 4.6 mm i.e. (Agilent Technologies, USA), operated at 25 °C.

The separation is achieved using a ternary linear elution gradient with (A) HPLC grade water 0.2% H_3PO_4 (v/v), (B) methanol and (C) acetonitrile. The injected volume was 20 μL. Detection: VWD detector set at 284 nm. Methanol and (C) acetonitrile. The injected volume was 20 μL. Detection: VWD detector set at 284 nm.

Sixteen standard phenolic constituents including:(gallic acid, catechol, p-hydro benzoic acid, Caffeine, Vanillic acid, Caffeic acid, Syringic acid, Vanillin, p-coumaric acid, Ferulic acid, Rutin, Ellagic acid, Benzoic acid, o-coumaric acid, Salicylic acid and Cinnamic acid) were used. Phenolic constituents in the samples were identified by comparing retention times and peak area with those of pure standards.

8.2.7   Statistical Analysis

The layout of the experiment was arranged in a randomized complete blocks design, with 4 replicates, and the resulted data were subjected to statistical analysis, employing F-test for significance at $P \leq 0.05$ and computing of LSD values to compare means in different statistical groups according to Gomez and Gomez (1984).

8.3   Results and Discussions

8.3.1   Establishment of Parkia biglobosa Cell Suspension Culture Under Salt Stress Conditions (NaCl)

Data presented in Table 8.1 and Fig. 8.1 showed the effect of NaCl on the growth of *P. biglobosa* cell suspension culture. The suspension cells were treated with different concentrations of NaCl (0, 100, 200, 300, 400, 500, 600 and 700 mM) and the viability (%), average number of cells/square and the concentration of viable cells/ml were estimated at different time intervals (3, 6, 9, 12, 15 and 18 days). The obtained results clearly showed that *P. biglobosa* cell viability has been affected by NaCl treatments comparing to the control treatment (zero NaCl).

Data indicated a lag phase from 0 to 3 days, but as the culture proceeded it showed increasing from day 6 and significantly produced the highest amount of cell number over a period of 9 days an exponential growth phase from 6 to 9 days (log phase).

Table 8.1 Effect of NaCl concentrations on *P. biglobosa* cell suspension culture during 18 days

Treatments (A)	Timing (B)	Viable	Non-viable	Total	Viability (%)	Average number of cells/square	Concentration of viable cells/ml
Control	Initial	12.75	8.50	21.25	59.88	3.19	63.75
	3rd day	18.00	5.75	23.75	76.03	4.50	90.00
	6th day	22.50	5.00	27.00	82.77	5.50	110.00
	9th day	25.25	3.75	29.00	87.76	6.31	126.25
	12th day	28.50	6.00	33.50	81.79	6.88	137.50
	15th day	23.50	7.75	31.25	74.88	5.88	117.50
	18th day	18.75	8.75	27.50	68.39	4.69	93.75
100 mM	Initial	15.00	7.75	22.75	65.27	3.75	75.00
	3rd day	19.50	6.00	25.50	76.12	4.88	97.50
	6th day	20.50	4.00	24.50	82.69	5.13	102.50
	9th day	20.75	1.75	22.50	91.75	5.19	103.75
	12th day	17.50	3.75	21.25	82.04	4.38	87.50
	15th day	11.75	5.00	16.75	70.64	2.94	58.75
	18th day	10.25	6.50	16.75	59.51	2.56	51.25
200 mM	Initial	18.75	8.00	26.75	70.05	4.69	93.75
	3rd day	19.50	6.00	25.50	76.69	4.88	97.50
	6th day	23.25	4.00	27.25	85.27	5.81	116.25
	9th day	27.25	2.75	30.00	90.74	6.81	136.25
	12th day	27.00	2.75	29.75	90.65	6.75	135.00
	15th day	18.75	4.00	22.75	82.06	4.69	93.75
	18th day	16.25	5.25	21.50	75.24	4.06	81.25
300 mM	Initial	24.50	7.75	32.25	75.59	6.13	122.50
	3rd day	27.75	5.75	33.50	82.74	6.94	138.75
	6th day	36.25	5.25	41.50	87.35	9.06	181.25
	9th day	40.75	4.00	44.75	91.15	10.19	203.75
	12th day	36.75	3.50	40.25	91.53	9.81	196.25

(continued)

Table 8.1 (continued)

Treatments (A)	Timing (B)	Viable	Non-viable	Total	Viability (%)	Average number of cells/square	Concentration of viable cells/ml
	15th day	34.00	6.50	40.50	83.72	8.50	170.00
	18th day	29.00	7.75	36.75	78.98	7.25	145.00
400 mM	Initial	28.50	5.50	34.00	83.80	7.13	142.50
	3rd day	33.00	4.25	37.25	88.52	8.25	165.00
	6th day	37.50	2.75	40.25	93.18	9.38	187.50
	9th day	43.75	0.75	44.50	98.42	33.44	218.75
	12th day	45.00	0.50	45.50	98.86	11.25	225.00
	15th day	39.25	3.00	42.25	93.08	9.81	196.25
	18th day	35.50	6.75	42.25	84.40	8.88	177.50
500 mM	Initial	25.00	6.75	31.75	78.87	6.25	125.00
	3rd day	32.00	6.75	38.75	82.58	8.00	160.00
	6th day	39.00	5.50	44.50	87.57	9.75	195.00
	9th day	42.00	3.50	45.50	92.48	10.50	210.00
	12th day	48.00	1.25	49.25	97.64	12.00	240.00
	15th day	43.50	5.50	49.00	89.02	10.88	217.50
	18th day	39.50	6.75	46.25	85.38	9.88	197.50
600 mM	Initial	29.25	8.00	37.25	78.42	7.31	146.25
	3rd day	31.75	7.50	39.25	81.03	7.94	158.75
	6th day	38.75	6.25	45.00	85.70	9.69	193.75
	9th day	43.50	2.50	46.00	94.52	10.88	217.50
	12th day	44.25	2.75	47.00	94.49	11.06	221.25
	15th day	43.25	6.25	49.50	87.73	10.81	216.25
	18th day	42.50	9.25	51.75	82.38	10.63	212.50
700 mM	Initial	12.00	10.25	22.25	54.04	3.00	60.00
	3rd day	19.50	8.25	27.75	70.21	4.88	97.50
	6th day	24.00	7.50	31.50	76.05	6.00	120.00

(continued)

Table 8.1 (continued)

Treatments (A)	Timing (B)	Viable	Non-viable	Total	Viability (%)	Average number of cells/square	Concentration of viable cells/ml
	9th day	28.75	6.25	35.00	82.23	7.19	143.75
	12th day	29.00	3.75	32.75	88.76	7.25	145.00
	15th day	26.75	5.50	32.25	83.21	6.69	133.75
	18th day	23.50	8.25	31.75	73.91	5.88	117.50
LSD 0.05	A	2.46	1.03	2.81	2.93	3.20	12.16
	B	2.30	0.96	2.63	2.74	2.99	11.37

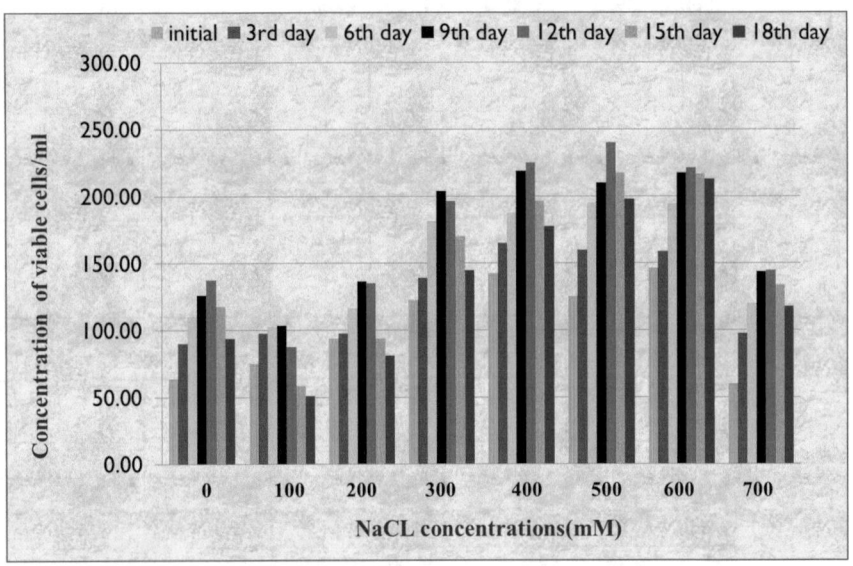

Fig. 8.1 Effect of NaCl concentrations on *P. biglobosa* cell suspension culture during 18 day of cultivation

The maximum increase in viable cell number reached on day 12. This means that lag phase from 0 to 3 days and exponential growth phase from 6 to 9 a day and stationary phase from 9 to 12 days, then decline from 15 to 18 days. During the 15th day under NaCl stress condition data showed that viable cell number in all treatments have a decline in the experiment.

Results obtained showed that in the 12th day, the best result was recorded using 500 mM NaCl shows a maximum increase in cell viability percentage, an average number of cells/square and concentration of viable cells/ml of *P. biglobosa* cell suspension culture.

Thought this experiment could be produced viable cells have tolerance for salinity until 500 mM NaCl in a good state. Therefore, it could be used 500 mM of NaCl to produce cells under saline conditions.

Generally, data indicates there was an inverse relationship between growth of cell suspension and salt concentrations in case of 600 and 700 mM NaCl in the culture media where the values decreased gradually with increasing the concentrations but, the relationship was absolutely using 500 mM NaCl in the culture media where cells were increased gradually with increasing concentrations.

It is observed that cell suspension culture could grow on liquid MS media till 500 mM NaCl. It appears from the obtained results that the cell of *P. biglobosa* exposed to different NaCl concentrations was unaffected by moderate NaCl levels of 100 or 200 mM NaCl coped with high salt exposure (300, 400, 500 and 600 mM NaCl). This result is suggesting that a salt-resistant cell culture of *P. biglobosa* plants have adapted to grow in an environment with high salts. Finally, it was found that at levels of (300, 400, 500 and 600 mM NaCl) treatments show higher viable cells comparing to the control (zero NaCl). While the levels of (100 and 200 mM NaCl) treatments showed results nearly the same as control.

The ability of *Parkia* cell culture grown in the presence of high salts indicates the unique adaptations and metabolism of *Parkia* cells. These results are in agreement with those obtained by Ferreira and Lima-costa (2006) reported that the citrus cell (*Carvalhal*) exposed to different NaCl concentrations was unaffected by moderate NaCl levels (50, 100 and 200 mM) and coped with high salt exposure (300–400 mM) NaCl suggesting that it is a salt-resistant cell culture.

On the other hand, Kumar et al. (2008) observed that NaCl provoked an inhibition of *Jatropha curcas* cell growth which may be due to natural imbalance, osmotic and metabolic disturbances.

8.3.2   Effect of Different Concentrations of NaCl on Ions Uptake (Na^+, K^+, Ca^{2+}, P^{3+} and N^{3+}) of P. biglobosa Cell Suspension Culture

Data in Table 8.2 and Fig. 8.2 illustrated that Na concentrations in stressed cells were higher than the control. Accumulation of Na^+ was shown to be higher in 300 and 600 mM concentrations as compared to 100, 200, 400 and 700 mM NaCl concentrations in cell cultures throughout the experiment. After 18 days of NaCl treatments exposure, *P. biglobosa* cells decreased the accumulation of Na^+ less than 700 mM NaCl which was similar to 100 mM NaCl concentration. The increasing in Na^+ content of cells accompanied by a decreasing in K^+ accumulation.

However, at 100, 200 and 300 mM stress cells of *P. biglobosa* exhibited higher K^+ content as compared to 500, 600 and 700 mM NaCl. Accordingly, Na/K ratio

Table 8.2 Effect of different NaCl concentrations on ion uptake (Na$^+$, K$^+$, Ca^{2+}, N^{3+} and P^{3+}) in *P. biglobosa* cell suspension culture after 18 days of culturing

Treatments (mM NaCl)	Mineral contents (%)					
	Na$^+$	K$^+$	Na/K	Ca^{2+}	P^{3+}	N^{3+}
Control	4.28	1.50	2.85	1.21	1.56	3.13
100	4.58	2.32	1.97	2.00	1.37	2.49
200	6.49	1.85	3.51	1.47	0.39	2.54
300	9.07	2.74	3.35	3.22	2.16	3.33
400	7.28	1.49	4.89	2.76	1.51	2.58
500	8.56	1.16	7.38	2.57	1.62	2.59
600	9.48	1.35	7.02	2.77	1.93	2.50
700	5.73	0.88	6.51	1.50	0.26	1.52
LSD 0.05	0.77	0.16	4.81	0.25	0.12	0.17

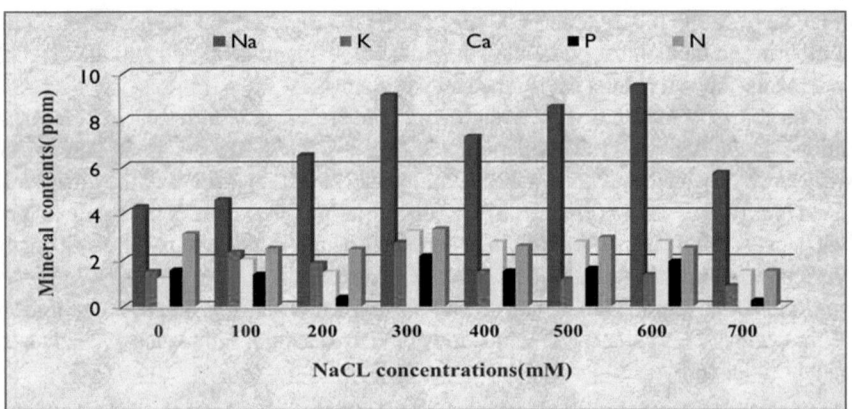

Fig. 8.2 Effect of different NaCl concentrations on the accumulation of (Na$^+$, K$^+$, Ca^{2+}, N^{3+} and P^{3+}) in *P. biglobosa* cell suspension culture after 18 days of culturing

at 100 mM remained lower than that of the rest NaCl concentration. In 500 mM Na/K ratio was 7.4%, 600 mM was 7 and 6.5% in 700 mM NaCl stressed cells after 18 days. While 100 mM NaCl exhibited 2% during the same period. This means that Na/K ratio increased as a function of external concentrations.

Ca^{2+} content showed high accumulation in 600, 300 mM NaCl and it decreased gradually till became low under 200 mM NaCl after 18 days of stress exposure. The control accumulation of Ca^{2+} was nearly similar to 200 mM. The data of *P. biglobosa* culture under NaCl stress showed that Na$^+$ and K$^+$ accumulation increased significantly under the lower concentration of NaCl then gradually a decrease was shown in K$^+$ content while an increase was observed in Na in higher concentrations of NaCl.

In general, Na^+ accumulation increased and K^+ content decline as a function of external NaCl treatment, in parallel with Na^+ accumulation and decline in K^+ content. Na/K ratio increased at all concentrations of NaCl. This finding was also reported by Cherian and Reddy (2003) found that the increase of Na^+ content in *Suaeda nudiflora* cells accompanied by a reduction of K^+ accumulation and differences in Na/K ratio under saline conditions. Hirschi (2004) concluded that the K^+ content of the salt-stressed cells at day 8 was approximately 50% lower than that of the control cells. The Na/K ratio in the control cells was approximately 19 times lower than that of the salt-stressed cells. Salt stress also, caused a decrease in the level of Ca^{2+} content particularly after 6 days of exposure to stress treatments. K^+ and Ca^{2+} have been reported to be the major cautions in cell organization as well as the major contributors to osmotic adjustment under stress condition in several plant *spp*.

P^{3+}content in data presented in Table 8.2 showed that there is high accumulation in 300, 600 mM NaCl treatment and it decreased gradually till became lesser under 700 mM NaCl after 18 days of stress exposure, accumulation of P in control was very similar to 500 mM NaCl.

N^{3+} content was found to be higher in 300 mM, and 500 mM NaCl stressed *P. biglobosa* cells as compared to 100, 200, 400, 600 and 700 mM NaCl. While in control accumulation of N^{3+} was similar to 300 mM NaCl. These results agree with Shibli et al. (2007) reported that Ca^{2+}, P^{3+} and N^{3+} contents decreased with the increased NaCl concentrations. This may be due to the displacement of these ions from cell membrane by Na^+.

8.3.3 Effect of Different Concentration of NaCl on Proline Content of Parkia biglobosa Cell Suspension Culture

Data in Table 8.3 and Fig. 8.3 confirm that there is no significant change was observed in proline content at all NaCl concentrations studied. The amount of proline was gradually decreased at treatments 500, 600 and 700 mM NaCl. While, there is an

Table 8.3 Effect of different NaCl concentrations on the accumulation of proline content in *P. biglobosa* cell suspension culture after 18 days of culturing

Treatments	Proline (mg/g fresh weight)
Control (zero NaCl)	1.03
100 mM NaCl	2.10
200 mM NaCl	3.41
300 mM NaCl	2.46
400 mM NaCl	2.04
500 mM NaCl	1.93
600 mM NaCl	1.45
700 mM NaCl	1.14
LSD 0.05	0.26

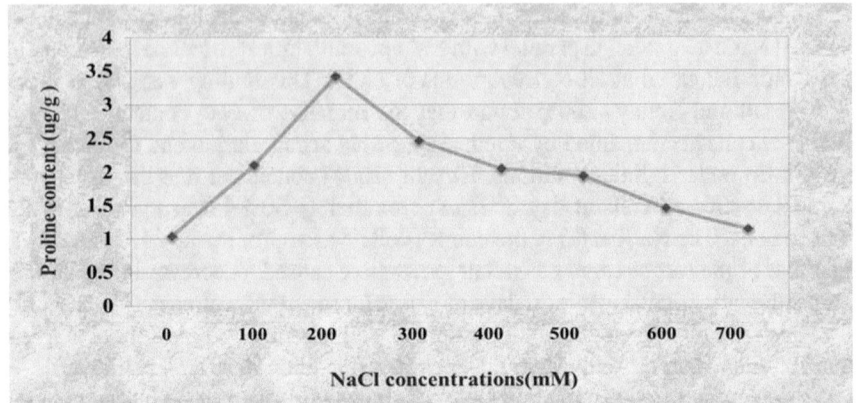

Fig. 8.3 Effect of different NaCl concentrations on the accumulation of proline in *P. biglobosa* cell suspension culture after 18 days of culturing

increment in proline content at treatments 100, 200 and 300 mM up to 400 mM NaCl compared to control. These results agree with Kumar et al. (2008) reported that there is no significant difference in proline content due to NaCl treatments. Also, Greenway and Munns (1980) mentioned that moderately salt resistant plants such as cotton showed a reduction of 60% in proline content and sensitive species such as soybean 200 mM NaCl had a lethal effect. However, El-Minisy et al. (2016) found that a positive relationship between salt tolerance and proline content in cassava suspension cultures up to 200 mM NaCl stress and the highest proline content compared to other treatments

On the other hand, the results obtained disagree with Celik and Atak (2012) illustrated that proline accumulation increased gradually with increasing concentrations of NaCl in both tobacco varieties Akhisar 97 and Izmir-ozbas, the largest increase in proline content compare to control plants was observed in plants of both varieties treated with 350 mM NaCl. The increase in proline content in the Izmir-ozbas variety was 9.4 fold compared to controls. It was 28.4 fold compared to controls in the Akhisar 97 variety. Finally, this result had been confirmed by Meloni et al. (2004) referred to the significance of proline accumulation in osmotic adjustment is still debated and varies according to species.

8.3.4   Electrophoretic Analysis

8.3.4.1   Characterization of Protein Patterns of *Parkia biglobosa* Cultures Using Sodium Dodecyl Sulphate Poly Acrylamide Gel Electrophoresis (SDS-PAGE)

The SDS-PAGE analysis of polypeptide banding patterns was used successfully as a biochemical and molecular marker for the detection of variations in tissue cultures and the investigation of biochemical makeup of plants in response to salt stress.

Accordingly, nine cell cultures of *P. biglobosa* plant including callus cultures and treated suspension cell cultures with NaCl was subjected to molecular analysis using The SDS-PAGE (Table 8.4).

The SDS-PAGE analysis revealed 21 protein bands with different molecular weights ranged from 11 to 170 KD as shown in Table 8.4 and Fig. 8.4. The number of protein bands was varied from 4 to 16, and number of protein bands was produced in lane (1) and lane (8) while less number of bands was observed in lane (3), lane (4) and lane(7).

These results illustrate that there are clear variations in protein banding patterns in different cell cultures. The most distinct variation is over 40 KDa protein band in all cell cultures. Among such protein bands, five bands with 130, 95, 43, 34 and 26 KDa were not appeared in the callus (lane 1) while six other protein bands with 130, 98, 95, 43 and 17 KDa were disappeared in salt-treated cell suspension culture (lane 8).

Also, it could be noticed that a fluctuation in the expression of protein bands is a common observation. Novel expression of some proteins, i.e. 40, 28 and 12 KDa in cell suspension cultures grown on salt stress (lane 8, lane 9) and callus (lane 1) as well as in other tissue cultures is the most striking observation.

In the present study types of modifications are shown on total soluble protein patterns of *P. biglobosa* cell cultures. Some protein bands were disappeared, other proteins were selectively increased, and synthesis of a new set of protein was induced. Some of these responses were shown under salinity treatments while others were induced by salinity. Sousa et al. (2003) showed that the synthesis of protein was affected by salt stress. In our study, several new proteins which are synthesized in response to interaction between environmental stress and growth substances applied have been reported as a stress protein in plants (Qasim et al. 2003).

Many of these proteins were suggested to protect the cell against the adverse effect of salt stress. Changes in protein synthesis under salinity treatments may be due to changes in the efficiency of mRNA translation of the regulation of RNA transcription transport and stability.

It was reported that in the seedlings of *Bruguiera parviflora* varying levels of NaCl lead to decrease in the intensity of several protein bands, which was proportional to NaCl concentrations (Parida et al. 2004). Salt stress lead to a difference in gene expressions were an alteration in protein could be due to alteration in the regulation of transcription, mRNA processing or due to altered rates of protein denaturation.

Table 8.4 SDS-PAGE analysis of total protein bands extracted from callus and salt treated cell suspension culture of *P. biglobosa*

Band No.	MW (KD)	Treatments								
		Callus	Salt treated cell suspension culture							
		1	2	3	4	5	6	7	8	9
1	170	+	−	−	−	−	−	−	+	−
2	130	−	−	−	−	−	−	−	−	−
3	98	+	−	−	−	−	−	−	−	−
4	95	−	−	−	−	−	−	−	−	−
5	80	+	+	−	−	−	−	−	+	−
6	72	+	+	−	−	−	−	−	+	−
7	56	+	−	−	+	−	+	−	−	−
8	50	+	+	+	+	+	+	−	+	−
9	43	−	−	−	−	−	−	−	−	+
10	40	+	+	+	+	+	+	+	+	+
11	34	−	−	−	−	−	−	+	+	−
12	32	+	+	−	−	−	+	−	+	−
13	28	+	−	+	+	+	+	+	+	+
14	26	−	+	−	−	+	+	−	+	+
15	24	+	+	+	−	+	−	−	+	+
16	22	+	+	−	−	−	−	−	+	−
17	20	+	−	−	−	−	−	−	+	−
18	18	+	−	−	−	−	+	−	+	−
19	17	+	+	−	−	−	−	−	−	−
20	14	+	−	−	−	−	−	+	+	+
21	12	+	−	−	−	−	+	−	+	+
Total no. of bands		16	9	4	4	5	8	4	15	7

Where 1 = callus, 2 = 0 NaCl, 3 = 100 mM NaCl, 4 = 200 mM NaCl, 5 = 300 mM NaCl, 6 = 400 mM NaCl, 7 = 500 mM NaCl, 8 = 600 mM NaCl, 9 = 700 mM NaCl. **(+) = detected, (−) = not detected

Bishnoi et al. (2006) suggested that the disappeared proteins in response to salt stress were a result of their denaturation.

A total number of bands in cell suspension culture of *P. biglobosa* treated with 100 mM NaCl (lane 3) was decreased as caused as compared with the respective control (lane 2). These results concluded that the decrease in the protein levels in salt-stressed plant might be attributed to a decrease in protein synthesis and the denaturation of enzymes involved in amino acids and protein synthesis.

Also, salinity stress caused the disappearance of some bands in *P. biglobosa* cell culture and some proteins completely disappear this may be due to the gene(s) responsible for certain proteins might be completely suppressed as the result of stress.

Lanes

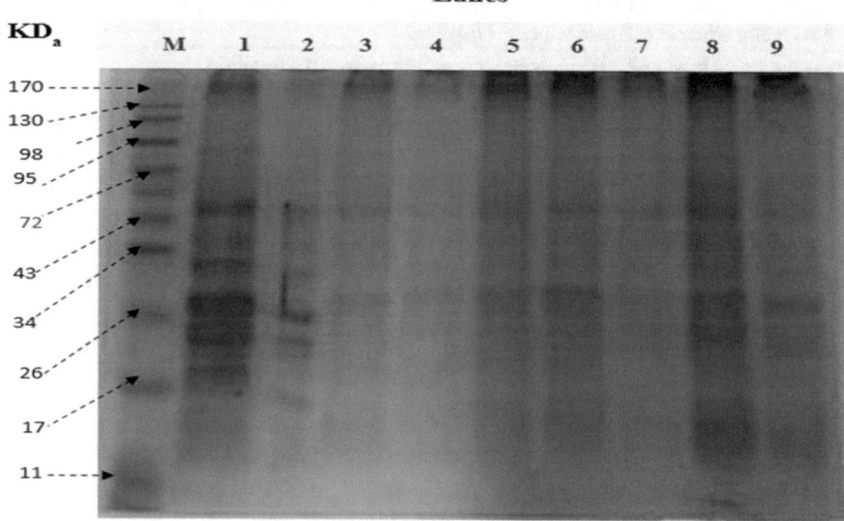

Fig. 8.4 SDS-PAGE analysis of total protein bands extracted from callus and salt stressed cell suspension cultures of *P. biglobosa* plant. Where lane: 1 = callus culture, 2 = 0 mM NaCl, 3 = 100 Mm NaCl, 4 = 200 mM NaCl, 5 = 300 mM NaCl, 6 = 400 mM NaCl, 7 = 500 mM NaCl, 8 = 600 mM NaCl, 9 = 700 mM NaCl

Therefore, the developed tissues had lost their ability to synthesize these proteins; proteins accumulated in salt-stressed plants might provide storage form of nitrogen for the utilization when stress is over (Qasim et al. 2003; Singh et al. 1987).

Our results obtained above agree with those obtained by Harrington and Aim (1988) observed that in the tobacco plant, polypeptide bands of molecular weights ranging from 15 to 115 KD in response to salt stress of cultured cells. El-Farash et al. (1993) found that in tomato plants the expression of 12 different protein bands induced salt stress was genetically regulated depending on salt concentration.

Also, Javid et al. (2004) stated that electrophoresis of proteins is a powerful tool for identification of genetic diversity and the SDS-PAGE is particularly being considered as reliable technology because seed storage proteins are highly independent of environmental fluctuations.

In contrast, the decrease in protein content observed at the higher NaCl concentration (100 mM) in the nutrient medium may be due to a decrease in the synthesis of protein. Also, Kumar et al. (2008) found that 83, 58 and 43 KDa proteins decreased with the increase in NaCl concentration in callus culture of *Jatropha curcas*.

Table 8.5 Distribution of the different forms of isozymes (peroxidase and glutathione peroxidase) bands among different cell cultures of *P. biglobosa*

Types of enzymes	Forms of enzymes	Distribution forms of isozymes bands(lanes)								
		Callus	Salt treated cell suspension culture							
		1	2	3	4	5	6	7	8	9
Peroxidase	POX1	+	+	+++	++	+++	++	+++	+	+
	POX2	+	+	+++	+++	+++	++	+++	++	++
Glutathione peroxidase	GPX1	+	+	+	+	++	+	++	+	ND
	GPX2	+	+	+	+	++	++	+++	++	+

Where ND = Not detected, (+) = Low expression, (++) = Medium expression, (+++) = High expression. **1 = callus culture, 2 = 0 NaCl, 3 = 100 mM NaCl, 4 = 200 mM NaCl, 5 = 300 mM NaCl, 6 = 400 mM NaCl, 7 = 500 mM NaCl, 8 = 600 mM NaCl, 9 = 700 mM NaCl

8.3.4.2 Characterization of Different in Vitro Cultures of *Parkia biglobosa* Using Isozyme Analysis

Isozyme analysis is considered one of the most important biochemical and molecular marker to the variations between the callus and cell suspension cultures of *P. biglobosa* produced in vitro. Peroxidase (POX) and Glutathione peroxidase (GPX) enzymes were chosen to differentiate between callus and cell suspension cultures for detection the biochemical changes or any changes in the enzyme form.

Peroxidase (POX) Isozymes

Expression of the POX isozyme was detected in *P. biglobosa* cell suspension culture treated with different concentrations of NaCl (0, 100, 200, 300, 400, 500, 600 and 700 mM) and callus using PAGE system.

The results showed that two bands were exhibited with different densities and intensities in treated and untreated cells. Data presented in Table 8.5 and Fig. 8.5a showed two forms of peroxidase enzymes (isozyme) appeared in callus and suspension cells. There are clear variations in peroxidase forms banding patterns of different cells. The first isomer was recorded in callus cells and cell suspension cultures (lanes 1, 2, 3, 4, 5, 6, 7, 8, 9, respectively).

These results indicated that salt stress increased the accumulation of the POX enzyme which may accelerate in response to salt stress in both treated and untreated cells, but the rate of acceleration was markedly higher in treated than untreated (callus and zero NaCl treated suspension cells). Our findings are in accordance with those obtained by Sreenivasulu et al. (1999) stated that high peroxidase isozyme activity was detected in a salt-tolerant cultivar of Foxtail millet which revelled to salt adaptation process.

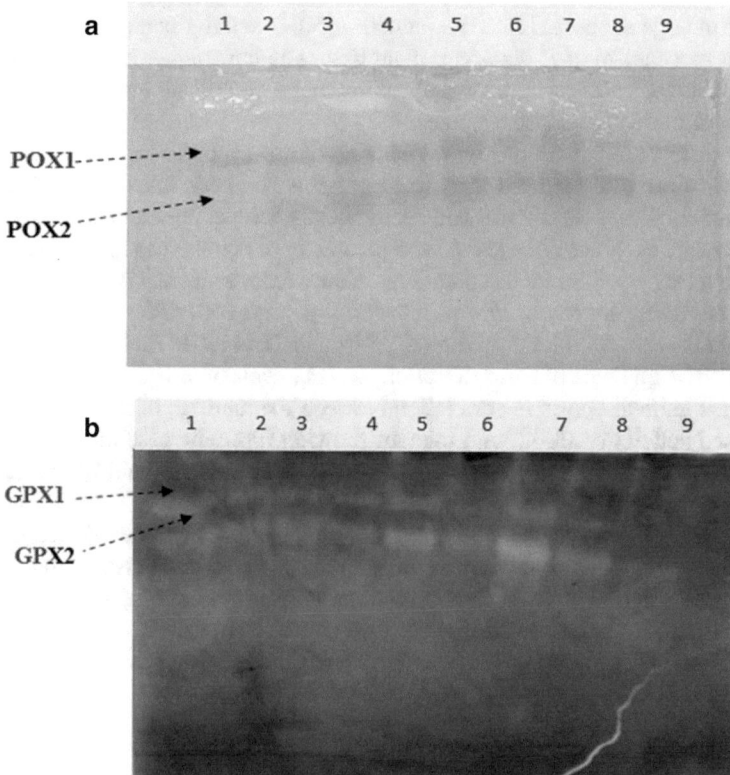

Fig. 8.5 **a** Electrophoretic patterns of peroxidase isozymes (POX) in callus and salt treated cell suspension culture of *P. biglobosa*. **b** Electrophoretic patterns of glutathione peroxidase (GPX) isozymes in callus and salt treated cell suspension culture of *P. biglobosa*. Where 1 = callus culture, 2 = 0 NaCl, 3 = 100 mM NaCl, 4 = 200 mM NaCl, 5 = 300 mM NaCl, 6 = 400 mM NaCl, 7 = 500 mM NaCl, 8 = 600 mM NaCl, 9 = 700 mM NaCl

Meanwhile, the second isomer was detected in callus and suspension cells. Isozyme banding pattern presented in Fig. 8.5a illustrates that fluctuation in peroxidase expression ranging from overexpression to mid expression was recorded among different treatments. It could be concluded that the two forms of peroxidase enzyme having different mobility and accordingly having different molecular weights were detected in callus and suspension cells.

In the present study, a highly significant elevation of antioxidant enzyme activity (POX) under salt stress condition was recorded in three suspension cells treatment (100, 300 and 500 mM NaCl) and medium elevation was shown in treatment (200 and 400 mM NaCl) while treatment (600 and 700 mM NaCl) recorded the lowest result. On the other hand, zero NaCl treated suspension cells showed the same result as callus.

So, the degree of elevation in (POX) activity of salt treated suspension cells were relatively high than untreated cells (callus and zero NaCl treated suspension cells).

These findings suggested that the increase (POX) activity could contribute to the oxidant mechanism of *P. biglobosa* plant tissues against higher NaCl concentration stress. This indicates that the level of salinity has a significant effect on the average activity of these peroxidase enzymes.

Our results are in harmony with Yun et al. (2002) reported that alteration in total peroxidase activities and iso-peroxidase patterns had been known to be related to environmental stresses such as salt, heavy metals, temperature and air pollutants.

Therefore POX has often served as a parameter of metabolic activity during cross alterations and environmental conditions. Also, Modarresi et al. (2014) observed that POX activity in *Aeluropus littoralis* shown higher at 250, 450 and 650 mM NaCl compared to the control. Furthermore, Yildiz and Terzi (2013) subjecting to tolerant and sensitive cultivars of barley to salinity stress (levels of 0, 100, 200 and 300 mM), found a significant positive correlation between the increase of salinity levels with increased activity of POX, with tolerant showing more increase than the sensitive one. They concluded that increased activity of the enzyme could be the cause of tolerance in the former.

In our study, the highest activity of POX was observed in NaCl treated cell suspension culture, and therefore we can conclude that it could be advisable to use tissue techniques in producing *P. biglobosa* tolerate NaCl up to a concentration of 500 mM, and higher concentrations cause eco-toxicity to the plant.

Glutathione Peroxidase (GPX) Isozymes

GPX isozymes are key enzymes of the sorbate-glutathione cycle, may be a potential mechanism of *P. biglobosa* adaptation to salinity. It was reported that in the stress conditions of intense light, high temperature, flood and salinity, free radicals and reactive oxygen molecules are also formed. Therefore, a scavenging system should very active.

In our study expression of the GPX isozyme was detected in *P. biglobosa* suspension cells treated with different concentrations of NaCl salt at levels of (0, 100, 200, 300, 400, 500, 600 and 700 mM) and callus cells using native page system. The results showed that two bands were exhibited with different densities and intensities in treated cell suspension culture and untreated cells.

Data tabulated in Table 8.5 and Fig. 8.5b represent two forms of GPX isozymes appeared in all cell cultures. There are clear variations in GPX forms banding patterns of different cells, the first isomer was recorded in callus cells and cell suspension cultures (lanes 1, 2, 3, 4, 5, 6, 7 and 8, respectively) and absent in treatment 700 mM NaCl (lane 9). These results indicated that salt stress increases the accumulation of GPX enzyme which may accelerate in response to salt stress, but the rate of the acceleration was little higher in treated than untreated.

Meanwhile, the second isomer was detected in callus and cell suspension culture. Isozyme banding pattern presented in Fig. 8.5b show clearly that GPX expression ranging from overexpression to mid expression was recorded among different treatments. For this reason, it could be concluded that the two forms of GPX isozymes having different mobility and accordingly having different molecular weight.

In this study, a highly significant elevation of GPX enzyme activity under salt stress condition was recorded only in cell suspension treatment at (level of 500 mM NaCl).The cell suspension treatments at (levels 300, 400 and 600 mM NaCl) showed a medium elevation.

The lowest results recorded at (levels 0, 100, 200 and 700 mM NaCl) treated suspension cells and in callus cells. These results could attribute to the increased activity of GPX enzyme of *P. biglobosa* plant tissue against high NaCl concentrations stress. The obtained results are in agreement with Mittler (2002) reported that environmental stresses such as salt stress cause oxidative damage to limit damage to cells by reactive oxygen species. Plant cells employ antioxidant enzymes including GPX enzymes. Increasing of GPX activity was also observed due to salinity in *Soybean* (Aghaleh and Niknam 2009; Weisanly et al. 2012) and *Pea* (Shahid et al. 2011). On the other hand, Hatamnia et al. (2013) mentioned that although both tobacco genotypes showed increases in GPX activity in leaves, their roots showed decreases in GPX activity under salt stress.

Our results suggested that GPX isozymes play a major role in estimating the salt stress tolerance of *P. biglobosa* plant which clarifies that using tissue technique in producing *P. biglobosa* tolerates NaCl salt up to a concentration of 500 Mm.

8.3.5   *Chemical Analysis*

8.3.5.1   **Determination of Total Phenols Content in Different in Vitro Cultures of *Parkia biglobosa***

This experiment was conducted to study the quantitative determination of total phenols content in different extraction from stem explant (mother plant), callus and salt-stressed suspension cells (0, 500, 600 mM NaCl) and its supernatant of *P. biglobosa* plant.

Data presented in Table 8.6 and Fig. 8.6 show the quantitative determination of total phenols in the extracts of different samples of the mentioned above. In case of the extract of mother plant (stem explant) show the highest amount of total phenols reached to (621.74 ug/g) followed by the extract of callus and untreated suspension cells (zero NaCl) that reached (401.88 and 112.49 ug/g), respectively. While the salt treated cell suspension cultures and their supernatant levels (500 and 600 mM NaCl), total phenols reached (70.65, 67.63, 32.23 and 2.32 ug/g) respectively.

It can conclude that there is an inverse relationship between the concentration of the used NaCl salt in suspension cells treatments and quantity of total phenols in the intact. It is observed that increasing the concentrations of NaCl salt in treatments decreased the total phenol content.

Table 8.6 Total phenolic compounds in mother plant, callus and salt treated cell suspension cultures of *P. biglobosa*

Code	Total phenol (ug gallic acid equivalent/g sample)
P1	621.74
P2	401.88
P3	112.49
P4	70.65
P5	67.63
P6	32.23
P7	2.32

Where P1 = Mother plant, P2 = callus culture, P3 = zero NaCl (untreated suspension cells), P4 = 500 mM NaCl treated suspension cells, P5 = 600 mM NaCl treated suspension cells, P6 = 500 mM NaCl supernatant, P7 = 600 mM NaCl supernatant

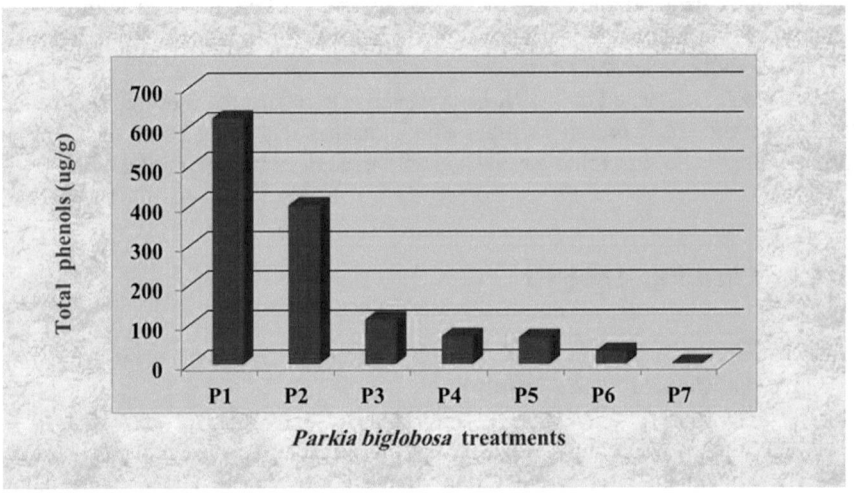

Fig. 8.6 Total phenolic compounds in the plant extracts of *P. biglobosa* expressed in terms of gallic acid equivalent (ug of GA/g of extract). Where P1 = Mother plant, P2 = callus culture, P3 = zero NaCl (untreated suspension cells), P4 = 500 mM NaCl treated suspension cells, P5 = 600 mM NaCl treated suspension cells, P6 = 500 mM NaCl supernatant, P7 = 600 mM NaCl supernatant

These results are in accordance with Telesinski et al. (2008) reported that the content of chlorides which increased in bean plant tissues with increasing NaCl concentration in the soil (0, 10, 30 and 50 mM) correlates negatively with total phenols content after 28 days of the experiment. Hence, Rao et al. (2013) reported that using different doses of salts causes a decrease in phenolic content of the wheat leaf. In contrast, Parida et al. (2004) reported that an increase in phenolic compound referred to moderate salinity in the mangrove. Also, Petridis et al. (2012) observed that high salinity leads to an increase in phenolic content up to 129% in leaves of the olive tree than the control sample.

8.3.5.2 Determination of Phenolic Compounds in Different in Vitro Cultures *Parkia biglobosa* Plant by HPLC

The objectives of this experiment were to separate and identify the main phenolic compounds by using HPLC then clarify the effects of salinity stress and changes in chemical compounds including phenolic compounds in *P. biglobosa* treatments.

Separation and Identification of the Main Phenolic Compounds

This experiment was carried out by extracting and acid hydrolysis the content of phenolic compounds was determined by HPLC quantitative analysis. A typical HPLC chromatogram of *P. biglobosa* samples was presented in Fig. 8.7. The amount of phenolic compounds detected in the sample are expressed in mg/g dry sample.

Data tabulated in Table 8.7 show the fractionation of phenolic compounds in *P. biglobosa* plant samples, the extracts of these samples indicated the presence of sixteen constituents. The present study indicated the occurrence of gallic acid in all sample extracts except treated cell suspension supernatant of (600 mM NaCl). The extract of the mother plant was found to contain a higher amount of gallic acid (30.07 ug/g), while in treated suspension cells BY 600 mM NaCl was 0.27 ug/g. These results are in harmony with Romaric et al. (2011) observed the accumulation of gallic acid in Spanish, cabbage and Chili pepper ranged from (0.49 to 3.3 mg/g).

The amount of salicylic acid content in extract of mother plant (15.34 ug/g) was much higher than all recorded results in other treatments, it is observed that the accumulation of salicylic acid was (1.89 and 0.41 ug/g) in treated suspension cells with (500 and 600 mM NaCl), respectively, while this acid not detected in the supernatant treatments. On the other hand, Singh et al. (2015) found that the amount of salicylic content increased significantly in the maize plant under salinity condition (18.14 mg/g).

Rutin is a phenolic compound with glycosidic- linkage, the amount of Rutin content in the extract of the mother plant and untreated suspension cells (zero NaCl) were recorded (14.47 and 3.50 ug/g), respectively. These results higher more than other treatments, while it is not detected in suspension cells supernatant at 600 mM NaCl. The obtained result was in agreement with those obtained by Moghaddasian et al. (2013) found that the amount of rutin in the leaves of *Fagopyrum esculentum* was 0.12 mg/g and in *Melisa officinalis* was 0.30 mg/g.

Ferulic acid is considered the most important phenolic due to its physiological functions. The concentration of this acid recorded (7.21, 1.83 ug/g) in mother plant and callus respectively. While the rest treatments its content ranged from (0.066–0.40 ug/g). Seal (2016) reported that the ethanol extract of *Oenanthe linearis* contained 0.023 mg/g of ferulic acid.

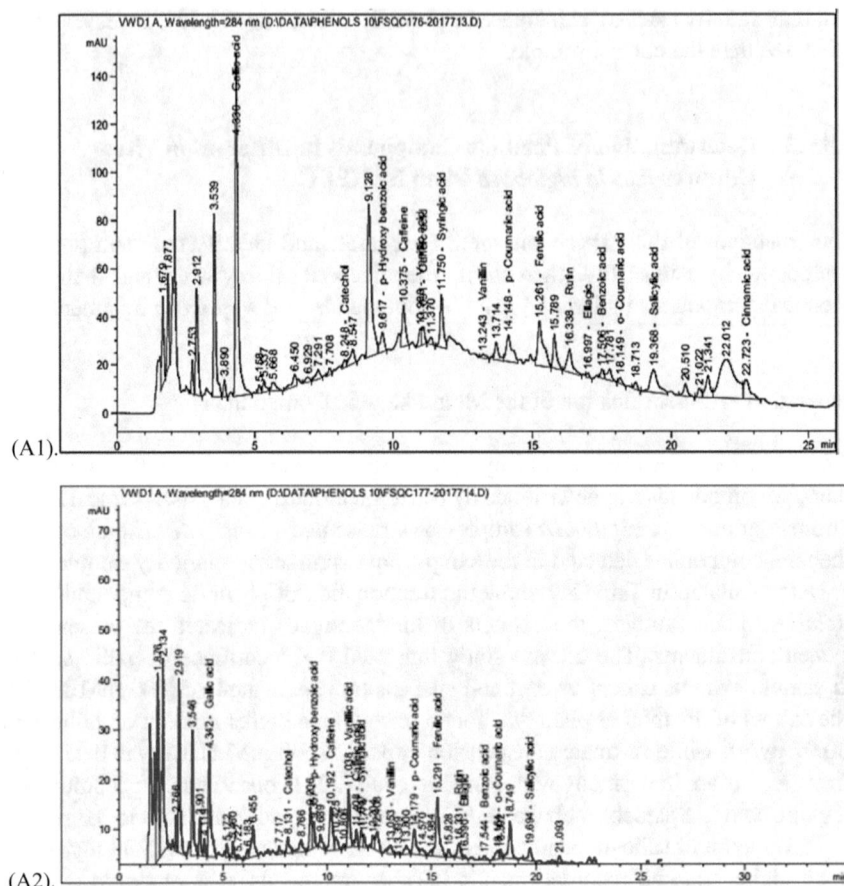

Fig. 8.7 HPLC chromatogram of methanol extract of *P. biglobosa* showing the presence of sixteen phenolic constituents. Where A1 = Mother plant, A2 = callus culture, A3 = zero mM NaCl (untreated suspension cells), A4 = 500 mM NaCl treated suspension cells, A5 = 600 mM NaCl treated suspension cells, A6 = 500 mM NaCl supernatant, A7 = 600 mM NaCl supernatant

Caffeic acid is one of the main hydroxyl cinnamic acid components found in the form of its ester in the extract of mother plant contained 1.05 ug/g more than other extracts of this experiment while the supernatant treatments (500 and 600 mM NaCl) recorded the lowest amount (0.02 mg/g). These results are in harmony with those obtained by Seal (2016) reported that all solvent extracts (acetic acid, ethanol, methanol and chloroform) of *Oenanthe linearis* contained a very good amount of caffeic acid ranging from 0.017 to 0.096 mg/g.

Syringic acid with hydroxyl benzoic acid skeleton is found in the extract of all *P. biglobosa* samples, the amount of this acid recorded 5.25 ug/g in mother plant, 0.34 ug/g in callus while its amount 0.46 ug/g in treated suspension cells (500 mM NaCl) and 0.11 ug/g in treated suspension cells (600 mM NaCl). These results

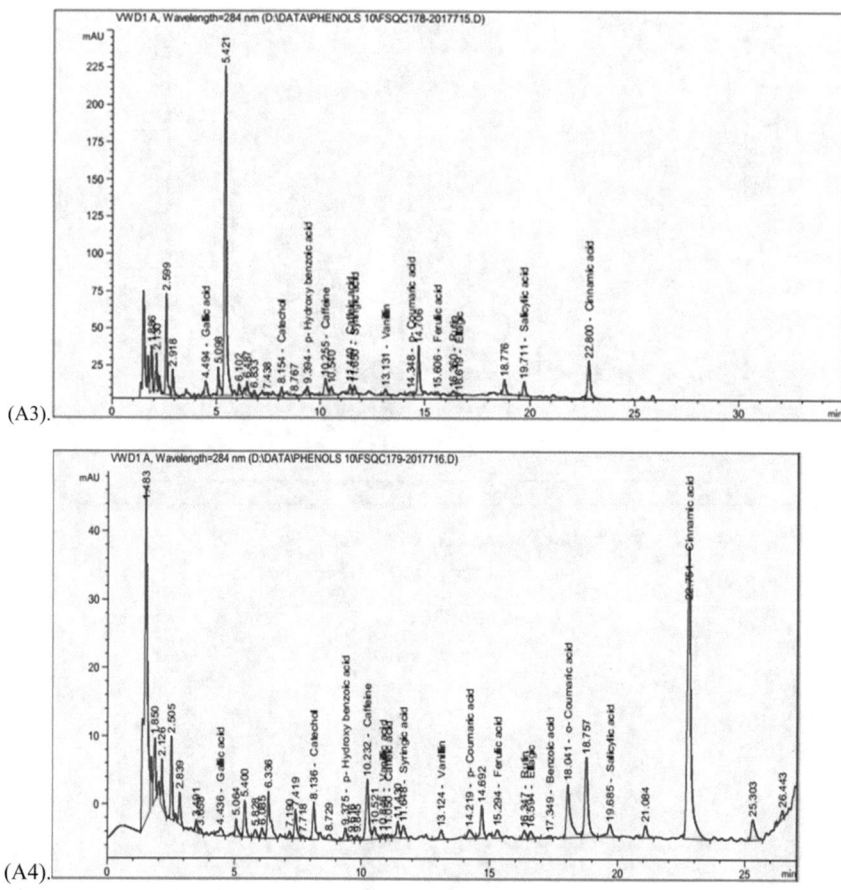

(A3).

(A4).

Fig. 8.7 (continued)

are in accordance with Kivilompolo et al. (2007) found the amount of this acid in different extracts such as *Salvia officinalis* (0.0335 mg/g) and *Origamum vulgare* (0.0375 mg/g).

The amount of p-coumaric acid (hydroxy cinnamic acid) in the extract of the mother plant was 2.39 ug/g and it is considered the higher amount compared to the remaining sample extracts. The least amount of this acid recorded (0.044 ug/g) in supernatant treatment (600 mM NaCl). In the same way, Seal (2016) revealed that P-coumaric acid (0.013 mg/g) was present only in the ethanol extract (80%) of *Oenanthe linearis*.

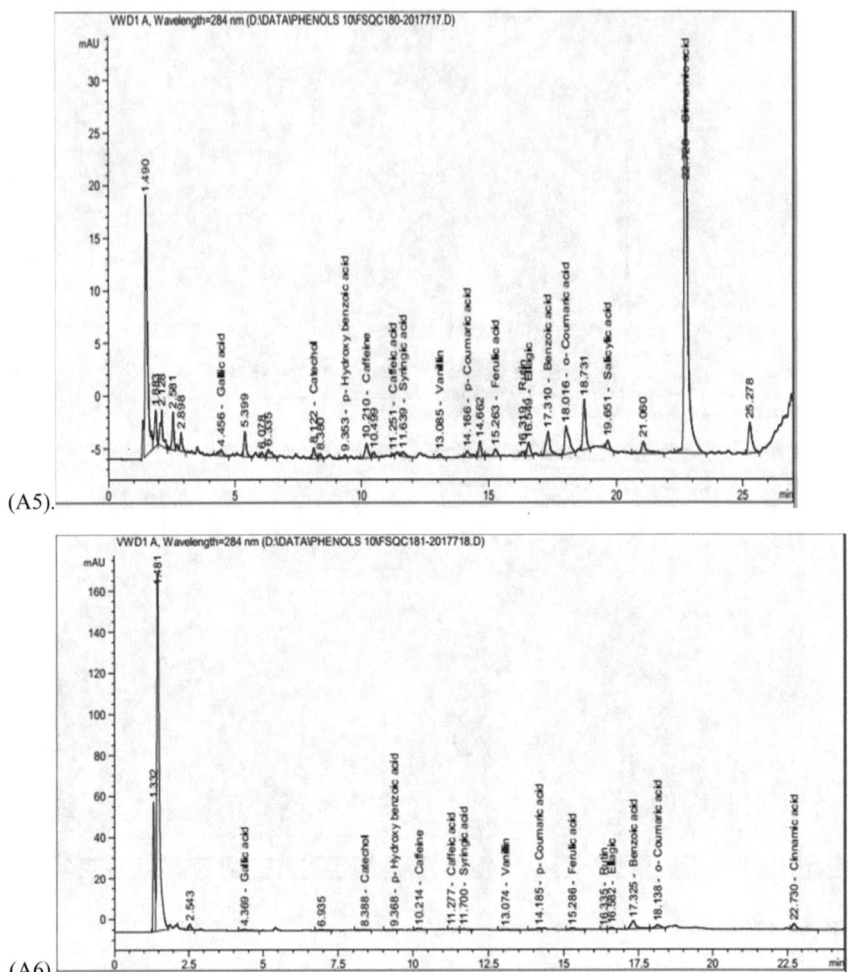

Fig. 8.7 (continued)

Effects of Salt Stress on Phenolic Compounds of *Parkia biglobosa* Cell Suspension Culture

Data tabulated in Table 8.7 showed that a high concentration of NaCl salt caused a highly significant decrease in phenolic compounds in the extract of treated suspension cells (500, 600 mM NaCl and their supernatant). While, the untreated suspension cells (Zero NaCl) caused a highly significant increase in all phenolic constituents except benzoic acid that not detected. This increase in phenolic acids in the untreated cells might be due to the increase in their biosynthesis.

The increase in phenolic compound levels has been reported in a number of plants grown under salinity stress condition (Ksouri et al. 2007). However, the exposure of

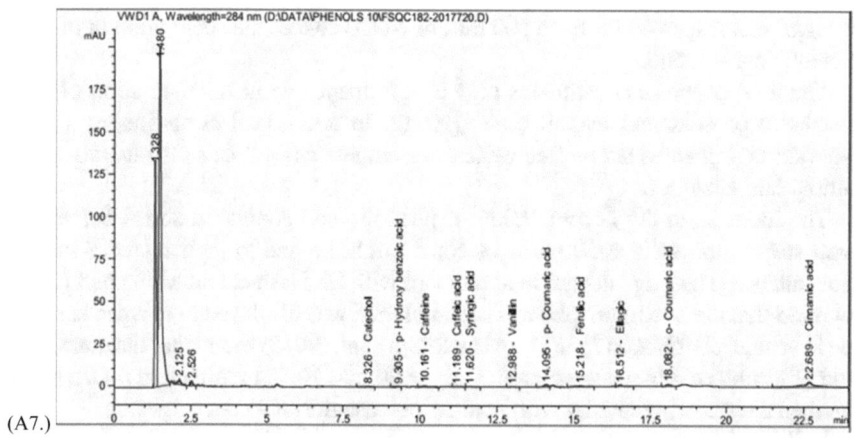

(A7.)

Fig. 8.7 (continued)

Table 8.7 Quantification of phenolic compounds in extracts of *P. biglobosa* cell cultures by using HPLC

Phenolic compound (ug/g)	Treatments						
	A1	A2	A3	A4	A5	A6	A7
Gallic acid	30.07	2.50	2.41	0.54	0.27	0.07	ND
Catechol	2.45	1.53	3.62	4.21	0.73	0.09	0.08
P-hydroxybenzoic acid	9.73	2.11	4.28	1.48	0.16	0.02	0.02
Caffeine	5.86	1.91	3.79	2.94	0.47	0.07	0.05
Vanillic acid	2.14	1.87	ND	0.19	ND	ND	ND
Caffeic acid	1.05	0.24	0.83	0.13	0.07	0.02	0.02
Syringic acid	5.25	0.34	1.08	0.46	0.11	0.02	0.013
Vanillin	0.51	0.29	0.93	0.33	0.08	0.02	0.014
p-coumaric acid	2.39	0.40	0.43	0.27	0.12	0.03	0.044
Ferulic acid	7.21	1.83	0.40	0.32	0.17	0.05	0.066
Rutin	14.47	1.76	3.58	1.13	0.29	0.03	ND
Ellagic acid	0.87	0.17	0.86	0.70	0.71	0.11	0.10
Benzoic acid	10.77	1.60	ND	1.80	8.48	4.09	ND
O-coumaric acid	0.66	0.32	ND	1.80	0.79	0.14	0.10
Salicylic acid	15.34	1.22	5.80	1.89	0.42	ND	ND
Cinnamic acid	1.92	ND	3.50	7.07	5.81	0.05	0.26

Where A1 = Mother plant, A2 = callus culture, A3 = zero NaCl (untreated suspension cells), A4 = 500 mM NaCl treated suspension cells, A5 = 600 mM NaCl treated suspension cells, A6 = 500 mM NaCl supernatant, A7 = 600 mM NaCl supernatant. ND = Not detected

P. biglobosa suspension cells to 600 mM of NaCl reduces the accumulation of these phenolic constituents.

The total phenolic constituents play a significant role in the regulation of plant metabolic process and overall plant growth. In addition, Martin-Tanguy (2001) reported that phenols act as free radical scavengers as well as substrates for many antioxidant enzymes.

The increase in the accumulation of phenolic compounds in stressed *P. biglobosa* suspension cells at 500 mM of NaCl might be due to their increase in their biosynthesis. These results are in agreement with El-Mashad and Mohamed (2012) reported that the maximum phenols accumulation was displayed in cowpea leaves at the lowest level of NaCl (25 mM). Also, Singh et al. (2015) found that the concentration of ferulic acid increased significantly at 50 mM NaCl treatment in maize plants but then decreased gradually with further increase in NaCl concentration.

In this study, the untreated cell suspension (zero NaCl) showed that the content of gallic, caffeic, vanillin, ferulic-acid, p-coumaric and salicylic acid increased compared to the treated cell suspension at the level of 500, 600 mM NaCl and their supernatant.

Our results are in line with results by Xuan et al. (2016) reported that under salinity stress cinnamic acid, vanillin, protocatechuic acid, ferulic acid and p- coumaric acid were detected as free phenolic acids and they play a direct role in salinity resistance in rice. Also, Singh et al. (2015) observed that caffeic acid level increased significantly at 50 mM NaCl treatment then decreased at higher concentrations of NaCl in the absence of salicylic acid, and they mentioned that the accumulation of gallic acid was also reduced up to 16.6%.

On the other hand, a significant increase in the accumulation of p-coumaric acid assists to decrease oxidative pressure because p-coumaric acid expresses high radical scavenging activity due to their hydroxyl nature (Jamalian et al. 2013). Also, Wakabayashi et al. (1997) reported that the presence of ferulic acid under osmotic stress might be related to the strengthening of the plant cell wall and the overall cell wall elongation. Moreover, Li et al. (2013) demonstrated that ferulic acid is matching with dehydration stress by decreasing of lipid peroxidation due to activation of antioxidant enzymes and accumulation of proline and soluble sugar in cucumber leaves.

In this study, we get supernatant from *P. biglobosa* cell suspension culture treated with 500 and 600 mM NaCl by running an experiment to obtain phenolic compounds in aqueous solutions of alcoholic extraction. After centrifuging, the combined supernatants were analyzed using HPLC.

Therefore, the higher buildup of phenolic and flavonoids in the plant under salt stress may assist the plant to tolerate salinity induced oxidative stress (Wahid and Ghazanfar 2006). The use of effective secondary metabolites promises development bioactive reagents to protect plant production under salt stress.

8.4 Conclusions

This chapter indicated that the cell suspension culture of *P. biglobosa* treated with NaCl could tolerate salt stress. Also, revealed that salinity caused highly significantly increased in viability percentage and concentration of viable cells/ml, increased in the Na content and decreased in the contents of K, Ca, P and N. The rate of proline accumulation in *P. biglobosa* suspension cells was higher at moderate NaCl stress. The electrophoretic analysis is considered one of the most important biochemical and molecular markers to the variations between the callus and NaCl treated cell suspension cultures of *P. biglobosa* produced in vitro. Also, phenolic constituents may have an antioxidant role during salt stress and could play a role in adaptation processes. By using HPLC could separate the phenolic constituents that responsible for *P. biglobosa* to tolerate salinity such as gallic, caffeic, vanillic, ferulic, p-coumaric and salicylic acids, it is used as a marker for salt tolerance.

8.5 Recommendations

From the present study, it can introduce African locust bean plant (*P. biglobosa*) as promising crop in Egyptian agriculture and expanding its cultivation and recognition of Egyptian farmers by this crop for its important uses in many industries especially pharmaceuticals, food, textile and other industries. By using genetic engineering can reveal the gene responsible for the ability of this legume plant to tolerate the salinity stress conditions and therefore transforming it to another legume plants to resist such conditions.

8.6 Acknowledgments

The authors wish to thank the Department of Natural Resources, Faculty of African Postgraduate Studies, Cairo University, and the Department of Plant Biotechnology, National Research Center, Cairo for providing financial support and laboratory facilities to carry out this investigation.

References

Abbas MS, El-Shabrawi HM, Soliman Ash, Selim MA (2018) Optimization of germination, callus induction, and cell suspension culture of African locust beans *Parkia biglobosa* (Jacq.) Benth. J Genet Eng Biotechnolo 16(1):191–201

Aghaleh M, Niknam V (2009) Effect of salinity on some physiological and biochemical parameters in explants of two cultivars of soybean (*Glycine max* L.). J Phytol 1(2):86–94

Bates LS, Waldren RP, Teare ID (1973) Rapid determination of free proline for water stress studies. Plant Soil 39:205–207

Bishnoi SK, Kumar B, Rani C, Datta KS, Kumari P, Sheoran IS, Angrish R (2006) Changes in protein profile of pigeon pea genotypes in response to NaCl and boron stress. Biol Plant 50:135–137

Celık Ö, Atak C (2012) The effect of salt stress on anti-oxidative enzymes and proline content of two Turkish tobacco varieties. Turk J Biol 36:39–356

Cherian S, Reddy MP (2003) Evaluation of NaCl tolerance in callus cultures of *Suaeda nudiflora*. Moq. Biol Plant 46:193–198

Daud Z, Keng CL (2006) Effects of plant growth regulators on the biomass of embryogenic cells of *Cyperus aromaticus* (Ridly) Mattf and Kukenth. Biotechnology 5:75–78

Djeridane A, Yousfi M, Nadjemi B, Boutassouna D, Stocker P, Vidal N (2006) Antioxidant activity of some Algerian medicinal plants extract containing phenolic compounds. Food Chem 97:654–660

El-Farash EM, El-Enany AE, Majen A (1993) Influence of genotype and NaCl on levels of growth, proteins, proline, free amino acids, viability and protein regulation in tomato callus cultures. Physiol Plant 74:345–352

El-kahoui S, Hernandez JA, Abdelly C, Ghrir R, Limam F (2005) Effects of salt on lipid peroxidation and antioxidant enzyme activities of *Catharanthus roseus* suspension cells. Plant Sci 168:607–613

El-Mashad AA, Mohamed HI (2012) *Brassinolide alleviates* salt stress and increases antioxidant activity of cowpea plants (*Vigna sinensis*). Protoplasma 249(3):625–635

El-Minisy AM, Abbas MS, Aly UI, El-Shabrawi HM (2016) In vitro selection and characterization of salt tolerant cell lines in cassava plant (*Manihot esculenta* Crantz). Int J ChemTech Res 9(5):215–227

Faithfull NT (2002) Methods in agricultural analysis: a practical handbook. CABI Publishing, New York, USA, pp 135–148

Fang Y, Dongyan Z, Fengwu BA, Lijia A (2005) The accumulation of iso camptothecin A and B in suspension cell cultures of *Camptotheca acuminate*. Plant Cell, Tissue and Organ Cult 81:159–163

Ferreira AL, Lima-Costa ME (2006) Metabolic response to salt stress in cell suspension cultures of sensitive and resistant citrus. J. Hortic. Sci. Biotechnol 81(6):983–988

Gomez KA, Gomez AA (1984) Statistical procedures for agricultural research. Wiley, New York, p 657

Greenway H, Munns R (1980) Mechanisms of salt tolerance in nonhalophytes. Ann Rev Plant Physiol 31:149–190

Harrington M, Aim DM (1988) Interaction of heat shock and salt shock in cultured tobacco cells. Plant Physiol 88:618–625

Hatamnia AA, Abbaspour N, Darvishzadeh R, Rahmani F, Heidari R (2013) Tobacco responds to salt stress by increased activity of antioxidant enzymes. Iran J Plant Physiol 3(4):801–808

Hirschi D (2004) The calcium conundrum, both versatile nutrient and Specific signal. Plant Physiol 136:2438–2442

Jamalian S, Gholami M, Esna-Ashari M (2013) Abscisic acid—mediated leaf phenolic compounds, plant growth and yield is strawberry under different salt stress regimes. Theor Exp Plant Physiol 25:291–299

Javid A, Ghafoor A, Anwar R (2004) Seed storage protein electrophoresis in groundnut for evaluating genetic diversity. Pak J Bot 36(1):25–29

Kho CW, Park SG, Lee DH, Cho S, Oh GT, Kang S, Park BC (2004) Activity staining of glutathione peroxidase after two-dimensional gel electrophoresis. Mol Cell 18:369–373

Kivilompolo M, Oburka V, Hyotylainen T (2007) Comparison of GC-MS and LC-MS methods for the analysis of antioxidant phenolic acids in herbs. Anal Bio Anal Chem 388:881–887

Koca M, Bor M, Ozdemir F, Turkan I (2007) The effect of salt stress on lipid peroxidation, antioxidative enzymes and proline content on Sesame cultivars. Environ Exp Bot 60:344–351

Ksouri R, Megdiche W, Debez A, Falleh H, Grignon C, Abdelly C (2007) Salinity effects on polyphenol content and antioxidant activities in leaves of the halophyte *Cakile maritima*. Plant Physiol Biochem 45:244–249

Kumar GS, Reddy MA, Sudhakar C (2003) NaCl effects on proline metabolism in two high yielding genotypes of mulberry (*Morus alba* L.) with contrasting salt tolerance. Plant Sci 165:1245–1251

Kumar N, Pamidimarri S, Kaur M, Boricha G, Reddy M (2008) Effects of NaCl on growth, ion accumulation, protein, proline contents and antioxidant enzymes activity in callus cultures of *Jatropha curcas*. Biologia 63(3):378–382

Laemmli UK (1970) Cleavage of structural protein during the assembly of the head of Bacteriophage T4. Nature 227:680–685

Larkindale J, Huang B (2004) Thermo tolerance and antioxidant systems in *Agrostis stolonifera*: involvement of salicylic acid, abscisic acid, calcium, hydrogen peroxide and ethylene. J Plant Physiol 161:405–413

Li DM, Nie YX, Zhang J, Yin JS, Li Q, Wang XJ, Bai JG (2013) Ferulic acid pretreatment enhances dehydration-stress tolerance of cucumber seedlings. Biol Plant 57:711–717

Maggio A, Reddy MP, Joly RJ (2000) Leaf gas exchange and solute accumulation in the halophyte *Salvadora persica* grown at moderate salinity. Environ Exp Bot 44:31–38

Martin-Tanguy J (2001) Metabolism and function of polyamine in plants. Plant Growth Regul 34:135–148

Meloni DA, Gulotta MR, Martinez CA, Olive MA (2004) The effect of salt stress on growth, nitrate reduction and proline and glycinebetaine accumulation in *Prosopis alba*. Braz J Plant Physiol 16:39–46

Millogo-Kone H, Guissou IP, Nacoulma O, Traore AS (2008) Comparative study of leaf and stem bark extracts of *Parkia biglobosa* against enterobacteria. Afr J Trad CAM 5:238–243

Mittler R (2002) Oxidative stress, antioxidants and stress tolerance. Trends in Plant Sci 7:405–410

Modarresi M, Moradian F, Nematzadeh GA (2014) Antioxidant responses of halophyte plant *Aeluro pulittoralis* under long term salinity stress. Biologia 69(4):478–483

Moghaddasian B, Eradatmand AD, Alaghemand A (2013) Simultaneous determination of rutin and quercetin in different parts of *Cappariss pinose*. Bull Env Pharmacol Life Sci 2:35–38

Murashige T, Skoog F (1962) A revised medium for rapid growth and bioassays with Tobacco tissue cultures. Physiol Plant 15:473–497

Nader BA, Karim BH, Ahmed D, Claude G, Chedly A (2005) Physiological and antioxidant responses of the perennial halophyte *Crithmum maritimum* to Salinity. Plant Sci 168:889–899

Naik HNK, Devaraj VR (2016) Effect of salinity stress on antioxidant defense system of niger (*Guizotia abyssinica* Cass.). Am J Plant Sci 7:980–990

Niknam V, Razavi N, Ebrahimzadeh H, Sharifizadeh B (2006) Effect of NaCl on biomass, protein and proline contents, and antioxidant enzymes in seedling and calli of two *Trigonella* Species. Biol Plant 50:591–596

Padgett PE, Leonard RT (1994) An amino from maize cell line for use in investigating. Nitrate assimilation. Plant Cell Rep 13:504–509

Parida AK, Das AB, Sanada Y, Mohanty P (2004) Effects of salinity on biochemical components of the mangrove, *Aegiceras corniculatum*. Aqua Bot 80:77–87

Petridis A, Therios I, Samouris G (2012) Genotypic variation of total phenol and oleuropein concentration and antioxidant activity of 11 Greek olive cultivars (*Olea europaea* L.). Hort Sci 47(3):339–342

Pourmorad F, Hosseinimehr SJ, Shahabimajd N (2006) Antioxidant activity, phenol and flavonoid contents of some selected Iranian medicinal plants. Afr J Biotech 5(11):1142–1145

Qasim M, Ashraf M, Ashraf MY, Rehman SU, Rha ES (2003) Salt-induced changes in two canola cultivars differing in salt tolerance. Biologia Plantrum 46:629–632

Ramanjulu S, Sudhakar C (2001) Alleviation of NaCl salinity stress by calcium is partly related to the increased proline accumulation in mulberry (*Morus alba* L.) callus. J Plant Biol 28:203–206

Rao A, Ahmad SD, Sabir SM, Awan SI, Shah AH, Abbas SR, Shafique S, Khan F, Chaudhary A (2013) Potential antioxidant activities improve salt tolerance in ten varieties of wheat (*Triticum aestivum* L.). Am J Plant Sci 4:69–76

Roberts MF (1988) Isoquinolines (Papaver alkaloids). In: Constabel F, Vasil IK (eds) Cell culture and somatic cell genetics of plants. Academic Press, San Diego, pp 315–334

Romaric GB, Fatoumata AL, Oumou HK, Mamounata D, Imael HNB, Mamoudou HD (2011) Phenolic compounds and antioxidant activities in some fruits and vegetables from Burkina Faso. Afr J Biotechnol 10:13543–13547

Schneider S (2014) Quality analysis of virgin olive oils—Part 6, Agilent Technologies Application Note, publication number 5991-3801EN 1-6

Seal T (2016) Quantitative HPLC analysis of phenolic acids, flavonoids and ascorbic acid in four different solvent extracts of two wild edible leaves, *Sonchus arvensis* and *Oenanthe linearis* of North-Eastern region in India. J Appl Pharm Sci 6(2):157–166

Sen A, Alikamanoglu A (2011) Effect of salt stress on growth parameters and antioxidant enzymes of different Wheat (*Triticum aestivum* L.) varieties on *in vitro* tissue culture. Fresen Environ Bull 20(2):489–495

Shahid MA, Pervez MA, Balal M, Mattson NS, Rashid A, Ahmad R, Ayub CM, Abbas T (2011) Brassinosteroid (24-epibrassinolide) enhances growth and alleviates the deleterious effects induced by salt stress in pea (*Pisum sativum* L.). Austr J Crop Sci 5(5):500–510

Shibli RA, Kushad M, Yousef GG, Lila MA (2007) Physiological and biochemical responses of tomato microshoots to induced salinity stress with associated ethylene accumulation. Plant Growth Regul 51:159–169

Singh NK, Bracken CA, Hasegawa PM, Handa AK, Buckel S, Hermodson MA, Pfankoch F, Regnier FE, Bressan RA (1987) Characterization of osmotin: a thaumatin-like protein associated with osmotic adjustment in plant cells. Plant Physiol 85:529–536

Singh PK, Shahi SK, Singh AP (2015) Effects of salt stress on physicochemical changes in maize (*Zea mays* L.) plants in response to salicylic acid. Indian J Plant Sci 4(1):69–77

Singleton VL, Ross JA (1965) Colorimetry of total phenolics with phosphomolybdic-phosphotungstic acid reagents. Am J Enol Viti Cult 16(3):144–158

Sousa MF, Campos FAP, Prisco JT, Enéas-Filho J, Gomes-Filho E (2003) Growth and protein pattern in cowpea seedlings subjected to salinity. Biol Plant 47:341–346

Sreenivasulu N, Ramanjulu S, Ramachandra-Kini K, Prakash HS, Shekar-Shetty H, Savithri HS, Sudhakar C (1999) Total peroxidase activity and peroxidase isoforms as modified by salt stress in two cultivars of fox tail millet with differential salt tolerance. Plant Sci 141:1–9

Telesiński A, Nowak J, Smolik B, Dubowska A, Skrzypiec N (2008) Effect of soil salinity on activity of antioxidant enzymes and content of ascorbic acid and phenols in bean (*Pheseolus vulgaris* L.) plants. J Elementol 13:401–409

Verpoorte R, Kraemer KH, Schenkel EP (2002) *Ilex paraguariensis* cell suspension culture characterization and response against ethanol. Plant Cell, Tissue Organ Cult 68:257–263

Wahid A, Ghazanfar A (2006) Possible involvement of some secondary metabolites in salt tolerance of sugarcane. J Plant Physiol 163:723–730

Wakabayashi K, Hoson T, Kamisaka S (1997) Osmotic stress suppresses cell wall stiffening and the increase in cell wall bound ferulic and diferulic acids in wheat coleoptiles. Plant Physiol 113:967–973

Weisanly W, Sohrabi Y, Heidari G, Siosemardeh A, Ghassemi-Golezani K (2012) Changes in antioxidant enzymes activity and plant performance by salinity stress and zinc application in soybean (*Glycine max* L.). Plant Omic J 5(2):60–67

Weisskopf L, Tomasi N, Santelia D, Martinoia E, Langlade NB, Tabacchi R, Abou-Mansourm E (2006) Isoflavonoid exudation from white lupin roots is influenced by phosphate supply, root type and cluster-root stage. New Phytology 171:657–668

Xuan TD, Minh LT, Khang DT, Ha PT, Tuyen PT, Minh TN, Quan NV (2016) Effects of salinity stress on growth and phenolics of rice (*Oryza sativa* L). Int Lett Nat Sci 57:1–10

Yang Y, Wei X, Shi R, Fan Q, An L (2010) Salinity-induced physiological modification in the callus from halophytes *Nitraria tangutorum* Bobr. J Plant Growth Regul 29:465–476

Yildiz M, Terzi H (2013) Effect of NaCl stress on chlorophyll biosynthesis, proline, lipid peroxidation and anti-oxidative enzymes in leaves of salt tolerant and salt sensitive barley cultivar. J Agri Sci 19:79–88

Yun BW, Huh GH, Lee HS, Kwon SY, Jo JK, Kim JS, Cho KY, Kwak S (2002) Differential resistance to methyl vio-logen in transgenic tobacco. J Plant Physiol 156:504–509

Zhang M, Fang YJ, Jiang Z, Wang L (2013) Effects of salt stress on ion content, antioxidant enzymes and protein profile in different tissues of *Broussonetia papyrifera*. South Afr J Bot 85:1–9

Part III
Recent Approaches for Biotic Stress Tolerance

Chapter 9
Varietal Differences and Their Relation to Brown Rot Disease Resistance in Potato

Yasser Hamad, W. I. Shaaban, S. A. G. Youssef, and N. M. Balabel

Abstract The current chapter was carried out in an attempt to study the transmission of *Ralstonia solanacearum*, the causal agent of potato brown rot, through field weeds, survey, isolation, identification and varietal differences in their relation to brown rot disease resistance in potato under Egyptian condition. The bacterial wilt problem is a major constraint for vegetable growers, especially potato and tomato farmers in many regions of the world. The number of weed varieties as hosts for *R. solanacearum* is very great. It causes disease at least of 200 different plant species, including herbaceous plants, shrubs and trees. There is increasing evidence of hosts which under certain condition act as latent or symptomless carriers of infection. Isolates of *R. solanacearum* from weeds and other plant species were identified as race 3 biovar 2 using morphological, physiological and biochemical tests. Also advanced Immunofluorescence Antibody Stains (IFAs), Polymerase Chain Reaction (PCR) and Real-time PCR (Taq-Man) techniques were applied and proved that the isolates are related to race 3 biovar 2 of R. *solanacearum*. Potato cultivars showed various levels of disease resistance *against R. solanacearum.* Spunta and Pyrnadet cvs. were more sensitive; Nicola cv. was moderate sensitive; Mondial cv. was moderate resistant, while Lady Roasita was resistant to the pathogen. Plant extracts from *Corchorus olitorius, Solanum. nigrum, Ricinus communis* and *Portulaca oleracea* appeared to be more effective against *R. solanacearum.*

Keywords Potato · *Ralstonia solanacearum* · Hosts · Immunofluorescence Antibody Stains · Polymerase chain reaction · Real-time PCR · Potato cultivars · Plant extracts

Y. Hamad (✉) · N. M. Balabel
General Administration of Plant Quarantine, Ministry of Agriculture, 12611 Giza, Egypt

W. I. Shaaban · S. A. G. Youssef
Agriculture Botany Department, Faculty of Agriculture, Suez Canal
University, 41522, Ismailia, Egypt

© Springer Nature Switzerland AG 2021
H. Awaad et al. (eds.), *Mitigating Environmental Stresses for Agricultural Sustainability in Egypt*, Springer Water,
https://doi.org/10.1007/978-3-030-64323-2_9

9.1 Introduction

Potato (*Solanum tuberosum* L.) is an important vegetable cash crop in Egypt. It is considered an important crop for exportation. Egypt is the largest potato producer in Africa, exporting annually more than 753,000 tons, during season 2016/2017 (Anonymous 2017). Potato plants in fields are attacked by a wide range of fungal, bacterial and viral diseases that cause serious losses in quantity and quality of potato tubers. Potato brown rot, caused by *R. solanacearum* (Yabuuchi et al. 1995) (Syn. *P. solanacearum*), has been reported in Egypt from many years ago (Briton-Jones 1925). The potato brown disease caused many quarantine problems during the export of table potatoes to Europe (Farag 2000).

Bacterial wilt caused by *Ralstonia solanacearum* (Yabuuchi et al. 1995) is an important soil-borne disease that affects potato production with a significant loss about 15% (Zehr 1969). Yield losses are mostly affected by tuber rotting. If weather environments are favorable for the development of the disease, the yield might decline by 50% (Anonymous 2015). The disease has been assessed to affect about 1.7 million hectares in nearly 80 countries worldwide, with global injury estimates of over USD 950 million per annum (Champoiseau et al. 2009). Direct yield losses triggered by *R. solanacearum* may vary FROM 33 to 90% dependent on the diverse reasons such as cultivar, climate, soil type, cropping pattern, strain of the bacteria etc. (Karim et al. 2018). Isolates of *R. solanacearum* in Egypt were identified as race 3 biovar 2 using morphological, physiological and biochemical tests. Also, advanced Immunofluorescence Antibody Stains (IFAs), Polymerase Chain Reaction (PCR) and Real-time PCR (Taq-Man) techniques were applied and proved that the isolates are related to race 3 biovar 2 of *R. solanacearum* (Hamad 2016).

High levels of varietal differences to brown rot disease resistance in potato were registered by many investigators of them Fahmy and Mohamed (1990), El-Didamony et al. (2003) and El Halag (2008) and Muthoni et al. (2014).

Many studies reported the antibacterial activity of plant extracts against pathogens (Gottlieb et al. 2002; Slusarenko et al. 2008; Guleria and Tiku 2009; Hamad 2016). Furthermore, Malafaia et al (2018) reported that the usage of antimicrobial agents that affect only the planktonic bacteria, an alternate method to the control of plant diseases. Plant extracts used for medicinal as *Croton heliotropiifolius*, *Eugenia brejoensis*, and *Libidibia ferrea* revealed to be favorable alternatives about the control of *R. solanacearum*. Therefore, this chapter has been prepared to focus light on the survey and identification of the pathogen. The importance of varietal differences in the resistance to *R. solanacearum* in potatoes, and the role of plant extracts in resistance to the pathogen.

9.2   Survey the Causal Organism of Potato Brown Rot Disease *Ralstonia solanacearum* in Weeds

Ralstonia solanacearum (Yabuuchi et al. 1995) is known to infect a wide range of economically important cash crops such as potato (*Solanum tuberosum* L.), tomato (*Lycopersicon esculentum* L.), eggplant (*Solanum melongena* L.), pepper (*Capsicum annuum* L.), tobacco (*Nicotiana tabacum* L), ginger (*Zingiber officinale* L.), banana (*Musae* sp), groundnut (*Arachis hypogea* L.) and other crops (Robinson-Smith et al. 1995; Atia et al. 2010) and many weeds, such as *Rumex dentatus* L. (sorrel) and *Solanum nigrum* L. (black nightshade) (Farag et al. 2004), *Portulaca oleracea* L. (purslane) (Elphinstone et al. 1998), *Polygonum capitata* and *Drymaria cordata* (Pradhanang et al 2000), *Dopatrium* sp. and *Monochoria vaginalis* (Pradhanang and Momol 2001) and many new plant species which have been identified as hosts. EL-Didamony et al. (2003) survied *R. solanacearum* race 3 biovar 2 from potato tubers cultivated in EL-Dakahlia and EL-Sharkia governorates, Egypt. They also surveyed the pathogen from aubergine, tomato and pepper. Whereas, Balabel (2006) collected different samples from potato fields at El-Minufiya governorate. Found that, the sever isolates were recovered from potato tubers and weeds while, soil, water and potato stem isolates were moderate in this regard. *Rumex dentatus* L. (sorrel) and *Solanum nigrum* L. (black nightshade) were found as alternative hosts for *R. solanacearum* race 3 biovar 2 in Egypt. Hamad (2008) collected potato samples, and indicated that, twelve samples of potato tubers showed positive reaction for *R. solanacearum* from total number of 50 samples, under El-Ismailia governorate. Three samples were positive towards *R. solanacearum* from total sample of 35 of different potato cultivars under El-Sharkia governorate. Spunta cultivar recorded the highest infection in both governorates. Survey results, showed that the infected potato in El-Ismailia were higher than in El-Sharkia. Infected weeds as *Solanum nigrum, Portulaca oleracea, Rumex dentatus, Arachis hypogea, Beta vulgaris* sub sp. martima, *Chenopodium album* and *Brassica kaber* are considered harbor and hosts for *R. solanacearum*, under both El-Sharikia and El-Ismailia governorates conditions.

Hamad et al. (2016) survey and isolation of the pathogen *R. solanacearum* from different potato fields in Egypt (El-Gharbia, El-Behera, El-Menoufia and El-Ismailia governorates) including different habitats. El-Ismailia governorate in pest-free areas (P.F.A.). It's important to note that, positive samples does not increase with through growing seasons form 2010/2011 to 2012/2013 at El-Gharbia, El-Behera, El-Menoufia and El-Ismailia governorates. Integrated pest management and applied the rules of plant quarantine are considered the main reasons to limiting infected plants.

9.3   Isolation of *R. solanacearum* from Different Potato, Weeds and Irrigation Water

Hamad (2008) isolated strains of *R. solanacearum* from potato cultivars and weed species during 2002–2005 at El-Sharikia and El-Ismailia governorates conditions. Whereas, Kago et al. (2013) isolated and cultured *R. solanacearum* from infected potato tubers, stems of capsicum and tomato collected from major solanaceous plant growing areas in Nairobi, Kenya. Isolation of the pathogen was made from different sources by plating on SMSA medium. All tested isolates were short-rod, non-spore formed and gram-negative, regarding cultural characteristics, colonies on NA medium were irregularly-round, convex with smooth surface, translucence and brownish in color. Meantime, colonies of isolates were fluidal white with pink center on SMSA or on TZC medium and whitish gray in color on King's B medium. Elphinstone et al. (1996) and Karim et al. (2018) came to similar findings. Furthermore, Hamad (2016) identified morphological characterization of *R. solanacearum* on SMSA medium. Isolated bacteria showed milky white, irregular and fluidal with blood red coloration in the center was identified as virulent (typical) colonies of *R. solanacearum*. On the other hand, bacterium forms developed less fluidal colonies which are completely pink to red on that medium identified virulent types (Fig. 9.1).

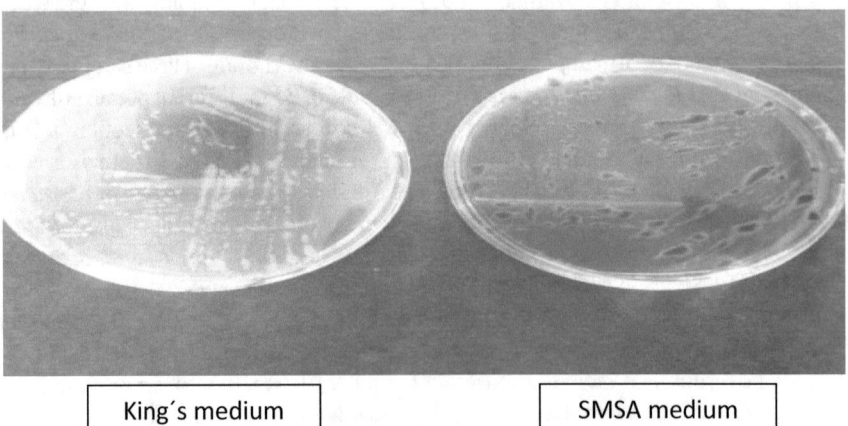

King's medium   SMSA medium

Fig. 9.1 Typical colonies of *R. solanacearum* are whitish gray in color on King's B medium and milky white, irregular and fluidal with blood red coloration in the center on Semi Selective Media of South Africa medium (Hamad 2016)

9.4   Identification of *R. solanacearum*

9.4.1   Traditional Methods

From the phytopathological point of view, the bacterium is found in nature into two forms: virulent (vi) and avirulent (av) or typical and atypical forms (Balabel 2006). Both forms may be recovered from diseased tissues on isolation, though the interrelation between them is not yet well understood.

9.4.1.1   Race and Biovar Determination of the *Ralstonia solanacearum*

Based on host range, five races were characterized according to He et al. (1983) as follows:

Race 1: Attacks solanaceous plants.
Race 2: Attacks musaceous plants.
Race 3: Attacks potatoes only.
Race 4: Attacks ginger.
Race 5: Attacks mulberry.

In this regard, five races were documented based on host range studies (He et al. 1983). Heather (2005) divided the pathogen *R. solanacearum* into biovars based on the utilization of the disaccharides cellobiose, lactose, and maltose and oxidation of the hexose alcohols dulcitol, mannitol, and sorbitol. Biovars classification of *R. solanacearum* is determined based on some biochemical properties i.e., their ability to produce acids from hexose alcohol and sugars (Hayward 1991), and the following biovars were identified as shown in Table 9.1.

Hamad et al. (2016) made identification and race determination of the most aggressiveness isolates (twelve isolates from different sources) based on pathological and bacteriological tests. They showed that physiological and bacteriological characteristics were found similar to those described for Race 3, biovar II of *R. solanacearum*. The biovars determination was made according to their ability to produce acids

Table 9.1 Identification of *R. solanacearum* biovars (Hamad 2016)

| Biovar | 1 | 2 | 3 | 4 | 5 |
Sugars test					
Mannitol	−	−	+	+	+
Sorbitol	−	−	+	+	−
Dulcitol	−	−	+	+	−
Maltose	−	+	+	−	+
Lactose	−	+	+	−	+
Cellobiose	−	+	+	−	+

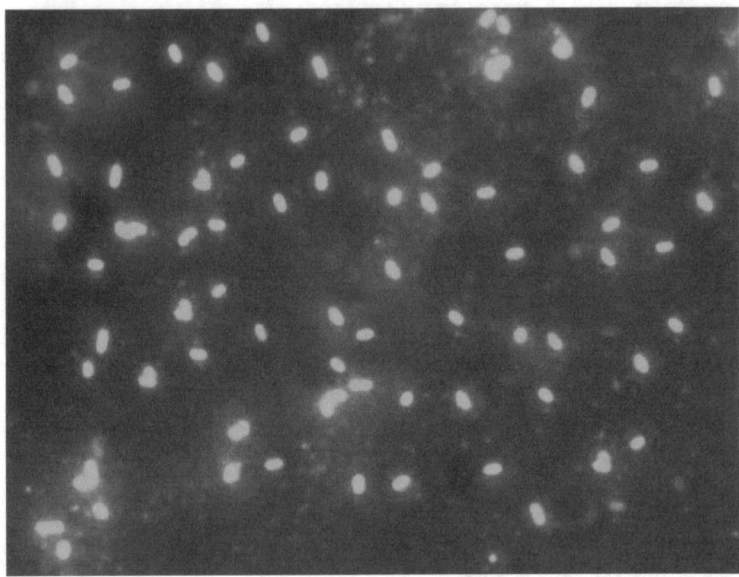

Fig. 9.2 Cell morphology of *R. solanacearum* in the serological immuno-fluorescent antibody staining (IFAS) test (Hamad 2016)

from hexose and alcohol sugars. The twelve isolates subject to identification were able to produce acids from maltose, lactose, and cellibiose. All isolates, however, were unable to produce acids from mannitol, sorbitol and dulcitol denoting that, the dominant race in Egypt is race 3, biovar II.

Immunofluorescent antibody staining test (IFAS) method is usually used in conjunction with the tomato bioassay test, Janse (1988) used a IFAS and obtained high levels of sensitivity for detection of *R. solanacearum* in potato tissue.). The detection level by the polyclonal antibodies in IFAS is 10^4 cells ml^{-1} and the false positive reaction due to cross-reacting bacteria was limited to only 2.3% using two polyclonal antisera against whole cells of *R. solanacearum*. Latent infections in seed potato tubers (Ciampi and Sequeira 1980) have led to the spread of the organism, both locally in Egypt and in other different countries. The effective control of brown rot is dependent on the reliability of detection of the pathogen. Serological techniques such as IFAS microscopy have been described for identification *R. solanacearum*. Hamad (2016) developed colonies on SMSA medium and tested by IFAS. The cells showed short rod-shaped morphology stained evenly as bright green fluorescent (Fig. 9.2).

9.4.2 Molecular Biology Techniques

Based on DNA sequence, Prior and Fegan (2005) applied phylogenetic analysis of sequence results produced from the 16S–23S internal transcribed spacer region, the *eg1* gene and the *mutS* gene for classification of *R. solanacearum*. Four phylo-types were distinguished corresponding with their geographic origins. The strains of bacteria within each phylotype originate from one location. Each phylotype can be more subdivided into sequevars based on variances in the sequence of a portion of the endoglucanase (*eg1*) gene (http://aem.asm.org/content/73/21/6790.full.pdf). They showed that race 3 biovar 2 (phylotype II, sequevar1) is pathogenic to potato.

9.4.2.1 PCR Amplification

Polymerase chain reaction (PCR) method has been applied to identify *R. solanacearum* strains (Balabel Balabel 2006; Ha et al. 2012). The advantage of PCR is the possible for discovery of a very small numbers of bacteria in the environment. However, detection of *R. solanacearum* from soil with PCR is quite problem. One reason for this is attributed to the property of PCR that needs DNA free from soil materials, mainly humic-materials that inhibit Taq polymerase in PCR (Tsai and Olson 1991). Another current difficult with PCR analysis is the inability to discriminate between dead cells and viable cells.

The Bio-PCR analysis developed by Schaad et al. (1995) seems to resolve the difficult of detecting viable cells of *R. solanacearum* in the soil. In this technique, an agar plating step before PCR technique offers benefit of biological amplification of the PCR target. These combinations between a selective medium for amplifying viable cells of *R. solanacearum* and PCR for amplifying specific fragments of *R. solanacearum* DNA has increased the effectiveness of the discovery of viable cells of *R. solanacearum* pathogen in the soil.

Grover et al. (2009) developed a reliable and sensitive procedure for detection of *R. solanacearum* by MDA-PCR (Multiple displacement amplification-PCR amplification). MDA-PCR procedure was performed on pure cell lysates and soil samples. Pure cell lysate and that of soil DNA was used as template in MDA reaction. The MDA amplified DNA was utilized for PCR amplification using *R. solanacearum*-specific PCR primers. MDA-PCR can detect as low as 1 (CFU ml^{-1}) of bacteria within 8 h containing DNA isolation. So, MDA followed by standard PCR facilitated the detection of pathogen from very low count samples. The detection methods play an important role in managing the brown rot disease of potato. Ha et al. (2012) distinguished *R. solanacearum* populations >1000 cells ml^{-1} by using RT-PCR in some plant and soil samples. They developed simple immunomagnetic separation (IMS) and magnetic capture hybridization (MCH) methods to purify *R. solanacearum* cells or DNA from PCR inhibitors. These tools permitted recognition the bacterium at levels >500 cells ml^{-1} in stem, tuber, and soil samples when direct RT-PCR failed, and reduced detection time from days to hours.

In general, DNA amplification for pathogen offers many advantages over traditional techniques; neither purification nor cultivation of the pathogen is necessary. The specificity, sensitivity and response time of tests are improved (Poussier et al. 2002). Nevertheless, the majority of laboratories became dependent mainly on PCR techniques. However, the inhibition of the amplification reaction by compounds

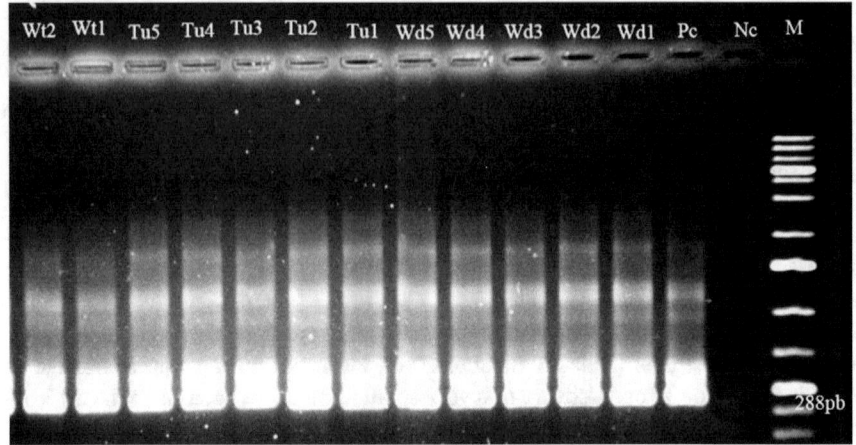

Fig. 9.3 Polymerase chain reaction (PCR) products amplified using primers OLI^{-1} and Y-2 from genomic DNA of different isolates from different habitats (Hamad 2016)

found in crude bacterial extracts might give false negative results (Farag et al. 2010) or low detection sensitivity (Picard et al. 1992). Polymerase chain reaction technique have been exploited to recognize *R. solanacearum* strains as illustrated in Fig. 9.3 in Egypt by Hamad (2016).

9.4.2.2   Real-Time PCR (Taq-Man) Assay

A fluorogenic (TaqMan) PCR assay was developed according to Weller et al. (2000) for identifying *R. solanacearum* strains. Two fluorogenic probes were used in a multiplex reaction; one broad-range probe (RS) discovered all biovars of *R. solanacearum*, and a second more specific probe (B2) identified only biovar 2. Amplification of the target was measured by the 5′ nuclease activity of *Taq* DNA polymerase on each probe, causing in emission of fluorescence. TaqMan PCR was completed with DNA extracted from *R. solanacearum* and genetically or serologically related strains to define the specificity of the assess. In pure cultures, recognition of *R. solanacearum* to $\geq 10^2$ cells ml^{-1} was succeeded. Sensitivity was reduced when TaqMan PCR was completed with inoculated potato tissue extracts, equipped by currently commended extraction processes. A third fluorogenic probe(COX), planned with the potato cytochrome oxidase gene sequence, was also developed to use as an internal PCR control and was revealed to recognize potato DNA in an RS-COX multiplex *Taq*Man PCR with infested potato tissue. The specificity and sensitivity of the assay, combined with high rapidity, robustness, reliability, and the opportunity of automating the technique, offer possible advantages in routine indexing of potato tubers and other plant genotypes for the presence of *R. solanacearum* (https://www.sigmaaldrich.com/cat alog/papers/10877778).

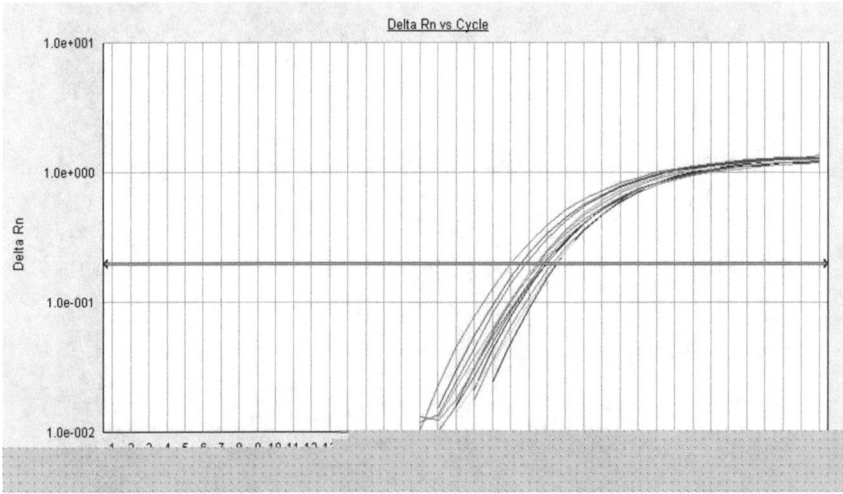

Fig. 9.4 Amplification of DNA extract of twelve selected isolates from different habitats by Real-time PCR. Negative and positive control was included. The curves exceed the threshold considered positive (Hamad et al. 2016)

Hamad et al. (2016) employed the RS primers and probe to detect all biovars and races of *R. solanacearum*. Positive results were noticed with all tested isolates indicating that, the twelve isolates are *R. solanacearum* biovar 2 race 3 (Fig. 9.4).

9.4.2.3 Phylotype Analysis of *Ralstonia solanacearum* by Multiplex- PCR

Phylotype-specific multiplex PCR analysis was applied on *R. solanacearum* by Prior and Fegan (2005) consisting of four phylotypes, each further divided into sequevars, using phylotype-specific multiplex PCR (Pmx-PCR). Based on Pmx-PCR product patterns, strains of *R. solanacearum* can be grouped into the four phylotypes. Phylotypes I (Asiatic origin) is characterized by production of 280 and 144 bp amplicons, Phylotypes II strains (American origin) gave 280 and 372 bp amplicons, Phylotypes III (mainly from Africa and nearby islands such as Reunion and Madagascar) generated 280 and 213 bp amplicons. The phylotyping scheme adds valuable evidences about the geographical origin and in some cases the pathogenicity of strains (Sagar et al. 2014).

Hamad et al. (2016) Phylotype specific multiplex PCR revealed that, all the twelve isolates of *R. solanacearum* belonged to phylotype II as 372-bp amplicon was observed with all tested isolates when phylotype multiplex-PCR (Pmx-PCR) products of these isolates were subjected to electrophoresis on 2% agarose gel (Fig. 9.5).

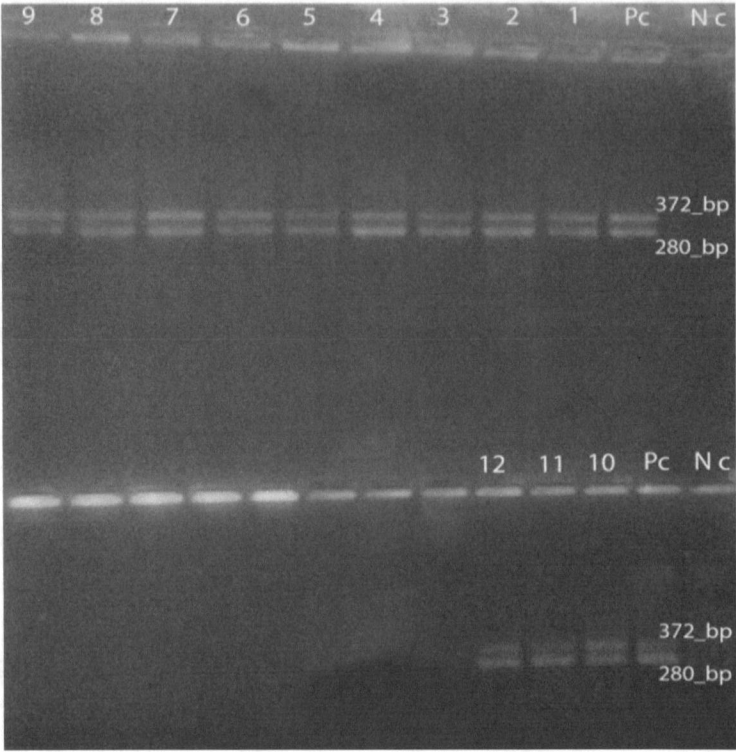

Fig. 9.5 PCR products of 12 isolates of *R. solanacearum* using different primers, 280-bp amplicon from *R. solanacearum* species and 372-bp amplicon from phylotype II strains (Hamad et al. 2016)

9.4.3 Pathogenicity Test

The bioassay with tomato seedling was revealed reliably to identify as few as 10^4 cells per ml infested potato extract (Janse 1988; Elphinstone et al. 1996; Atia et al. 2010). Tomato seedlings have also been widely used for pathogenicity testing. The typical wilting symptoms are generally apparent within a week of inoculation, based on the inoculum potential of the bacterium and the availability of optimum environmental circumstances (Elphinstone et al. 1996). Although tomato seedlings are easy to grow and are effective as indicator plants, they fail to develop wilt symptoms in case of the night temperature falls below 21 °C (Zehr 1970; Singh et al. 2014).

Hamad (2008) survey and isolation of the pathogen from different potato fields in Egypt (El-Gharbia, El-Behera, El-Menoufia and El-Ismailia governorates) including different habitats (tubers, different plant weeds and irrigation water). Several reports showed occurrence of potato brown rot bacteria in different locations of Egypt (Abd El-Ghafar et al. 1995; Elphinstone et al. 1996; Balabel 2006; Seaf Elyzal 2009).

Hamad (2016) applied pathogenicity test and showed that all tested isolates which obtained from different locations were able to infect the tomato seedlings 3–7 days after the inoculation and showing wilting symptoms under greenhouse conditions (Fig. 9.6). The highest number of the pathogenic isolates were recovered from the weed plants and potato tubers 104 and 55, respectively. Whereas, the lowest number of the pathogen was recorded from irrigation water (Table 9.2). It is interesting to note that, there was not any isolates can cause a bacterial wilt with tomato plants in Ismailia governorate.

Kumar et al. (2017) inoculation of *R. solanacearum* F1C1 on 6- to 7-day-old tomato seedlings by a simple leaf-clip strategy. They recorded a lethal pathogenic circumstance lead to kill the seedlings within a week post-inoculation. This prompted evaluating the effect of this inoculation procedure in seedlings from different cultivars of tomato. Colonization and spread of the bacteria through the diseased seedlings were proved using *gus*-tagged *R. solanacearum* F1C1.

Fig. 9.6 Pathogenicity test on tomato plants under greenhouse conditions (Hamad et al. 2016)

Table 9.2 Number of selected *R. solanacearum* isolates from different habitats (Hamad 2016)

Total	Irrigation water	Potato tubers	Weeds	Governorate
56	0	22	34	El-Gharbiather
54	1	17	36	El-Behera
52	2	16	34	El-Menoufia
0	0	0	0	El-Ismailia
162	3	55	104	Total

9.5   Varietal Differences in Relation to Brown Rot Disease Resistance

Potato varieties varied in their response to *P. solanacearum* infection. Cultivars Geno, Desiree and Alpha, were found to be susceptible, while cultivar Domina was moderately susceptible, whereas cultivar Prolina was highly resistant. Potato cultivation in clay and loamy soils resulted in an increase of disease incidence (Fahmy and Mohamed 1990).

Muthoni et al. (2014) reported that, the development of resistant cultivars is currently the best option for managing bacterial wilt; however, there are no known potato cultivars with resistance. Cultivars such as Cruza 148 and Molinera have been found to have some degree of tolerance to bacterial wilt but still transmit latent infection to their progeny tubers (French 1994). In this concern, El-Didamony et al. (2003) and El Halag (2008) reported that, when Spunta tested for its resistance to the infection by *R. solanacearum*, Spunta cultivar exhibited the highest percent of infection and disease severity. In contrast, Lady Rosetta exhibited moderate percentage of infection and disease severity, whereas Nicola cultivar achieved the lowest percentage of infection and disease severity. Furthermore, Hamad (2008) found that potato cultivars differed in their reaction to *R. solanacearum* isolate, race 3 biovar 2 and showed different degrees of sensitivity to the causal agent. Spunta cv. appeared to be very susceptible, Pernadet was susceptible. Nicola cultivar was classified as moderate susceptible, whereas Mondial was moderate resistant and Lady Roasita was resistant to *R. solanacearum*. These results may be due to the genetic variability existed in genes controlling brown rot resistance among the tested potato cultivars. Moreover, El-Didamony et al. (2003) and El Halag (2008) reported that, when Spunta tested for its resistance to the infection by *R. solanacearum*, Spunta cultivar exhibited the highest percent of infection and disease severity. In contrast, Lady Rosetta which exhibited moderate percentage of infection and disease severity, whereas Nicola cultivar achieved the lowest percentage of infection and disease severity. Moreover, thirty-six potato genotypes were planted on an infected field at the Kenya Agricultural Research Institute. Most of the potato cultivars were susceptible to bacterial wilt. Genotypes differed in resistance ranking among the three seasons. On average, the three most resistant genotypes were Kenya Karibu, Kenya Sifa and Ingabire. Eight potato genotypes (Meru, Ingabire, Kenya Karibu, Sherekea, Kihoro, Tigoni, Bishop Gitonga and Cangi) were identified to be used as promising parents for producing new crosses (Muthoni et al. 2014). Furthermore, Hamad (2016) in Egypt tested the reaction of five potato (Spunta, Hirms, Draga, Cara and Nicola cvs.) against *R. solanacearum* under the greenhouse conditions. Symptoms were recorded at 50, 60, 70, 80, 90, 100 and 110 days post sowing. Analysis of variance revealed significant increase of wilt severity with increase in age of potato cvs. (Table 9.3). He recorded different degrees of susceptibility to *R. solanacearum*. In this regard, Spunta and Hirmas cvs. were classified as sensitive to *R. solanacearum* infection, however Draga and Cara cvs., were classified as moderate sensitive and Nicola cv., appears to be

Table 9.3 The reaction of potato cvs. Spunta, Hirms, Draga, Cara and Nicola as a result of soil infested with *R. solanacearum* determined by percentage of wilt and disease severity index (DSI) (Hamad 2016)

Potato cultivars	Age (day) Characters	50	60	70	80	90	100	110	Negative control
Spunta	Wilt (%)	3.57ed	9.84d	11.29d	27.42c	51.61b	93.55a	100a	0.00
	DSI (%)	0.72	1.97	2.26	10.97	42.58	74.84	100	0
Hirms	Wilt (%)	6.52d	6.52d	7.55d	10.09d	26.42c	54.72b	100a	0.00
	DSI (%)	1.31	1.31	1.51	3.02	10.57	32.83	100	0
Draga	Wilt (%)	4.44e	6.38e	10.21de	18.36d	26.53c	50.02b	63.2a7	0.00
	DSI (%)	0.89	1.28	2.04	3.67	10.61	20.41	37.96	0
Cara	Wilt (%)	4.26f	6.12f	13.73e	19.61d	29.41c	49.02b	64.71a	0.00
	DSI (%)	0.85	1.23	2.75	3.92	11.77	19.61	38.82	0
Nicola	Wilt (%)	2.04d	3.64d	5.17d	11.11d	32.26c	43.55b	96.77a	0.00
	DSI (%)	0.41	0.73	1.03	2.26	12.9	17.42	77.42	0

moderate resistant to *R. solanacearum* infection. In this concern, motility is an important attribute for *R. solanacearum* to colonize potato roots in soil results in moving bacteria towards nutrient-rich regions is preferable (Chet and Mitchell 1976). One of the most nutrient- abundant regions of the soil is the root surface of actively growing plants which continuously secretes amino acids, sugars and other nutrients induced by microbes. The attractive force of different potato cultivar root exudates to *R. solancearum* showed that Spunta root exudates attract more number of bacteria than other cultivars because Spunta thought to have most of such nutrients and chemical signals secreted as root exudates around its root surface which attract the pathogen than in other potato cultivars. DeWeger et al. (1987) explained the important role of the chemotaxis mechanism in the colonization of microorganisms in potato roots by *R. solanacearum*. Therefore, Spunta potato cultivars may express secretion of some compounds that is considered as chemotatic signals between potato roots and the pathogen in soil which include amino acids, sugars and other nutrients induced by microbes (Chet and Mitchell 1976; Scher et al. 1985). Chemo attractive force of these exudates increase with time that increase after 40 days from potato cultivation than after 20 days and this is due to the amount of root exudates content of phenolics, organic acid and amino acid increase with time and that act as chemoattractive signals between potato roots and *R. solanacearum* (Bais et al. 2004).

9.6 Effect of Plant Extracts on *R. solanacearum* Growth in Vitro

The effectiveness of plant extracts against plant bacteria has been tested with certain degrees of success (Slusarenko et al. 2008). Control of bacterial wilt in infested soils is very difficult. It is generally considered that crop rotation with a non-host crop is of minimal value because of the wide range of crop and weed hosts of the pathogen (Hayward 1991). At the moment, no conventional bactericides are known to give operative control of *R. solanacearum*. The natural plant products derived from plant species has the ability to control diseases caused by different pathogens (Guleria and Tiku 2009). Many reports revealed that, plant metabolites and plant-based pesticides seem to be one of the best substitutes as they are identified to have insignificant environmental effect and risk to customers in difference to synthetic pesticides (Gottlieb et al. 2002). The usage of plant extracts and phytochemicals has significant results. Several studies have been investigated worldwide to prove antimicrobial activities from medicinal plants of them Alonso-Paz et al. (1995). Abdel-Sattar et al. (2008) isolated two furostane type saponin glycosides, protodioscin and pseudoprotodioscin from the unripe berries of *Solanum intrusum* (Soria). The constructions of both compounds were clarified using MALDI-TOF mass spectroscopy and one and two dimensional NMR procedures. Protodioscin and related glycosides are stated to have cytotoxic activity. Abo-Elyousr and Asran (2009) tested antibacterial activity of extract from datura, garlic and nerium in controlling *R. solanacearum* in vitro and in vivo. Garlic exhibited the strongest antibacterial activity against bacterial wilt in vitro and in vivo followed by datura and then nerium. Cold-water extracts of these plant species were more operative than hot water extract in the development of the disease in vivo.

Alemu et al. (2013) evaluated the antibacterial activities of aqueous and solvent (acetone and methanol) extracts from *Eichhorina crassipes, Mimosa diplotricha, Lantana camara* and *Prosopis juliflora* against *R. solanacearum*. In vitro antibacterial test was made in disc diffusion sensitivity test. Aqueous extract of *E. crassipes* gave the highest inhibition zone (26 mm), followed by *M. diplotricha* (14 mm). While, Seyed et al. (2015) reported that, the extracted hydro alcoholic of *P. oleracea* leave and seed had antibacterial effects on selected drug resistant bacterial strains. The leaves and seeds extract of *P. oleracea* has antibacterial effect and it can be a good alternative when we are faced with drug resistance bacteria. Furthermore, Hamad (2016) study the effect of some plant extracts on *R. solanacearum* growth in vitro. Results indicated that, crude of plant extracts from *C. olitorius, S. nigrum, P. oleracea* and *R. communis* gave inhibition growth of *R. solanacearum*. Significant differences were registered between the studied three isolates under different concentrations of solvents and for various plant extracts. Ethyl acetat extract had an in vitro potential antimicrobial activity against *R. solanacearum* as illustrated in Fig. 9.7 and Table 9.4. Moreovere, Malafaia et al. (2018) reported that the usage of antimicrobial agents that affect only the planktonic bacteria, an alternate method to the control of plant diseases. Plant extracts used for medicinal as *Croton heliotropiifolius, Eugenia brejoensis,*

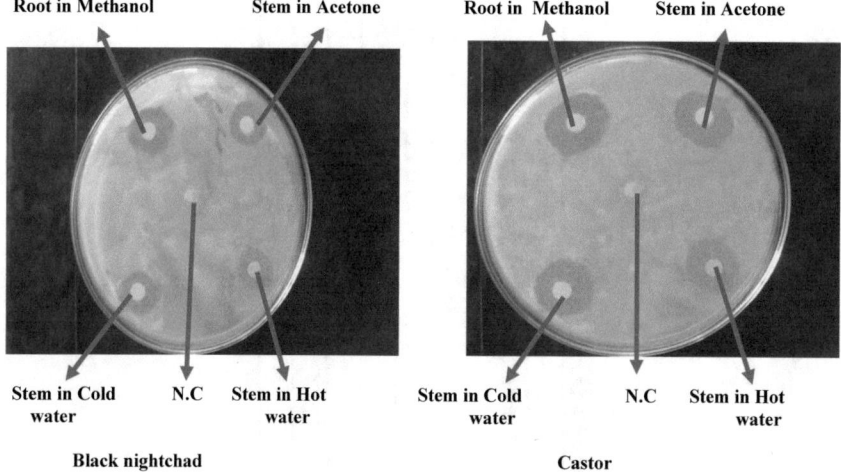

Fig. 9.7 Inhibition effect of different plant extracts against growth of *R. solanacearum* (mm) (Hamad 2016)

and *Libidibia ferrea* revealed to be favorable alternatives about the control of *R. solanacearum*.

9.7   Conclusions

Identification and race determination based on the physiological and bacteriological characteristics and PCR technique play an important role in the diagnosis of pathogens. These techniques proved that the pathogen is belong to race 3 biovar 2 of *R. solanacearum*. High degree of genetic variability has been recorded between potato cultivars in their reaction to *R. solanacearum*. Crud of plant extracts from *Corchorus olitorius*, *Solanum nigrum*, *Portulica oleracea* and *Ricinus communis* extracted were found to be more effective against the pathogen.

9.8   Recommendations

It is important to reduce the incidence of brown rot under Egyptian conditions, taking into account the following recommendations:

1. Detect the causal organism of potato brown rot bacterium in different hosts.
2. Study the cultivar reaction of potato brown rot.
3. Determine the effect of plant extracts on *R. solanacearum*.
4. Cultivate high-resistance varieties of the disease to reduce the causative damage
5. Integrated pest management and applied the rules of plant quarantine.

Table 9.4 Inhibition effect of different plant extracts on the growth of *R. solanacearum* in vitro (mm) (Hamad 2016)

Plant materials/inhibition zone (mm)

Black nightshade (*Solanum nigrum*)			Purslane (*Portulaca oleracea*)			Jews, mallow (*Corchorus olitorius*)			Castor (*Ricinus communis*)			Solvents
Stems	Roots	Seeds	Stems	Roots	Seeds	Stems	Roots	Seeds	Stems	Roots	Seeds	
17.0	–	–	15.0	–	8.0	–	–	–	13.0	9.0	–	Water extraction
–	11.0	–	–	–	8.5	10.0	29.3	–	14.2	–	–	Cold water
–	7.0	7.0	–	–	8.1	–	–	–	7.0	10.7	–	Hot water
–	6.0	7.0				–	13.0	–	7.0	–	–	Methanol extraction
												Acetone extraction

References

Abd El-Ghafar NY, Wafaa M, Abd El-Sayed SM, Abd El-Allah SM (1995) Identification of biovars and races of *Burkholderia* (*Pseudomonas*) *solanacearum* and their pathogenicity on tomato cultivars in Egypt. Egypt J Phytopathol 23:79–88

Abdel-Sattar E, Marwan MS, Sahar E (2008) Protodioscin and pseudoprotodioscin from *Solanum intrusum*. Res J Phytochem 2:100–105

Abo-Elyousr KA, Asran MR (2009) Antibacterial activity of certain plant extracts against bacterial wilt of tomato. Arch Phytopathol Plant Protect 42:573–578

Alonso-Paz E, Cerdeiras MP, Fernandez J, Ferreira F, Moyna P, Soubes M, Vazquez A, Veros S, Zunno L (1995) Screening of Uruguayan medicinal plants for antimicrobial activity. J Ethanopharm 45:67–70

Anonymous (2015) Brown rot disease mars export prospects of Indian potatoes.https://www.the hindubusinessline.com/economy/agri-business/brown-rot-disease-mars-export-prospects-of-ind ian-potatoes/article7185569.ece

Anonymous (2017) Ministry of Agriculture and Land Reclamation, Central Administration of Agriculture Economic and Statistics, Egypt

Atia MM, Tohamy MRA, Faiza FG, Hanna AM (2010) Integrated management of potato brown rot. In 6th International Conference of Sustainable Agricultural Development Faculty of Agriculture, Fayoum University, 27–29 December 2010, pp 275–294

Bais H, Park S, Weir T (2004) How plants communicate using the underground information superhighway. Trends Plant Sci 9:26–32

Balabel NM (2006) Persistence of *Rolstonia solanacearum* (Syn. *Pseudomonas solanacearum*) in different habitats in Egypt. Ph D Thesis, Microbiology Department, Faculty of Agriculture, Ain Shams University

Briton-Jones HR (1925) Mycological work in Egypt during the period 1920–1922. Egypt Min Agric Tech Sci Serv Bull 49:129

Champoiseau PG, Jones JB, Allen C (2009) Ralstonia solanacearum race 3 biovar 2 causes tropical losses and temperate anxieties [Online]. Madison: American Phytopathological Society. Available at http://www.apsnet.org/online/feature/ralstonia/. Accessed 25 June 2010

Chet I, Mitchell R (1976) Ecological aspects of microbial chemotactic behaviour. Ann Rev Microbiol 30:221–239

Ciampi L, Sequeira L (1980) Influence of temperature on virulence of race 3 strains of *Pseudomonas solanacearum*. Am Potato J 57:307–317

Derib A, Fikre L, Mulatu W, Gezahegn B (2013) Antibacterial activity of some invasive. *Alien* species extracts against tomato (*Lycopersicon esculentum* Mill) bacterial wilt caused by *Ralstonia solanacearum* (Smith). Plant Patholo J 12:61–70

Deweger L, Vander V, Wijfjes A, Bakker P, Schippers B, Lugternberg B (1987) Flagella of a plant growth-stimulating *Pseudomonas fluorescence* strain are required for colonisation of potato roots. J Bacteriol 169:2769–2773

El Halag KM (2008) Studies on the interaction between potato brown rot bacterium and root exudates in certain crops (MSc Thesis). Department of Botany, Faculty of Science, Benha University, pp 96–98

El-Didamony G, Ismail AEA, Sarhan M, Abdel-Azez SS (2003) Idenification and pathogenicity of bacterial wilt of potatoes in some types of Egyptian soil. Egypt J Microbiol 38(1):89–103

Elphinstone JG, Hennessy J, Wilson JK, Stead DE (1996) Sensitivity of different methods for the detection of *Ralstonia solanacearum* in potato tuber extracts. EPPO Bull/Bull OEPP 26:663–678

Elphinstone JG, Stanford HM, Stead DE (1998) Survival and transmission of *R. solanacearum* in aquatic plants of *Solanum dulcamara* and associated surface water in England. Bull OEPP/EPPO Bull 28:93–94

Fahmy FG, Mohamed MS (1990) Some factors affecting the incidence of potato brown rot. Assuit J Agric Sci 21:221–230

Farag NS (2000) Spotlights on potato brown rot in Egypt. In Proceeding of 9th Congress of the Egypt Phytopathology Society, May, pp 405–408

Farag NS, Eweda WE, Mostafa MI, Balabel Naglaa M (2004) Preliminary observations on the bacteriology and pathology of *Ralstonia solanacearum*. Egypt J Agric Res 82(4):1519–1523

Farag NS, Gomah AA, Balabel Naglaa M (2010) False negative multiplex PCR results with certain groups of antibiotics. Plant Pathol J 9(2):73–78

French ER (1994) Strategies for integrated control of bacterial wilt of potatoes. In: Hayward AC, Hartman GL (eds) Bacterial wilt: the disease and its causative agent, *Pseudomonas solanacearum*. CAB International, UK, pp 98–113

Gottlieb OR, Borin MR, Brito NR (2002) Integration of ethnobotany and phytochemistry: dream or reality? Phytochemistry 60:145–152

Grover A, Azmi W, Paul Khurana SM, Chakrabarti SK (2009) Multiple displacement amplification as a pre-polymerase chain reaction (pre-PCR) to detect ultra low population of *Ralstonia solanacearum* (Smith1896) Yabuchi *et al.* (1996). Lett Appl Microbiol 49(5):539–543

Guleria S, Tiku AK (2009) Integrated pest management: innovation-development process. Springer, Netherlands. Chapter 12, Botanicals in pest Management: Current Status and Future, Perspectives, pp 317–329

Ha Y, Kim JS, Timothy PD, Mark AS (2012) A rapid, sensitive assay for *Ralstonia solanacearum* race 3 biovar 2 in plant and soil samples using magnetic beads and real time PCR. Phytopathology 96(2):258–264

Hamad YI (2008) Studies on the transmission of potato brown rot causal organism through weeds in the Egyptian fields M.Sc. Thesis, Plant Pathology Department, Faculty of Agriculture, Zagazig University, Egypt

Hamad YI (2016) Pathological studies on potato brown rot under Egyptian conditions. PhD Plant Pathology, Agric Botany Department, Faculty of Agriculture, Suez Canal University, Egypt

Hamad YI, Balabel MN, Youssef SAG, Shaaban WI, Ahmed MI (2016) Detection the causal organism of potato brown rot bacterium in potato and water canals. Egypt J Appl Sci 31(6):90–109

Hayward AC (1991) Biology and epidemiology of bacterial wilt caused by *Pseudomonas solanacearum*. Ann Rev Phytopathol 29:65–87

He LY, Sequeira L, Kelman A (1983) Characteristics of *Pseudomonas solanacearum* from China. Plant Dis 67:1357–1361

Heather A (2005) *Ralstonia solanacearum*: requirement for soil borne plant pathogens. Springer, p 728

Janse JD (1988) A detection method for *Pseudomonas solanacearum* in symptomless potato tubers and some data on its sensitivity and specificity. Bull OEPP/EPPO Bull 18:343–351

Kago EK, Kinyua ZM, Okemo PO, Maingi JM (2013) Efficacy of brassica tissue and chalim TM on control of plant parasitic nematodes. J Biol 1(1):32–38

Karim Z, Hossain MS, Begum MM (2018) *Ralstonia solanacearum*: a threat to potato production in Bangladesh. Fundam Appl Agric 3(1):407–421

Kumar R, Barman A, Phukan T, Kabyashree K, Singh N, Jha G, Sonti RV, Genin S, Kumar Ray S (2017) *Ralstonia solanacearum* virulence in tomato seedlings inoculated by leaf clipping. Plant Pathol 66(5):835–841

Malafaia CB, Ana Cláudia SJ, Alexandre GS, de Souza EB, Alexandre JM, dos Santos Correia MT, MárciaVS (2018) Effects of Caatinga plant extracts in planktonic growth and biofilm formation in *Ralstonia solanacearum*. Microbial Ecol 75(3):555–561

Muthoni J, Shimelis H, Melis R, Kinyua ZM (2014) Response of potato genotypes to bacterial wilt caused by *Ralstonia solanacearum* Smith (Yabuuchi *et al.*) in the tropical highlands. Am J Potato Res (2014) 91:215–232

Picard C, Ponsonnet C, Paget E, Nesme X, Simonet P (1992) Detection and enumeration of bacteria in soil by direct DNA extraction and polymerase chain reaction. Appl Environ Microbiol 58:2717–2722

Poussier S, Cheron J, Couteau A, Luisetti J (2002) Evaluation of procedures for reliable PCR detection of *Ralstonia solanacearum* in common natural substrates. J Microbiol Methods 51:349–395

Pradhanang PM, Elphinstone JG, Fox RTV (2000) Sensitive detection of *Ralstonia solanacearum* in soil: a comparison of different detection techniques. Plant Pathol 49:403–413

Pradhanang PM, Momol MT (2001) Survival of *Ralstonia solanacearum* in soil under irrigated rice culture and aquatic weeds. J Phytopathology 149(11–12):707–711

Prior P, Fegan M (2005) Recent developments in the phylogeny and classification of *Ralstonia solanacearum*. Acta Hortic 695:127–136

Robinson-Smith A, Jones P, Elphinstone JG, Forde SM (1995) Production of antibodies to *Pseudomonas solanacearum*, the causative agent of bacterial wilt. Food Agric Immunol 7:67–79

Sagar V, Gurjar MS, Jeevalatha A, Bakade RR, Chakrabarti SK, Arora RK, Sharma S (2014) Phylotype analysis of *Ralstonia solanacearum* strains causing potato bacterial wilt in Karnataka in India. Afr J Microbiol Res 8(12):1277–1281

Schaad NW, Cheong SS, Tamaki S, Hatziloukas E, Panopoulas NJ (1995) A combined biological and enzymatic amplification (Bio-PCR) technique to detect *Pseudomonas syringae* pv. *phaseolicola* in bean seed extracts. Phytopathology 85:243–284

Scher F, Kloepper J, Singleton C (1985) Chemotaxis of fluorescent *Pseudomonas* spp. to soybean seed exudates *in vitro* and in soil. Can J Microbiol 31:570–583

Seaf Elyzal HA (2009) Studies on potato brown rot disease under Egyptian conditions. MSc Thesis, Agricultural Botany and Plant, Pathology Department, Faculty of Agriculture, Zagazig University

Seyed MM, Gholamreza B, Saeide S (2015) Antibacterial activities of the hydroalcoholic extract of *Portulaca oleracea* leaves and seeds in Sistan Region. Southeastern Iran Int J Infect 2(2):1–4

Singh D, Yadav DK, Sinha S, Choudhary G (2014) Effect of temperature, cultivars, injury of root and inoculums load of *Ralstonia solanacearum* to cause bacterial wilt of tomato. Arch Phytopathol Plant Prot 47(13):1574–1583

Slusarenko AJ, Patel A, Portz D (2008) Control of plant diseases by natural products: allicin from garlic as a case study. Eur J Plant Pathol 121:313–322

Tsai YL, Olson BH (1991) Rapid method for direct extraction of DNA from soil and sediments. Appl Environ Microbiol 57:1070–1074

Weller SA, Elphinstone JG, Smith NC, Boonham N, Stead DE (2000) Detection of *Ralstonia solanacearum* strains with a quantitative, multiplex, real-time, fluorogenic PCR (TaqMan) assay. Appl Environ Microbiol 66:2853–2858

Yabuuchi E, Kosaka Y, Yano I, Hotta H, Nishiuchi Y (1995) Transfer of two *Burkholderia* and *Alcaligenes* species. to *Ralstonia* gen. Nov., Proposal of *Ralstonia picketii* (Ralston, Palleroni and Doudoroff 1973) Comb. Nov., *Ralstonia solanacearum* (Smith, 1896) Comb. Nov. and *Ralstonia eutropha* (Davis 1969) Comb. Nov. Microbial Immunol 39(11):897–904

Zehr EI (1969) Bacterial wilt of ginger in the Philippines. Philippines Agriculturist 53(3 and 4):224–227

Zehr EI (1970) Strains of *Pseudomonas solanacearum* in the Philippines as determined by cross-inoculation of hosts at different temperatures. Philipp Phytopathol 6(1):44–54

Chapter 10
Effect of Soil Type and Crop Rotation on the Causal Agent of Potato Brown Rot Disease Under Egyptian Condition

Yasser Hamad, M. R. A. Tohamy, and G. A. El-Morsy

Abstract Soil type and crop rotation are the most important factors affecting on the causal agent of potato brown rot disease *Ralstonia solanacearum*. The effect of soil type, soil pH and previous crop in agricultural rotation are of fundamental importance on persistence of the causal organism. The reduction rate of the pathogen was higher in clay loam compared to sandy loam. Application of suitable agronomic practices which led to decreasing pH value lower than 7.2 or increase more than 7.7, might be useful for controlling the causal agent. Crop rotation is considered an important agricultural procedure in reducing the weed in potato fields and then decreasing the incidence of potato brown rot disease. The number of weed varieties as hosts for *R. solanacearum* is very great. It causes disease at least of 200 different plant species, including herbaceous plants, shrubs and trees. It has an important impact on disease management based on crop rotation. There is increasing evidence of hosts which under certain condition act as latent or symptomless carriers of infection. Previous researches indicated that Potato–Rice crop rotation was more effective in decline the population density of *R. solanacearum* rather than Potato–Maize and Potato–Peanut. This chapter focuses on the biotic capabilities of *R. solanacearum* in relation to dispersion, survival and effects which might be important in planning operative management against the pathogen.

Keywords Potato brown rot · Soil type · Soil pH · Crop rotation · Weed

Y. Hamad (✉)
General Administration of Plant Quarantine, Ministry of Agriculture, 12611 Giza, Egypt

M. R. A. Tohamy
Plant Pathology Department, Faculty of Agriculture, Zagazig University, 44511 Zagazig, Egypt

G. A. El-Morsy
Plant Pathology Research Institute, Agriculture Research Center, 12613 Giza, Egypt

© Springer Nature Switzerland AG 2021
H. Awaad et al. (eds.), *Mitigating Environmental Stresses for Agricultural Sustainability in Egypt*, Springer Water,
https://doi.org/10.1007/978-3-030-64323-2_10

10.1 Introduction

Potato plants in fields are attacked by an extensive range of fungal, bacterial and viral diseases that cause serious losses in quantity and quality of potato tubers. Potato plants are considered as sensitive crop to weed competition that causes harmful effects on quantity and quality of potato tubers. The bacterial wilt problem is a major constraint for vegetable growers, especially potato (Fig. 10.1) and tomato farmers in many regions of the world. Bacterial wilt caused by *R. solanacearum* is a distressing disease that often serious potato production and exportation. Estimation of yield losses due to the disease incidence ranged from 15 to 90% (Hayward and Hartman 1994). The members of the complex *R. solanacearum* that cause brown rot are greatly contagious and there are no known varieties with permanent pathogen resistance. Brown rot symptoms appear on potatoes, tomatoes, tree tomatoes, eggplant and sweet pepper only. Among the symptomatic plant species, potatoes, tomatoes and tree tomatoes were completely wilted (Uwamahoro et al. 2020).

Definition of pathogen during previous decades through biochemical and physiological characteristics has been detected. The causal agent of brown rot disease had many synonyms as *Bacillus solanacearum* (Smith); *B. nicotiana* (Uyeda); *B. sesami* (Malkoff); *Pseudomonas esmi* (Malkoff); *B. musae* (Roeve); *Bacterium solanacearum* (Smith); *B. musarium* (Zeman); *Erwinia necotianae* (Bergey); *Phytomonas solanacearum* (Bergey); *P. reicini* (Archibald); *Pseudomonas solanacearum* (Smith); *Burkholdaria solanacearum* and Recently *R. solanacearum* (Yabuuchi et al. 1995).

Fig. 10.1 Typical symptoms of the disease on tuber showing distinct brown vascular discoloration and ooze of bacterial masses (Hamad 2016)

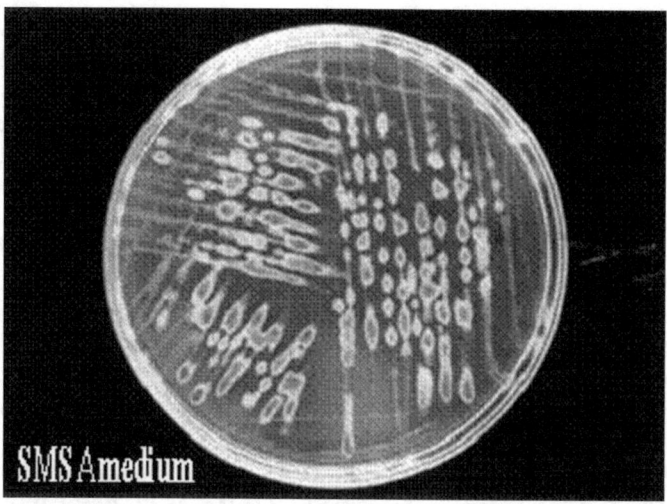

Fig. 10.2 Typical colonies of *R. solanacearum* on SMSA medium (Hamad 2008)

Ralstonia solanacearum is characterized as Gram-negative, non-fluorescent, non-spore forming, non-capsulating but slime forming rod shaped bacterium. The typical colonies on SMSA medium are milky white, irregular and fluidal with blood red coloration in the center (Fig. 10.2). The cells are motile with either a polar flagellum or tuft of polar flagella, though non-motile variants being recognizable under many circumstances. The cells accumulate poly β-hydroxy butyrate (PHB) as a cellular reserve inclusions. Cultures of *R. solanacearum* maintained in non-aerated liquid medium rapidly lose virulence and viability and shift from the fluidal wild type to a highly motile variants. Colonies of virulent wild types are irregularly rounded and are white with pink centers. Colonies of avirulent variants are uniformly round, butyrous, and deep red in media containing 2, 3, 5 triphenyltetrazolium chloride (Kelman 1954; Krieg and Holt 1984). These reactions included arginine (where ammonia is produced from arginine under an aerobic condition). Gelatin liquefecation, H_2S production tests, indol production, KOH 3%, oxidase reaction, starch hydrolysis and sugar tests, according to the methods described by Fahy and Persley (1983) and Lelliot and Stead (1987).

Balabel (2006) plating of bacterial suspensions from different sources revealed virulent and avirulent colony forms. The first is described as milky white, flat, irregular and fluidal with red coloration in the center. The so called avirulent form developed less fluidal colony which is completely pink to red. Both forms, however, induced wilting in tomato seedling thus the terms typical and atypical were suggested to replace and, respectively. Isolates in Egypt were identified as race 3 biovar 2 using morphological, physiological and biochemical tests. Also advanced Immunofluorescence Antibody Stains (IFAs) and Polymerase Chain Reaction (PCR) techniques were applied and proved that the isolates are related to race 3 biovar 2 of *R. solanacearum* (Hamad et al. 2008).

In some regions, obviously oppressive soils have been stated (Hayward 1991). The soil type effect on disease severity has been recurrently documented (Moffett et al. 1983; van Elsas et al. 2000; Elhalag et al. 2015). However, the actual effects are challenging and determined by geographical site, race, biovar of the pathogen, and crops. In some circumstances, bacterial wilt was more stark on well-drained sandy loams (French 1994; Hayward 1991), and the reduction rate of the pathogen was larger in clay loam compared to sandy loam (Moffett et al. 1983). On the other hand, the disease was most severe in heavy clay-loam soil (Kelman 1953). Thus, suppressiveness of soil to bacterial wilt might be associated with other reasons than soil texture, such as pH, organic matter content and microbial populations (van Elsas et al. 2005). Disease dominance was missing after soil dealing with methyl bromide or heat, demonstrating that a biological principle was the main reason. Resistance breeding and various cultural measures including the use of certified seed, crop rotations with non-host crops, soil type and careful water management has been practically successful against the pathogen (Hartman and Elphinstone 1994; French 1994; Elhalag et al. 2015). Potato–Rice crop rotation was more effective in decline the population density of R. solanacearum rather than Potato–Maize and Potato–Peanut (Hamad 2008). Therefore, many factors played importance role in persistence of the causal agent of brown rot R. solanacearum in potato such as soil type, soil pH and crop rotation.

10.2 Effect of Soil Type on Potato Brown Rot Disease Caused by *Ralstonia solanacearum*

Persistence of the causal agent R. solanacearum may be differed according to the soil type. Many investigators studied the effect of soil type on persistence of the pathogen. Among them, Engelbrecht (1994) calculated number of microorganisms under four soil types i.e. sandy loam, loamy sand, clay and black clay. He found that, the highest population density of *Pseudomonas solanacearum* have been counted on Semi Selective Medium of South Africa (SMSA) media from loam sand followed by sandy loam, black clay and then clay soil.

The effect of three soils, loamy sand and two different silt loam soils, besides the effects of temperature and soil moisture content on the population of R. solanacearum were investigated by van Elsas et al. (2000). They reported that soil type affected the population rate, which decline at 20 °C. The greatest declines were occurred in loamy sand soil. Also, they found occurring of viable but nonculturable strain 1609 cells in the loamy sand, and in silt loam soil at these conditions.

The survival of R. solanacearum in naturally infested sandy loam soil under irrigated rice culture in central Nepal was investigated by Momol (2001). The experimental plot had a preceding history of bacterial wilt at a range of 1.5×10^4–3×10^4 cfu/g soil. The presence of the causative agent R. solanacearum was monitored in roots of naturally growing aquatic weeds in plot of rice and in soil before and after rice harvested. The number of bacterial population in soil before rice harvest valued 1.5×10^4 cfu/g soil, their number was varied between $7.5 \times 10^2 – 1.9 \times 10^3$ cfu/g soil after rice harvest. Kehil (2002) collected soil samples from potato districts of Menufya, Minia and Nubaria governorates, Egypt. Among 195 soil samples, 17 contained *R. solanacearum* detected by using Semi Selective Medium of South Africa (SMSA). Results revealed that soil of Menufya has most *R. solanacearum* frequency (15.3%) from January 1999 to May 1999 and from September 1999 to January 2000. Under Minia conditions, 6.1% of samples showed presence of *R. solanacearum* through March and April 1999 than October and November. Nubaria soil exhibit least frequency (4.6%) from January 1999 to March. Moreover, the impact of soil amendment by addition of different organic matters, mineral fertilizers and bactericides on race 3, biovar 2 of *R. solanacearum* under artificial inoculation conditions was studied by Abd El-Ghafar et al. (2004). They indicated that organic matters and mineral fertilizers were more effective than bactericide treatments in reducing severity of potato bacterial wilt disease. Balabel (2005) studied the effect of mixed inoculum and persistence of the isolates of *R. solanacearum* recovered from symptomless *Rumex dentatus* and *Solanum nigrum*. All isolated produced severe wilt symptoms being 5% on tomato seedlings. On the other hand, the total microbial flora showed gradual decrease in densities, in Spunta and Diamont potato cultivars, up to the middle of June either in clay loam or loamy sand soil. In latter months, the brown rot pathogen showed a significant increase, being more pronounced in loamy sand soil.

Balabel (2006) studied the persistence of *R. solanacearum* in different habitats in Egypt. Unplanted clay loam and loamy sand soil maintained bare at 75% moisture showed a slight decrease in microbial densities in August, September and October. Infestation of soil with *R. solanacearum* showed considerable increase in microbial densities compared to the aforementioned uninoculated control. The slight decrease, however, was similarly observed in August and September in both soil types. Monitoring of microbial densities in both soil types, infested with brown rot pathogen, maintained under uncontrolled moisture conditions, showed similar trend of decrease in August and September. A noticeable increase was observed in October, November and December. Fluctuation in *R. solanacearum* densities in bare unplanted clay loam and loamy sand soil was noticed. Under controlled moisture conditions at 75% a remarkable decrease in densities in August and September was recorded followed by a pronounced increase in October, November and December. Under uncontrolled moisture condition the pathogen survived in loamy sand soil for 6 months with very high densities in December. On the other hand, densities of the pathogen in dry clay loam soil were extremely low after 5 months (November). The biofertilization with biosystemmicroorganisms product (EM) showed seasonal fluctuation in densities of *R. solanacearum*.

However, Hamad (2008) studied the persistence of *R. solanacearum* at different types of soil texture during different periods by using 10^{-3}, 10^{-4}, 10^{-5} dilutions. As showed in Table 10.1, there were significant differences for number of typical and atypical colonies among various types of soils. After one month from the start time of experiment, at 10^{-3} dilution, number of *R. solanacearum* typical colonies

Table 10.1 Effect of soil type on persistence (typical and atypical colonies) of *Ralstonia solanacearum* Ismailia isolate I_{SO} at different concentrations and periods (Hamad 2008)

Date after soil infestation	Dilution	Sandy soil		3/4 sandy 1/4 clay		1/2 sandy 1/2 clay		1/4 sandy 3/4 clay		Clay		L.S.D 0.05	
		Typical	Atypical	Typical	Atypical	Typical	Atypical	Typical	Atypical	Typical	Atypical	Typical	Atypical
After 30 days	10^{-3}	51	20	43	22	39	17	35	15	30	16	3.328	1.214
	10^{-4}	6	3	5	3	4	2	4	2	3	2	0.431	0.207
	10^{-5}	1	1	1	0	0	0	0	0	0	0	0.207	0.195
After 45 days	10^{-3}	32	11	26	13	20	11	18	7	17	11	2.628	0.169
	10^{-4}	4	2	3	1	2	1	2	1	2	1	0.274	–
	10^{-5}	1	0	0	0	0	0	0	0	0	0	0.142	–
After 60 days	10^{-3}	11	7	9	5	8	4	7	5	5	6	0.931	0.475
	10^{-4}	2	0	1	0	0	0	0	0	0	0	0.338	–
	10^{-5}	0	0	0	0	0	0	0	0	0	0	–	–
After 75 days	10^{-3}	7	4	3	5	1	2	0	2	1	5	1.163	0.510
	10^{-4}	1	0	0	0	0	0	0	0	0	0	0.207	–
	10^{-5}	0	0	0	0	0	0	0	0	0	0	–	–
After 90 days	10^{-3}	3	2	0	3	0	2	0	1	0	2	0.559	0.745
	10^{-4}	0	0	0	0	0	0	0	0	0	0	–	–
	10^{-5}	0	0	0	0	0	0	0	0	0	0	–	–
After 105 days	10^{-3}	0	1	0	0	0	0	0	0	0	0	–	0.372
	10^{-4}	0	0	0	0	0	0	0	0	0	0	–	–
	10^{-5}	0	0	0	0	0	0	0	0	0	0	–	–

in sandy soil reached the maximum population 51 with 20 colonies were atypical. After 45 days, 32 colony were typical and 11 colony were atypical colonies has been observed. Moreover, after 105 days, zero colony of typical with one atypical colony has been obtained. However, at 10^{-4} dilution after 30 days it was 6 typical colonies with 3 atypical. However, after 75 days, one typical without atypical colonies was detected. More reduction in cell population at 10^{-5} dilution was detected, after 30 days one typical with one atypical. The population density was decreased in the following periods until disappeared after 105 days. At 10^{-3} dilution, number of *R. solanacearum* typical colonies in (¾ sand: ¼ clay) soil reached the maximum population 43 with 22 colony of atypical. After 45 days, it was 26 colonies of typical and 13 atypical colonies. After 90 days, no typical colony with 3 atypical colonies were recorded. However, at 10^{-4} in the same periods, it was 5 typical colonies with 3 atypical. After 45 days, 3 colonies were typical and only one was atypical. Whereas, after 60 days, one typical without atypical colonies was recorded. More reduction in cell population under 10^{-5} dilution was detected, after 30 days with one typical and one atypical colony on TZC medium. At 10^{-3} dilution, number of *R. solanacearum* typical colonies in (½ sand: ½ clay) soil reached the maximum population after 30 days 39 colony of typical with 17 atypical colonies. After 90 days, no typical colonies with 2 atypical colonies were registered. However, at 10^{-4} in the same periods it was 4 colony of typical with 2 atypical colonies. After 45 days, 2 colonies were typical and only one was atypical. More reduction in cell population under 10^{-5} dilution was detected, after 30 days zero typical without atypical colonies on TZC medium. Population dynamics of the causal agent was rapidly declined from sandy soil to different mixtures until pure clay soil, with the time. Thus, the persistence of *R. solanacearum* was less than 105 days under sandy soil, but reached less than 90 days under clay soil. The faster decline rate of *R. solanacearum* recorded in clay as compared to sandy soil may be associated with the soil texture. It is well recognized that the fine texture of clay soil along with it is rich in organic matter permit or stimulate growing of antagonistic protoza, rhizobacteria (*Pseudomonas fluorescens* and *Bacillus cereus*), *Trichoderma harzianum* and etc. …which are previously affect the growth of bacteria (Minku and Bora 2000; Pardeep and Sood 2002). Whereas, in sandy soil the persistence of antagonistic organisms is scarcely done.

Concerning the preceding reports on suppression of the disease by organic amendments, and the extension of organic agriculture, Messiha Neven et al. (2007) compare the influence of organic and conventional management and different amendments on *R. solanacearum* race 3 biovar 2. They studied development of the causative agent of potato brown rot (bacterial wilt), in diverse soils (type: sand or clay; origin: Egypt or the Netherlands). Brown rot incidence was only slightly decreased in organic compared to conventional managed sandy soils of Egypt, however organic management increased significantly disease severity and pathogen persistence in Dutch sandy and clay soils. The absence of a disease oppressive influence of mineral and organic fertilization in Dutch clay soils might be associated with the great availability of inorganic and organic nutrients in the previous soils. Disease suppression mechanism of soil-borne plant pathogens can differ powerfully based on the soil type, especially if fairly diverse types of soil are used. Also, contents of Ca and K were also found to be greater in organic compared to in conventional Egyptian sandy soil. The percentage of *R. solanacearum*-infected plants was reduced in the organic soil, supporting former findings about improved resistance in plants high in Ca and/or K (https://link.spr inger.com/content/pdf/10.1007/s10658-007-9167-z.pdf) Under the Egyptian desert and the Netherlands, prevention of spread and quarantine remain main tools for combating potato brown rot. Thus instruments of disease suppression of soil-borne plant pathogens may differ intensely according to the soil type, particularly at quite various types of soil. In this respect, fertilization with potassium sulphate decreased the occurrence of potato brown rot (Fahmy and Mohamed 1990). The use of poultry manure will be an interesting option to be explored for the Nile Delta of Egypt (Islam and Toyota 2004). Furthermore, Farag et al. (2017) found that suppression of *R. solanacearum* after chitosan treatment was obviously detected in Wardan region, which was pronounced by a higher soil alkalinity compared to Talia area (silty clay soil) in Egypt.

10.3   Effect of Soil PH on Persistence of Potato Brown Rot Disease Caused by *Ralstonia solanacearum*

Persistence of the causal agent *R. solanacearum* may be differed according to the soil pH. Many investigators studied the effect of soil pH on persistence of the pathogen, among them Michel and Mew (1998) showed that a high pH value may have a harmful effect on longevity of *R. solanacearum*, whereas higher pH value restricts the viability of many nutrients necessary for the bacterium growth and longevity. Growth of the strains was suppressed at pH 3, 10 and 11 values and intensely reduced at pH 4 and 9. At pH 5 and 8 values, growth reduction was low.

Messiha Neven (2006) studied the effect of conventional soils with NPK and organic soils with compost or cow manure compared to non-amended controls along with soil pH. They found that the pathogen survived for longer periods in Dutch soil than in Egyptian soils, as well as in clay than in sandy ones of both countries. "Persistence has not been longer than 180 days and in several satuations much shorter, particularly in Egyptian sandy soil" (http://edepot.wur.nl/23415). The soil pH value was found to be negatively associated with both survival and disease suppression as the causative agent was severe in Egyptian soil with higher pH value.

Hamad (2008) showed that the optimum pH value for survival and persistence of *R. solanacearum* found to be 7.2–7.7 which gave the greatest number of colony i.e., 35 typical and 17 atypical as well as 40 typical and 19 atypical, respectively comparing with the low 5.8 and 6.2 pH values or the high 8.8 pH value (Table 10.2). Thus, application of suitable agronomic practices led to decreasing pH value lower than 7.2 or might increase soil pH value more than 7.7, might be useful for controlling the causal agent.

Table 10.2 The effect of soil pH on persistence (typical and atypical colonies) of *Ralstonia solanacearum* Ismailia isolate I_{SO} (Hamad 2008)

Date after soil infestation	Dilution	pH values												L.S.D 0.05	
		5.8		6.2		6.7		7.2		7.7		8.8			
		Typical	Atypical	Typical	Atypical	Typical	Atypical	Typical	Atypical	Typical	Atypical	Typical	Atypical	Typical	Atypical
After 30 days	10^{-3}	23	11	25	12	30	18	35	17	40	19	19	8	3.277	1.852
	10^{-4}	3	1	3	1	3	2	4	2	4	2	2	1	0.285	0.207
	10^{-5}	0	0	0	0	0	0	0	0	0	0	0	0	–	–
After 45 days	10^{-3}	11	4	13	7	18	11	21	12	28	15	12	4	2.731	1.889
	10^{-4}	1	0	1	0	2	1	2	1	2	1	1	0	0.207	0.207
	10^{-5}	0	0	0	0	0	0	0	0	0	0	0	0	–	–
After 60 days	10^{-3}	3	2	5	4	7	4	13	6	22	13	7	2	2.355	1.716
	10^{-4}	0	0	0	0	0	0	1	0	2	1	0	0	0.316	0.154
	10^{-5}	0	0	0	0	0	0	0	0	0	0	0	0	–	–
After 75 days	10^{-3}	0	0	0	0	2	5	6	4	19	6	2	2	2.752	1.067
	10^{-4}	0	0	0	0	0	0	0	0	2	0	0	0	0.308	–
	10^{-5}	0	0	0	0	0	0	0	0	0	0	0	0	–	–
After 90 days	10^{-3}	0	0	0	0	2	1	2	2	11	4	0	0	1.782	0.667
	10^{-4}	0	0	0	0	0	0	0	0	1	0	0	0	0.165	–
	10^{-5}	0	0	0	0	0	0	0	0	0	0	0	0	–	–
After 105 days	10^{-3}	0	0	0	0	0	1	2	3	8	3	0	0	1.211	0.613
	10^{-4}	0	0	0	0	0	0	0	0	0	0	0	0	–	–
	10^{-5}	0	0	0	0	0	0	0	0	0	0	0	0	–	–

10.4 Effect of Crop Rotation on Potato Brown Rot Disease

Crop rotation played an importance role in persistence of the causal agent of brown rot *R. solanacearum* in potato. Sohi et al. (1981) found that the crop rotation with cowpea–maize–cabbage, ladies finger (*Hibiscus esculentus*), cowpea–maize, *Eleusine coracana* and eggplants were effective in controlling the disease at Bangalore. Also, the effect of crop rotation on incidence of brown rot of potato in Plateau region of Bihar (India) was studied by Sinha et al. (1993). They found that the wilt incidence was reduced by 44.5–55% and by 56.0–67.5%, respectively in two (potato–wheat–maize–pea–potato) and (potato–wheat–maize–pea) and three years (potato–wheat–maize–cabbage–mung–cauliflower–potato) and (potato–wheat–maize–cabbage–mung–cauliflower) crop rotation.

Michel et al. (1996) studied the effect of preceding crop on soil populations of *Burkholderia solanacearum* and the incidence of tomato bacterial wilt in soil left fallow and following crops of cowpea, eggplant and rice. They detected that bacterial population decreased after cowpea and rice. The population was also reduced after soil left fallow, showing that an appropriate host plant is necessary to maintain the bacterial population. Lemaga et al. (2001) revealed that, one "season rotation in severely infested field (<90% wilt incidence) with wheat and maize did not significantly decrease wilt" (https://tspace.library.utoronto.ca/bitstream/1807/21909/4/cs01056.html), but improved tuber yield. Also, the highest wilt decrease was attained with potato–beans–maize–potato treatment, whereas the lowest decrease was with potato–maize–maize–potato. Whereas, Sharma and Kumar (2004) found that the reduction in bacterial population over the initial was 16.9% in rotation of maize–cabbage–cucumber. Increase in bacterial population over the initial population was 164.6% in tomato–tomato–tomato. Moreover, Alvarez et al. (2008) inoculated 20 crop species, most of them of potential importance in crop rotation with *R. solanacearum* biovar 2. Then after a month of inoculation, section of roots, and stems were examined to localize the pathogen on surface, cortex and or xylem. The results showed that host plants comprise cabbage, kidney bean and rutabaga cultivars, while alfalfa, barley, black radish, carrot, celery and maize were non-hosts.

The effect of previous crop through crop rotation on *R. solanacearum* was investigated by Hamad (2008). As given in Table 10.3, in potato–peanut crop rotation after one month at 10^{-3} dilution, number of *R. solanacearum* typical colonies reached the maximum population 75 typical with 20 atypical. After 45 days, typical colonies were 48 and 13 were atypical. However, at 10^{-4} of the same periods after 30 days, it was 8 typical colonies with 2 atypical. After 45 days, 5 colonies were typical and one only was atypical. More reduction in cell population at 10^{-5} dilution was detected; it was 2 typical with non-atypical colonies at TZC medium. The population density was decreased in the following periods until disappeared after 140 days. The obtained results reveal that survival of *R. solanacearum* was never longer than 150 days (under these conditions). Added that, in potato–maize crop rotation after one month from start time at the beginning of the experiment at 10^{-3} dilution, number of *R. solanacearum* typical colonies reached the maximum population 63 typical with

Table 10.3 Effect of previous crop (Potato–Peanut) on persistence (typical and atypical colonies) of causal organism of potato brown rot disease at different concentration and periods (Hamad 2008)

Days after soil infestation	Dilution					
	10^{-3}		10^{-4}		10^{-5}	
	Typical	Atypical	Typical	Atypical	Typical	Atypical
After 30 days	75	20	8	2	2	0
After 45 days	48	13	5	1	0	0
After 60 days	35	6	3	1	0	0
After 75 days	27	3	2	0	0	0
After 90 days	18	2	1	0	0	0
After 105 days	13	4	1	0	0	0
After 120 days	6	0	0	0	0	0
After 135 days	3	0	0	0	0	0
After 150 days	0	0	0	0	0	0
L.S.D 0.05	7.669	2.864	0.958	0.302	1.470	–

Table 10.4 Effect of previous crop (Potato–Maize) on persistence (typical and atypical colonies) of causal organism of potato brown rot disease at different concentration of periods (Hamad 2008)

Days after soil infestation	Dilution					
	10^{-3}		10^{-4}		10^{-5}	
	Typical	Atypical	Typical	Atypical	Typical	Atypical
After 30 days	63	22	7	4	1	0
After 45 days	45	15	5	2	0	0
After 60 days	32	15	4	1	0	0
After 75 days	18	11	3	1	0	0
After 90 days	9	8	1	0	0	0
After 105 days	3	7	0	0	0	0
After 120 days	1	3	0	0	0	0
After 135 days	0	0	0	0	0	0
After 150 days	0	0	0	0	0	0
L.S.D 0.05	8.033	3.118	0.925	0.568	0.117	–

22 atypical (Table 10.4). After 45 days typical colonies were 45, while 15 colonies were atypical. However, at 10^{-4} in the same periods it was 7 typical colonies with 4 atypical. After 45 days 5 colonies were typical and tow only was atypical. More reduction in cell population at 10^{-5} dilution was detected, it was 1 typical with non-atypical on TZC medium. The population density was decreased in the following periods until disappeared after 120 days. The obtained results reveal that survival of *R. solanacearum* was never longer than (under these conditions) 135 days.

Furthermore, for potato–rice crop rotation, significant differences were recorded for number of typical and atypical colonies between different periods after infestation. Table 10.5 indicates that, after one month at 10^{-3} dilution, number of *R. solanacearum* typical colonies reached the maximum population 52 typical with 30 atypical. After 45 days, typical colonies were 25 and 33 atypical. However, at 10^{-4} in the same periods it was 6 typical colonies with 3 atypical. After 45 days, tow colonies were typical and 2 only were atypical. More reduction in cell population at 10^{-5} dilution was detected, it was one typical with non atypical on TZC medium. The population density was decreased in the following periods until disappeared after 105 days. Therefore, survival of *R. solanacearum* was never longer than 120 days (under such conditions). The rice is not considered as host for the causal agent, as well as floating conditions of rice irrigation enhanced other anaerobic micro-organisms which may be decreased the population of the brown rot bacteria. On the other hand, the regarded causal agent is aerobic one, as it was proved experimentally, thus floating resulted in a reduction of it is activity by competing them. Also, this condition led to rapid decay of potato organic matter which is necessary for persistence of the causal agent. Surviving *R. solanacearum* from one potato-growing season to another depending on the pathogen is thought to survive in infected self-sown tubers or potato root debris (Graham and Lloyd 1979) or over wintering the roots of winter weeds or crops, or in perennial weeds, as reported for *Solanum dulcamara* plants (Olsson 1976; Elphinstone et al. 1996) and *Solanum cinereum* (Graham and Lloyd 1978). These hosts are not killed by the pathogen and so permit long-term pathogen persistence.

Elhalag et al. (2015) control potato brown rot, caused by *R. solanacearum* race 3 biovar 2, Phylotype II, sequevar 1 using different biocontrol approaches. They used the bacterial biocontrol agent *Stenotrophomonas maltophilia* (PD4560), under

Table 10.5 Effect of previous crop (Potato–Rice) on persistence (typical and atypical colonies) of causal organism of potato brown rot disease at different concentration of periods (Hamad 2008)

Days after soil infestation	Dilution					
	10^{-3}		10^{-4}		10^{-5}	
	Typical	Atypical	Typical	Atypical	Typical	Atypical
After 30 days	52	30	6	3	1	0
After 45 days	25	33	2	2	0	0
After 60 days	18	6	1	1	0	0
After 75 days	15	3	1	0	0	0
After 90 days	3	2	0	0	0	0
After 105 days	2	0	0	0	0	0
After 120 days	0	0	0	0	0	0
After 135 days	0	0	0	0	0	0
After 150 days	0	0	0	0	0	0
L.S.D 0.05	6.111	5.569	0.738	0.494	0.117	–

inoculated clay or sandy soils, grown with cowpea, maize or tomato in pots. The maximum survival of *S. maltophilia* in soil (more than 160 days) coincided with a major suppressing effect on *R. solanacearum* that expressed by wilt severity (up to 100% decrease), area under disease progress curve (AUDPC) (up to 99% decrease) and counts of the pathogen in soil (up to 75% reduction), rhizosphere (up to 80% decrease) and plant tissue (up to 97% decrease) of potato plants. Methionine is famous to improve the growth of *S. maltophilia*, therefore cowpea and maize are suitable for crop rotation with potato and will increase the sustainability of the biocontrol agent *S. maltophilia*.

10.5   Influence the Persistence of *Ralstonia solanacearum* by Different Plant Species and Habitats

Persistence of the causal agent *R. solanacearum* may be differed according to the plant species and habitats. Many investigators studied the effect of plant species and habitats on persistence of the pathogen, among them, Quimio and Chan (1979) surveyed *P. solanacearum* in the rhizosphere of several weed and economic plant species. They found that, some of the most susceptible weed species are members of the Solanaceae, Portulaceae and Asteraceae and being most susceptible. They would be expected to contribute most to soil populations of the pathogen, whereas, Moffett and Hayward (1980) found that the same weed species was the only one of various studied in tomato cropping soil to show symptoms of wilt. They isolated the pathogen from root system. Moreover, Vasse et al. (1995) showed that the bacteria ordinarily enter through the roots, either at wounds or sites of secondary root emergence, invading host plant xylem cells and spreading through the vascular system, leading to severe wilting that ultimately causing plant death. Kresten et al. (2001) showed that, *R. solanacearum*, a broadly distributed and economically important plant pathogen, invades the roots of diverse plant hosts from the soil and aggressively colonizes the xylem vessels, leading to a lethal wilting known as bacterial wilt disease. By testing bacteria from the xylem vessels of infested plants, they showed that *R. solanacearum* is essentially no motile in planta, while it can be highly motile in culture.

In this regard, Granda and Sequeira (1983) isolate the bacterium biovar 2 isolate UW270 from soil for up to a month after inoculation the soil. In this concern, *R. solanacearum* race 1 biovar 2 was isolated in Brazil from tomato, potato, sweet pepper and eggplant (Netto and Assis 2002); *R. solanacearum* race 3 biovar 2 was isolated in Japan from tomato, eggplant, pepper and tobacco (Horita and Tsuchiya 2001) and also isolated from cabbage, kidney bean and rutabaga cultivars (Alvarez et al. 2008).

Bagher and Taghavi (2000) collected samples of potato and tomato plants exhibiting yellow and sudden wilt symptoms from fields in Iran. The pathogen which caused these symptoms was isolated through the bacterial ooze extracted from stem and/or from vascular bundles of infested potato tubers. The bacterial ooze was streaked on sucrose peptone agar and triphenyl tetrazolium chloride media and the growth was examined. All strains were recognized as biovar II of *R. solanacearum*. At the same time, Sunaina et al. (2000) studied the persistence and distribution of *R. solanacearum* in naturally infested soil under changing temperature and moisture conditions during three years in the hills of Uttar Pradesh. The inoculums level in the infected soil was monitored by direct isolation and indirectly by assess based on wilted potato plants percentage growing in the soil. They recorded highest average bacterial population of 109 cfu/g soil in the month of July. Conditions favorable for high bacterial populations and wilt increase happened for a short period of six months between April and September. Coehlo-Netto et al. (2001) studied the effect of *R. solanacearum* race 1, biovar 1 and 3 on pathogenicity of some crop species. They observed that inoculation of healthy *Melanthera discoidea* plants reproduced the symptoms observed in the field. The isolated bacterium strains were also pathogenic to tomato, sweet pepper, potato and aubergine.

Saccardi et al. (2002) studied the effect of *R. solanacearum* race 3 on susceptibility of some cultivars of selected vegetable crops in an Italian experimental field contaminated with infected potatoes. None of the common radicchio (*Cichorium intybus*) types grown in the Veneto region, nor some cauliflower (*Brassica oleracea* var botrytis), savoy (*B. oleracea* var sabauda) and cabbage (*B. oleracea* var capitata) cultivars showed symptoms. No endophytic *R. solanacearum* was found.

Almeida et al. (2003) studied the effect of *R. solanacearum* on eleven commercial geranium cultivars (Avenida-Boogy, Brasilg, Opera, Rocky, Mountain Red, Rokoko, Rumba 98, Samba, Tang Hot Pink, Tango Orange and Tango Pink) under artificial inoculations by race 3 biovar 2 of *R. solanacearum*. The indicated that all cultivars were susceptible to the pathogen at different degrees. Castillo and Greenberg (2007) reported that *R. Solanacearum* is one of the most damaging bacterial pathogens, causes disease on at least 200 diverse plants species. It affects a wide range of crop plants, counting herbaceous plants, shrubs, and trees. Moreover, *R. Solanacearum* represented a main concern, given that it can seriously affect the yield of ornamental plants and valuable crops like tomato, potato, banana, peanut, eggplant and others. In this connection Robinson-Smith et al. (1995) reported that *Arachis hypogea* is considered as a host for *R. solanacearum*. Mwangi et al. (2007) reported that, adjustments in planting practices could contribute to reduce wilting incidence by the causal agent of brown rot (*R. solanacearum*) in banana and hence support efforts to crop production.

Hamad (2008) showed that in bare soil and rhizosphere the bacterium declined within 100 days from infested soil with *R. solanacearum* under natural and sterilized soil condition for *Solanum tuberosum* L. as well as within 90 days as inoculation for *Capsicum annum* L., *Lycopersicon esculentum* L., *Solanum melongena* L., *Solanum nigrum* L., *Arachis hypogea* L. and *Portulaca oleracea* L. at 10^{-4}, 10^{-5} and 10^{-6} dilutions. Isolation of the bacterium from surface roots showed an increase with increasing plant age up to 60 days then decreased in *Solanum nigrum* L., *Lycopersicon esculentum* L. and *Portulaca olerace* L. It showed an increase up to 75 days then decrease in *Solanum melongena* L. and *Capsicum annum* L. and up to 90 days then decrease in *Arachis hypogea* L. and *Solanum tuberosum* L. The results also indicated that populations of *R. solanacearum* released from symptomless infested weeds were not as great as those from wilted potato plants. On the light of these results, root is a favorable habitat for the bacterium compared with crown and stem which exhibited the lowest density with increasing the height of stem. Soil, rhizosphere and root habitats were higher in population density of bacterium than crown and stem, which may be due to that these habitats were more favorable for the pathogen propagation. For this reason, it is of importance to apply different agricultural practices of integration control for target pathogen. Chakraborty and Roy (2016) showed that potato brown rot disease could be controlled by use of fertilizers to alteration soil pH. The pathogen in the USA was eliminated by decreasing the soil pH to 4–5 in summer and rising it to pH 6 in the autumn. The disease is dangerous in sandy, loam, clay and peat soils, but it is never found in marl soils. They recommended treatment with stable bleaching powder (12–15 kg/ha) mixed with fertilizer in furrows while planting decreases the wilt occurrence by 80%.

10.6   Conclusions

Brown rot disease in potatoes is a serious disease that causes economic loss in the crop and affects the quality and export of the potato. Several reasons played significant role in persistence of the causal agent of brown rot *R. solanacearum* in potato such as soil type, soil pH and crop rotation. The decline rate of the pathogen was greater in clay loam compared to sandy loam. Soil pH value lower than 7.2 or more than 7.7, might be useful for controlling the causal agent. Potato–Rice crop rotation was more effective in decreasing the population density of *R. solanacearum* compared to Potato–Maize and Potato–Peanut.

10.7   Recommendations

To reduce the spread of brown rot disease caused by *R. solanacearum* in potatoes, the subsequent recommendations can be followed:

1. Follow the integrated pest management
2. Potato cultivation in pest free area of pathogen
3. Cultivation in clay loam soil with pH value ranged from 7.2 to 7.7.
4. Application of the crop rotation potato–rice.

References

Abd El-Ghafar NY, El-Zemaity MSS, Faiza Fawzi G, Haidi El-Henawy M (2004) Impact of soil amendment on population of *Ralstonia solanacearum* and development of potato bacterial wilt. J Environ Sci 9(2):649–669

Almeida IMG, Destefano SAL, Rodrigues-Neto J, Malavolta-Junior VA (2003) Southern bacterial wilt of geranium caused by *Ralstonia solanacearum* biovar 2/race 3 in Brazil. Revista de Agricultura Piracicaba 78(1):49–56

Alvarez B, Vasse J, Le-Courtois V, Trigalet-Demery D, Lopez MM, Trigalet A (2008) Comparative behaviour of *Ralstonia solanacearum* biovar 2 in diverse plant species. Phytopathology 98(1):59–68

Bagher A, Taghavi SM (2000) Characteristics of strains of the causal agent of bacterial wilt of potato and tomato in Fars province and reaction of some potato and tomato cultivars to the pathogen. Iran J Plant Pathol 36(3–4):233–243

Balabel NM (2005) Persistence of *Rolstonia solanacearum* (Syn. *Pseudomonas solanacearum*) in different habitats in Egypt. PhD Thesis, Microbiology Department, Faculty of Agriculture, Ain Shams University

Balabel, NM (2006) Persistence of *Rolstonia solanacearum* (Syn. *Pseudomonas solanacearum*) in different habitats in Egypt. PhD Thesis, Microbiology Department, Faculty of Agriculture, Ain Shams University

Castillo JA, Greenberg JT (2007) Evolutionary dynamics of *Ralstonia solanacearum*. Appl Environ Microbial 73(4):1225–1238

Chakraborty R, Roy TS (2016) Threats faced by brown rot of potato in Bangladesh. Microbiol Res 7(6258):1–12. www.pagepress.org/journals/index.php/mr/article/view/

Coehlo-Netto RA, Noda H, Boher B (2001) *Melanthera discoidea*: a new *Ralstonia solanacearum* host. Fitopatologia Brasiliera 26(4):781

Elhalag KM, Hassan ME, Messiha Nevein AS, Elhadad SA, Abdallah SA (2015) The relation of different crop roots exudates to the survival and suppressive effect of *Stenotrophomonas maltophilia* (PD4560), biocontrol agent of bacterial wilt of potato. J Phytopathol 163(10):829–840

Elphinstone JG, Hennessy J, Wilson JK, Stead DE (1996) Sensitivity of different methods for the detection of *Rolastonia solanacearum* in potato tuber extracts. Bull OEPP/EPPO, Bull 26:663–678

Engelbrecht MC (1994) Modification of a semi-selective medium for the isolation and quantification of *Pseudomonas solanacearum*. ACIAR Bacterial Wilt Newslett 10:3–5

Fahmy FG, Mohamed MS (1990) Some factors affecting the incidence of potato brown rot. Assuit J Agric Sci 21:221–230

Fahy PC, Persley GJ (1983) Plant bacterial disease. A diagnostic guide. Academic Press, New York

Farag SMA, Elhalag M, Hagag MH, Khairy AM, Ibrahim Heba M, Saker MT, Messiha Nevein AS (2017) Potato bacterial wilt suppression and plant health improvement after application of different antioxidants. J Phytopathol 165:522–537

French ER (1994) Strategies for integrated control of bacterial wilt of potatoes. In: Hayward AC, Hartman GL (eds) Bacterial wilt: the disease and its causative agent, *Pseudomonas solanacearum*. CAB International, Wallingford, UK, pp 98–113

Graham J, Lloyd AB (1978) *Solanum cinereum* R.Br., a wild host of *Pseudomonas Solanacearum* biotype II. J Aust Inst Agric Sci 44:124–126

Graham J, Lloyd AB (1979) Survival of potato strain (Race 3) of *P. solanacearum* in the deeper soil layers. Aust J Agric Res 30:489–496

Granda GA, Sequeira L (1983) Survival of *P. solanacearum* in soil, rhizosphere and plant roots. Can J Microbiol 29:433–440

Hamad YI (2008) Studies on the transmission of potato brown rot causal organism through weeds in the Egyptian fields M.Sc Thesis, Plant Pathology Department, Faculty of Agriculture, Zagazig University Egypt

Hamad YI (2016) Pathological studies on potato brown rot under Egyptian conditions. PhD Plant Pathology, Agric Botany Department Faculty of Agriculture, Suez Canal University, Egypt

Hamad YI, Tohamy MRA, El-Morsy GA (2008) Detection of *Ralstonia solanacearum* on some crops and weeds under Egyptian conditions. Zagazig J Agric Res 35(4):769–788

Hartman GL, Elphinstone JG (1994) Advances in the control of *Pseudomonas solanacearum* Race 1 in major food crops. In: Hayward AC, Hatman GL (eds) Bacterial wilt: the disease and its causative agent, *Pseudomonas solanacearum*. CAB International, Wallingford, UK, pp 157–177

Hayward AC (1991) Biology and epidemiology of bacterial wilt caused by *Pseudomonas solanacearum*. Ann Rev Phytopathol 29:65–87

Hayward AC, Hartman GL (1994) Bacterial wilt. The disease and its causative agent, *P. solanacearum*. "CAB" International, Wallingford, UK

Horita M, Tsuchiya K (2001) Genetic diversity of Japanese of *Ralstonia solanacearum*. Phytopathology 91(4):399–407

Islam TMD, Toyota K (2004) Suppression of bacterial wilt of tomato by *Ralstonia solanacearum* by incorporation of composts in soil and possible mechanisms. Microbes and Environ 19:53–60

Kehil YEI (2002) Studies on the pathological relationship between potato and brown rot bacteria *Ralstonia solanacearum* Ph.D. Thesis, Plant Pathology Department, Faculty of Agriculture, Minia University, Egypt

Kelman A (1953) The bacterial wilt caused by *Pseudomonas solanacearum*. A literature review and bibliography. North Carolina Agric Exp Stat Tech Bull 99:1–194

Kelman A (1954) The relationship of pathogenicity in *Pseudomonas solanacearum* to colony appearance on a tetrazolium medium. Phytopathology 44:693–695

Kresten J, Huang H, Allen C (2001) *Ralstonia solanacearum* needs motility for invasive virulence on tomato. J Bacteriol 183(12):3597–3605

Krieg NR, Holt JG (1984) *Pseudomonas solanacearum*. Bergey's Manual Syst Bacteriol 1:141–214

Lelliot RA, Stead DE (1987) Methods for diagnosis of bacterial diseases of plants. Methods in Plant Pathology, 2 ed. Blackwell Scientific Publication, 216 p

Lemaga B, Kanzikwera R, Kakuhenzire R, Hakiza JI, Maniz G (2001) The effect of crop rotation on bacterial wilt incidence and potato tuber yield. Afr Crop Sci 9(1):257–266

Messiha Neven AS (2006) Bacterial wilt of potato (*Ralstonia solanacearum* race 3, biovar 2). Disease Management, Pathogen survival and possible eradication. Ph D Thesis, Wayeningen University, The Netherlands

Messiha Neven AS, van Bruggen AHC, van Diepeningen AD, de Vos OJ, Termorshuizen AJ, Tjou-Tam-Sin NNA, Janse JD (2007) Potato brown rot incidence and severity under different management and amendment regimes in different soil types. Eur J Plant Pathol 119(4):367–381

Michel VV, Hartman GL, Midmore DJ (1996) Effect of previous crop on soil populations of *Burkholderia solanacearum*, bacterial wilt and yield of tomatoes in Taiwan. Plant Dis 80(12):1367–1372

Michel VV, Mew TW (1998) Effect of soil amendment on the survival of *Ralstonia solanacearum* in different soils. Phytopathology 88:300–305

Minku D, Bora LC (2000) Biological control of bacterial wilt of tomato caused by *Ralstonia solanacearum*. J Agric Sci Soc North-East India 13:52–55

Moffett ML, Hayward AC (1980) The role of weed species in the survival of *Pseudomonas solanacearum* in tomato cropping land. Australas Plant Pathol 9:6–8

Moffett ML, Giles JE, Wood BA (1983) Survival of *Pseudomonas solanacearum* biovars 2 and 3 in soil: effect of moisture and soil type. Soil Biol Biochem 15:587–591

Momol MT (2001) Survival of *Ralstonia solanacearum* in soil under irrigated rice culture and aquatic weeds. J Phytopathol 149(11–12):707–711

Mwangi M, Bandyopadhyay L, Ragama P, Tishemereirwe WK (2007) Assessment of banana planting practices and cultivar tolerance in relation to management of soil borne. Crop Prot 26(8):1203–1208

Netto RAC, Assis LAG (2002) *Coleus barbatus*: a new *Ralstonia solanacearum* host. Fitopatologia–Brasileira 27(2):226

Olsson K (1976) Experience of brown rot caused by *P. Solanacearum* (Smith) in Sweden. Bull OEPP/ EPPO Bull 6:199–207

Pardeep K, Sood AK (2002) Management of bacterial wilt of tomato with VAM and bacterial antagonists. Indian Phytopathol 55:513–515

Quimio AJ, Chan HH (1979) Survival of *Pseudomonans solanacearum* E.F. smith in the rhizosphere of some weed and economic plant species. Philippine Phytopathol 15:108–210

Robinson-Smith A, Jones P, Elphinstone JG, Forde SM (1995) Production of antibodies to *Pseudomonas solanacearum*, the causative agent of bacterial wilt. Food Agric Immunol 7:67–79

Saccardi A, Chillemi G, Pasqua-di-Bicceglie D, Lazzarin R, Traversa F, Mazzucchi U, Di-Bisc-Pasqua D, Brunelli A, Canova A (ed) (2002) Identification of vegetable crops which do not harbour *Ralstonia solanacearum* race 3 Atti, Giornate Fitopatologiche, Baselga di pine Trento, Italy, 7–11 April, 2:569–570

Sharma JP, Kumar S (2004) Effect of crop rotation on population dynamics of *Ralstonia solanacearum* in tomato will sick soil. Indian Phytopathol 57(1):80–81

Sinha SK, Mishra B, Verma SSP (1993) Effect of crop rotation on incidence of brown rot of potato in Plateau region of Bihar (India). Bacterial Wilt Newslett 9:7

Sohi HS, Rao MVB, Rawal RD, Kishun R (1981) Effect of crop rotation on bacterial wilt of tomato and eggplant. Indian J Agric Sci 51:572–573

Sunaina V, Kishore V, Shekawat GS, Kumar M, Khurana SMP, Shekhawat GS, Singh PB, Pandey SK (2000) Persistence of *Ralstonia solanacearum* in naturally infested soil under changing environment. Proceeding of the Global Conference on Potato. Held in New Delhi, India, 6–11 December 1999, 1:444–447

Uwamahoro F, Berlin A, Bucagu C, Bylund H, Yuen J (2020). *Ralstonia solanacearum* causing potato bacterial wilt: host range and cultivars' susceptibility in Rwanda. Plant Pathol 69:559–568

van Elsas JD, Kastelein P, Van Bekkum P, Van Der Wolf JM, De Vries PM, Van Overbeek LS (2000) Survival of *Ralstonia solanacearum* biovar 2, the causative agent of potato brown rot, in field and microcosm soil in temperate climates. Phytopathology 90(12):1358–1366

van Elsas JD, van Overbeek LS, Bailey MJ, Schönfeld J, Smalla K (2005) Fate of *Ralstonia solanacearum* biovar 2 as affected by conditions and soil treatments in temperate climate zones. In: Allen C, Prior P, Hayward AC (eds) Bacterial wilt disease and the *Ralstonia solanacearum* species complex. The American Phytopathological Society, Minnesota, USA, pp 39–49

Vasse J, Frey P, Trigalet A (1995) Microscopies studies of intercellular infection and protoxylem invasion of tomato roots by *Pseudomonas solanacearum*. Mol Plant Microbe Interact 8:241–251

Yabuuchi E, Kosaka Y, Yano I, Hotta H, Nishiuchi Y (1995) Transfer of two *Burkholderia* and *Alcaligenes* species. to *Ralstonia* gen. Nov., Proposal of *Ralstonia picketii* (Ralston, Palleroni and Doudoroff 1973) Comb. Nov, *Ralstonia solanacearum* (Smith, 1896) Comb Nov and *Ralstonia eutropha* (Davis 1969) Comb Nov Microbial. Immunology 39(11):897–904

Chapter 11
Advanced Methods in Controlling Late Blight Disease in Potatoes

Tamer Abd El-Azim Lotfy Ahmad

Abstract The aim of this chapter is to introduce recent information about the late blight disease in potatoes (*Solanum tuberosum* L.) which is considered one of the most famous diseases in agriculture. Late blight disease is caused by the oomycete pathogen *Phytophthora infestans*. This chapter focus also, assessment the economic importance of the disease. Also identification of the fungus using the most vital methods distinguishes between the new isolates of the fungus by numerous techniques, (a) the traditional method (b) using DNA markers and (c) bioinformatics and then control fungus by the best practices of chemical pesticides. Meanwhile, some chemicals are toxic and dangerous for both the environment and human health. Therefore, we will go to resistant cultivars and alternative methods such as plant oils and extracts as well as the use of nanotechnology and biocontrol compared to chemical fungicides to reduce the environmental problems on the plants, animals and then humans.

Keywords *Phytophthora infestans* · *Solanum tuberosum* · DNA markers · Bioinformatics · Nanotechnology · Biocontrol

11.1 Introduction

Potato is considered the most vital food crop worldwide after rice and wheat. Potato is distinguished by their high content of carbohydrates, microelements and vitamins, and very popular all over the world. However, potato plants are the host of numerous pathogens, such as fungi, bacteria, phytoplasmas, viruses, viroids and nematodes, which cause significant decreases in quantity and quality of the yield (Haverkort et al. 2009).

Furthermore, potato late blight disease caused by the oomycete *Phytophthora infestans* (Mont.) de Bary is considered the greatest known and remains among the most damaging plant diseases. Late blight affects humans because foliar phase limits

T. A. E.-A. L. Ahmad (✉)
Plant Pathology Research Institute, Agriculture Research Center, Giza 12613, Egypt

© Springer Nature Switzerland AG 2021   289
H. Awaad et al. (eds.), *Mitigating Environmental Stresses for Agricultural Sustainability in Egypt*, Springer Water,
https://doi.org/10.1007/978-3-030-64323-2_11

production of tubers that are a food source. Human suffering caused by dramatic epidemics was exacerbated in the 1840s by a lack of potatoes (Mizubtui and Fry 2006). The oomycete *Phytophthora infestans* that is causing the devastating late blight disease is one of the most important pathogens of potato. Recently, new strains with the ability to reproduce sexually are spreading that are associated with increased genetic variation and survival in several parts of the world (Fry 2008).

Phytophthora infestans is heterothallic, necessitating two mating types recognized as A1 and A2 for sexual reproduction and the creation of oospores. The occurrence of both mating types permits sexual reproduction that contributes to the formation in early infections and the acclimatization of the causal agent to some fungicides as well as to host resistance. Normally, sexual recombination leads to produce generation of mainly fit line-ages that have new combinations of troublesome characters (Smart and Fry 2001; Turkensteen et al. 2008). Alternative mechanism involved in genetic variability in the agricultural region is pathogen migration. This phenomenon seems to define the population dynamics of *P. infestans*. Population displacement by genotypes with improved fitness is a recurrent occurrence (Vleeshouwers et al. 2011). *Phytophthora infestans* unlike most Phytophthora species that causes soil borne root-rot diseases, *P. infestans* is a specialized pathogen, mainly causing disease on the foliage, stems, tubers of potato and tomato fruits. The most infection spread was done by airborne asexual sporangia through the growing season. Control of this disease is remains reliant on multiple uses of chemical fungicides during flowering and fruiting (Shattock 2002). A combination of high inoculum pressure, humid circumstances that favor pathogen growth and repeated pesticides uses have resulted in the emergence of resistant pathogen strains in some countries (Gisi and Cohen 1996). Further, the application of some synthetic chemicals to prevent fungal disease of food commodities is limited due to its possible carcinogenicity, great and serious toxicity, long periods of degradation, and environmental pollution. Therefore, cultivating high-tolerant varieties is essential in this respect. It is worth noting that, there are fears of the emergence of new physiological strains of the pathogen that lead to a loss of the effectiveness of traditional fungicides and the general unacceptability of the use of fungicides due to environmental hazards.

11.2   Backgrounds Economic

Solanaceae is a vital family which has a various set of plants varied from wild species to numerous imperative cultivated crops. The potato belongs to the previuos family, where it is considered a crop of economic importance that contributes to the global demand for food and commercial and economic advantages. More than one billion people worldwide depend on potato consumption (Anwar et al. 2015) which highlights its importance (Majeed et al. 2017). Potato has become an important vegetable crop for Egyptian growers as well as consumers and "one of the most vital economic crops in Egypt" (Mohamed and Abou-Hadid 2004). In Egypt, the potato is cultivated during the year in three growing seasons, summer, fall (Nili) and winter in the period from the early of September to the end of February and harvest time starts from the beginning of December until the end of May. Potato occupied an area of 183,990 feddans (Feddan = 4200 m^2) with a production of 4,800,000 tons (FAO 2013) and Egypt is ranked 15th among the world's most productive countries as shown in Table 11.1 and the first in Africa show (Fig. 11.1).

Table 11.1 Top 25 potato producing countries

Rank	Country	Potato production [tonnes]
1	China	95,987,500
2	India	45,343,600
3	Russian Federation	30,199,100
4	Ukraine	22,258,600
5	United States	19,843,900
6	Germany	9,669,700
7	Bangladesh	8,603,000
8	France	6,975,000
9	Netherlands	6,801,000
10	Poland	6,334,200
11	Belarus	5,913,710
12	United Kingdom	5,580,000
13	Iran (Islamic Republic of)	5,560,000
14	Algeria	4,928,030
15	Egypt	4,800,000

Year: 2013. *Source* FAOSTAT (Retrieval date July 3, 2015)

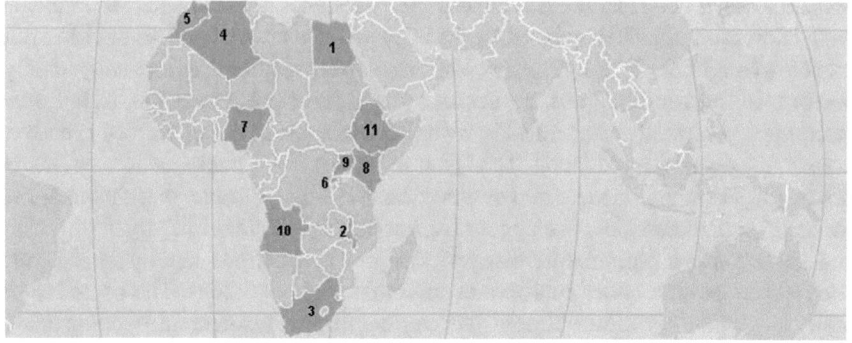

Fig. 11.1 Top producers in Africa. *Note* 1 Egypt; 2 Malawi; 3 South Africa; 4 Algeria; 5 Morocco; 6 Rwanda; 7 Nigeria; 8 Kenya; 9 Uganda; 10 Angola; 11 Ethiopia. *Source* 2008

Many factors can limit potato production. One of the most important factors is late blight disease. Unfortunately, the weather conditions allow the disease to start from the early of November and many outbreaks occur during December to the mid of April resulting in severe yield losses. Yield losses under favorable environmental conditions ranged from 50 to 70% potato yield (El-Ganainy 2013).

Potatoes in Egypt occupy the leading position for export vegetable crops, with annual production of more than 500,000 tons of fresh potatoes exported to European markets. In recent years, it has been possible to develop potato processing technology in Egypt and the methods of processing and preserving it to a large extent, leading to prolongation of the period of utilization and diversification to suit the wishes of consumers.

In Egypt, stark epidemics of late blight has been occurred in both 2000 and 2001, making the potato production process a major challenge. The outbreak of blight during the season depends on the period of adapting environmental circumstances such as temperature and relative humidity (Fahim et al. 2007).

11.3 The Late Blight Disease

Potato late blight is debatably one of the most notorious diseases in cultivation. It is premium identified as the reason for the Irish potato scarcity of the 1840s, caused in the death or migration of more than 2 million persons from Ireland (Martin et al. 2013). Also, potato late blight urged the advent of plant pathology field. To appreciate the reason for the disease, Anton de Bary confirmed that the fungal-like growth detected on blighted plants was the reason for late blight, compared to the result of the disease's existence. His results led further scientists to examine other fungi and bacteria related to plant disease, also through the progress of Koch's hypothesizes for founding pathogenicity, occasioned in the documentation of the causal agents of several crop diseases (Schumann 1991).

The causal agent *Phytophthora infestans* which is mainly of potatoes and tomatoes nonetheless has been identified to attack other associates of the Solanaceae family as well. Moreover, it is distinguished by the presence of a asexual and sexual life cycle https://usablight.org/lateblight. *Phytophthora infestans* reproduces asexually mainly, however in specific situations, the sexual cycle is principal. *Phytophthora infestans* reproduces principally asexually. However, other reports indicated the occurrence of occasional sexual reproduction (Danies et al. 2014). As a consequence, outbreaks of *P. infestans* might be identified as a portion of a clonal lineage, that characterized for certain characteristics, like aggressiveness and fungicide sensitivity.

Late blight is a "community disease", due to its capacity to rapidly spread from one field to another under appropriate environmental conditions. The bacterium is easily transmitted by wind with cold, wet weather to quickly infect surrounding fields. As such, understanding the disease symptoms and dealing with it when detected is necessary to prevent the disease from becoming into epidemic (https://usablight.org/lateblight).

11.4   The Pathogen

The sexual oospores of the causal organism was first discovered in the 1950s in Egypt and Mexico, and the existence of two mating forms, named A1 and A2, recognized (Al-Arousi et al. 1962; Gallegly and Gallindo 1957; Niederhauser 1956). Then, single R-genes governed resistance to late blight were recognized from the wild species *Solanum demissum*. Therefore, it is potential to discover R genes and their relation to virulence races. In the final, scientists succeeded in utilizing a series differential potato clones, each with a different single R-gene, to distinguish complementary avirulence/virulence races in groups of *P. infestans* (Malcolmson 1969; Malcolmson and Black 1966). Till the 1980s, both mating type and virulence to R-genes were the unique features for *P. infestans*, However lately a set of DNA markers were developed and used to study genetic variation in the pathogen population (Cooke and Lees 2004). The global population of *P. infestans* seemed to be asexual and to contain a single clonal ancestry of A1 mating type defined by a single genotype (one mitochondrial DNA haplotype, one multi-locus restriction fragment length polymorphism (RFLP) genotype and one di-locus isozyme genotype). However, in Egypt, the population was sexual and contained both genotypically diverse mating types A1 and A2 (Grünwald and Flier 2005). It is widely assumed that new strains of the pathogen migrated within shipments of imported potatoes (Niederhauser 1991), this led to following population variations counting the introduction of the A2 mating type (https://link.springer.com/content/pdf/10.1007/s11540-011-9187-0.pdf).

The mating type A2 was isolated from infested potatoes introduced from Egypt. The oospores created in the mating with welsh isolates grew readily (Abdel-Satter and Shaw 1985). In a study of 420 *Phytophthora infestans* isolates from potato, 93.1% were listed as A1 mating type, 5.3% belong to A2, and 1.6% were self-fertile. Isolates of potato were as pathogenic on tomato like the tomato isolates. Twenty-three virulence races were identified amongst the hundred verified isolates by nine potato differentials i.e. R1, R2, R3, R4, R7, R8, R10 and R11 (El-Korany 1994).

11.5   Diagnose the Disease

The diagnosis process is the basis on which the bases of disease control are based, especially if it requires chemotherapy or human intervention. Therefore, the accuracy and speed of the diagnosis are very important in the prevention of plant diseases and prevent or reduce the economic loss caused by infectious diseases (Youssef 2011).

11.5.1 Traditional Methods

11.5.1.1 Stages of Disease Diagnosis

First Stage: Field Diagnosis
Depending primarily on observations on the affected plants such as symptoms, signs of disease, distribution of disease in the field, the plant itself and its relationship to the ecosystem as well as information that can be obtained from farmers.

Second stage: laboratory diagnosis
There are some initial tests which may be useful in some cases to infer the disease, such as:

- Conduct a rapid microscopic examination.
- Place the sample in the humidifier chamber to encourage the growth of the causative.
- Use of isolation methods on different media.
- Reliance on accurate and sensitive methods (Youssef 2011).

11.5.1.2 Diagnostic Methods

Diagnostic can be divided into groups as follows:
First: Farm and physiological tests:
This step which is based on the biological processes carried out by the microbe such as the ability to analyze some compounds, starch, fat, gelatin and other respiratory breath test. These tests are useful in the case of fungi and bacteria and yeast, but not suitable for the diagnosis of viruses, viroid's and bacteria difficult in growth.

Second: Biological Tests:
This step depends on the relationship between the pathogen and the host. Include the pathogenicity tests, hypersensitivity, host range and ways of transmission of the pathogen, which is one of the most important tests that confirm the Koch's postulates to isolate and identification of pathogen.

Third: Microscopic Methods:
Depending on the use of optical microscopy or electronic types to identify the various pathogens, study their structural characteristics, the knowledge of their impact on the host or carrier, their location in the cells, increases the sensitivity of these microscopes use specialized dyes, chemical detectors and antibodies.

Fourth: Serological tests:
Which depends on the use of protein (antibody) in the detection of the cause of unknown pathogens and is suitable with several pathogens and the most famous viral infection is not suitable with the virus and divided into:

1. Sedimentation tests include the test of the deposition in the tubes and the testing of the assembly and the diffusion test in the agar which are intermediate sensitivity tests and consume a large quantity of antibodies. However, the propagation test during agar is still useful in information on the proximity of some viral isolates.

2. Tests of color and include Eliza tests by direct and indirect methods. These tests are carried out in special dishes. This is one of the most common methods of detecting viral diseases, fingerprints, and histology. It is also a color test used to detect viruses on a membrane of nitrocellulose for its ability to absorb protein, which is useful in the field of remote diagnosis.

The previous methods have pros and cons in detecting pathogens, which required the search for more sensitive techniques for rapid detection of the pathogen. So, the trend was to use genetic material is the perfect solution and has become the use of methods that depend on the genetic material of important technology. It gave high precision and sensitive results as the nucleic acids are complex chemical compounds with weights of molecular high cannot be the hives dispensed and so called for their concentration in the nucleus is divided into two types of DNA function carrying genetic information from one generation to another. The second type RNA function is the transmission of genetic material from the nucleus to the cytoplasm. The manufacture of proteins and nucleic acids are considered distinct qualities of each object from others and hence the importance in the process of diagnosis (Youssef 2011).

11.5.2   DNA Markers

A rapid and simple polymerase chain reaction was developed to recognize the pathogens of *Phytophthora infestans* oospores, causing the disease of potato late blight in the soil. This method included the disturbance of oospores by crushing dry soil, with abrasive characteristics in the presence of glass powder and skimmed milk powder in a short period. This leads to avoids loss of DNA through adsorption to soil particles and reductions the co-extraction of PCR inhibitors with the DNA. After phenol/chloroform extraction, the DNA is appropriate for PCR amplification without a precipitation step (https://www.ncbi.nlm.nih.gov/pmc/articles/PMC 4150230). This amplification leads to the discovery of the pathogen in infected soils before planting of the crop. The real-time PCR assay describes as highly sensitive and specific and has numerous advantages compared to conventional PCR analyzes used for detecting *P. infestans* to confirm positive inoculum level in seeds of potato and elsewhere (Hussain et al. 2014).

11.5.2.1   The Methods that Depend on DNA Molecules

PCR Test of Various Types

Polymerase chain reaction PCR is an operative and cheap tool to amplify minor segments of DNA or RNA. This technique combines between the principles of complementary nucleic acid hybridization with that of nucleic acid repetition which practical repeatedly through several cycles. Its consequences in the production of the definite target DNA/RNA sequences through a factor of 107 are done inside a little time. The amplification technique amplifies one copy of the DNA target through using two synthetic oligonucleotides "primers" that link to the target sequence, which has been prolonged by a Taq polymerase. Automated technique of repetitive cycles (commonly 25–40) of denaturation of the pattern DNA (at 94 °C), annealing of primers with their analogous sequences (50 °C), and extension of the primer (70 °C) is used to amplify the target sequence (Bartlett and Stirling 2003).

Applications of Polymerase chain reaction "PCR"

- Recognition and description of communicable agents.
- Determination of microbes in patient examples.
- Recognition of microorganisms in cultures.
- Detection of antimicrobial resistance.
- Study the association between the interest pathogens.
- DNA fingerprinting.
- Discovering mutations and genetic diseases.
- Cloning genes.
- PCR sequencing (https://microbeonline.com/polymerase-chain-reaction-pcr-steps-types-applications…)

In current centuries, PCR takes appeared as an influential device in the diagnosis of plant diseases, it does not need to isolate pure fungal cultures of infested tissues. It is more sensitive, accurate, faster and less expensive compared to traditional diagnosis (Wangsomboondee and Ristaino 2002). So, PCR is considered one of the first techniques that rely on the genetic material and the most accurate and sensitive in the detection of what is unknown, but it remains the most important aspect is to reach the genomic structure of the pathogen and study its expression.

Methods of Hybridization of DNA

A novel procedure was advanced for DNA extraction from *P. infestans* oospores. Baka (1997) study of 19 isolates from the Nile Delta, three distinct sets has been definite by isoenzyme investigation. Hussain et al. (2014) stated that oospores can be identified quantified by Real time PCR (both A1 and A2 mating type) and can detect their viability with MTT dye staining under microscope difference amongst dormant and germinating oospores. To discovery in the soil, where there is no noticeable injury tissue to sample from soil baiting has the improvement that it permits great quantities of soil (0.5–1 kg) to be verified while DNA extraction procedures are all founded on the extraction of small (1–10 g) samples.

So, when manipulating with samples of DNA extraction and analysis by polymerase chain reaction, the pathogen may be lost. The baiting also has the advantage that it detects only viable pathogens. Recently Hussain et al. (2015) established a novel adapted baiting technique for recognition *P.infestans* from potato fields. Regardless of the advantages, baiting is also slow and little throughput to be beneficial and could be subject to a great degree of untrue refusals too.

The quantities of inhibitory substances differ by soil type, vegetation type, and structure of the micro flora in the soil. For example, the micro flora differed even over minor distances (1 m scale) (Scala and Kerkhof 2000). PCR amplification efficiency is likely to differ widely even over small distances. So critical that an interior normal is used for PCR examination of soil samples (Hussain et al. 2005). They verified the positive discovery of *P. infestans* for up to twelve months of the soil in which infected leaf tissue were buried and for up to 24 months of soil having leaf tissue infested with both mating types.

A novel procedure was established for the extraction of DNA from *P. infestans* oospores. Use of DNA-based diagnostic techniques will help in international efforts to control the introduction of imported diseases into new geographic regions. Dependable and fast methods are necessary by different diagnostic laboratories to select genotypes for vital pathogens (Touseef 2018).

RFLP Test (Restriction Fragment Length Polymorphism)

Restriction Fragment Length Polymorphism (RFLP) is a useful technique for study differences in homologous DNA sequences and discriminate genotypes. Also to determine the sites of genes within DNA sequence. The term refers to polymorphism, which distinguished through the differing sites of restriction enzyme regions. Sample of DNA is digested into parts using specific restriction enzymes and the subsequent fragments are then isolated by gel electrophoresis based on the size (https://en.wikipedia.org/wiki/Restriction_fragment_length_polymorphism).

An RFLP probe is a branded DNA sequence which hybridizes with one or more fragments of the digested DNA sample after being separated by gel electrophoresis. Thus indicating a unique blotting arrangement characteristic to a specific genotype at a specific locus. Short, single- or low-copy genomic DNA or cDNA clones are exploited as RFLP probes. The RFLP probes are exploited in genome mapping and in variation analysis viz. genotyping, forensics, paternity assessments, diagnosis of genetic diseases, etc. RFLP analysis is useful in the work of DNA fingerprinting and characterization of genetic variability or breeding materials in plant populations and between pathogenic fungi (Saiki et al. 1985). Andrew et al. (2001) applied DNA-Markers on selected isolates of *Phytophthora infestans* collected from England and Wales and fingerprinted by multi-locus RFLP analysis.

RAPD Test (Random Amplified Polymorphic DNA)

Markers are DNA segments from PCR amplification of random fragments of genomic DNA with a single primer of an arbitrary nucleotide sequence (https://www.ncbi.nlm.nih.gov/probe/docs/techrapd/).

RAPD does not need any specific information on the target organism's DNA sequence compared to PCR analysis: the identical 10-mer primers may be amplify a segment of DNA, dependent on sites that are paired with the primers' sequence.

So, no segment is generated if primers annealed too far apart or if the ends of 3' of the primers are not fronting each other. Therefore, if a mutation has happened in the arrangement of DNA on the region that was complementary to the primer, a PCR product will not be formed, causing a diverse pattern of amplified DNA segments on the gel. Determining the correct sequence of the introductory material is important because the different sequences will gave different types of bands, and may allow a more specific perception of individual genotypes (Mbwana et al. 2006). Adillah and Nawsheen (2017) used RAPD fingerprinting with 40 random oligomers. They found that all the isolates of *Phytophthora infestans* were displayed to belong to mitochondrial type II and mating type A2.

Quantitative and Qualitative Testing (Real-Time PCR)

A real-time polymerase chain reaction (Real-Time PCR), identified as quantitative polymerase chain reaction (qPCR), is procedure of molecular biology according to the polymerase chain reaction (PCR). In this analysis it is amplified the targeted DNA molecule through the PCR, in real-time, and not at its end, as in PCR. Real-time PCR could be used quantitatively and semi-quantitatively, with a certain amount of DNA molecules (Bustin et al. 2009).

Two common approaches for the detection of PCR products in real-time PCR are non-specific fluorescent dyes which intercalate with any double-stranded DNA, and sequence-specific DNA probes containing oligonucleotides that are labeled with a fluorescent reporter. This permits discovery only after hybridization of the probe with its corresponding sequence (Julie et al. 2009).

Qualitative PCR is used to identify nucleic acids and recognized the infectious diseases, cancer, genetic abnormalities and to identify new strains (Espy 2006). The PCR assay was highly sensitive for identifying *Phytophthora infestans*, only 1 pg of purified DNA or DNA isolated from 500 sporangia was required to identify the pathogen. Applications of serological and molecular instruments were operative to distinguish the pathogen in infested symptomatic, asymptomatic tissues of leaves and tubers of potato and can also offer important information on *Phytophthora infestans* potato interaction and disease progress (El-Komy et al. 2010).

11.5.3 Diagnosis Remotely

Finally, the development of diagnostic methods and effective techniques that enable us to identify diseases is vital. The selection of various programs to monitor the health status of plants and the development of control programs and the production of crop plants free from disease represent the strategies to control the disease (Youssef 2011).

11.5.4 Bioinformatics

In recent periods, rapid progresses in genomics and proteomics have motivated the availability of a large amount of biological data. The conclusions drawn from these statements require accurate mathematical analysis. Bioinformatics is known as an interdisciplinary science for the interpretation of biological data via information technology and computer science (Alemu 2015; Xue et al. 2008).

The new field of investigation will become increasingly important as it continues to generate and integrate great amounts of genomic and proteomic data. There is parallel progress in the demand for methods in data management, visualization, integration, analysis, modeling, forecasting, and the increasing volume of data. Bioinformatics plays a vital role today in plant pathology in the improvement of new plant diagnostic tools. Pathogenic factors are among the traits considered in the main interest of plant bioinformatics (Alemu 2015).

Progress in bioinformatics has contributed to the possibility of mapping the genome of numerous organisms in more than a decade. These discoveries and current efforts to identify gene and protein functions have improved the capability to know the causes of plant diseases and to discover new treatments. This opens the way to the future of many innovations in the field of bioinformatics through data requirements and analysis of life sciences. Bioinformatics includes various practical applications in the management of current plant diseases, studying the interactions of host nurses, understanding the genetic factors of the pathogen of a nurse, ultimately helping to design the best management options (Koltai and Volpin 2003; Roca et al. 2004; Alemu 2015).

Genomic studies attentive on complete genome study, have opened up a novel range for biology in universal and for agriculture in specific. Lengthways with the usage of plant models and the advancement in sequencing agriculturally imperative organisms, the mixture of bioinformatics and functional genomics worldwide improve agricultural genomics. These information will lead the way towards a better understanding of the plant pathogenic biological network, and thus lead to the breaking of ideas in promoting plant resistance to pests (Koltai and Volpin 2003). Bioinformatics is acting an excessive role in controlling plant disease (https://www.iiste.org/Journals/index.php/ALST/article/viewFile/19040/19285).

11.6 Control Methods

Sustainable approaches to late blight management are required to reduce yield losses due to late blight problems caused by the use of fungicides. An integrated disease approach includes a set of different control practices used in suitable timing before and after the diseases emerge. This approach aims to reduce the incidence of pathogen, the effect of the source of inoculation and the time of interaction between host variety and pathogens (Nutter 2007). This is useful over single controlling repetition as joint methods have maximum effect on the pathogen and cover different feature of the disease management (https://journals.psmpublishers.org/index.php/biolres/article/download/71/43).

11.6.1 Chemical Methods

The use of chemical pesticides in the control of field crop pests and horticulture, both in the open fields and in the plastic hull, has a significant impact on the beneficial organisms and affects the natural enemies of plant pests, especially in the direct use of these pesticides. It is known that the use of pesticides is more harmful than the use of bribes for the difficulty of controlling the powders when fogging, especially during the season of flowering. The widespread use of chemical pesticides has a significant impact on environmental pollution and damage to the general health of humans and animals.

In the early 1990s, the World Health Organization (WHO) estimated that three million people a year suffer from the effects of pesticide use. And that most farmers were storing chemical pesticides in their homes in unsafe ways, especially in developing countries. The organization provides information and training to reduce the harmful effects of such pesticides.

In addition, many pesticides have become ineffective in resistance to plant pathogens in order to develop resistance status in these pathogens. FAO is also interested in issuing brochures that raise the level of safety when using pesticides. It is also useful to develop training programs aimed at reducing the harmful effects of plant pathogens and reducing environmental pollution.

The magnitude of crop losses is a major problem of late blight, particularly in regions with heavy rainfall and low temperatures. To reduce the loss of crop productivity, fungicides of various chemical properties are used as an aerosol spray on potato leaves to minimize the negative impacts of late blight. Whereas, based on the toxic effects of these chemicals, they can hurt public health, cause environmental pollution and incur heavy costs (Majeed et al. 2011; 2017). To reduce crop losses due to late blight and environmental harms posed by fungicides application, sustainable approaches for late blight management are required (https://journals.psmpublishers.org/index.php/biolres/article/download/71/43).

11.6.2 Resistant Cultivars

The use of resistant varieties is a vital aspect of sustainable management methods. Regarding its use, many plant features have been considered as traits related to host defense and determination of resistance genes (Bengtsson et al. 2014).

The resistant potato varieties to *P. infestans* infection were produced by commercial breeding companies, play a vital role in the sustainable management of the late potato disease. Cultivation of resistant cultivars started at the beginning of the twentieth century when the first resistance genes (R genes) were recognized in closely associated species (*Solanum demissum*) (Fry 2008). Unfortunately, after the production of resistance-bearing genes on a large scale, the R genes of *S. demissum* were broken down as a result of pathogen development. Now, numerous new resistance genes from diverse genetic resources are being used in traditional breeding programs to improve new resistant cultivars (Lammerts Van Bueren et al. 2008).

Garrett et al. (2001) showed that cultivation of various potato-resistant varieties and diverse genetic resources leads to reduce crop losses and to adopt somewhat fungal pesticides. Kirk et al. (2005) stated that host resistance to *P. infestans* can be helpful in reducing fungicides rates.

Hajianfar et al. (2014) showed the task of producing genotypes with resistant genes against late blight is faced by a new occurrence of a rapidly changing pathogen's genotypes capable of sexual reproduction throughout the worldwide. So, the resistant gene group plays a major role in the stability of the varieties. A number of minor genes and specific race-resistant genes have been known. Recently, resistant genes have been identified, isolated and introduced to sensitive species through breeding programs. So, it is important to study the genetic background of late blight resistance. Major resistant genes and other genes involved in resistance response as well as the identified QTLs are important in the development of molecular tools (http://real. mtak.hu/27954/) that can be used efficiently in breeding resistant varieties.

Fast alterations in the population of *P. infestans* can be managed by two alternate strategies comprising the use of more fungicides or cultivate of potato cultivars with horizontal resistance to the pathogen. The second approach could also reduce fungicide uses and reduce the crop production costs for farmers and less environmental pollutions. Hence, developing resistant cultivars is the focus of recent breeding programs (Inglis et al. 1996; Peters et al. 1999).

Releasing cultivars with resistant genes from wild relatives takes a long time. Therefore further agricultural practices are obligatory to protect new resistant genes in cultivars from breakdown of resistance (https://link.springer.com/article/10.1007/ s13593-016-0370-1).

11.6.2.1 The Role of Bioinformatics in Developing Disease Resistance Cultivars

Durale disease resistance is important, nevertheless the far-reaching goal in crop improvement programs. The study of the genome has a major effect on efforts to deal with plant diseases by identifying resistance genes available for crop improvement. Also, genomic studies on pathogens provide an appreciation of the molecular basis and an opportunity to choose more sustainable resistant cultivars (Michelmore 2003).

Plant genomics also aims to know the molecular basis of all biological practices in crop-associated species. This understanding is necessary to allow the effective utilization of plants as genetic resources in improving new cultivars with high quality and reducing economic and environmental costs. This knowledge is also essential for developing new plant diagnostic tools for resistance to pathogens, tolerating abiotic stress, improving plant quality characteristics, and reproductive features that determine yields (Vassilev et al. 2006). A genome programme can now be envisioned as a highly essential tool for plant improvement (Vassilev et al. 2005). Such an approach to recognize key genes and know their function will result in a quantum leap in crop improvement (https://www.iiste.org/Journals/index.php/ALST/article/viewFile/19040/19285). Moreover, the capability to study gene expression will permit us to identify how plants react with the physical environment and managing applies (Vassilev et al. 2006). Bioinformatics plays important roles in producing disease resistant cultivars. It will be vital for obtaining and establishing great quantities of information. Also, allows the supply of information from heterogeneous data sets (https://www.iiste.org/Journals/index.php/ALST/article/viewFile/19040/19285) to help the selection of promising genotypes.

The diseases resistance is one of several important features of a breeding program. Thus, bioinformatics will play a significant role in integrating the phenotypic and relative information of agricultural traits and resistant characters (Vassilev et al. 2006). Rendering to Vassilev et al. (2005) showed that the important role of bioinformatics for crop improvements include recorded all sequence information into the public domain through repositories to offer lucid annotation of genes, proteins and phenotypes, and to intricate relations both within and between crops and other organisms to provide outcomes counting sequence information, information for mutations, DNA markers, maps, functional discoveries, and others (https://www.iiste.org/Journals/index.php/ALST/article/viewFile/19040/19285).

11.6.3   Alternatives of Fungicides

11.6.3.1   Biological Control

The term biological control has a different meaning. Baker and Cook (1974) have proposed defining "biological control is reduction of inoculum or disease producing activity of a pathogen accomplished by or through one or more organisms other than man." This definition is broad adequate to encompass classic approaches to biocontrol that influence pest population as well as newly emerging biocontrol strategies. The biological control agents have two mechanisms of action, direct action on the pathogen and indirect mechanisms of activity. The direct mechanisms involve antibiosis, competition for nutrients and parasitism or predation. Indirect mechanisms involve alterations in plant physiology. Some of these indirect mechanisms include: the production of plant growth hormones that speed plant or root growth, improvement in water or nutritional status of the plant because of a more efficient root system, and increased nitrogen fixation. These mechanisms can lead to escape from disease or improve tolerance to stress (Jacobsen 1993).

Biological control of plant pathogens provides good opportunities for managing plant diseases. Isolation and identification of potential biological agents from the host itself and its use for managing of *P. infestans* is a prerequisite in this regard. Some studies have been performed in this direction (Kuznetsova et al. 1995, 1996; Webster 1997; Falconi and Benalcazer 1999; Gupta et al. 2004), but more needs to be done to take advantage of the full potential of biological control agents that occur in nature.

Now a day, procedures of biological control are gaining significance as these are non-toxic and also environment-friendly. While, *Phytophthora infestans* multiplies very fast; hence, the biological control method alone is not enough to manage late blight. Hence, the integrated pest management is essential. Lal et al. (2017) evaluated eleven treatments comprising biocontrol agents and fungicides against the late blight. The results revealed that the treatments are operative for dealing with the disease up to certain level. Whereas, lowest average disease severity was verified in treatment when *Bacillus subtilis* + *Trichoderma viride* was added before disease appearance followed by cymoxanil 8 + mancozeb 64%WP at the onset of late blight and one more spray of B5 + TV after seven days. These treatments could be integrated into farmer practices.

The efficacy of four bioagents, viz., *Penicillium viridcatum, Trichoderma harzianum, T. viride* and *Myrothecium verrucaria* were assessed for controlling potato late blight. The four biocontrol agents provided good control against late blight development both on detached leaves and whole plants of cultivar Kufri Chandra mukhi. *Trichoderma viride* and *Penicillium viridcatum* were better than the others (Gupta 2016).

The effect of *Trichoderma* isolates against *P. infestans*, and the possible modes of action involved were studied. In vitro assessments between *P. infestans* and *Trichoderma* isolates proved that the *P. infestans* colony was meaningfully repressed and overgrown through *Trichoderma* isolates. Antifungal products formed by the isolate HNA14 significantly prohibited the linear growth of the *P. infestans* community. Mycoparasitism leads to the aggressive nature of the *Trichoderma* isolate HNA14 in contradiction of *P. infestans*. The isolate HNA14 in planta bioassay reduced significantly disease index and enhanced the plant height as well as foliar fresh and dry weight. The *Trichoderma* isolate HNA14 under field conditions appeared to be more effective in controlling the pathogen out of all *Trichoderma* strains, and considerably decreased the disease severity rather than the control (Yao et al. 2016).

11.6.3.2   Plant Oils

Alternatives to synthetic fungicides now used to the control oomycete pathogen *Phytophthora infestans*, of potato late blight disease. The use of biologically based compounds in plant extracts might be an alternative to presently used fungicides to control fungi and bacteria because they virtually constitute a rich basis of bioactive substances such as phenols, flavonoids, quinons, tannins, alkaloids, saponins and sterols (Isman 2000; Burt 2004).

Since these extracts can be active against fungal and bacterial pathogens, are biodegradable to non-toxic products, and are potentially suitable for use in integrated pest management programs (http://www.entomoljournal.com/archives/2016/vol4issue1/PartA/3-5-121.pdf), they could become new class of safer disease control agents. Some phytochemicals of plant origin (i.e. azadirachtin, carvone, pyrethroids) have been formulated as botanical pesticides and are used successfully in integrated pest management programme (Shmutterer 1990).

Most of the efforts on antimicrobial impacts of essential oils have been conducted on human or food pathogens. To the best of our awareness, there is no information available on the antifungal action of essential oils from aromatic plants exploited against *P. infestans*. So, efforts have concentrated on the antimicrobial activities of essential oils, attained from aromatic plants growing in the Eastern Mediterranean Region of Turkey, against *P. infestans* (Soylu et al. 2006). The citrus essential oils *Citrus sinensis* Cadenera, *Citrus limon* Eureka and *Citrus bergamia* Castagnaro diverse "in their in vitro inhibitory effects on sporulation, mycelial growth and survivability of *P. infestans* isolates" (http://www.tandfonline.com/doi/abs/10.1080/10412905.2014.982877). However, the in vivo infection level of detached potato foliage in soaking treatment was appeared to be important and confirmed the fungicidal impact of citrus essential oils, which were in agreement with in vitro results. Bergamot essential oil was the best in both in vitro and in vivo assessments, followed by the orange and lemon essential oils (Moumene et al. 2015). Moreover, Quintanilla et al. (2002) studied the effect of up to 19 essential oils (EOs) against potato late blight caused by *Phytophthora infestans* in vitro level and under greenhouse circumstances. In a first screening exposed to an inhibition examination by the EOs in a standard plate assay assessing the diameter of the development fungal zone, they identified two isolates of *P. infestans*. Demonstrative oils designated from this first screening were applied to susceptible potato genotypes and examined the effects of both the fungicidal and phytotoxic. Whereas utmost of the EOs tried in the fungus plate assay displayed no desirable inhibitory action in contradiction of the pathogen, four of the EOs formed an inhibition zone of 80%. They observed that EOs from thyme gave an 89% inhibitory effect on the fungus isolates, however did not show defense in the potato, also caused phytotoxic symptoms. On the contrary, treatment with hyssop in the plate assay revealed only up to 45% of inhibition, while hyssop stimulated good fungus defense for the plants and enhanced the plant growth without any toxicity effects on plants

Soylu et al. (2006) evaluate antifungal activities of essential oils against *P. infestans* which extracted "from aerial organs of aromatic plants such as oregano (*Origanum syriacum* var. *bevanii*), thyme (*Thymbra spicata* subsp. *spicata*), lavender (*Lavandula stoechas* subsp. *stoechas*), rosemary (*Rosmarinus officinalis*), fennel (*Foeniculum vulgare*), and laurel (*Laurus nobilis*)" (https://link.springer.com/content/pdf/10.1007/s11046-005-0206-z.pdf). They found that various essential oils lead to inhibit the growth of *P. infestans* in a dose-dependent mode. Volatile phase effects of both oregano and thyme oils air were found to entirely prevent the growth of *P.*

infestans. Entirely growth inhibition of pathogen was achieved by essential oil of fennel, rosemary, lavender and laurel. To determinate of the contact phase effects of the tried essential oils, oregano, thyme and fennel oils were found to prevent the growth of *P. infestans* entirely. The effectiveness of the thyme oils inhibits the growth of the fungus. Rosemary, lavender and laurel essential oils were inhibitory at fairly higher doses. Whereas, volatile phase effects of essential oils appeared to be more active on fungal growth rather than contact phase effect. Also, inhibition of sporangial production by the essential oil has been observed. Results of light and scanning electron microscopic on pathogen hyphae, treated by both volatile and contact phase of oil, showed great morphological alterations in hyphae like cytoplasmic coagulation, vacuolations, hyphal shrivelling and protoplast leakage.

11.6.3.3 Plant Extracts

It is preferable to use alternatives to chemical pesticides for instance plant extracts (medicinal and aromatic plants as well as wild plants) to protect crop plants from fungal, bacterial, viral and nematode organisms that affect different field and horticultural crops and cause major losses to the national economy. Moreover, decrease the losses caused by the infection both during the planting season and in the post-harvest to keep pace with modern fashions in disease resistance. An important use of medicinal and aromatic plants as natural insecticides is to kill or repel insects such as neem, and also to combat fungus and plant-damaging bacteria as well as naturally harmful human and animal diseases. The importance of such plants in the safe use of plant diseases as an alternative to chemical pesticides is obvious.

Egypt is rich in medicinal and aromatic plants and wild plants due to its varied climate and suitable for the growth of these rare and important plants and is a huge source of wealth has registered more than 350 plant species with medicinal or aromatic use. There are many medicinal and aromatic plants cultivated in private farms spread in many provinces such as Faiyum and Beni Suef.

In Sinai, there are many medicinal and aromatic plants that grow wild and have medical therapeutic benefits. Medicinal and aromatic plants contain volatile oils and glycosides with a beneficial effect, which play a vital role in increasing the effectiveness of these plants.

The integrated control approaches are used for late blight control with sustainable production advantages of the crop and less injuries to the environment. Majeed et al. (2015) supported the use of plant extracts and natural combinations as another treatment to control late blight and reduce crop losses (https://journals.psmpublishers.org/index.php/biolres/article/download/71/43).

11.6.3.4 Nanotechnology

One of the most important objectives of agricultural policy in any country in the world is to improve productivity and increase the quantity of agricultural products to meet the needs of the growing population, what is the role of nanotechnology in achieving this objective?—The nanotechnology began in the manufacture of nanoparticles containing fertilizers added pesticides to fight all types of agricultural pests after studies have shown the negative effects of the current pesticides on human health and environment through pollution of the soil and water used to irrigate crops to reach time to groundwater and pollution.

Nanotechnology is defined as the treatment of the material on an atomic, molecular and supramolecular scale. The first, common explanation of nanotechnology has been described by Drexler (1986, 1992). This technology aims to accurately manipulate atoms and molecules to manufacture microscopic products, now denoted to as molecular nanotechnology.

Also, nanotechnology defines as the manipulation of material with at least one dimension sized from 1 to 100 nm. This reveals the fact that quantitative mechanical effects are important in a quantum-realm measure, and therefore the definition shifted from a technological aim to a research category that includes all types of research and techniques that deal with the specific properties of matter occurring less than the size limit. Therefore, the formula "nanotechnology" as well as "nanotechnology" refers to a wide range of research and applications that share volume. Given the variety of potential applications, governments have invested billions of dollars in nanotechnology investigation. During 2012, the USA has invested $3.7 billion using its National Nanotechnology Initiative, while the European Union has invested $1.2 billion, and Japan has invested $750 million (https://www.youtube.com/watch?v=doN3eQqo1So).

11.7 Conclusions

Late blight is one of the greatest serious fungal diseases affecting potatoes. The disease spreads almost everywhere grown in potatoes. The disease is of importance worldwide, particularly in traditional potato growing areas due to the disruption of potato production. Losses may be as high as 100% if the disease is not controlled, and low levels of infection affect the crop validity for storage. The disease can easily be controlled by the use of fungicides. In many developing areas, however, chemical control is hardly feasible for the farmer due to the high cost of fungicide applications. Furthermore, in nearly all developing countries, the fungicides or their active ingredients are imported, making the potato an expensive vegetable and at the same time costing the country valuable foreign exchange. The present finding revealed management approaches for its effective control to late blight include the use of alternatives of fungicides i.e. resistant cultivars, biological control, plant oils, plant extracts and nanotechnology.

11.8 Recommendations

In some countries such as Egypt, the use of fungicides excessively leading to the impact on human health and harm to the environment. This uses lead to the emergence of new strains of pathogens of the plant. All this makes us underestimate and guide the fungicides and use them only when necessary and then follow other ways to resist or combat pathogens.

Uses of resistant cultivars, biological control, plant oils, plant extracts and nanotechnology are alternatives to synthetic chemical pesticides in controlling the late blight disease. However, pesticides will continue to be required in many production systems if quality and competitiveness are to be maintained by producers. Therefore, within the foreseeable future resistance cultivars, biological control, plant oils and extracts and nanotechnology will be utilized in the management system as alternatives and supplement to pesticides.

References

Abdel-Sattar MA, Shaw DS (1985) Sexual production and oospore germination of Phytophthora infestans. In: First National Conference of Pests and Diseases of Vegetables and Field Crops in Egypt, 21–23 October 1985. Faculty of Agriculture, Suez Cana University, Ismailia, Egypt, pp 978–998

Adillah I, Nawsheen T (2017) Genotypic characterization of *Phytophthora infestans* from mauritius using random amplified polymorphic DNA (RAPD), mitochondrial haplotyping and mating type analysis. Plant Pathol J 16:121–129

Al-Arousi H, Michael S, Abdul Rahim MA (1962) Plant diseases. Knowledge facility in Alexandria

Alemu K (2015) The role and application of bioinformatics in plant disease management. Adv Life Sci Technol 28:28–33

Andrew IP, Nicholas DP, Jenny PD, Richard CS, David SS, Susan JA (2001) AFLP and RFLP (RG57) fingerprints can give conflicting evidence about the relatedness of isolates of *Phytophthora infestans*. Mycol Res 105(11):1321–1330

Anwar D, Shabbir D, Shahid MH, Samreen W (2015) Determinants of potato prices and its forecasting: A case study of Punjab. University Library of Munich, Germany, Pakistan

Baka ZAM (1997) Mating type, nuclear DNA content and isozyme analysis of Egyptian isolates of *Phytophthora infestans*. Folia Microbiol 42(6):613–620

Baker KF, Cook RJ (1974) Biological control of plant pathogens. W H Freeman Co, San Francisco

Bartlett JMS, Stirling D (2003) A short history of the polymerase chain reaction PCR protocols methods in molecular. Biology 226(2nd ed):3–6

Bengtsson T, Holefors A, Witzell J, Andreasson E, Liljeroth E (2014) Activation of defence responses to *Phytophthora infestans* in potato by BABA. Plant Pathol 63:193–202

Burt S (2004) Essential oils: their antibacterial properties and potential applications in foods. Rev Int J Food Microbiol 94:223–253

Bustin SA, Benes V, Garson JA, Hellemans J, Huggett J, Kubista M, Mueller R, Nolan T, Pfaffl MW, Shipley GL, Vandesompele J, Wittwer CT (2009) The MIQE guidelines: minimum information for publication of quantitative real-time PCR experiments. Clin Chem 55(4):611–622

Cooke DEL, Lees AK (2004) Markers old and new for examining *Phytophthora infestans* diversity. Plant Pathol 53:692–704

Danies G, Myers K, Mideros MF, Restrepo S, Martin FN, Cooke DEL, Smart CD, Ristaino JB, Seaman AJ, Gugino BK, Grünwald NJ, Fry WE (2014) An ephemeral sexual population of *Phytophthora infestans* in the northeastern United States and Canada. Plos one Online Journal. https://doi.org/10.1371/journal.pone.0116354

Drexler KE (1986) Engines of creation: the coming era of nanotechnology. Doubleday. ISBN 0-385-19973-2

Drexler KE (1992) Nanosystems: molecular machinery, manufacturing, and computation. Wiley, New York. ISBN 0-471-57547-X

El-Ganainy SMA (2013) Epidemiology and characterization of *Phytophthora infestans*, the cause of late blight disease in potato. Ph.D Thesis Menoufia University Egypt, 172 p

El-Komy MH, Abou-Taleb EM, Abo-Shosha SM, El-Sherif EM (2010) Serological and molecular detection of late blight pathogen and disease development in potato. Int J Agric Biol 12:161–170

El-Korany AE (1994) Pathological studies on late blight of potato caused by *Phytophthora infestans*. PhD Thesis Department of Agricultural Botany, Faculty of Agriculture Suez Canal University, Ismalia, Egypt 115 p

Espy MJ (2006) Real-time PCR in clinical microbiology: applications for routine laboratory testing. Clin Microbiol Rev 19(3):165–256

Fahim MA, Medany MA, Abou Hadid AF, Mosa AA, Mostafa MH (2007) Ecological studies and prediction of the onset of potato late blight under Egyptian climatic conditions. Acta Hortic 729:453–457

Falconi C, Benalcazar G (1999) Use of bacteria and fungus from phyllosphere as biological controllers of late blight. Proceedings of Global Initiative on Late Blight. A threat of global food security held on 16–19 March at Quito Ecuador

Fry W (2008) *Phytophthora infestans*: the plant (and R gene) destroyer. Mol Plant Pathol 9:385–402

Gallegly M, Gallindo J (1957) The sexual stage of *Phytophthora infestans* in Mexico. Phytopathology 47:13 (Abstr)

Garrett KA, Nelson RJ, Mundt CC, Chacon G, Jaramillo RE, Forbes GA (2001) The effects of host diversity and other management components on epidemics of potato late blight in the humid highland tropics. Phytopathol 91(10):993–1000

Gisi U, Cohen Y (1996) Resistance to phenylamide fungicides: a case study with *Phytophthora infestans* involving mating type and race structure. Annu Rev Phytopathol 34:549–572

Grünwald NJ, Flier WG (2005) The biology of *Phytophthora infestans* at its center of origin. Ann Rev Phytopathol 43:171–190

Gupta H, Singh BP, Mohan J (2004) Bio-control of late blight of potato. Potato J 31(1–2):39–42

Gupta J (2016) Efficacy of biocontrol agents against *Phytophthora Infestans* on Potato. Int J Eng Sci Comput 6(9):2249–2252. http://ijesc.org/

Hajianfar R, Polgar ZS, Wolf I, Takacs A, Cerank I, Taller J (2014) Complexity of late blight resistance in potato and its potential in cultivar. Improvement Acta Phytopathologica et Entomologica Hungarica 49

Haverkort A, Struik P, Visser RGF, Jacobsen E (2009) Applied biotechnology to combat late blight in potato caused by *Phytophthora infestans*. Potato Res 52:249–264

Hussain S, Lees AK, Duncan JM, Cooke DEL (2005) Development of a species-specific and sensitive detection assay for *Phytophthora infestans* and its application for monitoring of inoculum in tubers and soil. Plant Pathol 54:373–382

Hussain T, Singh BP, Anwar F (2014) A quantitative Real Time PCR based method for the detection of *Phytophthora infestans* causing Late blight of potato, in infested soil. Saudi J Biol Sci 21(4):380–386

Hussain T, Singh BP, Anwar F, Tomar S (2015) A simple method for diagnostic of *Phytophthora infestans* from potato agricultural fields of potato. Turk J Agric Food Sci Technol 3(12):904–907

Inglis D, Johnson D, Legard D, Fry W, Hamm P (1996) Relative resistances of potato clones in response to new and old populations of *Phytophthora infestans*. Plant Dis 80:575–578

Isman BM (2000) Plant essential oils for pest and disease management. Crop Prot 19:603–608

Jacobsen BJ (1993) Biological and cultural plant disease controls: Alternatives and supplements to chemicals in IPM systems. Plant Dis 77(3):311–315

Julie L, Kirstin E, Nick S (2009) Real-time PCR: current technology and applications, Logan J, Edwards K, Saunders N (eds). Caister Academic Press. ISBN 978-1-904455-39-4

Kirk WW, Abu-El Samen FM, Muhinyuza JB, Hammerschmidt R, Douches DS, Thill CA, Thompson AL (2005) Evaluation of potato late blight management utilizing host plant resistance and reduced rates and frequencies of fungicide applications. Crop Prot 24(11):961–970

Koltai H, Volpin H (2003) Agricultural genomics: an approach to plant protection. Eur J Plant Pathol 109:101–108

Kuznetsova MA, Filippov AV, Shcherbakova LA, Voinova JM (1996) Microbiological preparations for protection of potato from Phytophthorose, all Russia. Res Inst Phytopathol Moscow Russia 6:16–17

Kuznetsova MA, Shcherbakova LA, Il'Ins kaya LT, Filippov AV, Ozerets kovs kaya OL (1995) Mycelium extract of the fungus *Pythium ultimum* is an efficient preventive of potato *Phytophthora infestans* infection. Microbiol (New York) Phytopathol 64(4):422–424

Lal M, Yadav S, Sharma S, Singh BP, Kaushik SK (2017) Integrated management of late blight of potato. J Appl Nat Sci 19(3):1821–1824

Lammerts Van Bueren ET, Tiemens-Hulscher M, Struik PC (2008) Cisgenesis does not solve the late blight problem of organic potato production: alternative breeding strategies. Potato Res 51:89–99

Majeed A, Ahmad H, Chaudhry Z, Jan G, Alam J, Muhammad Z (2011) Assessment of leaf extracts of three medicinal plants against late blight of potato in Kaghan valley. Pakistan J Agric Technol 7(4):1155–1161

Majeed A, Chaudhry Z, Haq I, Muhammad Z, Rasheed H (2015) Effect of aqueous leaf and bark extracts of *Azadirachta indica* A. Juss, *Eucalyptus citriodora* Hook and *Pinus roxburghii* Sarg, on late blight of potato. Pak J Phytopathol 27(1):15–20

Majeed A, Muhammad Z, Ullah Z, Ullah R, Ahmad H (2017) Late blight of potato (*Phytophthora infestans*) I: Fungicides application and associated challenges. Turk J Agric Food Sci Technol 5(3):261–266

Malcolmson JF (1969) Races of *Phytophthora infestans* occurring in Great Britain. Trans Br Mycol Soc 53:417–423

Malcolmson JF, Black W (1966) New R. genes in Solanum demissum Lindl and their complementary races of *Phytophthora infestans* (Mont) de Bary. Euphytica 15:199–203

Martin MG, Cappellini E, Samaniego JA, Zepeda ML, Campos PF, Seguin-Orlando A, Wales N, Orlando L, Simon YWH, Dietrich FS, Mieczkowski PA, Heitman J, Willerslev E, Krogh A, Ristaino JB, Gilbert MTP (2013) Reconstructing genome evolution in historic samples of the Irish potato famine pathogen. Nat Commun Online J. https://doi.org/10.1038/ncomms3172

Mbwana J, Bölin I, Lyamuya E, Mhalu F, Lagergård T (2006) Molecular characterization of Haemophilus ducreyi isolates from different geographical locations. J Clin Microbiol Jan 44(1):132–137

Michelmore RW (2003) The impact zone: genomics and breeding for durable disease resistance. Curr Opin Plant Biol 6:397–404

Mizubtui ESG, Fry WE (2006) Potato late blight. The epidemiology of plant diseases, 2nd edn, pp 445–471

Mohamed AF, Abou-Hadid AF (2004) Potato late blight epidemiology and epidemic analysis in Egypt. Cairo 20 June 2004

Moumene SM, Li Y, Bachir K, Houmani Z, Bouznad Z, Chemat F (2015) Antifungal power of citrus essential oils against potato late blight causative agent. J Essent Oil Res 27(2):169–176

Niederhauser JS (1956) The blight the blighter and the blighted. Trans NY Acad Sci Ser II 19:55–63

Niederhauser JS (1991) *Phytophthora infestans*: The Mexican connection, Phytophthora JA, Lucas RC, Shattock, DS Shaw, LR Cooke (eds).Cambridge, Cambridge University Press, Cambridge, pp 25–45

Nutter FF (2007) The role of plant disease epidemiology in developing successful integrated disease management programmes. Gen Conc Integ Pest Dis Manag 45–79

Peters R, Platt H, Hall R, Medina M (1999) Variation in aggressiveness of Canadian isolates of *Phytophthora infestans* as indicated by their relative abilities to cause potato tuber rot. Plant Dis 83:652–661

Quintanilla P, Rohloff J, Iversen TH (2002) Influence of essential oils on *Phytophthora infestans*. J Eur Assoc Potato Res 45(2–4):225–235

Roca W, Espinoza C, Panta A (2004) Agricultural applications of biotechnology and the potential for biodiversity valorization in Latin America and the Caribbean. J Agro Biotechnol Manage Econ 7(1&2):13–22

Saiki R, Scharf S, Faloona F, Mullis K, Horn G, Erlich H, Arnheim N (1985) Enzymatic amplification of beta-globin genomic sequences and restriction site analysis for diagnosis of sickle cell anemia. Science 230(4732):1350–1354

Scala DJ, Kerkhof LJ (2000) Horizontal heterogeneity of denitrifying bacterial communities in marine sediments by terminal restriction fragment length polymorphism analysis. Appl Environ Microbiol 66:1980–1986

Schumann GL (1991) The Irish potato famine and the birth of plant pathology. In: Plant diseases their biology and social impact. APS Press

Shattock R (2002) *Phytophthora infestans*: populations, pathogenicity and phenylamides. Pest Manag Sci 58:944–950

Shmutterer H (1990) Properties and potential of natural pesticides from the neem tree Azadirachta indica. Annu Rev Entomol 35:271–297

Smart C, Fry W (2001) Invasions by the late blight pathogen: renewed sex and enhanced fitness. Biol Invasions 3:235–243

Soylu EM, Soylu S, Kurt S (2006) Antimicrobial activities of the essential oils of various plants against tomato late blight disease agent *Phytophthora infestans*. The Scientific and Technical Research Council of Turkey, 3104 Mycopathologia 161:119–128

Touseef H (2018) Diagnosis of phytophthora in soil samples by polymerase chain reaction. Curr Inves Agri Curr Res 1(1):8–10

Turkensteen L, Flier W, Wanningen R, Mulder A (2008) Production, survival and infectivity of oospores of *Phytophthora infestans*. Plant Pathol 49:688–696

Vassilev D, Leunissen JA, Atanassov A, Nenov A, Dimov G (2005) Application of bioinformatics in plant breeding. Wageningen University, The Netherland

Vassilev D, Nenov A, Atanassov A, Dimov G, Getov L (2006) Application of bioinformatics in fruit plant breeding. J Fruit Ornamental Plant Res 14:145–162

Vleeshouwers VGAA, Raffaele S, Vossen J, Champouret N, Oliva R, Segretin ME, Rietman H, Cano LM, Lokossou A, Kessel G (2011) Understanding and exploiting late blight resistance in the age of effectors. Annu Rev Phytopathol 49(1):507–531

Wangsomboondee T, Ristaino JB (2002) Optimisation of sample size and DNA extraction methods to improve PCR detection of different propagules of *Phytophthora* infestans. Plant Dis 86(3):247–253

Webster JM (1997) Antimycotic activity of *Xenorhabdus bovienii* (Enterobacteriaceae) metabolites against *Phytophthora infestans* on potato plants. Can J Plant Pathol 19:125–132

Xue J, Zhao S, Liang Y, Hou C, Wang J (2008) Bioinformatics and its Applications in Agriculture. In: IFIP International Federation for Information Processing, 259 Computer and Computing Technologies in Agriculture 2 Daoliang Li, Springer, Boston, pp 977–982

Yao Y, Li Y, Chen Z, Zheng B, Zhang L, Niu B, Meng J, Li A, Zhang J, Wang Q (2016) Biological control of potato late blight using isolates of *Trichoderma*. Am J Potato Res 93(1):33–42

Youssef SA (2011) Methods of diagnosis of plant diseases. Fifth scientific meeting, May 2011. http://ppathri.kenanaonline.com

Chapter 12
Developing Rust Resistance of Wheat Genotypes Under Egyptian Conditions

Hassan Auda Awaad and Doaa Ragheb El-Naggar

Abstract Wheat rusts constitute major threats which adversely effect on yield components and grain quality. Developing new genotypes having resistance genes is believed to be the most effective tool to overcome these challenges. This chapter is divided into three major categories to explain the Egyptian case of wheat rusts, i.e. yellow rust, leaf rust and stem rust. The used methods in breeding for resistance taking into account their economic impacts, during the last two decades. The most important findings are that yellow rust is able to attack most of the Egyptian commercial wheat cultivars causing severe infection and thereby high losses, particularly at the Northern Governorates. Recent studies evidence the responsibility of Yr1 resistant gene against yellow rust, at both seedling and adult stages. Breeding for rust resistance in wheat remains feasible and the most effective method for improving the level of cultivar resistance. Using molecular markers technique is one of the most effective methods in identifying genotypes carrying resistance genes and consequently transferring these genes to the commercial cultivars. On the other hand, several efforts have been done to obtain resistant genotypes using genetic engineering techniques.

Keywords Wheat · Genotypes · Stem rust · Yellow rust · Leaf rust · Resistance genes · Molecular markers · Genetic engineering yield loss

12.1 Introduction

Biotic stress is one of the most vital destructive factors affecting crop plants refer to infection with organisms such as viruses, fungi, bacteria, nematodes, insects and weeds. Biotic stresses are an actual threat to agricultural production and a threat to global food security. The origin of new physiological races due to environmental

H. A. Awaad
Crop Science Department, Faculty of Agriculture, Zagazig University, 44511, Zagazig, Egypt

D. R. El-Naggar (✉)
Wheat Disease Department, Plant Pathology Research Institute,
Agriculture Research Center, Giza 12619, Egypt

© Springer Nature Switzerland AG 2021
H. Awaad et al. (eds.), *Mitigating Environmental Stresses for Agricultural Sustainability in Egypt*, Springer Water,
https://doi.org/10.1007/978-3-030-64323-2_12

311

and genetic aspects is a major task for plant breeders for breeding resistant varieties. Wheat crop is exposed to three rust diseases, i.e. stripe rust, leaf rust and stem rust, caused by *Puccinia striiformis* Westend. f. sp. *tritici* Erikss., *Puccinia triticina* Eriks and *Puccinia graminis* f. sp. *tritici*, respectively, are severe foliar diseases in wheat growing regions of the world affecting imperative losses of yield.

Under Egyptian circumstances, stripe rust is considered the most common disease triggering significant damages in wheat production (Abu El-Naga et al. 2001) and can cause 100% yield loss, but often range from 10 to 70% (Chen 2005). Leaf rust of wheat attacks the leaf blades, sheath and glumes in greatly susceptible genotypes (Huerta-Espino et al. 2011), it reduced numbers of kernels per head and lower grain weights (Marasas et al. 2004; Kolmer et al. 2005). Early infection by leaf rust usually causes greater yield losses 60–70% based on the sensitivity of the wheat genotypes and the severity of epidemics (Appel et al. 2009). Bajwa et al. (1986) reported that losses in wheat grain weight due to leaf rust infection fluctuated between 2.0 and 41% in wheat varieties based on the level of resistance or susceptibility. Also, Kolmer et al. (2009) showed that leaf rust could cause yield losses of at least 25%.

Stem rust caused by *Puccinia graminis* f. sp. *tritici*, is the harshest disease to wheat. Under favorable circumstances, stem rust might cause yield losses up to 100% on the susceptible varieties (Roelfs 1985; Leonard and Szabo 2005). Whereas, yield injury in wheat from natural infections with *Puccinia graminis* f. sp. *tritici* varied from 10 to 45% in 3 experiments over 2 years (Loughman et al. 2005). The novel stem rust race designated as Ug99 in Uganda in 1999 has threatened wheat yield worldwide (Pretorius et al. 2000).

So, plant breeding for disease resistance is one of the most vital strategies for reducing the loss in wheat yield. The importance of disease depends principally upon the prevalence of virulent races and their compatibility with the genetic makeup of the host variety under specific environment. In breeding for disease resistance, plant breeders would not depend on the host only, but also the pathogen genetic makeup and environment, as the two important variables in the pathogen-host—environment schemes. Thus, the Egyptian wheat varieties have suffered from unexpected epidemics through the last decades as a result of climatic changes in relation to the genetic structure of both host and pathogen (El-Daoudi et al. 1987).

Nevertheless, wheat cultivars relying on race-specific resistance often mislay efficiency within a few years by impressive selection for virulent leaf rust races. Also, the cultivation of a great area of susceptible cultivars permits a great rust population to proliferate, producing a reservoir for mutation and selection (Kolmer et al. 2005).

New genetic resources of resistance might be joined into wheat to different the existing gene pools for resistance to rusts. The operative resistant genes against rust may be used singly or in combination with great yielding genes to improve high-yielding resistant cultivars in areas where wheat is grown. Genetic resistance is the greatest economic and operative means of decreasing yield harms due to leaf rust disease (Liu and Kolmer 1997). Singh et al. (1991) stated that loss in wheat yield owing to leaf rust could be decreased in levels similar to those of hypersensitive resistant genotypes through the use of partial resistance which gives long-lasting resistance at a small costing yield that is insufficient to justify treatment with fungicide.

Also, Herrera-Foessel et al. (2006) showed that losses in grain yield for susceptible, race-specific, as well as slow-rusting varieties were 51, 5, and 26%, respectively, at normal sowing date trial and 71, 11, and 44% when sown late. Though, breeding for resistance is a continuous procedure, and plant breeders need to increase new effective resources in breeding materials (Draz et al. 2015).

The current chapter provides detailed information's about genetic variability, looking for new sources of resistance, genetic system and genes conferring resistance, breeding efforts and biotechnology along with yield losses caused by the rust disease. These information's will serve as a foundation for developing durable rust-resistant wheat cultivars.

12.2 Types of Wheat Rusts

Wheat yellow rust caused by *Puccinia striiformis* f. sp *tritici* West., stem rust caused by *Puccinia graminis* f.sp. *tritici* and leaf rust caused by *Puccinia triticina* Eriks are considered a widespread disease of severe severity in Egypt and worldwide. Fungi which cause rusts are obligate biotrophic parasites. Yellow rust is a macro-cyclic, heteroecious fungus that needs both primary (wheat or grasses) and alternate (Berberis spp or *Mahonia* spp.) host plants to complete its life cycle. Strip rust seems as a mass of yellow to orange urediniospores erupt from pustules arranged in long, narrow strips on wheat leaves (usually between viens), leaf sheath glumes and awns on susceptible plants (see Fig. 12.1).

Wheat leaf rust symptoms are small brown pustules which are developed on the leaf blades in a random scatter distribution. They may group into patches in severe cases (Fig. 12.2).

Plants infected by stem rust contain red spores which are covered with fine spines. The pustules might be abundant and produced on both leaf surfaces and stems of wheat. Later in the season, pustules of black teliospores begin to appear in infected plants (Fig. 12.3).

Fig. 12.1 Wheat yellow rust caused by *Puccinia striiformis* f. sp *tritici* West

Fig. 12.2 Leaf rust caused by *Puccinia triticina* Eriks

Fig. 12.3 Stem rust caused *Puccinia graminis* f.sp. *tritici*

12.3 Economic Importance

12.3.1 Losses in Yield and Grain Quality

Biotic environmental stress is a limiting factor for wheat production. Rust diseases are the most important wheat diseases, that continued to destroy wheat crop since ancient periods.

Positive and significant associations were recorded between infection type and disease severity of stem rust in the four wheat crosses. It is important to notice that, negative and significant correlation coefficient has been observed between each of infection type and disease severity on the one hand, and grain yield/plant in 1st and 2nd crosses, on the other hand (Table 12.1). Thus, increasing infection type and disease severity resulted in a substantial reduction in wheal productivity (Salem et al. 2003).

Ashmawy et al. (2013) recorded significant differences for yield losses between protected and infected wheat genotypes as a result of the differences in the level of disease severity of stem rust. The loss % of grain yield per plot in 2011/2012 season varied from 2.47 to 6.29%. Wheat cultivars Sids 12 and Sids 13 displayed the highest values of yield loss % (6.29% and 5.89%) compared to the other genotypes. While, in 2012/2013 season, the yield loss % fluctuated from 1.96 to 8.21%. Misr 1, Misr 2, Sids 12 and Sids 13 gave the highest values of yield loss % valued 8.21, 6.81, 6.17 and 5.34%, respectively. Whereas, wheat cultivars Gemmeiza 7, Gemmeiza 11, Gemmeiza 10, Sakha 93, Gemmeiza 9, Sakha 61, Sakha 94 and Giza 168 recorded the lowest in yield loss %. This trend is in agreement with losses reported in previous studies attained by Loughman et al. (2005).

Ochoa and Parlevliet (2007) previously found that yield loss (%) was strongly correlated with AUDPC. Furthermore, Afzal et al. (2007) showed that the correlation

Table 12.1 Simple correlation coefficient among infection type, disease severity of stem rust and grain yield/plant in four wheat populations (Salem et al. 2003)

Character	Population	Disease severity	Grain yield/plant
Infection type	1	0.875*	−0.786**
	2	0.907**	−0.603*
	3	0.910**	0.246
	4	0.893**	0.025
Disease severity	1		−0.952**
	2		−0.643**
	3		−0.030
	4		−0.260

*, **significant and highly significant at 0.05 and 0.01 levels of probability

coefficient (r = − 0.67805) depicted a greatly significant effect of stripe rust in depressing wheat yield.

Stripe rust, or yellow rust, caused by *Puccinia striiformis* f.sp. *tritici*, is one of the most harmful rust diseases of wheat in many regions everywhere the world (Fu et al. 2008). The obligate parasitic fungus has destroyed wheat crop worldwide due to rapid semi systemic infection of infected plants causing shriveled kernels. At present, stripe rust of wheat has been registered in more than 27 countries of the world. In Egypt, wheat stripe rust was epidemic in 1967, 1995, 1997 and 2015 attacking bread wheat cultivars, i.e. Giza-144 and Giza-150, Gemmieza-1, Giza-163, Sakha-69, Sids-1, Sids-12, Sids-13, Gemmeiza-7 and Gemmeiza-11, respectively (Abd El-Hak et al. 1972; El-Daoudi et al. 1996; Abu El-Naga et al. 1997; Anonymous (2020). Putnik-Deliã (2008) found a significant association between the values of area under disease progress curve (AUDPC) and resistance characteristics of seedlings and adult stages.

Yield loss affected by stripe rust is determined by numerous factors, including the degree of sensitivity, infection time, the rate of disease development, and duration of disease (Chen 2005). Wheat yield and disease factors are influenced by environmental elements, among which temperature and moisture are the most vital in determining disease severity and yield loss (Chen 2007; Gladders et al. 2007). So, this disease is able to attack most of the Egyptian commercial wheat cultivars causing severe infection and consequently high losses, particularly at the Northern Governorates.

El-Daoudi et al. (1996) showed that average percentage loss in wheat grain yield of four Egyptian cultivars including Sakha-69, which was infected by stripe rust, was estimated by 20.5% at Delta region. Abu El-Naga et al. (1999a) indicated that Sids-7 and Sids-9 were dramatically affected by stripe rust infection since they exhibited the highest percentage loss in wheat grains and in yield components. Moreover, Menshawy and Najeeb (2004) indicated that all susceptible tested local cultivars are correlated by high percentage reduction in yield and 1000 kernel weight. Afzal et al. (2007) reported that the stripe rust can lead to 100% yield loss if the infection occurs very early and continues to develop during the growing season.

So, Omara et al. (2016) recorded significant differences between mean values of infected and protected plots for yield/fed. (Feddan = 4200 m^2). This would be due to the variances in the level of stripe rust severity. In 2013/2014 growing season, the estimated and actual percentage loss of the grain yield/fed. ranged from 18.9 to 55.4% and from 18.3 to 53.7%, respectively. Sids-12, Gemmeiza-11 and Gemmeiza-7 cultivars gave the highest values of estimated and actual percentage loss of the grain yield/fed., i.e. 55.4 and 53.7%, 53.4 and 51.8% and 42.8 and 41.5%, respectively. Accordingly, the cultivars Gemmeiza-10 and Gemmeiza-9 gave the lowest values of the estimated and actual percentage loss of the grain yield/fed. In 2014/2015 growing season, estimated and actual percentage loss of the grain yield/fed. fluctuated from 23.9 to 58.3% and from 23.0 to 56.3% of the estimated and actual percentage loss, respectively. Hence, Sids-12, Gemmeiza-11 and Gemmeiza-7 cultivars provided the highest values of estimated and actual percentage loss of the two parameters followed by Sids-1 and Sids-13. Irrespective of high susceptibility of the tested cultivars, the lowest level of percentage loss was recorded with Gemmeiza-10. This may be attributed to the process of reselection in such cultivars. Whereas, the significant difference between infected and protected plots in yield gave evidence to the justification of fungicide applications during critical times, i.e. times of epiphytotics especially when we know that the mean average percentage loss within the tested wheat cultivars in the two seasons was estimated by 18.9–58.3% (Table 12.2).

Omara et al. (2016) computed relationships between FRS (%), AUDPC, rAUDPC, r-value and estimated loss (%) of grain yield/fed. due to stripe rust through regression analysis during 2013/2014 and 2014/2015 growing seasons (Table 12.3). In season 2013/2014, they recorded strong positive relations between the four parameters under study and each of estimated loss (%) of 1000-kernel weight and grain yield/fed. Estimates of R^2 were 0.936 and 0.970; 0.897 and 0.982; 0.897 and 0.982 and 0.754 and 0.769 for FRS (%), AUDPC, rAUDPC, r-value, respectively. In the second season, the results were parallel to those obtained during the first season. Hence, the relation between FRS (%) and each of estimated loss (%) of 1000 kernel weight and grain yield/fed. was more stable than AUDPC, rAUDPC, r-value, during the two growing seasons of the study as previously reported by Ali et al. (2016).

Leaf rust infection of wheat causes a significant loss in grain yield and quality, due to the lack of translocation of nutrients towards the spike and the less effective of photosynthetic activities of leaf area. Awaad et al. (2003) recorded positive and significant associations between infection type and disease severity of leaf rust in three out of the four wheat crosses. It is important to notice that, negative and significant correlation coefficient has been observed between each of infection type and disease severity on one hand, and grain yield/plant in 1st, 2nd and 3rd crosses, on the other hand. Thus increasing infection type and disease severity led to substantial reduction in wheat grain yield (Table 12.4)

Ahmad et al. (2010) stated that the genotypes displayed 1–50, 55–100 and 100–200 AUDPC registered 3.43, 6.74 and 11.70% yield losses, respectively. However, genotypes showed medium range of AUDPC, i.e. 500–1000 suffered from 21.50% losses. Whereas the maximum area under leaf rust disease, i.e.1000–1500 and 1500–2000 caused 33 and 38% losses, respectively in different wheat genotypes. Draz et al. (2015) showed that leaf rust triggered by *Puccinia triticina* Eriks., is one of the main wheat diseases in Egypt, causing up to 50% of yield loss.

Shaheen and El-Orabey (2016) tested twelve wheat varieties for adult plant resistance and yield losses as a result of leaf rust under field conditions at Shibin El-Kom location. The experiment was surrounded by spreader area of highly susceptible varieties inoculated with a mixture of leaf rust pathotypes as a source of inoculum. The yield reduction ranged from 0.52% for Sids 13–17.58% for Sakha 93 during 2013/2014 growing season. Whereas in 2014/2015 it fluctuated from 0.48% for Sids 13–19.56% for Sakha 93. The yield injuries of the remaining genotypes were depends on leaf rust severity. The yield losses in 1000 kernel weight and plot weight were strongly associated with area under disease progress curve.

El-Naggar Doaa and Soliman (2015) assessment yield components for stripe rust of infected Sids-7, Giza-168 and Gemmeiza-9 wheat cvs. and their gamma ray induced mutant lines at M_4 generation. They showed that number of spikes and grain yield/m^2 of the eight selected mutants were surpassed their mother cvs. While, the gamma ray induced mutagenesis had alternative impact on number of

Table 12.2 Loss of grain yield for seven Egyptian wheat cultivars infected with stripe rust during 2013/2014 and 2014/2015 growing seasons (Modified after Omara et al. 2016)

Cultivar	Grain yield (ard./fed.)									
	2013/2014					2014/2015				
	Infected	Protected	Mean	Estimated loss (%)	Actual loss (%)	Infected	Protected	Mean	Estimated loss (%)	Actual loss (%)
Gem.-7	10.7	18.7	14.7	42.5	41.5	9.5	17.8	13.7	46.6	44.9
Gem.-9	15.1	19.9	17.5	24.1	23.4	14.7	20.1	17.4	26.8	25.8
Gem.-10	15.4	19.0	17.2	18.9	18.3	12.4	16.3	14.4	23.9	23.0
Gem.-11	10.4	22.3	16.4	53.4	51.8	8.6	20.0	14.3	57.0	55.0
Sids-1	11.5	17.7	14.6	35.0	34.0	10.0	16.5	13.3	39.4	38.0
Sids-12	9.5	21.3	15.4	55.4	53.7	8.5	20.4	14.5	58.3	56.3
Sids-13	13.7	19.8	16.8	30.8	29.9	11.3	17.5	14.4	35.4	34.2
Mean	12.3	19.8	–	–	–	10.7	18.4	–		
LSD 0.05: Cultivars Treatments C × T	2.35 0.61 1.61							1.84 0.65 1.73		

Ardab wheat = 150 kg Feddan = 4200 m^2

Table 12.3 Relationship between FRS (%), AUDPC, rAUDPC, *r*-value and estimated loss (%) of 1000 kernel weight (g) and grain yield/fed. (ard.) during 2013/2014 and 2014/2015 growing seasons (Omara et al. 2016)

Epidemiological parameter	Coefficient of determination (R^2)			
	2013/2014		2014/2015	
	1000 kernel weight	Estimated loss (%) of yield (ard./fed.)	1000 kernel weight	Estimated loss (%) of yield (ard./fed.)
FRS (%)	0.936	0.970	0.935	0.965
AUDPC	0.897	0.982	0.953	0.979
rAUDPC	0.897	0.982	0.953	0.979
r-value	0.754	0.769	0.914	0.893

Ardab wheat = 150 kg Feddan = 4200 m^2

Table 12.4 Simple correlation coefficient among infection type, disease severity of leaf rust and grain yield/plant in four wheat populations (Awaad et al. 2003)

Character	Population	Disease severity	Grain yield/plant
Infection type	1	0.663*	−0.847**
	2	0.992**	−0.877**
	3	0.944**	−0.717*
	4	0.388	−0.452
Disease severity	1		−0.795**
	2		−0.919**
	3		−0.763**
	4		−0.479

*, **significant and highly significant at 0.05 and 0.01 levels of probability

grains/spike and 1000-grain weight which differed from mother cv. to another and from mutant to another, at M_4 generation. Both mutants of Sids-7 wheat cv. increased the studied yield components with favorability to Sid-M_4-7 which recorded 316 spike/m^2, 84 grain/spike, 66.66 g/1000 grain and 1751.90 g grains/m^2 (Table 12.5). Results also indicated that plants of G-M_4-3 mutant recorded lower number of grains/spike it was characterized by the highest number of spikes/m^2, 1000-grains weight and grain yield/m^2, compared with its mother cv. Additionally, both mutants of Gemmeiza-9 at M_4 generation increased all studied yield components except number of spikes/m^2, compared with their mother cv. Statistical analysis revealed that significant differences were observed between all genotypes for all studied parameters.

At M_5 generation, they added that estimates of disease severity and AUDPC recorded significant differences between all the studied genotypes at M_5 generation (Table 12.6). The highest percentages of reduction in stripe, leaf and stem rust

Table 12.5 Assessment of some yield components for stripe rust of infected Sids 7, Giza 168 and Gemmeiza 9 wheat cvs. and their gamma ray induced mutant lines at M_4 generation (El-Naggar Doaa and Soliman 2015)

Genotypes	No. of spikes/m^2	No. of grains/spike	1000-grain weight (g)	Grain yield/m^2 (g)
Sids 7	253.00	68.00	56.66	1401.90
Sid-M_4-6	260.00	73.00	60.33	1450.00
Sid-M_4-7	316.00	84.00	66.66	1751.90
Giza 168	141.00	51.00	53.33	0381.12
G-M_4-1	283.00	42.00	46.66	0546.75
G-M_4-3	673.00	37.00	63.33	1568.76
G-M_4-5	280.00	45.00	43.33	0540.00
G-M_4-6	250.00	43.00	60.00	0636.17
Gemmeiza 9	357.00	55.00	41.00	750.00
Gem-M_4-1	337.00	59.00	43.33	0854.96
Gem-M_4-2	261.00	67.00	46.66	0804.40
L.S.D $_{0.05}$	4.267	2.066	1.677	39.26

severities were recorded by Sid-M_5-7 (88.00%), Gem-M_5-1 (87.14%) and G-M_5-3 (100%), respectively. The lowest percentages of leaf and stem rust severities were recorded by G-M_5-3 mutant line (1.00 and 0.00%, respectively) with the resulting lowest values of AUDPC (4.2 and 0.0, respectively), compared to the other genotypes. While, the lowest percentage of stripe rust severity was recorded by Sid-M_5-7 mutant line (2.00%), compared to the other genotypes. The responses of mother Sids-7, Giza-168 and Gemmeiza-9 wheat cvs. toward stripe rust infection were changed from susceptible (S) to moderately susceptible (MS), moderately resistant (MR) and moderately susceptible (MS) in their mutants (Sid-M_5-7, G-M_5-3 and Gem-M_5-1, respectively), at M_5 generation. Also, responses of these mother cvs. toward stem rust infection were changed from S to MS, stem rust free and MR in their mutants at M_5 generation, respectively. On the other hand, responses of these mother cvs. toward leaf rust infection did not changed except in Giza-168 wheat cv. in which the infection response was changed from MS (in mother cv.) to resistant (in its mutant, G-M_5-3). Thus, mutagenized Giza-168 wheat cv. at M_5 generation (G-M_5-3) can be considered as the promising resistant mutants toward the three rusts under study. This result can be strengthened by the values of AUDPC where this mutant (G-M_5-3) recorded the lowest values (18.67, 4.2 and 0.0) for stripe, leaf and stem rusts, respectively. Based on disease parameters and maturity duration as well as yield components, the during 2013/2014 season, grains of Sid-M_4-7, G-M_4-3 and Gem-M_4-1 were selected to be grown at M_5 generation during 2014/2015 growing season. Moreover as shown in Table 12.7 significant differences were observed between all genotypes for all studied parameters. The three tested mutants at M_5 generation increased the four examined yield components, compared with each mother cv. Among the three tested mutants, Gem-M_5-1 recorded the highest number (601.33) of spikes/m^2, Sid-M_5-7

Table 12.6 Evaluation of disease severities (%), infection responses and area under disease progress curve (AUDPC) of mutant lines of gamma irradiated grains of Sids-7, Giza-168 and Gemmeiza-9 wheat cvs. to stripe, leaf and stem rusts, during M_5 generation (El-Naggar Doaa and Soliman 2015)

Genotype	Stripe rust				Leaf rust				Stem rust			
	Disease severity (%)	% reduction in disease severity	Infection response	AUDPC	Disease severity (%)	% reduction in disease severity	Infection response	AUDPC	Disease severity (%)	% reduction in disease severity	Infection response	AUDPC
Sids-7	16.67	0.00	S	176.17	16.67	0.00	MS	210.0	10.00	0.00	S	118.77
Sid-M_5-7	2.00	88.00	MS	22.4	5.33	68.03	MS	53.2	6.67	33.30	MR	80.5
Giza-168	26.67	0/00	S	193.67	6.67	0.00	MS	92.4	3.00	0.00	S	49.7
G-M_5-3	4.33	83.76	MR/R	18.67	1.00	85.01	MR	4.2	0.00	100	–	0
Gemmeiza-9	16.67	0.00	S	208.83	23.33	0.00	MS	256.67	6.67	0.00	S	96.13
Gem-M_5-1	5.00	70.01	MS	80.27	3.00	87.14	MR	44.8	3.00	55.02	MR	15.4
L.S.D $_{0.05}$	7.545			48.54	6.951			41.29	3.177			23.51

Table 12.7 Number of spikes/m^2, number of grains/spike, weight of 1000 grain (g) and grain yield/m^2 (g) for stripe rust of infected Sids-7, Giza-168 and Gemmeiza-9 wheat cvs. and their gamma ray induced mutant lines at M$_5$ generation (El-Naggar Doaa and Soliman 2015)

Genotypes	No. of spikes/m^2	No. of grains/spike	1000-grain weight (g)	Grain yield/m^2 (g)
Sids-7	368.33	49.93	49.7	488.00
Sid-M$_5$-7	464.00	62.47	52.5	605.67
Giza-168	459.33	39.20	26.1	351.00
G-M$_5$-3	575.00	43.40	45.9	608.67
Gemmeiza-9	506.67	45.13	23.2	367.00
Gem-M$_5$-1	601.33	47.17	35.2	519.00
L.S.D $_{0.05}$	134.9	6.509	3.244	121.9

registered the highest number (62.47) of grains/spike and the highest weight (52.5 g) of 1000 grain and finally G-M$_5$-3 recorded the highest grain yield/m^2 (608.67 g) of grains/m^2.

12.4   Race Analyses

12.4.1   Yellow Rust

Identified the physiological races of strip rust (*Puccinia striiformis* f. sp *tritici* West.) during the period from 1999 to 2011 was performed by Ashmawy and co-workers (2012). Nine physiological races (0E0, 128E65, 194E69, 210E100, 226E109, 134E109, 236E 150, 450E45 and 454E128) were identified during 1999/2000. These races were determined based on of sum of high infection types for each of 17 wheat stripe rust monogenic differentials. Virulence frequencies were very high against Yr6, Yr7, Yr8, Yr9, Yr(6) and Yr(7), while they were very low against Yr1, Yr4, Yr5 and Yr10. Ten physiological races (0E0, 0E64, 230E150, 236E250, 435E117, 447E59, 497E85, 499E95, 505E71 and 505E119) were identified during 2000/2001. Virulence frequencies were very high against Yr1, Yr3, Yr4, Yr5, Yr6, Yr7, Yr8, Yr9, Yr10, YrSD, YrSU, YrSP, Yr(6) and Yr(7), and low against Yr2 and Yr (3). Also, they identified ten physiological races (0E0, 0E64, 4E2, 6E134, 142E182, 116E144, 230E148, 230E150 and 240E20) during 2001/2002. Virulence frequencies were very high against Yr2, Yr3, Yr6, Yr7, Yr8, Yr9, Yr10, YrSP, Yr (6) and Yr (7), and low against Yr1, Yr4, Yr5, YrSU and Yr (3). Also they identified eleven physiological races (0E0, 0E64, 4E16, 4E128, 6E16, 70E20, 70E26, 70E134, 70E182, 130E178 and 198E156) during 2002/2003. Virulence frequencies were very high against Yr2, Yr6, Yr7, Yr8, YrSU, YrSP, Yr(6) and Yr(7), and low against Yr1, Yr3, Yr4,Yr5, Yr9, Yr10, YrCV, YrSD and Yr(3). Nine physiological races (0E0, 0E64, 2E0, 4E2, 4E148, 6E1134, 70E20, 70E134 and 230E150) were recognized during 2003/2004.

Virulence frequencies were very high against Yr2, Yr6, Yr7, Yr8, YrSU, YrSP, Yr(6) and Yr(7), and low against Yr1, Yr3, Yr4, Yr5, Yr10, YrSD and Yr(3). Nine physiological races (0E0, 4E0, 4E2, 64E2, 68E2, 70E154, 134E18, 189E182 and 198E151) were known during 2004/2005. Virulence frequencies were very high against Yr2, Yr6, Yr7, Yr8, Yr9, YrSU and Yr(7), while the lowest frequencies were found against Yr1, Yr3, Yr4, Yr5, Yr10, YrCV, YrSD, YrSP and Yr(3). Nine physiological races (0E0, 0E64, 2E0, 4E0, 4E2, 6E134, 64E6, 68E2 and 70E142) were identified during 2005/2006. Virulence frequencies were very high against Yr6, Yr7, YrSU, Yr SP, Yr(6) and Yr(7), and low against Yr1, Yr2, Yr3, Yr4, Yr5, Yr8, Yr9, Yr10, YrCV, YrSD and Yr(3). Twelve physiological races (0E0, 0E64, 2E128, 32E20, 102E22, 102E128, 142E20, 198E144, 228E148, 230E158, 230E191, 238E0, 238E182 and 494E158) were identified during 2006/2007 season. Virulence frequencies were very high against Yr2, Yr3, Yr6, Yr7, Yr8, Yr9, YrSD, YrSU, YrSP, Yr(6) and Yr(7), and low against Yr1, Yr4, Yr5, Yr10 and YrCV. Six physiological races (0E0, 4E0, 6E153, 16E130, 102E128 and 230E150) were recognized during 2007/2008 season. Virulence frequencies were very high against Yr2, Yr4, Yr6, Yr7, Yr8, YrSD, YrSU, Yr(3), Yr(6) and Yr(7), and low against Yr1, Yr3, Yr5, Yr10, YrCV and YrSP. Eight races (0E0, 2E128, 102E128, 174E182, 228E148, 230E191, 238E0 and 494E158) were known during 2008/2009 season. Virulence frequencies were very high against Yr2, Yr3, Yr6, Yr7, Yr8, Yr9, YrSD, YrSU, Yr(6) and Yr(7), and low against Yr1, Yr4, Yr5, Yr10 and YrSP. Seven races (0E0, 4E0, 16E2, 16E128, 16E130, 60E153 and 60E177) were recognized during 2009/2010 season. Virulence frequencies were very high against Yr2, Yr6, Yr8, Yr10, Yr (6) and Yr(7), and low against Yr1, Yr3, Yr5, Yr7, YrSU, YrSP and Yr(3). Thirteen physiological races (0E0, 2E0, 2E16, 4E0, 6E0, 4E4, 6E4, 6E5, 6E20, 18E16, 34E20, 38E20 and 70E40) were recognized during 2010/2011 season. Virulence frequencies were very high against Yr6, Yr7 Yr8 and Yr(6) and low against Yr1, Yr2, Yr3, Yr4, Yr5, Yr9, YrCV, YrSP and Yr(3).

Shahin and co-workers (2015) identified thirteen and seven physiological races during 2012/2013 and 2013/2014, respectively. The pathotypes were recognized during two successive seasons. In 2012/2013 they were 0E0, 6E0, 2E0, 2E16, 4E0, 4E4, 6E5, 6E20, 18E16, 34E16, 34E20, 38E20, 70E4 and in 2013/2014 i.e. 0E0, 2E0, 2E8, 4E0, 6E116, 70E20 and 128E28 were known through growing seasons. Race 0E0 was the most frequent one followed by 2E0, 4E0 and 6E4. Results revealed that Yr's: 1, 5, and SP were the most energetic through growing seasons since no virulence were registered on either one (http://www.jofamericanscience.org/journals/am-sci/am110615/007_28630am110615_47_...) Whereas, Yr's 7, 6 and 6 were attacked by a great number of races. Concerning valuation of certain stripe rust wheat monogenic lines and Egyptian wheat varieties under greenhouse and field conditions, results showed that Yr1, 5,10,15, 17, 32 and SP were resistant at seedling and adult stages. Genes such as YrA and Yr18 were resistant only at adult plant resistance (APR), were testing the released wheat cultivars under natural conditions, Sakha 93 and Sids 12 were found infected (http://www.jofamericanscience.org/journals/am-sci/am110615/007_28630am110615_47_...). These findings would help as a useful device in wheat breeding program focused for disease resistance.

12.4.2 Leaf Rust

The availability of information on the geographical distribution of the physiologic races of *Puccinia triticina* and the discrimination of leaf rust resistance genes (Lr, s) in new Egyptian wheat cultivars is of great importance for improving resistance in the future genotypes. To identify the resistance in wheat cultivars in Egypt against wheat leaf rust and the most repeated race distributed Abdelbacki et al. (2013) collected infected samples from five Governorates, i.e., Dakahlia, Kafr El-Sheikh, Beheira, Sharqia and Sohag comprised the wheat growing area in Egypt. Samples were isolated, purified and recognized on the differential stes. Gene postulation was performed using fifteen known races on Egyptian wheat cultivars carrying Lr genes. Thirty-three races known through three seasons 2009/2010, 2010/2011 and 2011/2012. The most repeated race was TK (10%) followed by race BB (7.58%), PK (6.55%), TT (4.82%), PT (3.79%) and MT (3.44%). Races BB, TT and PT were existed through three seasons, but seemed in some Governorates and disappeared in others. Otherwise, the most repeatedly occurring gene in ten cultivars was Lr35 (70%), followed by Lr22 (60%), Lr27 (40%), Lr34 (30%), Lr19 (30%), Lr18 (10%), Lr36 (10%) and Lr46 (10%). Eight out of sixteen Lr genes were not found in the verified cultivars. Four genes; Lr28, Lr24, Lr34 and Lr19 were established using a molecular marker. There was a good difference in Lr genes carried by commercial Egyptian wheat cultivars. Therefore, strategies for using resistance genes to prolong operative disease resistance are proposed to resist wheat leaf rust disease (http://www.isaet.org/images/extraimages/phpU1.pdf). Furthermore, McVey et al. (2004) found three races, MBDLQ, MCDLQ, and TCDMQ, in Egypt in the years 1998, 1999, and 2000. Race MCDLQ presented at > 20% frequency each year. Virulence to wheat lines with Lr1, 3, 10, 14b, 15, 17, 23, and 26 occurred at > 45% each year.

Leaf rust samples were collected from diverse locations of Egypt during four seasons (2003/2004–2006/2007) by Nazim et al. (2010). They performed single pustule procedure of isolation for each sample. Rust data were recorded as infection types, and virulence frequencies were identified against 20 Lr genes, in monogenic lines, and some Egyptian cultivars. Virulence frequencies were very high against Lr's 1, 2a, 2c, 3, 16, 26, II, 30, 2b, 14b, 15 and 42, but were the lowest against Lr's 9, 24, 3 ka, 10, 12, 18, 21 and 36. They identified physiologic races based on their reactions on the 20 Lr's and virulence formula (virulence/avirulence) was recorded for each race. Shift, appearance and disappearance of the physiologic races were recognized. Race groups IT-, PT—and FT—were the most dominant being virulent mainly against Lr's 1, 2c, 3 and 9.

The annual survey of 2006/2007 wheat growing season revealed the presence of thrity-two physiologic races of wheat leaf rust fungus (*Puccinia triticina* Eriks.). The more frequent race was PTTS (23.07%) followed by race PTTT (12.30%), race TTTS (9.23%), race TTTT (6.15%), race PTTP (4.61%) and the two races PSTN and PTKS which were represented by (3.07%) each. The rest of the races were represented by (1.53%) each. However, the more effective leaf rust resistance genes were, i.e. Lr2a (73.84%), Lr2b (61.53%) and Lr36 (50.76%). On the other hand, the highest virulence was observed with Lr's, i.e., 10; 11. 14b, 17; 16, 21, 30; and 2c, 3, 24 which were represented by 100% susceptibility: 96.46; 96.92 and 95.38%, respectively. Leaf rust resistance gene(s) probably present in four Egyptian commercial wheat cultivars in comparison with nineteen leaf rust near-isogenic lines against 65 isolates, the cultivars Gemmeiza 10, Giza 168 and Sakha 94 probably have Lr's, i.e. 9, 16, 24, 11, 17, 30 and 14b. On the other hand, the tested Egyptian wheat cultivars were postulated to the lack of Lr2a, Lr2b and Lr36 which were proved to be the more effective genes in the present study. These results will remain an integral part of resistance breeding rust pathogen populations (Youssef et al. 2010).

Leaf rust, caused by *Puccinia triticina* Erikss, is the most prevalent and widespread disease in almost all the wheat growing areas In Egypt and worldwide. It is annually appears with different virulence frequencies of races within its populations. In this study, a total of 203 leaf rust samples were collected from different locations in Egypt, during the two successive growing seasons; 2009/2010 and 2010/2011. During the whole period of the study, 459 isolates were obtained and tested as infection types on a set of 16 single Lr gene differential wheat lines. Virulence frequency against 31 monogenic lines (Lr's) and some local wheat genotypes were also estimated against the collected isolates during the whole period of the study. Three races; PTTT, TTTT and TKTT were the most common races that occurred at consistently high frequencies throughout the two years of the current study. While, most races were rare, as they represented by only one or two isolates in the two pathogen populations tried. The frequency of race groups based on infection types (ITS) of the first 8 differential host lines was also detected. Race groups TT- and PT- were the most dominant, being occurred at high frequencies in both years collections. They also proved to have high virulent against Lr's 1, 2c, 3, 9, 16, 24 and 26. Otherwise, virulence occurred at relatively low frequency (less than 23%), against leaf rust resistance genes; Lr's 9, 18, 36, 42, 12, 18, 19, 32, 34, 46. Thus, these genes considered to be the most effective genes against a large number of pathogen isolates tested in the two years of study. Testing of virulence against 18 local wheat cultivars revealed that no or little virulence frequencies could be identified in wheat cvs.; Mirs-1, Misr 2, Sids-13, Gemmeiza-10, Gemmeiza-11 as well as Banisweif varities i.e. as characterized by their highest degree of resistant reaction baniswef 4. Banisweif-5 and Banisweif-6. As in consequence, these against a great number of leaf rust isolates, collected during the whole period of the study. Meanwhile, the three wheat cvs.; Sakha 93, Shandaweel 1 and Sids 1 showed the lowest levels of varietal resistance, as the occurrence of virulence corresponding to these genotypes were relative of high percentages of frequency in the two year's collections. Based on the obtained results in this investigation, it could be suggested that strategic control to wheat leaf rust that assists control by limiting the amount of genetic diversity in pathogen population and keeping population size small year round would be more effective, successful and fruitful. This ultimately leads to avoid a significant annual yield loss and prevent or avert any damaging leaf rust epidemics in the future (Negm et al. 2013).

In Egypt, during three growing seasons of 2011/2012, 2012/2013 and 2013/2014, El-Orabey et al. (2015) showed significant differences in pathotypes from season to another. They collected 50, 65 and 33 leaf rust samples in 2011/2012, 2012/2013 and 2013/2014, respectively from diverse wheat growing regions in eight governorates of Egypt, i.e. Beheira, Dakahlia, Gharbiya, Minufiya, Sharqiya, Damietta, Qalyubia and Bani Sweif. A total of 118, 166 and 61 physiologic races were recognized in 2011/2012, 2012/2013 and 2013/2014, respectively. The most frequent races comprised STTST and TKTTT (each with 2.54%) in 2011/2012; PKTST (6.63%), TTTTT (7.83%) and TTTST (10.24%) in 2012/2013 as well as FKTTT (4.92%) and PTTTT (11.47%) in 2013/2014. Race groups PT—and TK—were common at eight locations through the three growing seasons. Cluster analysis according to frequency of virulence of *P. triticina* race groups showed that in 2011/2012 and 2012/2013 growing seasons two main clusters were formed. While in 2013/2014 growing season the cluster analysis divided into six main clusters. Lines with Lr 1, Lr 2c, Lr 3, Lr 16, Lr 24 and Lr 26 were susceptible against most race groups, whereas, the leaf rust monogenic lines Lr 2a and Lr 9 displayed different reactions against the verified race groups (http://internationalscholarsjournals.org/download. php?id=127291490460539852.pdf…).

12.4.3 Stem Rust

Eight stem rust physiologic races of *Puccinia graminis* f. sp. *tritici* were identified in Egypt during 2003/2004 growing season according to the standard method of Stakman et al. (1962) which resulted in identifying physiologic race, i.e. 11, 14, 15, 17, 19, 39, 42 and 122. Race No. 11 displayed the higher frequency (53.49%) which followed by race No. 15 (17.83%), race No. 39 (9.3%), race No. 19 (6.98%), race No. 17 (6.2%) and both of races 42 and 122 (1.55%). Results indicated the presence of 129 isolates recognized using the 5 letter pathotype name according to MacCallum et al. (2002). Most of these isolates were frequented with the exception of Pgt PTTTT (4.6%), PPTTT (2.3%) and each of PTTIT TTTT, TTTT which were frequented by (1.6%). The total nets of Pgt pathotypes were 118 (Najeeb et al. 2004).

From our results, sixty-three virulence formulae (Physiologic race or pathotype) were derived from 72 isolates of (*Puccinia graminis* f.sp. *tritici*). Race TTTT was the prevalent one and comprised (6.9%) of all races followed by TKTP was represented by (5.52%) and TFKT and TRTT each of them was represented by (2.76%). While, the rest of races or pathotypes exhibited a frequency of (1.38%) (Youssef et al. 2012).

Puccinia graminis f.sp. *tritici* race TKTT was mostly spread, i.e. Denmark (2013), Ethiopia (2013), Germany (2013), Iran (2012 & 2013) and Turkey (2005 & 2013) making 2013 as an unusual year for stem rust in wheat. Because Pgt race TKTT was responsible for the stem rust endemic in Southern Ethiopia in 2013 (Hodson 2013), the question was whether this race could spread epidemiologically in Egypt or not? Pgt race TKTT was identified in Egypt, for the first time. During 2004 season by Najeeb et al. (2004) and during reappeared during 2012 and 2013 growing seasons. Till now Pgt race TKTT is not predominant in Egypt and did not cause rust epidemic. This may be attributed not only to the using of highly resistant wheat cultivars but also the unfavorable climatic circumstances which might be lead to the escape from wheat stem rust disease (Imbaby et al. 2014a).

From rust-infested wheat leaves and stems through the survey of wheat fields and rust trap in the most of Egyptian governorates, to define the virulence of wheat stem rust fungus in 2009/2010. Hermas et al. (2013) tested seventeen Egyptian wheat varieties and thirty-five stem rust resistance genes against seventy-four virulence formulae (pathotype or race) were obtained from ninety-five isolates of *Puccinia graminis* f.sp. *tritici*. The virulence phenotype (race) TTTS, which is virulent to each of the tested resistance genes used in race identification except Sr Tmp only, it represents the first most common virulent phenotype. Followed race TTTT was virulent to all tested stem rust resistance gene. On the other hand, stem rust resistance genes, i.e. Sr31 and Sr26 + 9 g displayed the highest efficacy (89.47% each of) followed by Sr29 was (81.05%). Conversely, the cultivar Gemmeiza 7 and Sids 1 postulate have 26 and 20 stem rust resistance gene, respectively. Whereas, the cultivar Giza 168 and Beni Sweif 5 were free from any gene at least against the studied pathotypes. The common gene probably found between Gemmeiza 7 and Misr 2, also Sakha 94 and Sakha 93.

Race UG99: a new race of stem rust pathogen represents a major threat to the wheat crop, which has become a problem due to the rapid development of new species or different forms of stem rust. Ug99 has evolved, commonly recognized a shifting enemy of wheat due to its rapid migration rate. Investigators have already expected Ug99 as the highest of threats ever confronted by wheat cultivation as it has the ability to break the strongest of resistant genes, like Sr31. A range of creativities and approaches have been occupied at the worldwide level, of which BGRI is the most prominent. Full cooperation must be made between scientific organizations in target countries and taking severe controlling procedures from governments along with enormous assurance of farmers on the scheme (Rautela and Dwivedi 2018).

In the year 1998, wheat nurseries implanted at CIMMYT in Uganda suffered from severe infection by stem rust of wheat. The sensitivity of plants was quite surprising because the planting genotypes were resistant to the pathogen. Pretorius et al. (2000) collected uredospore samples for race analysis, they detected of a new race with novel resistance was found against the resistance gene Sr31. The name for the new wheat stem rust race known as Ug99 comes from the country and the year of discovery. It was recognized as TTKS by Wanyera et al. (2006) by the North American nomenclature system (Roelfs and Martens 1988) and more recently as TTKSK after addition of the fifth set of differentials for further expansion of the classification (Jin et al. 2008).

East Africa highlands are a recognized "hot spot" for the development and continuation of new rust races. The pathogen is developing quickly, and till date thirteen known variants within Ug99 lineage were identified. The Ug99 race lineage is existent in 13 countries, with Egypt being the newest one. The latest variants are: TTKSF + which was discovered in South Africa and Zimbabwe (Pretorius et al. 2012), TTHST which was collected in Kenya in 2013; TTKTT, TTKTK, PTKTK and TTHSK (Patpour et al. 2015; Fetch and Zegeye 2016), all were collected from Kenya in 2014.

Since the first discovery of race TTKSK (Ug99) of *Puccinia graminis* f. sp. *tritici* in Uganda in 1998 (Pretorius et al. 2000), it has been importance to track its further spread to other regions of growing wheat. Up to now, 10 variants in the Ug99 race group were discovered in 12 countries, i.e., Uganda, Kenya, Ethiopia, Sudan, Tanzania, Eritrea, Rwanda, South Africa, Zimbabwe, Mozambique, Yemen, and Iran (Patpour et al. 2015).

In the 2014 growing season, the existence of virulence to Sr31 in Egypt was supposed according to preliminary field observations of great infection on sources of Sr31 grown as international stem rust trap nursery at (i) Sakha Agricultural Research Station in Kafrelsheikh (31.094059° N, 30.933899° E), (ii) Al-Sharqia (30.601400° N, 31.510383° E), and (iii) Nubaria (30.91464° N, 29.95543° E). At Sakha, wheat genotype PBW343 carrying Sr31 was recorded 30MS-S, and the monogenic line Benno Sr31/6*LMPG was registered 20MS-S at Al-Sharqia (https://experts.umn.edu/en/publications/first-report-of-the-ug99-race-group-of-w). Three samples from these lines were sent to the Global Rust Reference Center (GRRC, Denmark). At Nubaria, stem rust was detected on wheat cultivars Misr-1, Misr-2, Giza 168, and Giza 171, and infested samples were collected and sent under permit to the Foreign Disease-Weed Science Research Unit (Fort Detrick, MD). Urediniospores of each sample were recovered on both susceptible wheat cultivars Morocco and McNair 701. Twenty-three and eleven single pustule isolates were derived and examined at GRRC and USDA-ARS Cereals Disease Laboratory, respectively, using 20 North American stem rust differential lines following standard race-typing method and infection type (IT) criteria defining virulence and avirulence (Jin et al. 2008). Also, three additional tester lines of Siouxland (carrying Sr24 + Sr31), Sisson (carrying Sr31 + Sr36), and Triumph 64 (donor of SrTmp) were involved to approve virulence/avirulence to Sr24, Sr31, Sr36, and SrTmp. The experiments were replicated two to three times. They detected three races in the Ug99 race group; TTKST (four isolates, IT 3 + 4 for Sr24, Sr31, and cultivar Siouxland) from Al-Sharqia, TTKTK (13 isolates, IT 4 for Sr31, SrTmp, and cv. Triumph 64) from Sakha, and TTKSK (2 isolates, IT 4 for Sr31) from Nubaria. This is the first report of races in the Ug99 race group in Egypt, thus spreading the geographical distribution of Ug99-related races. Egypt plays the green-bridge for *P. graminis* f. sp. *tritici* between East and North African countries and the wheat belts in the Middle East and Mediterranean regions. In light of these hazards, rust control efforts must be joined between affected countries and neighboring areas (Patpour et al. 2016).

Stem rust collections were obtained from infected wheat stems throughout the survey of wheat fields and nurseries in three locations (Sids, El Sharkia and El Nubaria) during 2011/2012 growing season. Whereas during 2012/2013 growing season the samples collected from six locations (Giza, Sids, Tag El Aiz, Sakha, El Sharkia and El Nubaria). Based on the race analysis of stem rust populations, and race identification by inoculating stem rust differential hosts, the phenotypic characterization of *Puccinia graminis* f.sp. *tritici* during 2011/2012 growing season resulted in the identification of 86 races from 22 successful samples, all of them showed 1.16% frequency. Race BBBBC was avirulent on all the tested Sr genes, except Sr MCN, while race TTTTK was virulent on all the tested Sr except Sr

24. On the other hand, examination during the next growing season revealed that 123 races with a frequency fluctuated from 0.81 to 2.43% were recognized. About stem rust resistant gene efficiency during the study, Sr24, Sr38 and Sr31 display the greatest effectiveness percentage (95.34, 91.86 and 87.2%, respectively) throughout the two seasons. Therefore, deployment of active Sr genes such as Sr24, Sr38 and Sr31 in particular cultivar through gene pyramiding has paramount significance as the additive effects of numerous genes offers the cultivar a broader base stem rust resistance along with periodic race survey (El-Sherif et al. 2018).

12.5 Genetic Variability

The genetic differences existed among wheat genotypes in their resistant genes are the basics of the genetic improvement process. El-Naggar Doaa and Soliman (2015) showed that the mean disease severities of Sids 7, Giza 168 and Gemmeiza 9 Egyptian wheat cvs. and their eight selected promising mutants at M_4 generation during 2013/14 growing season were evaluated. They recorded a high degree of genetic variability between wheat genotypes (Table 12.8). Stripe, leaf and stem rusts disease severities on Sids-7 wheat cv. valued 16.67, 13.33 and 13.33, respectively, with infection responses of susceptible (S), moderately susceptible (MS) and susceptible (S), respectively. While M_4-7 mutagenized Sids-7 recorded the lowest values (2.67, 3.00 and 0.0 respectively) with the subsequent highest percentages of reduction in disease severities (83.98, 77.49 and 100.0%, respectively). These lowest values of rust severities were reflected in the lowest AUDPC values (28.00, 35.93 and 0.0) for stripe, leaf and stem rusts, respectively. Also, M_4-7 mutagenized Sids 7 recorded moderately susceptible infection responses for both stripe and leaf rusts and free from stem rust. Regarding Giza-168 cv. and their promising mutants showed very important findings about disease severity of the three rust diseases. All four mutants possessed zero stem rust disease severity. Moreover, M_4-3 promising mutant had zero stripe rust disease severity. Additionally, M_4-3 mutant had a moderately susceptible response to leaf rust with no change in this parameter compared with its mother cv. Concerning Gemmeiza 9 wheat cv. and their mutants, M_4-1 mutant recorded zero leaf rust disease severity and also registered the lowest values of stripe and stem rusts disease severities (4.33 and 3.00, respectively). These results might be attributed to that gamma-ray induced mutation and could be used to isolate beneficial genes related to harmful effects on characters through chromosome breaking and translocation. Accordingly, seeds of M_4-7 mutagenized Sids 7, M_4-3 mutagenized Giza 168 and M_4-1 mutagenized Gemmeiza 9 were selected for further investigations during 2014/2015 growing season. Where, significant differences were observed between all genotypes for all studied parameters.

Disease severity of stripe rust was recorded by Omara et al. (2016) on seven Egyptian wheat cultivars as the percentage of final stripe rust severity (%), AUDPC and r value during 2013/2014 and 2014/2015 growing seasons (Table 12.9). Results showed that the final rust severity fluctuated from 20 to 70% and from 20 to 90% of

Table 12.8 Evaluation of the percentage of disease severities, infection responses and area under disease progress curve (AUDPC) of mutant lines of gamma-irradiated grains of Sids-7, Giza-168 and Gemmeiza-9 wheat cvs. to stripe, leaf and stem rusts, at M_4 generation (El-Naggar Doaa and Soliman 2015)

Genotype	Stripe rust				Leaf rust				Stem rust			
	Disease severity (%)	% Reduction in disease severity	Infection response	AUDPC	Disease severity (%)	% Reduction in disease severity	Infection response	AUDPC	Disease severity (%)	% Reduction in disease severity	Infection response	AUDPC
Sids-7	16.67	0.00	S	159.13	13.33	0.00	MS	154.0	13.33	0.00	MS/S	161.0
Sid-M$_4$-6	4.33	74.03	MS	47.60	3.33	75.02	MS	36.63	2.00	85.00	R	6.3
Sid-M$_4$-7	2.67	83.98	MS	28.00	3.00	77.49	MS	35.93	0.00	100.0	free	0.00
Giza-168	16.67	0.00	MS	70.93	8.33	0.00	MS	73.73	3.00	0.00	S	59.5
G-M$_4$-1	3.00	82.00	R	7.70	0.00	100.0	free	0.00	0.00	100.0	free	0.00
G-M$_4$-3	0.00	100.00	Free	0.00	3.00	63.99	MS	36.4	0.00	100.0	free	0.00
G-M$_4$-5	3.67	77.98	MR	17.03	8.33	0.00	MS	106.4	0.00	100.0	free	0.00
G-M$_4$-6	2.00	88.00	MR	11.20	3.00	63.99	MS	30.8	0.00	100.0	free	0.00
Gemmeiza-9	20.00	0.00	S	189.0	16.67	0.00	MS	139.07	4.33	0.00	MS	72.8
Gem-M$_4$-1	4.33	78.35	MS	42.93	0.00	100.0	free	0.00	3.00	30.72	R	7.7
Gem-M$_4$-2	13.33	33.35	MS	101.07	8.33	50.03	MS	99.87	3.67	15.24	MS	43.87
L.S.D $_{0.05}$	5.15			36.49	4.70			30.04	3.08			13.37

Table 12.9 Final leaf rust severity (%), AUDPC, rAUDPC and r-value for 7 Egyptian wheat cultivars during 2013/2014 and 2014/2015 growing seasons (Omara et al. 2016)

Cultivar	Season/Epidemiological parameters							
	2013/2014				2014/2015			
	FRS (%)	AUDPC	rAUDPC	r-value	FRS (%)	AUDPC	rAUDPC	r-value
Gemmeiza 7	50S	1522.5	81.3	0.098	70S	2152.5	75.5	0.126
Gemmeiza 9	30S	631.4	33.7	0.095	40S	900.2	31.6	0.110
Gemmeiza 10	20S	462.0	24.7	0.060	20S	456.4	16.0	0.077
Gemmeiza 11	60S	1697.5	90.6	0.112	90S	2572.5	90.2	0.171
Sids 1	40S	1137.5	60.7	0.085	60S	1697.5	59.5	0.112
Sids 12	70S	1872.5	100.0	0.126	90S	2852.5	100.0	0.171
Sids 13	40S	924.0	49.3	0.092	50S	1270.5	44.5	0.116
$LSD_{0.05}$	–	15.07	6.56	0.006	–	17.11	7.94	0.002

the verified wheat cultivars during the two seasons, respectively. The highest final rust severity in the two seasons was registered by Sids 12 (70S, 90S), Gemmeiza 11 (60S, 90S) and Gemmeiza 7 (50S, 70S). While, the remaining cultivars exhibited final rust severity varied from 20 to 40% and 20 to 60% during the two seasons, respectively. Percentage of disease severity was different among the cultivars. Sids 12, Gemmeiza 11 and Gemmeiza 7 were the fast cultivars for the disease development followed by Sids-1 and Sids-13. Otherwise, stripe rust development was slow with Gemmeiza 10 and Gemmeiza 9 cultivars during 2013/2014 and 2014/2015 growing seasons (Fig. 12.4). AUDPC values have taken the same direction and went in parallel line with the final rust severity. The highest estimates of AUDPC were detected by Sids-12 (1872.5, 2852.5), Gemmeiza 11 (1697.5, 2572.5) and Gemmeiza 7 (1522.5, 2152.5) followed by Sids 1 (1137.5, 1697.5) and Sids-13 (924.0, 1270.5) for the two seasons 2013/2014 and 2014/2015, respectively. Whereas, Gemmeiza 10 and Gemmeiza 9, attained low values of AUDPC, i.e. (462.0, 456.4) and (631.4, 900.2) for the two seasons, respectively. Therefore, Gemmeiza 10 and Gemmeiza 9 may be exhibited partial resistance or tolerance against the disease. But under the prevalent circumstances of the elapsed season, the situation was completely different.

The differential sets and isogenic lines indicated a wide range of rust response during the four-year investigation by Shahin (2017). The field data obtained in four cropping seasons indicated that resistance genes, Yr2 + , Yr5, Yr10, Yr15, YrSD and YrND were effective. The genotypes with Yr2, Yr6, Yr7, Yr9, Yr17, Yr27, YrSU, and YrA were susceptible. The genotype with Yr 29 was moderately resistant while those with Yr18 were moderately susceptible and Anza (YrA + Yr18) was moderately susceptible. The genotypes with Yr4, Yr5, Yr70, and Yr75 exhibited 0-type or R-type reaction, advising these genes to be immune or resistant against the

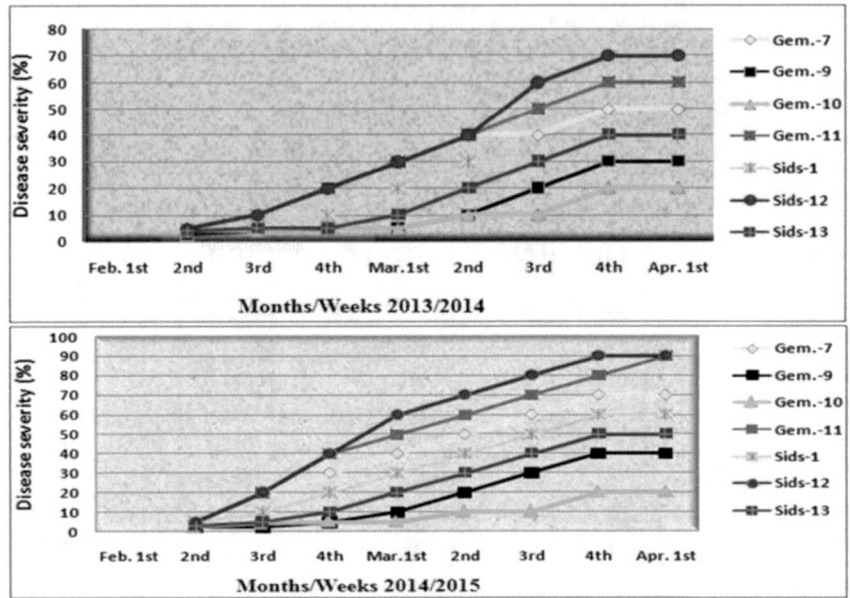

Fig. 12.4 Maximum percentage of disease severity for commercial wheat cultivars during 2013/2014 and 2014/2015 growing seasons) Omara et al. 2016)

pathogen populations at four growing seasons. Sakha Station considered to be stripe rust hot spot in Egypt. The responses of genotypes were compared across the four seasons, the responses at 2015/2016 were more different, while those at three other growing seasons were relatively similar. Based on the reaction of the isogenic lines, no virulence was recorded at three growing seasons. Partial virulence was recorded at four seasons for YrA, Yr18 (Anza). The Egyptian wheat genotypes had more variation in reaction, during four seasons. The FRS of susceptible control, Morocco was 70 to 100%, indicating that high adult plant infection type was established over four-year field experiment. The genotypes Misr 3 and Sakha 95 were resistant, one variety (Gemmeiza 7) showed moderate resistance, three varieties Gemmeiza 3, 10 and 12) were moderately susceptible and the remaining were susceptible. The mean rust severity during 2015/2016 was the highest across the years. Twenty varieties that had different reactions may be categorized into three classes. Class 1 included fifteen varieties (Misr 2, Gemmeiza 3, 9, 10, 11, 12, Sids 1, 12, 13, Sakha 8, 61, 69, 94, Giza 168 and Giza 171) showing susceptible to moderately susceptible reaction. Class 2 contained moderately resistant to resistant varieties (Misr 1 and Shandaweel 1) were the least affected during the four years, relative estimates for AUDPC and ACI were also the minimum for these two entries and were the most resistant. Varieties in class 3 included (Misr 3 and Sakha 95), displaying R-type or 0-type reaction. Most Egyptian varieties were susceptible, and their susceptibility levels were higher at the 2015 growing seasons than in the other seasons of experimentation under field condition at Sakha during 2014–2017. Among the Egyptian varieties, Gemmeiza 9 and Misr 2 showed partial virulence at 2013/2014 season, while complete virulence was registered at 2016/2017 season. Furthermore, differences in virulence pattern detected in consecutive years. Especially variety, Sakha 61 showed immunity in 2013/2014 and 2014/2015 while complete virulence was recorded at 2016/2017. With regard to mean of rAUDPC and coefficient of infection, the least area of disease progression were observed with Mist 1 while the highest values were recorded with Sids 12.

Eisa and El-Naggar Doaa (2015) assessed the leaf rust in four Egyptian bread wheat cultivars i.e. Gemmeiza 7, Misr 1, Sids 1 and Sids 12 with their subsequent histological correlations. Assessments were carried out through 2012/2013 and 2013/2014 growing seasons under realistic field circumstances of Kafr El-Hamam Agricultural Research Station in El-Sharkia governorate, ARC, Egypt. Based on their responses, Misr 1 and Sids 12 were expressed as resistant cultivars. Meanwhile, Gemmeiza 7 and Sids 1 were susceptible. Different associations between disease parameters and some histological features were achieved. Constitutive, pre-formed defenses are thought to comprise a major part and first step in resistance against pathogens. They include the stomata, epidermal reinforcements such as the epidermal leaf hairs (trichomes), as well as sclerenchyma tissue and leaf bundles. These pre-formed barriers are believed to delay the initial colonization by microbes, allowing the plant sufficient time to mount effective inducible defense responses. Stomata on the leaf surfaces serve as entrance pathways for the majority of the urediospores of rust fungi. The emerging germ tubes should adhere first to the leaf surface, and next they will grow in a direction where they may encounter a stoma. This is called directional growth. Figure 12.5 shows the stomatal features on the upper surface of the flag leaf of adult plants. The studied wheat cultivars can be divided into two groups. The first group (Fig. 12.5a and b) included the resistant cvs. (Misr 1 and Sids 12) which characterized by the lowest average numbers of stomata (33.33 and 36.32/mm^2, respectively) with the highest distances between stomatal rows (193.78 and 173.23 μm, respectively), compared to the second group (Fig. 12.5c and d) included the susceptible cvs. (Sids 1 and Gemmeiza 7) which characterized by the highest average numbers of stomata (51.52 and 42.62/mm^2, respectively) with the lowest distances between stamatal rows (107.11 and 148.37 μm, respectively). Hairs are surface structures originating from the epidermal cells which usually protrude from the leaf and/or the stem. They act as physical barriers providing protection against various abiotic and biotic factors of the external environment, including pathogen attack. In recent years, there has been growing interest in the outermost layers of the plant. The epidermal hairs of the tested wheat cultivars were examined and illustrated in Fig. 12.6 and Table 12.10. Transactions showed in Fig. 12.6a and results in Table 12.10 indicated that Misr-1 wheat cultivar was characterized by the highest number of epidermal hairs. This result coincides with those obtained in the aforementioned disease assessment section where Misr 1 was the highly resistant cultivar among the studied ones. Sclerenchyma cells are characterized by having uniformly thick lignified secondary walls. Loss of their protoplasm at maturity gave mechanical supporting to plants making them resistant to different pathogenic

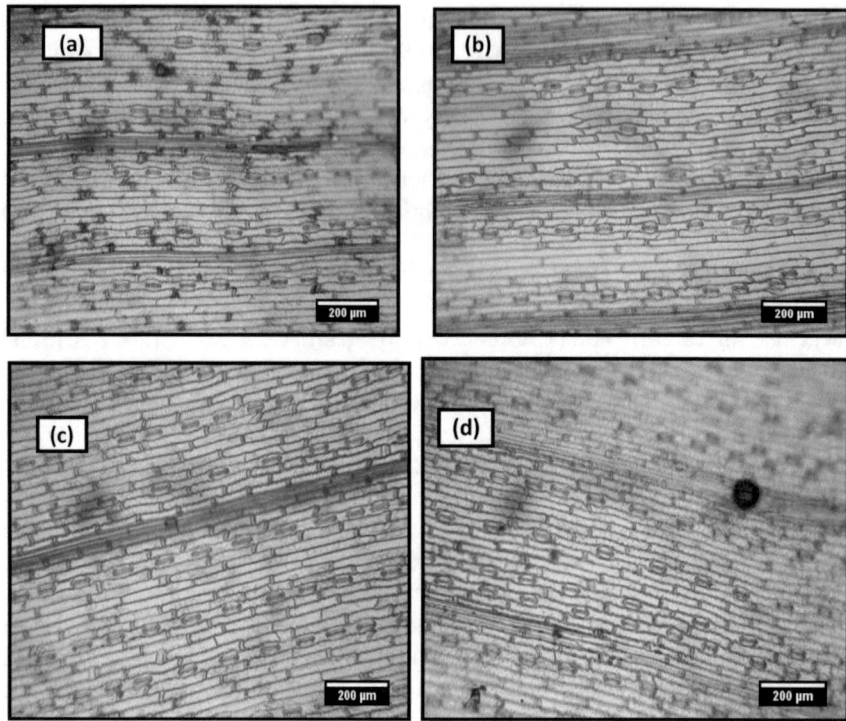

Fig. 12.5 Stomata distribution on the upper surface of flag leaf of adult plants of **a** Misr 1, **b** Sids 12, **c** Gemmeiza 7 and **d** Sids 1 wheat cultivars, at the age of 70 days during 2013/2014 growing season (Eisa and El-Naggar Doaa 2015)

attacks. In susceptible examined wheat cultivars (Gemmeiza 7 and Sids 1), the sclerenchyma tissue did not extend up the laminar (lateral or veinlets) vascular bundles (Fig. 12.6c and d). Moreover, data in Table 12.10 indicated that leaves of Gemmeiza-7 and Sids-1 wheat cvs. recorded the lowest values of sclerenchyma tissues thickness (13.96 and 9.22 μm, respectively), the lowest average values of laminar vascular bundles thickness (40.05 and 36.94 μm, respectively) and the lowest average values of laminar vascular bundles width (50.47 and 47.71 μm, respectively). Meanwhile leaves of resistant wheat cvs. (Misr-1 and Sids-12), recorded the highest values of sclerenchyma tissues thickness (38.11 and 26.66 μm, respectively), the highest values of the average thickness of laminar vascular bundles (46.04 and 41.83 μm, respectively) and the highest average values of laminar vascular bundles width (76.53 and 57.64 μm, respectively).

Awaad et al. (2003) evaluated six wheat populations of four crosses (Sakha 69 × Sahel I, Sakha 69 × ACS.W 945, Gemmeiza 9 × PAT 10/ALD"S" and Gemmeiza 5 × Giza 168) derived from seven wheat genotypes varied in their resistance to leaf rust. Wheat plants were artificially infested with urediospores mixture of eleven physiological races of *Puccinia recondite* Rob. Ex Desm. f.sp. *tritici* Erikes and

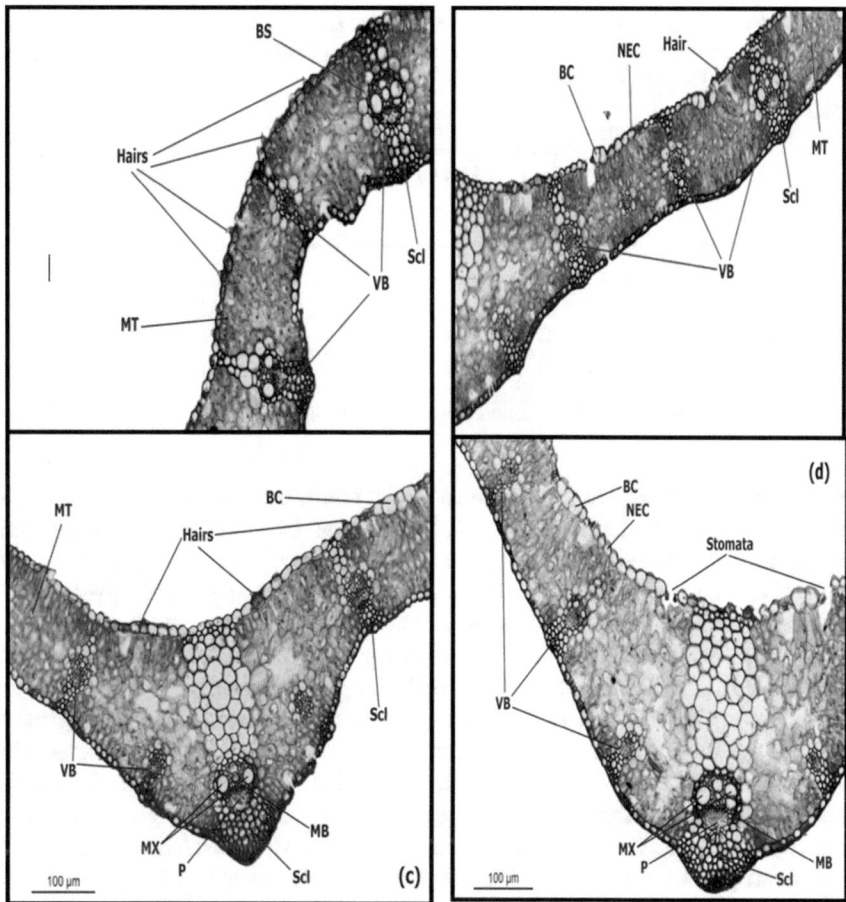

Fig. 12.6 Epidermal structural features and internal tissue patterns of flag leaf blade of wheat cultivars of **a** Misr 1, **b** Sids 12, **c** Gemmeiza 7 and **d** Sids 1, at the age of 70 days during 2013/2014 growing season. BC = Bulliform cell, BS = Bundle sheath, MB = Midvein bundle, MT = Mesophyll tissue, MX = Meta xylem, NEC = Normal epidermal cell, P = Phloem, Scl = Sclerenchyma tissue and VB = Veinlet (laminar or lateral) bundles (Eisa and El-Naggar Doaa 2015)

Henn and evaluated for their resistance. The results revealed various levels of resistance among the four cross populations. The most resistant wheat parents were Giza 168 followed by PAT10/ALD"S", ACSAD 945, Gemmeiza 9, where, Gemmeiza 5, Sakha 69 and Sahel 1 expressed as susceptible ones. The F_1 displayed highly resistant in 1st cross, complete dominance in 2nd cross and partial dominance towards their respective resistance parent in 3rd one. On the same genetic materials, Salem et al. (2003) registered diverse levels of resistance against stem rust in the evaluated cross populations. The most resistant parents were Giza 168 followed by Gemmeiza 9, Sakha 69, PAT10/ALD"S" and Gemmeiza 5, while, Sahel 1 and ACSAD 945

Table 12.10 Flag leaf blade structural features (stomata, hairs, sclerenchyma tissue, laminar bundles and bulliform cells) of four adult wheat cultivars, at the age of days during 2013/2014 growing season (Eisa and El-Naggar Doaa 2015)

Cultivar	Stomata		Hairs		Laminar bundles	
	Average no./mm^2	Distance between stomata rows (μm)	Average no./microscopic field	Scl.	Average thickness (μm)	Average width (μm)
Misr 1	33.33	193.78	9	38.11	46.04	76.53
Sids 12	36.32	173.23	1	26.66	41.83	57.64
Gemmeiza 7	42.62	148.37	3	13.96	40.05	50.47
Sids 1	51.52	107.11	1	9.22	36.94	47.71

Scl. = Sclerenchyma tissue thickness under upper epidermis and up lower epidermis (μm)

were susceptible ones. The F1 displayed partial dominance towards stem rust resistance in most cases. Moreover, Minaas, Sallam et al. (2014) evaluated thirty-seven monogenic wheat lines carrying single leaf rust genes for their stability resistance to leaf rust caused by *Puccinia triticina* Eriks. at adult plant stage under diverse field conditions comprising four environments, i.e. Italy El-Baroud, Sakha and Nubariya Agric. Res. Stations and Shibin El-Kom during three successive growing seasons, i.e. 2011/2012, 2012/2013, and 2013/2014. All genetic materials were chosen as derivatives to Thatcher variety. The tested Lr genes were classified into three groups according to the infection type. The first group contained the effective genes that included Lr 9, Lr 18, Lr 19 and Lr 28. The second group involved genes differentiated ineffectiveness, whereas, the third group comprised ineffective genes. Stability parameters revealed that wheat monogenic line Lr 33 was considered as stable and broadly adapted in its resistance. Moreover, Minaas, Sallam et al. (2016) tested thirteen Egyptian wheat genotypes against leaf rust resistance at seedling stage under greenhouse condition as well as adult plant stage under field conditions during three growing seasons i.e. 2011/2012, 2012/2013 and 2013/2014 in three locations, Itay El-Baroud and Nubariya Agricultural Research Stations as well as the Farm of the Faculty of Agriculture, Minufiya University, Shibin El-Kom. The verified wheat genotypes were divided into three groups based on their reaction. The first group, race-specific resistant genotypes containing Shandweel 1, Misr 1, Misr 2, Sids 12 and Sids 13, exhibited the lowest values of final rust severity (FRS %) and area under disease progress curve (AUDPC). The second group, slow-rusting or partially resistant genotypes comprising Sakha 94, Gemmeiza 9, Giza 168, Sakha 95, Gemmeiza 10 and Gemmeiza 11, exhibited low level of FRS and AUDPC. The third one contains Gemmeiza 7 and Sids 1, showed the highest estimates of FRS and AUDPC. Whereas, Draz et al. (2015) screened 42 Egyptian wheat genotypes for leaf rust resistance, and found only 9 varieties (Sakha 94, Giza168, Gemmeiza 9, Gemmeiza10, Gemmeiza 11, Sids 12, Sids13, Misr1 and Misr2) displayed seedling and adult plant resistance during 2010/2011 and 2011/2012 seasons. Out of 41 tested monogenic line (Lr genes), only 13 Lr genes (Lr9, Lr10, Lr11, Lr16, Lr18, Lr19, Lr26, Lr27, Lr29, Lr30,

Lr34, Lr42 and Lr46) showed seedling resistance whereas, 9 Lr genes (Lr19, Lr20, Lr21, Lr24, Lr29, Lr30, Lr32, Lr34 and Lr44) revealed adult plant resistance during both seasons. Therefore, they might be utilized as good sources for resistance. Also, 12 varieties (Sids12, Misr 2, Sakha94, Misr1, Sids13, Giza168, Gemmeiza 9, Sids 7, Beniswef 4, Sakha 93, Gemmeiza11 and Sids 6) were presented partial resistance of wheat seedlings, registering the longest incubation and latent period. However, 10 varieties (Sakha 8, Sakha 93, Giza144, Giza155, Giza156, Giza157, Sids4, Sids5, Sids8 and Beniswef4) were obvious as having a high level of partial resistance of an adult plant, giving ACI less than 20%, AUDPC less than 332.5 and r-value less than 0.101. The highest significant loss % has been registered in susceptible wheat cultivars i.e. Gemmeiza 7, Sakha 61 and Giza 164 valued 12.24, 12.10 and 9.08%, respectively. Whereas, insignificant loss % was observed in resistant cultivars i.e. Giza168 (1.87%), Misr 2 (2.44%) Sakha 94 (2.46%). Inverse relation was found between the disease level and grain yield (https://www.sciencedirect.com/science/article/pii/S0570178315000020). Furthermore, Shaheen and El-Orabey (2016) tested twelve wheat genotypes in respect to adult plant resistance and yield losses as a result of leaf rust under field conditions at Shibin El-Kom location, Egypt during 2013/2014 and 2014/2015 growing seasons. The field trial was surrounded by spreader area of highly susceptible genotypes inoculated with a combination of leaf rust pathotypes as an inoculum source. Final leaf rust severity (%) varied from Tr MR to 80S in 2013/2014 growing season, while it varied from Tr MR to 70S in 2014/2015.

Both leaf and stem rust were evaluated by Darwish et al. (2018) using six parents, i.e. Gemmeiza 9, Sids 12, Misr 1, Misr 2, Sids 1 and Sham 4, with corresponding 15 F_2 crosses. The phenotypic variances in the F_2 crosses were varied significantly from the environmental differences in the corresponding parents in most cases. The parents Gemmeiza 9, Sids 12, Misr 1, Misr 2 were resistant to leaf rust and Gemmeiza 9, Sids 12 and Sids 1 were resistant to stem rust. Among the crosses, three cross combinations, i.e. Misr 2 × Sids 1, Misr 1 × Sids 1 and Gemmeiza 9 × Sids 1 produced the highest grain yield. Means of F_2 crosses were greater than those of the parents for most evaluated characters. The ranges of the F_2 values exceeded the ranges of the two parents in most situation as a result of transgressive segregation. According to the resistance to leaf and stem rusts, suitable plant height (90–100 cm) and grain yield greater than the highest parent, then 8–17 distinctive plants were selected from seven crosses.

Cluster analysis was carried out using the manner proposed by Ball and Hall (1967) to identify natural groupings or clusters of genotypes. El-Naggar Doaa and Soliman (2015) showed that the dendrogram based on the similarities of disease severity of the three tested rusts at M_4 generation (Fig. 12.7) clustered the 11 genotypes into two main groups. The first group contained the three mother cultivar (Sids-7, Giza-168 and Gemmeiza-9) in addition to Gem-M_4-2 mutant line. While, the second group included the rest seven mutant lines. The second group clustered the seven mutant lines into six subgroups. Jointing between G-M_4-3 and G-M_4-6 mutant lines gave the highest similarity (91.56) followed by between this subgroup and Sid-M_4-7 (91.12). Also, the dendrogram based on the similarities of disease severity of the three tested rusts at M_5 generation (Fig. 12.8) clustered the 6 genotypes into two main groups. The

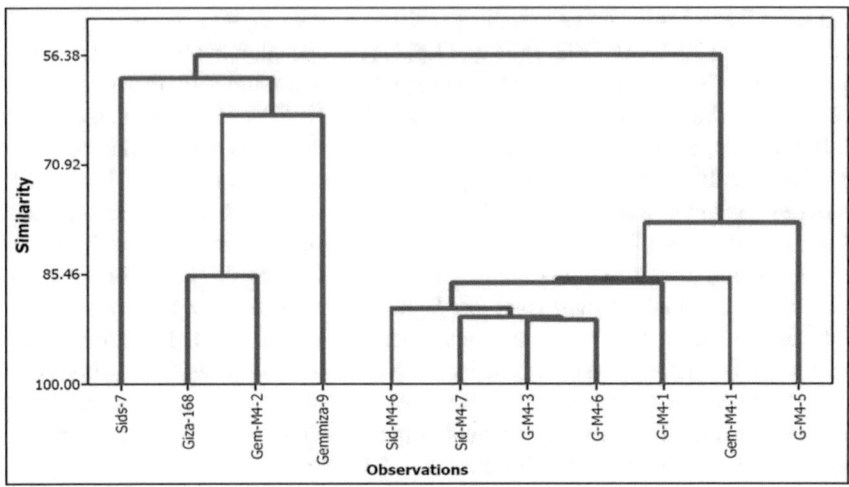

Fig. 12.7 Dendrogram of 11 wheat genotypes developed from stripe, leaf and stem rust severities at M$_4$ generation (El-Naggar Doaa and Soliman 2015)

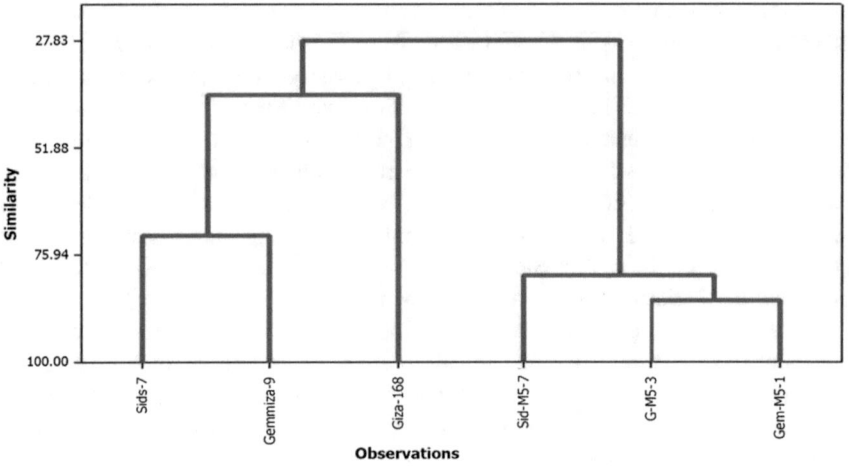

Fig. 12.8 Dendrogram of 6 wheat genotypes developed from stripe, leaf and stem rust severities at M$_5$ generation (El-Naggar Doaa and Soliman 2015)

first group included the three mother cultivars, while the second group included the three mutant lines. Jointing between G-M$_5$-3 and Gem-M$_5$-1 mutant lines gave the highest similarity (86.09) followed by jointing between this subgroup and Sid-M$_5$-7 (80.51).

Genetic differences between parents and their progeny in histological characters related to stem rust resistance were detected by as illustrated in Figs. 12.9 and 12.10

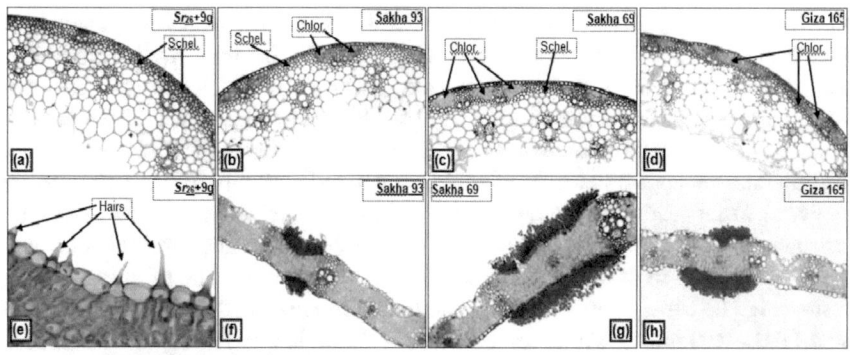

Fig. 12.9 Histological characteristics of three susceptible wheat varieties and parent $Sr_{26} + 9$ g: **a** $Sr_{26} + 9$ g; **b** stem section of Sakha 93; **c** stem section of Sakha 69; **d** stem section of Giza 165; **e** hairs of $Sr_{26} + 9$ g; **f** pustule of Sakha 93; **g** pustule of Sakha 69 and **h** pustule of Giza 165 (Swelam et al. 2010)

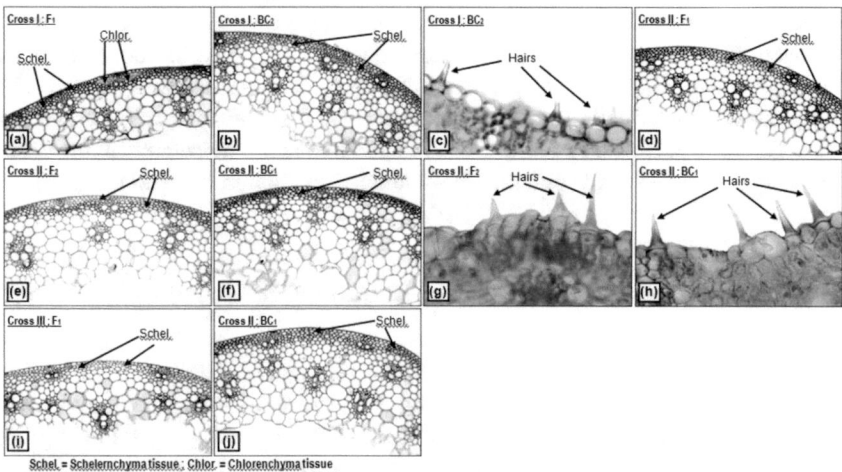

Fig. 12.10 Histological characteristics of **a** stem section of F_1 of cross I; **b** stem section of BC_2 of cross I; **c** hairs of BC_2 of cross I; **d** stem section of F_1 of cross II; **e** stem section of F_2 of cross II; **f** stem section of BC_1 of cross II; **g** hairs of F_2 of cross II; **h** hairs of BC_1 of cross II; **i** stem section of F_1 of cross III and **j** stem section of BC_1 of cross III (Swelam et al. 2010)

(Swelam et al. 2010). As illustrated in Fig. 12.10 clearly showed that hybridization of susceptible wheat variety Sakha 93 with the resistant parent Sr26 + 9 g (cross 1) increased sclerenchyma tissue percentage in F1 by 83.48% (Fig. 12.10a) which led to the appearance of the epidermal hairs in order of 6.7/mm and conse-quently the absence of uridiopustule. In F2 generation, the uridiopustule was reap-peared, and schlerenchyma tissue percentage was decreased while chlorenchyma tissue percentage was increased and epidermal hairs were decreased. The resistance character was highly increased in BC2 by increasing sclerenchyma tissue percentage to 100% and the appearance of epidermal hairs in order 10/mm (Fig. 12.10b). Sakha 69 wheat variety acquired the resistance to rust disease by its hybridization with the resistant parent (Sr26 + 9 g) (cross II). This improved by increasing sclerenchyma tissue percentage to 100% in F1, F2 and BC1 with the absence of chlorenchyma tissue (Fig. 12.10d–f) and a noticeable increase in epidermal hairs number in leaf Sect. (10.0, 10.0 and 13.4/mm, respectively) as shown in Fig. 12.10g and h. Although the urediopustule was absent in BC2 plants but the resistance decreased because of decreasing schlerenchyma tissue percentage to 77.5 and No. of epidermal hairs in leaf section to 3.4/mm. Resistance to rust disease appeared in Giza 165 wheat variety when hybridized with the resistant parent (Sr26 + 9 g) (cross III). The resistance showed its maximum in F1 and BC1 by increasing sclerenchyma tissue percentage to 94.05 and 100%, respectively, and appearance of epidermal hairs in order 6.7 and 13.4/mm of leaf section, respectively. These results might be due to the fact that, chlorenchyma tissue has a thin cell wall which became easy to be penetrated by the pathogen, while the sclerenchyma tissue has a thick cell wall. The resistance character was presented in Fig. 12.10i and j. On the other hand, the resistance disappeared in F2 by increasing chlorenchyma tissue percentage to 40.33% and absence of epidermal hairs. The resistance still found in BC2, in spite of decreasing sclerenchyma tissue percentage to 69.58% and epidermal hairs number to 3.4/mm compared to BC1.

12.6 Genetic System and Genes Responsible Resistance

The study of genetic behavior, the nature of the gene action and the number of gene pairs controlling the resistance are imperative to be aware before starting the breeding program. It is useful to identify the line of the suitable breeding program to improve resistance in the promising genotypes. As yet, more than 187 rust resistance genes (80 leaf rust, 58 stem rust and 49 stripe rust) were derivative from various bread wheat or durum wheat cultivars and the related wild species using various molecular procedures (Aktar-Uz-Zaman et al. 2017).

In this respect, a field test was carried out by Shahin and Ragab (2015) to study the inheritance of stripe rust resistance in wheat at adult plant stage under field conditions during 2010/2011 growing season at Sakha Agricultural Research Station. Giza160 crossed with the four Yr monogenic lines showed adult plants susceptibility to yellow rust except for the cross Giza160/Compair. The F2 generation segregated into 30R:180S, 0R:205S, 10R:195S and 20R:189S plants with expected ratios 13:3, 0:1, 1:15 and 7:57, respectively. The F1 plants of the cross Giza168/Giza160 was resistant to yellow rust, while the F2 population segregated into 40R:120S with expected ratio 3:13. Adult plant response for stripe rust; observed hypothetical ratios and chi-square and probability values for nine wheat F2 populations inoculated with Pst under field conditions during 2010/2011. Regarding variance estimates, environmental (VE), phenotypic (VP) and genotypic (VG) variances ranged from 0.3, 7.9 and 7.6 for the cross Giza 168/Jupateco R to 10.8, 235.9 and 225.1 for the cross Giza 160/Lee, respectively. Broad-sense heritability ($h_b^2\%$) estimates ranged from 64.0% for cross Giza 168/Kalyansona to 97.7% for the cross Giza 160/Compair. The genetic advance from selection ranged low for cross Giza168/Jupateco R to high for cross Giza160/Kalyansona. Meanwhile, the genetic coefficient of variation ranged from 4.0 for cross Giza 160/Lee to 23.0 for cross Giza 168/Compair. Regarding resistant by susceptible crosses, the F1 plant's field response were closed to the resistant parent in the crosses Giza 168/Jupateco R and Giza160/Compair); closed to the susceptible parent in the cross Giza168/Lee; closed to mid parent in the cross Giza168/Kalyansona.

Hasan et al. (2014) studied crosses between the nine tested cvs. and five monogenic lines. As shown in Table 12.11 and summarized in Table 12.13. All of the 394, 410 and 371 F2 seedlings of the crosses between monogenic line Lr 10 and the wheat cultivars Gemmeiza 9, Gemmeiza 11 and Shandweel 1 were resistant and showed no segregation when tested with the race QBLBN. These results indicated that the wheat cultivars Gemmeiza 9, Gemmeiza 11 and Shandweel 1 possess the gene Lr 10. Moreover, the F2 seedlings of the crosses between Lr 10 and the wheat cultivars Misr 1, Misr 2, Sakha 94, Gemmeiza 7, Gemmeiza 10 and Sids 12 segregated to 305 Resistant:100 Susceptible, 287 R:93 S, 299 R:103 S, 274 R:97 S, 304 R:99 S and 295 R:95, respectively. These segregations fit the ratio 3 R:1 S, indicated that these cultivars do not have Lr 10 gene. All of the 399 and 408 F2 seedlings of the crosses between Lr 13 and the wheat cultivars Gemmeiza 11 and Shandweel 1 were resistant and showed no segregations when tested with the race QBLBN. These results indicated that the wheat cultivars Gemmeiza 11 and Shandweel 1 possess the gene Lr 13. The F2 seedlings of the crosses between Lr 13 and the wheat cultivars Misr 1, Misr 2, Sakha 94, Gemmeiza 7, Gemmeiza 9, Gemmeiza 10 and Sids 12 segregated to 298 R:103 S, 281 R:91 S, 303 R:96 S, 278 R:89 S, 304 R:105 S, 294 R:99 S and 302 R:98, respectively. These segregations fit the ratio 3 R:1 S, indicated that these cultivars do not have Lr 13 gene. Moreover, all of the 415, 397, 405, 389 and 405 F2 seedlings of the crosses between Lr 19 and the wheat cultivars Misr 1, Misr 2, Gemmeiza 9, Gemmeiza 10 and Shandweel 1, respectively were resistant and showed no segregation when tested with the race NSHLQ. These results revealed that these cultivars possess the gene Lr 19. The F2 seedlings of the crosses between Lr 19 and the wheat cultivars Sakha 94, Gemmeiza 7, Gemmeiza 11 and Sids 12 segregated to 281 R:96 S, 277 R:101 S, 305 R:103 S and 298 R:99 S, respectively. These segregations fit the ratio 3 R:1 S, showed that these cultivars do not have Lr 19 gene. Furthermore, all of the 394, 381, 403, 379, 405, 391 and 400 F2 seedlings of the crosses between Lr 35 and the wheat cultivars Misr 1, Gemmeiza 7, Gemmeiza 9, Gemmeiza 10, Gemmeiza 11, Sids 12 and Shandweel 1, respectively were resistant and showed no segregations when tested with the race MHCQQ. These results indicated that these cultivars possess the gene Lr 35. The F2 seedlings of the crosses between Lr 35 and the wheat cultivars Misr 2 and Sakha 94 segregated to

Table 12.11 Segregations and Chi-square analysis of F_2 plants of the crosses between five leaf rust monogenic lines and nine bread wheat cultivars at seedling stage under greenhouse condition during 2013/2014 growing season (Hasan et al. 2014)

Cross	Leaf rust race	No. of F_2 plants		Expected ratio	χ^2	P value
		Resistant (R)	Susceptible (S)			
Gemmeiza 7 × Lr 10	QBLBN	274	97	3:1	0.260	0.610
Gemmeiza 9 × Lr 10		394	0	No segregation	–	–
Gemmeiza 10 × Lr 10		304	99	3:1	0.041	0.840
Gemmeiza 11 × Lr 10		410	0	No segregation	–	–
Misr 1 × Lr 10		305	100	3:1	0.021	0.886
Misr 2 × Lr 10		287	93	3:1	0.056	0.813
Sakha 94 × Lr 10		299	103	3:1	0.083	0.773
Sids 12 × Lr 10		295	95	3:1	0.085	0.770
Shandweel 1 × Lr 10		371	0	No segregation	–	–
Gemmeiza 7 × Lr 13	QBLBN	278	89	3:1	0.110	0.740
Gemmeiza 9 × Lr 13		304	105	3:1	0.099	0.753
Gemmeiza 10 × Lr 13		294	99	3:1	0.008	0.930
Gemmeiza 11 × Lr 13		399	0	No segregation	–	–
Misr 1 × Lr 13		298	103	3:1	0.101	0.751
Misr 2 × Lr 13		281	91	3:1	0.057	0.811
Sakha 94 × Lr 13		303	96	3:1	0.188	0.665
Sids 12 × Lr 13		302	98	3:1	0.053	0.817
Shandweel 1 × Lr 13		408	0	No segregation	–	–
Gemmeiza 7 × Lr 19	NSHLQ	277	101	3:1	0.596	0.440
Gemmeiza 9 × Lr 19		405	0	No segregation	–	–
Gemmeiza 10 × Lr 19		389	0	No segregation	–	–

(continued)

Table 12.11 (continued)

Cross	Leaf rust race	No. of F_2 plants		Expected ratio	χ^2	P value
		Resistant (R)	Susceptible (S)			
Gemmeiza 11 × Lr 19		305	103	3:1	0.013	0.909
Misr 1 × Lr 19		415	0	No segregation	–	–
Misr 2 × Lr 19		397	0	No segregation	–	–
Sakha 94 × Lr 19		281	96	3:1	0.043	0.835
Sids 12 × Lr 19		298	99	3:1	0.001	0.977
Shandweel 1 × Lr 19		405	0	No segregation	–	–
Gemmeiza 7 × Lr 35	MHCQQ	381	0	No segregation	–	–
Gemmeiza 9 × Lr 35		403	0	No segregation	–	–
Gemmeiza 10 × Lr 35		379	0	No segregation	–	–
Gemmeiza 11 × Lr 35		405	0	No segregation	–	–
Misr 1 × Lr 35		394	0	No segregation	–	–
Misr 2 × Lr 35		297	102	3:1	0.068	0.795
Sakha 94 × Lr 35		291	97	3:1	0.000	1.000
Sids 12 × Lr 35		391	0	No segregation	–	–
Shandweel 1 × Lr 35		400	0	No segregation	–	–
Gemmeiza 7 × Lr 37	LMSDB	382	0	No segregation	–	–
Gemmeiza 9 × Lr 37		276	94	3:1	0.032	0.857
Gemmeiza 10 × Lr 37		298	99	3:1	0.001	0.977
Gemmeiza 11 × Lr 37		411	0	No segregation	–	–

(continued)

Table 12.11 (continued)

Cross	Leaf rust race	No. of F$_2$ plants		Expected ratio	χ^2	P value
		Resistant (R)	Susceptible (S)			
Misr 1 × Lr 37		401	0	No segregation	–	–
Misr 2 × Lr 37		393	0	No segregation	–	–
Sakha 94 × Lr 37		303	105	3:1	0.118	0.732
Sids 12 × Lr 37		302	98	3:1	0.053	0.817
Shandweel 1 × Lr 37		309	106	3:1	0.065	0.799

P values higher than 0.05 indicate non-significant of χ^2

297 R:102 S and 291 R:97 S, respectively. These segregations fit the ratio 3 R:1 S, showed that these cultivars do not have Lr 35 gene. Also, all of the 401, 393, 382 and 411F2 seedlings of the crosses between Lr 37 and the wheat cultivars Misr 1, Misr 2, Gemmeiza 7 and Gemmeiza 11, respectively were resistant and showed no segregations when tested with the race LMSDB. Hereby, these cultivars possess the gene Lr 37. The F2 seedlings of the crosses between Lr 37 and the wheat cultivars Sakha 94, Gemmeiza 9, Gemmeiza 10, Sids 12 and Shandweel 1 segregated to 303 R:105 S, 276 R:94 S, 298: 99 S, 302 R:98 S and 309 R:106 S, respectively. These segregations fit the ratio 3 R:1 S, indicated that these cultivars do not have Lr 37 gene. On the other hand, crosses between Lr 13 and the wheat cultivars Misr 1, Misr 2, Sakha 94, Gemmeiza 7, Gemmeiza 9, Gemmeiza 10 and Sids 12 segregated to 298 R:103 S, 281 R:91 S, 303 R:96 S, 278 R:89 S, 304 R:105 S, 294 R:99 S and 302 R:98, respectively. These segregations fit the ratio 3 R:1 S, indicated that these cultivars do not have Lr 13 gene. All of the 415, 397, 405, 389 and 405 F$_2$ seedlings of the crosses between Lr 19 and the wheat cultivars Misr 1, Misr 2, Gemmeiza 9, Gemmeiza 10 and Shandweel 1, respectively were resistant and showed no segregation when tested with the race NSHLQ. These results indicated that these cultivars possess the gene Lr 19. The F$_2$ seedlings of the crosses between Lr 19 and the wheat cultivars Sakha 94, Gemmeiza 7, Gemmeiza 11 and Sids 12 segregated to 281 R:96 S, 277 R:101 S, 305 R:103 S and 298 R:99 S, respectively. These segregations fit the ratio 3 R:1 S, indicated that these cultivars do not have Lr 19 gene. While, all of the 394, 381, 403, 379, 405, 391 and 400 F$_2$ seedlings of the crosses between Lr 35 and the wheat cultivars Misr 1, Gemmeiza 7, Gemmeiza 9, Gemmeiza 10, Gemmeiza 11, Sids 12 and Shandweel 1, respectively were resistant and showed no segregations when tested with the race MHCQQ. Therefore, these cultivars possess the gene Lr 35. The F$_2$ seedlings of the crosses between Lr 35 and the wheat cultivars Misr 2 and Sakha 94 segregated to 297 R:102 S and 291 R:97 S, respectively. These segregations fit the ratio 3 R:1 S, revealed that these cultivars do not have Lr 35 gene. Finally, all of

the 401, 393, 382 and 411F_2 seedlings of the crosses between Lr 37 and the wheat cultivars Misr 1, Misr 2, Gemmeiza 7 and Gemmeiza 11, respectively were resistant and showed no segregations when tested with the race LMSDB. Thus, these cultivars possess the gene Lr 37. The F_2 seedlings of the crosses between Lr 37 and the wheat cultivars Sakha 94, Gemmeiza 9, Gemmeiza 10, Sids 12 and Shandweel 1 segregated to 303 R:105 S, 276 R:94 S, 298. Furthermore, Hasan and co-workers also evaluated genetic analysis of the five-leaf rust monogenic lines and nine bread wheat cultivars at the adult plant stage in F2 under field conditions (Table 12.12 and summarized in Table 12.13). All of the 451, 423 and 439 F2 plants of the crosses between the adult plant resistance gene Lr 10 and the wheat cultivars Gemmeiza 9, Gemmeiza 11 and Shandweel 1, respectively were resistant under field conditions and showed no segregations, indicating that the wheat cultivars Gemmeiza 9, Gemmeiza 11 and Shandweel 1 have the leaf rust resistance gene Lr 10. The F2 plants of the crosses between Lr 10 and the rest of the tested cultivars, i.e. Misr 1, Misr 2, Sakha 94, Gemmeiza 7, Gemmeiza 10 and Sids 12 segregated to 344 R:118 S, 373 R:120 S, 366 and the rest of the tested cultivars, i.e. Misr 1, Misr 2, Sakha 94, Gemmeiza 7, Gemmeiza 9, Gemmeiza 10 and Sids 12 segregated to 362 R:115 S, 350 R:119 S, 373 R:126 S, 338 R:113 S, 358 R:121 S, 377 R:125 S and 368 R:121 S, respectively. These segregations fit the ratio 3 R:1 S, revealed that these cultivars do not have Lr 13 gene. All of 454, 491, 457, 451 and 479 F2 plants of the crosses between Lr 19 and the wheat cultivars Misr 1, Misr 2, Gemmeiza 9, Gemmeiza 10 and Shandweel 1, respectively were resistant and therefore expressed Lr 19. The crosses between Lr 19 and the wheat cultivars Sakha 94, Gemmeiza 7, Gemmeiza 7, Gemmeiza 11 and Sids 12 segregated at a ratio of 366 R:124 S, 368 R:121 S, 366 R:130 S and 369 R:118 S, respectively which were a good fit to the ratio 3 R:1 S, representing that these cultivars do not have Lr 19 gene

The inheritance of leaf rust resistance was studied under field condition at Sakha Agric. Res. Station during 1998/1999 and 1999/2000 growing seasons by Boulot and El-Sayed (2001). They constructed 21 crosses among four leaf rust resistant wheat cultivars Giza 155, Giza 168, Sakha 69 and Sakha 93 and three susceptible ones i.e. Giza 139, Chenab 70 and Thatcher. The measured average coefficient of infection (ACI), final rust severity (FRS), receptivity (pustules/cm), rate of disease increase FRS value) and area under disease progress curve (AUDPC), as components of leaf rust resistance. Results revealed that the inheritance of leaf rust resistance best fit an additive and/or dominance gene effects. However, the additive genetic component (D) was found to be greater in magnitude than its corresponding dominance (H1 and H2) values, reflecting its importance in the expression of wheat resistance to leaf rust. The quantitative polygenic analysis evidenced resistance is a quantitatively inherited and governed by at least three gene pairs. Positive values of F revealed that resistant cultivar in the diallel crosses carry more dominant alleles than recessive ones. Moreover, both general and specific combining ability variances were highly significant for leaf rust resistant measurements. Results showed that delay selection to late segregating generations would be more effective.

Table 12.12 Segregations and Chi square analysis of F2 plants of the crosses between five leaf rust monogenic lines and nine bread wheat cultivars at adult stage under field conditions at Itay El-Baroud location during 2013/2014 growing season (Hasan et al. 2014)

Cross	No. of F2 plants		Expected ratio	χ^2	P value
	Resistant (R)	Susceptible (S)			
Gemmeiza 7 × Lr 10	372	126	3:1	0.024	0.877
Gemmeiza 9 × Lr 10	451	0	No segregation	–	–
Gemmeiza 10 × Lr 10	353	113	3:1	0.140	0.708
Gemmeiza 11 × Lr 10	423	0	No segregation	–	–
Misr 1 × Lr 10	344	118	3:1	0.072	0.788
Misr 2 × Lr 10	373	120	3:1	0.114	0.735
Sakha 94 × Lr 10	366	117	3:1	0.155	0.694
Sids 12 × Lr 10	369	118	3:1	0.154	0.695
Shandweel 1 × Lr 10	439	0	No segregation	–	–
Gemmeiza 7 × Lr 13	338	113	3:1	0.001	0.978
Gemmeiza 9 × Lr 13	358	121	3:1	0.017	0.895
Gemmeiza 10 × Lr 13	377	125	3:1	0.003	0.959
Gemmeiza 11 × Lr 13	480	0	No segregation	–	–
Misr 1 × Lr 13	362	115	3:1	0.202	0.653
Misr 2 × Lr 13	350	119	3:1	0.035	0.852
Sakha 94 × Lr 13	373	126	3:1	0.017	0.897
Sids 12 × Lr 13	368	121	3:1	0.017	0.896
Shandweel 1 × Lr 13	477	0	No segregation	–	–
Gemmeiza 7 × Lr 19	368	121	3:1	0.017	0.896
Gemmeiza 9 × Lr 19	457	0	No segregation	–	–
Gemmeiza 10 × Lr 19	451	0	No segregation	–	–
Gemmeiza 11 × Lr 19	366	130	3:1	0.387	0.534
Misr 1 × Lr 19	454	0	No segregation	–	–
Misr 2 × Lr 19	491	0	No segregation	–	–
Sakha 94 × Lr 19	366	124	3:1	0.024	0.876
Sids 12 × Lr 19	369	118	3:1	0.154	0.695
Shandweel 1 × Lr 19	479	0	No segregation	–	–
Gemmeiza 7 × Lr 35	477	0	No segregation	–	–
Gemmeiza 9 × Lr 35	481	0	No segregation	–	–
Gemmeiza 10 × Lr 35	458	0	No segregation	–	–
Gemmeiza 11 × Lr 35	479	0	No segregation	–	–
Misr 1 × Lr 35	483	0	No segregation	–	–
Misr 2 × Lr 35	374	126	3:1	0.011	0.918
Sakha 94 × Lr 35	372	133	3:1	0.481	0.488

(continued)

Table 12.12   (continued)

Cross	No. of F$_2$ plants		Expected ratio	χ^2	P value
	Resistant (R)	Susceptible (S)			
Sids 12 × Lr 35	493	0	No segregation	–	–
Shandweel 1 × Lr 35	475	0	No segregation	–	–
Gemmeiza 7 × Lr 37	491	0	No segregation	–	–
Gemmeiza 9 × Lr 37	375	120	3:1	0.152	0.697
Gemmeiza 10 × Lr 37	377	122	3:1	0.081	0.776
Gemmeiza 11 × Lr 37	479	0	No segregation	–	–
Misr 1 × Lr 37	366	131	3:1	0.489	0.484
Misr 2 × Lr 37	342	111	3:1	0.060	0.807
Sakha 94 × Lr 37	362	115	3:1	0.202	0.653
Sids 12 × Lr 37	350	113	3:1	0.087	0.768
Shandweel 1 × Lr 37	363	125	3:1	0.098	0.754

P values higher than 0.05 indicate non-significant of χ^2

Table 12.13   Resistance genes for leaf rust identified in nine bread wheat cultivars at seedling and adult plant stages (Hasan et al. 2014)

No.	Cultivar	Plant stage/leaf rust resistance genes (*Lr*,s)	
		At seedling stage	At adult plant stage
1	Misr 1	19, 35, 37	19, 35
2	Misr 2	19, 37	19
3	Sakha 94	–	–
4	Gemmeiza 7	35, 37	35, 37
5	Gemmeiza 9	10, 19, 35	10, 19, 35
6	Gemmeiza 10	19, 35	19, 35
7	Gemmeiza 11	10, 13, 35, 37	10, 13, 35, 37
8	Sids 12	35	35
9	Shandweel 1	10, 13, 19, 35	10, 13, 19, 35

For leaf rust, high narrow sense heritability values have been reported by Ageez and Boulot (1999). Moreover, Awaad et al. (2003) showed that additive and dominance genetic variances were significant, with the predominant of the additive type in governing infection type in all the evaluated four crosses and disease severity in 1st, 3rd and 4th crosses, resulting in dominance ratio was less than unity, reinforcing the efficiency of phenotypic selection for improving leaf rust resistance. The dominance genetic variance was the prevailed type controlling the inheritance of disease severity in 2nd cross as well as grain yield/plant in all crosses, resulting in dominance ratio was more than unity. This indicating the effectiveness of using hybrid-breeding method when commercial seed production of wheat is feasible. Narrow sense heritability "Tn" (Table 12.14) reflects the fixable type of gene action transmissible from

Table 12.14 Genetic components of variance and heritability for Infection type, disease severity (%) of leaf rust and grain yield/plant in four wheat cross populations (Awaad et al. 2003)

Genetic parameters	Infection type				Disease severity (%)				Grain yield/plant			
	1	2	3	4	1	2	3	4	1	2	3	4
VA	0.179**	0.735**	0.172**	0.585*	11.434**	3.474*	15.368***	15.669**	4.202*	2.124*	4.338*	2.720*
VD	0.018*	0.059*	0.031*	0.024*	1.045*	18.173**	0.571*	0.287*	5.985**	3.914**	4.996**	3.084**
VAD	0.0025	0.061	0.046	0.046	0.500	3.894	4.11	2.257	0.950	0.462	1.943	0.950
VE	0.115	0.174	0.160	0.081	2.271	2.495	1.817	1.772	1.114	0.449	0.778	1.606
Dominance ratio	0.448	0.400	0.600	0.286	0.428	3.235	0.273	0.191	1.688	1.919	1.518	1.506
Heritability (Tn)	57.37	75.93	47.38	84.78	77.52	14.38	86.55	88.39	37.18	32.74	42.89	36.70

*, **significant and highly significant at 0.05 and 0.01 levels of probability
VA: Additive genetic variance
VD: Dominance genetic variance
VAD: Dominance genetic variance
VE: Environmental variance

Table 12.15 Segregation of reaction to *Puccinia recondita* f. sp *tritici* in F2 progenies of four wheat cross populations (Awaad et al. 2003)

Cross populations	Number of plants		Expected ratio	χ^2
	Resistant	Susceptible		
I. Resistant × Moderate Susceptible				
1. ACSAD 945 × Sakha 69	163	47	13:3	1.817
2. Giza 168 × Gemmeiza 5	125	75	9:7	3.174
II. Resistant × Resistant				
Gemmeiza 9 × PAT10 ALD"S"	191	19	15:1	2.805
III. Moderate susceptible × Moderate Susceptible				
1. Sakha 69 × Sahel 1	62	148	1:3	2.292

For 1 d.f, the value of χ^2 is 3.84 ($P = 0.05$) R: Resistant, S: Susceptible

the parents to the progeny, was high for infection type and disease severity on three out of four crosses studied. This result is allowing for considerable progress from selection. Whereas, low (14.38%) "Tn" estimate was reported for disease severity in 2nd cross and moderate (47.38%) for infection type in 3rd cross, also moderate for grain yield/plant in all crosses. This result supported those obtained from adequacy genetic model which indicated that additive and dominance, as well as epistasis were involved in the expression of these characters in abovementioned crosses.

In continuous, for leaf rust, most of 60 leaf rust resistance genes give race-specific resistance in a gene-for-gene fashion were identified by McIntosh et al. (2007). Awaad et al. (2003) revealed that genetic analysis using x^2 test indicated that leaf rust resistance was conditioned by two interacting gene pairs in the 1st cross, two complementary gene pairs in the 2nd cross, two double dominant gene pairs in the 3rd cross as well as two recessive complementary gene pairs in the 4th one (Table 12.15).

Abou-Elseoud et al. (2014) postulated twenty leaf rust resistance genes (Lr genes) in nine Egyptian wheat cultivars according to infection types (ITs) on the verified cultivars by 72 *Puccinia triticina* pathotypes compared to the ITs expressed on the monogenic lines. The best carrier genes cultivars were Giza168 and Misr1 both might have five genes, i.e. Lr2c, 10, 18, 24, 41 and Lr3, 10, 19, 22b, 24, respectively. Five cultivars, Sakha94, Gemmeiza9, Gemmeiza10, Sids12 and Misr2 each probably include four genes i.e. Lr9, 19, 29, 37; Lr18, 21, 24, 41; Lr3, 9, 19, 29; Lr9, 19, 26, 29 and Lr3, 10, 19, 26, respectively. Gemmeiza11 carries two genes, i.e. Lr24 and Lr41. The most postulated genes were Lr19 and Lr24, both postulated within five cultivars followed by Lr41 within four cultivars. Five Lr genes, Lr3, Lr9, Lr10, Lr26 and Lr29 each within three cultivars. The Lr9 gene was recognized in cultivar Sids12, Lr10 in cultivar Misr1, Lr19 in two cultivars, Misr1 and Misr2, while Lr24 and Lr26 were absent in all the tested Egyptian cultivars. Moreover, Minaas, Sallam et al. (2016) evaluated thirteen Egyptian wheat varieties for leaf rust resistance at seedling stage under greenhouse environment and adult plant stage under field environment during three growing seasons i.e. 2011/2012, 2012/2013 and 2013/2014 under three regions, Itay El-Baroud and Nubariya Agricultural Research Stations as well as the Farm of the Faculty of Agriculture, Minufiya University, Shibin El-Kom. Postulation of leaf rust resistance genes was varied among the tried genotypes. Sakha 95 and Sids 12 might have seven resistance genes. Moreover, Gemmeiza 10 may have five genes, and Misr 1 may have three genes. Whereas, Giza 168, Sids 1, Misr 2 and Shandweel 1 may have two genes. Wheat cultivars Gemmeiza 11 and Gemmeiza 12 may have only single gene. Furthermore, wholly the tested wheat genotypes might contain some additional genes. Otherwise, wheat genotypes Sakha 94, Gemmeiza 7, Gemmeiza 9 and Sids 13 did not carry any of the verified genes.

For stem rust, Salem et al. (2003) evaluated four crosses (Sakha 69 × Sahel I, Sakha 69 × ACSAD 945, Gemmeiza 9 × PAT 10/ALD"S" and Gernmeiza 5 × Giza 168) derived from seven wheat genotypes for stem rust resistance. They found that additive and dominance gene effects as well as all their interaction types, i.e. additive × additive and additive × dominance were important in the genetics of infection type in 3rd cross. Duplicate type of epistasis has been recorded for infection type in the 4th cross and disease severity in the 1st, 2nd and 4th crosses and grain yield/plant in 2nd and 3rd crosses. Whereas the complementary type was identified for infection type in the 2nd cross. Narrow sense heritability for disease severity was high (< 50%) in all crosses, except in the 4th cross which was moderate (34.93%). Also, it was low (23.2%) for infection type in the 3rd cross, and moderate in the 2nd (38.8%), and 4th (35.38%) crosses. Low to moderate narrow sense heritability for grain yield/plant was noticed. Genetic analysis using x^2 test indicated that stem rust resistance was controlled by a single dominant gene pair in the 2nd cross, two complementary gene pairs in the 1st cross and three dominant gene pairs in both 3rd and 4th crosses (Table 12.16).

Genetic nature for stem rust resistance was studied by Omara et al. (2017) in seven parents and their wheat crosses, i.e. Misr-1 × Morocco, Misr-2 × Morocco, Gemmeiza 11 × Morocco, Gemmeiza 12 × Morocco, Shandaweel 1 × Morocco, Giza 171 × Morocco and Sakha 94 × Morocco in F1 and F2 crosses, at Kafr El-Hamam Agriculture Research Station—Sharkia Governorate, Egypt. Qualitative analysis showed that the observed ratios, fitted to the expected rations 1:15, 3:13, 9:7,

Table 12.16 Segregation of reaction to *Puccinia graminis* f. sp *tritici* in F2 progenies four wheat cross populations (Salem et al. 2003)

Cross populations	Number of plants		Expected ratio	χ^2
	Resistant	Susceptible		
I. Resistant × Susceptible				
1. Sakha 69 × Sahel 1	120	80	9:7	1.143
2. Sakha 69 × ACSAD 945	148	52	3:1	0.107
II. Resistant × Resistant				
Gemmeiza 9 × PAT10 ALD"S"	195	5	63:1	1.149
Gemmeiza 5 × Giza 168	200	2	63:1	0.428

For 1 df, the value of χ^2 is 3.84 ($P = 0.05$) R: Resistant, S: Susceptible

Table 12.17 Heritability in the broad sense (T_{bs}%) for a percentage of disease severity and area under disease progress curve (AUDPC), under field conditions at M_4 and M_5 generations (El-Naggar Doaa and Soliman 2015)

	M_4 generation		M_5 generation	
	Disease severity (%)	AUDPC	Disease severity (%)	AUDPC
Stripe rust	82.15	88.82	83.61	91.12
Leaf rust	77.53	89.81	83.53	95.06
Stem rust	81.42	97.63	79.29	92.69

3:1, 13:3, 1:3, and 7:9 for the aforementioned seven crosses, respectively. Hereby, stem rust resistance inherited as a simple trait, as it was controlled by only one or two gene pairs in most cases at adult plant stage. While quantitative analysis revealed partial dominance mode of inheritance was more pronounced in its genetic expression. Heritability in broad-sense was high and varied from 79.5 to 96.3%. Thus selection for stem rust resistant genotypes in an early generation was possible.

Heritability also helps in predicting the performance of succeeding generations by devising the suitable selection criteria and measuring the level of genetic enhancement (Ajmal et al. 2009). Stripe, leaf and stem rusts, El-Naggar Doaa and Soliman (2015) represented the percentages of heritability in the broad sense (T_{bs}) for a percentage of disease severity and area under disease progress curve (AUDPC), under field conditions at M_4 and M_5 generations. They obtained higher values of heritability indicated the presence of a high degree of variability which may be due to the genetic effects (not changes under different environmental conditions from season to another). Heritability estimates in a broad sense showed an increase from M_4 to M_5 generations in stripe and leaf rust (Table 12.17), suggesting that selection would be more effective in M_5 generation for isolating promising mutants more resistant to both rust diseases.

Wheat stripe and stem rusts were evaluated by Shahin and Abu Aly (2015) through a set of tester wheat lines carrying stripe Yr's, stem Sr's rust genes and selected Egyptian varieties for their reaction to Pst and Pgt at adult plant stage under field circumstances in Sakha Agriculture Research Station, during the 2011 to 2014 seasons. They revealed that, the new stripe rust race Yr27-virulence to Pst. Also, pathotypes were virulent for Yr2, Yr6, Yr7, Yr8, Yr9, Yr27, while Yr18 revealed moderate susceptibility. Whereas, Yr1, Yr5, Yr10, Yr15, Yr17, Yr32 and YrSP displayed high levels of resistance (https://www.globalrust.org/rgenes/yr27). Concerning resistance genes sources of stem rust of ICARDA, CIMMYT wheat germplasm, and Egyptian wheat varieties released i.e. Misr1 and Misr2 that having Ug99-resistance genes Sr2 and Sr25 were found susceptible to Pgt, also Sr31, recorded infection moderately susceptible to susceptible at the adult stage. Genes Sr2 complex, Sr24, Sr26, Sr27, and Sr32 were resistant at adult plant stage. The combination of Sr26 with Sr2 and Sr25 giving stem rust resistance in several CIMMYT wheat germplasm. Egyptian cultivars Misr 1 and Misr 2 were found to be susceptible rated 10S-20S and Sr31 rated MSS. This indicates presence a new Sr31-virulence. However, genes Sr2 complex,

Sr24, Sr26, Sr27 and Sr32 were resistant, and the combination of Sr26 with (Sr2 and Sr25) produced resistance to stem rust in some CIMMYT wheat germplasm.

Genetic analysis of both leaf and stem rust, as well as grain yield, were evaluated by Darwish et al. (2018) using six parents, i.e. Gemmeiza 9, Sids 12, Misr 1, Misr 2, Sids 1 and Sham 4, with corresponding 15 F_2 crosses. Moderate to high broad sense heritabilities were registered for most characters. The tested plants in the F_2 's crosses segregated and provided ratios fitted to 9:7, 9:7, 3:1, 1:3, 13:3 and 3:13 for leaf rust and 9:7, 7: 9, 3:1, 1:3, 3:13 and 1:15 for stem rust with non-significant χ^2 values. Hereby the resistant parents for leaf and stem rusts exhibited one or two genes and were complimentary dominance, recessive or independent in their expressions.

Leaf rust monogenic lines, i.e. Lr27, Lr29 were crossed with four Egyptian bread wheat cultivars (*Triticum aestivum* L.). Also Lr46 was crossed with 3 wheat cultivars, and Lr2a was crossed with one cultivar (Sakha 93). Parents, F1's and F2's were evaluated at adult plant stage under field conditions against race mixtures of the pathogen (*Puccinia triticina* Eriks.) under artificial infection. The parents of monogenic lines Lr2a, Lr46, Lr29 and Lr27 exhibited low disease severity. While, Gemmeiza 7, Sakha 61, Sakha 93 and Sids 1, presented high disease severity against leaf rust disease. The F1's tested plants displayed low disease severity with most of the crosses excluding three crosses, i.e. Lr27 + Sakha 61, Lr27 + Sakha 93 and Lr27 + Sids 1, which displayed high disease severity. The F2 plant populations were segregated into two gene pairs. The dominance tends to the direction of low disease severity (partial leaf rust resistance) with eight crosses. Though, the dominance tends to the side of high disease severity with three crosses that were previously mentioned with F1's. But no segregates were noted with the cross Lr2a + Sakha 93 and dominance tends to the aside of low disease severity (partial leaf rust resistance). The cultivar Sakha 93 showed the adult plant resistance to the gene Lr2a under field conditions. These conclusions may demonstrate that this gene is effective under the Egyptian environmental circumstances. The partial leaf rust resistance in the tested wheat cultivars was supposed to be controlled in adult plant stage by digenic pairs. The selection for partial leaf rust resistant genotypes in the early generations was possible but delaying it to late ones is more operative, due to the significant role of dominance effect in the expression of the resistance (Youssef 2011).

Swelam et al. (2010) made three crosses viz, Sakha 93 × Sr26 + 9 g, Sakha 69 × Sr26 + 9 g and Giza 165 × Sr26 + 9 g among four diverse genotypes of bread wheat (*Triticum aestivum* L.). Populations P_1, P_2, F_1, F_2, BC_1 and BC_2 were grown during 2008/2009 in a randomized complete block design in a field experiment at Kafer El-Hamam Agric. Res. Station, ARC, Egypt, to test their resistance at adult plant stage. Also, the studied populations were grown in the greenhouse to evaluate the infection type at the seedling stage. Statistical analysis of variance showed significant differences among the studied genotypes for all traits. Results indicated the importance of additive, dominance and their interactions in the inheritance of most studied traits and stem rust reaction. Significant positive or negative heterosis based on better parent values were attained for all crosses and traits except for No. of kernels/spike in the first cross. All crosses showed significant inbreeding depression except for plant height and spike length in the first cross, infection type-seedling stage and disease severity-adult stage in all crosses. Broad sense heritability fluctuated from 53.19 to 97.40 for No. of kernels/spike in the third cross and No. of spikes/plant in the first cross, respectively, However, narrow sense heritability ranged from 40.24 to 71.53 for 100-kenels weight in the third cross and infection type-seedling stage in the second cross, respectively. Genetic advance varied between the studied traits. Resistance to stem rust disease was determined in the current study by the presence and thickness of sclerenchyma tissue in stem and presence of epidermal hairs on leaves of wheat plants. The stem rust resistant parent (Sr26 + 9 g) contained sclerenchyma layer without chlorenchyma tissue and epidermal hairs in order 20/mm of leaf section. The susceptible wheat varieties (Sakha 93, Sakha 69 and Giza 165) exhibited sclerenchyma layer of about 43.04, 39.31 and 23.32%, respectively, from the stem circuit and 56.96, 60.69 and 76.68%, respectively for chlorenchyma tissue.

12.7 Effect of Climatic Factors on Wheat Rusts

A significant effect of environmental conditions has been registered concerning wheat rusts through many investigators. Stripe rust infected by *Puccinia striiformis* f. sp. *tritici* is an important disease in the world. Low temperate and high relative humidity are suitable to the wide distribution of the disease (Stubbs 1988; Danial et al. 1994). Grabow et al. (2016) explained the Simple environmentally based stripe rust models at the crop reporting districts (CRD) level might be combined with disease notes at the field level and understanding the various reactions to strip rust as part of the Knesset disease forecasting system.

In Egypt, stripe rust is an intermittent disease because it seems in same years in Near and Middle East areas. However starting from the 1990s (http://www.jofame ricanscience.org/journals/am-ci/am110615/007_28630am110615_47_), it became familiar due to its continuous occurrence (Abu El-Naga et al. 2001). The first yellow rust epidemic was recorded in 1967/1968 season on wheat variety Giza 144 at Manzala district, however, the most important stripe rust epidemic occurred in 1995/1996, 1997/1998 on wheat cultivars Sakha 69, Giza 163, Gemmeiza1 of particularly in the Northern and Southern Delta areas (El-Daoudi, et al. 1996; Abu El-Naga et al. 1997, 1999b, 2001).

Roelfs et al. (1992) reported that the optimum temperature required to wheat leaf rust development were 20 °C, 15–20 °C, 20 °C, 25 °C for germination, germling appressorium, penetration and growth sporulation, respectively. Besides genetic resistance, that affects the individual crop hazard, weather variables effect the incidence and severity of leaf rust (Moschini and Perez 1999). Several environmental factors affect the production, spreading and survival of urediniospores (Eversmeyer and Kramer 1995; Vijaya Kumar 2014).

Diab (1994) stated that results of field experiments in Egypt proved high levels of distribution of wheat leaf rust at Sakha province compared with Giza province. The clear variance in meteorological factor between such province gave a reasonable explanation, where Sakha province recorded higher RH values and a greater amount of rainfall and wind speed, but on the other hand, a relative less degrees of temperature. The successive records of temperature were 17.9, 20.0 and 24.3 °C for Sakha and were 19.1, 22.6 and 28.3 °C. for Giza at February, March and April, respectively. The disease intensity on susceptible Giza 160 and Giza 155 varieties were highly correlated with temperature degree and rainfall.

In Australia, 10–20 °C daily average temperature was determined as the optimal temperature range for the incidence of leaf rust in wheat (Anonymous 2005). It's also a decision support system, based on temperature, leaf wetness period and rainfall was developed (Steinberg 2000). Weeks 7–9 of the crop growing season at Ludhiana, Faizabad and Sabour and weeks 10–12 at Kanpur were identified as critical periods for relating weather variables to disease (https://link.springer.com/content/pdf/10.1007/s10658-014-0478-6.pdf).

Tohamy (2004) explained the percentage of disease severity of wheat leaf rust during 2000/2001 season, was ranged between 40% at Beni-Suef governorate to 90% of disease severity at Kafr el-Sheikh. The record of disease severity at both Minufiya and Beheira governorates was 80%, while it was 60% at Sharkiya governorate. This variation in wheat rust severity at such location was more related to the decrease in over minimum temperature in March and April, the period of initial disease infection and appearance of symptoms. The highest severity (90%) of wheat leaf rust was at Kafr el-Sheikh, where minimum temperature was 4.5 °C in March and increased up to 9.0 °C at April. However, the lowest disease severity was 40% at Beni-Suef where the minimum temperature in the same province was 10.5 and 17.0 °C in March and April, respectively. It is clear that the minimum temperature, when increased over 10.0 °C during March or over 17 °C, during April has lowered the infection development during these two months. On the other hand, the maximum temperature when ranged between 20–27 °C during March and over 30 °C April, has resulted in the same effect on decreasing the disease development. The average daily temperature during two months which was clearly related with high development of leaf rust was (4.5–16.7 °C), and (9.0–23.7 °C) during March and April 2001 at Kafr el-Sheikh governorate. Moreover, the relative humidity (R.H. %) at 5 governorates and their effect on wheat leaf rust disease incidence indicated that although no clear variance in the percentage of R.H. was recorded during March 2001 at Kafr el-Sheikh (90%) and both Minufiya and Beheira (88%), however, the disease severity record at the first province was 90%, and it was 80% at the two later provinces. Meanwhile, when the percentage of R.H. was decreased to 80% at Sharkiya and to 75% at Beni-Suef during the same month (March), the disease record was also decreased to 60 and 40% at both or the two governorates, respectively. The maximum level of R.H. % during March 2001 was ranged from 29% at Beni-Suef and 40% at Minufiya and Beheira governorates and near values 37.6% was recorded at Kafr el-Sheikh. It is clear that the minimum level of R.H. % seemed to be effective on the development of wheat leaf rust disease when R.H. was recorded during April 2001. The range was between 85-33.2% at Sharkiya, 82-29% at Kafr El-Sheikh, 75-31% at Beheira and Minufiya, and 65-20% at Beni-Suef. This last record that was clearly inferior have any other location and the disease development at such location was also the lowest compared with others. Also, the R.H. values during March were higher than at April, and it seems that initiation of infection with leaf rust disease at March was accordingly enhanced with higher levels of R.H. No records of rainfall was received in any province of wheat fields except that in Kafr el-Sheikh governorate. The maximum values of rainfall in this location was 2 and 3 mm., while the minimum was 1.0 mm. during March and April, respectively. The infection of wheat leaf rust (disease severity %) at Kafr el-Sheikh, reached the maximum value (90%) which was higher than any other location at the same location. Here again, the relation between rainfall and disease development was clear in one location which recorded rain deposition.

Abd El Malik (2011) stated that, temperature degrees, average minimum and maximum temperatures were relatively higher during the growth period and rust incidence of the late sowing date which enhanced the development of the disease.

Calendar-month summaries of weather were made from the hourly series of time from August (from the year of planting of winter wheat) to the following July (year of harvest). These included descriptive factors such as average temperature, average RH, and the total number of hours a specified condition was satisfied per month (e.g., hours with temperatures between 5 and 12 °C). Several favorable or optimal temperature ranges (i.e., 5–12, 7–12, 10–15, 10–18, and 2–23 °C) have precipitation. Other factors were constructed to reflect known associations as pronounced above between weather and stripe rust etiology by summarizing been described for stripe rust and are conditional on the constituent of the disease cycle (spore germination, infection, etc.) and adaptations inside the *P. striiformis* population (Coakley et al. 1982; de Vallavieille-Pope et al. 1995; Eddy 2009; Milus et al. 2009).

Gebril et al. (2018) explained that the sowing dates (1st November, 15th November, 1st December and 15th December) on stripe, stem and leaf rust of nine wheat cultivars. (Giza-171, Shandaweel-1, Misr-1, Misr-2, Sakha-94, Sakha-95, Sids-12, Gemmeiza-11 and Gemmeiza-12) were evaluated during 2016/2017 growing season. Among sowing dates, rust severity was lower in early sowing (1st November and 15th November), where the temperature in November was 14.8 (minimum) and 28.3 (maximum). Sever infection was in late sowing (1st December and 15th December) where the temperature in December was 11.3 (minimum) and 23.6 (maximum). In the early sowing has paramount importance to combat rust severity effectively. There was a significant effect of sowing date on the average coefficient of infection ACI. ACI was higher in late sowing conditions comparing to normal sowing condition. Meanwhile, the lowest values of ACI had obtained from sowing on 15th November. Regarding ACI of stripe rust, revealed that Sids-12 and Gemmeiza-11 were the most affected cvs. with ACI 60 and 40 respectively. Sakha-95 was the resistant cv. showing ACI of 30. The least values of ACI for leaf rust and stem rust were those of Giza-171 = 0. On the other hand, ACI of stem rust was less than those of stripe and leaf rust. Also, at sowing date 1st of December ACI of stripe rust showed intermediate values ranged from 30 to 80. While, ACI for leaf rust, it was less than that of stripe rust on all tested cultivars.

12.8 Breeding Efforts and Biotechnology

12.8.1 Classical Methods

Breeding against wheat rusts is a well-established strategy. Growing resistant cultivars to rusts is considered the most sustainable, environmentally friendly and cost-effective approach for controlling wheat rusts. The choice of procedures and strategies for developing plant resistance principally depends upon the availability of genetic resources for resistance. The breeding approaches can be divided into classical and modern methods. Classical breeding approaches like introduction of exotic germplasm, hybridization, composite crossing, multiline varieties and backcross breeding were exploited for this purpose (Swelam et al. 2010).

Through the application of breeding programs, the National Wheat Improvement Program in Egypt over the last 25 years has succeeded in improving wheat production from 3.3 tons per hectare in 1981 to 5.1 tons per hectare in 1991 to about 6.9 tons per hectare in 2016, an increase of 109% (Anonymous 2017). Plant breeders in Egypt developed a collection of bread wheat cultivars more resistant to rusts such as Giza 168, Giza 171, Gemmeiza 7, Gemmeiza 9, Gemmeiza10, Gemmeiza 11, Gemmeiza12, Sids 12, Sids 13, Sids 14, Sakha 94, Sakha 95, Misr 1, Misr 2, Shandweel 1, and durum wheat cultivars i.e. Sohag 3, Sohag 4, Sohag 5, Beni Suef 1, Beni Suef 4, Beni Suef 5 and Beni Suef 6 (Anonymous 2020).

In another direction, Hickey et al. (2015) suggested a new technique for rapid generation improvement known as 'speed breeding', has significant advantages over doubled haploids technology for wheat. This technique offers increased recombination during line improvement and allows selection in early generations for economic traits. The technique was developed over 8 years at Queensland University, using controlled temperature systems and 24 h light to accelerate plant growth and development. The low-cost management method allows produce up to 6 wheat generations annually. Presently, three of the six Australian wheat breeding companies are exploiting speed breeding, and promising genotypes developed by this technology are in the last stages of yield evaluation. Recently, they developed the speed breeding system, and exploited for evaluating adult plant resistance APR to wheat rusts. The protocols enables selection for APR against 'Triple rust' pathogens and crossing of selected plants in a single plant generation. The technique was applied to introgression rusts resistance genes into numerous Australian cereal cultivars. This technique also benefits in accelerating research efforts to resist wheat rusts. New genetic recombinations designed for mapping novel adult plant resistant genes can be improved within 12–18 months. This technique is also useful in understanding the function of genes in terms of temperature and their effect on resistance.

12.8.2   Mutation Breeding

Molecular genetics methodology like mutation was improved by breeders to improve effective resistance in crop genotypes in a shorter time. Mutagenic agents can be divided into physical and chemical mutagens. The use of physical mutagens, like gamma rays, is well recognized for inducing useful mutants has been used in many crops such as wheat, rice, barley, maize, etc. (Abdel-Hady et al. 2008; Sharma et al. 2011; Marcu et al. 2013). These induced mutation help breeders to develop several agronomical important traits such as a shorter growing period increased resistance to biotic stresses (Maluszynski and Kasha 2002; Kenzhebayeva et al. 2013).

Of the total 3222 accessions of FAO/IAEA Mutant Varieties Database (http://www-mvd-iaea.org), wheat is represented by 250 officially released mutant cvs. from which 153 mutants were developed by gamma irradiation. Among these 153 mutants cvs., induction against biotic stresses was represented by 49 mutants from which 45 cvs. were developed to improve the resistance against rust diseases (FAO/IAEA 2015). Njoro-BW1 is the only mutant variety which officially released in Africa in Kenya during 2001 (Njau et al. 2006). This variety was developed by gamma irradiation with the main improved attributes of tolerance to drought, resistance to wheat rusts and high yielding (FAO/IAEA 2015). El-Naggar Doaa and Soliman (2015) used gamma rays to produce mutant lines from Sids-7, Giza-168 and Gemmeiza-9 wheat cultivars. They selected promising mutants according to the earliness of maturation, disease parameters, and improved quality of yield components. At M_4 generation, earliness of maturation was observed in one mutant from each mother cultivar (Sid-M_4-7 of Sids-7, G-M_4-3 of Giza-168 and Gem-M_4-1 of Gemmeiza-9). Furthermore, Sid-M_4-7 mutant line recorded the lowest percentages of disease severity for the three rusts (2.67, 3.00 and 0.0% respectively). Also, G-M_4-3 was considered as promising mutant line due to zero stripe and stem rust disease severity with moderately susceptible response to leaf rust. While Gem-M_4-1 mutant line recorded the lowest percentages of the disease severity for the three rusts. Thus, seeds of Sid-M_4-7, G-M_4-3 and Gem-M_4-1 mutant lines were selected for further investigations at M_5 generation in which G-M_5-3 mutant line can be considered as the promising resistant mutant line toward the three rusts with the highest yield (608.67 g) of grains/m^2. Moreover, Yin et al. (2018) showed that the R39 wheat induced mutant has adult-plant resistance (APR) to Pst. Genetic analysis showed that a single recessive gene YrR39, was responsible for the (APR) of R39 to Pst.

12.8.3   Recent Approaches

12.8.3.1   Molecular Marker Technology

Using a molecular marker technology is imperative in ascertaining rust resistance genes in wheat genotypes, particularly when used in combination with multipatho-types test at the beginning of the breeding program. This method may help into understand the wheat—pathogen interaction and offer information to build an oper-ative management program for rust resistance. Molecular markers are an important tool in improving diseases resistance of varieties that can lead to increased crop productivity.

Recent developments in the molecular marker technique have brought operative tools for resolving many difficulties and have numerous benefits over classical pheno-type trait selection (Todorovska et al. 2009). Marker-assisted selection MAS tech-nology has also been widely castoff to target rust resistance genes. These procedures can improve selection efficacy in plant breeding, where, Anderson (2003) stated that to assist breeding for durable resistance to wheat rust, molecular markers are suit-able for improving resistant cultivars, especially to pyramiding of numerous disease-resistance genes. Also, MAS can be exploited at an early stage of plant growth, with several DNA markers being used to recognize several genes simultaneously (Babar et al. 2010).

Molecular markers are used for two purposes to improve resistance, first to monitor the incorporation of specific resistance genes or QTLs into the best wheat genotypes (i.e., MAS, marker-assisted selection) and to detect resistance genes in genotypes. A large amount of information on leaf rust resistance genes has been identified from different genotypes. In this regard in Mexico, Lr68 gene was recognized using SSR and CAPS in *T. aestivum* wheat cultivar Arula1/Arula2. In Egypt, Abou-Elseoud et al. (2014) recognized five Lr genes, Lr9, Lr10, Lr19, Lr24 and Lr26 by PCR-based molecular marker. The Lr9 gene was recognized in cultivar Sids12, while Lr10 was recognized in cultivar Misr1. The Lr19 was present in two cultivars, Misr1 and Misr 2. The Lr24 and Lr26 were absent in the screened Egyptian wheat cultivars. The obtained results for Lr9, Lr10, Lr19, Lr24 and Lr26 marker are in agreement with their identification by gene postulation. Markers for Lr9, Lr10 and Lr19 might be beneficial in marker-assisted breeding. Furthermore, Imbaby et al. (2014b) utilized PCR amplification of 15 Egyptian cultivars genomic DNA using ten leaf rust *Lr* molecular markers. Ten genes, Lr13, Lr19, Lr24, Lr26, Lr34, Lr35 Lr36, Lr37, Lr39, and Lr46, were identified in fifteen wheat cultivars. The most frequently occurring genes in fifteen Egyptian wheat cultivars were Lr13, Lr24, Lr34, and Lr36 recognized in all the used cultivars, followed by Lr26 and Lr35 (93%), Lr39 (66%), Lr37 (53%), and Lr4 (26.6%) of the cultivars, and Lr19 was found in 33.3% of cultivars. High variation in Lr genes was recorded among commercial wheat cultivars in Egypt. Therefore, this is useful for developing and sustain durable resistance.

Abu Aly et al. (2014a) screened thirteen genotypes with three DNA markers aimed at identifying the presence of Yr9, Yr17 and Yr18. Figure 12.11 showed

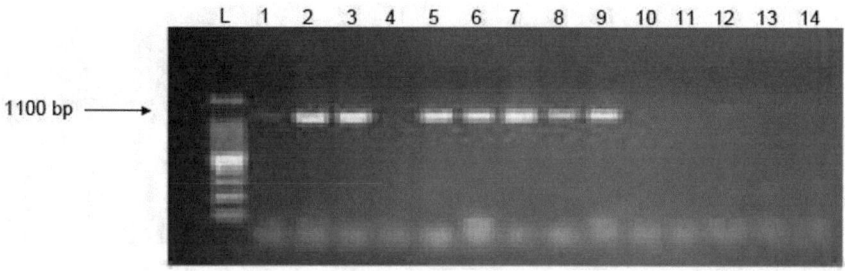

Fig. 12.11 Amplification products of Yr9 marker using PCR in the tested wheat genotypes running on agarose gel. L: DNA Ladder, Lane (1): monogenic Yr9, (2): Misr-1, (3): Misr-2, (4): Sakha-93, (5): Sids-12, (6): Sids-13, (7): Gemmeiza-9, (8): Gemmeiza-10, (9): Gemmeiza-11, (10): Giza-171, (11): Line-6043, (12): Line-6085, (13): Line-6086 and (14): Line-6107. The arrow shows the fragment which is associated with Yr9 at 1100 bp (Abu Aly et al. 2014a)

the polymorphic survey of Yr9 gene marker which was identified as a fragment of the 1100 bp band in 7 genotypes (Misr-1, Misr-2, Sids-12, Sids-13, Gemmeiza-9, Gemmeiza-10 and Gemmeiza-11). The only indicative band for Yr17 was observed at 252 pb fragment Line-6043 (8STEMRRS), as shown in Fig. 12.12. While, Yr18 was identified as a fragment of 517 pb in all tested Egyptian and CIMMYT genotypes, Fig. 12.13. Data obtained from Figs. 12.11–12.13 were summarized in Table 12.18.

Abu Aly et al. (2014b) tested seven Egyptian wheat cultivars (Sakha-93, Sids-12, Sids-13, Gemmeiza-10, Gemmeiza-11, Misr-1, and Misr-2) and four lines (Line-6043, line-6086, line-6107 and line-6085) from the CIMMYT and four monogenic lines (Sr2, Sr24, Sr26 and Sr31) were chosen as resistant plant materials (Sr genes) to detect stem rust resistance genes using molecular markers. They used Sr specific primer to detect the presence/absence of Sr gene. The polymorphic survey revealed that the marker for Sr2 was recognized as a fragment of 310 bp in eleven genotypes as

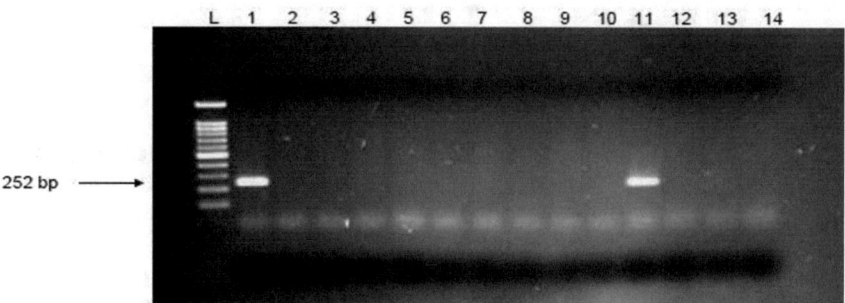

Fig. 12.12 Amplification products of Yr17 marker using PCR in the tested wheat genotypes running on agarose gel. L: DNA Ladder, Lane (1): monogenic Yr9, (2): Misr-1, (3): Misr-2, (4): Sakha-93, (5): Sids-12, (6): Sids-13, (7): Gemmeiza-9, (8): Gemmeiza-10, (9): Gemmeiza-11, (10): Giza-171, (11): Line-6043, (12): Line-6085, (13): Line-6086 and (14): Line-6107. The arrow shows the fragment which is associated with Yr17 at 252 bp. (Abu Aly et al. 2014a)

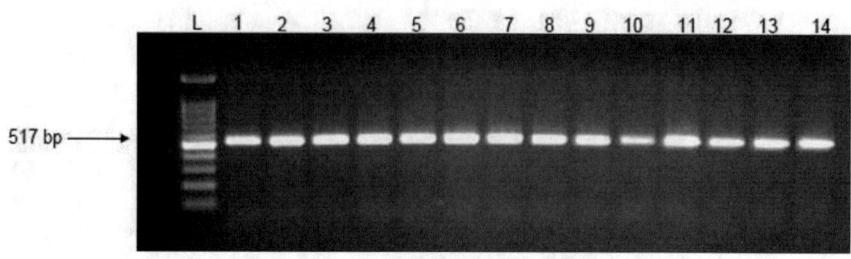

Fig. 12.13 Amplification products of Yr18 marker using PCR in the tested wheat genotypes running on agarose gel. L: DNA Ladder, Lane (1): monogenic Yr9, (2): Misr-1, (3): Misr-2, (4): Sakha-93, (5): Sids-12, (6): Sids-13, (7): Gemmeiza-9, (8): Gemmeiza-10, (9): Gemmeiza-11, (10): Giza-171, (11): Line-6043, (12): Line-6085, (13): Line-6086 and (14): Line-6107. The arrow shows the fragment which is associated with Yr18 at 517 bp (Abu Aly et al. 2014a)

Table 12.18 Yr genes detected with PCR based markers in nine Egyptian wheat cultivars and four CIMMYT wheat lines (Abu Aly et al. 2014a)

No.	Genotypes	Yr9	Yr17	Yr18
1	Misr-1	+	−	+
2	Misr-2	+	−	+
3	Gemmeiza-9	+	−	+
4	Gemmeiza-10	+	−	+
5	Gemmeiza-11	+	−	+
6	Sakha-93	−	−	+
7	Sids-12	+	−	+
8	Sids-13	+	−	+
9	Giza-171	−	−	+
10	Line-6043	−	+	+
11	Line-6068	−	−	+
12	Line-6107	−	−	+
13	Line-6085	−	−	+

shown in Fig. 12.14. The marker for Sr24 was recognized as a fragment of 500 bp in three genotypes only (Sakha-93, Misr-1 and line-6085), while other listed genotypes did not show the presence of Sr24 Fig. 12.15. The polymorphic screening of the eleven genotypes revealed that the marker for Sr26 was known as a fragment of 250 bp in nine cultivars, i.e. Sakha-93, Sids-12, Sids-13, Gemmeiza-10, Gemmeiza-11, Misr-1, Misr-2 and two lines, i.e. line-6043 and line-6107. While two tested genotypes line-6068 and line-6085 did not show the presence of Sr26 Fig. 12.16. The marker for Sr31 was identified as a fragment of 1100 bp in all Egyptian cultivars except cv. Sids-13. While all CIMMYT lines did not show the presence of Sr31 except line 6043 (Fig. 12.17 and Table 12.19).

250 bp

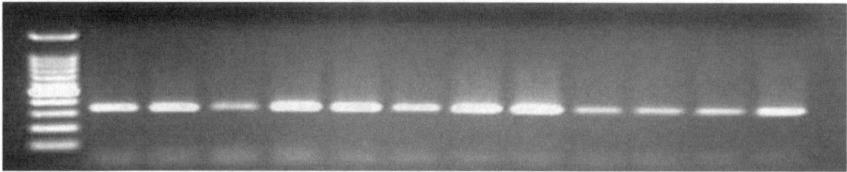

Fig. 12.14 Marker csSr2 tested on diverse wheat genotypes and run on an agarose gel. Lanes: (1) monogenic Sr2, (2) Sakha-93, (3) Sids-12, (4) Sids-13, (5) Gemmeiza-10, (6) Gemmeiza-11, (7) Misr-1, (8) Misr-2, (9) Line-6043, (10) Line-6068, (11) Line-6107 and (12) Line-6085. The arrow showed the fragment which is associated with Sr2 (Abu Aly 2014b)

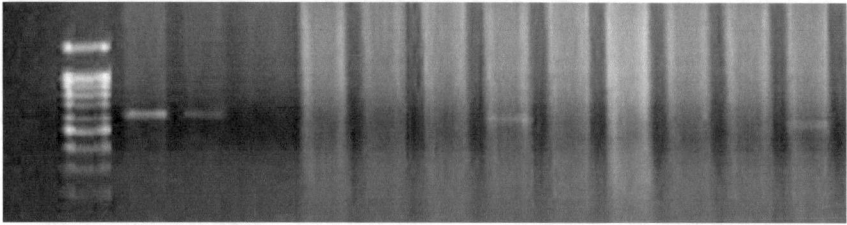

Fig. 12.15 Marker Sr24#12 tested on diverse wheat genotypes and run on an agarose gel. Lanes: (1) monogenic Sr24, (2) Sakha-93, (3) Sids-12, (4) Sids-13, (5) Gemmeiza-10, (6) Gemmeiza-11, (7) Misr-1, (8) Misr-2, (9) Line-6043, (10) Line-6068, (11) Line-6107 and (12) Line-6085. The arrow showed the fragment which is associated with Sr24 (Abu Aly 2014b)

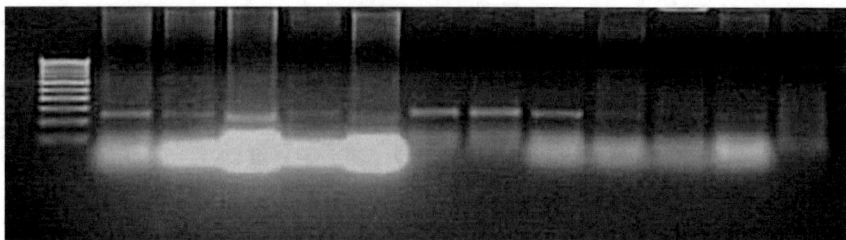

Fig. 12.16 Marker Sr26#43 tested on diverse wheat genotypes and run on an agarose gel. Lanes: (1) monogenic Sr26, (2) Sakha-93, (3) Sids-12, (4) Sids-13, (5) Gemmeiza-10, (6) Gemmeiza-11, (7) Misr-1, (8) Misr-2, (9) Line-6043, (10) Line-6068, (11) Line-6107 and (12) Line-6085. The arrow showed the fragment which is associated with *Sr26* (Abu Aly 2014b)

Elkot et al. (2017) in Egypt, used four SSR markers linked to genes controlling stem rust resistance as well as two SSR and one STS marker linked to yellow rust resistance genes for detecting the presence of rust resistance genes in Egyptian wheat. They exploited molecular marker for characterization and identification of candidate lines and cultivars for predicting stem rust and yellow rust resistance genes (Sr2, Sr28, Sr35, Sr40; Yr10, Yr15 and Yr18), respectively. The analysis showed that four Egyptian cultivars and three Libyan local cultivars and three advanced breeding

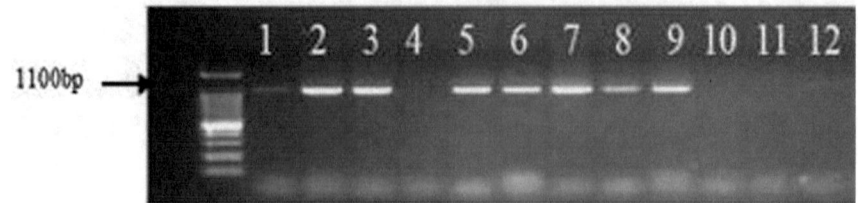

Fig. 12.17 Marker iag95 tested on diverse wheat genotypes and run on an agarose gel. Lanes: (1) monogenic Sr31, (2) Sakha-93, (3) Sids-12, (4) Sids-13, (5) Gemmeiza-10, (6) Gemmeiza-11, (7) Misr-1, (8) Misr-2, (9) Line-6043, (10) Line-6068, (11) Line-6107 and (12) Line-6085. The arrow showed the fragment which is associated with *Sr31* (Abu Aly 2014b)

Table 12.19 Sr genes detected with PCR based markers in 7 Egyptian wheat cultivars and four CIMMYT wheat lines (Abu Aly 2014b)

No.	Genotypes	Sr2	Sr24	Sr26	Sr31
1	Sakha-93	+	+	+	+
2	Sids-12	+	−	+	+
3	Sids-13	+	−	+	−
4	Gemmeiza-10	+	−	+	+
5	Gemmeiza-11	+	−	+	+
6	Misr-1	+	+	+	+
7	Misr-2	+	−	+	+
8	Line-6043	+	+	+	+
9	Line-6068	+	−	−	−
10	Line-6107	+	−	+	−
11	Line-6085	+	+	−	−

(+) = presence of Sr gene in wheat cultivars and (−) = absence of Sr gene in wheat genotype

genotypes from CIMMYT and ICARDA for a marker linked to an individual gene, Sr2 was present in all test genotypes, Sr28 was present in six genotypes, Sr35 was present in four genotypes, Sr40 was present in two genotypes. Likewise, Yr10 was present in two genotypes, Yr15 in six genotypes and Yr18 wasn't found in any of the tested genotypes. These markers would be beneficial in marker-assisted pyramiding genes of stem rust and yellow rust resistance to improve new cultivars with numerous gene resistance against stem rust, yellow rust races in Egyptian wheat.

PCR and loop mediated isothermal amplification of DNA (LAMP) were utilized by Anonymous (2018) for diagnostic *Puccinia triticina*, the causal against of leaf rust. Based on in silico analysis of *P. triticina* genome, PTS68, they determined a simple sequence repeat highly specific to leaf rust fungus. A marker (PtRA68) generates a unique and sharp band of 919 bp in *P. triticina* pathotypes. A novel gene amplification technique LAMP allows visual detection for leaf rust pathogen by naked eye. A set of six primers was designed from a specific region of *P. triticina* and detected presence of *P. triticina* on wheat at 24 hpi (pre-symptomatic stage) which was much former than PCR. The sensitivity of LAMP assay was 100 fg more than conventional PCR (50 pg) and equivalent to qPCR (100 fg).

12.8.3.2   Gene Transfer Technology

Several achievements have been made in the field of wheat genetic transformation. Particle bombardment, electroporation and co-cultivation with *Agrobacterium*, microinjection, direct imbibition, permeabilization, silicon carbide fiber-mediated and pollen tube pathway have also been applied for the introduction of foreign DNA into wheat genotypes (Patnaik and Khurana 2001).

The early efforts in Egypt by Ali et al. (1992) showed that the transformation with Bacillus DNA increases the resistance of cells and wheat plants to leaf rust caused by *Puccinia recondite tritici*. However, Feng et al. (2013) used a target gene of a wheat miRNA (tae-miR408), designated TaCLP1, to offer vital roles in wheat response to increase stripe rust resistance. Transcript accumulation of TaCLP1 and tae-miR408 showed contrasting divergent expression patterns in wheat response to stripe rust and high copper ion stress. Silencing of individual cDNA clones in wheat challenged with stripe rust indicated Electronic supplementary material. The results show that the target of tae-miR408, TaCLP1, play an significant role in regulating resistance to stripe rust in host wheat plants. Whereas, Eissa Hala et al. (2017) obtained genetically modified wheat plants harboring the barley chitinase gene (*chi26*), driven by maize *ubi* promoter using biolistic bombardment. Whereas the herbicide resistance gene, *bar,* driven by the *CaMV 35S* promoter was involved as a selectable marker. Genetically modified wheat with barley *chi26* was appeared to be resistant even after five generations under conditions of artificial infection. One genetically modified line was proved to be equivalent in comparison to the non-genetically modified control. The plant expression vector pBarley/chi/bar with barley *chi26* gene was transferred to the immature embryos of wheat cultivar Hi-Line through particle bombarding. Transgenic plants were selected using the herbicide Basta (1 g/L). Fourteen genetically modified lines out of 72 were recognized, and their transgenic seeds were grown individually to obtain T1 generation (https://plantmethods.biomedcentral.com/articles/10.1186/s13007-017-0191-5). According to the field data, some genetically modified wheat families with *chi26* gene exhibited high resistance to leaf rust, yellow rust and powdery mildew.

12.9   Conclusions

Wheat yield and disease factors are affected by environmental conditions, among which temperature and moisture are the utmost vital in determining disease harshness and yield loss. Yield loss due to rusts is determined by numerous aspects, including the degree of sensitivity, infection time, rate of disease development and duration of disease, causing severe infection and consequently high yield losses. Genetic resistance is the most economical and effective methods to reduce yield losses caused by rust diseases. The effective resistance genes against rust can be deployed individually or in combination with high yield genes to develop high-yield resistant wheat cultivars in wheat-growing regions.

Breeding against wheat rusts is a well-established strategy. Growing resistant cultivars to rusts is considered the most sustainable, cost-effective and environmentally friendly trend in rust diseases control. The select of procedures and strategies for enhancing plant resistance principally depends upon the availability of genetic resources for resistance. At present, more than 187 rust resistance genes were derived from diverse bread or durum wheat cultivars and the related wild species using diverse molecular methods. These 187 consisted of 80 leaf rust, 58 stem rust and 49 stripe rust.

Classical breeding methods like the introduction of exotic germplasm, hybridization, composite crossing, multiline, and backcross breeding were exploited for this target. Speed breeding has substantial advantages over doubled haploids technology for wheat. Current advances in molecular marker technology and gene transfer have created operative tools for resolving many difficulties and have numerous advantages over traditional phenotypic trait selection in improving the resistance of varieties

Through breeding programs, the National Wheat Improvement Program in Egypt over the last 25 years has succeeded in increasing wheat productivity from 3.3 tons per hectare in 1981 to 5.1 tons per hectare in 1991 to about 6.9 tons per hectare in 2016, an increase of 109%. Plant breeders were able to develop a collection of bread wheat cultivars that resistant to rusts such as Giza 168, Giza 171, Gemmeiza 7, Gemmeiza 9, Gemmeiza10, Gemmeiza 11, Gemmeiza12, Sids 12, Sids 13, Sids 14, Sakha 94, Sakha 95, Misr 1, Misr 2, Shandweel 1, and durum wheat cultivars i.e. Sohag 3, Sohag 4, Sohag 5, Beni Suef 1, Beni Suef 4 Beni Suef 5and Beni Suef 6.

12.10   Recommendations

– Improving the genotypes for disease resistance is an ongoing process, and plant breeders need to add new effective sources to their breeding materials
– Cultivating of resistant cultivars are recommended to escape yield losses caused by the rusts diseases
– Plant breeders will not only depend on the host variety, but will take into account the pathogen genetic makeup and the environment.
– It is importance to know the geographical distribution for physiological rust races and recognition leaf rust resistance genes in the modern Egyptian wheat cultivars to maximize resistance in future-bred cultivars.
– It is of prominence to identify genetic differences between germplasm and the study of genetic behavior, the nature of the gene action and the number of gene pairs controlling the resistance before starting the breeding program.

- It is useful to identify the method of the suitable breeding program to improve resistance to produce the promising genotypes.
- It can be utilized Marker-assisted selection breeding as widely castoff to target rust resistance genes to improve selection effectiveness at early stage of plant development, to screen numerous genes simultaneously
- Applied gene transfer technologies such as particle bombardment, electroporation and co-cultivation with *Agrobacterium*, microinjection, direct imbibition, permeabilization, silicon carbide fiber-mediated and pollen tube pathway have also been tried to enter foreign DNA with various degrees of success.

References

Abd El Malik NI (2011) Effect of environmental conditions and virulence dynamics on wheat leaf rust incidence and losses in grain yield. Ph D Thesis, Institute of Environmental Studies & Research, Ain Shams University

Abd El-Hak TM, Stewart DM, Kamel AH (1972) The current rust situation in the Near East countries. Regional Wheat Workshop, Beirut, Lebanon, pp 1–29

Abdelbacki A, Soliman A, Najeeb M, Omara R (2013) Postulation and identification of resistance genes against *Puccinia triticina* in new wheat cultivars in Egypt using molecular markers. International J Chem Environ Biol Sci (IJCEBS) 1(1):2320–4087

Abdel-Hady MS, Okasha EM, Soliman SSA, Talaat M (2008) Effect of gamma radiation and gibberellic acid on germination and alkaloid production in Atropa Belladonna L. Aust J Basic Appl Sci 2(3):401–405

Abou-Elseoud MS, Kamara AM, Alaa-Eldein OA, El-Bebany AF, Ashmawy NA, Draz IS (2014) Identification of leaf rust resistance genes in Egyptian wheat cultivars by multipathotypes and molecular markers. J Plant Sci 2(5):145–151

Abu Aly AA, Shahin AA, El-Naggar DR, Ashmawy MA (2014a) Identification of stripe rust resistance genes Yr's in candidate Egyptian and CIMMYT wheat genotypes by molecular markers. J Plant Prot Pathol Mansoura Univ 5(6):717–727

Abu Aly AAM, Shahin AA, El-Naggar DR, Hermas GA (2014b) Identification and evaluation of resistance genes sources of stem rust in different Egyptian and CIMMYT wheat genotypes using conventional and molecular techniques. J Plant Prot Pathol Mansoura Univ 5(6):729–740

Abu El-Naga SA, Khalifa MM, Alaa El-Dien OA, Youssef WA, Embaby IA (1999a) Effect of stripe rust *Puccinia striiformis* West infection on grain yield in certain wheat cvs. and control application in Egypt during (1996/1997). J Agric Sci Monsoura Univ 24(10):5228–5308

Abu El-Naga SA, Khalifa MM, Bassiouni AA, Youssef WA, Sheheb El-Din TM, Abd-Latif HA (1999b) Revised evaluation for Egyptian wheat germplasm against physiologic pathotypes of stripe rust. J Agric Sci Mansoura Univ 24(2):477–488

Abu El-Naga SA, Khalifa MM, Sherif S, Youssef WA, El-Daoudi YH, Shafik I (2001) Virulence of wheat stripe rust pathotypes identified in Egypt during 1999/2000 and sources of resistance. First Regional Yellow Rust Conference for Central & West Asia and North Africa 8–14 May, SPH, Karj, Iran

Abu El-Naga SA, Khalifa MM, Youssef WA, Imbaby IA, El-Shamy MM, Amer E, Shehab El-Din TM (1997) Effect of stripe rust infection on grain yield in certain wheat cultivars and the economic threshold of chemical control application in Egypt during 1996/1997 growing season. National Annual Coordination Meeting, NVRSRP/Egypt, pp 81–90

Afzal SN, Haque MI, Ahmedani MS, Bashir S, Rattu AR (2007) Assessment of yield losses caused by *Puccinia striiformis* triggering stripe rust in the most common wheat varieties. Pak J Bot 39(6):2127–2134

Ageez AA, Boulot OA (1999) Quantitative determination of the gene action of leaf rust resistance. Egypt J Appl Sci 14(6):216–226

Ahmad S, Khan MA, Haider MM, Iqbal Z, Iftikhar Y, Hussain M (2010) Comparison of yield loss in different wheat varieties/lines due to leaf rust disease. Pak J Phytopathol 22(1):13–15

Ajmal SU, Zakir N, Mujahid MY (2009) Estimation of genetic parameters and character association in wheat. J Agric Biological Sci 1(1):15–18

Aktar-Uz-Zaman M, Tuhina-Khatun Mst, Mohamed MH, Mahbod S (2017) Genetic analysis of rust resistance genes in global wheat cultivars: An overview. Biotechnol Biotechnol Equip 31(3):431–445

Ali AH, El-Hennawy MA, El-Kholy HK, Hagran A (1992) Genetically engineered sodium chloride and *Puccinia recondite tolerant* wheat cells and plants. Egypt J Appl Sci 7(8):675–690

Ali RG, Omara RI, Zinap A (2016) Effect of leaf rust infection on yield and technical properties in grains of some Egyptian wheat cultivars. Menoufia J Plant Prot, 19–35

Anderson JA (2003) Plant genomics and its impact on wheat breeding. Plant Mol Breed, 184–215

Anonymous (2005) Managing stripe rust and leaf rust of wheat, farm note no. 43. Department of Agriculture, Western Australia

Anonymous (2017) Wheat production and consumption. Economic Affairs Sector. ARC, Giza, Egypt

Anonymous (2018) Rapid detection of *Puccinia triticina* causing leaf rust of wheat by PCR and loop mediated isothermal amplification. PLoS ONE13(4):e0196409. https://doi.org/10.1371/journal.pone.0196409. eCollection, 26 Apr 2018

Anonymous (2020) Recommendation techniques of field crops. ARC, Giza, Egypt

Appel JA, DeWolf E, Bockus WW, Odd TT (2009) Preliminary 2009 Kansas wheat disease loss estimates. Kansas cooperative plant disease survey report

Ashmawy MA, Abu Aly AAM, Youseef WA, Shahin AA (2012) Physiologic races of wheat yellow rust *Puccinia striiformis* f. sp. *tritici* in Egypt during 1999–2011. Minufiya J Agric Res 37(2):297–305

Ashmawy MA, El-Orabey WM, Mohamed N (2013) Effect of stem rust infection on grain yield and yield components of some wheat cultivars in Egypt. ESci J Plant Pathol 2(3):171–178

Awaad HA, Salem AH, Atia MMM, Sallam ME (2003) The genetic system controlling leaf rust resistance in bread wheat (*Triticum aestivum* L.). Zagazig J Agric Res 30(4):1151–1167

Babar M, Mashhadi AF, Mehvish A (2010) Identification of rust resistance genes Lr10 and Sr9a in Pakistani wheat germplasm using PCR based molecular markers. Afr J Biotech 9(8):1144–1150

Bajwa MA, Aqil KA, Khan NI (1986) Effect of leaf rust on yield and kernel weight of spring wheat. RACHIS 5:25–28

Ball G, Hall D (1967) A clustering technique for summarizing multivariate data. Behav Sci 12:153–155

Boulot OA, El-Sayed EAM (2001) Quantitative genetic studies on leaf rust resistance and its components in a 7-parents diallel cross of wheat. J Agricul Sci Mansoura Univ 26(1):133–146

Chen XM (2005) Epidemiology and control of stripe rust (*Puccinia striiformis* f. sp. *tritici*) on wheat. Can J Plant Pathol 27(3):314–337

Chen XM (2007) Challenges and solutions for stripe rust control in the United States. Aust J Agric Res 58:648–655

Coakley SM, Boyd WS, Line RF (1982) Statistical models for predicting stripe rust on winter wheat in the Pacific Northwest. Phytopathology 72:1539–1542

Danial DL, Stubbs RW, Parleyflit JE (1994) Evolution of virulence patterns in yellow rust races and its implication for breeding for resistance in wheat in Kenya. Euphytica 80(3):165–170

Darwish MAH, Farhat WZE, El-Sabagh A (2018) Inheritance of some agronomic characters and rusts resistance in fifteen F_2 wheat populations. Cercetări Agronomice în Moldova 1(173):5–28

De Vallavieille-Pope C, Huber L, Leconte M, Goyeau H (1995) Comparative effects of temperature and interrupted wet periods on germination, penetration, and infection of *Puccinia recondita* f. sp. *tritici* and *P. striiformis* on wheat seedlings. Phytopathology 85:409–415

Diab MH (1994) Epidemiology of wheat leaf rust in Egypt in relation to ecological conditions. Ph.D. Thesis, Institute Environment. Science Studies and Research. Ain Shams University, 114 p

Draz ISMSA bou-Elseoud, Kamara AM, Alaa-Eldein OA, El-Bebany AF (2015) Screening of wheat genotypes for leaf rust resistance along with grain yield. Annals of Agric Sci 60(1):29–39

Eddy R (2009) Logistic regression models to predict stripe rust infections on wheat and yield response to foliar fungicide application on wheat in Kansas. Thesis, Kansas State University, Manhattan

Eisa GSA, El-Naggar Doaa R (2015) Leaf rust assessment in four Egyptian bread wheat cultivars with their subsequent physio-histological correlations. Minufiya J Agric Res 40(3):609–622

Eissa Hala F, Hassanien SE, Ramadan AM, El Shamy MM, Saleh OM, Shokry AM, Abdelsattar M, Morsy YB, El Maghraby MA, Alameldin HF, Hassan SM, Osman GH, Mahfouz HT, Gad El-Karim GA, Madkour MA, Bahieldin E (2017) Developing transgenic wheat to encounter rusts and powdery mildew by overexpressing barley chi26 gene for fungal resistance. Plant Methods 13:41

El-Daoudi YH, Shafik I, Ghamem EH, Abu El-Naga SA, Sherif SO, Khalifa MMO, Mitkees RA, Bassiouni AA (1996) Stripe rust occurrence in Egypt and assessment of grain yield loss in 1995. In: Proceedings Du Symposium Regional Sur les Maladies des Ceraleset des Legumineuses Alimentaries. Rabat Morocco, pp 341–351

El-Daoudi YH, Shenoda Ikhals S, Bassiouni AA, Sherif SE, Khalifa MM (1987) Genes conditioning resistance to wheat leaf and stem rust in Egypt. In: Proceedings 5th The Egypt Phytopathological Society Giza, pp 387–404

Elkot AFA, Abd El-Aziz MH, Aldrussi IA, El-Maghraby MA (2017) Molecular identification of some stem rust and yellow rust resistance genes in egyptian wheat and some exotic genotypes. Assiut J Agric Sci 47(4):124–135

El-Naggar Doaa RM, Soliman SSA (2015) Evaluation of some mutant lines in three Egyptian bread wheat cultivars for resistance to biotic stress caused by wheat rusts. Egypt Journal of Applied Science 30(8):254–269

El-Orabey WM, Sallam Minaas E, Omara RI, Abd El-Malik Nagwa I (2015) Geographical distribution of *Puccinia triticina* physiologic races in Egypt during 2012-2014 growing seasons. Afr J Agric Res 10(45):4193–4203

El-Sherif NA, Hermas GA, Hasan MA, Tohamey Somaya (2018) Physiological races and virulence diversity of *Puccinia graminis* PERS. f. sp. *tritici* ERIKS. & E. HENN. during 2011/2012 and 2012/2013 growing seasons in Egypt. Menoufia J Plant Prot 3:71–84

Eversmeyer MG, Kramer CL (1995) Survival of *Puccinia recondita* and *P. Graminis* urediniospores exposed to temperatures from subfreezing to 35°C. Phytopathology 85:161–164

FAO/IAEA (2015) Mutant varieties database. http://www-mvd-iaea.org, IAEA, Vienna

Feng H, Zhang Q, Wang QL, Wang XJ, Liu J, Li M (2013) Target of tae-miR408, a chemocyaninlike protein gene (TaCLP1), plays positive roles in wheat response to high-salinity, heavy cupric stress and stripe rust. Plant Mol Biol Reporter 83:433–443

Fetch T, Zegeye T (2016) Detection of wheat stem rust races TTHSK and PTKTK in the Ug99 race group in Kenya in 2014. Plant Dis 100:1495

Fu D, Uauy C, Distelfeld A, Chen X, Fahima T, Dubcovsky J (2008) High density map of wheat stripe rust resistance gene Yr36. Plant and Animal Genome XVI Conference, San Diego CA, USA

Gebril EEMA, Gad MA, Kishk AMS (2018) Effect of sowing dates on potential yield and rust resistance of some wheat cultivar Mansoura Journal of Plant. Production 9(4):369–375

Gladders P, Langton SD, Barrie IA, Hardwick NV, Taylor MC, Paveley ND (2007) The importance of weather and agronomic factors for the overwinter survival of yellow rust (*Puccinia striiformis*) and subsequent disease risk in commercial wheat crops in England. Annals of Appl Biol 150:371–382

Grabow BS, Shah DA, DeWolf ED (2016) Environmental conditions associated with Stripe Rust in Kansas winter wheat. Plant Dis 100(11):2306–2312

Hasan MA, El-Orabey WM, El-Naggar Doaa R, Ashmawy MA (2014) Genetic analysis of seedling and adult plant resistance genes to leaf rust in nine bread wheat cultivars. Zagazig J Agric Res 41(6):1217–1229

Hermas GA, Youssef IAM, Shahin AA, Diab Hoda M (2013) Physiological specialization in *Puccinia graminis* f.sp. *tritici* and probable genes for resistance in 13 Egyptian wheat varieties during 2009/2010 growing seasons

Herrera-Foessel SA, Singh RP, Huerta-Espino J, Crossa J, Yuen J, Djurle A (2006) Effect of leaf rust on grain yield and yield traits of durum wheats with race-specific and slow-rusting resistance to leaf rust. Plant Dis 90:1065–1072

Hickey L, Rutkoski J, Riaz A, Singh WNGD, Godwin I, Aitken E, Platz G, Dieters M (2015) Using 'speed breeding' to harness rust resistance: faster, cheaper and easier. BGRI 2015 Technical Workshop. 17–20 September. Sydney Australia, Borlaug Global Rust Initiative

Hodson D (2013) Stem rust epidemic in southern Ethiopia. Extreme caution and vigilance need in east Africa and middle east region. (http://rusttracker.cimmyt.org/?p=5473)

Huerta-Espino J, Singh RP, German S, McCallum BD, Park RF, Chen WQ, Bhardwaj SC, Goyeau H (2011) Global status of wheat leaf rust caused by *Puccinia triticina*. Euphytica 179:143–160

Imbaby IA, El-Sherif Nabila, Hermas GA, El-Naggar Doaa R, Diab Hoda (2014a) Could TKTT race of *Puccinia graminis* f.sp. *tritici* cause epidemic disease in Egypt? Egypt J Appl Sci 29(8):169–171

Imbaby IA, Mahmoud MA, Hassan MEM, Abd-El-Aziz ARM (2014b) Research article identification of leaf rust resistance genes in selected Egyptian wheat cultivars by molecular markers. Sci World J, Article ID 574285, 7 pages. http://dx.doi.org/10.1155/2014/574285

Jin Y, Pretorius ZA, Singh RP, Fetch T (2008) Detection of virulence to resistance gene Sr24 within race TTKS of *Puccinia graminis* f. sp. *tritici*. Plant Dis 92:923–926

Kenzhebayeva S, Alybayeva R, Atabayeva S, Doktrbai G (2013) Developing salt tolerant lines of spring wheat using mutation techniques. Proceeding International Conference. Life Science and Biological Engineering, Japan, pp 2318–2376

Kolmer J, Chen XM, Jin Y (2009) Diseases which challenge global wheat production. The wheat rusts. 89–124. In: BF Carver (ed) Wheat: science and trade. Wiley-Blackwell, Wiley, Ames, Iowa, p 569

Kolmer JA, Long DL, Hughes ME (2005) Physiological specialization of *Puccinia triticina* on wheat in the United States in 2003. Plant Dis 89:1201–1206

Leonard KJ, Szabo LJ (2005) Stem rust of small grains and grasses caused by *Puccinia graminis*. Molecular Plant Pathol 6:99–111

Liu JQ, Kolmer JA (1997) Genetics of leaf rust resistance in Canadian spring wheats AC Domain and AC Taber. Plant Dis 81:757–760

Loughman R, Jayasena K, Majewski J (2005) Yield loss and fungicide control of stem rust of wheat. Aust J Agric Res 56:91–96

MacCallum RC, Zhang S, Preacher KJ, Derek DR (2002) On the practice of dichotomization of quantitative variables. Psychol Methods 7(1):19–40

Maluszynski M, Kasha KJ (2002) Mutations, in vitro and molecular techniques for environmentally sustainable crop Improvement (eds). Kluwer Academic Publishers, Dordrecht/Boston/London. ISBN 1-4020-0602-0

Marasas CN, Smale M, Singh RP (2004) The economic impact in developing countries of leaf rust resistance breeding in CIMMYT-related Spring bread wheat. International Maize and Wheat Improvement Center, Mexico, DF

Marcu D, Damian G, Cosma C, Cristea V (2013) Gamma radiation effects on seed germination, growth and pigment content, and ESR study of induced free radicals in maize (Zea mays). J Biol Phys 39:625–634

McIntosh RA, Yamazaki Y, Devos KM, Dubcovsky J, Rogers J, Appels R (2007) Catalogue of gene symbols for wheat. 2007 Supplement. KOMUGI Integrated Wheat Science Database. http://www.shigen.nig.ac.jp/wheat/komugi/genes/symbolClass-List.jsp

McVey DV, Nazim M, Leonard KJ, Long DL (2004) Patterns of virulence diversity in *Puccinia triticina* on wheat in Egypt and the United States in 1998–2000. Plant Dis 88:271–279

Menshawy AM, Najeeb MAA (2004) Genetical and pathological studies on certain Egyptian wheat genotypes as affected both leaf and stripe rust. J Agric Sci Mansoura Univ 29(4):2041–2051

Milus EA, Kristensen K, Hovmoller MS (2009) Evidence for increased aggressiveness in a recent widespread strain of *Puccinia striiformis* f. sp. *tritici* causing stripe rust of wheat. Phytopathology 99:89–94

Moschini RC, Perez BA (1999) Predicting wheat leaf rust severity using planting date, genetic resistance, and weather variables. Plant Dis 83:381–384

Najeeb MA, Hermas GA, Youssef EA, El-Shamy MM (2004) Physiologic specialization in *Puccinia graminis tritici* and genes conferring wheat resistance in Egypt. Annals of Agric Sci Moshtohor J 42(4):1603–1612

Nazim M, Aly MM, Ikhlas Shafik, Abed-Almalek Nagwa I (2010) Frequency of virulence and virulence formula of wheat leaf rust identified in Egypt during 2004/05–2007/08. Egypt J Phytopathol 38(1–2):77–88

Negm SS, Boulot OA, Hermas GA (2013) Virulence dynamics and diversity in wheat rust (*Puccinia triticina*) populations in Egypt during 2009/2010 and 2010/2011 growing seasons. Egypt J Appl Sci 28(6):183–212

Njau PN, Kimurto PK, Kinyua MG, Okwaro HK, Ogolla JBO (2006) Wheat productivity improvement in the drought prone areas of Kenya. Afr Crop Sci J 14(1):49–57

Ochoa J, Parlevliet JC (2007) Partial resistance to barley rust, *Puccinia hordie* on the yield of the barley cultivars. Euphytica 153:309–312

Omara RI, Abd El_malik Nagwa, Abu Aly AA, I, Abu Aly AA (2017) Inheritance of stem rust resistance at adult plant stage in some Egyptian wheat cultivars. Egypt J Plant Breed 21(2):261–275

Omara RI, El-Naggar DR, Abd El Malik NI, Ketta HA (2016) Losses Assessment in some Egyptian Wheat Cultivars caused by Stripe Rust Pathogen (*Puccinia striiformis*). Egypt J Phytopathol 44(1):191–203

Patnaik D, Khurana P (2001) Wheat biotechnology: A minireview. Electron J Biotechnol 4(2):1–29. ISSN: 0717-3458

Patpour M, Hovmoller MS, Justesen AF (2015) Emergence of virulence to SrTmp in the Ug99 race group of wheat stem rust, *Puccinia graminis* f. sp. *tritici*, in Africa. Plant Dis 100:522

Patpour M, Hovmøller MS, Shahin AA, Newcomb M, Olivera P, Jin Y, Luster YD, Hodson D, Nazari K, Azab M (2016) First report of the Ug99 race group of wheat stem rust, *Puccinia graminis* f. sp. *tritici*, in Egypt in 2014, 100(4):863–864

Pretorius ZA, Singh RP, Wagoire WW, Payne TS (2000) Detection of virulence to wheat stem rust resistance gene *Sr31* in *Puccinia graminis* f. sp. *tritici* in Uganda. Plant Disease 84:203

Pretorius ZA, Szabo LJ, Boshoff WHP, Herselman L, Visser B (2012) First report of a new TTKSF race of wheat stem rust (*Puccinia graminis* f. sp. *tritici*) in South Africa and Zimbabwe. Plant Disease 96:590

Putnik-Deliã M (2008) Resistance of some wheat genotypes to *Puccinia triticina*. Proc Nat Sci Matica Srpska Novi Sad 115:51–57

Rautela A, Dwivedi M (2018) Wheat Stem Rust Race Ug99: a Shifting Enemy. Int J Curr Microbiol Appl Sci 7(1):1262–1266

Roelfs AP (1985) Wheat and rye stem rust. In: The cereal rusts, Vol. II: diseases, distribution, epidemiology and control. Academic Press, Orlando, FL, pp 3–37

Roelfs AP, Martens JW (1988) An international system of nomenclature for *Puccinia graminis* f. sp. *tritici*. Phytopathology 78:526–533

Roelfs AP, Singh RP, Saari EE (1992) Rust diseases of wheat, concept and methods of disease management. CIMMYT, Mexico, D.F., p 81

Salem AH, Awaad HA, Sallam Minaas E, Atia MMM (2003) Inheritance of stem rust resistance in bread wheat (*Triticum aestivum* L.) Tenth Congress of Phytopathology, Giza, Egypt, December, pp 39–52

Shaheen S, El-Orabey WM (2016) Assessment of grain yield losses caused by *Puccinia triticina* in some Egyptian wheat genotypes. Article in Span J Agric Res 41(1):1–9

Sallam Minaas E, El-Orabey WM, Ashmawy MA, Omara RI (2014) Stability of some specific genes of wheat leaf rust resistance in near-isogenic lines of Thatcher variety. Can J Plant Prot 2(2):44–54

Sallam Minaas E, El-Orabey Walid M, Omara Reda I (2016) Seedling and adult plant resistance to leaf rust in some Egyptian wheat genotypes. Afr J Agric Res 11(4):247–258

Shahin A, Abu Aly, AA (2015) Wheat stripe and stem rust situation in Egypt: Yr27 and Sr31 virulence. In: APS Annual Meeting, 1–5 Aug, Pasadena, CA, US, © 2018 Borlaug Global Rust Initiative

Shahin A, Ragab KHE (2015) Inheritance of adult plant stripe rust resistance in wheat cultivars Giza160 and Giza186. Mansoura J Plant Prot Pathol 6(4):587–596

Shahin AA (2017) Effective genes for resistance to wheat yellow rust and virulence of *Puccinia striiformis* f.sp. *tritici* in Egypt. Egypt Acad J Biol Sci 8(2):1–10

Shahin AA, Abu Aly AA, Shahin SI (2015) Virulence and diversity of wheat stripe rust pathogen in Egypt. J Am Sci 11(6):47–52

Sharma A, Singh S, Joshi N, Meeta M, Sharma I (2011) Induced resistance for stripe rust and leaf blight in barley (*Hordeum vulgare* L.) using chemical mutagen. J Wheat Res 3(2):56–58

Singh RP, Payne TS, Figueroa P, Valenzuela S (1991) Comparison of the effect of leaf rust on the grain yield of resistant, partially resistant, and susceptible spring wheat cultivars. AM J Altern Agric 6(3):115–121

Stakman EC, Stewart DM, Loegering WQ (1962) Identification of physiologic races of *Puccinia graminis* var. *tritici*. Agricultural Research Service E617. Washington, DC, United States Department of Agriculture

Steinberg D (2000) Modeling the basis for rational disease management. Crop Prot 19:747–752

Stubbs RW (1988) Pathogenicity analysis of yellow (stripe) rust of wheat and its significance in a global context breeding strategies for persistence to the rust of wheat N,W. Simmonds and S. Rajaram (CIMMYT) ISBN 96-127

Swelam AA, El-Naggar Doaa RM, Eisa GSA (2010) Genetic studies to stem rust and subsequent historical examinations in three bread wheat (*Triticum aestivum* L.) crosses. Ann Agric Sci Moshtohor J 48(1):81–102

Todorovska E, Christov N, Slavov S (2009) Biotic stress resistance in wheat breeding and genomic selection implications. Biotechnol Biotechnol Equip 23(4):1417–1426

Tohamy S (2004) Studies on the dynamics of *Puccinia triticina* races the cause of leaf rust on wheat under dominant environmental conditions in Egypt. M Sc Thesis, Institute of Environmental Studies & Research, Ain Shams University

Vijaya Kumar P (2014) Development of weather-based prediction models for leaf rust in wheat in the Indo-Gangetic plains of India. Europ J Plant Pathol 140:429–440

Wanyera R, Kinyua MG, Jin Y, Singh, RP (2006) The spread of stem rust caused by *Puccinia graminis* f. sp. *tritici,* with virulence on Sr31 in wheat in Eastern Africa. Plant Dis 90:113

Yin JL, Fang ZW, Sun C, Zhang P, Zhang X, Lu C, Wang SP, Ma DF, Zhu YX (2018) Rapid identification of a stripe rust resistant gene in a space-induced wheat mutant using specific locus amplified fragment (SLAF) sequencing. Sci Rep 8:1–9

Youssef IAM (2011) Inheritance of adult plant resistance to leaf rust in four Egyptian wheat cultivars and their crosses with four leaf rust resistance monogenic lines. J Agric Chem Biotechnol Mansoura Univ 2(12):305–315

Youssef IAM, El-Naggar Doaa R, Hermas GA, Negm SS (2010) Identification of physiologic races of *Puccinia triticina* and postulated genes of resistance in certain Egyptian commercial wheat cultivars. J Plant Prot Pathol 1(2):75–85

Youssef IAM, Hermas GA, El-Naggar Doaa R, El-Sherif Nabila A (2012) Virulence of *Puccinia graminis* f.sp. *tritici* and postulated resistance genes for stem rust in thirteen wheat varieties during 2008/2009 growing seasons in Egypt. Egypt J Appl Sci 27(11):326–344

Chapter 13
The Importance of Faba Bean (*Vicia faba* L.) Diseases in Egypt

Said Ahmed Mohamed Omar

Abstract Faba bean (*Vicia faba* L.) is a major food crop in Egypt. The cultivated area was decreased considerably, and subsequently, yield production was greatly reduced. Moreover, faba bean plants are attacked by fungal diseases (chocolate spot, rust, root rot and wilt) and viral diseases (Faba Bean Necrotic Yellow Virus, Bean Yellow Mosaic Virus and Bean Leaf Roll Virus). These pathogens considered as a main constrains affected growth of the plant and contributed significantly to causing great yield loss both in quantity and quality. Yield loss caused by diseases is estimated by about (25–80%) and reduced protein content by about 4%. It is worth mention that combined infections with more than one pathogens on faba bean plant caused substantial damage and significant yield loss than infection by a single pathogen (85%). However, some control measures for minimizing these diseases are available. These involve: breeding for disease resistance, fungicides treatment, induce disease resistance, biological control, plant extracts, growth regulators, and agricultural practices.

Keywords Faba bean · Botrytis · Rust · Disease · Control

13.1 Introduction

Faba bean (*Vicia faba* L.) is considered as one of the oldest crops and was known to the ancient Egyptian (Abdalla 1979; Hegab et al. 2014). It is believed to be originated in Mediterranean regions and spread all over the world. The plant belonged to family Fabaceae and classified generally to large, intermediate and small-seeded size (Bond 1979). Different names were given to the plant according to their seed size and growing locations. These names are faba bean, broad bean, field bean, tick bean, pigeon bean, horse bean, garden bean, fava, spring, winter, and common bean. In Egypt, the popular name is fool misery. The crop more often uses as human food and animal feed due to its valuable nutritional contents, i.e. protein, carbohydrate,

S. A. M. Omar (✉)
Plant Pathology Research Institute, Agriculture Research Center, 12619 Giza, Egypt

© Springer Nature Switzerland AG 2021
H. Awaad et al. (eds.), *Mitigating Environmental Stresses for Agricultural Sustainability in Egypt*, Springer Water,
https://doi.org/10.1007/978-3-030-64323-2_13

fatty acids, lipids, phenols and elemental composition N, P, K, Ca, Mg, S. (Brown 1977; Hill-Coltinghan 1983; Baseony and Abd El-Moneim 2012). Nevertheless, anti-nutritional factors were also recognized such as tannins, lactin, tyrosin, vicine and convincing (Hossain and Mortuza 2006). It is well established that faba bean plants increase soil fertility by nitrogen fixation process through different Rhizobium strains. Also, improve soil textures by their root traits (Roughley et al. 1983). Faba bean cultivated in 56 countries, the global area was 2.4 million ha and the total production was 4.46 million tons of dry grains (FAOSTAT 2018). The plant ranked as the fourth crop regarding total yield production after beans, pisum, and chickpea. This presenting 7.5% of the total yield production of food crops in the world (FAOSTAT 2016).

In Egypt, the grown areas of faba bean were decreased gradually and remarkably during the last two decades from about 220.00 to 80.00 feddan (feddan = 4200 m^2) (Anonymous 2017a). The shortages of the grown areas with numerous increasing of Egyptian populations as well as other factors (plant diseases, pests, weeds, parasitic plants, nutrition's and cost of production materials, etc.), these constraints led to imported a considerable amounts of faba bean seeds to fulfill the requirements and demands of the consumers (El-Metwally et al. 2013). This chapter aims to provide some knowledge is about the impact of the major plant diseases as a factor affecting faba bean yield production.

13.2 Fungal Diseases

Despite its importance, the crop productivity can be reduced due to several biotic and abiotic factors. Fungal diseases are considered one of the most important biotic factors causing faba bean yield reduction.

13.2.1 Shoot Diseases

Among them, chocolate spot and rust which consider the most important diseases in the Mediterranean region (Honounik and Bisri 1991). Other foliar fungal diseases also could contribute in reduction the productivity level of the crop, alternaria leaf spot, Ascochyta blight, downy mildew and white mold/Sclerotinia stem rot diseases are best examples (Biddle and Cattlin 2007; Nasraoui 2008).

13.2.1.1 Chocolate Spot

The main causal fungus is *Botrytis fabae* Sard. Other botrytis species (*Botrytis cinerea* and *Botrytis fabiopsis* were also reported with less virulence on faba bean. Chocolate spot disease was first recorded in Egypt by EL-Helaly (1939). Different strains of

B. fabae was existed and varied in their morphological characters and pathological capabilities. Mohamed et al. (1996) found 17 strains of *B. fabae* isolates differed in their morphological and pathological features in the north of Egypt. The fungus attacks all above ground parts of the plants involving leaves, stems, flowers, pods and seeds inside the pods (see Fig. 13.1).

Lesions of infected leaf vary from the small brown spot (1–3 mm in diameter) to conspicuous well-defined lesions with reddish brown margins and tan colored centers. Under favorable conditions (high humidity 85–100% associated with low temperature 10–15%), the disease becomes aggressive, and lesions may coalesce causing blackening and partial defoliation. Disease severity of chocolate spot was greatly high in north parts of Egypt especially at Kafr EL-Sheikh, Dakhalea, Domitta, and Behara governorates. The environmental conditions of these regions favor growth and development *B. fabae* during January and February. However, disease severity

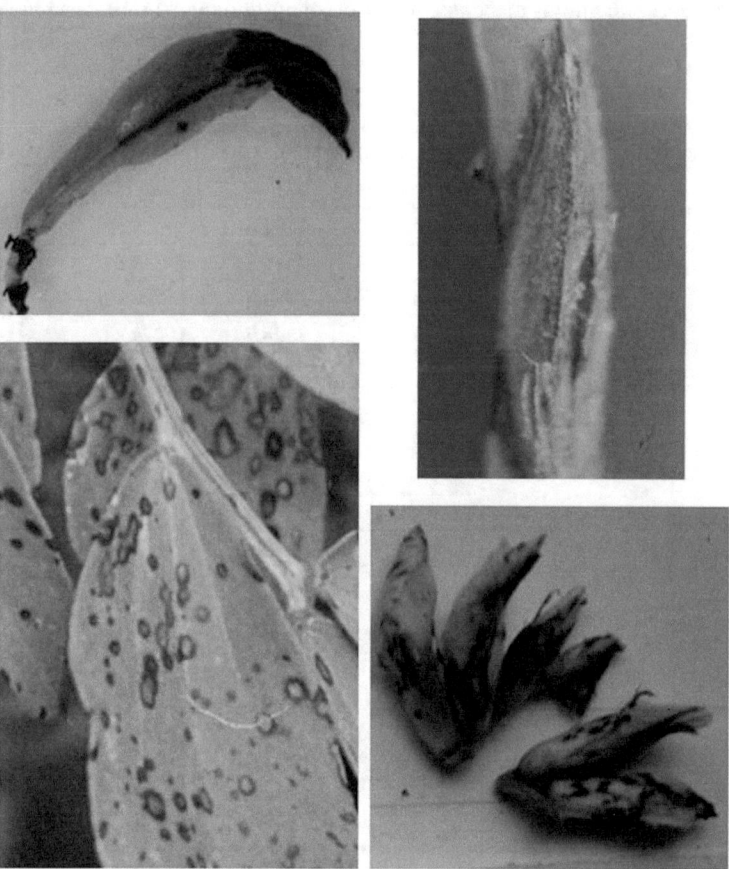

Fig. 13.1 Symptoms of chocolate spot disease on the faba bean plant (Omar 1984)

reduces significantly in the middle and almost absent in the south of Egypt where the temperature is high, and relative humidity is low.

Impact of chocolate spot disease on yield loss varied greatly according to the severity of infection, aggressiveness of Botrytis isolate(s), the sensitivity of plant cultivars and stage of plants at which infection tack place. Different methods have been used to estimate yield loss include artificial inoculation and/or leaf defoliation (Williams 1975; Morsy and Tarrad 2005; Negash et al. 2015). It was found that leaves damaged caused by *B. fabae* reduced the efficacy of the leaves in the photosynthesis process which in turn decreased their efficiency to form pods on the plants. Mohamed (1996) found that yield loss of faba bean ranged from 25–85% as a result of chocolate spot disease (Table 13.1). Awaad et al. (2005) registered a negative and highly significant correlation between faba bean seed yield and infection type of chocolate spot disease in F2 (−0.675**) and F4 (−0.773**) generations.

The work conducted by Morsy and Tarrad (2005) revealed that artificial inoculation with *B. fabae* and/or mechanical leaf defoliation reduced faba bean yield 80% than the control and lowered protein content by 4% than the check (Table 13.2).

Table 13.1 Area under disease progress curve (AUDPC), seed yield and losses percentage in seed yield of three faba bean cultivars sown at three different dates in two seasons (Mohamed 1996)

Season	Main-treat. (dates)	Sub-treat. (cvs.)	AUDPC	Yield (Kg/fed.)	Loss%
1990/1991	D1 Nov. 11, 90	Giza 461	1292.6	906.39	0
		Giza 3	1599.8	794.61	0
		Giza 402	2269.5	568.62	0
	D 2 Nov. 25, 90	Giza 461	1128.9	374.22	58.71
		Giza 3	1441.8	508.36	36.02
		Giza 402	2143.1	394.15	30.68
	D3 Dec. 9, 90	Giza 461	624.0	134.14	85.20
		Giza 3	1007.6	245.43	69.11
		Giza 402	1335.8	173.99	69.40
1991/1992	D1 Nov. 14, 91	Giza 461	573.8	2020.68	0
		Giza 3	740.6	1899.35	0
		Giza 402	1064.3	1626.34	0
	D2 Nov. 28, 91	Giza 461	411.28	1225.00	39.38
		Giza 3	578.20	1416.34	25.43
		Giza 402	743.20	1129.34	30.56
	D3 Dec. 12,91	Giza 461	396.2	644.94	68.08
		Giza 3	544.2	678.07	64.30
		Giza 402	657.2	730.34	55.09
L.S.D between: 2 Cvs. at each date 1990/1991 1991/1992			5% 1% 144.0 197.3 83.9 115.0	5% 1% 61.7 84.1 133.0 182.5	

Table 13.2 Effect of artificial infection and leaf defoliation with a chocolate spot on some yield parameters of faba bean (Morsy and Tarrad 2005)

Treatment	Rate/conc.	2000/2001		Mean	2000/2001		Mean
		Giza 40	Giza 461		Giza 40	Giza 461	
Leaf defoliation %	100%	3.03	10.27	6.65	3.43	10.43	6.93
	75%	3.70	10.23	6.97	3.83	13.23	8.53
	50%	7.30	24.13	15.72	7.63	23.50	15.57
	25%	19.53	41.77	30.65	18.13	39.80	28.97
Infection with *Botrytis fabae*	10^8	9.20	22.03	15.62	10.40	20.80	15.60
	10^4	10.10	40.50	25.30	11.77	25.30	18.54
	10^2	21.33	36.30	28.82	22.63	30.07	26.35
	10	25.00	42.47	33.74	26.50	41.17	33.84
Control		27.13	43.87	35.50	29.10	45.73	37.42
Dithane M 45		29.13	46.10	37.62	31.10	46.47	38.79

13.2.1.2   Control

Several disease control measures have been adopted for controlling chocolate spot disease includes:

1. Breeding for disease resistance with high yielding ability using traditional breeding programs (Khalil et al. 1993) as well as advanced molecular studies (EL-Badawy 2008). Results of detached leave technique, greenhouse and field experiments revealed resistant cultivars to the disease, i.e. Sakha 1, Sakha 3, Sakha 4, Giza 3, and Giza 716 (see Fig. 13.2).

It was found that faba bean resistant cultivars to B.fabae infection due probably to its contains phytoalexins compounds (Omar et al. 1992 and Tarrad et al. 1993). Antimicrobial phytoalexin derivatives especially wyerone acid is responsible for resistance against chocolate spot disease (Table 13.3).

1. Fungicides treatment, spraying Mancozeb and Tridex 80% W.P. at the recommended rate (250 g/100 L water) four times with 15 days intervals gave good protection against the disease (Mohamed 1982; Anonymous 2017b).
2. Application of other treatments i.e. induce disease resistance, growth regulators, biological control, plant extracts, as well as some agricultural practices offered a degree of resistance to the disease, and increased yield (Eisa et al. 2006; Baraka et al. 2008; Salem et al. 2012; EL-Wakil et al. 2016).

13.2.1.3   Rust Disease

The disease caused by *Uromyces vicia fabae* (Pers.), rust ranked as the second shoot fungal disease of faba bean plants (Sarhan 2006). The pathogen infects leaves, stems and pods forming brown pustules (Fig. 13.3).

Fig. 13.2 Screening methods for chocolate spot disease resistance (Khalil et al. 1993)

Table 13.3 Wyerone acid concentration (ug/g fresh weigh) of faba bean cultivars (Omar et al. 1992)	Cultivars (tissues)	Reaction type	Lesion sites + Peripheries
	ILB 938	Resistant	334
	Giza 461	Resistant	30.9
	Giza 402	Susceptible	22.1

Two types of pustules were observed; either as a single one or surrounded by small pustules. The pustules contain numerous amounts of uredospores which infect the leaves through stomata during the growing seasons. It is worth mention that, there are several physiological strains of faba bean rust varied in their morphological and pathological capabilities (Conner and Bernier 1980). They reported the existence of three dominant genes for resistance to two isolates of rust disease.

Moreover, rust has a wild host range, can infect other legume plants such as Pisum, Lens and Lathyrus. The favorable conditions for growth and development the fungus are relatively hot weather 25–30 °C associated with heavy dew. Such ideal conditions are present in North and Delta regions were rust disease did occur. While a negligible rust infection was noted in the middle or south of Egypt. Since rust infection occurs relatively at a late stage of plant growth were most of the pods formed, the influence of rust disease on the yield, therefore, was limited. In this concern, Awaad et al. (2005) recorded a negative and highly significant association between faba bean seed yield and infection type of rust disease in F2 (−0.688**) and F4 (−0.664**) generations.

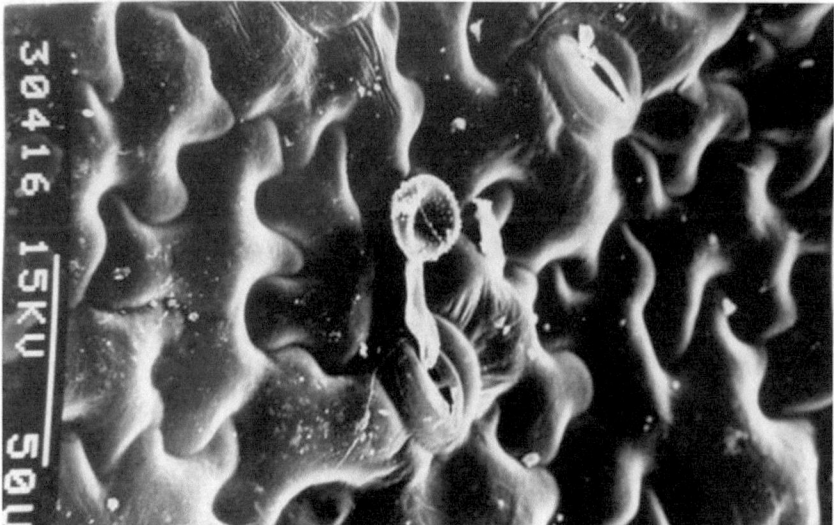

Fig. 13.3 Rust infection and symptoms in faba bean leaves (Omar 1984)

Control

1. Spraying fungicide Mancozeb at the dose of 250 g./100 L water three times reduce rust disease severity

Table 13.4 Influence of soaking seed with ethephon and foliar application of Dithane M 45 on growth, yield and yield components of faba bean (Salem et al. 1992)

Treatment	Plant height (cm)	No. branches	No. pods/plant	No. seeds/pod	100-seed weight(g)	Seed yield/plant (g)
Ethephon (600 ppm)	60.0ab[a]	2.6a	18.7a	3.2a	61.9a	36.6a
Ethephon (800 ppm)	57.9c	3.1a	19.0a	3.3a	63.4a	39.9a
Ethephon (1000 ppm)	55.5c	3.0a	16.2a	3.2a	60.0a	31.3a
Dithane M-45 (0.3%)	68.6ab	3.2a	19.6a	3.3a	59.8a	38.1a
Control	72.1a	3.4a	18.1a	3.2a	57.8a	32.6a

[a]Treatments with the same letter are not significantly different

2. Application of some agricultural practices especially (sowing date) might escape and lower rust disease.
3. Salem et al. (1992) found that ethephon treatment as seed soaking (600–1000 ppm) or foliar spry of Dithane M-45 reduced disease severity and improved growth and increased yield component in faba bean plants (Table 13.4).

13.2.1.4 Other Shoot Diseases

Faba bean plants can also attack with less economical important diseases which occur occasionally (Fahim et al. 1975; Mohamed 1982). These diseases are:

Stemphylium leaf spot caused by *Stemphylium botryosum* Wall.
Alternaria leaf spot caused by *Alternaria solani.*
Downy mildew caused by *Peronospora viciae* (Berk).

13.2.2 Root Diseases

Root diseases have been recognized as serious diseases affecting faba bean plants in Egypt (EL-Gantiry et al. 1994; Atwa 2016; Khalifa 2016). Increasing numbers of rooted plants (disease incidence) reduced subsequently total yield production of faba bean. Yield loss caused by root diseases estimated with 40% by (Salt 1981). The diseases have been reported to be caused by various soil-born fungi rather than a particular distinct fungus. Therefore, it is often described as complex pathological conditions (Root-rot/wilt disease complex).

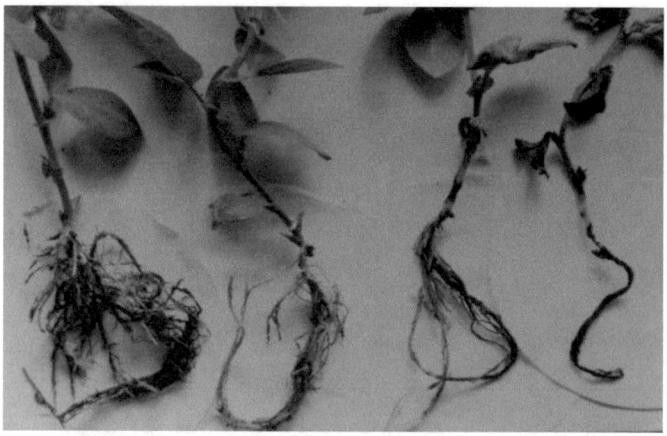

Fig. 13.4 Symptoms of faba bean root rot disease (Omar 1984)

13.2.2.1   Root Rot

The fungal pathogens associated with the disease were identified as *Rhizoctonia solani, Fusarium solani, Fusarium avenacerum, Fusarium oxysporum, pythium sp, Microphomina phasolina, Sclerotium rolfsii and Sclerotinia sclerotiorum* (Salt 1981; Mohamed 1982; Khalifa 2016). The survey conducted by EL-Gantiry et al. (1994; Atwa 2016) revealed that the disease recognized in most governorates grown faba bean in Egypt.

The fungal pathogens attack the plants at seedlings stage causing pre- and post-emergence damping-off. Also, infect adult plants showing a progressive necrotic breakdown of the cortex of tap root and laterals (Fig. 13.4).

The environmental conditions predispose the plants to root infection can be summarized as follows: high temperature, imbalance of water, plant nutrient deficiencies and adverse soil conditions (Salt 1981).

13.2.2.2   Wilt

The causal organism that causes wilt disease in faba bean is *Fusarium* oxysporium *f. sp. fabae* Yu and Fang. The fungus interplant roots through wounds, the presence of nematode in soil may facilitate interring the pathogen (Hassanein et al. 2001). The fungus grows rapidly and extensively in vascular cylinder preventing the flow of water and nutrient from the root to the shoot system. The leaves turn yellow then brown and finally damaged. Cross and/or longitudinal section in the wilted plants (root and stem) show clear distinguish brown pigments in the vascular tissues (Fig. 13.5).

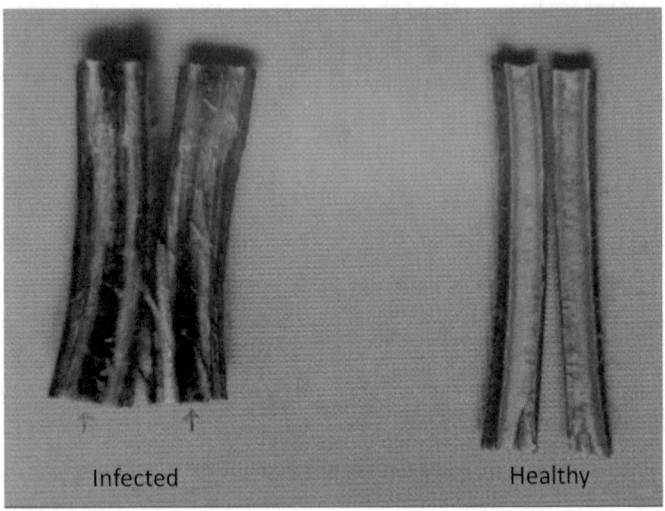

Fig. 13.5 Longitudinal section in the faba bean wilted plant (Omar 1984)

Control

1. The most effective control measures to root fungal diseases (seedling damping off, root rot and wilt) is using fungicides seed treatment. Rizolex-T, Vitavax-200 and Benlate at the recommended dose (2 g/kg seeds) before sowing gave sufficient protection against the pathogenic soil-borne fungi (Khaled et al. 1995; Anonymous 2017a).
2. Growth regulators "Jasmonic acid" application has a positive effect on the defense of faba bean against Fusarium wilt disease (Ahmed et al. 2002).
3. Omar (1993) found that use of biological control "*Pseudomonas cepacia*" offered significant effect against *Rhizoctonia solani* and *Pythium ultimum* on some legume crops.
4. Rouging the infected seedlings and rooted as well as wilted plants at the early growth stage from the field reduce spread infection units to neighboring healthy plants.
5. Removal of plant residues at the end of the season.
6. Atwa (2016) found that induction of resistance using salicylic acid (5 mM) reduced disease incidence of damping-off and root rot in faba bean plants.
7. Application of the recommended agricultural practices (low sowing rates) might minimize the percentage of infected faba bean root diseases.
8. Crop rotation is the most practical solution for the management of root diseases.

New Disease

More recently, Teferi et al. (2018) reported a new disease emerged in Tigray, Ethiopia infected faba bean. The disease is "gall" (*Olpidium viciae* kusano). Further details studies about this disease are undertaken especially in controlling the disease using fungicidal treatments.

13.3 Viral Diseases

Faba bean is known as highly susceptible to viral diseases. Yield loss caused by viral diseases considerable high. These viruses transmitted by different ways, i.e. insect, mechanically, parasitic plant or by seeds. It is well established that, early virus infection reduces significantly growth and yield of the plants Mazyad et al. (1975) found that broad bean inoculated with BBTMV at the seedling stage yielded 38% fewer seeds than uninoculated plants. Survey of faba bean viral diseases in Egypt reported about 20 viruses distributed in most cultivated areas with differences in their disease severity (Tolba 1980; Makkook et al. 1994; Abdel-Salam and EL-Sharkawy 1996; Khattab 2002; Fegla et al. 2008; Fath-Allah 2010). Differentiations among these viruses based fundamentally on: symptoms, host range, serology, methods of transmission, virus particles features and molecular characterizations. The most major viral disease problems of faba bean in Egypt are reported here in are Faba Bean Necrotic Yellow Virus, Bean Yellow Mosaic Virus, and Bean Leaf Roll Virus, since these viruses are economically important and causing a notable reduction in the yield.

13.3.1 Faba Bean Necrotic Yellow Virus (FBNYV)

FBNYV particles about 18 nm. The virus translated by aphids in a persistent manner. At early virus infection, the symptoms described as yellowish, reduction in chlorophyll, necrosis, and rolling of the leaves as well as remarkable plant stunt. The survey conducted by Makkok et al. (1994) indicated that the hot spot of the virus was in the middle of Egypt (Fayoum, Bani Souf and Minia governorates) caused an epidemic status during 1991–1993 growing seasons. The sever FBNYV infection was responsible for reducing dramatically faba bean grown plants in these regions and probably elsewhere. Extensive work has been done by Fath-Allah (2010) for purification, serological and molecular studies on an Egyptian isolate of FBNYV in order to find out a way(s) for controlling the virus on faba bean plants.

13.3.2 Bean Yellow Mosaic Virus (BYMV)

BYMV widely spread in almost faba bean cultivated areas. Virus particles about 750 nm long, 15–20 nm wide. The symptoms produced can describe as transient vein chlorosis followed by obvious green or yellow mosaic, but usually, there is no leaf distortion. The virus is transmitted rapidly by sap inoculation and by many aphid species in a non-persistent manner. Yield loss caused by BYMV infection on faba bean plants ranged between (17–59%) less than uninfected plants.

The main principle factor which contributed to reducing pods production per plant is possible due to the decline in a number of mature flowers (Senanayake 1983)

13.3.3 Bean Leaf Roll Virus (BLRV)

BLRV belongs to the luteo virus group and has isometric particles about 25 nm in diameter. The virus is persistently transmitted by several aphid species manly *Acyrthosiphon pisum*. BLRV symptoms are a faint intervenes chlorosis in the youngest leaves developed to bright yellow, leaves become thickened and brittle, the margins are rolled upward and inward, the stunt becomes obvious in all infected faba bean plant parts (Fig. 13.6).

Early faba bean infection with BLRV reduces significantly plants and yield components. BLRV infected faba bean reduced yield by about 50–90% compared to virus-free plant.

Fig. 13.6 Symptoms of bean leaf roll virus (BLRV) on faba bean plant (Omar 1984)

13.3.3.1 Control of Viral Diseases

1. In general, controlling virus diseases in the plants, to some extent, is difficult. Sources of resistance in faba bean plants against virus pathogens are limited. However, controlling insect transmitting viruses using insecticides might be an indirect way of checking the spread of the viruses. Cockbain (1983) found that spraying insecticide Dimethoate intervals on faba bean plants decreased notable BYMV disease incidence from 19 to 6%.
2. Early rouging by removing infected seedling plants that show clear virus symptoms may be advisable.
3. For those faba bean viruses transmitted by seeds, ensure that seed crops are kept as a virus-free as possible.
4. Spraying faba bean plants with some plant extracts, inducers and nanoparticle materials may reduce or delay virus infection (EL-Shazly et al. 2017).

13.4 Disease Interaction

Within a field situation for faba bean one, and several possible diseases (viral and fungal pathogens) are endemic on the individual plant (Fig. 13.7).

These combined infections (virus–fungus); (virus–virus); (fungus–fungus); (nematode–fungus); (bacteria–fungus) are more likely to cause heavy yield reduction than one particular pathogen. The intensive studies conducted by Omar (1984; Omar et al. 1986; Hilal et al. 2016) indicated that faba bean previously infected with

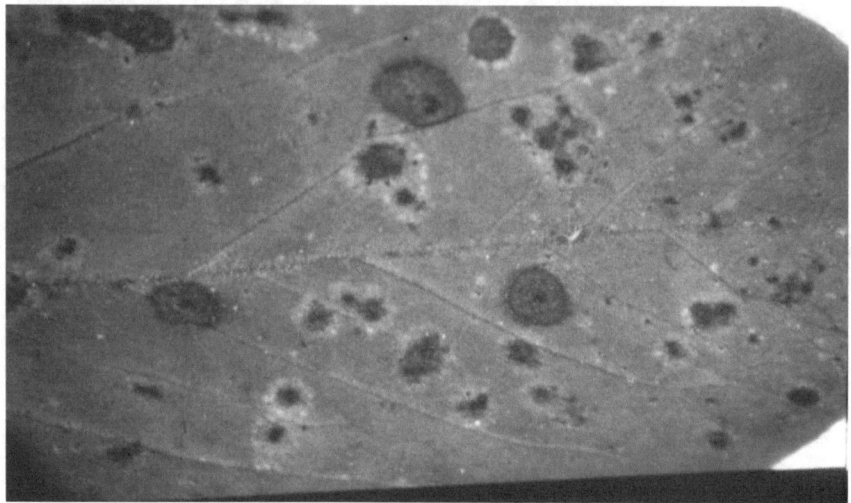

Fig. 13.7 Combined of different fungal pathogens on faba bean leaves (Omar 1984)

BYMV or BLRV render the plants to be more susceptible to *Botrytis fabae* infection (Table 13.5). This synergistic effect between viral and fungal diseases (BYMV and/or BLRV with *B.fabae*) led to severe disease severity causing remarkable damage to the plant and subsequently cause a high significant reduction in plant growth and yield components than each pathogen alone. Moreover, the quality of yield was also affected as a result of the diseases manifested by reducing protein content in the seeds by about 4%.

Omar (1984) demonstrated that presence of rust fungus in half leaflets enhanced the development of chocolate spot fungus on the other half as manifested by increasing lesions grade and much more spreading compared with the control (Table 13.6).

Also, Cockbain (1980) found that field bean inoculated with BBSV or BBTMV yielded 44% and 41% less than uninoculated ones; plants inoculated with both viruses yielded 62% less. The synergism effect between BYMV and root rot fungus (*Rhizoctonia solani*) was observed by Salt (1981) and confirmed by Omar (1984) on faba bean plant. He found that double infection with BYMV and root rot diseases reduced plant growth than virus-free plants (Table 13.7).

The interaction between root-knot nematode (*Meloidogyne incognita*) and *Fusarium oxysporum* in faba bean was studied by Hassanein et al. (2001). They

Table 13.5 Lesion development of *B.fabae* and *B.cinerea* on detached leaves of healthy or virus—infected faba bean (Omar et al. 1986)

	Infection grade after 3 days		Lesion spread after 7 days (mm)	
Leaves	*B. faba*	*B. cinerea*	*B. faba*	*B. cinerea*
BYMV experiment				
Virus-free	48.1	11.2	2.1	0
BYMV-infected	66.5**	21.4**	3.0*	0
BLRV experiment				
Virus-free	53.2	10.5	3.2	0
BLRV-infected	84.5**	19.3**	5.3*	0

*, **Values significantly greater than virus-free controls at $p \leq 0.05$ and $p \leq 0.01$, respectively

Table 13.6 Placement of *B. fabae* spores from rust infection on faba bean leaves (Omar 1984)

Treatment	Infection grades 3 days after inoculation	Lesion spread (mm) 7 days after inoculation	Rate of lesion spread per day (mm)
Un-rusted half leaflet (control)	[a]56.8	3.1	1.2
Rusted half leaflet	71.5	4.4	1.9
LSD 0.05	8.9	0.8	0.3

[a]values are average of 36 infection sites

Table 13.7 Effect of BYMV and *R. solani* individually and in a combination with field bean growth (Omar 1984)

Treatments	Shoot length (cm)	Shoot fresh weight (g)	Shoot dry weight (g)	Root length (cm)	Root fresh weight (g)	Root dry weight (g)	Nodulation number
Healthy	[a]28.8	7.4	2.3	26.0	6.4	2.1	28.2
BYMV infected	25.9	5.5	1.5	23.2	4.7	1.3	25.3
R. solani infection	21.4	3.5	0.3	21.0	2.8	0.3	10.4
Plant infected with both pathogens	17.6	2.9	0.1	16.6	1.9	0.1	7.1
L.S.D.0.05	2.9	2.1	1.7	3.1	1.7	1.0	1.3

[a]Average of 8 replicates

found synergistic interaction occurred between the Pathogens, resulting in a significant reduction in plant growth. On the other hand, the antagonistic effect between microorganisms can also detect. The work of Trabelsi et al. (2017) revealed that nodules and roots of *Vicia faba* are inhabited by quite different populations of associated bacteria

13.5 Conclusions

It can be concluded that faba bean grown in the field came under the bombardment of wild arrays of the pathogen(s) causing various diseases to the plants. These diseases are affecting greatly productivity of the yield both quantity and quality. Such an effect depends mainly on the aggressiveness of the pathogens, the susceptibility of the host associated with favorable environmental conditions. It is worth mention that some disease control measures are available to diminish disease severity.

13.6 Recommendations

The safe, economical and stable mean of controlling faba bean pathogens is breeding for disease resistance. The breeding program showed involving single and multiple disease resistance (since the plant can infect by more than one pathogen). This achieved by corroboration between plant breeders and pathologists of faba bean plants.

Other control measures, i.e. fungicides, biological control, induce disease resistance, growth regulators, plant extracts, and agricultural practices also to be considered.

Acknowledgements The author is indebted to Dr: Maali S. M. Soliman and Dr: Saieda S. Abd EL-Rahman of Plant Pathology research Institute ARC, for their helpful assistance during which this chapter was completed.

References

Abdalla MMF (1979) The origin and evolution of *Vicia faba* L. In Proceedings 1st Mediterranean Conference of Genetics, pp 713–746

Abdel-Salam AM, El-Sharkawy AM (1996) The use of monoclonal and polyclonal antibodies for the detection of an Egyptian isolate of faba bean necrotic yellow virus (FBNYV) in faba bean tissues. Bull Fac Agric Cairo Univ 47:355–368

Ahmed HFS, EL-Araby M, Omar SA (2002) Differential effect of Jasmonic acid on the defense of faba bean against Fusarium wilt. Modulation of other phytohormones and simple phenols. Int J Agric Biol 4:447–453

Anonymous (2017a) Field Crops. Statistics Year Book. Egypt, Ministry of Agriculture and Land Reclamation

Anonymous (2017b) Fungicides recommendations book. Ministry of Agriculture and Land Reclamation, Egypt

Atwa MA (2016) Induction of resistance against damping off and root rot diseases in faba bean. Arab Univ J Agric Sci, Ain Shams 24:555–578

Awaad HA, Salem AH, Mohsen AMA, Atia MMM, Hassan EE, Amer MI, Moursi AM (2005) Assessment of some genetic parameters for resistance to leaf miner, chocolate spot, rust and yield of faba bean in F2 and F4 generations. Egypt J. Plant Breed 9:1–15, Special Issue, Proceed. Fourth Plant Breeding Conference, 5 March 2005 (Ismailia)

Baraka MA, Omar SA, Mazen MM, Soltan HH (2008) Enzymatic activities in faba bean plants induced by some biotic and abiotic agents. Egypt J Basic Appl Sci 23:59–68

Baseony A, Abdel Moneim M (2012) Nutritional values of faba bean and its utilization. In Hamdy AH (ed) The faba bean. Agricultural Research Center, Giza, Egypt, pp. 381–416

Biddle AJ, Cattlin ND (2007) Pests, Diseases and Disorders of Peas and Beans: A Colour Handbook. CRC Press, Boca Raton

Bond DA (1979) English names of *Vicia faba* in UK. FABIS 1:5–6

Brown GD (1977) Field bean (*Vicia faba* L) as a potential human food. In: Proceedings Symposium on the Production, Processing and Utilization of the Field Bean (*Vicia faba* L.), pp. 80–87

Cockbain AJ (1980) Viruses of spring sown field bean (*Vicia faba*) in Great Britain. In Bond DA (ed) *Vicia faba*; Feeding value, Processing and viruses, pp 297–308. EEC. Seminar Cambridge 1979 Marting Nijhoff. The Hague, 422 pp

Cockbain AJ (1983) Virus and virus—like diseases of *Vicia faba* L. In. PD Hebblethwaite (ed) The Faba Bean (*Vicia faba* L.). Butterworth, London, pp 421–462

Conner RL, Bernier CC (1980) The inheritance of rust resistance in inbred lines of *Vicia faba*. Phytopathology 72:1555–1557

Eisa NA, EL-Habbaa, Omar SA, EL-Sayed SA, GM, Omar SA, EL-Sayed SA (2006) Efficacy of antagonists, natural plant extract and fungicides in controlling wilt, root rot and chocolate spot pathogens of faba bean in vitro. Egypt J Appl Sci 44(4):1547–1570

EL-Badawy NFA (2008) Biochemical and molecular studies on some genes that induce after infection of faba bean plants by chocolate spot disease. Ph.D. Thesis, Faculty of Agriculture, Cairo University

EL-Gantiry SM, Omar SA, Salem DE, Rahhal MM (1994) Survey, host response and fungicides treatment of root rot wilt disease complex in faba bean. Egypt J Appl Sci 9(7):366–375

EL-Helaly AF (1939) Further studies on the control of bean rust with some references to the prevention of chocolate spot of beans. Technical Bulletin, Ministry of Agriculture. Cairo, Egypt 236 p

EL-Metwally IM, El-Shahawy TA, Ahmed MA (2013) Effect of sowing dates and some broomrape control treatments on faba bean growth and yield. J Appl Sci 9(1):197–204

EL-Shazly MA, Attia YA, Kabil FF, Anis E, Haziman M (2017) Inhibitory effects of salicylic acid and silver nanoparticles on potato virus Y-infected potato plants in Egypt. Middle East J Agric Res 6:835–848

EL-Wakil MA, Abass MA, EL-Metwally MA, Mahmoud MS (2016) Green chemistry for inducing resistance against chocolate spot disease of faba bean. J Environ Sci Technol 9(1):170–187

Fahim MM, Abdou YA, Eisa NA (1975) Effect of some systemic and contact fungicides in controlling Stemphylium leaf spot of broad bean. Annals of Agric Sci Moshtohor J 3:63–69

FAOSTAT (2016) Statistical database of The United Nation Food and Agriculture Organization (FAO), Rome. http://www.Fao.org/faostat/en/#data/QC

FAOSTAT (2018) Statistical database of The United Nation Food and Agriculture Organization (FAO), Rome. http://www.fao.org/faostat/en/#data/QC

Fath-Allah MM (2010) Purification, serological and molecular studies on an Egyptian isolate of faba bean necrotic yellows virus (FBNYV) infected faba bean plants. Egypt J Phytopathol 38(1–2):185–199

Fegla GI, Younes HA, Atta Alla SIM, Efaisho EEM (2008) Host range, insect and seed transmission of two viruses isolated from faba bean in EL-Beheira. Egypt J Viroids 5(2):67–83

Hassanein AM, EL-Hamawi MH, EI-Sharkawy T, Solaiman MSI (2001) Interaction between Root-knot nematode and *Fusarium oxysporum* in faba bean. Egypt J Agric Res 79(2):395–405

Hegab ASA, Fayed MTB, Hamada MM, Abdrabbo MAA (2014) Productivity and irrigation requirements of faba-bean in North Delta of Egypt in relation to planting dates. Icel Agric Sci 59(2):185–193

Hilal AA, Shafie RM, EL-Sharkawy HH (2016) Interaction between Bean Yellow Mosaic Virus and *Botrytis fabae* on faba bean and the possibility of its control by plant growth promoting rhizobacteria. In 13th Congress of Phytopathology, 10–11 May 2016, Giza, Egypt

Hill-Coltingham DG (1983) Chemical constituents and Biochemistry. In Hebblethwait PD (ed) The Faba Bean (*Vicia faba* L.). Butterworth, London, pp 159–180

Honounik SB, Bisri M (1991) Status of diseases of faba bean in the Mediterranean region and their control. Options Mediterraneennes. Serie A: Seminaires Mediterraneens (CIHEAM)

Hossain MS, Mortuza MG (2006) Chemical composition of kalimator, locally grown strain of faba bean (*Vicia faba* L.). Pak J Biol Sci 9:1817–1822

Khaled AA, Abdel-Moity SM, Omar SA (1995) Chemical control of faba bean diseases with fungicides. Egypt J Agric Res 73(1):45–55

Khalifa NAM (2016) Pathological studies on controlling wilt and root—rot diseases on faba bean plants in Egypt and Sudan. Ph.D Thesis Institute Of African Research and Studies. Cairo University

Khalil SA, EL-Hady MM, Dissouky RF, Amer MI, omar SA (1993) Breeding for high yielding ability with improved level of resistance to chocolate spot (*Botrytis fabae*) disease in faba bean (*Vicia faba* L). J Agric Sci, Mansoura Univ 18(5):1315–1328

Khattab EAH (2002) Recent technique to study some broad bean viral diseases. Ph D Thesis Faculty of Agriculture Zagazig University, Egypt

Makkook KM, Rizkalla L, Madkour M, EL-Sherbeeny M, Kumare SG, Amriti AW, Solh MB (1994) Survey of faba bean (*Vicia faba* L) for viruses in Egypt. Phytopathologia Mediterr 3:207–211

Mazyad H, EL-Hamady M, Tolba MA (1975) The broad bean mosaic disease in Egypt. Annals Agricu Sci Moshtohor J 4:87–94

Mohamed HA (1982) Major disease problems of faba bean in Egypt. In G. Hawtin and C. Webb (ed) Faba bean Improvement. Proceeding of the faba Bean Conf. held in Cairo, Egypt 7–11, 1981, pp 213–225, ICARDA, Aleppo, Syria. Martinus Nijhoff/Dr Junk, for the ICARDA/ IFAD Nile Valley Project

Mohamed HA, Omar SA, EL-Gantiry SMM (1996) Interaction between Botrytis isolates and faba bean strains with special references to the effect of diffusion from leaves on the fungus conidia germination. Agricu Res Rev 64(2):233–243

Mohamed NAM (1996) Studies on chocolate spot disease of broad bean and loss occurrence. Ph.D. Thesis Faculty of Agriculture Minufiya University, Egypt

Morsy KM, Tarrad AM (2005) Effect of infection with Botrytis fabae Sard. and mechanical leaf defoliation on yield loss in faba bean. Egypt J Basic Appl Sci 20(11 B):443–454

Nasraoui B (2008) Main fungal diseases of cereals and legumes in Tunisia. Centre de Publication Universitaire, Tunisia, p 186

Negash TT, Azanaw A, Tilahun G, Mulat K, Woldemariam SS (2015) Evaluation of Faba bean (Vicia faba L.) varieties against chocolate spot (Botrytis fabae) in North Gondar, Ethiopia. Afr J Agric Res 10(30):2984–2988

Omar SAM (1984) Disease interaction and host response in Vicia faba L. Ph D Thesis London Univ

Omar SA (1993) Use of Pseudomonas cepacia biocontrol agent against Rhizoctonia solani and Pythium ultimum on some legume crops. Egypt J Appl Sci 8(3):265–274

Omar SAM, Bailiss KW, Chapman GP (1986) Virus induce changes in the response of faba bean to infection by Botrytis. Plant Pathol 35:86–92

Omar SA, Khalil SA, EL-Hady MM (1992) Phytoalexin production in resistant and susceptible faba bean cultivars to chocolate spot disease (Botrytis fabae Sard.). Egypt J Appl Sci 7(1):42–47

Roughley RJ, Sprent JI, Day JM (1983) Nitrogen fixation. In: PD Hebblethwait (ed) The Faba Bean (Vicia bean L.). Butterworth, London, pp 233–260

Salem DE, Omar SA, Aly MM (1992) Induction of resistance using ethephon seed treatments. FABIS 1:29–33

Salem MF, Rizk NM, Omar SA, Abdel Hamed AS (2012) The Potential utilization of alfalfa root saponins in controlling chocolate spot disease in faba bean. J Biol Chem Environ Sci 7(3):157–171

Salt GA (1981) Factors affecting resistance to root rot and wilt diseases. International Conference on Faba bean, Cairo, Egypt 6–11 Mar 1981

Sarhan EAD (2006) Pathological studies on faba bean rust in Egypt. Ph D Thesis. Faculty of Agriculture, Cairo University

Senanayake AHS (1983) Virus infection and the development and yield indeterminate and determinate lines of field bean (Vicia faba L.). MPhil Thesis University of London

Tarrad AM, EL-Hyatemy YY, Omar SA (1993) Wyerone derivatives and activities of peroxidase and polyphenoloxidase in faba bean leaves as induce by chocolate spot disease. Plant Sci 89:161–165

Teferi TA, Weldemichael GB, Wakeyo GK, Mindaye TT (2018) Fungicidal management of the newly emerging faba bean disease "gall"(Olpidium viciae Kusano) in Tigray, Ethiopia. Crop Prot 107:19–25

Tolba MA (1980) A note on bean yellow mosaic virus and other viruses in Egypt. FABIS 2:42

Trabelsi D, Chihaoui SA, Mhamdi R (2017) Nodules and roots of Vicia faba are inhabited by quite different populations of associated bacteria. Appl Soil Ecol 119:72–79

Williams PF (1975) Growth of broad bean infected by Botrytis fabae. J Horti Sci 50:415–424

Part IV
Advanced Procedures in Improving Crop Productivity

Chapter 14
Role of Helium-neon Laser in Improving Wheat Grain Yield Potentiality

Hassan Auda Awaad

Abstract Agriculture is a fertile area for the application of modern technologies to improve the productivity of crop plants. The advantages of laser in agriculture include many aspects which include encouraging plants to tolerate environmental effects and safely mitigate the effects of environmental changes, such as drought, high temperature, salinity and pollution. Also, it biostimulate plant resistance to biotic stresses of disease injuries. Moreover, the application of laser technology is considered a sustainable, secure and clean means to improve growth, yield and quality of crop plants. In wheat, laser led to improve morph physiological, biochemical, enzymatic activity, yield and grain quality. This chapter presents the state-of-the-art on various benefits of helium-neon laser technology when applied to agriculture.

Keywords Wheat · Helium-neon laser · Biostimulation · Plant resistance · Environmental stresses · Yield · Quality

14.1 Introduction

A helium–neon laser or He–Ne laser is defined as a gas laser consists of a mixture of 85% helium and 15% neon inside a small bore capillary tube, which is motivated by a DC electrical discharge. The laser (Light Amplification by Stimulated Emission of Radiation) was discovered by Maiman, 50 years ago, using a flash lamp pumped ruby crystal as the environment (Maiman 1960). The helium-neon laser operates at a wavelength of 632.8 nm, in the red part of the visible spectrum.

Sustainability of agricultural and food security are main challenges to continued population growth. The integration of existing and recent technologies to induction and exploit genetic variability is an important objective in the development of healthy, nutritious and productive crops. Laser is a proven technology for improving crop genotypes with desirable qualities (Suprasanna and Jain 2017).

H. A. Awaad (✉)
Crop Science Department, Faculty of Agriculture, Zagazig University, 44511, Zagazig, Egypt

© Springer Nature Switzerland AG 2021   391
H. Awaad et al. (eds.), *Mitigating Environmental Stresses for Agricultural Sustainability in Egypt*, Springer Water,
https://doi.org/10.1007/978-3-030-64323-2_14

Since the discovery of the laser in the last century, this has made great advances in biological science. Agriculture has had a good deal of laser contribution in promoting and improving productivity and quality. Laser applications in the field of agriculture have contributed to increased tolerance to a biotic and biotic stresses and enhancing yield of crop plants. Joanna et al. (2016) showed that genetic variation is considered as a source of phenotypic diversity and is a main engine of evolutionary diversification. Heritable variation was observed and utilized thousands of years ago in the improving plants and animals. The laser-induced differences were widely used to study the genome of plant species and improve many crop geniuses like Arabidopsis, barley, soybean, tomato and wheat.

The use of physical mutagens like laser beams for inducing variation is well established and helps to develop of many agronomical important traits.

It should be noted that the use of safe physical methods to promote physiological processes and improve plant growth has become more widespread due to the least harmful effect on the environment. Govindaraj et al. (2017) stated that, physical factors can be used to get positive biological changes in crop plants without affecting the ecology. It initiates physiological and biochemical changes, which reflect the plant growth and development processes and ultimately improve the yield and quality of products.

Laser irradiation is considered as a new approach in agriculture. Laser is one of the physical sources for encouraging a biostimulation effect and genetic changes in crop plants. Lower doses of laser stimulate plants and increasing bioenergetical ability of the cell and stimulation their biological and physiological processes. Higher doses of the laser affect the genetic makeup leading to genetic changes in plant traits. It is successfully used in the field of molecular biology, plant genetics and crop breeding (Rybiński 2000). Helium-neon laser applied under field conditions to improve morphological parameters and the yield of wheat crop (Chen and Han 2014). Therefore, this chapter was prepared to focus more light on the helium-neon laser and its advantages and applications on crop plants, especially wheat in improving yield and quality characteristics.

14.2 Laser Light Properties

Svelto (2010) explained the properties of lasers in the following points:

1. The light is extremely monochromatic and characterized by wavelength of 632.8 nm.
2. The light has a high coherence property. The distance in which the waveform still analogous to a sine-wave is named the coherence-length of the beam. It is commonly around 10–30 cm in case of the helium-neon laser.
3. The light is unidirectional and affiliated.
4. Light is characterized by being spatially coherent, and the phase of radiation is constant over the cross-sectional width of the beam.

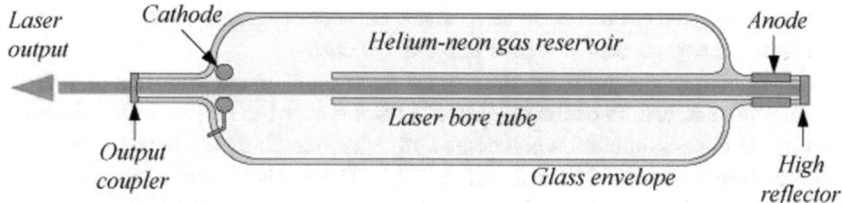

Fig. 14.1 Schematic diagram of a helium-neon laser. *Source* https://upload.wikimedia.org/wikipe dia/commons/a/af/Hene-1.png, viewed 14 July 2016

5. Manufacturers often dearly insert a Brewster window into the laser for light flux with a specific case of linear polarization (LASERS.web.physics.ucsb.edu/~phys128/experiments/laser/LaserFall06.pdf University of California, Santa Barbara Physics Department, Senior Lab).

14.3   Helium-neon Laser Design and Operation

A helium–neon laser is one of the most common and inexpensive lasers types in production (Andrews and Phillips 2005). The gas reservoir consists of a mixture of 85% helium and 15% neon at 1/300 atmospheric pressure inside of a small bore capillary tube, regularly excited by a DC electrical discharge. The most famous and most used helium-neon laser operates at a wavelength of 632.8 nm, in the red portion of the visible spectrum. Figure 14.1 shows the basic construction of a helium-neon laser. The collimation of the laser beam is accomplished by the reflectors at each end of the laser.

14.4   Mode of Action of Helium-neon Laser Biostimulation

The procedure of generating He and Ne in specific excited situations is famous as pumping. In helium-neon laser, pumping happens through collision of electrons and atoms throughout an electrical release. The 632.8 nm optical light is discharged via the laser when the Neon atoms degeneration from an excited state to an normal state. The light or photons released from the neon atoms will transfers back and forth between two mirrors or output coupler to create laser light (Anonymous 2018). The rest of the radiation remains in the tube to continue the release and laser production. Neon atoms are excited by collisions with helium atoms, which in turn are induced by electrical discharge inside the tube and so on. This procedure of photon release is named stimulated emission of energy. The neon excited electrons continue on to the ground state through radioactive and nonradioactive transitions, it is important for the continuous wave operation. The light emitted from the helium-neon laser is very monochromatic, unidirectional, spatially coherent and aligned to be parallel to the laser body. Red helium-neon laser have a number of technical uses in the field of agriculture due to the ease of operation and low cost.

The laser has a significant impact on agronomic science as a tool for biostimulation. The results of the reference studies showed the effect of low intensity laser light in biostimulation when treated on seeds, seedlings and plants (Vasilevsky 2003; El Tobgy et al. 2009; Perveen et al. 2010). The mechanism of biostimulation of the laser in the plant is explained by the synergy among the polarized monochromatic laser beam and the photoreceptors, leading to the activation of many biological reactions (Bielozierskich and Zolotariewa 1981; Karu 1989; Koper et al. 1996). Where plants absorb light through their photoreceptors, which stimulates the physiological and chemical processes of the plant cells and increase their bioenergetics potential (Vasilevski et al. 2001).

14.5   Advantages and Disadvantages of Helium-neon Laser

14.5.1   Advantages

Helium-neon laser emits laser light in the visible portion of the spectrum.
High stability level
Relatively low cost
Works without injuries at the higher levels of temperatures.

14.5.2   Disadvantages

Low efficiency
Low gain
Helium-neon laser is limited to low energy tasks.

14.6   Objectives of Using Helium-neon Laser in Agriculture

Laser irradiation is considered as a recent physical source for inducing a stimulation effect and changes in the genetic makeup of crop plants. Laser applications in the agricultural field should not be ignored, especially with the increasing importance of agriculture in the current decades and the tendency of most countries towards food security. It was necessary to pay attention to the importance of laser in agriculture to solve many of the problems related to agricultural production. Laser applications in agriculture can be classified as follows (Anonymous 2007):

1. Accelerate seed germination and stimulate growth and early flowering.

2. Used as pesticide, it kills insects, bacteria, viruses and others at any stage of its growth and development. It penetrates plants or seeds without causing any harm. At the same time, it deals with insects, diseases and others easily and sterilized and eliminates the risk of pathogens.
3. Identification of weeds in the fields and their control by short pulses of lasers in the field under sustainable agriculture activities aimed to avoid the use of pesticides as much as possible.
4. Improving seed quality and produce pure and sterile seeds.
5. Improving the ability of crop varieties to adapt with climate changes and environmental stresses *i.e.* drought, salinity and UV-B stress.
6. Increasing agricultural production without high costs.
7. Improving the quality of crops such as protein, fat and vitamins.

14.7   Applications of Helium-neon Laser in Improving Crop Plants

The use of physical mutagens like laser beams in crop plants was found to improve germination and growth of the plants. It was also noted that the laser leads to induce genetic variability and development of many important agronomic traits. In barley, Rybiñski (2001) used laser beams combined with chemo mutagen (MNU) for inducing mutation frequency in spring barley. Irrespective of the variety, a short period exposure to laser light (30 min) caused a bio stimulation effect on the yield attributes traits in the M_1 plants. They recorded effects of laser and MNU on the percentage of chlorophyll mutants frequency in the M_2 population. Prolonging irradiation period to 120 min caused somatic damage of plants expressed by decrease of the traits. Chemo mutagen-N-methyl-N-nitrosourea (MNU) induced a great degree of somatic damages, but this effect was mitigated after exposure for a longer period to laser light. He added that laser light can cause genetic variation in plants with lower levels of somatic injury than MNU.

Furthermore, the influence of laser light on photosynthetic parameters activity and growth characters was detected in doubled haploid lines of spring barley by Rybiñski and Garczyñski (2004). They irradiated barley seeds with helium-neon laser (helium-neon laser) in red light spectrum at wavelength of 632 nm. Laser light increased the flag leaves significantly rather than the control, and exposure to 180 min of irradiation was more effective rather than 60 min. The photosynthetic coefficient of water use was greater for laser light combinations, it was improved with longer exposure to laser light, especially for flag leaf.

In rice, Abo-Hegazi (2007) showed that laser technique could help the breeders and geneticists in selection breeding programs for the important characters in field crops. He utilized laser as a good non-destructive tool for screening rice genotypes for the color of endosperm of the paddy kernel without dehulling as a specific trait for quality of marketing. On the other hand, Zhang et al. (2014) detected enhancement of chlorophyll content and Rubisco enzyme activity in leaves of rice group treated by helium-neon laser, hereby improve the photosynthetic efficiency than the control. Helium-neon laser can repair damage of enhanced UV-B to rice.

In oat, Drozd and Szajsner (2007) applied three doses of semi-conductor laser radiation on two bare-grained oat genotypes, Akt and Polar as well as STH 6102 and STH 6503 lines. The investigated doses induced significantly the stimulation of coleoptile length over 20% rather than control. Also significant increase in germ root length of Akt cultivar was noticed after laser irradiation.

Legume crops played an important role as protein source in sustainable agriculture of different countries. Laser had encouraging effect on the growth, development and yield of legumes. Appropriate doses of He-Ne and CO_2 accelerate plant growth and metabolism. Qi et al. (2000) exposed the embryos of broad bean seeds to He–Ne laser or CO_2 laser radiation. Then subjected to 1.02, 3.03 or 4.52 kJ m^{-2} UV-B radiation, respectively, at 70 μmol m^{-2} s^{-1} photosynthetically active radiations (PAR) under growth cabinet. Laser pretreatment of embryos improved UV-B stress resistance in the epicotyls of the broad bean through increasing the content of Abscisic acid and UV-B absorbing compounds. Thus increase in stress resistance achieved in the broad bean. When comparing the effect of the laser on faba bean and lupine crops, Podleśny et al. (2012) exposed seeds of faba bean var. Albus and white lupine var. Katon to three levels of a helium–neon laser, D0-no irradiation, D3-triple irradiation, D5-fivefold irradiation. Irradiated seeds showed significant increase in amylolytic enzymes activity of both crop plants. Treatment of 120 h from the time of sowing produced the greatest differentiation of the enzymatic activity. Indole-3-acetic acid IAA content was found to be greater in irradiated seeds than the control. The activity of IAA was slightly higher in faba bean than in white lupine seeds. The irradiated seeds of white lupine and faba bean exhibited greater fresh weight at the time of imbibition than the control. Treated white lupine seeds with laser beams was induced enlarged their mass after 12, 24, 48, 72, 96, 120 and 144 h from the time of sowing by 1.7, 2.4, 3.3, 3.8, 4.6, 5.24 and 5.9 g $\times$ 10 seeds^{-1} and the faba bean seeds by 2.9, 3.7, 4.6, 5.4, 6.1, 6.4 and 7.0 g $\times$ 10 seeds^{-1}, respectively (Fig. 14.2).

In lentil, Drawish et al. (2013) treated dry seeds of two Egyptian lentil cultivars Sinai 1 and Giza 370 with three wave lengths of Diode Laser 408 (UV), 650 (VIS) and 980 (IR) nm. Each diode laser type was used with 6 exposure periods. The UV and VIS were used for 1, 5, 10, 15 and 20 min and the control. However, IR irradiation was applied for 0.5, 2, 3, 4, 5 min and control. They registered significant differences between the two cultivars for all traits, and responded variably from exposure period to another of the three laser types. Both seedlings emergence and survived plants % significantly influenced by laser types. Exposure periods (EP) affected all traits significantly, except survived plants %. Laser types x EP interaction had significant effect on seedlings emergence and survived plants %. Sinai 1 treated with the three investigated laser types produced higher growing and survived plants than untreated ones, and Giza 370 only on survived plants with UV and VIS. All laser treatments improved seed yields of both lentil cultivars.

Podleśna et al. (2015) showed that exposure seeds of pea cultivar Ramrod to helium-neon laser resulted in improving the germination rate and uniformity and lead to modified growth stages. Thus laser acceleration of flowering and maturity.

Fig. 14.2 Dynamics of mass change of faba bean (solid lines) and lupine (dashed lines) imbibing seeds treated (filled circles) and non-treated (empty circles) by laser light (Podles´ny et al. 2012)

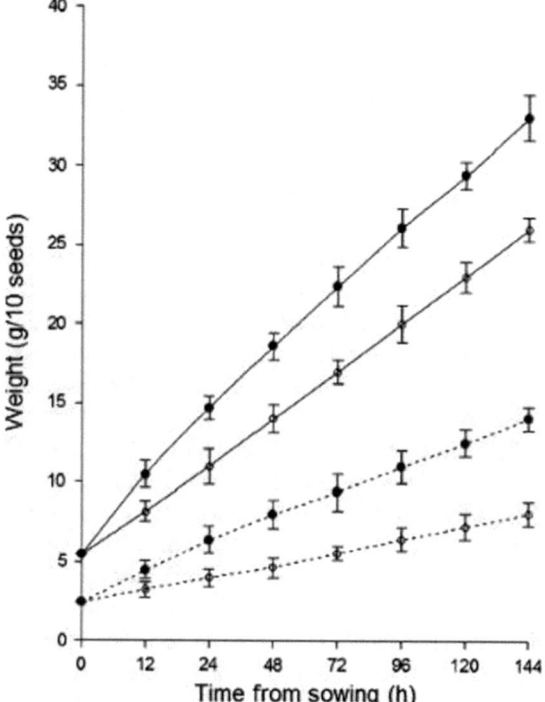

Helium-neon laser light improved the amylolytic enzymes activities and indole-3-acetic acid (IAA) content in seeds and seedlings of pea. They also found that laser light improved plant height, leaf surface area, vegetative and reproductive organs yield. Stimulation effect by low-intensity laser irradiation in pre-sowing on crop seeds has been established in several studies using promoting growth and yield of sweet basil plant (El-Kereti et al. 2013).

In cotton, Delibaltova and Ivanova (2006) irradiated seeds of some cotton varieties by helium-neon laser of 623.8 nm wavelength and power 20 mW. at 4, 6, 8 and 10 times and control (C). They showed that helium-neon laser pre-sowing irradiation of cotton seeds did not affect the plant development stages. A stimulation effect has been observed on the rate of growth of high stem length and flowering dynamic of cotton cultivars. Maximum stimulating effect was attained by 8 times treated seeds. Irradiated genotypes were characterized by high rate of growth of high stem length and flowering dynamic.

In soybean, Khalifa and El Ghandoor (2011) exposed Egyptian soybean cultivar Giza 111 of pre-sowing laser treatment at wave length $\Lambda = 532$ nm. for 5, 10, 30, 60 and 120 min. They registered gradually increase in growth potential of 4-week old seedlings with prolonging laser treatment. Protein patterns appeared comparative increased levels of a ~ 55 KD protein in alliance with the pre sowing exposure time to laser irradiation, as the large subunit of the enzyme 1, 5 ribulose bisphosphate carboxylase/oxygenase (RubisCO). They also observed enhancement of growth vigor of soybean seedlings with high level of RubisCO enzyme.

In canola, laser had a positive effect on yield components and led to increase tolerance plants to stress environmental conditions. Mohammadi et al. (2012) showed that laser radiation applied for 45-min had the greatest effect on yield and yield traits and significantly reduces the negative effects of salinity. Whereas, application of laser radiation for 60 and 75 min, resulted in a dramatic reduction in yield and its components.

In sugar beet, Prośba-Białczyk et al. (2013) found that germination, germinal roots and hypocotyls length were increased after seed stimulation by laser light. A positive stimulation was observed on the morphological characters for laser irradiation applied in a three-fold dose. The irradiation of seeds with laser light has a positive change in chlorophyll a to b ratio. Seed stimulation also had a positive effect on yield and sucrose content.

In medical and aromatic plants, El Tobgy et al. (2009) exposed dry and wet seeds of anise and cumin to helium-neon laser at 5, 10 and 20 min with power density of 95 mW/cm. They observed enhancement in growth parameters, yield and its attributes i.e., plant height, number of branches/plant, number of umbels/plant and fruits yields/fed. (Feddan $= 4200$ m^2), essential oil content and oil yield/fed of both anise and cumin plants due to exposure time for 20 min especially in wet case. Maximum phosphorus content was recorded at 20 min in wet case

In tall fescue, helium-neon laser illumination was found to be alleviated the inhibitory effect of cadmium sulfide nanoparticles in tall fescue seedlings. Helium-neon laser also improved seedling development through activation antioxidant antioxidants biosynthesis and antioxidant enzyme. Moreover, phytochelatins and glutathione induced by laser was also proved to be involved in reducing the cytotoxicity of CdSNPs (Mei et al. 2016).

Furthermore in radish, helium-neon laser light irradiation caused significantly effects on the final germination percentage and germination index when compared to the untreated seeds. These results depended on the germination temperature (Muszyñski and Gadyszewska 2008).

14.8   Applications of Helium-neon Laser in Improving Wheat Grain Yield Potentiality

Germination
Plant physiology
Plant Morphology
Grain yield potentiality
Quality
Tolerance to environmental stresses.

Wheat is often called as the king of cereals. It is considered the main source of food crops in the world and Egypt. Raising wheat production through improving the local cultivars by conventional and mutation breeding methods is the most important national target to minimize the gap between the production and consumption in Egypt.

Where, the area allotted to wheat crop in Egypt was 1.32 million hectares gave total production 8.45 million metric tons in 2017/2018 (USDA 2019). However, the total production is far than that required for local consumption which is up to 15 million tons per year. Great efforts have been made to increase food and agricultural production, mainly through developing new cultivars characterized by high yielding potentiality and better quality to fill the gap in wheat production. This is the target of wheat breeders where they use biotechnological techniques besides the traditional breeding methodology.

The major purpose of using laser has been to induce genetic variability, which is the principal step in any breeding programme. Since wheat crop is a polyploidy plant, duplication of genes permits a higher number of induced variants to be preserved and transferred to the subsequent generation. So, the segregating populations have been exploited to improve wheat genotypes have proper combination of pest resistance, plant height, fertile tiller, spike length and grain weight. These characteristics could be genetically manipulated using helium-neon laser to produce maximum yield stability under stress conditions. Where, the induction of mutations from the point of view of plant breeders has been an important means for over 70 years to increase genetic diversity and produce new promising mutants of improved traits (IAEA 2017).

So, the application of helium-neon laser in wheat improvement could be discussed in the following respects.

14.8.1   Germination

The impact of Laser treatment on germination dynamics has been recorded by Nenadic et al. (2008). They found that laser treatment was operative in improving sprout efficacy and decreasing fungus survival, while less effective compared to fungicide treatment. This reinforcing the importance of using laser as an environmentally safe alternative to fungicide treatment for various seed genotypes.

Seeds pre-sowing helium-neon laser irradiation was found to be improved the germination parameters in wheat. Abu-Elsaoud and Tuleukhanov (2013) exposed seeds of wheat genotypes from Kazakhstan (Aksay, kazakhstanskaya-10 and eretrospermum-350) and from Egypt (Giza-168) to helium-neon laser irradiations with 5 mW.mm^{-2} power intensity for to IR laser at 0, 1, 3, 10, 30, 60, 180, 600, 1200 and 1800s exposition time. Wheat seed pre-treatment by helium-neon laser gave significant increase in germination percentage, germination rate, relative germination rate coefficient and improving various growth parameters. Various doses of helium-neon laser enhanced germination to 100%. Helium-neon laser caused major variations in levels of hydrogen peroxide, lipid peroxidation and Cu-Zn superoxide dismutase enzyme activity.

14.8.2 Plant Physiology

Under Egyptian conditions, an increase in Cu/Zn superoxide dismutase activity, germination percentage, shoot and root lengths was registered after irradiated wheat seeds of Egyptian common wheat cultivar with different doses of Helium-neon laser ranging from 2 to 3600s (Abu-Elsaoud 2013). Whereas, Chen and Han (2014) explored the effects of continuous Helium-neon laser irradiation (632 nm, 5 mW mm^{-2}, 2 min d^{-1}) on the physiological parameters of wheat seedlings treated with an enhanced UV-B radiation (10 KJ m^{-2} d^{-1}) at the early growth stages. Wheat seedlings were exposed to enhanced UV-B, helium-neon laser treatment or a combination of the two. Chlorophyll content and chlorophyll fluorescence were improved in UV-B-exposed wheat exposed to helium-neon laser irradiation. The parameters of treated seedlings appeared to be near the control levels and the enzyme activities improved. This result revealing that helium-neon laser treatment partly alleviates the damage caused by enhanced UV-B irradiation. Furthermore, the use of helium-neon laser alone had a favorable effect on seedling photosynthesis rather than the control.

14.8.3 Plant Morphology

Guo et al. (2017) developed hexaploid winter wheat mutants through high phenotypic variations, included by mutants. They added that this mutated genetic variation included spike, leaf and seed morphology, plant architecture, and days to heading. It can be used as a resource of broad-spectrum phenotypic and genotypic differences that may be useful in wheat improvement, gene discovery, and functional genomics.

When Abaza Ghada et al. (2017) exposed three Egyptian bread wheat cultivars namely Sids 12, Sakha 94 and Gemmeiza 9 to 0.0, 1.0, 1.5 and 2 h laser (Table 14.1), Sids 12 produced the shortest plant height (76.96 cm), otherwise Gemmeiza 9 was the tallest (89.26 cm) one, whereas, Sakha 94 located in between them with value of (83.07 cm). Spike characteristics i.e. spike length, No. of fertile spikelets/spike and No. of sterile spikelets/spike exhibited significant differences between bread wheat cultivars. The maximum and the minimum values for such characters were 13.79 cm (Sids 12) and 12.13 cm (Sakha 94) for spike length; 21.00 (Gemmeiza 9) and 17.90 (Sakha 94) for No. of fertile spikelets/spike as well as 3.29 (Sids 12) and 2.47 (Gemmeiza 9) for No. of sterile spikelets/spike.

Significant increase in plant height was observed due to laser treatments compared to the control. The highest effect was observed at 1 h (85.75 cm) followed by 1.5 h and 2 h with values of (82.32 cm) and (81.93 cm) in descending order, respectively. All tested treatments caused significant increase in spike length rather than the control

Table 14.1 Mean performance of plant height and spike characteristics for Sids 12, Sakha 94 and Gemmeiza 9 wheat cvs. as influenced by laser irradiation treatments and their interaction in M_2 generation (Abaza Ghada et al. 2017)

Main effects and interaction	Plant height (cm)	Spike length (cm)	No. of fertile spikelets/spike	No. of sterile spikelets/spike
Cultivars				
Sids 12	76.96	13.79	19.00	3.29
Sakha 94	83.07	12.13	17.90	2.87
Gemmeiza 9	89.26	13.17	21.00	2.47
F. test	*	*	*	*
L.S.D $_{0.05}$	1.166	0.8701	1.619	0.0979
Laser irradiation treatments				
Control	78.34	12.45	19.00	2.83
1 h	85.75	13.15	23.30	2.45
1.5 h	82.32	13.28	20.00	2.95
2 h	81.93	13.26	20.00	3.22
F. test	*	*	*	*
L.S.D.$_{0.05}$	1.947	0.5818	1.1570	0.1100
Interaction A × B	*	*	N.S	*

*, Significant at 0.05 level of probability N.S., not significant

without significant differences between all laser treatments (Table 14.1). Whereas, 1 h treatment resulted in pronounced effect on No. of fertile spikelets/spike with value of (23.30) compared to both 1.5 and 2 h period which gave the same effect valued (20.00). In contrast, 1 h treatment reduced significantly No. of sterile spikelets/spike compared to the control. While, it was increased from 2.95 to 3.22 with prolonged exposure periods from 1.5 to 2 h, respectively. An increase in spike length and No. of fertile spikelets/spike after mutagen treatments has been was also obtained by Drozd and Szaysner (1999).

Interaction between cultivar × laser treatments was significant for plant height and spike characteristics, except for No. of fertile spikelets/spike which was insignificant. At this moment, these cultivars had differential responses to laser irradiation treatments. Furthermore, Chen and Han (2014) showed that single helium-neon laser irradiation yielded improved morphological parameters, while the combined helium-neon laser and improved UV-B radiation group displayed rehabilitation effects.

14.8.4 Grain Yield and Its Components

It is interest to mention that, the laser offers a pure environmental source of energy that ensures high production, hence the laser started to be used as a biostimulator in plant yield. High level of genetic variability between wheat genotypes under laser treatments has been recorded by Nenadic et al. (2008) and showed the impact of laser treatment on dynamics of germination and wheat characters. In continuous, Abaza Ghada et al. (2017) subjected seeds of three wheat cultivars (Sids 1, Sakha 94 and Gemmeiza 9) to laser treatments viz 0.0, 1 h, 1.5 h and 2 h. They recorded a significant variability between wheat cultivars for all studied yield characters under laser treatments. This result revealing the great effect of the laser on performance of wheat cultivars. Wheat cultivar Sakha 94 produced the biggest number of spikes/plant (10.75), otherwise, Sids 12 gave the smallest (3.31) number as well as Gemmeiza 9 attained value of (9.25). Wheat cultivar Sids 12 produced the highest number of grains/spike (75.16) and 1000-grain weight (49.64 gm). Otherwise, Sakha 94 produced the lowest number of grains/spike and 1000-grain weight with values of 56.42 and 38.68 gm, respectively. Whereas, Gemmeiza 9 exhibited relatively moderate values of 60.97 and 41.66 gm for the same studied characters, in the same respective order. Meanwhile, Gemmeiza 9 produced the heaviest grain yield/plant (18.00 gm), otherwise Sids 12 gave the lowest productivity (10.32 gm), while Sakha 94 was moderate between them (14.59 gm) as given in Table 14.2.

Table 14.2 Mean performance of yield and its components for Sids 12, Sakha 94 and Gemmeiza 9 wheat cvs. as influenced by laser irradiation treatments and their interaction in M_2 generation (Abaza Ghada et al. 2017)

Main effects and interaction	No. of spikes/plant	No. of grains/spike	1000-grain weight (gm)	Grain yield/plant (gm)
Cultivars				
Sids 12	3.31	75.16	49.64	10.32
Sakha 94	10.75	56.42	38.68	14.59
Gemmeiza 9	9.25	60.97	41.66	18.00
F. test	*	*	*	*
L.S.D $_{0.05}$	0.709	3.336	0.953	1.577
Laser irradiation treatments				
Control	7.26	63.00	40.56	13.67
1 h	7.08	65.78	44.22	11.94
1.5 h	8.47	63.03	45.07	15.78
2 h	8.37	64.20	43.40	15.57
F. test	*	*	*	*
L.S.D $_{0.05}$	1.106	2.036	1.488	1.694
Interaction A × B	*	*	*	*

*, Significant at 0.05 level of probability

Laser irradiation treatments induced an increase in No. of spikes/plant due to exposure periods of 1.5 h and 2 h valued (8.47) and (8.37) compared to the control (7.26). While, No. of grains/spike showed an increase due to 1 h, followed by 2 h with values of (65.78) and (64.20), respectively compared to the control valued (63.03). An increase in 1000-grain weight and grain yield/plant was observed due to laser treatments with highest effect by 1.5 h exposure followed by 1 h, then 2 h valued (45.07), (44.22) and (43.40 gm), respectively rather than the control (40.56 gm). Meanwhile grain yield/plant showed significant increase with 1.5 h and 2 h valued (15.78 gm) and (15.57 gm), respectively, but decreased significantly (11.94 gm) at 1 h rather than the control (13.67 gm). Hexaploid wheat cultivars had differential responses to laser irradiation treatments as revealed by the significant interaction of cultivar × laser treatments. Kannan et al. (2007) used physical mutagens, laser ray (0.5, 1.0 and 1.5 min) and UV ray (10, 20 and 30 min). They showed that mean values of number of tillers/plant, grains/spike, grain yield/plant and 100-grain weight were shifted in the negative direction. PBW 226 cv. was more sensitive to the mutagens compared to WH 542. The higher nuclear DNA content explained the differential mutagenic sensitivity between the two cultivars in PBW 226 cv.

Awaad et al. (2018) induced genetic variability in Sids 12, Sakha 94 and Gemmeiza 9 wheat cultivars using Gamma ray, LASER beams and Ethyl Methan Sulphonate. Generally, maximum assessments of phenotypic (PCV), genotypic (GCV) and environmental (ECV) coefficients of variability were detected for No. of spikes/ plant followed by grain yield/plant and then No. of grains/spike in both Sids 12 and Sakha 94 as well as No. of grains/spike followed by grain yield/plant and flag leaf area in Gemmeiza 9. Heritability estimates in the broad sense in M3 generation varied from moderate to high for grain yield/plant and its contributing characters. The genetic advance was high for No. of grains/spike and ranged from low to moderate for the remaining yield contributing characters in the three mutant cultivars. They added that the most promising mutant populations derived from mother cultivar Sakha 94 were Sk-94 350 Gy, Sk-94 400 Gy and Sk-94 2 h LASER for grain yield and its components, flag leaf area, flag leaf chlorophyll content, spike length. Furthermore, those obtained from Sids 12 was Sd-12 2 h LASER for flag leaf area, flag leaf chlorophyll content, spike length, plant height, No. of grains/spike, 1000-grain weight with moderate grain yield/plant.

14.8.5   Quality

Jia and Duan (2013) revealed that some proteins like catalase, DNA polymerase and decreased glutathione declines in their expression under the ultraviolet-B radiation, while increases at the UV-B + helium-neon laser. They added that helium-neon laser leads to up-regulation of the certain initial responsive proteins. Fourteen of protein spots show protein expression rise after helium-neon laser radiation such as 1602, 2204, 2502, 3205, 3502, 3503, 3506, 3507, 4505, 5405, 5503, 6202, 6401and 8302.

There are 4 differential individual proteins are found to be expressed after helium-neon laser radiation such as 4204, 4603, 7305, 8601. The number of up-regulated proteins on 2-DE gels progressively rises during UV-B + He–Ne laser group. Four helium-neon laser responsive proteins appear. Furthermore, Chen and Liu (2015) showed that exposed wheat seeds to laser resulted in higher the soluble protein content in the seedling than those the untreated seeds.

14.8.6   Tolerance to Environmental Stresses

Lasers have a positive physiological effect on wheat growth under abiotic stress conditions.

Helium-neon laser treatment can mitigate and repair the destruction effects of UV-B radiation on wheat chlorophyll and expression of peroxidase isozymes as well as photosynthesis process (Yang et al. 2012). Chen and Han (2014) explained that helium-neon laser irradiation activated the enzymes and altered different photosynthesis parameters involved in wheat seedling *i.e.* chlorophyll content, Hill reaction, chlorophyll fluorescence parameters and yield. It also stimulates enzyme activity in the dark reaction *i.e.* phosphoenolpyruvate carboxylase, carbonic anhydrase, malic dehydrogenase and chlorophyllase. Single helium-neon laser irradiation had an encouraging effect on physiological parameters of seedling. Helium-neon laser treatment partly alleviates the damage caused by UV-B irradiation.

Laser radiation improves plant tolerance to disease and abiotic stress environmental conditions. Chen et al. (2010) exposed wheat seeds to CO(2) laser radiation for 300 s. After being cultivated for 48 h at 25 °C, the seedlings of wheat were exposed to chilling stress for 24 h. They found that when chilling stress was preceded by CO(2) laser irradiation, the contents of malondialdehyde and oxidized glutathione were reduced whereas the activities of nitric oxide synthase, catalase, peroxidase, superoxide dismutase and the contents of nitric oxide and glutathione increased. Therefore an appropriate dose of CO(2) laser can stimulate the physiological tolerance of wheat seedlings to chilling stress. Also, Qiu et al. (2013) showed that laser radiation increases plant tolerance to heavy metal of cadmium stress.

14.8.7   Increasing Water Use Efficiency

Helium-neon laser play a positive physiological role in increasing wheat plants to water stress tolerance by decreasing the concentrations of MDA and H_2O_2, also increase nutrient uptake and transport (Qiu et al. 2018). Helium-neon laser treatment of wheat under UV-B radiation increased chlorophyll content and photosynthetic efficiency. Chen and Han (2014) observed that this may reflect the expression of five genes encoding proteins know to regulate photosynthetic activity. These genes were found to be up-regulated in helium-neon laser pretreated wheat seedlings under water stress.

On the other side, the use of laser in leveling of the wheat field is a useful indirect measure to improve productivity. Laser leveling achieve the following benefits, (1) Laser has the potential to level unevenly sloping fields and improve drainage, (2) Saving up to 25% of irrigation water and (3) It enhances the effectiveness of salt leaching, which is considered an important priority for soil health under the conditions of irrigated agriculture (Bekheit and Latif 2015). Under scarce of water resources in Nubarya region, Egypt. Abdelraouf et al. (2014) revealed that it can be saved 40% of irrigation water (IR) by adding 60% IR with laser land leveling method to irrigate wheat under Nubarya region as sandy soil environment.

14.9   Conclusions

Laser irradiation is considered a new branch in agriculture proved to be very beneficial in enhancing the yield of plants. The effect of helium-neon laser on the seed germination and acceleration of the growth rate of crop plants has been observed. Also, laser effect was noticed to improve early maturity and increase leaf area, plant height, yield, yield components, and quality characters of crop plants. Laser also has positive physiological effect on wheat growth and yield and enhances its tolerance to stress environmental conditions.

14.10   Recommendations

1. The safety of helium-neon laser in improving productivity is recommended as new technology in agriculture.
2. Application of laser technology is considered a sustainable, secure and clean means to improve growth, yield and quality of crop plants.
3. Application of helium-neon laser has shown to improve morph physiological, biochemical, enzymatic activity, yield and grain quality in wheat.
4. Through this chapter, it is recommended to use helium neon laser in the field of agriculture, and investigate the criteria to produce desirable effects of biostimulation conditions
5. Set up specialized research centers more in Egypt.

References

Abaza Ghada MShM, Gomaa MA, Awaad HA, Atia ZMA (2017) Performance and breeding parameters for yield and its attributes in M2 generation of three bread wheat cultivars as influenced by Gamma and LASER ray. Zagazig J Agric Res 44(6B):2431–2444

Abdelraouf RE, Mohamed MH, Kh Pipars S, Bakry BA (2014) Impact of Laser Land Leveling on Water Productivity of Wheat under Deficit Irrigation Condations. Curr Res Agric Sci 1(2):53–64

Abo-Hegazi AMT (2007) Laser as a good non-destructive tool for selection of rice kernel color. Egypt J Plant Breed 11(2):647–650

Abu-Elsaoud AM (2013) Double-Pulse Laser Light Treatment Stimulate Germination and Changes the Oxidative Stress and Antioxidant Activities of Wheat (Triticum aestivum). J Ecol Health Environ 1(1):1–9

Abu-Elsaoud AM, Tuleukhanov ST (2013) Can He-Ne laser induce changes in oxidative stress and antioxidant activities of wheat cultivars from Kasakhstan and Egypt? Sci Int 1:39–50

Andrews LC, Philips RL (2005) Laser Beam Propagation through Random Media, 2nd edn. SPIE Press, Bellingham, Washington

Anonymous (2007) Laser applications in the field of agriculture. Specialized Physics Forums. http://www.hazemsakeek.org/vb/archive/index.php/t-6197.html

Anonymous (2018) Helium-Neon (HeNe) Lasers: Properties and Applications. https://www.azoopt ics.com/Article.aspx?ArticleID=464

Awaad HA, Attia ZMA, Abdel-lateif KS, Gomaa MA, Abaza Ghada MShM (2018) Genetic improvement assessment of morphophysiological and yield characters in M3 mutants of bread wheat. Menoufia J Agric Biotechnol 3:25–41

Bekheit H, Latif M (2015) Improvement of Wheat Productivity in Egypt through Farmer Field School. Plant Protection Research Institute, ARC Egypt, and Food and Agriculture Organization, Regional Office for the Near East and North Africa, pp 1–26. file:///C:/Users/Dear/Downloads/Documents/Rakin_upload_FFS_149.pdf https://upload.wik imedia.org/wikipedia/commons/a/af/Hene-1.png, viewed 14 July 2016

Bielozierskich MP, Zolotariewa TA (1981) Laser treatment of seeds (in Russian). Sugar Beet 2:32–33

Chen HZ, Han R (2014) He-Ne laser treatment improves the photosynthetic efficiency of wheat exposed to enhanced UV-B radiation. Laser Phys 24:10–17

Chen YP, Liu Q (2015) Effect of laser irradiation and ethylene on chilling tolerance of wheat seedlings. Russ J Plant Physiol 62(3):299–306

Chen YP, Jia JF, Yue M (2010) Effect of CO_2 laser radiation on physiological tolerance of wheat seedlings exposed to chilling stress. Photochem Photobiol 86:600–605

Delibaltova V, Ivanova R (2006) Impact of the pre-sowing irradiation of seeds by helium-neon laser on the dynamics of development of some cotton varieties. J Environ Prot Ecol 7(4):909–917

Drawish DS, Abo-Hegazy SRE, Harb AH (2013) Effects of Laser Irradiation on Two Lentil Cultivars. Egyptian J Plant Breed 17(4):29–39

Drozd D, Szajsner H (1999) Influence of presowing laser radiation on spring wheat characters. Int Agrophysics 13:79–85

Drozd D, Szajsner H (2007) Effect of application of presowing laser stimulation on bare-grained oat genotypes. Acta Agrophysica 148:583–589

El Tobgy KMK, Osman YAH, Sherbini S (2009) Effect of laser radiation on growth, yield and chemical constituents of anise and cumin plants. J Appl Sci Res 5:522–528

El-Kereti MA, El-Feky SA, Khater MS, Osman YA, A El-Sherbini SA (2013) ZnO nanofertilizer and He-Ne laser irradiation for promoting growth and yield of sweet basil plant. Recent Patents Food, Nutr Agric 5:169–181

Govindaraj M, Masilamani P, Alex Albert V, Bhaskaran M (2017) Effect of physical seed treatment on yield and quality of crops: a reveiw. Agric Rev 38(1):1–14

Guo H, Yan Z, Li X, Xie Y, Xiong H, Liu Y, Zhao L, Gu J, Zhao S, Liu L (2017) Development of a High-Efficient Mutation Resource with Phenotypic Variation in Hexaploid Winter Wheat and

Identification of Novel Alleles in the *TaAGP.L-B1* Gene. Front Plant Sci 8:1404. http://doi.org/10.3389/fpls.2017.01404

IAEA (2017) Mutation induction. International Atomic Energy Agency, Vienna International Centre, PO Box 100 A-1400 Vienna, Austria 100. https://www.iaea.org/topics/mutation-induction

Jia Z, Duan J (2013) Protecting effect of He-Ne laser on winter wheat from UV-B radiation damage by analyzing proteomic changes in leaves. Adv Biosci Biotechnol 4:823–829

Joanna JC, Mba C, Bradley JT (2016) Mutagenesis for Crop Breeding and Functional Genomics. Biotechnologies Plant Mutat Breed 3–18

Kannan R, Reddy VRK, Thamayanthi K (2007) Improvement of Indian hexaplantoid wheat. I. Induced mutations. Res on Crops 8(1):147–150

Karu T (1989) Photobiology of low power laser effects. Health Phys 56:691–704

Khalifa NS, El Ghandoor H (2011) Investigate the effect of Nd-Yag laser beam on soybean (Glycine max) leaves at the protein level. Int J Biol 3:135–144

Koper R, Wójcik S, Kornas-Czuczwar B, Bojarska U (1996) Effect of the laser exposure of seeds on the yield and chemical composition of sugar beet roots. Int Agrophysics 10:103–108

Maiman T (1960) Stimulated optical radiation in ruby. Nature 187:493–494

Mei LG, Yong FL, Rong H (2016) The detoxification effects of He-Ne laser irradiation on cytotoxicity of cadmium sulfide nanoparticles (CdSNPs) in tall fescue seedlings. Canadian J Plant Sci 96(4):539–550

Mohammadi SK, Shekari F, Fotovat R, Darudi A (2012) Effect of laser priming on canola yield and its components under salt stress. Int Agrophysics 26:45–51

Muszyñski S, Gadyszewska B (2008) Representation of He-Ne laser irradiation effect on radish seeds with selected germination indices. Int Agrophysics 22:151–157

Nenadic K, Franjo J, Stjepan P (2008) An investigation of automatic treatment of seeds with low power laser beam. Automatika 49:127–134

Perveen R, Ali Q, Ashraf M, Al-Qurainy F, Jamil Y, Ahmad MR (2010) Effects of different doses of low power continuous wave He-Ne laser radiation on some seed thermodynamic and germination parameters, and potential enzymes involved in seed germination of sunflower (*Helianthus annuus* L.). Photochem Photobiol 86:1050–1055

Podles´ny J, Stochmal A, Podles´na A, Misiak LE (2012) Effect of laser light treatment on some biochemical and physiological processes in seeds and seedlings of white lupine and faba bean. Plant Growth Regul 67:227–233

Podleśna A, Gładyszewska B, Podleśny J, Zgrajka W (2015) Changes in the germination process and growth of pea in effect of laser seed irradiation. Agrophysics 29:485–492

Prośba-Białczyk U, Szajsner H, Grzyś E, Demczuk A, Sacała E, Bąk K (2013) Effect of seed stimulation on germination and sugar beet yield. Int Agrophys 27:195–201

Qi Z, Yue M, Wang XL (2000) Laser pretreatment protects cells of broad bean from UV-B radiation. J Photochem Photobiol B Biol 59:33–37

Qiu ZB, Jin TL, Man MZ, Zhen ZB, Zhen LL (2013) He-Ne laser pretreatment protects wheat seedlings against cadmium-induced oxidative stress. Ecotox Environ Safe 88:135–141

Qiu ZB, Yuan M, He Y, Li Y, Zhang L (2018) Physiological and transcriptome analysis of He-Ne laser pretreated wheat seedlings in response to drought stress. Sci Rep 7:6108

Rybiñski W (2000) Influence of laser beams on the variability of traits in spring barley. Int Agrophysics 14(2):227–232

Rybiñski W (2001) Influence of laser beams combined with chemomutagen (MNU) on the variability of traits and mutation frequency in spring barley. Int Agrophysics 15:115–119

Rybiñski W, Garczyñski S (2004) Influence of LASER light on leaf area and parameters of photosynthetic activity in DH lines of spring barley (*Hordeum vulgare* L.). Int Agrophysics 18:261–267

Suprasanna P, Jain SM (2017) Mutant Resources and Mutagenomics in crop plants. Emirates J Food Agric 29(9):651–657

Svelto O (2010) Principles of Lasers, Springer US, pp 9–15.

USDA (2019). World Agricultural Production. United States Department of Agriculture, Foreign Agricultural Service, Circular Series WAP 6–19 June 2019, Office of Global Analysis, FAS, USDA, Foreign Agricultural Service/USDA

Vasilevski G, Bosev D, Bozev Z, Vasilevski N (2001) Biophysical methods as a factor in decreasing of the soil contamination. Int. Workshop Assessment of the Quality of Contaminated Soils and Sites in Central and Eastern European Countries (CEEC) and New Independent States (NIS), Sept 30–Oct 3, Sofia, Bulgaria

Vasilevsky G (2003) Perspectives of the application of biophysical methods in sustainable agriculture. Bulgarian J Plant Physiol 179–186

Yang LY, Han R, Sun Y (2012) Damage repair effect of He-Ne laser on wheat exposed to enhanced ultraviolet-B radiation. Plant Physiol Biochem 57:218–221

Zhang M, Liang Z, Liang Z, Shan Y (2014) Effects of he-ne laser and enhanced UV-B radiation on the photosynthesis on rice flags. https://www.researchgate.net/publication/287023618

Chapter 15
Seed Technology and Improvement Productivity of Field Crops

Ahmed Abd-El-Ghany Ali

Abstract Applying the new technology of seed production, multiplication, storage facilities and distribution could be the ways to improve the productivity and quality of field crops. Since planting high seed quality with genetical purity, high vigor, free from weed seeds and infested or borne diseases needs should be used for planting seed crops with applying specific seed treatments that will be stimulated, protected seed during field emergence which led to grow strong seedlings and faster plant growth under different field conditions. Perfect harvesting, seed processing and suitable storage facilitates would be required for saving and protect the seed for next seed sowing. Seed multiplying of new improved varieties with much care through different grades of multiplications should be taken to keep the new varieties from mixture or losses their pure genetical or vigor during the cycle processes. Finally, seed inspection and testing must be occurred to evaluate the planting seed crops before sowing in order to planting high seed quality and attained good stand in the field with better plant performance during growth and reproductive stages.

Keywords Seed quality · Seed vigor · Seed processing · Storage facilitates · Seed multiplication · Seed inspection

15.1 Introduction

Seed quality of field crops plays an essential role in speed of germination, field emergence capacity, seedling vigor and subsequent growth of crop plants. Source of variation in seed quality from year to another is difficult to pin point. Weather, fertilization, cross pollination and agronomic practices all influence final seed quality of field crops.

A major factor contributing to the present crisis in agricultural production is the inability of many developing countries to increase crop yields by using improved varieties and good seed quality. Under the traditional farming system, the farmer

A. A.-E.-G. Ali (✉)
Crop Science Department, Faculty of Agriculture, Zagazig University, 44511, Zagazig, Egypt

© Springer Nature Switzerland AG 2021   409
H. Awaad et al. (eds.), *Mitigating Environmental Stresses for Agricultural Sustainability in Egypt*, Springer Water,
https://doi.org/10.1007/978-3-030-64323-2_15

usually saves a part of each crop as seed for planting in the next season. This will no longer suffice if a substantial increase in crop productions is to be obtained. To stimulate improved crop production, an essential prerequisite is the production, distribution and utilization of quality seed of improved cultivars.

Quality seed is produced by specialized farming and does not tend itself to the type of controls applicable to other agricultural inputs. Crop cultivars must be replaced quickly on the farm with saving their quality from season to another.

The main characteristics which determine seed quality are: genetical quality, health, purity, weed seed content, germination, moisture content, 1000-seed weight, mechanical damage and pests infection. As these characteristics cannot be evaluated by visual inspection many countries have established official seed testing stations for proper evaluation. Thus, with using improved and high quality varieties could be improved productivity of growing field crops.

In light of the climate changes, the international seed companies are working to adapt their products to combat the impact of environmental change and meet the needs of nutrition through the production of high-quality seeds (Anonymous 2019).

15.2 Principles of Seed Science

Seed is a mysterious source of plant life of more seeds. Deep within the "seed" are its development forces, nutritive elements, place and time mechanisms which assign the next growth stage. The seed itself has the ability to disperse and scatter using different forces of surround nature such as wind, water, insects, birds, and/or animals (Copeland and McDonald 1995).

Seed-producing plant may be grouped into three taxa: Angiospermae, Gymnospermae, and Pteriodospermae. The angiosperms with 220,000 extant species comprise a more successful group, both numerically and in area occupied, than the gymnosperms with 520 extent species (Kozlowski 1972a). The basic differences between the seeds of the two taxa are that seeds of angiosperms are the product of double fertilization within ovary (included field crops), whereas seeds of gymnosperms are the product of the single fertilization and the ovary is absent.

Thus, the study of seed is a kind of study the life itself. That the seed unit provides us with a neatly wrapped package, inside of which a living organism is capable of displaying all the processes that found in the mature plant. Furthermore, studying the seed or germinating seedling, man has the opportunity to gain much of his knowledge about growth regulators, respiration, morphogenesis, cell division, photosynthesis and further metabolic processes (Copeland and McDonald 1995).

15.2.1 Seed as a Basic for Development

In the seed are hidden all the characters which later manifest themselves in the behavior of the plant. The practical value of a seed—provided that it is viable and healthy—depends entirely on these hidden characters that it is to mention, on the genetical background of the seed. All the characters are hereditary, and through the genes—the carriers of the characters—they transmitted from generation to another. The genes can be combined in unlimited number ways, and because of this we find enormous variations among plants, not only between species but also between individual varieties of the same species. For the same reason also it is possible to find plants suitable for almost all environmental conditions in different parts of the world.

Natural selection resulted in presented suitable and competitive types of plants have developed in the various ecological regions. Natural selection has also been effective in cultivated plants, and different agro-ecotypes, mostly referred to as local varieties, have been developed which are well adapted to the environmental conditions offered to them. Through plant breeding it is, however, possible to bring about a more systemic recombination of species and to develop variety highly superior to the old local ones in the way to improvement of the productivity of grown field crops.

Varieties can be improved through plant breeding mainly in three different respects, namely, yield capacity, cultural reliability and quality of products. Then, the improved varieties should be multiplicities through different grades of multiplication breeder's seed, foundation seed, registered seed and certified seed. The certified seed must be produced in amount required for growing the area of each crop. More information about the seed multiplication would be discussed in seed propagation Sect. 15.6.

15.2.2 Seed Quality

Seed quality means differing seed characters including genetic structure, physical appearance, germination, seed vigor, storability and health condition (without seed infect by diseases or insects).

Increasing seed quality is very important in several ways to obtain the good stand and results in the field. Grower, farmers and others stated that high quality seed produced by different organizations (seedsmen, centers, seed companies or imported from outside) are generally needed for improved seed germination, growth and speed of development.

Producing good quality seed would be useful for either seed consumers or producers. Use good quality seed affords the seed consumers through various projects such as better tolerate stressful planting condition. Thus, good quality seed can be able to germinate under adverse condition with perfect field emergence, speed of growth and high production. In other cases, using of good quality seed will help to produce a seedling for every seed planted that would resulted in saving seed rate which could be used and help in reducing seed cost. Also, planting a high quality seed lot often results in more uniform stand and field homogeneity which allows best tillage practices and weed and pest control (Ali 2017).

Therefore, high quality seed is an essential basic input in agriculture production, that the successful agriculture production could be identified based on the quality of seeds that used for sowing. So, the seed producers could hold greater responsibility in maintaining and enhancing genetically pure seed to preserve the seed quality through different stages and periods to be free from various contaminants and in healthy condition.

15.2.3   Seed Formation and Structure

The seed is defined as "mature fertilized ovule". The seed is produced through the double fertilization, which happens in angiosperms and valuable crops. This result in form an endosperm and the embryo. After the fertilization, the endosperm started its development before the embryo (Kozlowski 1972b).

The first step, which approximately 80% of the growth process happens, is initiation by cell division and elongation as seed weight enhances and increases through supplying nutrients from the mother plant. However, the second step is the seed separated from the mother plant. Hence, the time that this process is completed, the seed got its maximum dry weight, that is defined as the physiological maturity and the seed is reaches maximum quality (McDonald and Copeland 1997).

The over all, double fertilization resulted in zygote (2n) and primary endosperm (3n) which developed into embryo and endosperm or food storage which surrounded by ovule wall or "testa" (seed coat). The embryo axis contains plumule, hypocotyl and radicle.

The ovary which developed to fruit may contain one ovule or more depend on dicotyledon families which resulted in one seed such as some fruits of lentil or peanuts or numerous seeds such as pea and some beans. However, monocotyledon plants such as cereals crops, the flower has one ovary which contain one ovule that their walls associated together in form indehiscent one-seed fruit in which the integument and the pericarp are fused and defined as caryopsis either covered caryopsis (rice and barley) or naked caryopsis (wheat and corn). Thus, the caryopsis stricture included the embryo, endosperm and pericarp which formed from several layers (ovary and ovule walls).

Generally, seed development which called the "critical period" significantly affected the seed quality reached at physiological maturity through the interaction between genetic structure and environmental condition (edaphic and climatic factors). Thus, the high quality seed could be obtained by supply the mother plants by enough nutrients required, water, space under suitable climatic factors by chosen the optimum sowing date. Also, harvesting should take place as soon as possible after physiological maturity to avoid storage on the mother plant that leads to deterioration and reduced seed quality.

15.2.4 Seed Variability

Despite seeds have a number of general characters, biologists will also find a number of interesting structural, chemical and functional differences between seeds of different genes, species and varieties. Such different can be classified mainly in four groups.

A. External variation (physical properties):

These variations included mostly seed size, weight and density, shape, color, and surface texture, are several and are very important in seed identification and seed separation (cleaning and upgrading) through seed processing after harvesting (Sect. 15.4). Some kinds have little seed variability, while others have much variability. Some idea of the extent of variation in seed weight may be gained from fact that 100-seed of large faba been have a dry weight of 140 g whereas 1000-seed of Egyptian clover weight about 2.5 g. Also, the variation in seed size or weight could be observed between species and varieties under the same genes. Furthermore, such variation in seed size could be noticed among the seeds of the same varieties which gave a chance to select the large and big one to used for planting crops to improve and increase crop yields in many cases especially small grains, lentil, flax and clover. Seed shapes have not been adequately studied. As Duke (1969) mused, "which seed shapes are frequent and which are rare?" Common seed shapes are ellipsoid globes, lenticular, oblong ovoid, rein form, and sectored.

The most common colors of seeds are brown and its derivatives (light and dark and yellows). Browns and blacks make up almost of seed colors. Conspicuous colors such as red, green and white are infrequent. The color separator usually using to separate peanut seeds into classer depend on seed coat color to separate the seeds without seed coat or part of it.

The surface textures of seed coats vary from highly polished to markedly roughen. Although the surface topography have been used to advantage (Murley 1951), it is the appendages which have naturally drawn most attention. These appendages include wings, arils, carbuncles, spines, tubercles, hairs, and elaisones (Kozlowski 1972b). Such variation in seed surface texture has been used to separate weeds and other contaminants from pure seed such as roll machine, magnetic and inclined drapers separators.

B. Internal variation:

Internal variation is based on seed anatomy which provide by Martin (1946). His comprehensive study was based on embryo type, size and placement food reserve quantity and quality; and seed size. Also, seed coats, cotyledons, and endosperms have been studied in detail by several investigators which are not needed to illustrate in this chapter.

C. Chemical variation:

The biochemical data are usually recorded under percent ash, oil, and protein, fraction of alcohol-soluble nitrogen, and trichloroacetic acid-soluble nitrogen, starch, and tannin, etc. In a supplement, Jones and Earle (1966) tabulated oil and protein content for seeds of 759 species. Variations in chemical composition of seeds are required to explain their relation and its affecting on seed longevity as well as seed deterioration in this chapter.

D. Physiological variation:

Variability exists among plant species in different types included the source of carbohydrates and other growth requirements of seed germination and early development of seeding. Food reserves of angiosperms seeds are stored primary in the endosperm, and in the cotyledons.

The patterns of utilization of endosperm food also vary among species of plant. In some seeds the reserve foods may be absorbed by the embryo before the seed is mature (e.g., beans, peas, and cotton); in seeds (e.g., cereal grain) endosperm foods still until after the mature seeds are planted and imbibe water.

Also, the amount of food storage present in seeds varies greatly among species of angiosperms. Such different should be taken in mind when planting small size seed.

Type of germination another source of variation, whereas some species germinate with pushing the cotyledons by the active of division hypocotyle cues through the surface of ground which identified as epigous germination (e.g., cotton, peas and phaseolus), or it may be hypogeous with the fleshy cotyledons remaining below ground (e.g. faba bean and most all grasses) (Woodstock 1969).

15.2.5   Seed Vigor and Field Emergence

Seed vigor reflect that condition of better good quality seed which permits field emergence and good stand, healthy, rapid growth and development under extreme environmental conditions (Woodstock 1969). Seed Vigor is also defined as "potential of rapid germination and speed of growth in extreme environmental conditions" (May Ching 1973). Many characters identify the meaning of seed vigor in seedling and plant performs including better stand, seedling growth rate, uniformity, germinate and development under adverse conditions of edaphic and climatic factors pathogen-infested soil, as well as high crop yield and quality, storability under such conditions (Copeland and McDonald 1995).

Numerous factors affecting seed vigor (e.g., genetic structure, seed maturity, environmental conditions, seed size or weight, mechanical damage, age and deterioration). Thus, should be taken in mind such concepts to produce high vigor seeds, and using in planting crops such vigorous seed to maintain uniform, vigor and good seedlings which would be resulted in higher yield productivity.

Furthermore, using vigor seeds in planting crops will gave a good stand under adverse condition and reduces the amount of seed rates used with no need to replanting or thinning practice after planting.

Evaluate of seed and seedling vigor tests taken place in most seed testing laboratory, especially those have strong relationship with field emergence. Such test expose the seed to various stresses through its germination or measurement seedling growth or speed of germination and evaluate the seed vigor depend on the results obtained. The big different between seed viability and seed vigor is the first evaluated under optimum conditions with reflect relationship with field emergence, while the second evaluated under some kind of stresses conditions with relatively a good relationship with field emergence (Fig. 15.1).

The most promising seed vigor tests included germination speed and first count (range grasses and cotton, Whalley et al. 1966); cold test (corn, cotton, soybean and rice Duane Isely 1960); seedling growth rates, glutamic acid decarboxylase activity GADA test (corn and wheat, Grabe 1967); electrical conductivity test (peanut, soybean cotton and maize (Ali 1979; Hopper and Hinton 1980; Joo et al. 1980); respiration level for corn, soybean and wheat (Woodstock and Pollock 1965; Woodstock

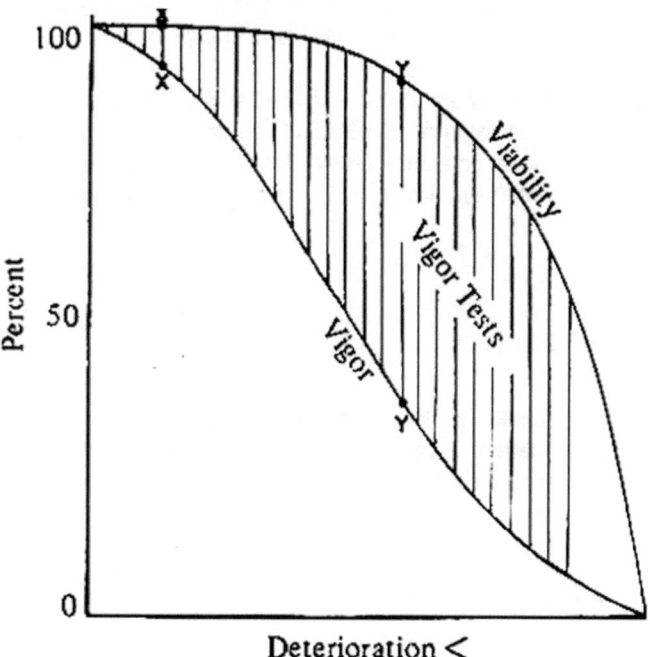

Fig. 15.1 Relationship between seed viability and vigor under seed deterioration; the shaded area between viability and vigor indicates the area where vigor testing is valuable in indicating seed quality (From: Delouche and Caldwell 1960)

1968; Woodstock and Grabe 1967), tetrazolium test for vigor (Kittock and Law 1968) reported that Tz-vigor results correlated by 0.91 with wheat seeds and field emergence, also other several investigators came to the same relationship with other crops, peanut, corn, cotton, soybeans etc.); and accelerated aging test (soybean, primarily for measuring seed storability, Delouche and Baskin (1973).

15.2.6 Seed Longevity

Seeds are unique to survive as valuable regenerative organisms till the time and place that is relevant for the beginning of the new generation like any other form of life, however they cannot retain their viability indefinitely and so deteriorate and die (Copeland and McDonald 1995). The longevity of seeds periods had been classified into three groups, long-lived seeds (macrobiotic seeds), middle-lived seeds (measobiotic) and short-lived seeds (microbiotic seeds) that ranged from 10,000, 1000, 160 years to a few months (Porsid and Harrington 1967; Libby 1951).

The life-span of seeds are influence by numerous factors included internal factors such as physical conditions and physiological states of seeds (anatomy structure, chemical components, dormancy, mechanical damages and environmental stresses during seed development); relative humidity and temperature and storage conditions. These influence factors will be in more details in storage of seed (Sect. 15.5 of this chapter).

Thereafter, should be aware about each kind of seed species and avoiding any stresses during seed development, mechanical damage, high moisture content, pests infections and mixture by contaminations. Also, storage under optimum conditions to remain the seed viable and healthy to be able to germinate without deterioration occurrence.

15.3 Seed Treatments

The term seed treatments has been used to describe ways of treating to increase their vigor of germination using seed stimulation, protection, breaking dormancy and other specific seed treatments which can be summarized in the following titles.

15.3.1 Seed Protection

Treated seed either through seed processing or in storage and planting practices is very important and recommended for protect the seeds from fungi or some other diseases and insects. Such practices needed for improve the performance by growth and developing high vigorous and good seedling.

Ali and Abd-Alla Maha (1992) showed that treated cotton seed with vitavax or captan fungicides improved seed and seeding vigor. Moreover, seed treatment with Amistra (azoxystrolain), Bion (Benzothiadiazole), and Vitavax-200 (Carboxin) resulted in better performance of peanut emerged seedlings (Sultan et al. 2012).

Furthermore, Rafael et al. (2013) indicated that, the fungicide treatments, thiram, carbenduzim, fludioxonil, thiabendazole and mixure of them could control fungi that associated with peanut seeds and consequently the seedling emergence and growth enhanced. In addition, Conceicao et al. (2016) reported that the use of fungicides would associate or not with polymers and/or insecticides in the control of fungi preset in the soybean seeds would reduce the rate of infestation of the seeds.

Seed treatments for protection have been used with the most of field crops either through seed processing or before planting by using various suitable fungicide for each; or using other protect facilities such as hot-air (cotton-55–58 °c for 2–6 min), soaking in formalin or hot-water (wheat, cotton and barley) to control nematode, fusarium and rot diseases (Ali 2017).

15.3.2   Seed Stimulation

Planting high quality seeds is the first basic store in increasing and improving productivity of field crops. Thus better knowledge of the improving seed quality or increasing seed vigor of planting seeds is essential.

The term "seed stimulation" has been used to describe ways of treating seeds to increase the vigor of germination. The obvious dream of growers for any easy way of seed treatment to stimulate more vigorous germination and greater yields and the willingness of certain persons to persons to promote dubious methods of seed stimulation have created a general distrust of the entire concept (Copeland and McDonald 1995).

Seeds soaked in few amount of water and slowly dried at ordinary temperature imbibe water and could develop more quickly when allowed to absorb water and germinate than untreated seeds (common practices with rice seeds in Egypt), (Keller and Bleak 1968 and Keller et al. 1970 on wheatgrass and forage species).

Seed stimulation by substances other than water has been tried with varying success. Flax yield has been increased by treating seeds with a boric acid solution before planting (Sokolora 1956). Pretreatment with growth substances such as kinetin, gibberellic acid, naphthalene and indoleacetic acid, potassium nitrate, potassium phosphate and even sodium chloride have been used to increase the vigor and speed of germination. Zink and other micronutrients (Rasmussen and Bowan 1969 with beans). Ali et al. (1982b) on cotton, Ramadan et al. (1986) on rice. Using alternative temperature improved of seedling rate, good seedling and germination index of maize cultivars (Ali et al. 1982a).

More recently, treated seed of maize with water or Zink increased germination %, field emergence and improved seedling growth (Mir-Mahmoodi et al. 2011). Soaking maize seeds in salicylic acid, ascorbic acid, and hydrogen peroxide improved speed of germination rate and field emergence (Ahmed et al. 2012). Soaking maize seeds in gibberellic acid and ascorbic acid increased good seedling %, germination index and improvement seedling growth (Eraky and Hania 2015).

15.3.3 Seed Coating

Produce high seed quality is great target to supply the growers and other using the seed in various ways. Thus, different companies and organization produce seed at moisture-safe storage, freed from weed seeds, with perfect physical purity. Larger healthy vigorous seed produced good seedlings with higher speed and growth seedlings (Kozlowski 1972a; Ali et al. 1986).

Several companies and researchers started to produce either coated or primed seeds for production of high quality seeds. Different types of seed coating usually used in the production process include; seed pellet which produced by treating the seed with natural materials such as montmorillonite clay. This treatment making up the seed in specific size to be easier planting. Second seed coat practice applied to using materials such as mixtures of fungicides, micronutrients, and/or insecticides to protect and stimulate the seed performance. Also, using applying plastic film coating to better hold chemicals of the seed, and protect the seed from losing of elements and improve the seed appearance. Many seeds are being marketed after priming such as polyethylene glycol or vermiculite or clay-based substance as indicated by McDonald and Copeland 1997).

15.3.4 Breaking Dormancy

Seed dormancy might be a puzzling challenge to the seeds men, nevertheless it is the method that the plants survive and adapt to their surround environment (Copeland and McDonald 1995).

Different reasons appear to be responsible to seed dormancy such as impermeable seed coat, rudiment embryo dormancy, embryo dormancy, chemical inhibition, secondary dormancy, after-ripening and physiological dwarfing. Therefore, a specific method has been used to breaking dormancy of seeds according to the natural of dormancy reasons.

Hard seeds of agricultural species are often scarified before planting to permit faster, more uniform water uptake, germination, and stand establishment. Mechanical scarification machines use a rotating, tumbling, or flailing action which rubs the seeds together and against an abrasive surface. Chemical scarification with sulfuric acid, hydrochloric acid, sodium hydroxide, acetone, and alcohols has been used. Radio frequency radiation, an intriguing method of scarification, has been developed and is commercially available (Nelson 1961).

Seeds with embryo dormancy usually require a low-temperature pretreatment before germination can occur. Such a treatment is known as stratification. The moistened seeds are usually pre conditioning at temperatures from 3 °C to 10 °C, even the specific temperature and length of exposure could be varied.

In addition, Ali et al. (1982a) mentioned that the germination rate of seeds may be increased if exposed to daily alternating temperature cycles. That Seeds which needed light to germinate at a constant temperature could be germinated in the dark when alternating temperatures would be used (Asakawa 1959).

Light wavelength, intensity, and photo period are main parameters that affect the germination of seeds that have embryo dormancy. Dormancy is broken by exposure the seed of lettus and Virginia pine to red light (700 to 760 nm) (from Copeland and McDonald 1995).

Many growth regulators, including gibberellins, auxin, kinins and some herbicides, can act as inhibitors as well as promoters of germination, particularly at high concentrations. Mechanical and chemical scarification and growth regulator increased germination % and growth germination period (Kar-Ling and Tao 1982).

Generally, it is very important to applied the suitable method to broken seed dormancy in planting craps to avoid replanting practice and its effect on stand uniformity and saving seeding rate required, especially for the improved varieties, high-value seeds and multiplication procedures.

15.3.5 Specific Seed Treatments

Specific technic or treatment may be used for improvement, germination, seedling emergence and growth, plant growth and development, as well as yield and quality of planting crops from them:

1. Legume seed inoculation with especially species of Rhizobium for each legume crop by mixture with seeds before planting which resulted in increasing yield and seed protein contents. Such seed treatment would be much needed when the legume crops will be planting the first time in new soil.
2. Cotton seed delinting by either methods, acid, or mechanical delinting to remove the residual lint or fuzz from cotton seed coat to make it easier in mechanical planting and mixture well with fungicides used. Also, acid delinting able to cleaning the seeds from all microorganisms covered the seed coat beside make the hard seed coat soft to germinate after that.
3. "Vernalization" treatment. In reference to flowering, it is defined as the process by which the floral induction is promoted. The spring cereal types moisted seed (30–35 moisture %) exposed to low temperature (0–4 °C) for 4–8 weeks, then, the seeds will be started the sporting. These vernalized seeds when planting would be germinated well and being flowering without need to exposure to low or frozen temperature. Such process must be applied when planting winter wheat types in

moderate region. Also, it could be applied to seeds to identify the stratification to break dormancy and enabling them to germinate.

4. Soaking the seeds in light concentrate solution of manitol, cycocel or water (20–35 °C) for 24–48 h to make the seeds able to germinate under salinity, cold, high temperature or drying soil condition.
5. Seed pellet and plastic film coating former illustrated in seed coating Sect. 15.3.3.
6. Divide poly-germ fruit of sugar beet by especial machine to separate into small parts included one germ, which well be similar to monogerm seed and easier for planting with saving seeds and good emergence without need to thinning practice which may affect the growth and development of the survive seedlings.

15.4 Seed Processing

Seeds are processed to clean the seeds from contaminants by upgrading the seed quality through removing the damaged or deteriorated seeds, broken seed, off type seed and treat the seeds with specific materials.

Produce and improve planting varieties are very important objective to increase crop yields and national input. Such varieties should be pure in a viable condition, free of contaminating weed seeds, and adequate quantizes at the quiet time. In addition, seed processing is the way of transform the imported varieties with high quality to growers in various places and countries.

Thus, seed processing is an integral part of the technology that involved in transforming the genetic engineering of the plant breeder and geneticist into improved seed.

15.4.1 Before Harvesting

Specific agronomy practices should be considered through seed development stages and seed filling periods included.

a. **Nutrition and irrigation**

In the nutrition of seed crops nitrogen, phosphorus, and potassium all play an important role. The amount required from each element must be applied at perfect time by the right application method. A part from these, three major elements, a deficiency of certain other elements such as calcium, sulpher, magnesium, boron, iron, copper, zinc, manganese and molybdenum may occur and some cases will need to be made good. The foliar application of some mineral elements such as copper, zinc and molybdenum with low dose of nitrogen is regular practice applied during later stages of maturity to improve quality and increasing yield and protein content of wheat, Ries et al. (1970). Also, gypsum application tended to produce heavier and larger seeds, in addition to improve seed quality of peanuts (Sullivan et al. 1974; Sullivan et al. 1977; Sullivan 1989; Abd-Alla Maha and Ali 1992).

Furthermore, over irrigation in the late stages of seed crops is to be avoided, because it can cause excessive growth, delayed maturity, plant lodging, loss yield and decrease quality of seed. Also if moisture tension increased just a little during the late stages resulted from low water or drying soil will affect seed filling period and decrease seed yields and its quality.

b. **Control of weeds, pests and diseases**

Weed control is a very important practice must be occurred during late mature stages and before maturity to remove all following weeds and other varieties or plant crops to avoid present such contaminants with the seed of planted crop as well as prevent weed plants from spread their seeds in the soil. Also, the removed of weed plants may be make the harvesting easier and produced cleaned seeds through threshing and shelling processes.

The use of resistant varieties offers the best means of control of pests and diseases. However, for many diseases and pests direct application of chemical sprays to the seed crop is necessary to protect the seeds or fruit from infection or borne diseases and insects and transfer this infection through processing and storage. Also, the application of defoliant before harvesting is necessary in some planting crop such as cotton to dropping the leaves from plants for mechanical harvest occur and allowing close bolls to open and increase the cotton yield.

c. **Harvesting and threshing**

It is of great importance to harvest a seed crop at the time that will allow both the greatest yield and best quality seed. Therefore, the mature seeds should be harvested as soon as possible to avoiding losses from shattering of the seed and lodging or falling down of plants. Also, the seed deterioration would be started when stored on mother plant. Furthermore, the losses resulting from harvesting immature seed crops are shown in smaller seed yields and a reduction in seed quality. The smaller yields are due to the failure of the seed to develop fully and to shrinkage after harvest.

The choice of the combine harvester threshing machine with ideal speed at suitable time for process should be with greatest case for each crop for remove the seeds from straw with prevent cracking or skinning the seed and impairing the germination.

15.4.2 After Maturity

Seed processing after maturity is an essential part of the total technology that involved in making valuable high quality seed of planting crops that include seed drying, seed separation (cleaning and upgrading), seed treatment for protection, and packing before storage.

15.4.3   Seed Drying

It is better to allow the seed drying naturally on mother plant before harvest practices in the field crops. This is normally in seed production. Otherwise, some crop plants needed to harvest before drying in the field because of climatic conditions and some other reasons to avoid the field deterioration. Then, the artificial drying required to reduce seed moister tell safe storage.

In addition, It should be considered and taken in mind two main principal parameters that control seed moisture content, both relative humidity and/or temperature. These parameters mainly differed to seed deterioration that starting from physiological maturity at mother plant till they are planted (these two factors will be discuss their effects on seed quality in Sect. 15.5.3, concerning seed deterioration)

Drying technology included two ways; natural drying and/or sun drying that represent the most method for drying seeds in the larger area of the world. Natural drying usually did by spread the seeds in then layers on floors, on some kinds of screen or pieces of cloth where they exposed to the sun. The seed layer should be periodically stirred many times per day to obtain uniform drying seed and rapidly draying. The problem facing natural drying is that environmental conditions may be unstable through seed drying and that method depending on environmental especially climatic factors. Also, the seed can infested by diseases and insects or damage by sun-cracking occurs.

The second method, artificial drying by using heated air. Two concepts should be taken to avoid "over drying" or/and damaged seeds and losing viability of seeds during seed drying. The first one is the air temperature which must be around 43 °c (110 °F) that has been considered as the safe upper limit for drying seeds without damage (McDonald and Copeland 1997). The second concept is depth of the seeds in bin which must be related to seed size and rate of airflow that should be in thin layers with small seeded crops (Table 15.1) and low-rate of airflow to avoiding over-drying of seeds located in the bottom of the bin.

Table 15.1 showing the maximum depth and recommended temperature for drying seeds of some field crops (Agrawal 1980)

Recommended temperature °C	Maximum depth	Seed crops
43.3 (110 °F)	50.8 (20 in)	Shelled corn
43.3 (110 °F)	50.8 (20 in)	Wheat
40.4 (105°F)	50.8 (20 in)	Barley
43.3 (110 °F)	91.4 (36 in)	Oats
43.3 (110 °F)	45.7 (18 in)	Rice
43.3 (110 °F)	50.8 (20 in)	Soybean
32.2 (90 °F)	152.4 (60 in)	Peanut
43.3 (110 °F)	50.8 (20 in)	Grain sorghum

15.4.4 Seed Separation (Cleaning an Upgrading)

Seed quality appeared to be good through using two directions. The first is cleaning the seed from other contaminants included other varieties, crop seeds, weeds, damaged, off-type and inert matters. The second is upgrading or separate the seed to a few grader seeds. Such processing applied to reach the maximum quality with high germination potential and good uniform stand in the field. This indicates to the pure live percentage can be calculated by multiplying the percent of germination (80%) by the percent of purity (90%). Then the pure life seed $= 90 \times 80/100 = 72\%$ which it may be called as the agricultural value of seed and can be used for determinate the seeding rate required for planting unit area of soil.

Seed separation can be occur by mechanical means only when the physical characteristics are differed than pure seeds and contaminants mixed which can be separated by mechanical, electric and light and magnetic process. Then, the processor user in external variation of the seed crop being cleaned and the seed of other crop types, weeds, varieties, off-type and damaged seed as well as inert matter. The physical properties are the basic for separate seeds are size, weight, density, shape, color, surface texture, affinity of liquids, electrical properties and powers of airflow.

Numerous equipments had been used for cleaning and upgrading seeds of planting crops according to their physical properties included air-screen cleaner, width and thickness separator, air separators, specific gravity separators, magnetic separator, air separators (seed blowers), electric color sorters and spiral separators. Thus the choice of a machine or sequence of machines for processing seed depended mainly on the nature and kinds of contaminants, the kind of seed being processed, the quality standards that must be met, and the quantity of each in the raw seed (Vaughan and Delouch 1968).

15.4.5 Treated Seed

Seed treatment for protection through seed processing has become a widely accepted practice by both seeds men and their customers as a recognized part of seed processing for many species.

Protection seed treatment benefits the seed in different ways. It disinfects the seed of seed-borne diseases and renders them harmless. It protects the seed against soil-borne organisms, including insects, prior to and during germination and seedling establishment. And a few act as systemic protection against disease or insect infection for the seedling (more details were aforementioned in seed treatments Sect. 15.3.1).

The ideal seed treatment chemical should be: highly effective; relatively nontoxic to plants, harmless to humans and livestock; stable for long periods of time during seed storage; easy to applied; and economically competitive. Also, treated seed must be identified by incorporating a dye into the treatment that will give the seed a contrasting color and by labeling the seed as poison-treated with a special tag showing the outline of a skull and crossbones and indicating the name of the materials used and the antidote in case the chemical are taken internally.

15.4.6 Packing

When the seed has been thoroughly processed, it should be properly bagged before entering the distribution channels. Since seed is such an important and valuable commodity, a primary consideration of the dealer should be a bag which will carry the seed safely to its final destination without breakage or loss of seed. Meanwhile, each container should be designated for a particular kind and quantity of seed. When selecting a bag for a particular use, major two points should be taken in mind, the value and the fineness of the seed.

Among other things to be considered are: the material of which the bag is to be made, its size, the printing desirable, and how it will be stored, shipped and handled. Seed bags are usually closed by sewing, this being the most economical and meatiest method. Also it gives them a better appearance and makes them easier to handle. Other seed bags is the self-closing bags which closes itself after filling. A part from bags, cartons, tins packets as well as hermetically sealed and waterproof containers are also used.

Field seed packaged in burlap, cloth, or paper and synthetic fibers are normally used and closed by industrial sewing machines. Seed packed in plastic bags are closed with a heat sealer. Boxes of seed are closed by tape or glue. The dealer usually packs seed in bags containing 50 kg or 100 lb, which may be satisfactory with certain crops.

Finally, the sequence of the all seed processings either before or after harvesting are showing in Fig. 15.2.

15.5 Storage of Seed

15.5.1 The Role of Seed Storage in Agricultural Development

The main purpose of storing seeds of economic plants is to reservation planting stocks from a season until the next. As agriculture developed, man expanded his knowledge regarding both the requirements of seed for maintenance of viability and methods of providing suitable storage conditions (Justice and Bass 1978). The storage of seed stocks for planting the following season remnants the most significant cause for storing seeds, farmers and seedsmen have found it beneficial to carry over seeds for diverse years (https://naldc.nal.usda.gov/download/CAT87208646/PDF). This practice results in accumulation supplies of desired genetic stocks for use in years following periods of low production.

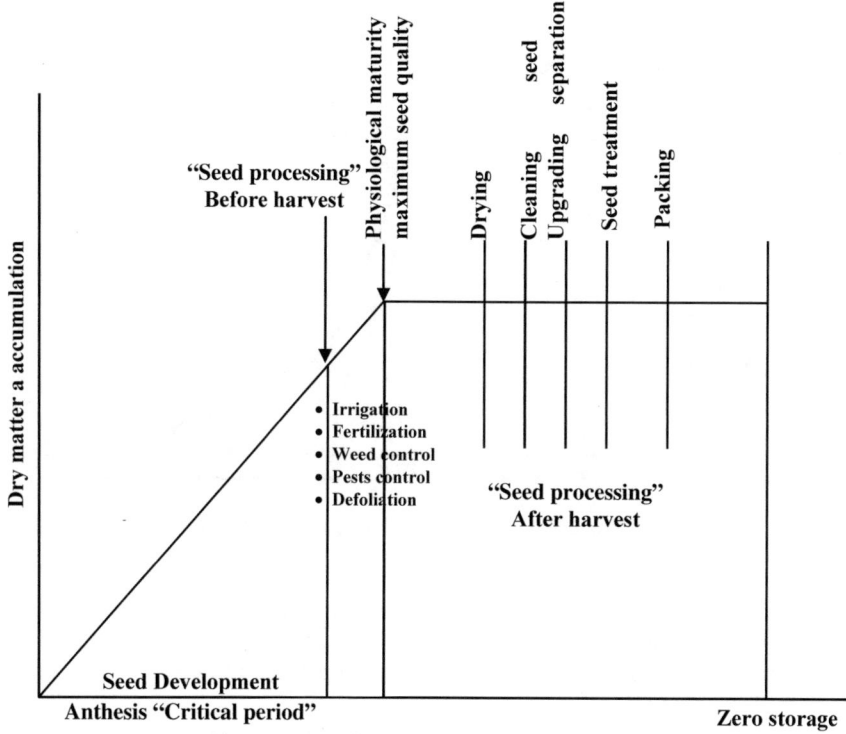

Fig. 15.2 Illustrate seed processing before and after harvest

With increased information and knowledge of plant genetics and plant breeding, the need for long-time storage of small quantities of the various cultivars becomes apparent (germplasm bank). Facilities for storing genetic stocks now exist at several countries. Research workers can obtain genetic stocks considered useful in their breading programs from three germplasm banks (Justice and Bass 1978).

Furthermore, facilitates for storing grades of propagation seeds of improved varieties (breeder's seeds; and foundation, registered and certified seeds in addition to "carry over seeds") that can be used through seed multiplication of improved varieties or regenerate the deteriorated cultivars. Thus, these improved cultivars should be distributed in several locations and regions to obtain the greatest productivity of cultivated crops.

15.5.2   Seed Physiology and Storage

At physiological maturity, the seed reached its highest quality and there after the deterioration processes would be started under adverse condition in the field (field

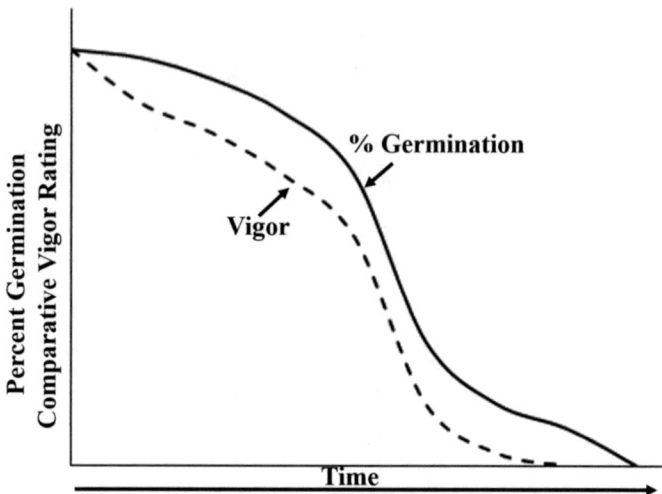

Fig. 15.3 The decline in vigor and germination of seed lot with time (From: Kozlowski 1972a)

deterioration) or through harvesting and processing. Thus, the aging of a quantity or lot of seed is not uniform but is a function of the history of each seed in the lot. Figure 15.3 illustrates, in a generalized manner, the loss of vigor and germination with time of a seed lot.

The decline in seed vigor, and finally death, of an individual seed, and the loss in vigor and germination of seed lot, have not yet been stepped by investigator. However, we know much about the environmental factors that influence aging and these can slow the aging process with proper storage conditions. Two most important factors of the environment that influence the speed of seed aging are the relative humidity of the air, which control seed moisture content, and temperature, which affects the rates of biochemical processes in seed. Harrington (1959) gives two rulers of thumbs which are of general validity and are useful guides to the effect of seed moisture and ambient temperature on the rates of seed aging.

1. For each 1% increase in seed moisture the life of the seed is halved. This role applied when seed moisture content is between about 5 and 14%.
2. For each 5 °C increase in seed temperature the life of the seed is halved. This role applies between at least 0 °C and 50 °C. There two roles apply independently. Thus, seeds having 10% moisture and stored at 20 °C will survive about as long as those with 8% moisture stored at 30 °C. Therefore, the major requirements for storage of most seeds are to create storage conditions of low relative humidity and low temperature.

Also, seed longevity is also affected by the composition of the storage atmosphere. High oxygen tends to hasten loss in viability, while high Co_2, N or a vacuum may retard deterioration under some circumstances (Kozlowski 1972a).

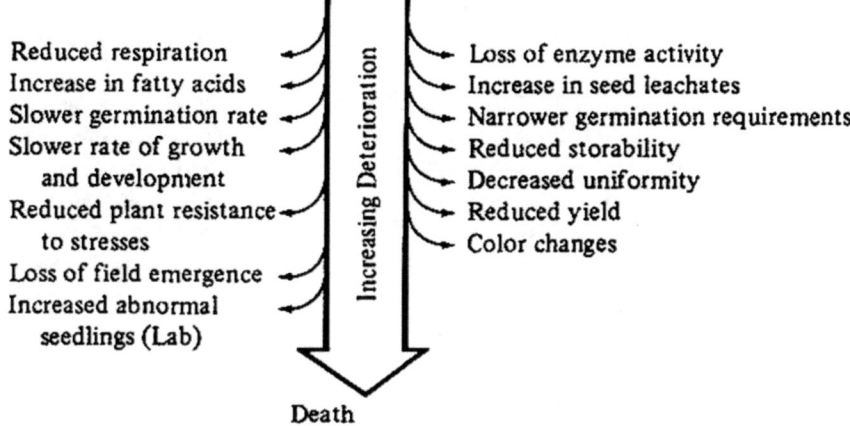

Reduced respiration → Loss of enzyme activity
Increase in fatty acids → Increase in seed leachates
Slower germination rate → Narrower germination requirements
Slower rate of growth → Reduced storability
    and development → Decreased uniformity
Reduced plant resistance → Reduced yield
    to stresses → Color changes
Loss of field emergence →
Increased abnormal
    seedlings (Lab) →

Increasing Deterioration

Death

Fig. 15.4 Probable sequence of changes in symptoms of deterioration seeds (From: Delouche and Baskins 1973)

15.5.3 Seed Deterioration

Two main groups affected the deterioration of seeds, the first are internal factors which indicate to physical condition and physiological state of seed that greatly influence their life-span. Seeds that have been broken, crashed, or even bruised deteriorate more rapidly than undamaged seeds. The second are the relative humidity and temperature which are the principal external factors that influence seed longevity and both groups aforementioned and discussed before in Sects. 15.2.6 and 15.5.2.

Advanced stages of used desecration are evidenced by visible symptoms during germination and seeding growth. However, there are processed by more subtle physiological changes whose symptoms can be detected only by sophisticated measurement techniques. Figure 15.4 shows the probable sequence of changes in deteriorating seeds, some of which can be measured by growth tests, other can be detected only by biochemical analysis (Copeland and McDonald 1995).

15.5.4 Losses and Damaged of Seeds During Storage

The rate of seed losses and damage of crop seeds either for commercial storage (food, feed and other purposes) or seed planting through harvesting, threshing, drying and storage as well as marketing and distribution amounted to 30–40% or more.

The rate of losses varied according to kinds, varieties, diseases, insects, climatic condition and seed processing and storage facilitate. International Academic Science in U.S.A, estimated the losses in cereal crops in developing countries which ranged from 2–40% in rice; 1–100% in corn; 2–42% in wheat, 1–68 in pulses crops. In

Egypt, it is common stated that the losses of wheat grains ranged from 25 to 35% annually (Ali 2017). Such losses included the reduction in quantity and reducing quality.

The main reasons affected the losses of seeds and damaged seeds included pre-harvest factors, harvesting process, biological factors (microorganisms, birds, rodents, and insects) and after harvesting (threshing and drying losses) as well as storage losses. Thus, it is of great important to protect the crop seeds through development; harvesting; processes; storage; handling and distribution for reducing those kind of losses which it will be reflected in increasing productivity of field crops with higher quality seeds.

15.5.5   Storage Methods

Much amount of seeds is losses annually due to storage under adverse conditions. The seed should be stored dry and saved dry through storage period. The longevity of seed is mainly controlled by seed moisture content and temperature of storage methods. Nevertheless, in most cases, most of seeds produced each year required to be stored only from harvest to next planting. Such seed may not need this storage process under these specific conditions other than the normal air temperature and relative humidity considering the seed types and local environment condition. Seed storage facilities required basic features for storage seeds generally included storage structure which must be protection from water, contamination, insects, rodents, fungi and fire (Justice and Bass 1978).

The types of storage in general can be classified into: farm storage; seed processor storage; country elevator storage; retail storage, research storage and germplasm storage.

Thus, it should be clear the longevity of a seed lot is mainly intimately related to its storage facilities. There should be protected against birds, rodent, fungi, and insects that might get into the storage and destroy the seed from outside. There must also be controlled of relative humidity and temperature to minimize destruction of the seeds.

15.6   Seed Propagation

Plant breeders looking always to improve productivity and quality of crop varieties and resistance to insects and diseases as well as adaptable to different environmental conditions. These advances can be obtained just under seed production programs through genetically superior and makes such seeds available to the user in enough amounts with reasonable prices. Better seed production needed to be free from other mixture, genetically pure, and able to produce better seedling with uniform stand and rabidly emerged under the wide-range of environmental conditions.

15.6.1 Production and Seed Quality Use

Availability of new seed varieties required a strong seed producer with the ability for rapid seed increase with adequate sulfured for different varietals purity. Also, there are circumstance where the introduction of a safeguard for varietal purity. Also, there are circumstances where the introduction of a cultivar with resistance to drought or salinity, or to pest and diseases, could be itself making a significant contribution to improving crop production.

The use of quality seed of improved cultivars has been mainly responsible for the success of the maximize field crops productivity, especially food grain in the developing countries.

The production of improved varieties through multiplication by following planting of multiplication degree, started by breeds seeds which produced foundation seed, then registered seeds and certified seed which released to farmers. Such procedures must be protect the multiplying seed from mixture or crossing with other varieties or volunteers plants in addition to be free from weed seeds and other contaminates and planting under propriety conditions and recommended agronomy practice as well as harvesting, processing and storage conditions.

15.6.2 Seed Certification

Seed certification the way to produce enough amounts of improved varieties to distributed on growers and other companies and centers. Certification program is the only method for attachment high improved seed quality. Thus, it becomes very important for field crop production, for which most varieties have traditionally been publically released as indicated by Copeland and McDonald (1995).

The certification concept is a progeny concept, that the pedigree of better crop varieties is achieved through using steps of seed production program. Mostly, three or four generation steps (Fig. 15.5) has been devised to maintain certified seeds. The four multiply generation called breeders seed, foundation seed, registered and certified seed as could be shown in Fig. 15.5.

15.6.3 The Continues Need to Produce New Varieties

The agriculture development concerning increasing productivity and improve quality of planting seed crops required plant breeders to continue to make significant advances in developing, adaptable, diseases-free varieties of growing field crops.

Also, the new improved varieties are badly needed for replacement the farmer distributed or local cultivars which might be deteriorated through its planting from

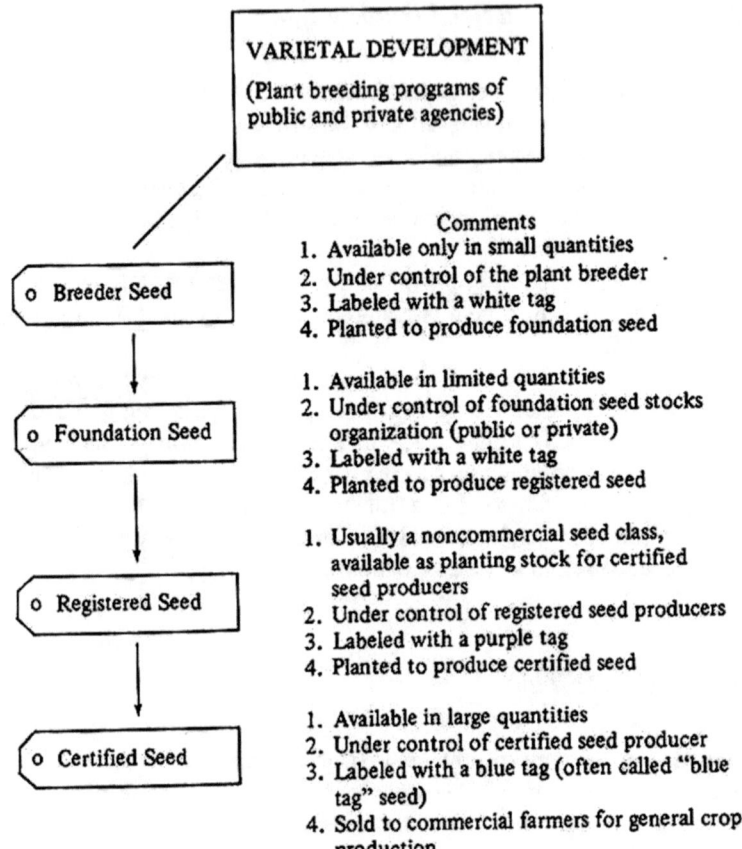

Fig. 15.5 Scheme of the overall seed certification program, that developed from of a new variety to its availability to local farmers (From: Copeland and McDonald 1995)

season to another as a resulted from several reasons; genetically mixture, isolation, varieties mechanical mixture, mutations, more stresses on mother plant during development, and appearance of new generations of diseases and insects types.

Thus, it is from great important to be continuing in create and produce new improving varieties forever to reaching the maximum productivity and quality of field crops on the long road.

15.7 Seed Inspection and Testing

Seed testing could be identified as the science of evaluating the seed quality for agricultural purposes. Even this concept initially was developed for evaluating the planting quality of different field crops and other planting seeds included vegetables, lawn, flower and tree seeds (Kozlowski 1972a).

15.7.1   Seed Sampling

Seed sampling is the first prerequisite to good seed testing. Tests can only relate to the sample as received and results of test can only be a measure of the quality of the bulk if the sample is a truly representative one. It has to be recognized that true uniformity is not always attained in commercial bulks of seed and sampling procedure has to take cognizance of this fact.

Bulk samples received by the seed testing station are entered in a register where they are listed in consecutive order, and during all tests each sample is only known under its species name and registered number until the seed analysis certificate or report is finally made out.

As it is neither possible nor necessary to test the whole bulk sample, it is reduced to a so-called working sample which is then used. The working sample can be obtained by a mechanical mixer and divider or it may be drawn by one of the methods described in the ISTA rules. These rulers also prescribe the weight of the working samples for different kinds of seed.

15.7.2   Seed Purity

Purity analyses consists of separating the sample into pure seed of kind under consideration, other variety seed, other crop seed, weed seed, broken and immature seeds and inert matter.

When a sample has been divided into its parts, each fraction is weighed and these weights from the basis for calculation of percentages. However, in cases a clear picture is needed for weed seeds, number of weed seeds in a specific weight of the seed and expressing the result on a count basis. The occurrence of noxious weed seeds is reported in number of seeds per pound or ounce. Then, the purity report included purity percent (pure seed per sample weight) and number of noxious weed seeds per specific weight.

15.7.3   Seed Viability

Seed viability as expressed by seed analysts indicated to ability of a seed to germinate and producing normal seedling under optimum conditions according to the A.O.S.A and ISTA rulers. Standard germination test is the main viability test through the wholly world at least for the commercial using. Other viability tests also can be used such as TZ-potential test; vital coloring methods (sulfuric acid and indigo carmine and other aniline dyes); enzyme activity method, conductivity test; X-ray test and free fatty acidity test.

Standardized methods have been developed by which reliable germination tests can be made in the laboratory at any time of the year. It is not possible here to go into details of these methods, or to describe various types of equipment used.

In certain cases it has been found that ordinary germination tests under more or less optimal conditions in the laboratory do not always give sufficient information on the seed for sowing purposes. A sample of old seed, or seed harvested under unfavorable weather conditions, or seeds with cracks in the seed coat due to threshing damage, may produce an acceptable percentage of normal seedlings in the laboratory but give a poor field emergence. Thus, it is very important to find out some other tests to evaluate seed and seedling vigor, as it is usually called, to obtain better field emergence even under adverse condition.

15.7.4 Seed and Seedling Vigor Tests

The principle of seed vigor is reaching seed having potential to germinate strong, healthy and vigor seedlings under wide range of environmental conditions rather than weakened performance potential that result of deterioration, damage or other reasons. May Ching (1973) defined that vigor, under field conditions, is the potential for rapid uniform germination and seedling growth. The principal included seedling emergence, speed of uniformity germination, plant performance, ability to germinate under wide-range adaption of edaphic and climatic conditions, crop yield and storability under optimum or adverse conditions as illustrated by Copeland and McDonald (1995).

Some factors affecting seed vigor included: genetic structure; seed maturity, environmental during seed development, seed size and density; mechanical damage; age and deterioration, and attack by microorganisms, that former discussed through different sections of this chapter especially Sects. 15.2.5, 15.2.6, and 15.5.3.

The most vigor tests have been correlated with laboratory or field germination included: seedling growth rate; speed of germination; cold test; GADA test, birch gravel test; respiration level; TZ-test for vigor, measurement of leachates (EC-test), strength of emergence test and accelerated aging test. It is not possible here to go into details of those tests or to describe various types of equipment used.

15.7.5 Specific Seed Tests for Seed Quality

In spite of general tests occurred on seed "lots" that include: purity, germination and common and noxious weed evaluation, several addition tests also reflect seed quality such as:

- **Variety tests**

  It is notably difficult for variety tests to distinguish between crop varieties due to their similarity in morphological characteristics. However there are varies methods to compare and identify between crop varieties including: Visual observation of the seed; visual observation of seedlings, greenhouse and/or field grow-out tests; ultraviolet light tests; chemical tests; chromosome count and other cytological methods.

- **Testing for pathological quality**

  Seed Pathological testing has not been developed to the level that needed and is almost never a part of routine seed laboratory analysis in several countries. The methods used for pathogen quality tests included; visual examination; agar media tests for identifying pathogenic; bacteriophage technique; plant injection tests … etc.

- **Effectiveness of seed treatments determination:**
- To identify the seed effectiveness treatment with chemical pesticides.
- Effectiveness of legume seed inoculation.
- Seed tolerance to define acceptable limits within which different test results may be vary
- Tolerance for purity, florescence, germination, and noxious weed seed examinations.

15.8 Conclusions

High quality seed of improved crop varieties, with modern equipment, good quality fertilizers and better method for pests and weed control, has motivated the farming resulted in enhancing the field crops productivity and quality. Availability of new varieties required a strong seed industry with the ability for rapid seed distribution with adequate safeguards for seed varietal purity. Seed vigor and longevity would be able to establishment good seedlings and stand per field under environmental conditions. Improved varieties and local ones might be deteriorated when grown for generation after generation in general cultivation due to several factors during production cycles. Therefore, breeding programmer must be continuing in order to produce new improved varieties to be replaced the deteriorated varieties. It is of great important to be continue to produce high quality seeds with safety storage; protected from mixture and deterioration; as well as better agronomic practices and protection from pests, with much care at harvesting, handling and processing to reaching the maximum productivity and quality from growing field crops.

15.9   Recommendations

The use of seed with high quality will increases the productivity and quality of field crops. Thus, the development of new varieties characterized by seed vigor and viability with high genetic purity will contribute to increase production. The storage of high quality seed under controlled suitable conditions should be applied in order to keep such seeds in the best case till next planting. Also, application of seed pre-sowing treatments will improve field germination and obtained good healthy stand which led to faster plant growth, flowering and maturity and an increase in productivity. It is important to treat seeds whichever before or after harvesting to obtain clean, up-graduated, safe moisture content and treated seed before storage.

15.10   Buzz Words

See Table 15.2.

Table 15.2 Buzz words within Seed Technology

Buzz Word	Quick explanation
A.O.S.A	Association of Official Seed Analysts, organization of state and federal seed analysis of the united states and Canada
I.S.T.A	Initials of the International Seed Testing Association
G.A.D.A	Glutamic Acid Decarboxylase Activity test which used to evaluate seed and seedling vigor
E.C.	Electrical conductivity test used to evaluate seed and seedling vigor
T.Z.	Tetrazoluim test which used to evaluate seed viability and seed vigor (quick test)
Cultivar	Crop variety
Inoculation	A seed treatment prepared using nitrogen fixing bacteria that added to legume seed before planting to assure nitrogen fixation ability (Glossary, Copeland and McDonald 1995)

References

Abd-Alla Maha M, Ali A-GA (1992) Seed quality of two peanut cultivars as influenced by sowing dates and gypsum application in newly reclaimed soil. Egypt J Appl Sci 7(12):52–67

Agrawal RL (1980) Seed Technology. General Principles of seed production. Part (2):43–66. Oxford and IBH Publishing CO., New Delhi

Ahmed I, Khaliq TAA, Basra SMA, Hasnain Z, Ali A (2012) Effect of seed primary with ascorbic acid, salicylic acid and hydrogen peroxide on emergence, vigor and antioxidant activities of maize. African J Biotechnol 11(5):1127–1132

Ali A-GA (1979) Electrical and chemical evaluation of planting seed quality. PhD Thesis, Faculty of Agriculture North Carolina State University, USA

Ali A-GA, Abd-Alla Maha M (1992) Effect of acid delinting and fungicide on viability and vigor of some Egyptian cotton cultivars. Zagazig J Agric Res 19(1):1–10

Ali A-GA (2017) The seed and technological of seed processings and storage. Seed Phycol Storage (8) 183–258. Fac Agric. Zagazig Univ. Egypt. 2017/ 23761

Ali A-GA, Ramadan IE, Saleh ME (1982a) Germination rate and seedling vigor of some maize varieties at constant and alternating temperature. Egyptian Crop Science J No 10

Ali A-GA, Ramadan IE, Saleh ME, El-Nagar EM (1982b) Response of cotton to growth regulator seed treatment. Response of cotton seed to gibberellic acid. Wiss Z Univ Halle XXXII 83 M, H.5, 5:101–112

Ali A-GA, Ramadan IE, Saleh ME (1986) Relation of seed size to germination, vigor and field emergence in lentil seed. Egyptian J Appl Sci 1:8–14

Anonymous (2019) The Access to Seeds Index 2019—Global Seed Companies. https://knowledge4food.net/the-access-to-seeds-index-2019-global-seed-companies/

Asakawa S (1959) Germination behavior of several coniferous seeds. J Japanese For Soc 41:430–435

Conceicao GM, Lucio AD, Mertz-Henning LM, Henning FA, Beche M, Andrade FF (2016) Physiological and sanitary quality of soybean seeds under different chemical treatment during storage. Revista Brasileria de Engenharia Agricolae Ambiental 20(11):1020–1024

Copeland LO, McDonald MB (1995) Principles of seed science and Technology. Burgess Publishing, Minneapolis, MI

Deloch JC, Baskin CC (1973) Accelerated aging techniques for predicting the relative storability of seed lots. Seed Sci Technol 1(2):427–452

Delouche JC, Caldwell WP (1960) Seed vigor and vigor tests. Proceedings of the AOSA 50(1):136

Dl Kittock, Law AG (1968) Relationship of seedling vigor to respiration and tetrazolium chloride reduction by germinating wheat seeds. Agron J 60:286–288

Duan Isely (1960) The cold test for corn. Proceedings of the ISTA, 1950. 16:299–311; WN Rice, Development of the cold test for seed evaluation. Proceedings of the AOSA 50:118–123

Duke JA (1969) On tropical tree seedling. 1. Seed, seedling systems, and systematics. Ann Mot Bot Gard 56:125

Jones Q, Earle, FR (1966) Chemical analysis of seeds. II Oil and protein content of 759 species. Econ Bot 20:127

Eraky Hania, AME (2015) Influence of specific seed treatments on field emergence and seedling growth of maize (Zea mays, L.) varieties. PhD Thesis, Fac of Agric Zagazig University Egypt

Grabe DF (1967) Glutamic acid decarboxylase activity as a measurement of seedling vigor. Proc AOSA 54:100–109

Harrington JF (1959) Drying, storage, and packing seeds to maintain germination and vigor. Proc. Short Course Seedsmen, State College, MS, pp 89–108

Hopper NW, Hinton HR (1980) The use of electrical conductivity as a measure of cotton seed quality. Agron Abs 1980:109

Joo PK, Orman BA, Moustafa AM, Hafdahl M (1980) Can leachate electro-conductivity be a useful tool for corn seed emergence potential evaluation? Agron Abs 1980:109

Justice OL, Bass Louis N (1978) Principles and practices of seed storage. Agric Handbook No 506 U.S.A. Det of Agric

Keller W, Bleak AT (1968) Preplanting treatment to hasten germination and emergence of grass seed. J Range Manag 21(4):213–216

Keller W, Bleak AT, Hansen AA (1970) Preplanting seed treatment may reduce failures in range seedling. Proceedings of the 11th International Grassland Congress, 1970, 166–119

Kozlowski TT (1972a) Seed Biology, Insects, and Seed Collection, storage, testing, and certification, vol III. Academic Press, New York and London

Kozlowsky TT (1972b) Seed Biology, Importance, Development, and Germination, vol I. Academic Press, New York and London

Libby WF (1951) Radiocarbon dates II. Science 114:291–296

Martin AC (1946) The comparative internal morphology of seeds. Amer Midl Nature 36:513

May Ching, Te (1973) Biochemical aspects of seed vigor. Seed Sci Technol 1:73–88

McDonald MB, Copeland Lawrence O (1997) Seed Production (Principles and Practices). International Thomson Publshing, New York and Washington

Mir-Mahmoodi TM, Golezani KG, Habibi D, Paknezhad F, Ardekani MR (2011) Effect of priming techniques on seed germination and seedling emergence of maize (*Zea mays* L.). J Food Agric Environ 9(2):200–202

Murley M (1951) Seeds of cruciferae of northeastern North America. Amer Midl Nature 46:1

Nelson SO (1961) Radio-frequency electric seed treatment. Seed World 88(12):6–7

Porsild AE, Harrington CR (1967) Lupinus articus wats, grown from seeds of the Pleistocene Age. Science 158:113–114

Rafael MB, Faria SJ, Lopes MD, Panzizzi RD, Vieira RD (2013) Chemical control of pathogens and physiological performance of peanut seeds. J Food Agri Environ 11(2):322–326

Ramadan IE, Ali AA-G, Saleh ME, Mohamed MA (1986) Effect of pregermination treatment and temperature on germination and seedling growth of some rice cultivars. Zagazig J Agric Res 13(2):278–309

Rasmussen PE, Bowan LC (1969) Zink seed treatment as a source of Zink for beans (*Phaselous vulgaris*). Agron J 61:674–676

Ries SK, Moreno O, Meggitt WF, Schewizer GJ, Ashkar SA (1970) Wheat seed protein-chemical influence on and relationship to subsequent growth and yield in Michigan and Mexico. Agron J 62:746–748

Sokolora GP (1956) Preplanting treatment of flax seeds with solutions of micronutrients. Zapiski Leningrad sel'skokhoz Institute, 11:412–416. Chemical Abstracts, 1958, 52:2041 8a

Sullivan GA (1989) Peanut-Year Book, Peanut production practices. Crop Science Dept North Carolina State University, Raleigh, NC

Sullivan GA, Jones GL, Moore RP (1974) Effects of dolomitic limestone, gypsum, and potassium on yield and seed quality of peanuts. The Journal series of the North Carolina State University Agriculture Extension service, No 140 Raleigh, NC

Sullivan GA, Ali AA-G, Hube K, Celeste Dye (1977) Quality of seed from the peanut variety and quality evaluation trials, 1977. North Carolina State University Agriculture Extension service, Raleigh, NC

Sultana N, Hossain I, Akhter K (2012) Comparatine effect of seed treatment with bion, amistar and vitavax-200 in controlling tikka disease of peanut var. Jhinga Badam. J Exp Biosci 3(1):37–44

Tao Kar-Ling J (1982) Improving the germination of Johnson grass seeds. J Seed Technol 7(1):1–9

Vaughan C, Delouch JC (1968) Seed Processing and Handling. Mississippi State University, Starkville, MS

Whalley RDB, Mckell CM, Green LR (1966) Seedling vigor and the early nonphotosynthetic stage of seedling growth in grasses. Crop Sci 6:147–150

Woodstock LW (1968) Relationship between respiration during imbibition and subsequent growth rates in germinating seeds. In 3rd international symposium on Quantitative Biology of Metabolism, ed. A Looker, pp 136–146

Woodstock LW (1969) Seedling growth as a measure of seed vigor. Proceeding of the ISTA 34(2):273–280

Woodstock LW, Grabe DF (1967) Relationships between seed respiration during imbibition and subsequent seedling growth in *Zea mays* L. Plant Physiol 42:1071–1076

Woodstock LW, Pollock BM (1965) Physiological predetermination: imbibition, respiration, and growth of lima bean seeds. Science 150:1031–1032

Chapter 16
Identification of Salt Tolerant Genotypes Among Egyptian and Nigerian Peanut (*Arachis hypogaea* L.) Using Biochemical and Molecular Tools

Mohamed S. Abbas, Amani M. Dobeie, Clara R. Azzam, and Amira Sh. Soliman

Abstract The present chapter aims to define salt tolerance of Egyptian and Nigerian peanut at morphological, biochemical and molecular levels. Growth parameters were estimated, and Na^+, K^+ and proline were measured by biochemical methods. The polymerase chain reaction was used to amplify the KAT1 gene. The results showed that growth parameters reduced under high salt conditions. Statistical analysis revealed a significant increase in Na^+ and proline accumulation in Ismailia1 and Samnut 22 genotypes, while K^+ decreased significantly in peanut under salinity conditions in all genotypes. Sequence analysis confirmed presences of KAT1 gene in peanut and showed that Giza 6 and Ismailia1 shared in one SNP (A) in codon 195, while Ismailia1 and Samnut 22 shared in SNPs (T) at codon 258, 259, 263 and 406. The present study reported that Ismailia1 and Samnut 22 are more salt tolerant genotypes than others. Ismailia1 and Samnut 22 possess SNP in KAT1 gene due to increase salt tolerance levels. Therefore, this study recommends that Ismailia1 and Samnut 22 could be involved in peanut salt tolerance breeding programs.

Keywords Peanut · Salt stress · Proline · NaCl · Morphological · Minerals · Biochemical and KAT1 gene

16.1 Introduction

Peanut (*Arachis hypogaea* L.) is an important legume crop grown worldwide. It contains high oil content and protein (Reddy et al. 2003). Salinity is abiotic stress that restricts of legumes production and determines it's spreading through environments (Giannakoula et al. 2012). Several plant growth stages were significantly affected

M. S. Abbas (✉) · A. M. Dobeie · A. Sh. Soliman
Natural Resources Department, Faculty of African Postgraduate Studies,
Cairo University, 12613 Cairo, Egypt

C. R. Azzam
Cell Research Department, Field Crops Research Institute,
Agricultural Research Centre, 12619 Giza, Egypt

© Springer Nature Switzerland AG 2021
H. Awaad et al. (eds.), *Mitigating Environmental Stresses for Agricultural Sustainability in Egypt*, Springer Water,
https://doi.org/10.1007/978-3-030-64323-2_16

by salt stress. Roots are the first organ that interacts with salt stress (Delgado and Sanchez-Raya 2007). Mensah et al. (2006) discovered reductions in the height of plant and vigour with rises in salinity. Moreover, there was a decline in the number of leaves/plant with an increase in the salt.

Salinity is one of the supreme problems to affect plant crop growth production (Sibole et al. 1998). NaCl is the most familiar accumulated salt in soils suffering from salinity. Shrivastava and Kumar (2015) showed that salinity is one of the most brutal environmental factors limiting the productivity of crop plants because most of the crop plants are sensitive to salinity caused by high concentrations of salts in the soil, and the area of land affected by it is increasing day by day. However, Negrao et al. (2017) discuss how to quantify the impact of salinity on different traits, such as relative growth rate, water relations, transpiration, transpiration use efficiency, ionic relations, photosynthesis, senescence, yield and yield components. A proposed experimental route from design to data analysis should include a classical screening of germplasm for breeding purposes, characterization of genes, and discovery of new quantitative trait loci (QTL) and, ultimately, genes using forward genetics.

16.1.1   Effect of Salinity on Plant Characterizations

Adaptation to all these stresses is associated with metabolic adjustments that lead to the modulation of different enzymes (Ehsanpour and Amini 2003).

16.1.1.1   Effect of Salinity on Seed Germination

Seed germination is one of the most fundamental and vital phases in the growth cycle of a plant that determines the yield. However, the salinity adversely affects the process of germination (Khodarahmpour et al. 2012; Fernández-Torquemada and Sánchez-Lizaso 2013). Salinity affects the germination process many-folds. It alters the imbibition of water by seeds due to lower osmotic potential of germination media (Ashraf and Orooj 2006), causes toxicity which changes the activities of enzymes of nucleic acid metabolism (Gomes-Filho et al. 2008) alters protein metabolism (Dantaset al. 2007) disturbs hormonal balance (Khan and Rizvi 1994), and reduces the utilization of seed reserves (Othman et al. 2006). Lauchli and Grattan (2007) proposed a generalized relationship between percent germination and time after adding water at different salt levels. In *Solanum lycopersicum*, Kaveh et al. (2011) found a significant negative correlation between salinity and the rate and percentage of germination which resulted in delayed germination and reduced germination percentage. Bordi (2010) reported that Compared with control, germination percentage, and germination speed were decreased by 38 and 33%, respectively, at 200 mM NaCl in *Brassica napus*. Khodarahmpour et al. (2012) observed a drastic reduction in germination rate, length of radicle and plumule, seedling length and seed vigor in maize seeds exposed to 240 mM NaCl. Moreover, the germination reaction, seedling growth, photosynthetic pigments and antioxidants parameters indicated that peanut had a reaction to different levels of salinity and under salinity stress. Increasing salinity reduced the size of the morphological traits of the seedlings (Nokandeh et al. 2015). Peanut is most sensitive to salt stress in the germination and seedling stages (Liu et al. 2012). Soil salinity may influence the germination of seeds either by creating an osmotic potential external to the seed preventing water uptake, or the toxic effects of Na^+ and Cl^- ions on germinating seed (Khajeh-Hosseine et al. 2003). Salt and osmotic stresses are responsible for both inhibition or delayed seed germination and seedling establishment (Almansouri et al. 2001). Contrary, Gebreegziabher and Qufa (2017) found that seed priming with NaCl and $CaCl_2$ had a significant increase in germination, early growth, number branches, number of cobs and grain yield. Also, Aflaki et al. (2017) found that with an increase in salinity levels, indices including germination percentage, germination rate, seed vigor, the coefficient of germination rate, coleoptile to plumule ratio, and daily germination mean decreased.

16.1.1.2   Effect of Salinity on Vegetative Growth

Salinity is affected all physiological and growth in plants, which showed a decrease with increasing electrical conductivity of the irrigation water (Oliveira et al. 2016). The mechanism by which salinity inhibits plant growth depends on whether the exposure to salt is short-term (hours to days) or long-term (weeks). Short-term effects usually represent an osmotic effect (water deficit) with little or no ionic effect. Long-term effects allow the ion toxicity symptoms to appear on the plant due to the accumulation of toxic ions as (Na^+ and Cl^-) in the cytoplasm (Rengel 1992). Also, Sharma (1997) indicated that under salinity stress, plants tend to record low yields because of adverse effects of salinity on such parameters as relative water content, total dry weight, plant height and a number of leaves per plant.

   Salinity reduces shoot growth by suppressing leaf initiation, expansion, as well as internodes growth, and accelerating leaf abscission (Zekri 1991). Salt stress causes a rapid and potentially lasting reduction in the rate of leaf growth (Feng et al. 2002). A leaf cell of a NaCl-treated plant can expand at reduced rates because of reduced uptake rates of water or osmolytes, stiffened cell walls, or because of lowered turgor (Cosgrove1997). For moderate salinity stress, inhibition of lateral shoot development becomes apparent over weeks, and over months there are effects on reproductive development, such as early flowering or a reduced number of florets. During this time, a number of older leaves may die. However, the production of younger leaves continues. The reduction in leaf development is mainly due to the osmotic effect of the salt (Sumer et al. 2004).

16.1.2   Mechanism of Salt Tolerance

Salt tolerance is depending on plant species and/or environmental factors. Some plant species are very sensitive to salt stress during germination stage and others during reproduction (Howat 2000).

16.1.2.1   Osmotic Tolerance

Osmotic tolerance includes the plant tolerance for drouth and to maintain leaf expansion and stomatal conductance (Rajendran et al. 2009). So, increase osmotic tolerance caused increase number and area of the leaf, and higher stomatal conductance resulting increase in plant production generally (Munns and Tester 2008).

16.1.2.2   Sodium Exclusion

Sodium exclusion from leaves is associated with salt tolerance in cereal crops (James et al. 2011). This exclusion occurs resulting the reduction of cell uptake of Na^+ in the root cortex as well as the tight control of net loading of the xylem by parenchyma cells in the stele (Davenport et al. 2005).

16.1.2.3   Tissue Tolerance

Segmentation of Na^+ and Cl^- at the cellular and intracellular level to avert toxic concentrations within the cytoplasm of leaf (Munns and Tester 2008) and synthesis and accumulation of compatible solutes within the cytoplasm. However, compatible solutes play a role in plant osmotolerance in various ways, protecting enzymes from denaturation, stabilizing membrane or macromolecules or playing adaptive roles in mediating osmotic adjustment (Ashraf and Foolad 2007). These solutes protect the cellular structures through scavenging ROS (Hasegawa et al. 2000).

16.1.3   Biochemical Studies

16.1.3.1   Effect of Salinity on Minerals Accumulation

The large amounts of accumulated inorganic ions may cause ion toxicity. The uptake of ions needed for the synthesis of organic osmolytes and other metabolites is affected by external salt stress. This is due to competition and nutrient imbalances (Kinraide 1999). The presence of excessive soluble salts in the soil competes with the uptake and metabolism of mineral nutrient that is essential to plants. The appropriate ion ratios provide a tool to the physiological response of a plant in relation to its growth and development (Wang et al. 2003).

Salinity enhances the Na^+ content in *Vicia faba* while the Na^+/K^+ ratio was decreased Gadallah (1999) thus suggesting a negative relationship between Na^+ and K^+. In addition, many of the deleterious effects of Na^+ seem to be related to the structural and functional integrity of membranes (Kurth et al. 1986).

Désiré et al. (2010) revealed that K^+, Mg^{++}, Ca^{++}, P, N, K^+/Na^+ and Ca^{++}/Na^+ uptake of peanut plant organs were significantly reduced with increasing salinity. There was selective absorption of minerals by salinity tolerant cultivars under saline conditions. Salinity causes accumulation of Na^+ and Cl ions in seedling roots, shoots, and leaves. Further, Chakraborty et al. (2016) evaluated the ameliorative role of potassium in salinity tolerance of peanut and found that external K^+ application resulted in improved salinity tolerance in terms of plant water status, biomass produced under stress, osmotic adjustment and better ionic balance. This study identified Na^+ exclusion as a key strategy for salt-tolerance.

16.1.3.2   Effect of Salinity on Proline Accumulation

Proline is one of the most studied compatible solute playing a predominant role in protecting plants from osmotic stress (Rai et al. 2010). The mitigation of toxicity may be caused by regulating Na^+/K^+ ratio and increasing accumulation of proline (Wu et al. 2017). Compatible solutes, proline and glycine betaine increase highly under stresses of salt and drought (Munns et al. 2002) and constitute the major metabolites found in durum wheat under salt stress, as in other Poaceae (Carillo et al. 2005). Also, proline is a proteinogenic amino acid with an exceptional conformational rigidity, essential for primary metabolism, which normally accumulates in large quantities in response to drought or salinity stress (Szabados and Savouré 2010).

Under salt stress, several functions are proposed for the accumulation of proline in tissues which include osmotic adjustment, carbon and nitrogen reserve for growth after stress resistance, detoxification of excess ammonia, stabilization of membranes, protecting photosynthetic activity and mitochondrial functions, and scavenging of free radicals (Kavi-Kishore et al. 2005). However, the significance of proline accumulation in osmotic adjustment is still controversial and varies according to the species (Silveira et al. 2003; Meloni et al. 2004). Hossain et al. (2011) showed a significant increase in proline content after 7 days of exposure to NaCl stress and the magnitude of proline accumulation was positively associated with NaCl in the culture solution on peanut seedlings. However, Nithila et al. (2013) reported that higher levels salinity and sodicity stresses resulted in an increase in proline accumulation. Therefore, it was revealed that proline synthesis might have accelerated at the sublethal level of stress rather than severe stress.

Proline accumulation occurs rapidly after the onset of stress, rapid breakdown of proline during stress may introduce enough reducing agents that support mitochondrial oxidative phosphorylation and production of ATP for recovery from stress and repairing of stress-induced damages (Carillo et al. 2008). So, proline induces expression of salt stress-responsive genes, which possess proline responsive elements in their promoters (Ashraf and Foolad 2007). Proline plays a highly helpful role in plants exposed to several stresses. Besides acting as an excellent osmolyte, proline plays three major roles during stress, i.e., as an antioxidative defense molecule, a metal chelator and a signaling molecule (Hayat et al. 2012; Szabados and Savouré 2010).

Musa et al. (2015) found that proline level and the activity of glutathione reductase (GR) appeared to be only components that play an important part in salt stress protection in two cultivars of peanut.

As a result of competition and nutrient imbalances under salt stress, ion toxicity may be affected by a large amount of inorganic ions accumulation (Kinraide 1999). Proline as compatible solute acting as an important role in protecting plants from damaging effects of salt stress (Rai et al. 2010). The K^+ genes play an efficient role of plant K^+ uptake channels in order to avoid K^+ drip under disapproving environments (Geiger et al. 2009). KAT1 gene is expressed in guard cells and faces there variations of external K^+ (Hertel et al. 2005). The family of voltage-gated (Shaker-like) K^+ channels involves nine associates (KAT1, KAT2, AKT1, AKT2, AKT4, AKT5, SKOR, GORK and SPIK) in *Arabidopsis thaliana* (Gajdanowiez et al. 2009).

So, the objectives of this chapter were screening of eleven peanut genotypes from Egypt and Nigeria for salt stress in terms of morphological, biochemical and molecular genetic level. Also, to design specific primers for salt tolerance gene for detecting and sequencing salt tolerance gene in Egyptian and Nigerian peanut genotypes for identify SNPs of salt tolerance gene and comparing between tolerant and susceptible genotype for salinity tolerance on SNPs level.

16.2 Materials and Methods

16.2.1 Plant Materials and Growth Conditions

This study was carried out during the summer season of the year 2016–2018 in the greenhouse and laboratories of the Institute of African Research and Studies, and Agricultural Research Center, Giza, Egypt.

Eleven peanuts (*Arachis hypogea* L.) genotypes from two different origins and gene pools were used in this study, six Egyptian peanut genotypes (Sohag 110, Giza 6, Giza 5, R-92, Ismailia 1 and Sohag 104) which were kindly obtained from Oil Crops Research Department, Field Crops Res., Inst., Agricultural Res. Center, Giza, Egypt and five Nigerian peanut genotypes (Ex-Dakar, Samnut 24, Samnut 23, Samnut 22 and Samnut 21) were kindly obtained from Ahmadu Bello University, Zaria, Nigeria. Levels of salinity *i.e.*, 0, 1000 and 3000 mgl^{-1} of NaCl were used.

16.2.2   Greenhouse Experiment

The experiment was conducted under greenhouse conditions at the Greenhouse of Cell Research Department, Field Crops Research Institute, Agricultural Research Center, Giza, Egypt during the summer season of the year 2016. A total of 20 l PVC pots (50 cm diameter) were filled with 2:1 sand and clay soil, respectively. Seeds of the eleven peanut genotypes were sterilized for 5 min in 10% sodium hypochlorite and then rinsed several times with distilled water before planting. Ten seeds were planted per pot. Then irrigation of these pots by tap water in order to germinate seeds and form the crop. Seedlings began to appear at the sixth day after sowing. Salinity treatment started 20 days after sowing. The growth parameters were measured under salinity and normal conditions after germination in greenhouse such as plant height (cm), number of shoots/plant, number of leaves/plant and leaf area (cm^2).

16.2.2.1   Minerals and Proline Determination

Samples of leaves and roots of eleven peanut genotypes were collected after 90 days of planting for sodium and potassium, measurements according to Page et al. (1982). On the other hand, leaves were collected for proline determination according to the protocol of Bates et al. (1973).

16.2.2.2   Molecular Genetic Level

DNA was extracted by Gen Elute™ plant DNA extraction kit (Sigma-Aldrich) as follow:

a. **Disrupting cells**

   Plant tissues were grinded into a fine powder in liquid nitrogen using a mortar and pestle. 100 mg of the powder was transferred to a microcentrifuge tube. The samples were kept on ice for immediate use or were frozen at $-70\ °C$.

b. **Lysing cells**

   350 μl of lysis solution [Part A] and 50 μl of lysis solution [Part B] were added to the tube; then were mixed well by vortexing and inverting. A white precipitate was formed upon the addition of a lysis solution [Part B]. The mixture was incubated at 65 °C for 10 min with occasional inversion to dissolve the precipitate. 50 units of RNAase were added to the lysis mixture just prior to incubation at 65 °C.

c. **Precipitating Debris**

   130 μl of precipitation solution was added to the mixture, and then the mixture was mixed completely by inversion, and the sample was placed on ice for 5 min. The sample was centrifuged at maximum speed (12,000–16,000 $\times$ g) for 5 min to pellet the cellular debris, proteins and polysaccharides.

d. Filtering Debris

The supernatant from the previous step was carefully pipetted on to a Gen Elute filtration column (blue insert with a 2 ml collection tube), and then was centrifuged at maximum speed for 1 min. This removes any cellular debris not removed in step 3 (previous step). The filtration column was discarded, but the collection tube was kept.

e. Preparing for Binding

700 µl was added to the binding solution directly to the flow through the liquid from step (4). Then it was mixed thoroughly by inversion.

f. Preparing the Binding Column

A Gen Elute Miniprep Binding Column (with a red O-ring) was inserted into a provided microcentrifuge tube if it not already assembled. 500 µl of the Column Preparation Solution was added to each miniprep column and was centrifuged at $12,000 \times g$ for 30 s to 1 min. The flow-through liquid was discarded.

g. Loading Lysate

700 µl of the mixture from step 5 was pipetted carefully onto the column prepared in step 6 and was centrifuged at maximum speed for 1 min. The flow-through liquid was discarded; while the collection tube was kept. The remaining lysate from step 5 was applied onto the column. The centrifugation was repeated as above, and the flow-through liquid was discarded, and the collection tube was kept.

h. First Column Washing

Prior to first-time use, ethanol was added to the wash solution concentrate. The binding column was placed into a 2 ml fresh collection tube and 500 µl was applied to the diluted wash solution to the column. Then it was centrifuged at maximum speed for 1 min. The flow-through liquid was discarded, but the collection tube was kept.

i. Second Column Washing

Another 500 µl of diluted wash solution was applied to the column and centrifuged at maximum speed for 3 min to dry the column. Care was taken not to allow the flow-through liquid to contact the column, as well as, any fluid that was adhered to the outside of the column was wiped off.

j. Eluting DNA

The binding column was transferred to a fresh 2 ml collection tube. 100 µl of pre-warmed (65°C) Elution Solution was applied to the column and centrifuged at maximum speed for 1 min. DNA was stored at $-20\ °C$ until it was used.

16.2.2.3   Primer Designing

The primer was designed to detect the KAT1 gene at 543 bp in Egyptian and Nigerian peanut genotypes based on the genome sequence of KAT1 gene of *Arabidopsis thaliana* as deposited in GeneBank (accession number: NM_123993). The sequence of the specific primer as follows:

F: 5'-GGACTAGCCGAACCAGAACC-3'
R: 5'-CATGAGCATGCATCGCACTC-3'

16.2.2.4   Polymerase Chain Reaction Condition

PCR amplification reaction was performed in a total reaction volume of 25 μl containing 100 ng of template DNA, 10 pM/μl specific primers and 12.5 μl of commercially available PCR master mix (Promega, UK). The PCR amplifications were carried out in a Biometra thermocycler. Cycling conditions were as follows: pre-denaturation 3 min. at 94 °C, followed by 35 Cycles at 94 °C for 30 s. (denaturation), 55°C for 1 min. (annealing) and 72 °C for 50 s. (polymerization) with a final extension at 72 °C for 5 min. After amplification, 10 μl of PCR product was loaded in a 2% agarose gel; electrophoresis was done at 120 V for 1 h, with 100 bp DNA ladder (LAROVA®, Germany) as a standard molecular weight marker. Gels were stained with ethidium bromide and photographed under UV-light for examination.

16.2.2.5   Sequence Analysis

The three PCR products of amplified specific nucleotide sequence gene were purified and sequenced in Macrogen, Inc. (Seoul, South Korea). The nucleotide sequences were aligned with existing sequences of KAT1 gene in the GenBank databases using BLAST programs and databases of the NCBI (National Center for Biotechnology Information, Bethesda, MD, USA) (www.blast.ncbi.nlm.nih.gov/Blast.cgi). Two sets of nucleotides sequence at Giza 6 and Ismailia 1 (the Egyptian peanut genotypes) and one set nucleotide sequence at Samnut 22 (the Nigerian peanut genotype) at the Molecular mass of 543 bp were separately aligned against homologous sequences reference in GenBank using ClustalW method in the Bioedit software version 7.2.5.

16.3   Statistical Analysis

Data were exposed to proper statistical analysis of a completely randomized design (CRD) and were subjected to two-factors of analysis of variance by using two-way ANOVA. The treatment means were compared by Duncan multiple range test at 0.05 confidence level by the use of Co-Stat software (Gomez and Gomez 1984).

16.4 Results

16.4.1 Growth Parameters Under Greenhouse Conditions

16.4.1.1 Plant Height (cm)

Means of plant height of peanut genotypes as affected by NaCl concentrations are presented in Table 16.1 and Fig. 16.1. Data show that increasing NaCl concentration from 0 up to 1000 and 3000 mgl^{-1} led to a significant decrease in plant height overall genotypes with values of 67.00, 57.55 and 48.88 cm, respectively.

Among Nigerian genotypes, Samnut 22 recorded the highest plant height (72.89 cm), while Samnut 21 recorded the least plant height overall the NaCl concentrations (41.56 cm). On the other hand, among the Egyptian genotypes, Giza 6 recorded 78.11 cm (Table 16.1 and Fig. 16.2).

Giza 6 recorded the highest plant height (79.00 cm) under 1000 mgl^{-1} NaCl and R 92 recorded the highest plant height (68.00 cm) under 3000 mgl^{-1} NaCl among Egyptian genotypes. On the other hand, the highest plant height among Nigerian genotypes was 62.67 and 52.67 cm in Samnut 22 under 1000 and 3000 mgl^{-1} NaCl, respectively (see Table 16.1).

Table 16.1 Interactive effects of peanut genotypes and NaCl levels on plant height (cm) under greenhouse conditions

Name	NaCl (mgl^{-1})			Mean
	0	1000	3000	
Sohag 110	76.00^c	61.33^i	63.33^{ghi}	66.89^c
Giza 6	88.00^a	79.00^b	67.33^{ef}	78.11^a
Giza 5	77.33^{bc}	74.33^{cd}	67.00^{ef}	72.89^b
R92	71.67^d	75.00^c	68.00^{ef}	71.56^b
Ismailia 1	65.00^{fgh}	68.67^e	55.33^{jk}	63.00^d
Sohag 104	62.33^{hi}	42.00^m	37.33^n	47.22^g
Ex-Dakar	72.00^d	44.00^m	36.00^n	50.67^f
Samnut 24	66.00^{efg}	31.67^o	35.67^n	44.44^h
Samnut 23	56.00^j	44.67^m	34.67^n	45.11^h
Samnut 22	48.00^l	62.67^{hi}	52.67^k	54.44^e
Samnut 21	54.67^{jk}	49.67^l	20.33^p	41.56^i
Mean	67.00^a	57.55^b	48.88^c	

Means within a column showing the same letters are not significantly different ($P \leq 0.05$)

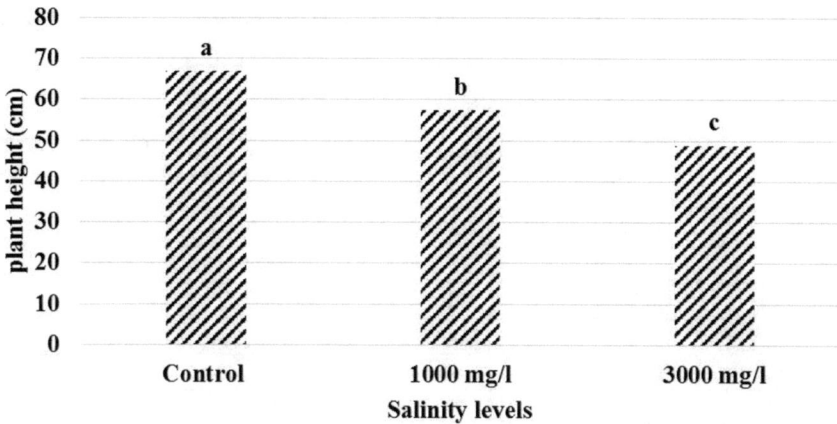

Fig. 16.1 The effect of different salinity levels on plant height (cm) under greenhouse conditions

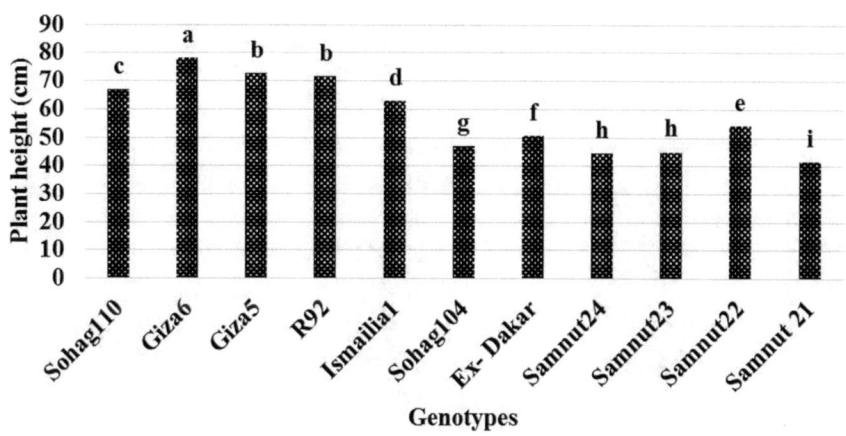

Fig. 16.2 The effect of genotypes on plant height (cm) under greenhouse conditions

16.4.1.2 Number of Shoots per Plant

Results in Table 16.2 and Fig. 16.3 reveal that there were significant differences among different salt concentrations in a number of shoots per plant. The highest number of shoots plant^{-1} was obtained from control (3.88) and followed by 1000 and 3000 mgl^{-1} NaCl (3.15 and 2.42, respectively).

Overall the NaCl concentrations, the highest number of shoots plant^{-1} was recorded on Egyptian genotype Sohag 110 and Giza 5 without significant difference with Giza 6 (4.67) and Nigerian genotype Samnut 21 (2.67), while the least number of shoots plant^{-1} was observed in Sohag 104 (3.67) among the Egyptian genotypes and Samnut 22 among the Nigerian ones (1.33) (Table 16.2 and Fig. 16.4).

Table 16.2 Interactive effects of peanut genotypes and NaCl levels number of shoots plant-1 under greenhouse conditions

Name	NaCl (mgl^{-1})			Mean
	0	1000	3000	
Sohag 110	5.00^{ab}	5.30^a	3.70^{cdef}	4.67^a
Giza 6	5.00^{ab}	5.30^a	3.30^{def}	4.56^a
Giza 5	5.00^{ab}	4.70^{abc}	4.30^{abcd}	4.67^a
R92	4.67^{abc}	4.00^{bcde}	3.70^{cdef}	4.12^{ab}
Ismailia 1	4.30^{abcd}	3.30^{def}	3.70^{cdef}	3.77^b
Sohag 104	5.30^a	3.00^{efg}	2.70^{fgh}	3.67^b
Ex-Dakar	3.30^{def}	1.30^i	1.00^i	1.89^d
Samnut 24	2.70^{fgh}	1.70^{hi}	1.00^i	1.78^d
Samnut 23	1.70^{hi}	2.00^{ghi}	1.00^i	1.56^d
Samnut 22	1.70^{hi}	1.00^i	1.30^i	1.33^d
Samnut 21	4.00^{bcde}	3.00^{efg}	1.00^i	2.67^c
Mean	3.88^a	3.15^b	2.42^c	

Means within a column showing the same letters are not significantly different (P ≤ 0.05)

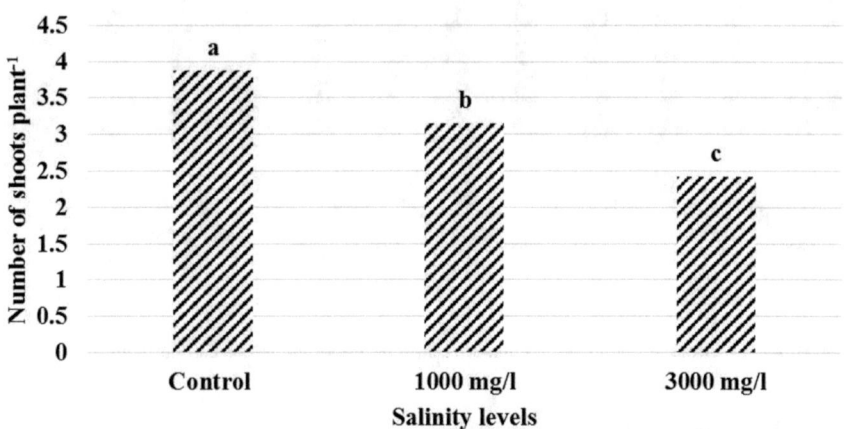

Fig. 16.3 The effect of different salinity levels on a number of shoots under greenhouse conditions

Effect of the interaction between genotypes and NaCl concentrations showed that the highest number of shoots plant^{-1} was recorded on Samnut 21 and Sohag 104 among Nigerian and Egyptian genotypes were 4.00 and 5.30, respectively under control condition. While under salinity conditions, the highest shoots plant^{-1} was 3.00 and 5.30 under 1000 mgl^{-1} of NaCl in Sohag 110 and Giza 6, respectively among Egyptian genotypes and 3.00 for samnut 22 among Nigerian genotypes. At

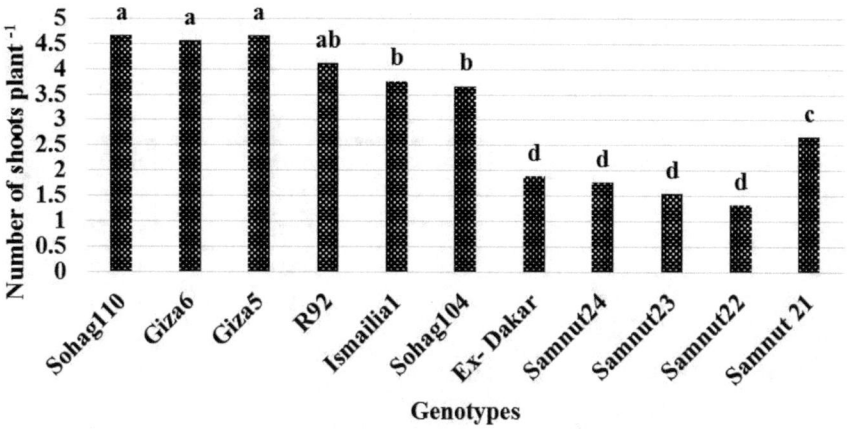

Fig. 16.4 The effect of genotypes on a number of shoots under greenhouse conditions

3000 mgl^{-1} NaCl, Giza 5 and Samnut 22 recorded 4.30 and 1.30, respectively, as shown in Table 16.2.

16.4.1.3   Number of Leaves per Plant

NaCl concentrations had a significant effect on a number of leaves per plant. Data in Table 16.3 and Fig. 16.5 show that the highest reduction was observed when salinity concentration increased from 1000 and 3000 mgl^{-1} NaCl (3.79 and 26.88), respectively.

A number of leaves plant^{-1} of all evaluated peanut genotypes significantly varied in response to salinity concentrations. A number of leaves plant^{-1} ranged from 39.22 to 52.78 (Egyptian genotypes) and 20.44 to 13.11 (Nigerian genotypes). Egyptian genotype Ismailia 1 showed the highest number of leaves plant^{-1}, as well as, Nigerian genotype Samnut 21 (Table 16.3 and Fig. 16.6).

Giza 5 showed the highest number of leaves plant^{-1} (55.00) under 3000 mgl^{-1} NaCl among Egyptian genotypes while Samnut 22 gave the highest number of leaves plant^{-1} (17.67) among Nigerian genotypes under the same salt concentration (see Table 16.3).

16.4.1.4   Leaf Area (cm^2)

Means of leaf area of peanut genotypes as affected by NaCl concentrations are presented in Table 16.4 and Fig. 16.7. Data show that increasing NaCl concentration from 0 up to 1000 and 3000 mgl^{-1} led to a significant decrease in leaf area overall genotypes valued 18.78, 15.76 and 13.94 cm^2, respectively.

Table 16.3 Interactive effects of peanut genotypes and NaCl levels on a number of leaves plant-1 under greenhouse conditions

Name	NaCl (mgl^{-1})			Mean
	0	1000	3000	
Sohag 110	56.70^c	54.00^{cde}	40.33^i	50.33^{bc}
Giza 6	53.33^{de}	60.00^b	35.33^{jk}	49.56^c
Giza 5	48.67^{fg}	51.33^{ef}	55.00^{cd}	51.67^{ab}
R92	48.00^g	45.00^h	44.67^h	45.89^d
Ismailia 1	70.00^a	37.00^j	51.33^{ef}	52.78^a
Sohag 104	72.33^a	21.67^m	23.67^m	39.22^e
Ex-Dakar	28.67^l	16.00^{nop}	5.30^u	16.67^g
Samnut 24	28.67^l	10.67^{rs}	7.67^{tu}	15.67^g
Samnut 23	12.67^{qr}	17.30^{no}	9.33st	13.11^h
Samnut 22	14.67^{opq}	14.00^{pq}	17.67^n	15.44^g
Samnut 21	33.33^k	22.67^m	5.33^u	20.44^f
Mean	42.45^a	31.79^b	26.88^c	

Means within a column showing the same letters are not significantly different (P ≤ 0.05)

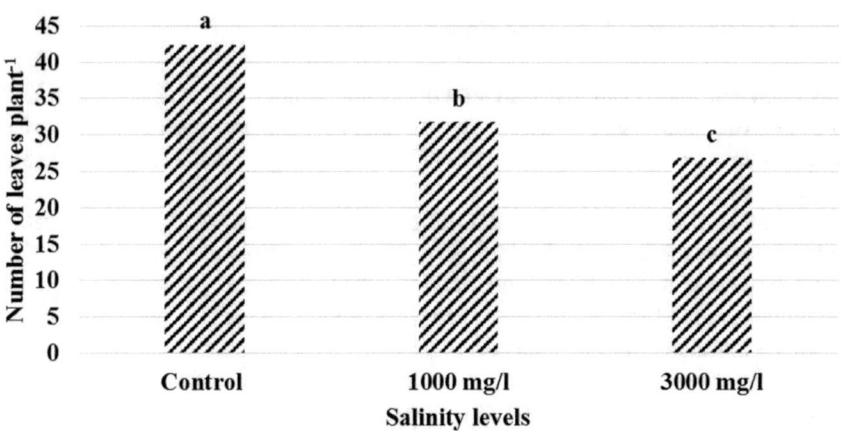

Fig. 16.5 The effect of different salinity levels on a number of leaves under greenhouse conditions

Among Nigerian genotypes, Ex-Dakar recorded the highest leaf area (25.68 cm^2), while Samnut 21 recorded the least leaf area overall the NaCl concentrations (8.71 cm^2). On the other, among the Egyptian genotypes, Sohag 104 recorded the highest leaf area (19.13 cm^2), while Giza 5 recorded the least leaf area overall the NaCl concentrations (11.91 cm^2) (Table 16.4 and Fig. 16.8).

Sohag 104 recorded the highest leaf area (18.33 and 16.67 cm^2) under 1000 and 3000 mgl^{-1} NaCl, respectively among Egyptian genotypes. On the other hand,

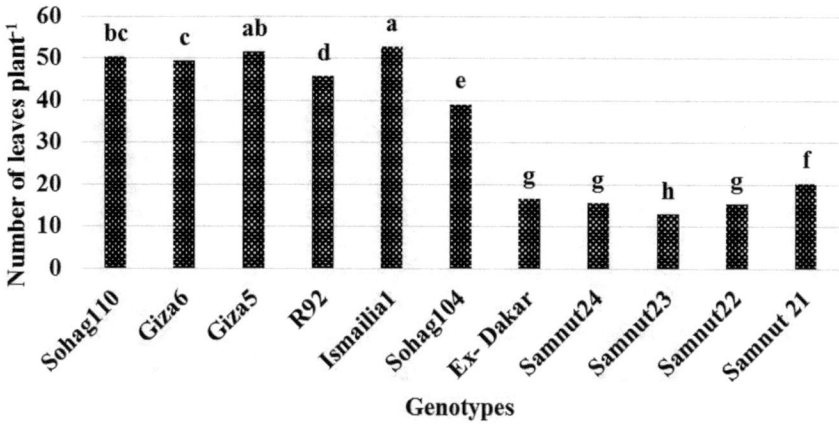

Fig. 16.6 The effect of genotypes on a number of leaves per plant under greenhouse conditions

Table 16.4 Interactive effects of peanut genotypes and NaCl levels on leaf area (cm^2) under greenhouse conditions

Name	NaCl (mgl^{-1})			Mean
	0	1000	3000	
Sohag 110	15.67^{ijk}	12.57^{mno}	11.17^{nop}	13.13^g
Giza 6	16.17^{hij}	12.63^{mno}	8.60^q	12.47^{gh}
Giza 5	13.13^{lmn}	11.83^{no}	10.77^{op}	11.91^h
R92	16.67^{ghij}	14.83^{jkl}	13.83^{klm}	15.11^f
Ismailia 1	16.80^{ghij}	17.50^{fghi}	15.50^{ijk}	16.60^e
Sohag 104	22.40^c	18.33^{efg}	16.67^{ghij}	19.13^c
Ex-Dakar	31.37^a	19.17^{def}	26.50^b	25.68^a
Samnut 24	22.43^c	20.50^d	19.67^{de}	20.87^b
Samnut 23	22.80^c	20.50^d	9.70^{pq}	17.66^d
Samnut 22	18.03^{efgh}	16.17^{hij}	15.33^{jk}	16.51^e
Samnut 21	11.13^{nop}	9.33^{pq}	5.67^r	8.71^i
Mean	18.78^a	15.76^b	13.94^c	

Means within a column showing the same letters are not significantly different ($P \leq 0.05$)

the highest leaf area among Nigerian genotypes was 20.50 cm^2 in Samnut 24 and samnut 23 under 1000 NaCl. In addition under 3000 mgl^{-1} NaCl Ex-Dakar verified the highest leaf area 26.5 cm^2 (see Table 16.4).

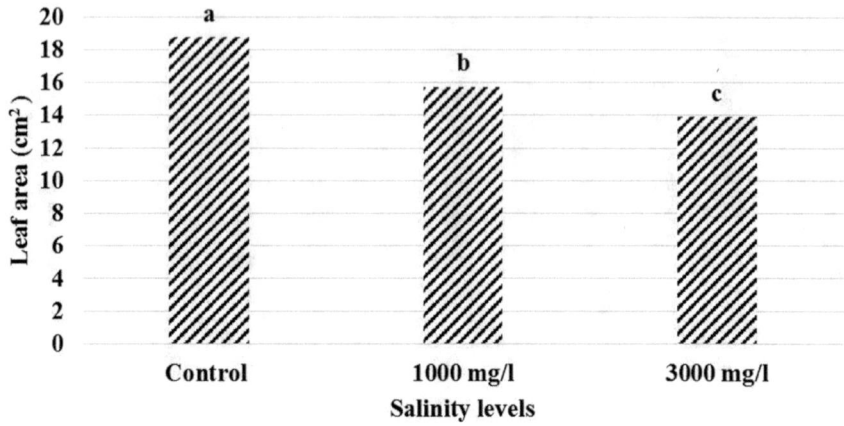

Fig. 16.7 The effect of different salinity levels on leaf area under greenhouse conditions

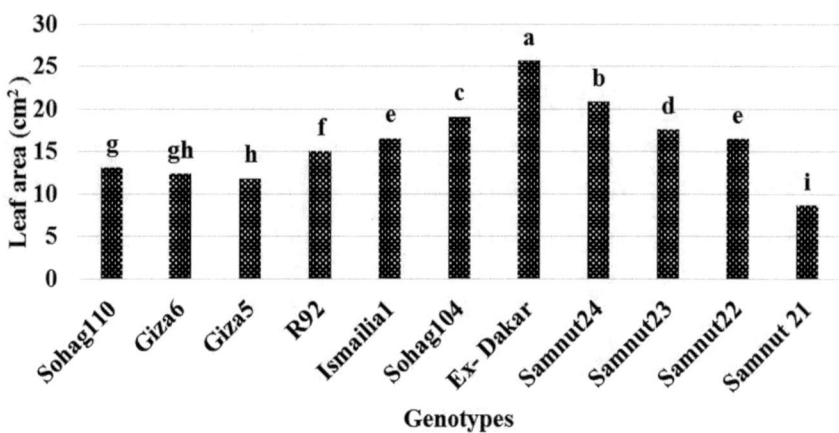

Fig. 16.8 The effect of genotypes on leaf area under greenhouse conditions

16.4.2 Minerals Accumulation

16.4.2.1 Sodium Accumulation

Sodium accumulation in leaves was affected by the salinity level (see Fig. 16.9). The lowest Na^+ concentration in the leaves were observed under control (0 mgl^{-1} NaCl non-saline conditions), it recorded 415.45 mgl^{-1}, while the highest Na^+ concentration in the leaves were observed from (3000 mgl^{-1} NaCl) was 1218.18 mgl^{-1}. Sodium concentration increased in response to salt treatment (see Table 16.5).

The effects of genotypes on the adsorption of sodium in leaves are presented in Fig. 16.10 and Table 16.5. The highest adsorption for sodium was recorded on

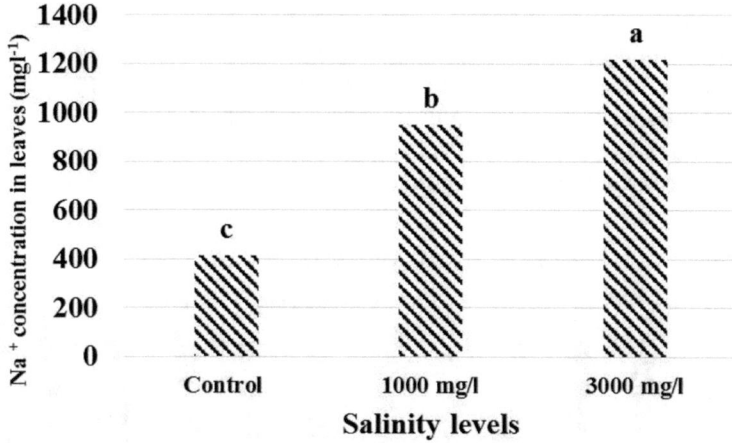

Fig. 16.9 The effect of salinity levels on Na$^+$ concentration in the leaves under greenhouse conditions

Table 16.5 Interactive effects of peanut genotypes and NaCl levels on Na$^+$ accumulation in leaves under greenhouse conditions

Name	NaCl (mgl^{-1})			Mean
	0	1000	3000	
Sohag 110	440.00^i	850.00^{gh}	950.00^{efgh}	746.67^d
Giza 6	450.00^i	950.00^{efgh}	1150.00^{bcd}	850.00^{bc}
Giza 5	430.00^i	926.00^{efgh}	1050.00def	802.00^{cd}
R92	435.00^i	800.00^h	1000.00^{defg}	745.00^d
Ismailia 1	350.00^i	1100.00^{cde}	1500.00^a	983.33^a
Sohag 104	470.00^i	950.00^{efgh}	1250.00^{bc}	890.00^{abc}
Ex-Dakar	400.00^i	980.00^{efg}	1150.00^{bcd}	843.33^{bc}
Samnut 24	410.00^i	900.00^{fgh}	1250.00^{bc}	853.33^{bc}
Samnut 23	420.00^i	1000.00^{defg}	1300.00^b	906.67^{ab}
Samnut 22	345.00^i	1100.00^{cde}	1500.00^a	981.67^a
Samnut 21	420.00^i	900.00^{fgh}	1300.00^b	873.33^{bc}
Mean	415.45^c	950.54^b	1218.18^a	

Means within a column showing the same letters are not significantly different (P ≤ 0.05)

Ismailia 1 (983.33) among Egyptian genotypes and Samnut 22 (981.00) among Nigerian genotypes.

The highest contents of sodium ion in leaves (1100.00 and 1500.00 mgl^{-1}) were recorded on Ismailia 1 and Samnut 22 at 1000 and 3000 mgl^{-1} NaCl, respectively. While control recorded 470.00 mgl^{-1} on leaves for Sohag 104 of Egyptian genotypes and 420.00 mgl^{-1} for Samnut 23 of Nigerian genotypes (see Table 16.5).

Results of Na$^+$ accumulation in root tissues indicated that the reactions of peanut genotypes were significantly differed by increasing NaCl concentrations under 0, 1000 and 3000 mgl^{-1} NaCl, they recorded 415.45, 1182.97 and 1218.18 mgl^{-1} sodium ion, respectively as shown in Table 16.6 and Fig. 16.11.

Accumulation of sodium in roots of peanut genotypes showed that the highest accumulation for sodium was recorded on Ismailia 1 (1225 mgl^{-1}) among Egyptian

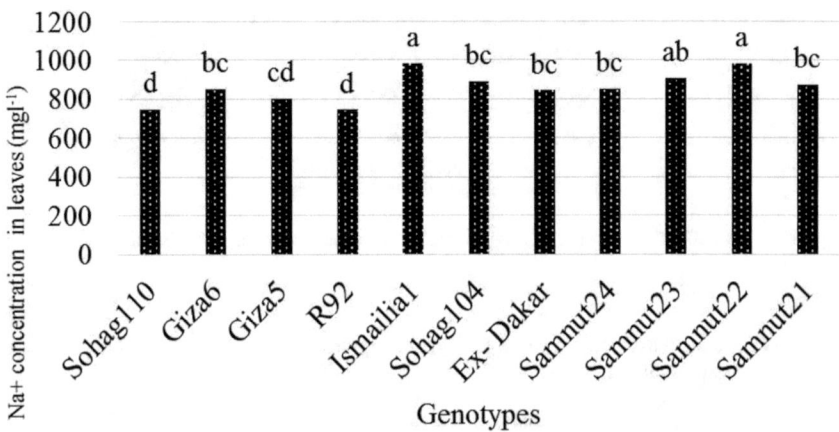

Fig. 16.10 The effect of genotypes on Na$^+$ concentration in the leaves under greenhouse conditions

Table 16.6 Interactive effects of peanut genotypes and NaCl levels on Na$^+$ accumulation in roots under greenhouse conditions

Name	NaCl (mgl^{-1})			Mean
	0	1000	3000	
Sohag 110	520.00^h	920.00^g	1000.00^{fg}	813.33^d
Giza 6	510.00^h	1000.00^{fg}	1500.00^{bc}	1003.33^{bc}
Giza 5	550.00^h	940.00^g	1400.00^{bcd}	963.33^{bc}
R92	500.00^h	950.00^{fg}	1200.00^{def}	833.33^{cd}
Ismailia 1	475.00^h	1200.00^{def}	2000.00^a	1225.00^a
Sohag 104	470.00^h	1000.00^{fg}	1500.00^{bc}	990.00^{bc}
Ex-Dakar	490.00^h	1050.00^{fg}	1300.00^{cde}	946.67^{bc}
Samnut 24	520.00^h	900.00^g	1500.00^{bc}	973.30^{bc}
Samnut 23	510.00^h	1000.00^{fg}	1600.00^b	1036.67^b
Samnut 22	470.00^h	1200.00^{def}	1900.00^a	1190.00^a
Samnut 21	485.00^{lmh}	1100.00^{efg}	1600.00^b	1061.6^b
Mean	415.45^b	1182.97^a	1218.18^a	

Means within a column showing the same letters are not significantly different (P ≤ 0.05)

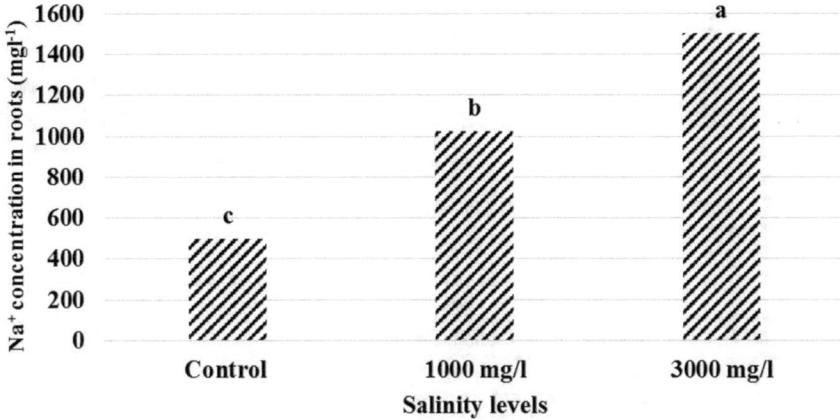

Fig. 16.11 The effect of salinity levels on Na$^+$ concentration in the roots under greenhouse conditions

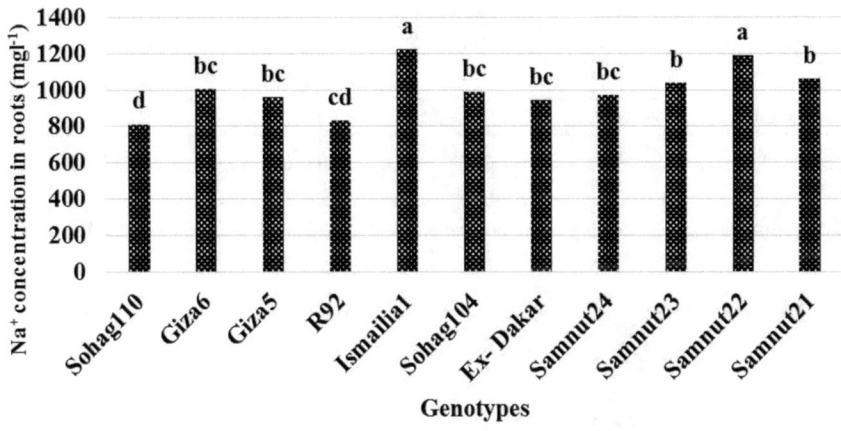

Fig. 16.12 The effect of genotypes on Na$^+$ concentration in the roots under greenhouse conditions

genotypes and Samnut 22 (1190 mgl^{-1}) among Nigerian genotypes (Table 16.6 and Fig. 16.12).

Concerning of Egyptian genotypes, Giza 5 recorded the highest contents of sodium ion in roots (550.00 mgl^{-1}) under control, while, the highest contents of sodium ion (1200.00 and 2000.00 mgl^{-1}) were recorded on Ismailia 1 under 1000 and 3000 mgl^{-1} NaCl, respectively as shown in Table 16.6. Regards to Nigerian genotypes, Samnut 24 recorded the highest contents of sodium ion in roots (520.00 mgl^{-1}) under control, while, the highest contents of sodium ion (1200.00 and 1900.00 mgl^{-1}) were recorded on Samnut 22 under 1000 and 3000 mgl^{-1} NaCl, respectively (see Table 16.6).

16.4.2.2 Potassium Accumulation

In this study, K^+ concentration decreased with increasing salinity levels (Table 16.7 and Fig. 16.13). The effect of salinity levels on K^+ accumulation in the leaves of

Table 16.7 Interactive effects of peanut genotypes and NaCl levels on K^+ accumulation in leaves under greenhouse conditions

Name	NaCl (mgl^{-1})			Mean
	0	1000	3000	
Sohag 110	100.00^{cde}	60.00^{efgh}	35.00^{gh}	65.00^{abc}
Giza 6	110.00^{cd}	55.00^{efgh}	20.00^{h}	61.67^{abc}
Giza 5	80.00^{defg}	50.00^{efgh}	25.00^{h}	51.67^{c}
R92	90.00^{def}	40.00^{fgh}	30.00^{gh}	53.33^{bc}
Ismailia 1	210.00^{a}	30.00^{gh}	10.00^{h}	83.33^{a}
Sohag 104	200.00^{a}	40.00^{fgh}	20.00^{h}	86.67^{a}
Ex-Dakar	190.00^{a}	50.00^{efgh}	25.00^{h}	88.33^{a}
Samnut 24	170.00^{ab}	55.00^{efgh}	20.00^{h}	81.67^{ab}
Samnut 23	140.00^{bc}	40.00^{fgh}	15.00^{h}	65.00^{abc}
Samnut 22	200.00^{a}	30.00^{gh}	10.00^{h}	80.00^{abc}
Samnut 21	200.00^{a}	35.00^{gh}	15.00^{h}	83.33^{a}
Mean	153.64^{a}	44.09^{b}	20.45^{c}	

Means within a column showing the same letters are not significantly different ($P \leq 0.05$)

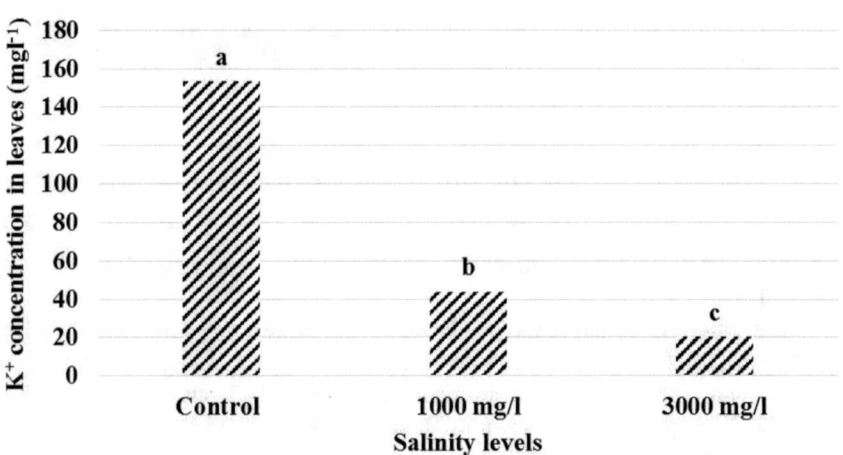

Fig. 16.13 The effect of salinity levels on K^+ concentration in the leaves under greenhouse conditions

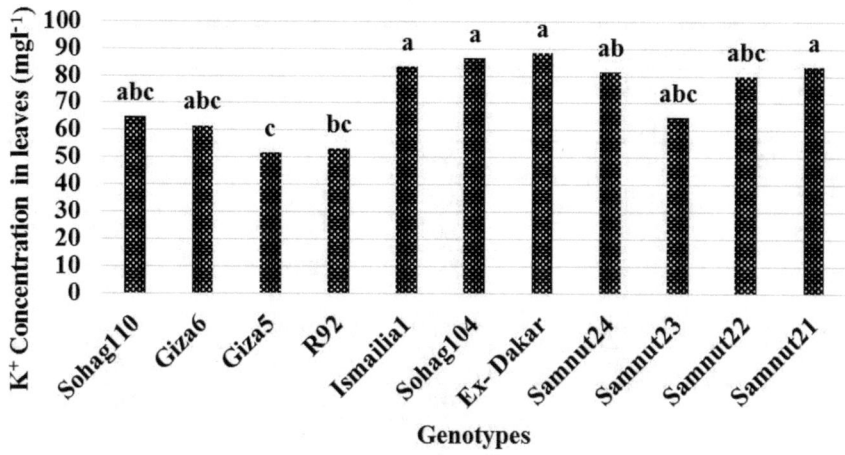

Fig. 16.14 The effect of genotypes on K$^+$ concentration in the leaves under greenhouse conditions

peanut genotypes was significant. Control application had the highest K$^+$ concentration in leaves (153.63 mgl^{-1}), while 3000 mgl^{-1} NaCl treatment had the lowest K$^+$ concentration in leaves (20.45 mgl^{-1}).

The effects of genotypes on the accumulation of potassium in leaves confirmed that the highest accumulation for potassium was recorded on Sohag 104 (Egyptian genotype), as well as, Ex-Dakar (Nigerian genotype) with values of 86.67 and 88.33 mgl^{-1} K$^+$, respectively (Table 16.7 and Fig. 16.14).

Concerning of Egyptian genotypes, Ismailia 1 recorded the highest contents of potassium ion in leaves (210.00 mgl^{-1}) under control. While, the highest contents of potassium ion (60.00 and 35.00 mgl^{-1}) were recorded on Sohag 110 under 1000 and 3000 mgl^{-1} NaCl, respectively (Table 16.7). Concerning of Nigerian genotypes, Samnut 21and Samnut 22 recorded the highest contents of potassium ion in leaves (200.00 mgl^{-1}) under control. However, Samnut 24.00 mgl^{-1} recorded 50 mgl^{-1} under 1000 mgl^{-1} NaCl, and Ex-Dakar recorded 25.00 mgl^{-1} under 3000 mgl^{-1} NaCl (Table 16.7).

Under the control conditions, the highest K$^+$ accumulation in peanut roots was 92.73 mgl^{-1}, while 3000 mgl^{-1} NaCl concentration recorded the lowest K$^+$ accumulation in roots (18.18 mgl^{-1}), as in Table 16.8 and Fig. 16.15.

The effects of genotypes on the accumulation of potassium ions in roots presented in Table 16.8 and Fig. 16.16 show that the highest accumulation for potassium was recorded on Giza 6 among Egyptian genotypes and Samnut 21 among Nigerian genotypes, 58.33 and 55.00 mgl^{-1}, respectively.

Concerning of Egyptian genotypes, Ismailia 1 recorded the highest contents of potassium ion in roots (110.00 mgl^{-1}) under control. While, the highest contents of potassium ion (60.00 and 24.00 mgl^{-1}) were recorded on Sohag 110 under 1000 and 3000 mgl^{-1} NaCl, respectively (Table 16.8).Concerning of Nigerian genotypes, Samnut 22 and Samnut 21 recorded the highest contents of potassium ion in roots

Table 16.8 Interactive effects of peanut genotypes and NaCl levels on K$^+$ accumulation in roots under greenhouse conditions

Name	NaCl (mgl^{-1})			Mean
	0	1000	3000	
Sohag 110	90.00^{bc}	60.00^{de}	24.00^{gh}	58.00^a
Giza 6	100.00^{ab}	55.00^{ef}	20.00^h	58.33^a
Giza 5	80.00^c	50.00^{ef}	21.00^{gh}	50.33^{ab}
R92	90.00^{bc}	40.00^{fg}	22.00^{gh}	50.67^{ab} bc
Ismailia 1	110.00^a	25.00^{gh}	10.00^h	48.33^{ab}
Sohag 104	90.00^{bc}	40.00^{fg}	20.00^h	50.00^{ab}
Ex-Dakar	85.00^{bc}	50.00^{ef}	23.00^{gh}	52.67^{ab}
Samnut 24	80.00^c	40.00^{fg}	20.00^h	46.67^b
Samnut 23	75.00^{cd}	55.00^{ef}	15.00^h	48.33^{ab}
Samnut 22	110.00^a	25.00^{gh}	10.00^h	48.33^{ab}
Samnut 21	110.00^a	40.00^{fg}	15.00^h	55.00^{ab}
Mean	92.73^a	43.64^b	18.18^c	

Means within a column showing the same letters are not significantly different (P $\leq$ 0.05)

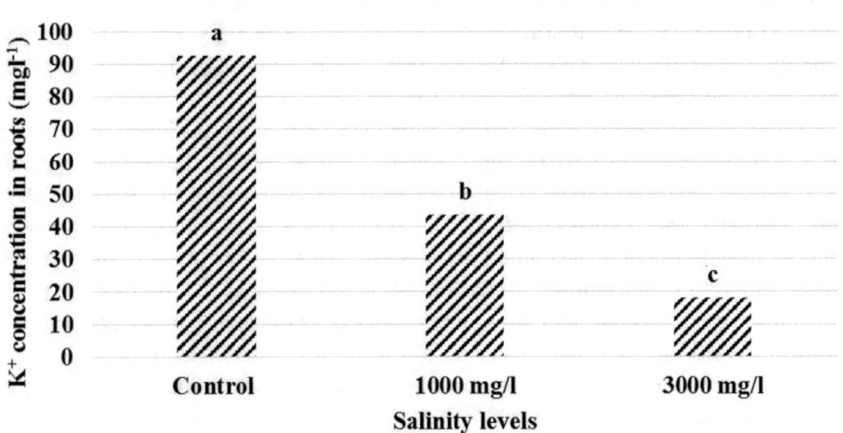

Fig. 16.15 The effect of salinity levels on K$^+$ concentration in the roots under greenhouse conditions

(110.00 mgl^{-1}) under control. Whereas, the highest contents of potassium ion (55 mgl^{-1}) was recorded on Samnut 23 under 1000 and Ex-Dakar recorded 23 mgl^{-1} under 3000 mgl^{-1} NaCl, as in Table 16.8.

16.4.2.3 Proline Accumulation

Significant proline accumulation was observed in 3000 mgl^{-1} NaCl treatment (555.90 mgl^{-1}) compared with control (266.70 mgl^{-1}) as shown in (Table 16.9) and Fig. 16.17.

The effects of genotypes on proline accumulation in leaves are presented in Table 16.9 and Fig. 16.18. The highest proline accumulation was recorded on Ismailia

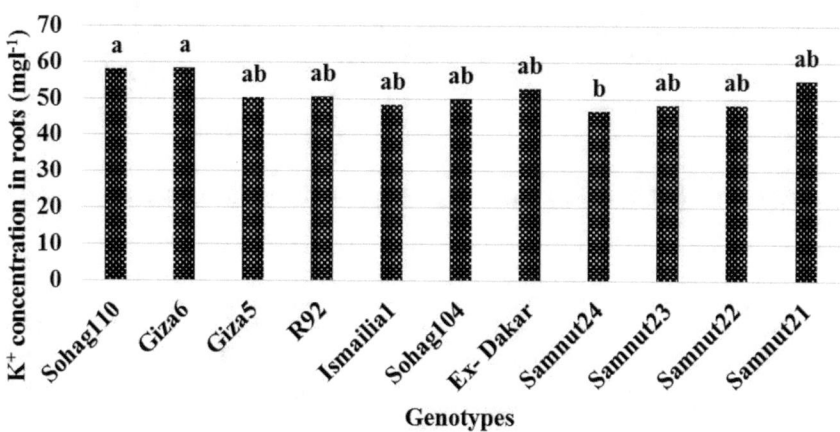

Fig. 16.16 The effect of genotypes on K$^+$ concentration in the roots under greenhouse conditions

Table 16.9 Interactive effects of peanut genotypes and NaCl levels on proline in leaves under greenhouse conditions

Name	NaCl (mgl^{-1})			Mean
	0	1000	3000	
Sohag 110	233.70^r	187.40^t	436.30^{jk}	285.80^g
Giza 6	121.70^v	208.80^s	494.10^h	274.86^h
Giza 5	307.40^p	347.40^{no}	550.90^g	401.90^d
R92	288.80^q	282.60^q	706.40^d	425.90^c
Ismailia 1	349.20^{mno}	588.20^f	753.50^c	563.70^a
Sohag 104	159.90^u	631.80^e	358.10^{mn}	383.30^e
Ex-Dakar	203.50st	386.50^l	276.30^q	288.80^g
Samnut 24	149.20^u	359.90^{mn}	795.30^b	434.80^c
Samnut 23	333.20^o	245.20^r	450.50^j	343.00^f
Samnut 22	367.90^m	399.00^l	825.50^a	530.80^b
Samnut 21	419.40^K	311.90^p	468.30^i	399.80^d
Mean	266.70^c	359.00^b	555.9^a	

Means within a column showing the same letters are not significantly different (P ≤ 0.05)

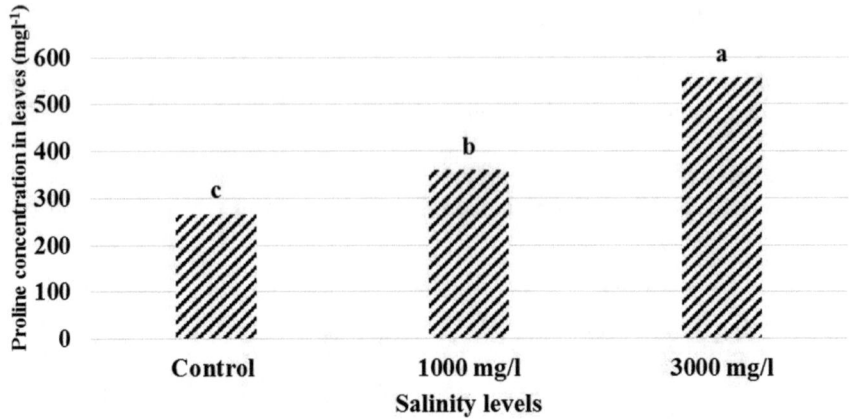

Fig. 16.17 The effect of salinity levels on proline accumulation in leaves under greenhouse conditions

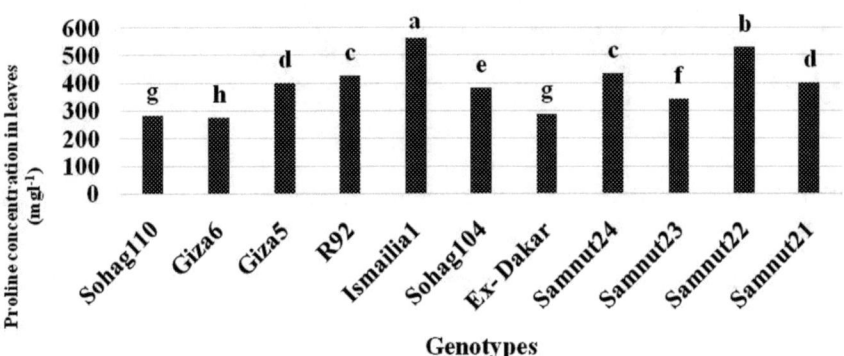

Fig. 16.18 The effect of genotypes on proline accumulation in leaves under greenhouse conditions

1 (563.70 mgl^{-1}) among Egyptian genotypes and Samnut 22 (530.00 mgl^{-1}) among Nigerian genotypes.

Concerning of Egyptian genotypes, Ismailia 1 recorded the highest proline accumulation in leaves (349.20, 588.50 and 753.50 mgl^{-1}) at control, 1000 and 3000 mgl^{-1} NaCl, respectively (Table 16.9). Concerning of Nigerian genotypes, Samnut 22 recorded the highest proline accumulation in leaves (367.90, 399.00 and 825.50 mgl^{-1}) at control, 1000 and 3000 mgl^{-1} NaCl, respectively (see Table 16.9).

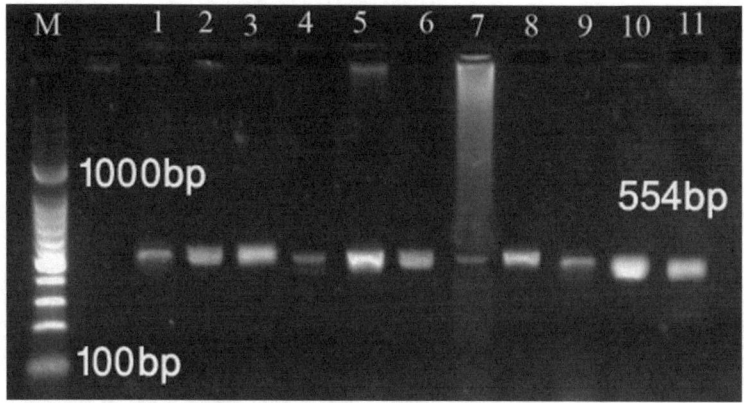

Fig. 16.19 Detection of KAT1 gene at 554 bp, M refers marker ladder (from 1000 to 100 bp) (1—Sohag 110; 2—Giza 6; 3—Giza 5; 4—R 92; 5—Ismailia 1, 6—Sohag 104, 7—Ex-Dakar, 8—Samnut 24, 9—Samnut 23, 10—Samnut 22 and 11—Samnut 21)

16.4.3   Molecular Genetics Studies

16.4.3.1   KAT1 Gene in Egyptian and Nigerian Peanut

In this study, the specific primer was designed to detect the KAT1 gene in Egyptian and Nigerian peanut based on sequence database of KAT1 gene of *Arapidopsis thaliana* in GenBank. Figure 16.19 shows amplified specific band of KAT1 gene at 554 bp in peanut. Samples from 1 to 6 represent Egyptian peanut and samples from 7 to 11 represent Nigerian peanut. Comparative genomic studies were investigated between *Arachis hypogaea* and four crops (*Arabidopsis, Glycine, Lotus* and *Medicago*). The maximum number of matches was with sequences from the *Lotus* Gene Indices (581 hits), followed by 491 hits against the *Arabidopsis* Gene Indices, 358 hits against the *Glycine* Gene Indices and only 221 hits against *Medicago* Gene Indices (Jayashree et al. 2005).

16.4.3.2   Sequence analysis

Sequence Alignments of KAT1 gene among Egyptian and Nigerian peanut generated under salinity conditions yielded 554 base pair with six stop codons and no gaps. The results of sequencing reported the existence of SNPs between the three different samples, where Giza 6 has one SNP, 4 and 5 SNPs for Samnut 22 and Ismailia 1 compared with KAT1 gene of *Arapidopsis thaliana* (Fig. 16.20), respectively. BLAST data showed identical 100% between the sequence of the Egyptian peanut genotype Giza 6 and sequence of KAT1 gene of *Arapidopsis thaliana*, and 99% between Ismailia 1 and Samnut 22 (accession number: NM_123993.3). Giza 6 and

Ismailia 1 share in one SNP (A) in codon 195, while, Ismailia 1 and Samnut 22 share in four SNPs (T) in codon 258, 259, 263 and 406 (see Table 16.10).

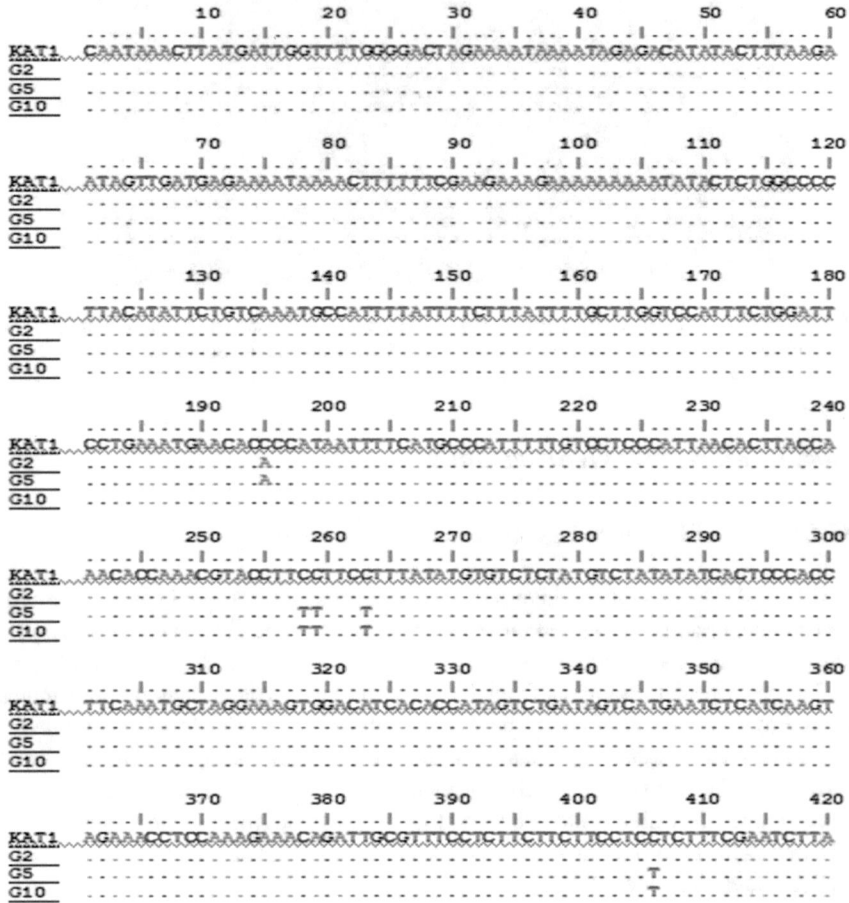

Fig. 16.20 ClustalW multiple alignments of KAT1 gene in *Arapidopsis thaliana* and two Egyptian peanut genotypes (G2 = Giza 6, G5 = Ismailia 1) and one Nigerian genotype (G10 = Samnut 22)

Table 16.10 SNPs of KAT1 gene of *Arachis hypogaea* compared with *Arapidopsis thaliana*

Position of SNPs	Giza5	Ismailia 1	Samnut 22
195	A	A	....
258	....	T	T
259	....	T	T
263	....	T	T
406	....	T	T

Fig. 16.21 Phylogenetic tree of KAT1 gene in Egyptian and Nigerian peanut generated under salinity conditions, where G2 = Giza 6, G5 = Ismailia 1 and G10 = Samnut 22 genotypes

16.4.3.3   Phylogenetic tree

The phylogenetic tree shows the evolutionary relationship of the sequences in which the length of the branch was proportional to the estimated genetic distance between the sequences (see Fig. 16.21). Giza 6 is more close to Ismailia 1 than Samnut 22, while Ismailia 1 and Samnut 22 very close, although Ismailia 1 and Giza 6 are Egyptian peanut genotypes and Samnut 22 is Nigerian one. Evolutionary tree emphasized the data of screening of all investigated genotypes for their salt tolerant, which reported that Ismailia 1 (G5) and Samnut 22 (G10) were tolerance for salinity while Giza 6 (G2) was a sensitive one.

16.5   Discussion

Sodium chloride causes decreasing in plant growth as a result of metabolic stress (Dodd and Donavan 1999). Salt tolerant plants may continue its growth in NaCl presence as a result of cellular responses such as compartmentation and osmotic adjustment (Volkmar et al. 1998). However, plant height is affected negatively by salt stress because cytokinesis and growth of the cell are inhibited due to salts toxic effect. Additionally, the reduction in hormones that stimulate the growth and rise in hormones that inhibit growth can cause less development in plant heights (Taiz and Zeiger 1998). Rising in osmotic pressure around the roots as a result of saline conditions can also stop water uptake by roots, causing shorter plant height (Graciano et al. 2011). Concerning a number of shoots per plants revealed a decline with increasing level of salinity (Lacerda et al. 2006). Shoot number may be related to a reduction rate of leaf production which led to a decrease in photosynthesis and accumulation of dry matter (Mansour et al. 2005). Leaf area reduction in saline environments is an adaptation to decrease roots ion uptake. Therefore, the plant reduced the surface of transpiration and the unprotected area in order to intercept photosynthetically active radiation, perhaps owing to the rise of the xylem ABA

(Abcisic Acid) concentration. This causes stomatal closure in the leaf and reduced leaf expansion (Mensah et al. 2006)

The results in the present chapter showed that K^+ uptake of peanut plant organs was significantly reduced with increasing salinity. On the other hand, Na^+ uptake is significantly increased. Similar observations were reported by Chakraborty et al. (2016). Under salt stress, greater accumulation of Na^+ in plants could as well serve to increase the cell solute potential and hence increase their osmotic adjustment (2015). NaCl in the plant environment decreased K^+ concentrations in the shoot and root in according with antagonism between Na^+ and K^+ [Kamoker et al. 2008]. Excess of sodium ions as a result of NaCl in soil depolarize cell membrane, causing loss of potassium ions. Sodium and potassium ions have similar chemical nature. Consequently, sodium inhibits potassium uptake by roots (Weimberg 1987).

Many studies also reported that salinity resulted in substantial proline accumulation in leguminous plants, including alfalfa, soybean, pea, peanut, green gram and chickpea (Garcia-Sanchez et al. 2002; Girija et al. 2002; Amirjani et al. 2010; Jain et al. 2006; Panda and Khan 2009; Hayat et al. 2012; Mansour and Ali 2017) the amount depending on the species, cultivar, tissue type and age (Celikkol et al. 2010). Many investigators found an accumulation of amino acids especially proline in plants exposed to salt stress. Proline accumulation may contribute to osmotic adjustment at the cellular level (Perez-Alfocea et al. 1993). Increase in proline content under NaCl stress may be due to the breakdown of a proline-rich protein or de novo synthesis of proline (Tewari and Singh 1991). Proline may act as an enzyme protectant, stabilizing the structure of macromolecules and organelles. Proline also acts as a major reservoir of energy and nitrogen for utilization upon exposure to salinity (Girija et al. 2002). Proline accumulation is reported as a symptom of stress in less salinity-tolerant plants, and it plays multiple roles in stress tolerance as a mediator of osmotic adjustment (Yoshiba et al. 1997).

The potassium is trasfered into and within the plant by membrane protein which differs with respect to their affinity for potassium (Schachtman and Schroeder 1994) and perhaps their mode of energization (Maathuis and Sanders 1993). KAT1 gene is expressed in guard cells and faces there fluctuations of external K^+ in the millimolar range (Fernández-Torquemada and Sánchez-Lizaso 2013).

KAT1 gene is expressed in guard cells and faces there fluctuations of external K^+ in the millimolar range (Fernández-Torquemada and Sánchez-Lizaso 2013). The results of alignment showed very similarity in sequence of KAT1 gene between applied peanut and *Arapidopsis thaliana*. Comparative genomic studies were investigated between *Arachis hypogaea* and four crops (*Arabidopsis*, soybean, *Lotus* and *Medicago*). The maximum number of matches was with sequences from the *L. japonicus* Gene Indices (581 hits), followed by 491 hits against the *Arabidopsis* Gene Indices, 358 hits against the soybean Gene Indices and only 221 hits against Medicago Gene Indices (Jayashree et al. 2005). The present study reported existence SNPs between the three different samples. Single amino acid changes in the pore region of a channel or transporter can change the selectivity for cations, so a single base-pair mutation can have a large effect (Munns 2005).

16.6   Conclusions

- Salinity reduces vegetative growth in all Egyptian and Nigerian Peanut genotypes under greenhouse conditions.
- High significance increase in proline accumulation and Na^+ concentration of Ismailia1 and Samnut 22, while K^+ decreased significantly in all peanut genotypes treated with salinity compared with control.
- KAT1 gene was detected by specific primer and amplified using PCR at 554pb.
- BLAST data showed identical 100% between the sequence of the Egyptian peanut genotype (Giza 6) and the sequence of KAT1 gene of *Arapidopsis thaliana*, and the similarity was 99% between Ismailia1 and Samnut22. Giza 6 and Ismailia1 share in one SNP (A) in codon 195, while Ismailia1 and Samnut22 share in four SNPs (T) in codon 258, 259, 263 and 406.
- The phylogenetic tree shows that Giza 6 is more close to Ismailia 1 than Samnut 22, while Ismailia 1 and Samnut 22 very close, although Ismailia 1 and Giza 6 are Egyptian peanut genotypes and Samnut 22 is Nigerian one.
- Evolutionary tree emphasized the data of screening of all investigated genotypes for their salt tolerant, which reported that Ismailia 1 and Samnut 22 were tolerance for salinity while Giza 6 was a sensitive one.

16.7   Recommendations

Ismailia1 (for Egyptian genotypes) and Samnut 22 (for Nigerian genotypes) are more tolerance for salinity than other varieties.

References

Abd El-RheemKh M, Zaki SS (2015) Effect of soil salinity on growth, yield and nutrient balance of peanut plants. Int J Chemtech Res 8(12):564–568

Aflaki F, Sedghi M, Pazuki A, Pessarakli M (2017) Investigation of seed germination indices for early selection of salinity tolerant genotypes: a case study in wheat, Emirates. J Food Agric 29(3):222–226

Almansouri M, Kinet JM, Lutts S (2001) Effect of salt and osmotic stresses on germination in durum wheat (*Triticum durum* Desf.). Plant Soil 231:243–254

Amirjani MR (2010) Effect of salinity stress on growth, mineral composition, proline content, antioxidant enzymes of soybean. Am J Plant Physiol 5:350–360

Ashraf M, Foolad MR (2007) Roles of glycine betaine and proline in improving plant abiotic stress resistance. Environ Exp Bot 59(2):206–216

Ashraf M, Orooj A (2006) Salt stress effects on growth, ion accumulation and seed oil concentration in an arid zone traditional medicinal plant ajwain (*Trachyspermum ammi* [L.] Sprague). J Arid Environ 64(2):209–220

Bates LS, Waldren RP, Teare ID (1973) Rapid determination of free proline for water stress studies. Plant Soil 39:205–207

Bordi A (2010) The influence of salt stress on seed germination, growth and yield of canola cultivars. Not Bot Horti Agrobo 38:128–133

Carillo P, Mastrolonardo G, Nacca F, Fuggi A (2005) Nitrate reductase in durum wheat seedlings as affected by nitrate nutrition and salinity. Funct Plant Biol 32(3):209–219

Carillo P, Mastrolonardo G, Nacca F, Parisi D, Verlotta A, Fuggi A (2008) Nitrogen metabolism in durum wheat under salinity: accumulation of proline and glycine betaine. Funct Plant Biol 35(5):412–426

Celikkol AU, Ercan O, Kavas M, Yildiz L, Yilmaz C, Oktem HA, Yucel M (2010) Drought-induced oxidative damage and antioxidant responses in peanut (*Arachis hypogeal* L.) seedlings. Plant Growth Regul 61:21–28

Chakraborty K, Bhaduri D, Meena HN, Kalariya K (2016) External potassium (K^+) application improves salinity tolerance by promoting Na^+-exclusion, K^+-accumulation and osmotic adjustment in contrasting peanut cultivars. Plant Physiol Biochem 103:143–153

Cosgrove DJ (1997) Assembly and enlargement of the primary cell wall in plants. Annu Rev Cell Dev Biol 13:171–201

Dantas BF, DeSa RL, Aragao CA (2007) Germination, initial growth and cotyledon protein content of bean cultivars under salinity stress. Rev Bras de Sementes 29:106–110

Davenport R, James R, Zakrisson-Plogander A, Tester M, Munns R (2005) Control of sodium transport in durum wheat. Plant Physiol 137:807–818

Delgado IC, Sanchez-Raya AJ (2007) Effects of sodium chloride and mineral nutrients on initial stages of development of sunflower life. Commun Soil Sci Plant Anal 38:2013–2027

Désiré TV, Liliane MT, Le prince NM, Jonas PI, Akoa A (2010) Mineral nutrient status, some quality and morphological characteristics changes in peanut (*Arachis hypogaea* L.) cultivars under salt stress. African J Environ Sci Technol 4(7):471–479

Dodd GL, Donavan LA (1999) Water potential and ionic effects on germination and seedling growth of two cold desert shrubs. Am J Bo 86:1146–1153

Ehsanpour AA, Amini F (2003) Effect of salt and drought stress on acid phosphatase activities in alfalfa *Medicago sativa* l. Explants under *in vitro* culture. Afr J Biotech 2:133–135

Feng G, Zhang FS, Li XL, Tian CY, Tang C, Rengel Z (2002) Improved tolerance of maize plants to salt stress by arbuscular mycorrhiza is related to higher accumulation of soluble sugars in roots. Mycorrhiza 12:185–190

Fernández-Torquemada Y, Sánchez-Lizaso JL (2013) Effects of salinity on seed germination and early seedling growth of the Mediterranean seagrass *Posidonia oceanica* (L.) Delile. Estuarine. Coastal Shelf Sci 119:64–70

Gadallah MAA (1999) Effects of proline and glycine betaine on *Vicia faba* response to salt stress. Biol Plant 42:249–257

Gajdanowicz P, Garcia-Mata C, Gonzalez W, Morales-Navarro SE, Sharma T, González-Nilo FD, Gutowicz J, Mueller-Roeber B, Blatt MR, Dreyer I (2009) Distinct roles of the last transmembrane domain in controlling Arabidopsis K + channel activity. New Phytol 182(2):380–391

Garcia-Sanchez F, Jifon J, Carvajal M, Syvertsen JP (2002) Gas exchange, chlorophyll and nutrient contents in relation to Na^+ and Cl^- accumulation in 'Sunburst' mandarin grafted on different rootstocks. Plant Sci 162:705–712

Gebreegziabher BG, Qufa ChA (2017) Plant physiological stimulation by seeds salt priming in maize (*Zea mays*): prospect for salt tolerance. Afr J Biotech 16(5):209–223

Geiger D, Becker D, Vosloh D, Gambale F, Palme K, Rehers M, Anschuetz U, Dreyer I, Kudla J, Hedrich R (2009) Heteromeric AtKC1 /AKT1 channels in *Arabidopsis* roots facilitate growth under K^+-limiting conditions. J Biol Chem 284:21288–21295

Giannakoula A, Ilias IF, Maksimović JJD, Maksimović VM, Živanović BD (2012) Does overhead irrigation with salt affect growth, yield, and phenolic content of lentil plants. Arch Biol Sci 64(2):539–547

Girija C, Smith BN, Swamy PM (2002) Interactive effects of sodium chloride and calcium chloride on the accumulation of proline and glycinebetaine in peanut (*Arachis hypogaea* L.). Environ Exp Bot 47:1–10

Gomes-Filho E, Machado Lima CRF, Costa JH, Da Silva AC, Da Guia Silva Lima M, De Lacerda CF, Prisco JT (2008) *Cowpea ribonuclease*: properties and effect of NaCl-salinity on its activation during seed germination and seedling establishment. Plant Cell Rep 27:147–157

Gomez KA, Gomez AA (1984) In Statistical Procedures for Agricultural Research, 2nd edn. Wiley, New York, p 680

Graciano ESA, Nogueira RJMC, Lima DRM, Pacheco CM, Santos RC (2011) Crescimentoe capacidade fotossintética da cultivar de amendoim BR 1 sob condições de salinidade. Rev Bras Eng Agric Ambient 15(8):794–800

Hasegawa PM, Bressan RA, Zhu JK, Bohnert HJ (2000) Plant cellular and molecular responses to high salinity. Annu Rev Plant Physiol Plant Mol Biol 51:463–499

Hayat Sh, Haya Q, Alyemeni MN, Wani AS, Pichtel J, Ahmad A (2012) Role of proline under changing environments. Plant Signaling Behav 7(11):1456–1466

Hertel B, Horvath F, Wodala B, Hurst A, Moroni A, Thiel G (2005) KAT1 inactivates at submillimolar concentrations of external potassium. J Exp Bot 56:3103–3110

Hossain MA, Ashrafuzzaman M, Ismail MR (2011) Salinity triggers proline synthesis in peanut leaves. Maejo Int J Sci Technol 5(01):159–168

Howat D (2000) Acceptable salinity, sodicity and pH values for Boreal forest reclamation: Alberta Environment, Environmental Sciences Division, Edmonton Alberta. Report # ESD/LM/00-2

Jain S, Srivastava S, Sarin NB, Kav NNV (2006) Proteomics reveals elevated levels of PR 10 proteins in saline-tolerant peanut (*Arachis hypogaea*) calluss. Plant Physiol Biochem 44(4):253–259

James RA, Blake C, Byrt CS, Munns R (2011) Major genes for Na$^+$ exclusion, Nax1 and Nax2 (wheat HKT1; 4 and HKT1; 5), decrease Na$^+$ accumulation in bread wheat leaves under saline and waterlogged conditions. J Exp Bot 62(8):2939–2947

Jayashree B, Morag F, Doyle DIJ, Crouch JH (2005) Analysis of genomic sequences from peanut (*Arachis hypogaea*). Electron J Biotechnol 8(3):226–237

Karmoker JL, Farhana S, Rashid P (2008) Effects of salinity on ion accumulation in maize (*Zea Mays* L. CV. BARI-7). Bangladesh J Bot 37:203–205

Kaveh H, Nemati H, Farsi M, Jartoodeh SV (2011) How salinity affect germination and emergence of tomato lines. J Biol Environ Sci 5:159–163

Kavi-Kishore PB, Sangam S, Amrutha RN, Sri Laxmi P, Naidu KR, Rao KRSS, Rao S, Reddy KJ, Theriappan P, Sreenivasulu N (2005) Regulation of proline biosynthesis, degradation, uptake, and transport in higher plants: its implications in plant growth and abiotic stress tolerance. Curr Sci 88:424–438

Khajeh-Hosseini M, Powell AA, Bingham IJ (2003) The interaction between salinity stress and seed vigour during germination of soybean seeds. Seed Sci Technol 31:715–725

Khan MA, Rizvi Y (1994) Effect of salinity, temperature and growth regulators on the germination and early seedling growth of *Atriplex griffithii* var. Stocksii Can J Bot 72:475–479

Khodarahmpour Z, Ifar M, Motamedi M (2012) Effects of NaCl salinity on maize (*Zea mays* L.) at germination and early seedling stage. Afr J Biotechnol 11:298–304

Kinraide TB (1999) Interactions among Ca^{2+}, Na$^+$ and K$^+$ in salinity toxicity: quantitative resolution of multiple toxic and ameliorative effects. J Exp Bot 50:1495–1505

Kurth E, Cramer GR, Lauchli A, Epstein E (1986) Effects of NaCl and CaCl$_2$ on cell enlargement and cell production in cotton roots. Plant Physiol 82:1102–1106

Lacerda CF, Morais MMM, Prisco JT, Filho EG, Bezerra MA (2006) Interaction between salinity and phosphorus in forage sorghum. Agronomic Sci 37(3):258–263

Lauchli A, Grattan SR (2007) Plant growth and development under salinity stress. In: Jenks MA, Hasegawa PM, Mohan JS (eds) Advances in molecular breeding towards drought and salt tolerant crops. Springer, Berlin, pp 1–32

Liu YH, Shen Y, Wang ZF, Yan W (2012) Identification of salt tolerance in peanut varieties/lines at the germination stage. Chin J Oil Crop Sci 34(2):168–173

Maathuis FJM, Sanders D (1993) Energization of potassium uptake in *Arabidopsis thuliana*. Planta 191:302–307

Mansour MMF, Ali EF (2017) Evaluation of proline functions in saline conditions. Phytochemistry 140:52–68

Mansour MMF, Salama KHA, Ali FZM, Abou Hadid AF (2005) Cell and plant responses to NaCl in *Zea mays* L. cultivars differing in salt tolerance. Gen Appl Plant Physiol 31:29–41

Meloni DA, Gullota MR, Martinez CA, Oliva MA (2004) The effects of salt stress on growth, nitrate reduction and proline and glycine betaine accumulation in *Prosopis alba*. Braz J Plant Physiol 16:39–46

Mensah JK, Akomeah PA, Ikhajiagbe B, Ekpekurede EO (2006) Effects of salinity on germination, growth and yield of five groundnut genotypes. Afr J Biotech 5(20):1973–1979

Munns R (2005) Genes and salt tolerance: bringing them together. Tansely Reviewer 167:645–663

Munns R, Husain S, Rivelli AR, James RA, Condon AG, Lindsay MP, Lagudah ES, Schachtman DP, Hare RA (2002) A venues for increasing salt tolerance of crops, and the role of physiologically based selection traits. Plant Soil 247:93–105

Munns R, Tester M (2008) Mechanisms of salinity tolerance. Annu Rev Plant Biol 59:651–681

Musa K, Oya EA, Ufuk CA, Begüm P, Seçkin E, Hüseyin AÖ, Meral Y (2015) Antioxidant responses of peanut (*Arachis hypogaea* L.) seedlings to prolonged salt-induced stress. Arch. Biol Sci Belgrade 67(4):1303–1312

Negrao S, Schmockel SM, Tester M (2017) Evaluating physiological responses of plants to salinity stress. Ann Bot 119:1–11

Nithila S, Durga DD, Velu G, Amutha R, Rangaraju G (2013) Physiological evaluation of groundnut (*Arachis hypogaea* L.) varieties for salt tolerance and amelioration for salt stress. Res J Agric For Sci 1(11):1–8

Nokandeh SE, Mohammadian MA, Damsi B, Jamalomidi M (2015) The effect of salinity on some morphological and physiological characteristics of three varieties of (*Arachis hypogaea* L.). Int J Adv Biotechnol Res 6(4):498–507

Oliveira DS, Dias TJ, Oliveira EP (2016) Growth and physiology of peanut (*Arachis hypogaea* L.) irrigated with saline water and biofertilizer application times. African J Agric Res 11(44):4517–4524

Othman Y, Al-Karaki G, Al-Tawaha AR, Al-Horani A (2006) Variation in germination and ion uptake in barley genotypes under salinity conditions. World J Agric Sci 2:11–15

Page AL, Miller RH, Keeney DR (1982) Methods of chemical analysis. Part 2: Chemical and microbiological properties (Second Edition). American Society of Agronomy, Inc. and Science Society of America, Inc., Madison, WI

Panda SK, Khan MH (2009) Growth, oxidative damage and antioxidant responses in greengram (*Vigna radiata* L.) under short-term salinity stress and its recovery. J Agron Crop Sci 195:442–454

Perez-Alfocea F, Estan F, Caro M, Balarin MC (1993) Response of tomato cultivars to salinity. Plant Soil 150:203–211

Rai MK, Jaiswal VS, Jaiswal U (2010) Regeneration of plantlets of guava (*Psidium guajava* L.) from somatic embryos developed under salt-stress condition. Acta Physiol Plant 32:1055–1062

Rajendran K, Tester M, Roy SJ (2009) Quantifying the three main components of salinity tolerance in cereals. Plant, Cell Environ 32(3):237–249

Reddy TY, Reddy VR, Anbumozhi V (2003) Physiological responses of groundnut (*Arachis hypogea* L.) to drought stress and its amelioration: a critical review. Plant Growth Regul 41:75–88

Rengel Z (1992) Disturbance of cell Ca^{++} homeostasis as a primary trigger of Al toxicity syndrome. Plant, Cell Environ 15:931–938

Schachtman D, Schroeder JI (1994) Structure and transport mechanism of a high-affinity potassium uptake transporter from higher plants. Nature 370(65):54–58

Sharma SK (1997) Plant growth, photosynthesis and ion uptake in chickpea as influenced by salinity. Indian J Plant Physiol 2:171–173

Shrivastava P, Kumar R (2015) Soil salinity: A serious environmental issue and plant growth promoting bacteria as one of the tools for its alleviation. Saudi J Biol Sci 22:123–131

Sibole JV, Montero E, Cabot C, Poschenrieder C, Barcelo J (1998) Role of sodium in the ABA mediated long-term growth response of bean to salt stress. Physiol Pl 104:299–305

Silveira JAG, Viegas RA, Rocha IMA, Moreira ACOM, Moreira RA, Oliveira JTA (2003) Proline accumulation and glutamine synthetase activity are increased by salt-induced proteolysis in cashew leaves. J Plant Physiol 160:115–123

Sumer A, Zorb C, Yan F, Schuber S (2004) Evidence of sodium toxicity for the vegetative growth of maize (*Zea mays* L.) during the first phase of salt stress. J Appl Bot Food Qual 78:135–139

Szabados L, Savouré A (2010) Proline: a multifunctional amino acid. Trends Plant Sci 15(2):89–97

Taiz L, Zeiger E (1998) Plant Physiology, 2nd edn. Sunderland, MA, Sinauer Associates Ins

Tewari TN, Singh BB (1991) Stress studies in lentil (*Lens esculenta* Moench). II. Sodicity-induced changes in chlorophyll, nitrate, nitrite reductase, nucleic acids, proline, yield, and yield components in lentil. Plant Soil 135:225–250

Volkmar KM, Hu Y, Steppuhn H (1998) Physiological responses of plants to salinity: a review. Can J Plant Sci 78:19–27

Wang W, Vinocur B, Altman A (2003) Plant responses to drought, salinity and extreme temperatures: towards genetic engineering for stress tolerance. Planta 218:1–14

Weimberg R (1987) Solute adjustments in leaves of two species of wheat at two different stages of growth in response to salinity. Physiol Plant 70:381–388

Wu GQ, Feng RJ, Li SJ, Du YY (2017) Exogenous application of proline alleviates salt-induced toxicity in sainfoin seedlings. J Animal Plant Sci 27(1):246–251

Yoshiba Y, Kiyosue T, Nakashima K, Yamaguchi-Shinozaki K (1997) Regulation of levels of proline as an osmolyte in plants under water stress. Plant Cell Physiol 38:1095–1102

Zekri M (1991) Effects of NaCl on growth and physiology of sour orange and *Cleopatra mandarine* seedlings. Sci Hortic 47:305–315

Chapter 17
Importance of Mycorrhizae in Crop Productivity

Mahmoud Fathy Seleiman and Ali Nasib Hardan

Abstract Arbuscular mycorrhizal fungi symbioses can pose a beneficial impact for sustaining agroecosystem functioning, for instance, improving plant nutrient uptake, plant growth under water deficit and crop productivity and quality. Arbuscular mycorrhizal fungi are known to receive photosynthetic carbon from the host plants; in return, they provide host plants with some nutrients. Organic fertilizers and related sources of nutrients, as well as slow-release mineral fertilizers, can inspire arbuscular mycorrhizal fungi activity in the rhizosphere, while many investigations revealed that most of the chemical fertilizers suppress the activity of mycorrhiza and their colonization with host plant roots. Arbuscular mycorrhizal fungi mainly play a vital role in plant phosphorus nutrition, and consequently, increase plant uptake of phosphorus. Moreover, arbuscular mycorrhizal fungi can play an important in the plant uptake for inorganic phosphate as well as some other immobile nutrients in the soil and moving them into the host plants. Thus, the main function of arbuscular mycorrhizal fungi is to supply colonized plant roots with phosphorus. Recently, some investigations reported that arbuscular mycorrhizal fungi could provide the host plants with nitrogen from organic sources via converting it into inorganic nitrogen. Also, arbuscular mycorrhizal fungi hyphae can directly take up ammonium, nitrate and amino acids from the rhizosphere and translocate them into their host plants in inorganic form. In the current chapter, the focus will include background about the arbuscular mycorrhizal fungi, the importance of arbuscular mycorrhizal fungi in plant nutrient and crop productivity under water deficit and salinity stress.

Keywords Arbuscular mycorrhizal · Agroecosystem · Plant nutrition · Water deficit · Crop productivity · Sustainability

M. F. Seleiman (✉)
Plant Production Department, College of Food and Agriculture Sciences, King Saud University, 2460, Riyadh 11451, Saudi Arabia
e-mail: mahmoud.seleiman@agr.menofia.edu.eg; mseleiman@ksu.edu.sa

M. F. Seleiman · A. N. Hardan
Department of Crop Sciences, Faculty of Agriculture, Menoufia University, 32514, Shibin El-kom, Egypt

© Springer Nature Switzerland AG 2021
H. Awaad et al. (eds.), *Mitigating Environmental Stresses for Agricultural Sustainability in Egypt*, Springer Water, https://doi.org/10.1007/978-3-030-64323-2_17

471

17.1 Introduction

In the rhizosphere, there are six types of mycorrhiza fungi, viz. arbuscular, arbu-toid, ecto-, orchid, ericoid and monoptropoid, which are categorized based on their morphological properties (Garg et al. 2006; Wang and Qiu 2006). The majority of terrestrial plant species (250,000) have almost symbiosis process with mycorrhizal fungi (Harley and Harley 1987; Peterson et al. 2004; Wang and Qiu 2006; Smith and Read 2008; Helgason and Fitter 2009). Generally, plant species belong to *Brassicaceae* and *Chenopodiaceae* families do not have a symbiotic relationship with mycorrhizal fungi (Newman and Reddell 1987; Peterson et al. 2004). The arbuscular mycorrhizal fungus is considered the most vital type among those six types of mycorrhizae fungi that can enhance and improve plant growth and soil fertility (Gosling et al. 2006; Smith et al. 2011). Furthermore, arbuscular mycorrhizal fungi can enhance the host plant resistance to drought or diseases (Smith and Read 2008). Plant roots are the location where the carbohydrate and mineral nutrient exchange between arbuscular mycorrhizal fungi and host plant cells can occur. Arbuscular mycorrhizal fungi receive photosynthetic carbon from the host plant species, in return they provide the host plants with the acquisition and transport of nitrogen (Fellbaum et al. 2012; Liu et al. 2013).

The symbiosis process between arbuscular mycorrhizal fungi and its host plants begins from the colonization of roots by the hyphae and asexual spores (Requena et al. 1996; Smith et al. 2011). The hyphae start to enter the root cortical cell walls of host plants and form morphologically different structures, such as arbuscules and vesicles (Smith and Read 2008; Smith et al. 2010; Smith and Smith 2011). In addition, arbuscular mycorrhizal fungi spores start to grow in the soil and germinate spontaneously and freely of plant-derived signals. Following root colonization of host plants, the mycelium grows out of the root exploring the soil, up-taking different nutrients and water, and then translocating them to the roots, and it can colonize other susceptible roots (Smith and Read 2008; Smith et al. 2010; Smith and Smith 2011). The arbuscules of mycorrhizal fungi are considered the location of the nutrient transfer between the mycorrhizal fungus and the host plant (Taiz and Zeiger 2006). However, vesicles are considered storage organs for lipids, which indicate that they can act as propagules for arbuscular mycorrhizal fungi (Peterson et al. 2004; Smith and Read 2008).

Root arbuscular mycorrhizal fungi colonization has many benefits such as plant uptake for nitrogen, phosphorus, potassium and manganese (Smith and Read 2008; Smith et al. 2011; Seleiman et al. 2013), and copper and zinc (Li et al. 1991; Azaizeh et al. 1995; Taiz and Zeiger 2006). The host plant provides up to 20% of the organic carbon for the formation and function of arbuscular mycorrhizal fungi (Smith and Read 2008; Smith and Smith 2011).

17.2 Effects of Organic and Inorganic Fertilizers on Activity of Arbuscular Mycorrhizal Fungi in Rhizosphere

Organic sources of different nutrients and slow-release chemical fertilizers can enhance arbuscular mycorrhizal fungi activity (Joner 2000; Alloush and Clark 2001; Smith et al. 2011; Seleiman et al. 2013), however other several chemical fertilizers can suppress mycorrhizal colonization activity with the roots of host plants (Liu et al. 2000; Seleiman et al. 2013). Arbuscular mycorrhizal fungi can enhance the plant uptake and use efficiency of phosphorus (Koide et al. 2000), and consequently can result in an improvement in growth and yield of plant species (Koide et al. 2000; Smith et al. 2011; Seleiman et al. 2013). The phosphorus depletion zone is closed to root hair length in the non-mycorrhizal plant (Marschner and Dell 1994), whereas it exceeds the root hair zone in mycorrhizal plants, which shows that unavailable phosphorus to the plant is linked to the fungal hyphae (Garg and Chandel 2010). The routes of symbiotic phosphorus process begin from the absorption of inorganic phosphorus through the fungal high-affinity transporters (Maldonado-Mendoza et al. 2001). Inside the arbuscular mycorrhizal fungi, inorganic phosphate is condensed in polyphosphate. The polyphosphate becomes depolymerized into inorganic phosphorus before it is released into the periarbuscular interface (Solaiman et al. 1999; Ohtomo and Saito 2005). Then, phosphorus is translocated from the interface via phosphate transporters to the plant root cells (Solaiman et al. 1999).

Arbuscular mycorrhizal fungi can uptake the nitrogen from different organic sources, change it into mineral nitrogen, and transfer it to the host plant through roots (Hodge et al. 2001). The hyphae can uptake the NH_4^+ and NO_3^- (Johansen et al. 1996; Smith and Read 2008) and amino acids (Hawkins et al. 2000; Hodge et al. 2001) from their surroundings and transfer the nitrogen in inorganic form into their host plants roots (Hawkins et al. 2000; Azcón et al. 2001; Vazquez et al. 2001). From Fig. 17.1, NH_4^+ can be taken by the extraradical mycelium of arbuscular mycorrhizal fungi, assimilated into amino acids, transfer from extra- to intraradical fungal structures as arginine and then transported as NH_4^+ to the host plant roots (Govindarajulu et al. 2005; Jin et al. 2005; Chalot et al. 2006).

There are two pathways (direct and indirect pathways) of phosphorus uptake by plants from the rhizosphere (Fig. 17.1). In the direct pathway, plant roots uptake the phosphorus directly from the rhizosphere (root zones) through the transporters in the epidermis and root hairs (green circles). In the mycorrhizal or indirect pathway, fungal phosphorus transporters (blue circles, behind the root apex) can uptake the phosphorus and transfer it into hyphae and translocate it into arbuscules and hyphal coils in root cortical cells. Then, plant phosphorus transporters (black circle) can easily translocate the phosphorus from the interfacial apoplast to plant cortical cells (Fig. 17.1).

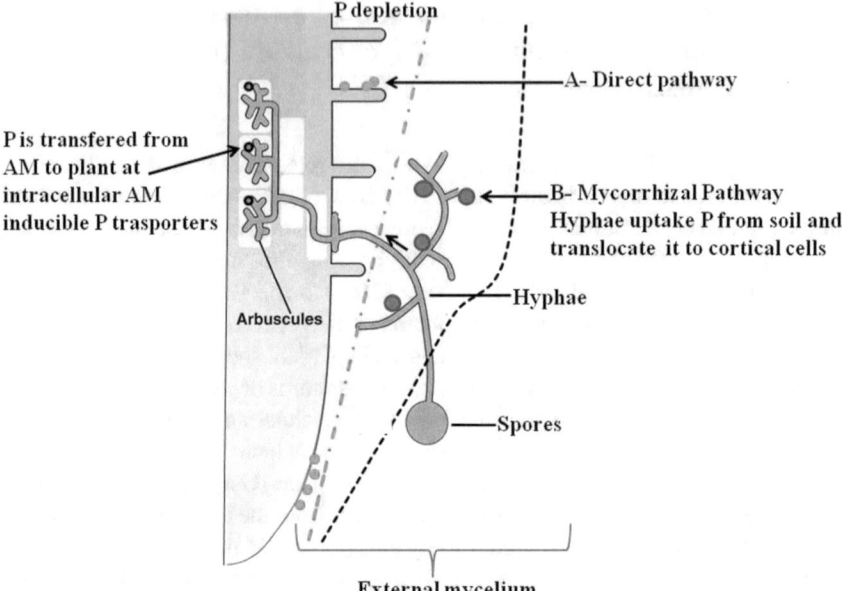

Fig. 17.1 Mycorrhizal pathways of plant P uptake (Modified from Smith et al. 2011). *Source* Seleiman (2014)

17.3   Role of Arbuscular Mycorrhizal Fungi on Phosphorus and Nitrogen Uptake by Host Plants

In most cases, arbuscular mycorrhiza uptake the nitrogen in the form of nitrate and ammonium, and therefore different investigations on different plants have shown that arbuscular mycorrhizal fungi can play a beneficial role in nitrogen uptake and transport (Blanke et al. 2005; Smith and Read 2008; Seleiman et al. 2013; Seleiman 2014). In most cases, mycorrhizal plants contain higher nitrogen than non-mycorrhizal plants (Tobar et al. 1994; Barea et al. 2002; Jemo et al. 2007; Seleiman et al. 2013) whereas, in some cases there is no significant difference in the nitrogen content of inoculated and untreated plants with mycorrhizal fungi (Johansen 1999; Hawkins et al. 2000). Arbuscular mycorrhizal fungi prefer to uptake the nitrogen in the form of NH_4^+ over NO_3^-, although the export of nitrogen from the hyphae of arbuscular mycorrhizal fungi to the host root and shoot can be higher in case of NO_3^- uptake than NH_4^+ uptake (Ngwene et al. 2013). Govindarajulu et al. (2005) have shown that arbuscular mycorrhizal fungi can provide the host plants with 20–50% of the total plant nitrogen uptake. Also, Seleiman et al. (2013) reported that arbuscular mycorrhizal fungi enhanced the uptake and accumulation of nitrogen in maize and hemp were grown with digested sludge in comparison to the synthetic chemical fertilizers.

Phosphorus, which is a vital and important mineral nutrient for plant growth, is one of the most three important mineral nutrients applied to agricultural land for plant nutrition. Rock phosphate sources are limited, and most of the phosphate mines will be depleted during the next 100 years (Herring and Fantel 1993). The consumption of triple-phosphate has been lessened in developed countries by about 35% during 2000–2006. However it increased by about 35% in the developing countries. The excessive application of synthetic phosphorus fertilizers is a critical concern and can cause eutrophication, and for this reason, the improvement of phosphorus uptake and it use efficiency through plant species is an important issue. It was noted that a mycorrhizal plant root is more efficient in phosphorus uptake than a non-mycorrhizal plant root (Smith and Read 2008) (Figs. 17.2 and 17.3). Mycorrhizal fungi colonization with plant roots in the soil can act as extensions of the root system and are considered more effective for nutrients and water absorption in comparison to the roots themselves. In additions, they also can explore the soil and arrive to places inaccessible for plant roots. Taffouo et al. (2014) investigated the effects of phosphorus application (i.e. 0.1, 0.5 or 1.0 mM) and arbuscular mycorrhizal fungi (arbuscular mycorrhizal fungi; *Funneliformis mosseae*) on growth, foliar nitrogen mobilization, and phosphorus partitioning in cowpea (*Vigna unguiculata*) grown in

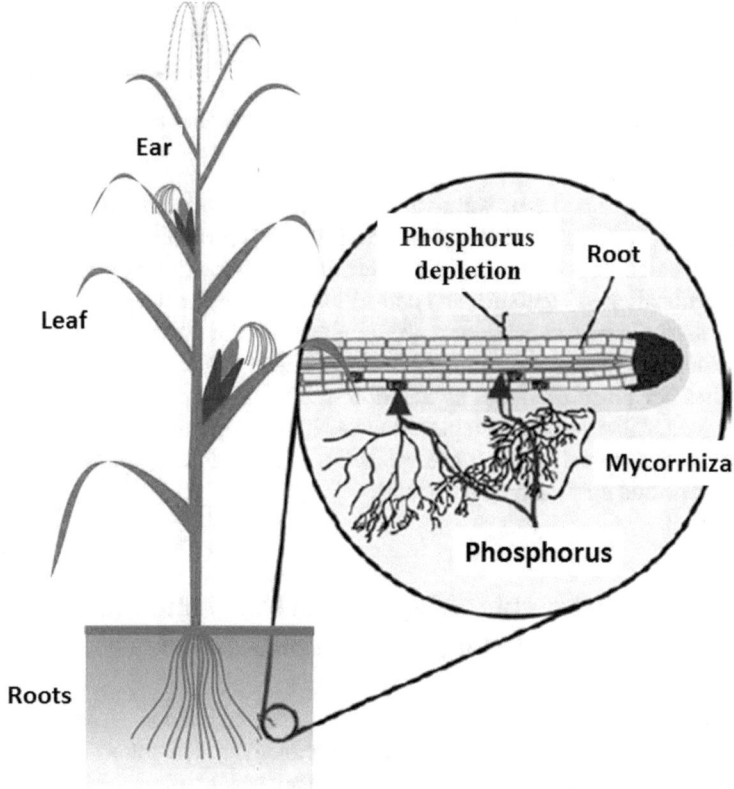

Fig. 17.2 Mycorrhized plant showing extraradical hyphae of AMF

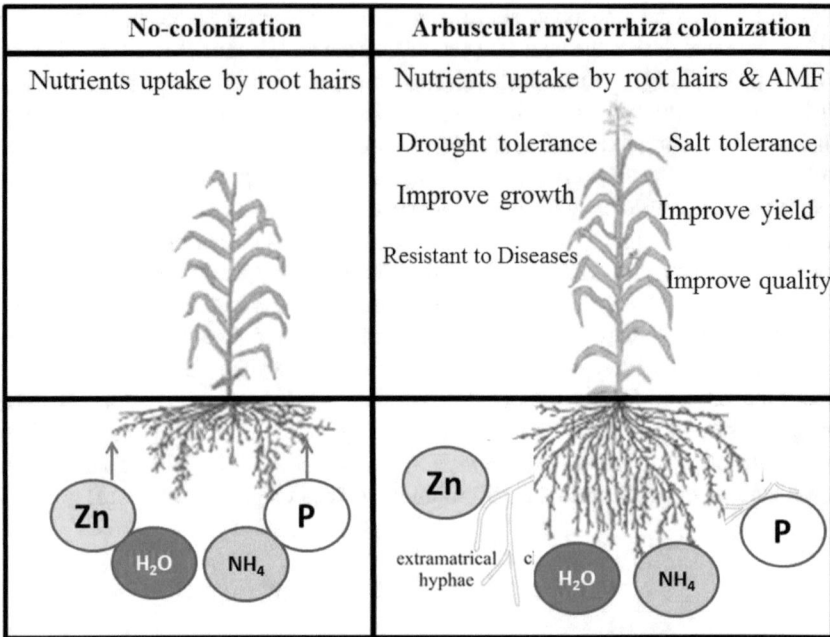

Fig. 17.3 The beneficial effects of arbuscular mycorrhiza colonization on plant roots

pots containing a mixture of vermiculite and sterilized quartz sand. They reported that the lowest application of phosphorus fertilization resulted in a significant higher root colonization of cowpea than those fertilized with the other two level of phosphorus fertilization at vegetative and pod-filling stages. Also, the phosphorus uptake and growth traits of cowpea plants were enhanced when mycorrhizal inoculation was applied with the medium phosphorus fertilization at the vegetative stage. Moreover, it has been estimated that a reduction of 80% of the recommended synthetic phosphorus fertilizer could be replaced through the inoculation of plant seeds with arbuscular mycorrhizal fungi (Jakobsen 1995), and this is an important outcome for the economic and environmental impacts.

17.4   Role of Arbuscular Mycorrhizal Fungi on Plants Grown Under Drought and Salinity Stress

Abiotic stresses such as drought, salinity, heavy metals, mineral depletion and heat stress can cause high reduction of agricultural productivity. The colonization of arbuscular mycorrhizal fungi with plant roots can enhance plant tolerance against abiotic stress (Smith and Read 2008). The manipulation of arbuscular mycorrhizal fungi in sustainable agricultural systems will be of marvelous vital for soil fertility and crop yield under severe edapho-climatic conditions (Lal 2009). Arbuscular mycorrhizal fungi living symbiotically with host plant species can improve plant growth through improvement the acquisition of different nutrients and water relations as well as due to the hyphal extraction of soil water (Al-Karaki 1998). He added that mycorrhizal plants were more water-use efficient than non-mycorrhizal plants. The root of host plants associations with arbuscular mycorrhizal fungi appeared to be more resistance for drought stress through enhancing water relations (Ruiz-Lozano et al. 1995). Marulanda et al. (2006) reported that indigenous drought-tolerant strain of *Glomus intraradices* associated with a native bacterium resulted in a reduction in the water required for the production of *Retama sphaerocarpa* by 42%.

The negative salt effects on plant growth and productivity might include a lessening in the soil solution osmotic potential that can reduce plant-available water, and toxicity of excessive Na^+ or Cl^- towards the plasma membrane (Feng et al. 2002). Effects of osmotic are linked with inhibition of plant cell wall extension and cellular expansion, resulting in a reduction in plant growth (Staple and Toenniessen 1984). The salinity of soil can affect the formation and function of mycorrhiza fungi (McMillen et al. 1998), however different investigations have shown that soil treated with arbuscular mycorrhizal fungi improved plants growth when grown under salinity stress environments (Al-Karaki and Hammad 2001). Consequently, arbuscular mycorrhizal fungi have been considered as important bio-ameliorators in saline soils (Singh et al. 1997). Also, arbuscular mycorrhizal fungi can alleviate the stress of salinity in the olive tree where palm grove yields are can be affected by drought and soil salinity (Porras-Soriano et al. 2009).

Elhindi et al. (2017) investigated the role of arbuscular mycorrhizal fungi [Glomus deserticola and non-inoculated in mitigating salt-induced [control (0.64 dS m^{-1}), low salinity (5.0 dS m^{-1}), and high salinity (10.0 dS m^{-1})] adverse effects on growth and productivity of sweet basil (*Ocimum basilicum* L.). They reported that the arbuscular mycorrhizal fungi mitigated the reduction of potassium, phosphorus and calcium uptake by sweet basil due to salinity impact. Arbuscular mycorrhizal fungi enhanced the growth and productivity of sweet basil plants grown under salinity conditions. The application of arbuscular mycorrhizal fungi significantly improved chlorophyll content and water use efficiency of sweet basil plants. In conclusion, there was an indication that arbuscular mycorrhizal fungi colonization on roots of sweet basil plants can lessen the harmful impact of salinity stress on growth and productivity of sweet basil plants. Also, Feng et al. (2002) conducted a pot experiment on maize (*Zea mays* L. cv. Yedan 13) grown in sand with 0 or 100 mM NaCl and at two phosphorus (0.05 and 0.1 mM) levels for 34 days, following 34 days of non-saline pre-treatment. They reported that the contents of chlorophyll, phosphorus and soluble sugars were higher in maize grown with mycorrhizal fungi than in non-mycorrhizal plants under the different treatment of NaCl and phosphorus. They also found that maize plants grown with a low level of phosphorus plus arbuscular mycorrhizal fungi treatment and those grown with high phosphorus minus arbuscular mycorrhizal fungi had almost phosphorus content without a significantly different, the plants treated with mycorrhiza still had higher root dry weights, soluble sugars and electrolyte content. The same trend was obtained in the presence or absence of salt stress. Therefore they concluded that the improved tolerance of maize plants against salt stress by arbuscular mycorrhiza was associated with the higher accumulation of soluble sugars in roots.

Generally, the mechanism of arbuscular mycorrhizal fungi on the improvement of plant growth in saline soils is associated with the physiological processes such as increasing exchange rate of carbon dioxide, stomatal conductance, transpiration rate and water use efficiency instead of the nutrient uptake particularly nitrogen or phosphorus (Ruiz-Lozano et al. 1996).

17.5 Role of Arbuscular Mycorrhizal Fungi on Quality and Productivity of Plants

Arbuscular mycorrhizal fungi root colonization can affect the quality of plant seeds in terms of improving protein and lipid contents through prompting phosphorus nutrition or prompting other metabolic reactions in the host plant species. Different types of proteins (Gianinazzi-Pearson and Gianinazzi 1989), sugars reduction and fractions of amino acid (Tawaraya and Saito 1994) and lipids (Bethlenfalvay et al. 1997) are yielded by host plant species when arbuscular mycorrhizal fungi root colonization is applied. This indicates that seeds of mycotrophic plants can be changed and modified, not only in terms of whole biomass production but also in the comparative plenty of storage products such as proteins, starch and lipids (Lu and Koide 1991; Bethlenfalvay et al. 1997). Although wheat grains are considered to be a starchy food, they contain additional valuable nutritive compounds such as proteins (Abdel-Aal et al. 2009; Seleiman et al. 2010a, b, 2011) and lipids in significant amounts (Al-Karaki and Ereifej 1998).

Al-Karaki and Clark (1999) conducted a greenhouse experiment to determine the impact of arbuscular mycorrhizal fungi and soil phosphorus on quality and yield of two genotypes of durum wheat (*Triticum durum* L.). They concluded that arbuscular mycorrhizal fungi root colonization increased with the reducing application of phosphorus into the soil. Also, arbuscular mycorrhizal fungi application, without soil phosphorus application, resulted in a significant increase in the content of lipid and seed dry weight, and a significant reduction in the protein content of seed in comparison to untreated plants. Wheat seed protein and lipid contents were highly associated with phosphorus content in plants.

On the other hand, about 30% of the world's agricultural lands have zinc deficient, mainly in tropical regions (Cavagnaro 2008) and this can lead to a significant reduction in productivity and zinc content of different crop products, causing an inadequate dietary zinc intake for many human populations as well as causing a negative impact on human health. Some investigations have shown that arbuscular mycorrhizal fungi can enhance zinc uptake by plants under field conditions (Cavagnaro 2008; Seleiman et al. 2013). In this respect, the zinc content in shoots and fruits of mycorrhizal wild-tomato plants was higher by 50% than those obtained in a mutant with a reduction in mycorrhizal colonization for roots (Cavagnaro et al. 2006).

17.6 Role of Arbuscular Mycorrhizal Fungi in Increasing Soil Stability

The fungal symbioses grow out from the mycorrhizal plant root during the development of arbuscular mycorrhizal fungi to improve a complex though a network of branches (i.e. 30 m of fungal hyphae per g of soil) into the surrounding rhizosphere (Cavagnaro et al. 2005; Wilson et al. 2009). Such network of fungal hyphae can form about 50% of fungal mycelium in the rhizosphere (Rillig et al. 2002) thus can form a major share of the microbial biomass in soil (Leake et al. 2004). The mycelium network can result in a binding action on the soil and consequently can improve soil structure. Furthermore, the excretion by arbuscular mycorrhizal fungi of hydrophobic can improve the soil stability and water retention (Bedini et al. 2009). The mixture of the extensive arbuscular mycorrhizal fungi hyphal network and the ooze of the glomalin is a significant key for improving the stabilize soil aggregates (Rillig and Mummey 2006), thus can increase the stability of soil structure and quality (Bedini et al. 2009). However, some agricultural and agronomical practices for instance ploughing process, monoculture cropping and/or chemical fertilization can result in a negative influence on the activity and the diversity of arbuscular mycorrhizal fungi in the zones of soil rhizosphere (Helgason et al. 1998; Oehl et al. 2005). Therefore, the lessening of fungal biomass of arbuscular mycorrhizal fungi can result in an undesirable impact on soil stability and accordingly can increase the potential risk of soil erosion.

17.7 Conclusions

Arbuscular mycorrhizal fungi are considered the most vital type among different types of mycorrhizae fungi that can enhance and improve plant productivity and soil fertility. They uptake the nitrogen in the form of nitrate and ammonium as well as they can uptake the N from different organic sources and change it into mineral nitrogen for plant uptake. Arbuscular mycorrhizal fungi enhance plant uptake and use efficiency of phosphorus and some micronutrients, and consequently can result in an improvement in growth and yield of plants. Moreover, they can improve protein and lipid contents of plants through prompting phosphorus nutrition or prompting other metabolic reactions in the host plant species.

Organic sources of different nutrients and slow-release chemical fertilizers can enhance arbuscular mycorrhizal fungi activity, while synthetic fertilizers can suppress mycorrhizal colonization activity with the roots of host plants. A reduction of more than 50% of the recommended synthetic phosphorus fertilizer can be replaced through the inoculation of plant seeds with arbuscular mycorrhizal fungi, and this is an important outcome for the economic and environmental impacts.

Arbuscular mycorrhizal fungi can increase the host plant resistance for drought stress through enhancing the water relations. Also, they are important bio-ameliorators in saline soils. The mechanism of arbuscular mycorrhizal fungi on the improvement of plant growth in saline soils is associated with the physiological processes such as increasing exchange rate of carbon dioxide, stomatal conductance, transpiration rate and water use efficiency instead of the nutrient uptake particularly nitrogen or phosphorus.

Therefore, inoculation of plant seeds with arbuscular mycorrhizal fungi and reducing synthetic fertilizer is highly recommended for suitability of agriculture.

17.8 Recommendations

- Arbuscular mycorrhizal fungi application should be encouraged in Egypt since they can enhance and improve plant productivity, quality and soil fertility, plant nutrients uptake.
- Arbuscular mycorrhizal fungi application should be used with less synthetic fertilizers, because synthetic fertilizers can cause an adverse effect on the roots colonization with mycorrhiza fungi.
- Application of arbuscular mycorrhizal fungi enhances plant tolerance against abiotic stress such as drought and salt stress, so it is highly recommended in arid and semi-arid regions.

References

Abdel-Aal SM, Ibrahim ME, Seleiman MF (2009) Effect of nitrogen fertilization on yield, technological and rheological characters of wheat (*Triticum aestivum* L.). Egypt J Agron 31:95–107

Al-Karaki GN (1998) Benefit, cost and water-use efficiency of arbuscular mycorrhizal durum wheat grown under drought stress. Mycorrhiza 8:41–45

Al-Karaki GN, Clark RB (1999) Mycorrhizal influence on protein and lipid of durum wheat grown at different soil phosphorus levels. Mycorrhiza 9:97–101

Al-Karaki GN, Ereifej KI (1998) Seed yield and chemical composition of durum wheat under arid and semiarid Mediterranean environments. In: Jaradat AA (ed) Triticeae III. Science, Enfield, NH, pp 439–444

Al-Karaki GN, Hammad R (2001) Mycorrhizal influence on fruit yield and mineral content of tomato grown under salt stress. J Plant Nutr 24:1311–1323

Alloush GA, Clark RB (2001) Maize response to phosphate rock and arbuscular mycorrhizal fungi in acidic soil. Commun Soil Sci Plant Anal 32:231–254

Azaizeh HA, Marschner H, Römheld V, Wittenmayer L (1995) Effects of a vesicular-arbuscular mycorrhizal fungus and other soil microorganisms on growth, mineral nutrient acquisition and root exudation of soil grown maize plants. Mycorrhiza 5:321–327

Azcón R, Ruiz-Lozano JM, Rodriguez R (2001) Differential contribution of arbuscular mycorrhizal fungi to plant nitrate uptake of ^{15}N under increasing N supply to the soil. Canad J Bot 79:1175–1180

Barea JM, Toro M, Orozco MO, Campos E, Azcón R (2002) The application of isotopic (^{32}P and ^{15}N) dilution techniques to evaluate the interactive effect of phosphate-solubilizing rhizobacteria, mycorrhizal fungi and rhizobium improve the agronomic efficiency of rock phosphate for legume crops. Nutr Cycl Agroecosyst 63:35–42

Bedini S, Pellegrino E, Avio L, Pellegrini S, Bazzoffi P, Argese E, Giovannetti M (2009) Changes in soil aggregation and glomalinrelated soil protein content as affected by the arbuscular mycorrhizal fungal species Glomus mosseae and Glomus intraradices. Soil Biolog Biochem 41:1491–1496

Bethlenfalvay GJ, Schreiner RB, Mihara KL (1997) Mycorrhizal fungi effects on nutrient composition and yield of soybean seeds. J Plant Nutr 20:581–591

Blanke V, Renker C, Wagner M, Füllner K, Held M, Kuhn AJ, Buscot F (2005) Nitrogen supply affects arbuscular mycorrhizal colonization of Artemisia vulgaris in phosphate-polluted field site. New Phytol 166:981–992

Cavagnaro TR (2008) The role of arbuscular mycorrhizas in improving plant zinc nutrition under low soil zinc concentrations, a review. Plant Soil 304:315–325

Cavagnaro TR, Smith FA, Smith SE, Jakobsen I (2005) Functional diversity in arbuscular mycorrhizas: exploitation of soil patches with different phosphate enrichment differs among fungal species. Plant, Cell Environ 28:642–650

Cavagnaro TR, Jackson LE, Six J, Ferris H, Goyal S, Asami D, Scow KM (2006) Arbuscular mycorrhizas, microbial communities, nutrient availability, and soil aggregates in organic tomato production. Plant Soil 282:209–225

Chalot M, Blaudez D, Annick B (2006) Ammonia: a candidate for nitrogen transfer at the mycorrhizal interface. Trend Plant Sci 11:263–266

Elhindi KM, El-Din AS, Elgorban AM (2017) The impact of arbuscular mycorrhizal fungi in mitigating salt-induced adverse effects in sweet basil (Ocimum basilicum L.). Saud J Biolog Sci 24:170–179

Fellbaum CR, Gachomo EW, Beesetty Y, Choudhari S, Strahan GD, Pfeffer PE, Kiers ET, Bucking H (2012) Carbon availability triggers fungal nitrogen uptake and transport in arbuscular mycorrhizal symbiosis. Saud J Biolog Sci 109:2666–2671

Feng G, Zhang FS, Li XL, Tian CY, Tang C, Rengel Z (2002) Improved tolerance of maize plants to salt stress by arbuscular mycorrhiza is related to higher accumulation of soluble sugars in roots. Mycorrhiza 12:185–190

Garg N, Chandel S (2010) Arbuscular mycorrhizal networks: process and functions. A review. Agron Sust Develop 30:581–599

Garg N, Geetanjali Kaur A (2006) Arbuscular mycorrhiza: Nutritional aspects. Arch Agron Soil Sci 52:593–606

Gianinazzi-Pearson V, Gianinazzi S (1989) Cellular and genetic aspects of interactions between hosts and fungal synbionts in mycorrhizae. Genome 31:336–341

Gosling P, Hodge A, Goodlass G, Bending GD (2006) Arbuscular mycorrhizal fungi and organic farming. Agric Ecosyst Environ 113:17–35

Govindarajulu M, Pfeffer PE, Jin HR, Abubaker J, Douds DD, Allen JW, Bucking H, Lammers PJ, Shachar-Hill Y (2005) Nitrogen transfer in the arbuscular mycorrhizal symbiosis. Nature 435:819–823

Harley JL, Harley EL (1987) A check-list of mycorrhiza in the British Flora. New Phytol 105:1–102

Hawkins HJ, Johansen A, George E (2000) Uptake and transport of organic and inorganic nitrogen by arbuscular mycorrhizal fungi. Plant Soil 226:275–285

Helgason T, Fitter A (2009) Natural selection and the evolutionary ecology of the arbuscular mycorrhizal fungi (Phylum Glomeromycota). J Exper Bot 60:2465–2480

Helgason T, Daniell TJ, Husband R, Fitter AH, Young JPW (1998) Ploughing up the wood-wide web. Nature 394:431

Herring JR, Fantel RJ (1993) Phosphate rock demand into the next century: impact on world food supply. Nat Resour Res 2:220–246

Hodge A, Campbell CD, Fitter AH (2001) An arbuscular mycorrhizal fungus accelerates decomposition and acquires nitrogen directly from organic material. Nature 413:297–299

Jakobsen I (1995) Transport of phosphorus and carbon in VA mycorrhizas. In: Varma A, Hock B (eds) mycorrhiza. SpringerVerlag, Berlin, pp 297–324

Jemo M, Nolte C, Nwaga D (2007) Biomass production, N and P uptake of Mucuna after Bradyrhizobia and arbuscular mycorrhyzal fungi inoculation, and P application on acid soil of Southern Cameroon. In: Bationo A (ed) Advances in integrated soil fertility management in Sub-Saharan Africa: challenges and opportunities. Springer, Dordrecht, pp 855–864

Jin H, Pfeffer PE, Douds DD, Piotrowski E, Lammers PJ, Shachar-Hill Y (2005) The uptake, metabolism, transport and transfer of nitrogen in an arbuscular mycorrhizal symbiosis. New Phytol 168:687–696

Johansen A (1999) Depletion of soil mineral N by roots of *Cucumis sativus* L. colonized or not by arbuscular mycorrhizal fungi. Plant Soil 209:119–127

Johansen A, Finlay RD, Olsson PA (1996) Nitrogen metabolism of external hyphae of the arbuscular mycorrhizal fungus *Glomus intraradices*. New Phytol 133:705–712

Joner EJ (2000) The effect of long-term fertilization with organic or inorganic fertilizers on mycorrhiza-mediated phosphorus uptake in subterranean clover. Biol Fert Soils 32:435–440

Koide RT, Goff MD, Dickie IA (2000) Component growth efficiencies of mycorrhizal and nonmycorrhizal plants. New Phytol 148:163–168

Lal R (2009) Soil degradation as a reason for inadequate human nutrition. Food Sec 1:45–57

Leake JR, Johnson D, Donnelly D, Muckle G, Boddy L, Read D (2004) Network of power and influence: the role of mycorrhizal mycelium in controlling plant communities and agroecosystem functioning. Canad J Bot 82:1016–1045

Li XL, Marschner H, George E (1991) Acquisition of phosphorus and copper by VA-mycorrhizal hyphae and root-to-shoot transport in white clover. Plant Soil 136:49–57

Liu A, Hamel C, Hamilton RI, Ma BL, Smith DL (2000) Acquisition of Cu, Zn, Mn and Fe by mycorrhizal maize (*Zea mays* L.) grown in soil at different P and micronutrient levels. Mycorrhiza 9:331–336

Liu ZL, Li YJ, Hou HY, Zhu XC, Rai V, He XY, Tian CJ (2013) Differences in the arbuscular mycorrhizal fungi-improved rice resistance to low temperature at two N levels: aspects of N and C metabolism on the plant side. Plant Physiol Biochem 71:87–95

Lu X, Koide RT (1991) *Avena fatua* L. seedling nutrient dynamics as influenced by mycorrhizal infection of the maternal generation. Plant, Cell Environ 14:931–939

Maldonado-Mendoza IE, Dewbre GR, Harrison MJ (2001) A phosphate transporter gene from the extra-radical mycelium of an arbuscular mycorrhizal fungus *Glomus intraradices* is regulated in response to phosphate in the environment. Mol Plant-Microbe Interact 14:1140–1148

Marschner H, Dell B (1994) Nutrient uptake in mycorrhizal symbiosis. Plant Soil 159:89–102

Marulanda A, Barea JM, Azcon R (2006) An indigenous drought tolerant strain of *Glomus intraradices* associated with a native bacterium improves water transport and root development in Retama sphaerocarpa. Microbial Ecol 52:670–678

McMillen BG, Juniper S, Abbott LK (1998) Inhibition of hyphal growth of a vesicular-arbuscular mycorrhizal fungus in soil containing sodium chloride limits the spread of infection from spores. Soil Biol Biochem 30:1639–1646

Newman EI, Reddell P (1987) The distribution of mycorrhizas among the families of vascular plants. New Phytol 106:745–751

Ngwene B, Gabriel E, Eckhard G (2013) Influence of different mineral nitrogen sources (NH_4^+-N vs NO_3^--N) on arbuscular mycorrhiza development and N transfer in a *Glomus intraradices*–cowpea symbiosis. Mycorrhiza 23:107–117

Oehl F, Sieverding E, Ineichen K, Ris EA, Boller T, Wiemken A (2005) Community structure of arbuscular mycorrhizal fungi at different soil depths in extensively and intensively managed agroecosystems. New Phytol 165:273–283

Ohtomo R, Saito M (2005) Polyphosphate dynamics in mycorrhizal roots during colonization of an arbuscular mycorrhizal fungus. New Phytol 167:571–578

Peterson RL, Massicotte HB, Melville LH (2004) Mycorrhizas: anatomy and cell biology. NRC Research Press, Ottawa/CABI Publishing, Wallingford, UK

Porras-Soriano A, Soriano-Martin ML, Porras-Piedra A, Azcon R (2009) Arbuscular mycorrhizal fungi increased growth, nutrient uptake and tolerance to salinity in olive trees under nursery conditions. J Plant Physiol 166:1350–1359

Requena N, Jeffries P, Barea JM (1996) Assessment of natural mycorrhizal potential in a desertified semiarid ecosystem. Appl Environ Microbiol 62:842–847

Rillig MC, Mummey D (2006) Mycorrhizas and soil structure. New Phytol 171:41–53

Rillig MC, Wright SF, Nichols KA, Schmid WF, Torn MS (2002) The role of arbuscular mycorrhizal fungi and glomalin in soil aggregation: comparing effects of five plant species. Plant Soil 238:325–333

Ruiz-Lozano JM, Azon R, Gomez M (1995) Effects of arbuscular-mycorrhizal Glomus species on drought tolerance: physiological and nutritional plant responses. Appl Environ Microbiol 61:456–460

Ruiz-Lozano JM, Azcon R, Gomez M (1996) Alleviation of salt stress by arbuscular mycorrhizal Glomus species in *Lactuca sativa* plants. Physiol Plant 98:767–772

Seleiman MF (2014) Towards sustainable intensification of feedstock production with nutrient cycling. PhD Thesis, University of Helsinki, Finland

Seleiman MF, Abdel-Aal SM, Ibrahim ME, Monneveux P (2010a) Variation of yield and milling, technological and rheological characteristics in some Egyptian bread wheat (*Triticum aestivum* L.) cultivars. Emirat J Food Agric 22:84–90

Seleiman MF, Ibrahim ME, Abdel-Aal SM, Zahran GA (2010b) Effect of seeding rates on productivity, technological and rheological characteristics of bread wheat (*Triticum aestivum* L.). Inter J Curr Res 4:75–81

Seleiman MF, Abdel-Aal S, Ibrahim M, Zahran G (2011) Productivity, grain and dough quality of bread wheat grown with different water regimes. J Agron Crop Sci 2:11–17

Seleiman MF, Santanen A, Kleemola J, Stoddard FL (2013) Improved sustainability of feedstock production with sludge and interacting mycorrhiza. Chemosphere 91:1236–1242

Singh RP, Choudhary A, Gulati A, Dahiya HC, Jaiwal PK, Sengar RS (1997) Response of plants to salinity in interaction with other abiotic and factors. In: Jaiwal PK, Singh RP, Gulati A (eds) Strategies for improving salt tolerance in higher plants. Science Publishers, Enfield, NH, pp 25–39

Smith SE, Read DJ (2008) Mycorrhizal Symbiosis. Elsevier, London, UK

Smith SE, Smith FA (2011) Roles of arbuscular mycorrhizas in plant nutrition and growth: new paradigms from cellular to ecosystem scales. Ann Rev Plant Biol 62:227–250

Smith SE, Christophersen HM, Pope S, Smith FA (2010) Arsenic uptake and toxicity in plants, integrating mycorrhizal influences. Plant Soil 327:1–21

Smith SE, Jakobsen I, Grønlund M, Smith FA (2011) Roles of arbuscular mycorrhizas in plant phosphorus nutrition: Interactions between pathways of phosphorus uptake in arbuscular mycorrhizal roots have important implications for understanding and manipulating plant phosphorus acquisition. Plant Physiol 156:1050–1057

Solaiman MZ, Ezawa T, Kojima T, Saito M (1999) Polyphosphates in intraradical and extraradical hyphae of an arbuscular mycorrhizal fungus, *Gigaspora margarita*. Appl Environ Microbiol 65:5604–5606

Staple RC, Toenniessen GH (1984) Salinity tolerance in plant: strategies for crop improvement. Wiley, New York

Taffouo VD, Ngwene B, Akoa A, Franken P (2014) Influence of phosphorus application and arbuscular mycorrhizal inoculation on growth, foliar nitrogen mobilization, and phosphorus partitioning in cowpea plants. Mycorrhiza 24:361–368

Taiz L, Zeiger E (2006) Plant Physiology. Sinauer Associates, Sunderland, MA

Tawaraya K, Saito M (1994) Effect of vesicular-arbuscular mycorrhizal infection on amino acid composition in roots of onion and white clover. Soil Sci Plant Nutr 40:339–343

Tobar R, Azcón R, Barea JM (1994) Improved nitrogen uptake and transport from ^{15}N-labelled nitrate by external hyphae of arbuscular mycorrhiza under water-stressed conditions. New Phytol 126:119–122

Vazquez MM, Barea JM, Azcón R (2001) Impact of soil nitrogen concentration on Glomus spp.-*Sinorhizobium* interactions as affecting growth, nitrate reductase activity and protein content of *Medicago sativa*. Biol Fert Soil 34:57–63

Wang B, Qiu YL (2006) Phylogenetic distribution and evolution of mycorrhizas in land plants. Mycorrhiza 16:299–363

Wilson GWT, Rice CW, Rillig MC, Springer A, Hartnett DC (2009) Soil aggregation and carbon sequestration are tightly correlated with the abundance of arbuscular mycorrhizal fungi: results from long-term field experiments. Ecol Lett 12:452–461

Part V
Sustainability of Environmental Resources from a Crop Production Perspective

Chapter 18
Optimizing Inputs Management for Sustainable Agricultural Development

Mahmoud Fathy Seleiman and Emad Maher Hafez

Abstract Sustainable agriculture has become very popular in recent years and it refers to the plant production without relying on toxic chemical pesticides or herbicides, synthetic fertilizers, genetically modified grains or agricultural practices that pollute soil, water, or other natural resources. In the current chapter, we will discuss the importance of sustainable agriculture for protecting environment and humans from potential risks of some agricultural practices such as using excessive synthetic fertilizer as well as using toxic chemical pesticides or herbicides in crop productivity. We also will discuss the importance of organic (biochar) and bio-fertilizers and their importance in agricultural system. The attention will also be paid into the advantages of cropping rotation systems for better health of soil, environment and human.

Keywords Sustainable agriculture · Biofertilizers · Synthetic fertilizer · Crop rotation · Organic farm · Biochar

18.1 Introduction

Agriculture is the main source of livelihood for human being over hundreds of years (Paull 2011). With reference to the Food and Agricultural Organization (FAO), people in the developing countries are increasing rapidly, leading to a starvation if the food output does not grow by 50–60%. Population of the whole world is about 7.4 billion. It is expected to increase for over than 9.0 billion by 2025. The starvation should be disclaimed from the earth's surface, whereas the most important objective of any

M. F. Seleiman (✉)
Plant Production Department, College of Food and Agriculture Sciences, King Saud University, 2460, Riyadh 11451, Saudi Arabia
e-mail: mseleiman@ksu.edu.sa; mahmoud.seleiman@agr.menofia.edu.eg

Department of Crop Sciences, Faculty of Agriculture, Menoufia University, Shibin El-Kom 32514, Egypt

E. M. Hafez
Agronomy Department, Faculty of Agriculture, Kafrelsheikh University, Kafr El-Sheikh 33516, Egypt

© Springer Nature Switzerland AG 2021
H. Awaad et al. (eds.), *Mitigating Environmental Stresses for Agricultural Sustainability in Egypt*, Springer Water,
https://doi.org/10.1007/978-3-030-64323-2_18

organization to supply enough food for the populace who are under poorness line. A fundamental increase in the crop production was achieved through new cultivars and application of high fertilization rates (Francis and Daniel 2004; Hafez and Geries 2018). High productivity of crops can cause a decline in soil fertility and maintain the agricultural sustainable. Also, roughly 80% of the agricultural land suffers from excessive use of irrigation water and synthetic chemical fertilizers. Faulty agricultural practices which can lead to soil deterioration and environmental degradation can be declaimed by organic farming (Pagliai et al. 2004).

18.2 Sustainable Agriculture

The main principles of sustainability in agriculture are (a) integration biological and ecological processes for instance N fixation, different nutrient cycling and soil regeneration, (b) reducing the usage of pollutants inputs that can pose harm effect to the environment, animals and humans, (c) solving common agricultural issues as well as natural resource problems for instance pest, irrigation and forest management.

Sustainable of agriculture has become very popular recently. It means production of crops without relying on toxic chemical pesticides or herbicides, synthetic fertilizers, genetically modified grains or agricultural practices that pollute soil, water, or other natural resources. Sustainable of agriculture can be achieved by cultivating high-yielding cultivars and using sustainable techniques such as crop rotation and tillage (DeLonge et al. 2016). Using farming techniques that can protect the environment and public health from hazardous pesticides and herbicides are very important to grow crops that are safe for consumers and farmers (Gallandt 2014). Sustainable farmers are also able to protect humans from exposure to pathogens, toxins, and other hazardous pollutants. This form of agriculture enables us to produce healthful food without compromising future generations (Ryals et al. 2014).

To increase the crop production and provide the requirements of the overpopulation, it is time to change agricultural production methods. These included the use of high-productivity cultivars and high fertilization levels; increasing the irrigated zone and intensive agriculture; merging large areas to grow one crop; crop cultivation in non-traditional spaces and changing the agricultural cycle (Magdoff and van Es 2000; Hafez and Kobata 2012). Excessive using of chemical fertilizers is devastated the environmental ecosystem. In the past, our predecessors consumed food free of chemicals, but nowadays there are huge amounts of chemicals and pollutants in food which can cause much anxiety. The use of modern agricultural technology such as productivity of new seeds, fertilizers, irrigation systems and suitable management strategies has saved such serious expectations (Abou El Hassan et al. 2014). Recently, reliance on agricultural production technology has led to a significant increase in the yield; however it has been increased deterioration of agricultural land as well as the increase of chemical residues in the agricultural production, water pollution and environmental degradation (Ryals et al. 2014). The increase in growing population can cause harmful environmental impacts and can degrade natural resources that are necessary to agriculture. Some of these harmful impacts are soil erosion and decrease in organic matter, hence modern agriculture is not sustainable in some cases (Gallandt 2014).

18.3 Synthetic Fertilizer

Agriculture system is one of the most critical human activities that can increase the chemical pollutants through the excessive use of different synthetic chemical fertilizers and pesticides. Such activities can result in a further environmental risk with a potential risks to the human health (Seleiman et al. 2012, 2017). Nitrous oxide is one of the agricultural chemical pollutants that produced through the excessive application of synthetic nitrogen fertilizer (Vejan et al. 2016). Application of synthetic nitrogen fertilizers with excessive rates can also reduce the activity of mycorrhizal root colonization (Seleiman et al. 2013) biological nitrogen fixation in the soil. In cases of high application of synthetic fertilizers, in particular ammonium nitrate form, plant species do not need for the symbiotic microbes to deliver ammonium, consequently this can reduce the symbiosis process for nutrients release (Hafez and Kobata 2012). Additionally, nitrifying bacteria take the benefit of such high application of ammonium and use it to yield the nitrate. The yielded nitrate can then utilized through denitrifying bacteria to result the nitrous oxide and the rest of nitrate can be leached into the groundwater (Galloway et al. 2008).

Synthetic fertilizers contain the most important nutrients such as N, P, K, Zn, Ca and Mg that are needed for plants. The world production of synthetic chemical fertilizers was over 200 Mt in 2014, Fig. 18.1. Nitrogen was accounted for almost 60% of the total production of synthetic chemical fertilizers (FAOSTAT 2017). Globally, the production of synthetic N fertilizers that added into crops has increased about 7 fold, whereas the yield production of crops has increased by almost 2.4 times (Tilman et al. 2002). The production of P synthetic fertilizer was about 38 Mt in 2009, while the production of K synthetic fertilizers (about 18 Mt) has been moderately stable since 1977 (IFA 2010). Globally, nearby 5.0% of the total commercial energy is used in agriculture practices, and 40.0% of that is used for manufacturing the synthetic nitrogen fertilizers (Isherwood 2000). Manufacturing synthetic N fertilizer consumes higher energy (78 MJ Kg N^{-1}) than synthetic P (18 MJ Kg $P_2O_5^{-1}$) and K (14 MJ Kg K_2O^{-1}) fertilizers due to the huge input that needed for reducing N_2 to ammonia in the Haber-Bosch process (Helsel 1992).

18.4 Organic Farming

Organic farming is referred to an alternative farming system which endeavors sustainable agricultural development, enrichment the soil fertility, crop diversity without

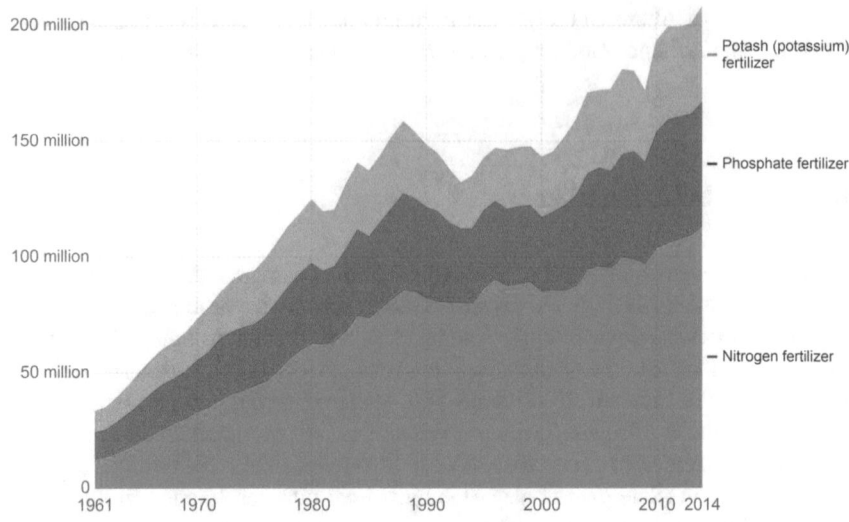

Fig. 18.1 Total production of synthetic chemical fertilizer by tones (FAOSTAT 2017)

damaging its products by prohibiting synthetic pesticides, chemical fertilizers, nano-materials, human sewage sludge, hormones and genetically modified grains & plant growth regulators (Gallandt 2014). It is a natural agricultural system which relay on organic fertilizers such as farmyard manure, compost, bio-fertilizer, green manures, crop residues recycling and biochar (Ryals et al. 2014). Organic farming includes crop diversity and encourages integrated pest and weed management (Pagliai et al. 2004). Hafez and Abou El-Hassan (2015) showed that gypsum application improved clay saline-sodic soil properties, enhanced water use efficiency, improved growth and yield of rice and reduced the effect of water stress. In general, organic farming is designed to reduce the use of inputs to ensure sustainable of agricultural development for a long time and to maintain soil productivity for better water management and biological resource (Hafez et al. 2015; Wu et al. 2012). It was created to supply different nutrients into plants for increasing sustainable of agricultural production in an eco-friendly and to produce free food of contaminants with high nutritional value (Ryals et al. 2014).

18.5   Rotating Crops and Embracing Diversity

Crop rotation is the practice of growing different crops in succession on the same area in sequenced seasons. It can reduce the application of herbicide, and fertilizers, reduce water pollution and avoid depleting soil, increase soil fertility and yields of crops. These systems can improve soil structure, reduce carbon and nitrogen losses, increase soil biomass and fertility, control diseases and weeds (Anderson 2010), increase the beneficial microorganisms and reduce synthetic chemical application. Perennial crops that aren't rotated may rely on other practices, such as cover crops to reduce soil erosion and improve soil productivity as well as can reduce nitrogen and phosphorus losses (Montgomery 2007).

Some reports indicated that nutrients availability as a result of crop rotation under organic farming conditions can increase the beneficial microbes in soil organic matter compared to traditional farming (Pagliai et al. 2004). Agronomists explained the different benefits of crop rotation systems: a crop grown in rotation led to increase 10–25% in yield versus monoculture (Gallandt 2014) as a result of enhancing the availability of nutrients, improve soil fertility and control of pest and weeds (Ryals et al. 2014). Also, crop rotation system alleviates the negative impact of monoculture cropping systems.

18.5.1 Benefits of Cropping Rotation Systems

18.5.1.1 Soil Organic Matter

The use of crop rotation improves soil organic matter, soil health and maintains water use. Soil organic matter is a mix of decaying material from biomass with active microorganisms (Ryals et al. 2014). Crop rotation, by nature, increases exposure to biomass from green manure. The decreased request for aggravating tillage under crop rotation helps biomass aggregation for high nutrient efficiency, declining the request for applied nutrients (Baldwin 2006). For tillage, disturbance and oxidation of soil makes a low conducive environment for variety and proliferation of microorganisms in the soil (Gallandt 2014). He added that these microorganisms are very important for nutrients availability to plants. Soil microorganisms also can decline pathogen and pest activity meanwhile competition. Furthermore, crops output root exudates and other chemicals which manipulate their soil environment and their weed environment.

18.5.1.2 Carbon Sequestration

Crop rotations can enhance soil organic carbon. No-till methods along with crop rotations have been reported an increase in soil organic carbon (Hobbs et al. 2008). Furthermore, increasing crop productivity and carbon sequestration from atmosphere has high impacts on improving the climate by removing carbon dioxide from the atmosphere (Cornejo et al. 2008).

18.5.1.3 Nitrogen Fixing

Crop rotation provides the soil with nutrients. Legumes, for example, have nodules on their roots which include nitrogen-fixing bacteria called rhizobia (Ryals et al. 2014).

Through modulation, the rhizobia bacteria use nutrients and water coming from the plant to change atmospheric nitrogen into ammonia, which is then changed into an organic compound that the plant can use as its nitrogen source (Gallandt 2014). For that, it is logically to rotate them with grains (i.e. Poaceae) and other crops that need nitrates (Naeem et al. 2016).

18.5.1.4   Pathogen and Pest Control

Crop rotation is used to pests, weeds and diseases control which can appear in the soil on the long term. The cropping system declines the high levels of pests by interrupting pest life cycles and pest habitat (Baral 2012). Crop rotation and cultivation soil with cover crops can break the life cycle of pests and restricted (Cornejo et al. 2008). For instance, root-knot nematode is a dangerous for crops in warm climates and sandy soils, where it can decrease crop productivity by cutting off the stream from the plant roots (Gallandt 2014). This principle is especially used in organic farming, where pest control must be achieved without synthetic pesticides (Cornejo et al. 2008).

18.5.1.5   Weed Management

Weeds has a negative impact on crop quality and yield, weeds can slow down the harvesting process (Duffy 2008). Cover crops into crop rotations have a great importance in weed control through competition (Galloway et al. 2008). Moreover, the compost from cover crops and green manure control the weeds growing in the soil, furthermore a competitive advantage to the crop (Hobbs et al. 2008). Cover crops into crop rotations have advantages compared to other common practices for weeds management through tillage (Gallandt 2014). Tillage is aimed to prevent spreading of weeds by stirring the soil; and decrease the viable seeds of weeds in the soil (Duffy 2008).

18.5.1.6   Preventing Soil Erosion

Crop rotation greatly increased soil quality and decreased soil degradation due to crop residue left after harvesting. Crop rotation that includes legume crops and perennial grasses enable to organic matter production in the soil by decomposition of canopy. Plant roots enhance soil organic matter. Soil organic matter content affects soil bulk density, soil water holding capacity, cation exchange capacity, soil aggregation, and nutrient availability (Ryals et al. 2014). Crop rotations thorough remnant cereal crops (e.g. wheat or rice), while used in maintenance tillage systems, can increment soil water (by boosting rainfall/irrigation infiltration and decreasing soil water evaporation) and maintain soils cooler (Karlen et al. 1994). Rotations that increment soil organic matter boost useful bacteria and fungi populations in the soil. Rotations can almost eliminate soil erosion. Crop rotation including legumes (e.g. alfalfa) minimizes the need to add nitrogen fertilizer for next seasons (Gallandt 2014). Crop rotation including legumes may enrich physical and chemical soil properties through N-fixation. Legume cover crops increase soil nitrogen amounts and decrease nitrogen application to the next crops (Galloway et al. 2008).

18.6 Bio-Fertilizers

Towards a sustainable agricultural system, growing of crops require to have a tolerance for diseases, drought, heavy metal and salt stress as well as have a better nutritional properties. The possibility to achieve the above favorite properties of crop is to apply soil microorganisms such as bacteria, fungi and algae that can enhance the nutrient uptake and improve the water use efficiency (Lugtenberg and Kamilova 2009; Seleiman et al. 2012, 2013; Armada et al. 2014; Seleiman and Abdel-Aal 2018). Biofertilizers are fertilizers that contain living different microorganisms, which can enhance microbial activity in soil. They can convert the ambient nitrogen into available forms for plants use such as nitrate and ammonia, improve soil porosity and improve the resistance of plants against pathogens. *Rhizobacteria* are vital microorganisms and can play essential roles in promoting plant growth and its productivity. Such microorganism can straight promote plant growth through producing phytohormones, solubilise phosphate, or fix nitrogen. Ultimately, they can also improve plant growth via inhibition of pathogen development (Lugtenberg and Kamilova 2009).

Modern agriculture is increasingly relying on the application of continuous synthetic chemical fertilizers. Intensive agriculture using chemical fertilizers in huge quantities has undoubtedly led to multiple increases in agricultural production, but the negative impact of the chemical fertilizers is remarkable on soil structure and health, water quality, food quality and environment (Hegde 2008). For these reasons, bio-fertilizers are beneficial complement to chemical fertilizers. Organic agriculture has been confirmed as the only way to sustainable agriculture and the environment. Bio-fertilizers are one of the methods used in the organic agriculture practices. They are the most important biotechnology needed to enhance the expansion of organic farming, sustainable agricultural development and alternative clean agriculture. Bio-fertilizers are natural and organic fertilizers that help to maintain soil with all nutrients and living microorganisms needed for crop health (Venkatashwarlu 2008). Nowadays, the biotechnology in plant nutrition such as bio-fertilizers has shown that biological control is widely considered as a desirable technique for controlling insects and pests because of its environmental effects and obviate of resist the problems caused by agricultural pests. Bio-fertilizers are energy-efficient, free-pollution and rely on exploiting the ability of some microorganisms such as bacteria, algae, and fungi. Bio-fertilizers fix atmospheric nitrogen in the soil and root nodules of legume crops and make it available into the plant. Bio-fertilizers dissolve the insoluble forms of phosphates such as tri-calcium, iron and aluminium phosphates into available forms, decompose organic matter and help mineralization in soil or oxidize sulfur in soil. Their application into soil can enhance N and P, organic matter cornet, as well as they can promote root and growth and yield of plant species (Baral 2012). In continues, Bio-fertilizers are low-cost plant nutrient source, eco-friendly and has a complementary role with synthetic fertilizers. They are classified into two groups viz. fixation of nitrogen such as *Rhizobium, Azotobacter, Azospirillum*, Blue Algae Green, Azola, Phosphorus solubilizers/mobilizers such as Bismuth & Mycorrhizae. Bio-fertilizers excrete several growth hormones (auxins and ascorbic acid) and vitamins which enhance seed germination and growth of crop plants. They improve physical and chemical properties of soil such as water holding capacity, buffer capacity etc. in addition to excrete antibiotics and thus act as pesticides. Zinc and sulfur solubilizers such as thiopasylos and manganese soluble species such as fungal culture *Pencilium citronium* has also been identified for commercial operations. These new strains will also address the issue of "Fertilizer Use Efficiency" and would improve efficiency of bio-fertilizers. Figure 18.2 shows the performance of wheat plants under chemical nitrogen fertilizer and bio-fertilizer. Bio-fertilizers can improve the yield by 10–25% and minimize the use of inorganic N by 25–50 kg/ha without adversely affecting the soil and environment. The global market for bio-fertilizers was estimated to be about five billion USD in 2012 and is expected to be the double by 2020.

Different investigations were conducted under the Egyptian condition to investigate the enhancement of growth, yield and quality of some crops when fertilized with biofertilizers. Seleiman and Abdel-Aal (2018) studied the effect of organic, inorganic and biofertilizer on growth, yield and quality of two cultivars of chickpea (*Cicer arietinum* L.), Menoufia Governorate, Egypt. They reported that bio-fertilizer as a single application should be encouraged for the best productivity and quality of chickpea (cv. Giza 195), since it enhanced and improved yield and quality as well as lessened the amount of mineral fertilizer and consequently the environmental pollution in comparison to inorganic fertilizer (Table 18.1). Sary et al. (2009) investigated the response of wheat (cv. Sakha-93) to weed control and bio-organic fertilization in Kalubia, Egypt. They reported that the application of 25% bioorganic + 75% mineral NPK resulted in the best reduction of fresh and dry broad leaved, grassy and total weeds at 75 and 105 DAS as well as resulted in the highest wheat yield, yield components, protein, P and K yields. Also, Abd El-Lattief (2013) studied the impact of integrated use of bio (*Azotobacter* and/or *Azospirillum*) and mineral nitrogen (230 kg N ha^{-1}) fertilizers and their combination on productivity and profitability of wheat grown in sandy soil, Upper Egypt. He stated that the application of 75% mineral N + bio-fertilizer with *Azotobacter* and *Azospirillum*) resulted in the highest growth and yield of wheat, followed by the application of 50% mineral N + biofertilizer with *Azotobacter* and *Azospirillum*. In this study, bio-fertilizers could save about 25–50% of the recommended dose of mineral N. Megawer and Mahfouz (2010) conducted a field experiment to investigate the effects of bio-fertilizers (*Azotobacter* + *Azospirillum*, free nitrogen fixers and *Trichoderma* a phosphate solubilizing fungi) in combinations with mineral N fertilizer on yield quantity and quality of canola (*Brassica napus* L.) in newly reclaimed soil, Faiyoum, Egypt. They found that application of bio-fertilizer in combination with half of recommended dose of mineral N resulted

Fig. 18.2 Wheat grown with chemical nitrogen fertilizer (left spikes) and bio-fertilizer (right spikes) (Picture: Prof. Dr. Mohamed Salem)

in the highest yield and the best quality of canola. EL-Shamy et al. (2016) investigated the effect of combinations of mineral N with biofertilizers on sugar beet and faba bean productivity under intercropping system in Kafer EL-Sheikh, Egypt. The application of 90 or 72 kg mineral N as individual or 72 kg mineral N in combination with bio-fertilizer. *Cerealine* or *Rizobacterien* recorded the highest significant values of plant height, no. of pods and seeds/ plant, 100-seed weight, seeds yield, protein % of faba bean, and no. of leaves/plant, dry leaves weight/plant, dry root weight/plant, root yield/fed, sugars%, purity% and TSS% of sugar beet.

In conclusion, the application of bio-fertilizers became nowadays inescapable to reduce the environmental pollution that can be caused by the chemical fertilizers. In addition, they improve the yield quality of different crops. About 25–50% of mineral N can replaced by bio-fertilizers.

Table 18.1 Effect of organic, mineral and bio-fertilization on growth, yield and quality of chickpea

Traits/Treatments	Number of branches/plant		Leaves area/plant (cm^2)		Grain yield (ton ha^{-1})		Crude protein (%)	
	S1	S2	S1	S2	S1	S2	S1	S2
Control	6.78^d	6.58^d	263.16^d	239.82^e	0.779^d	0.810^d	17.97^c	17.49^c
Macro elements	7.91^c	7.90^c	356.44^c	395.61^c	1.493^c	1.542^c	20.74^b	20.86^b
Micro elements	7.60^{cd}	8.23^c	348.37^c	350.96^{cd}	1.568^c	1.741^c	20.18^b	20.23^b
Biofertilizer	11.72^a	12.90^a	408.04^b	543.42^b	2.927^a	3.038^a	22.88^a	21.82^a
Humic acid	6.82^d	8.11^c	341.49^c	341.88^d	1.467^c	1.563^c	20.70^b	20.25^b
Combination	9.01^b	10.00^b	493.73^a	597.98^a	2.140^b	2.210^b	23.53^a	22.41^a
LSD$_{0.05}$	0.87	0.73	51.31	48.30	0.250	0.242	0.72	0.79

S1 = First season; S2 = Second season; LSD = Least significant difference
(*Source* Seleiman and Abdel-Aal [2018])

18.7 Compost

Composting of organic residues and the use of compost in agriculture can enhance nutrients and organic matter in the soil (Vázquez and Soto 2017). The remarkable advantages of compost utilize are the enhancement of soil organic matter, soil humus, soil microbial biomass, enzyme activity, resistance against pests and diseases. Compost application into soil can increase carbon sink in the soil, consequently it can reduce the global warming effect (Alejandra et al. 2018). Also, application of compost into lands can improve the phosphorus and potassium availability for the absorption by the plant as well as can increase the soil cation exchange capacity. It can also improve soil quality in terms of bulk density, porosity and available water capacity leading to the increase of plant resistance to the water deficit. Most of reports stated that the application of compost has a positive impact on crop productivity (Boldrin et al. 2009).

Many researchers conducted different investigations on the role of compost application on the growth and yield of different plant species as well as its effect on soil properties. Rady et al. (2016) investigated the role of compost (i.e. 10, 20 and 30 ton ha^{-1} with 50% of the recommended NPK) on growth and yield of *Phaseolus vulgaris* grown in saline soil, Beni Sueif Governorate, Egypt. 100% NPK was also included treatment as the control. Compost application at the rate of 20 ton ha^{-1} improved the soil chemical and physical properties as well as significantly reduced the Cd and NO$_3^-$ in *Phaseolus vulgaris* leaves, pods and seeds. Also, the application of 20-ton compost ha^{-1} showed the same level of growth, and pod and seed yields traits as the control treatment (100% of mineral-NPK fertilizers). They concluded that compost can be used as a partial alternative to NPK-chemical fertilizers; consequently it can increase the possibility of sustainable agronomic performance of common bean using the available recycled organic materials. Therefore, application of compost tea and filtrate biogas slurry liquid at two different applications (i.e. 50 and 100%)

and control treatment as foliar application on sixteen-year-old navel orange trees in Kafer El-Sheikh Governorate, Egypt. Application of compost tea at 100% resulted in the highest yield, fruit weight, and vitamin C of orange trees, while the highest % of fruit set, fruit number, and soluble solid content, lowest fruit drop, and highest reducing and total sugars were obtained from trees treated with 100% filtrate biogas slurry liquid. To sum up, Omar et al. (2012) reported that using the foliar application of compost tea and filtrate biogas slurry liquid at the rate of 100% as food nutrients can be recommended to enhance the productivity and fruit quality of navel orange fruits. In addition, Uzoma et al. (2011) studied the effects of organic-waste compost application (i.e. biosolids, municipal solid waste and water hyacinth compost, incorporated with or without shale deposits [tafla] at the rate of 8%) on growth, yield and quality of sesame grown in the sandy soil, Nuclear Research Center, Egypt. In this study, all compost treatments enhanced the sesame growth and its pigment, carbohydrate and elements. The concentration of metals in sesame seeds were below the phytotoxicity levels and were not exceed 226 Fe, 12.5 Mn, 81.7 Zn, 23.7 Cu, 3.8 Co, 24.6 Ni, 5.4 Pb and 1.72 Cd mg kg^{-1}.

18.7.1   Beneficial Effects of Compost Application

18.7.1.1   Soil Organic Matter

Compost usually contains 15–50% organic matter and 15–45% humic acid. Therefore, compost contains high humus-reproduction value (Abdel-Sabour and Abo El-Seoud 1996). Most arable soils contain 2–4% organic matter, and as it is known that organic matter can enhance the soil's capacity to store nutrients and water. Soil humus can enhance the stable aggregate structure, which gets better water-holding capacity, buffer soil pH, soil temperature, soil tillage without compaction under moist condition and plant growth. The high content of organic carbon in the soil can reduce the adverse effects of global warming (Campbell and Zentner 1993). Application of 6–7 t ha^{-1} year^{-1} (dry weight.) of compost application in medium-textured soils under temperate climate conditions is usually sufficient for the maintenance of the soil humus level (Abdel-Sabour and Abo El-Seoud 1996). Also, application of compost regularly enhances microbial biomass in the soil and activates enzymes (Aranyos et al. 2016), resulting in an increment of organic matter mineralization and biotic (pests and disease) stress tolerance (Abdel-Sabour and Abo El-Seoud 1996). This property enables soils to hold nutrients in the soil-plant cycle or at least can post their leaching or lost into adjacent ecosystems such as lakes and rivers or groundwater.

18.7.1.2   Soil pH

Application of compost (soil organic amendment) fertilization such as alfalfa meal into the soil can decrease pH by increasing bacterial populations. As soil acidity increases, the mineral elements such as phosphorous, iron, and zinc become more available (Aranyos et al. 2016). Compost acts in such cases as a buffer to the soil's pH modifying. The suitable pH needed for most crops range from 6.0–7.0.

18.7.1.3 Nitrogen

Nitrogen content in composts generally ranges from 11.5–16.4 g kg^{-1}. It is not completely available in compost for plants at beginning of application, but it can be mineralized and thus absorbed by the plant or volatized, denitrified or leached over the time (Aranyos et al. 2016). Mineralization of N from composts and soil are affected by almost same factors. The biochemical composition, such as soluble C, cellulose and lignin, has a vital role in the mineralization process (Gabrielle et al. 2004). Mineralization process is highly affected by plant-soil relations, so it is necessary to understand such mechanism to regulate the date of compost application. Also, the N absorption of plants relies on the crop N demands and N-absorption dynamics (Abdel-Sabour and Abo El-Seoud 1996). Subsequently, N mineralization accounted from the consequences of field experiments differed from −14% to +15%. High N-mineralization and N-recovery rates are connected with well mature composts that contain high N, while low N-mineralization and N-recovery rates are connected with immature composts that contain low N. This indicates that the application of different types of agricultural wastes which differ in biochemical composition can greatly enhance soil C and reduce plant N requirements. Otherwise, application compost to pulses cover crops might buffer increase nitrogen to diminish the problem of N-leaching. On the other hand, compost application can lead to N-leaching higher than application of mineral fertilization under different soils and environmental conditions. N-leaching in drainage water as a result of compost application at rate of 40–80 ton ha^{-1} was accounted for 5.2 ppm, while N-leaching in drainage water as a result of mineral N fertilization at rate of 400 kg ha^{-1} was account for 41.5 ppm.

18.7.1.4 Phosphorus

Phosphorus content in composts generally ranges from 2.7 to 4.0 g kg^{-1}. Also, composts can provide 20–40% phosphorus to the soil which can be uptake by plants. Completely decomposed compost can increase the available P in soils through mineralization with a support of the microbial biomass. Compost may diminish the capacity of acid soils by P fixation. High P flux is a vital for high N fixation by legumes (Oehl et al. 2002).

18.7.1.5 Potassium

Potassium contents in composts generally range from 8.4 to 12.5 g kg^{-1}. Composting process has a significant effect on K flux in the soil. Because of the excess water solubility of K, leaching losses may take place when the compost is exposed to rainfall. P-enrich compost application was almost better than P mineral fertilization for the soil properties and plant growth (Abdel-Sabour and Abo El-Seoud 1996). Thus, it can be recommended that the K content in compost could be considered as an alternate for mineral K fertilization.

18.7.1.6 Micronutrients

Iron (Fe), Mn, Cu, Zn, B, and Mo are necessary nutrition's for plant growth, production and quality. Application of compost into soil may either decline or increment element flux, solubility, and plant absorption. The impact of compost on elements availability may be adjusted by the solution pH. It can be recommended that application of high-quality compost (EU regulation 2092/91); only a little change in the available micronutrient status of the soil is to be predictable. As for crop plants, different varieties state differing micronutrient-absorption/exclusion systems and micronutrients are not partitioned regularly between plant roots, stem, leaves, and fruits (Abdel-Sabour and Abo El-Seoud 1996).

18.7.1.7 Soil Structure

Soil structure is one which clears the coming properties: aggregate stability, which improves soil physical properties such as water and air movement into the soil; available water capacity and rate of water infiltration (Garcia-Orenes et al. 2005). Compost application can compensate problems of soil structure more than chemical fertilization by increasing aggregate stability (Wortmann and Shapiro 2008). Lower bulk density and higher porosity enhance soil aeration and drainage regard to compost application.

Crop rotation, cover crops, mulches, organic fertilization, and tillage practices affect root growth, organic carbon oxidation and increase aggregate stability. Compost application causes a slow, however long-standing increment in aggregate stability as its organic matter fundamentally consists of humic substances (Wortmann and Shapiro 2008). It usually affects aggregate stability after a short time and reduces soil bulk density (Garcia-Orenes et al. 2005).

Also, adding compost leads to an increase soil pore volume, soil aeration and warming, and thus for root growth, and for soil water infiltration and earthworm abundance. Application of compost is considered an important in soil–water–plant relationships and in maintaining good soil structure conditions (Kowaljow et al. 2017). They also stated that increasing soil organic matter content has a positive connection with available water capacity. Hudson (1994) documented that higher soil water contents connected with compost fertilization. Compost use increases transpiration and gas exchange of plant due to the improvement in crop growth and leaf area index which increase potential water use by transpiration. Conversely, high water use for transpiration and photosynthesis increases dry matter production and leaf area index. Compost increases transpiration rate of maize crops. Adamtey et al. (2010) reported that compost application increased both photosynthesis and stomatal conductance, this positive impact greater with higher compost applications. Compost application encourages root growth which increase water uptake in plants.

18.7.1.8 Plant-Disease Suppression

Nowadays, composts are known to be as effective as fungicides for root rots inhibition such as *Phytophthora* and *Pythium*. Composts are specific concern in cultivation systems which are depending on fungicides is prohibited or not permitted, as in organic farming (Christopher 2008). Compost does not remove diseases organisms or prevent infections. However, compost does turn the microbe environment in the soil and this turn will influence the microbe's number and types in the soil. This may affect the interaction of microbes and their capability to infect plants. Compost can influence plant diseases by indirect way (Bruggen and Finch 2016). Also, it ameliorates soil structure which makes it beneficial for plant growth and a perfect root system. Both of these contribute to a healthier plant, and indirectly fight plant diseases.

18.8 Biochar

Biochar is the carbon-rich organic matter that remains after heating biomass (crop residues) in little or no oxygen conditions during a process called pyrolysis of various types of vegetable biomass, such as corn, rice or wheat waste (Naeem et al. 2014). The potential for biochar to increase soil fertility and health could lead to higher crop productivity from previously deteriorated soils for farmers (Kamara et al. 2015).

18.8.1 Agricultural and Environmental Benefits of Biochar

Biochar ameliorates soil physical and chemical properties and increases the availability of mineral nutrients in soil leading to increase crop yields and soil improvement (Naeem et al. 2014; Seleiman and Kheir 2018). Many reports stated that biochar application to soils significantly affected crop productivity taking into account soil health and germination rate in addition to water stress tolerance. Biochar as soil conditioner reduces nutrient loss by decrease leaching. It increases water-holding capacity due to its porous structure (Seleiman and Kheir 2018). Biochar sustains soil organic matter content (Seleiman and Kheir 2018). It can increase soil fertility of acidic soils (low pH soils), protects the plants from diseases, promotes growth soil microbes and increase the soil's carbon content. Biochar can be used for long-term storage of carbon in the soil thus decreasing the amount of CO_2 released in the atmosphere (Viger et al. 2014; Seleiman et al. 2019). Figure 18.3 shows biochar application in Experimental Farm of Kafrelsheikh University, Egypt.

Many different investigations were carried out in Egypt to study the effect of biochar on growth and yield of plant species as well as on soil properties. Ibrahim et al. (2015) evaluated the role of biochar (2, 4, 6 ton/fed, fed = 4200 m^2) on wheat productivity grown in sandy soil conditions under two levels of water requirements (2000 and 2500 m^3/fed). They found that biochar application significantly improved plant fresh weight, chlorophyll a, chlorophyll b and proline content during growth stage as well as improved number of grain/spike, grain weight/spike, biological, grain and straw yields at maturity. On the other hand, Ghoneim and Ebid (2013) studied the effect of rice-straw biochar on some soil properties and productivity of rice (*Oryza sativa* L.), Kafr El-Sheikh, Egypt. They stated that application of biochar resulted in a reduction in NH$_4$$^+$–N and NO$_3$$^-$–N in soil under flooding and intermittent conditions. Also they found that rice yield was the highest when biochar application was applied

Fig. 18.3 Biochar application in Experimental Farm of Kafrelsheikh University, Egypt by Dr. Emad Hafez during winter of 2017

at the rate of 30 g/kg soil as well as it enhanced the total soil N, soil organic carbon and soil pH under flooding and intermittent conditions.

Mousa (2017) investigated the influence of biochar (2 and 5%) and seaweed extract (1 and 2 g/ L) applications and their combinations on growth, yield and mineral composition of wheat (*Triticum aestivum* L.) grown in sandy soil, Cairo, Egypt. Results revealed that application of biochar (2%) individually or in combination with sprayed seaweed extract treatment has more promotion enhancement on growth and yield components and resulted in the highest contents of macro- and micronutrients in roots, leaves and grains of wheat in comparison to the control. Also, Mousa (2017) conduct a field experiment on the effect of using soil conditioners (biochar 5, 10 and 20 ton/ fed, and humic acids 5, 10 and 20 kg/fed) on salt-affected soil properties and its productivity at El-Tina Plain Area, North Sinai, Egypt. Data indicated that different treatments significantly improved grain yields of wheat and maize. The application of 20 kg humic acids/fed was the best treatment. The results concluded that using biochar and humic acids as organic amendments is great strategy to improve soil characterization, compensate the deficiency of different plant nutrients in soil and their significant impact on the yield of wheat and maize.

18.9   Nano-Fertilizers for Sustainable Crop Production

It seems that nano-fertilizers can have valuable potential for attaining the sustainable crop production in agriculture systems. Application of nanotechnology is considered an advanced and favorable technology to feed the increase in world population for sustainability of agriculture. Nanotechnology has not revolutionized agriculture with innovative nutrients in the form of nano-fertilizers, but has also supported the plant protection via the development of efficient water management system, increasing the efficiency of plant to utilize the sun's energy and improvement of nano-pesticides (Ditta et al. 2015). The low use efficiency (20–50%) of synthetic chemical fertilizer, high cost, application rate, and high energy used in the production process have encouraged scientists to improve and promote the technology of nano-fertilizers (Aziz et al. 2006). Nano-fertilizers are the nano-materials (i.e. 1–100 nm) that could work as macro- or/and micronutrients for the different plant species or work as carriers of the synthetic chemical fertilizers–nano-carriers for effectual use of the different nutrients. Nano-fertilizers have been showed to be moderately effective over the synthetic chemical fertilizers because of their new mechanisms of actions, lessen nutrient loss, enhance use efficiency, and reduce worsening of the environment. Furthermore, nano-fertilizers have the capability of entering the plant cells straight due to their small size, and this decreases the required energy of their uptake and transportation into the plant cells (DeRosa et al. 2010; Brackhage et al. 2013). Nano-fertilizers are similar to the synthetic chemical fertilizers in terms of their dissolving in the rhizosphere and can be directly taken by the plants. Application of nano-fertilizers can reduce N loss through leaching into ground-water. Lately, biodegradable, polymeric chitosan (78 nm) has been as carrier for NPK fertilizer sources for instance urea, calcium phosphate, and potassium chloride. In addition other nano-materials such as kaolin and polymeric biocompatible might be used for similar purpose (DeRosa et al. 2010).

18.10 Conclusions

Principles of sustainability in agriculture include the integration biological and ecological processes for instance N fixation, different nutrient cycling, crop rotation, high-productivity cultivars as well as reducing the use of pollutants inputs that can pose harm effect to the environment, animals and humans. Agriculture system is one of the most critical human activities that can increase the chemical pollutants through the excessive use of different synthetic chemical fertilizers and pesticides. Excessive use of chemical fertilizer can reduce the biological nitrogen fixation in the soil and increase the chemical pollutants of nitrous oxide. Thus, organic farming including organic and bio-fertilizers, crop rotations, high yield cultivars and nano-fertilizer are highly recommended for sustainability in agriculture.

18.11 Recommendations

- Excessive use of chemical fertilizer should not be used, because it reduces the biological nitrogen fixation in the soil and increases the chemical pollutants of nitrous oxide.
- Crop rotation is highly recommended in Egyptian agriculture because it results in a reduction in the application of herbicide, fertilizers, growth of weeds and water pollution as well as improves soil fertility, micro-organisms activity and productivity of field crops.
- Bio-fertilizers are highly recommended as single or as a complementary with synthetic fertilizer, because they are energy-efficient, free-pollution, low-cost plant nutrient source and eco-friendly.
- Biochar and compost are also highly recommended particularly in poor soils with 50% of the recommended NPK, because their application improves the soil properties and fertility as well as enhance the yield and quality of field crops.
- Nano-fertilizers is not used in Egypt on the wide scale, however it is recommended because of their new mechanisms of actions, lessen nutrient loss, enhance use efficiency, and reduce worsening of the environment. In addition, nano-fertilizers have the capability of entering the plant cells straight due to their small size, and this decreases the required energy of their uptake and transportation into the plant cells

References

Abd El-Lattief SA (2013) Impact of integrated use of bio and mineral nitrogen fertilizers on productivity and profitability of wheat (*Triticum aestivum* L.) under Upper Egypt conditions. Int J Agron Agric Res 3(12):67–73

Abdel-Sabour MF, Abo E1-Seoud MA (1996) Effects of organic-waste compost addition on sesame growth, yield and chemical composition. Agric Ecosyst Environ 60:157–164

Abou El Hassan WH, Hafez EM, Ghareib AA, Ragab MF, Seleiman MF (2014) Impact of nitrogen fertilization and irrigation on N accumulation, growth and yields of Zea mays L. J Food Agric Environ 12(3&4):217–222

Adamtey N, Cofi O, Ofosu-Budu KG, Ofosu-Anim J, Laryea KB, Forester D (2010) Effect of N-enriched co-compost on transpiration efficiency and water-use efficiency of maize (*Zea mays* L.) under controlled irrigation. Agric Water Manag 97:995–1005

Alejandra C, Adriana A, Xavier F, Raquel B, Teresa G, Antoni S (2018) Composting of food wastes: status and challenges. Bioresour Technol 248:57–67

Anderson RL (2010) Rotation design to reduce weed density in organic farming. Renew Agric Food Syst 25(3):189–195

Aranyos JT, Tomócsik A, Makádi M, Mészáros J, Blaskó L (2016) Changes in physical properties of sandy soil after long-term compost treatment. Int Agroph 30:269–274

Armada E, Portela G, Roldan A, Azcon R (2014) Combined use of beneficial soil microorganism and agrowaste residue to cope with plant water limitation under semiarid conditions. Geoderma 232:640–648

Aziz T, Maqsood RMA, Tahir IA, Cheema MA (2006) Phosphorus utilization by six brassica cultivars (*Brassica juncea* L.) from tri-calcium phosphate, a relatively insoluble P compound. Pak J Bot 38:1529–1538

Baldwin KR (2006) Crop rotations on organic farms (PDF) (Report). Center for Environmental Farming Systems. Retrieved May 4, 2016, Available at: https://www.carolinafarmstewards.org/wp-content/uploads/2012/12/7-CEFS-Crop-Rotation-on-Organic-Farms.pdf

Baral KR (2012) Weeds management in organic farming through conservation agriculture practices. J Agric Environ 13:60–66

Boldrin A, Andersen J, Møller J, Christensen T (2009) Composting and compost utilization: accounting of greenhouse gases and global warming contributions. Waste Manage Res 27:800–812

Brackhage C, Schaller J, Bäucker E, Dudel EG (2013) Silicon availability affects the stoichiometry and content of calcium and micro nutrients in the leaves of common reed. Silicon 5:199–204

Bruggen AH, Finckh MR (2016) Plant diseases and management approaches in organic farming systems. Annu Rev Phytopathol 4(54):25–54

Campbell C, Zentner R (1993) Soil organic matter as influenced by crop rotations and fertilization. Soil Sci Soc Am J 57:1034–1040

Christopher AG (2008) Sustainable agriculture and plant diseases: an epidemiological perspective. Philosophical Transactions of the Royal Society of London. Ser B, Biol Sci 363(1492):741–759

Cornejo P, Meier S, Borie G, Rillig MC, Borie F (2008) Glomalin-related soil protein in a Mediterranean ecosystem affected by a copper smelter and its contribution to Cu and Zn sequestration. Sci Total Environ 406:154–160

DeLonge MS, Miles A, Carlisle L (2016) Investing in the transi-tion to sustainable agriculture. Environ Sci Pollut 1:266–273

DeRosa MC, Monreal C, Schnitzer M, Walsh R, Sultan Y (2010) Nanotechnology in fertilizers. Nat Nanotechnol 5:91–94

Ditta A, Arshad M, Ibrahim M (2015) Nanoparticles in sustainable agricultural crop production, applications and perspectives. Nanotechnology and plant sciences. Springer International Publishing, Switzerland, pp 55–75

Duffy M (2008) Estimated costs of crop production in Iowa-2009. FM1712. Iowa State University Extension

EL-Shamy MA, Hamadny MK, Mohamed AA (2016) Effect of faba bean sowing distance and some combinations of mineral nitrogen levels with bio fertilizers on sugar beet and faba bean productivity under intercropping system. Egypt J Agron 38(3):489–507

FAOSTAT (2017) Fertilizers. FAO Statistical Databases & Data-sets. Available from: http://faostat. fao.org/site/291/default.aspx. Accessed 8 Dec 2018

Francis CA, Daniel H (2004) Organic farming, p 77–84. Encyclopedia of soils in the environment. Elsevier, Oxford, UK

Gabrielle B, Da-Silveira J, Houot S, Francou C (2004) Simulating urban waste compost effects on carbon and nitrogen dynamics using a biochemical index. J Environ Qual 33:2333–2342

Gallandt E (2014) Weed management in organic farming. In: Chauhan B, Mahajan G (eds) Recent advances in weed management. Springer, New York, NY

Galloway JN, Townsend AR, Erisman JW, Bekunda M, Cai Z, Freney JR, Martinelli LA, Seitzinger SP, Sutton MA (2008) Transformation of the nitrogen cycle: Recent trends, questions, and potential solutions. Science 320:889–892

Garcia-Orenes F, Guerrero C, Mataix-Solera J, Navarro-Pedreno J, Gomez I, Mataix-Beneyto J (2005) Factors controlling the aggregate stability and bulk density in two different degraded soils amended with biosolids. Soil Till Res 82:65–76

Ghoneim AM, Ebid AI (2013) Impact of rice-straw biochar on some selected soil properties and rice (*Oryza sativa* L.) grain yield. In J Agron Agric Res 3(4):14–22

Hafez EM, Abou El-Hassan WH (2015) Nitrogen and water utilization efficiency of barley subjected to desiccation conditions in moderately salt-affected soil. Egypt J Agron 37(2):231–249

Hafez EH, Abou El Hassan WH, Gaafar IA, Seleiman MF (2015) Effect of gypsum application and irrigation intervals on clay saline-sodic soil characterization, rice water use efficiency, growth, and yield. J Agric Sci 7(12):208–219

Hafez EM, Geries L (2018) Effect of N fertilization and biostimulative compounds on onion. Cercetari Agron Moldova 1(173):75–90

Hafez EM, Kobata T (2012) The effect of different nitrogen sources from urea and ammonia sulfate on the spikelet number in irrigated Egyptian spring wheat. Plant Prod Sci 15(4):332–338

Hegde SV (2008) Liquid bio-fertilizers in Indian agriculture. Bio-Fertilizer News Letter, pp 17–22

Helsel ZR (1992) Energy and alternatives for fertilizer and pesticide use. In: Fluck, RC (ed) Energy in farm production, vol. 6. Elsevier, New York, USA, pp 177–201

Hobbs PR, Sayre K, Gupta R (2008) The role of conservation agriculture in sustainable agriculture. Philos Trans R Soc B 363:543–555

Hudson BD (1994) Soil organic matter and available water capacity. J Soil Water Conserv 49:189–194

Ibrahim OM, Bakry AB, El kramany MF, Elewa TA (2015). Evaluating the role of bio-char application under two levels of water requirements on wheat production under sandy soil conditions. Int J Adv Res 2(2):411–418

IFA (2010) Statistics: Fertilizer Industry Association. Available from: http://www.fertilizer.org/ifa/ HomePage/STATISTICS. Accessed 8 Mar 2018

Isherwood KF (2000) Mineral fertilizer use and the environment. International Fertilizer Industry Association (IFA) and United Nations Environment Programme (UNEP). Revised edition. Paris, France, 53 p

Kamara A, Hawanatu SK, Mohamed SK (2015) Effect of rice straw biochar on soil quality and the early growth and biomass yield of two rice varieties. J Agric Sci 6:798–806

Karlen DL, Varvel GE, Bullock DG, Cruse RM (1994) Crop rotations for the 21st century. Adv Agron 53:1–45

Kowaljow E, Gonzalez-Polo M, Mazzarino MJ (2017) Understanding compost effects on water availability in a degraded sandy soil of Patagonia. Environ Earth Sci 76:255

Lugtenberg B, Kamilova F (2009) Plant-growth-promoting rhizobacteria. Annu Rev Microbiol 63:541–556

Magdoff F, van Es H (2000) Building soils for better crops. Sustainable Agriculture Network, National Agriculture Library, Beltsville, MD

Megawer EA, Mahfouz SA (2010) Response of Canola (Brassica napus L.) to biofertilizers under Egyptian conditions in newly reclaimed soil. Int J Agric Sci 2(1):12–17

Montgomery DR (2007) Soil erosion and agricultural sustainability. Proc Nat Acad Sci 104(33):13268–13272

Mousa AA (2017) Effect of using some soil conditioners on salt affected soil properties and its productivity at el-tina plain area. North Sinai. Egypt J Soil Sci 57(1):101–111

Naeem MA, Khalid M, Ahmad Z, Naveed M (2016) Low pyrolysis temperature biochar improves growth and nutrient availability of Maize on Typic Calciargid. Commun Soil Sci Plant Anal 47:41–51

Naeem MA, Khalid M, Arshad M, Ahmad R (2014) Yield and nutrient composition of biochar produced from different feedstocks at varying pyrolytic temperatures. Pak J Agric Sci 51:75–82

Oehl F, Tagmann HU, Oberson A, Besson JM, Dubois D, Mäder P (2002) Phosphorus budget and phosphorus availability in soil under organic and conventional farming. Nutr Cycl Agroecosyst 62:25–35

Omar A, Belal E, El-Abd A (2012) Effects of foliar application with compost tea and filtrate biogas slurry liquid on yield and fruit quality of Washington navel orange (Citrus sinenesis Osbeck) trees. J Air Waste Manag Assoc 62(7):767–772

Pagliai M, Vignozzi N, Pellegrini S (2004) Soil structure and the effect of management practices. Soil Till Res 79:131–143

Paull J (2011) Nanomaterials in food and agriculture: The big issue of small matter for organic food and farming, Proceedings of the Third Scientific Conference of ISOFAR (International Society of Organic Agriculture Research), 28 September–1 October, Namyangju, Korea, 2:96–99

Rady M, Semida W, Hemida K, Abdelhamid M (2016) The effect of compost on growth and yield of Phaseolus vulgaris plants grown under saline soil. Inter J Recyl Org Waste Agric 5:311–321

Ryals R, Kaiser M, Torn S, Berhe AA, Silver WL (2014) Impacts of organic matter amendments on carbon and nitrogen dynamics in grassland soils. Soil Biol Biochem 68:52–61

Sary GA, El-Naggar HM, Kabesh MO (2009) Effect of bio-organic fertilization and some weed control treatments on yield and yield components of wheat. World J Agric Sci 5(1):55–62

Seleiman MF, Abdel-Aal MSA (2018) Effect of organic, inorganic and bio-fertilization on growth, yield and quality traits of some chickpea (cicer arietinum l.) varieties. Egypt J Agron 40(1):105–117

Seleiman MF, Kheir AS (2018) Maize productivity, heavy metals uptake and their availability in contaminated clay and sandy alkaline soils as affected by inorganic and organic amendments. Chemosphere 204:514–522

Seleiman MF, Refay Y, Al-Suhaibani N, Al-Ashkar I, El-Hendawy S, Hafez EM (2019). Integrative effects of rice-straw biochar and silicon on oil and seed quality, yield and physiological traits of Helianthus annuus L. grown under water deficit stress. Agron 9(10):637

Seleiman MF, Santanen A, Kleemola J, Stoddard FL (2013) Improved sustainability of feedstock production with sludge and interacting mycorrhiza. Chemosphere 91:1236–1242

Seleiman MF, Santanen A, Stoddard FL, Mäkelä P (2012) Feedstock quality and growth of bioenergy crops fertilized with sewage sludge. Chemosphere 89:1211–1217

Seleiman MF, Selim S, Jaakkola S, Mäkelä P (2017) Chemical composition and in vitro digestibility of whole-crop maize fertilized with synthetic fertilizer or digestate and harvested at two maturity stages in boreal growing conditions. Agric Food Sci Finland 26:47–55

Tilman D, Cassman KG, Matson PA, Naylor R, Polasky S (2002) Agricultural sustainability and intensive production pratices. Nature 418:671–677

Uzoma KC, Inoue M, Andry H, Fujimaki H, Zahoor Z, Nishihara E (2011) Effect of cow manure biochar on maize productivity under sandy soil condition. Soil Use Manag 27:205–212

Vázquez MA, Soto M (2017) The efficiency of home composting programmes and compost quality. Waste Manage 64:39–50

Vejan P, Abdullah R, Khadiran T, Ismail S, Boyce AN (2016) Role of plant growth promoting rhizobacteria in agricultural sustainability—a review. Molecules 21:573–590

Venkatashwarlu B (2008) Role of bio-fertilizers in organic farming: organic farming in rain fed agriculture: Central institute for dry land agriculture. Hyderabad, 85–95

Viger M, Robert DH, Miglietta F, Taylor G (2014) More plant growth but less plant defence? First global gene expression data for plants grown in soil amended with biochar. GCB Bioenerg 7:658–672

Wortmann CS, Shapiro CA (2008) The effects of manure application on soil aggregation. Nutr Cycl Agroecosyst 80:173–180

Wu W, Yang M, Feng Q, Mcgrouther K, Wang H, Lu H, Chen Y (2012) Chemical characterization of rice straw-derived biochar for soil amendment. Biomass Bioenergy 47:268–276

Chapter 19
Maize Productivity in the New Millennium

Abdelrehim A. Ali

Abstract In plant genetics, maize (*Zea mays L.*) is an exciting model plant and also the most essential crop global for nutrition, animal feed and bioenergy. The most significant reason maize has extent wider than other major crops because of its greater adaptability and its high yielding ability under a wide range of environmental conditions. Although, maize is a lowly source of protein for both humanity and monogastric animals, it is considered a principal food for numerous millions of people worldwide, mainly in the developing countries. Maize grain yields fluctuate between 2.5 and 9 t/ha; production in excess of 20 t/ha have locally been recorded, representing an excellent transformation of photosynthesis-related solar radiation into exploitable biochemical energy which is considered among the highest crops. Climate change and ecological degradation overall the world threaten cereal productivity and food security. In Egypt, the expected climate changes, according to climate change scenarios will cause an increase in evapotranspiration (ET) between 2.4% to 16.2% in the Delta region, between 5.9% to 21.1% in the Middle Egypt region and 5.8% to 22.5% in the Upper Egypt region up to the year 2100 as compared to the current situation. So, moving from the traditional to the advanced breeding methods became necessary to overcome the different limitations faced maize production, particularly in climate change conditions. Also, the actual challenge under climate change environments is to use adaptation approaches, which are used to develop agricultural management applies. Therefore, by using phenotyping an infra-red thermography and other advanced tools could be a beneficial tool in selecting tolerant genotypes. The objectives of this chapter are to summarize the historical trend of maize production in the world (Developed countries, USA, and developing countries, Egypt), discuss the main constraints that affect maize productivity. Also, it provides a perspective on the challenges faced in maize agriculture which must be met both in terms of increased productivity and quality.

Keywords Maize · Production · Consumption · Heterosis · Grain yield · Abiotic and biotic stresses · Genetic improvement · Infrared approach · Maize modern breeding

A. A. Ali (✉)
Crop Science Department, Faculty of Agriculture, Suez Canal University, 41522, Ismailia, Egypt

© Springer Nature Switzerland AG 2021
H. Awaad et al. (eds.), *Mitigating Environmental Stresses for Agricultural Sustainability in Egypt*, Springer Water,
https://doi.org/10.1007/978-3-030-64323-2_19

19.1   Introduction

19.1.1   Maize Production Directions

Maize (*Zea mays* L.) is the most important cereal crop worldwide after wheat and rice. Based on production volume, in 2016/2017, corn is the most important grain in the world (1.07 billion metric tons) comparing with other grain crops (Fig. 19.1). The United States in that year was the biggest corn producer worldwide (388 million metric tons). This represents over one third (36.67%) followed by China 20.93% of the global corn production.

The average of the area cultivated with maize reached approximately 161.6 million hectares during the period (2005–2012). Besides, nearly half of the cultivated area is concentrated in two countries which are China and the United States, as they produce about 63.5% of the total global production. The global production reached about 713.7 million tons in 2005 and then it rose to about 872.1 and 1033.74 million tons in 2012 and 2017 at a rate of increase reached about 18.16 and 30.96%, respectively. The United States is the first country producing maize crops in the world, as its production represents about 36% of the total global production, while its cultivated area represented about 20.2% of the total cultivated area at the level of the world during the period (2005–2017). More than 80% of maize production is concentrated

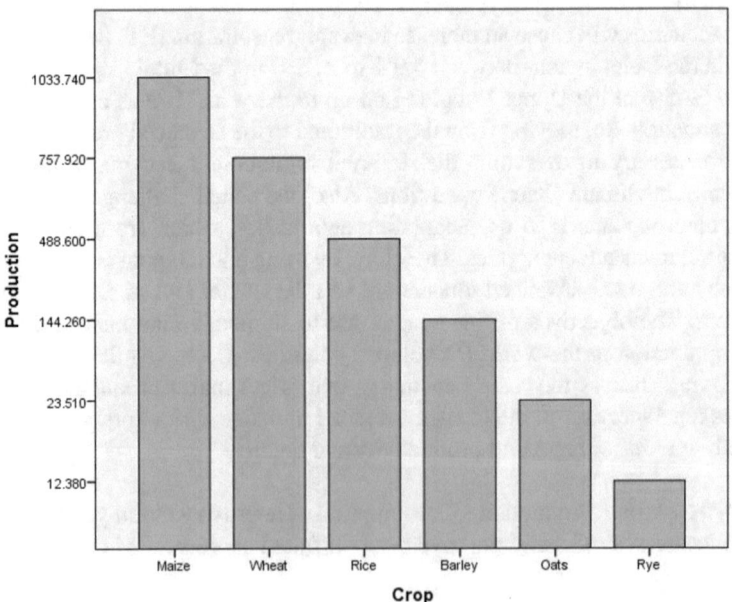

Fig. 19.1 Worldwide production of grain in 2017/2018, by type (in million metric tons) (*Source* https://www.statista.com/statistics/263977/world-grain-production-by-type/)

Table 19.1 Total domestic corn production and growth of production in the largest corn producing countries (2012–2017). The data obtained from FAOSTAT's estimates 2012–2017

Rank	Country	Production (million tons)					
		2012	2013	2014	2015	2016	2017
1	USA	273.19	351.27	361.09	345.49	384.78	370.96
2	China	205.7	218.6	215.8	224.8	231.8	215.89
3	Brazil	71.10	80.30	79.90	85.30	64.10	94.50
4	India	22.26	24.26	24.17	22.57	26.26	27.15
5	Argentina	21.20	32.10	33.10	33.80	39.80	36.00
6	Ukraine	20.96	30.95	28.50	23.33	28.07	24.12
7	Mexico	22.07	22.66	23.27	24.69	28.25	26.50
8	Indonesia	19.39	18.51	19.01	19.61	20.37	11.35
9	France	15.39	15.04	18.34	13.72	12.13	13.90
10	Russia	8.21	11.63	11.33	13.17	15.31	13.23
11	South Africa	12.12	11.81	14.25	9.96	7.78	13.00
12	Egypt	8.10	7.96	8.60	7.80	8.00	6.00

(*Source* https://www.statista.com/search)

in the Americas (53%) and Asia (28%), followed by Europe (15%). In the world ranking, the USA is the first in maize production followed by China, Brazil, India, Argentina, Ukraine and Mexico (Table 19.1).

In developed countries, such as USA, France and Italy, maximum yield per unit area could be attained by cultivating hybrids that can withstand high plant density up to 90,000 plants ha^{-1} (Gozubenli et al. 2003). Also, Maize production in these developed countries, is greatly mechanized and based on profitable production procedures using selected hybrids and agrochemicals. These greatly mechanized production situations are, however, in clear difference with those in several developing countries. Maize is often cultivated by medium and small scale farmers with cultivation procedures are less advanced (Verheye 2010).

Average maize production per unit area in the USA increased intensely through the second half of the twentieth century, due to improvement in crop agricultural practices and greater tolerance of new hybrids to high plant densities (Duvick and Cassman 1999; Tollenaar and Wu 1999).

As indicated in Fig. 19.2, the areas cultivated with corn in Egypt, have fluctuated from 1960 to 2018 as a result to fluctuate cultivated area of rice and cotton. As a result of an increase in rice and cotton area in 2011/2012, the area of corn significantly decreased. The production has increased significantly, during the same time. The domestic production was raised from about 5.6 million in 2000 to about 6.9 million tons in 2014 (Heba et al. 2015). In 2015 and 2016, according to Food and Agriculture Organization (FAO) reports, Egypt ranks between the fifth and the sixth place in the world with regard to average productivity after Germany, France, Canada, Italy and

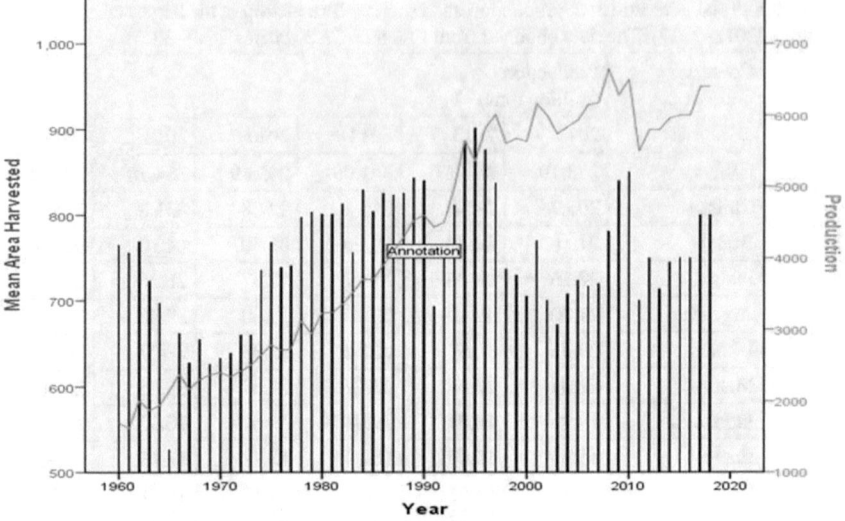

Fig. 19.2 Production (1000 MT) and area harvested (ha) of maize in Egypt, 1960–2018 (*Source* United States Department of Agriculture)

USA, where average production of these developed countries reached 10–11 ton ha^{-1}.

19.1.2 Maize: Consumption and Gap

Maize has numerous uses, it is used for three major purposes: as an essential food crop for humans, a forage for livestock, as raw material for various industrial uses, as well as bio-fuel production. Prasanna (2015) stated that by 2050, the demand for corn in the developing countries is expected to be twice. A study by FAO and the Organization for Economic Cooperation and Development (OECD) estimates that universal consumption of cereals will growth by 390 million tons in the period from 2014 to 2024. Operable ethanol plants increased by 5% per year (EIA report 2018)-over 700 million gallons within January 2017 and January 2018. Snapp et al. (1998) mentioned that the use of cereals as biofuel feedstock has been expected to grow to 182 million tons by 2020; Also, according, Vanlauwe et al. (2010) and Nyamangara et al. (2013), it could reach almost 450 million tons by 2050 (Fig. 19.3).

In Egypt, the domestic consumption was increased from about 10.1 million tons in 2000 to about 14.3 million tons in 2013. In the same period, the domestic production was raised from about 5.6 million in 2000 to about 8.6 million tons in 2014. The amount of imports was raised from about 4958.2 million tons in 2000 to about 10805.6 in 2014 (Heba et al. 2015). So, there is a gap between production and consumption

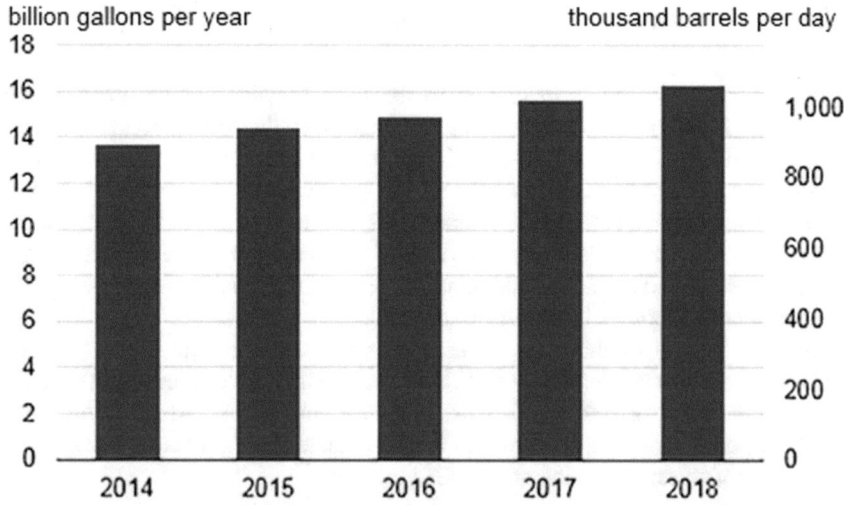

Fig. 19.3 U.S. fuel ethanol production capacity (2014–2018) (*Source* U.S. Energy Information Administration, *U.S. Fuel Ethanol Plant Production Capacity*) https://www.indexmundi.com/agriculture/?country=eg&commodity=corn&graph=area-harvested

of maize of about 45%. This gap is compensated by importation, which put a burden on the country's budget (Zohry et al. 2016).

Generally, In order to balance the global population increase with the global food demand, it is a must to use improved agricultural technologies that help in boosting crop production and productivity (FAO 2005).

19.2 History, Origin and Distribution

As most historians believe, domestication maize maybe from a wild teosinte form (*Euchlaena mexicana*), since 6000 to 7500 years ago in the Tehuacan Valley of Mexico (Piperno 2011). Teosinte look like corn in habit, but produces a number of basilar tillers. It crosses freely with maize, and the crosses are fruitful. In Africa, maize has been introduced by Portuguese and Arab explorers then to other tropical countries, from where it expanded inland via the slave trade routes, and finally to Asia (Piperno 2011).

Maize variation exceedingly could be found in the Meso-American and the northern portion of South America. The considerable diversity of conditions and environments have formed the base for the improvement of maize varieties well adapted to severe conditions of climate and soil and to biotic stresses (Louette and Smale 1998).

19.3   Taxonomy and Classification

Corn is an annual crop is belonging to the tribe Maydeae of the grass family Gramineae. *Zea mays* Mexicana is the closest vegetal relative to maize and considered annual teosinte variety (Smith 2013). Using numerical taxonomy methods based on morphological characters, genetic effects, and genotype x environment interactions were detected by Sanchez and Goodman (1992). In most Mexican territory considerable amount of maize landraces are widely distributed. In some scattered areas devilishly associated to maize, wild relatives, particularly teosinte, are growing. The number of chromosomes in *Z. mays* is 2n = 20.

There is variety of the color of kernels maize ranging from white to black. According to National Chamber of Milling, South Africa (2008), most of the maize grown in the USA is yellow, whereas in several countries of Africa and Central America, people are used to consume a white maize grains; while, yellow maize grains mostly consumed in the animal feed industry.

Composition of the endosperm is used as a base of maize classification, resulting in an artificially definition by the type of kernel such as Flour corn *(Zea mays var. amylacea)*, Popcorn: *Zea mays* var. *everta*, Dent corn: *Zea mays* var. *indentata* and other types

19.4   The Botany of Maize

Maize is a determinate annual plant. It produces large, narrow, opposite leaves on a solid single stem 3–4 cm in diameter, 2–3 m high. The maize plant would have reached its full height by tasseling initiation stage. The total leaves number may vary between different crosses, seasons, planting times and locations. The maize root system contains no tap-root, and its feathery roots spread out in all sides, mostly in the top soil (Fig. 19.4). Maize is a monoecious grass generally protandrous, i.e. the male flowers matures earlier than the female flowers. Tasseling stage is the stage at which the tassels (male flowers) appear as a terminal panicle, up to 40 cm in long. Silking stage starts 2–3 days after tasseling when any silks are initiating outside the husk after developing the female inflorescence on a short side-branch. Pollination in maize mainly is cross-pollination, but there may be about 5% self-pollination. It takes place when the new moist silks catch the falling pollen grains. Pollen grains can be carried by wind up to 500 m distances to grow as a pollen tube inside of the silk. A single one nucleus fertilizes the haploid egg to produce a diploid embryo inside the ovule, the other one combine with the diploid central cell to produce the triploid endosperm.

Ear of maize commonly carry 600 grains. The maximum size of grains is apparently 2.5 cm (Stevenson and Goodman 1972). Average kernel volume of 0.29 cm^3 (U.S. Grains council 2017/2018). The maize seed comprises two major compartments, the embryo (representative 10–13% of the grain), and the endosperm, both

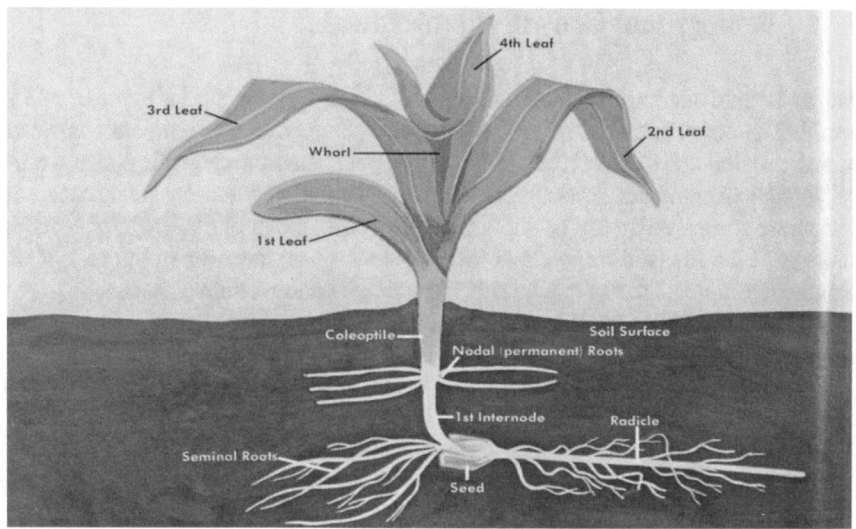

Fig. 19.4 Parts of a young maize plant (Hanway 1966)

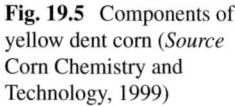

Fig. 19.5 Components of yellow dent corn (*Source* Corn Chemistry and Technology, 1999)

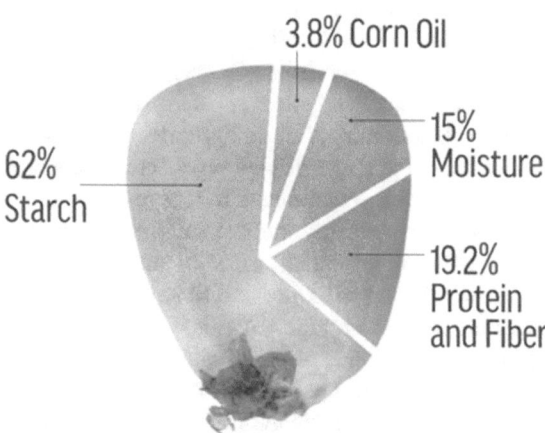

resulting from the double fertilization process and of a seed coat, the pericarp, of maternal origin. In absence of the testa the endosperm is combined with the pericarp (Fig. 19.5).

19.5 Ecology and Growth Requirements

Maize is cultivated in tropical, subtropical and temperate climatic zones of the world. The majority (70%) of maize production happens under temperate environments. Aquino and Calvo (1999) identify five mega-environments for maize worldwide: tropical highlands, lowland tropics, subtropics, mid-altitude tropical zones, and temperate zones. Maize can be cultivated in spring-summer and fall-winter growing seasons at the lowland tropics and lowland subtropics, however in the rest of the mega-environments maize can be cultivated in the spring-summer season.

Maize is cultivated in all agro-climatic zones in Egypt from north to south. The cultivated area with maize is irrigated using surface irrigation, as it is the most prevailing irrigation system in Egypt expansion, which is a limited by limited water resource (Zohry et al. 2016).

Maize foremost is considered a warm season crop and it is grown in wide-ranging of climatic conditions. The main maize production areas are existing in temperate regions of the globe. Temperature is one of the most significant factors in the germination process, the initial temperature to germinate maize seed is about 10°C. Maize is comparatively sensitive to cold conditions, and it does not adapt to lower temperatures like most cool-season crops. Temperatures of 5–7°C might be led to photo-inhibited physiological harm that may decrease photosynthetic rates for several days. Maize as a C4 plant, respond well to high temperatures and intensive sunlight. At midday temperatures of 32–35°C, well-watered maize plants reach highest leaf photosynthesis rates. Photosynthetic rates of sun-adapted maize do not drench until the intensity of light oncoming, full sunlight. Because photosynthesis take of sunlight energy is considered the major driving vigor for maize growth and yielding ability, extreme cloudiness and short days tend to reduce maize production (Fahad et al. 2017).

Maize grows healthy in most soils, but deep fertile soils rich in organic matter, have a good water holding capacity and well-drained are the most preferred ones. Maize is less adapted to the conditions of compact clay and sand soils. The main soil factor is soil water storage, and this is determined by texture and building. The major factor of success under rain-fed maize production is available soil moisture, mainly where the rains are not equivalently distributed over the season. In this situation, maize plant has to rely on the stored moisture in the root zone to overcome the water shortage. Sandy soils have a low moisture retention capacity. Sands hold less than 5% water, while clay loam soils have available water up to 20% of their volume. The ability of the soil to hold few weeks of water in situation of dry periods through the season is a most important factor of the potential of such soil to produce maize.

Maize is a comparatively exacting crop regarding to plant nutrients. It demands a great quantity of nitrogen (N). In supplying the essential nutrients for the crop, at least the same quantity of these elements should be incorporated in the soil. The soil analysis should be the basis for determining and adding the particular amounts of fertilizers to the soil. Adequacy of supplies of nutrients as needed by the plants can be assured by adding of nutrients in organic or inorganic (chemical fertilizers) forms (http://www.eolss.net/Sample-Chapters/C10/E1-05A-13-00.pdf).

In the tropics, soil temperature is rarely considered a dangerous limiting factor for maize growing. There are two cases, however, in which soil temperatures can be restricted: extremely high temperatures in sandy soils specially, and cold temperatures in the equatorial highlands (Lal 1973). Maximum soil temperatures 41°C reduced the rate of leaf elongation by 22% and the decrease was 44% with maximum soil temperatures of 43°C. Root tip elongation rate was significantly influenced by short-term exposure to high soil temperature (Maize production manual 1982).

The maize crop is greatly sensitive to excess soil moisture circumstances and therefore it is important that fields where maize is grown are adequately drained. The most critical growth period is from tasseling to dough stage of grain development so far as the availability of water to the maize crop is procedures of increasing water use efficiency under rainfed situations.

In regions receiving an annual rainfall of 60 cm, maize can successfully be grown in condition that these quantities should be well distributed throughout the stages of crop growth (ICAR 2006). Sufficient moisture is necessary for germination, and the rate of water use increases after germination as the leaves number increases and through the advanced stages. Generally, the crop utilizes about 2.5 mm of water daily up to about 25–30 cm tall. The average daily use increased to 6–8 mm during the vigorous period of growth from silking until dough stage of grain growth. In the high temperatures zones and low relative humidity, the average may be quite higher. The evapotranspiration ratio from the maize is about: 0.35 in the seedling stage and raises up to 0.80 at the silking stage, then decreasing again (Nafziger 2010).

19.6   Maize Production Under Egyptian Conditions

Maize is one of the most strategically important crops for the Egyptian national economy because it is main source of human food. Production of maize in Egypt was increased significantly over the past three decades. Under Egyptian conditions, both loamy and light clay soils are ideal for corn production and it is grown after clover, faba bean and wheat or barley and followed by cotton in fertile soil or by clover and faba bean. Agriculture in Egyptian is entirely depends on irrigation supply of water from the Nile, and to a lesser degree from groundwater. With the limited water resources in Egypt, there is need to efficient use of water through modern irrigation schemes and growing drought-resistant genotypes. Hybrid varieties, developed by the National Maize Breeding Program (NMBP) in Egypt are selected at 21,000 plants fed^{-1} (ca. 52,000 plants ha^{-1}), i.e. almost half of the density applied in developed countries. Therefore, maize hybrids released by NMBP of Egypt are not capable of producing high grain yield under higher plant density, because of increased height, non- prolificacy, non- erect leaves and large plant size (Al-Naggar et al. 2011).

19.7 Challenges to Maize Production

Maize production, particularly, in the tropical zones, is affected by a several of restrictions, including an range of abiotic stresses such as poor soil fertility, lack of access to key inputs, specially quality seed and fertilizers, low levels of mechanization and poor post-harvest managing and biotic stresses like diseases, pests and lack of suitable cultivars as stated by Salasya et al. (1998). The average maize production in Africa and many developing countries stood at 1.3 t/ha which could be attributed to such constraints (FAO 2011). Also, essential amino acids deficiency such as lysine, tryptophan, and methionine (Bantte and Prasanna 2003; Huang et al. 2006) are the main nutritional limitations of normal grain of maize.

Besides improving elite maize germplasm for tolerance to various abiotic stresses, particularly drought, heat, acidity, another important challenge of adapting crops to climate changes will be to preserve their genetic resistance to diseases and insect-pests. So, improving adapted maize germplasm for climate change in the tropical and subtropical -growing areas has become one of the major priorities to reduce climate change impacts through increasing use of preservation agriculture practices. Also, identifying and developing new crosses of maize appropriate for conservation agriculture-based procedures and achieving increased adaptation actions (Prasanna 2015).

To prioritize of the more appropriate action among different possible solutions, operative adaptation of agriculture to climate change will require knowing about the relative hazards posed by climate change across diverse locations and crop-ping systems, to prioritize the use of the rare resources dedicated to adaptation, and the likely mechanisms of potential harm from climate change (Araus et al. 2008; Barnabas et al. 2008)

19.7.1 Abiotic Stresses

Drought, salinity, and heat are main abiotic stresses that reduce crop production and reduce global food security, particularly given the current situation of impacts of climate change and increases in the incidence and severity of abiotic stress factors. Several independent studies have shown that increased drought and temperature can reduce crop production by as much as 50% (Mouna et al. 2018).

Although the maize has wide adaptability, among cereal crops it is the least tolerant to abiotic stresses. Among the major abiotic stresses drought, salinity, and high temperatures, are that negatively influence maize production in most maize produc-tion zones worldwide. The effects of stress differ with its severity and duration, and the stage of crop development. A wide range of plant reactions to these stresses could be due to morphological, physiological, and biochemical responses. So for better crop management, it is very significant to understand the physiological, biochemical, and ecological interferences regarding to these stresses.

Drought is the main abiotic stress restricting maize production. It's obviously a major reason of maize yield loss in worldwide (Bänziger and Araus 2007). Approximately 16% loss of maize productivity in lowland tropics and 17% in lowland tropics due to drought stress (Edmeades et al. 1992 and Edmeades et al. 2006) and up to 60% in rigorously drought-affected areas/seasons (Rosen and Scott 1992). From early vegetative stages till about two weeks pre-pollination, maize is fairly tolerant of moisture stress. While it is very sensitive to drought during two weeks pre, post and during pollination. The severe drought during this period can cause significant yield losses, that are mostly due to failure of pollination, and the most widespread, cause is the failure of silks to appear. Drought in grain-fill period has a less serious influence on yield, but the kernels may not fill completely. Generally, Maize production loss due to drought is expressed as top firing, tassel blast and poor seed set which mostly due to embryo sac abortion or non-availability of viable pollen and drying of silk (Dhillon and Prasanna 2001).

Using modern maize crosses tolerate or avoid water stress besides some practices such as reducing tillage, especially on lighter soils, adjusting planting dates, and using lower plant densities lead to mitigation the severe influence of the drought and avoid complete loss of the yield in dry areas (Nafziger 2010).

High temperatures are a dangerous problem for maize because drought and heat usually occur together. Higher temperatures negatively effect on maize ordinarily at anthesis, silking, and grain filling reproductive phenophases. High-temperature stress during grain-filling stage in spring maize (*Zea mays* L.) is the major obstacle to increasing yielding ability. There is evidence that hybrids differ in their tolerance to heat and drought.

Increasing temperature as a result of future climatic changes would be the main challenge facing agricultural development because it will increase the water requirements of crops. In Egypt, in the light of the limited water resources, this will affect future plans for agricultural development. So, adaptation strategies indicate that the most suitable sowing date is through 10th to 20th May in the Gemmeiza and Sids regions and through July in the Mallawy region (in El-Minia, Governorate, Egypt). Most suitable cultivars in future climate are SC10 (Gemmeiza), SC129 (Sids) and SC166 (Mallawy) as reported by Samia et al. (2012).

Poor soil fertility, inadequate mineral nutrition and low nutrient use efficiency are major challenges for crop production and rank among the most important elements limiting yield stability (Abu et al 2011 and Achiri (2017). Also, plants often take up less than half the applied nitrogen, while the remainder is lost as a nitrate leaching and gaseous nitrous oxide emissions which reduces Nitrogen use efficiency (Liu et al. 2013; Zhang et al. 2015). Low soil nitrogen has been recorded as one of the most widely spread problems between small-scale farmers in the tropics. Nitrogen deficiency affects various yield-determining factors. It reduces plant leaf area and leaf stay-green, due to reducing the photosynthetic rate and increasing ear abortion (Dhillon and Prasanna 2001).

There are main constraints such as minimum tillage and direct seeding equipment availability that needs to be addressed to ensure sustainable agriculture and developing locally connected options for conservation agriculture (Liu et al. 2013; Zhang et al. 2015).

There is clear variation in nitrogen use efficiency shows that this trait may be genetically determined and could be developed by different breeding methods. A suitable breeding strategy could be used to improve genotypes that tolerate low nitrogen stress and give high yielding ability in each of the shortage of Nitrogen in the soil and optimal conditions (Mi et al. 2005).

One of the great challenges on maize productivity on acid soils is Aluminum (Al) toxicity which decreases maize yields on around 8 million hectares, generally in tropical developing countries (Granados et al. 1993). So, soil acidity is considered the main constraint limiting maize growing in several countries over the world. Firstly Aluminium toxicity affects root growth followed by reducing the water and nutrients uptake affecting the physiological process and, consequently, accumulation of aboveground biomass (Roy et al. 1988). Application of lime has been suggested to increase the productivity of crops at such conditions. Several studies have revealed that selection maize for tolerance to acidic soils has been operative in increasing root length of seedlings, improve grain yield and other characteristics of maize (Stockmeyer et al. 1978). So, Aluminum resistance became the main aim of maize breeding programs for these conditions (Bushamuka and Zobel 1998).

In Egypt, one million ha in the irrigated regions suffer from salinization problems, mostly, are located in the northern-central region of the Nile Delta and on its eastern and western regions. To improve grain yield per unit area, breeding programs of maize in Egypt should pay more attention to develop new hybrids, characterized by adaptive traits to high plant density and more tolerance than older ones, such as prolificacy, short stature and leaf erectness (Al-Naggar et al. 2011).

19.7.2   Biotic Stresses

Insects and diseases cause a significant reduction in grain yield over all the world. However, 10% reduction of global food production may be due to diseases (Christou and Twyman 2004).

Climate change take into consideration as one of the most substantial causes that will play a role in diseases spreading. Rapid changing of insects and pathogens races is making crop resistance a non-durable process. So, the improvement of crop plants resilient to biotic stresses a difficult job (Strange 2005). High temperatures and humidity remarkably effect on pathogens and insect-pests responsiveness, and may lead to new and unpredictable epidemiologies Gregory et al. (2009). In several maize production regions, average annual yield reduction of 18, 80, and 44–55.9% was stated due to stem borers, grain weevils and ear rots (Shiferaw et al. 2011).

The majority of corn diseases are caused by fungi. However, some are caused by bacteria. Most of the major corn diseases are foliar—affect the leaves—which vary from year to year because they are intensely influenced by weather conditions. Major challenge of adapting field crops to climate change is maintaining their genetic resistance to diseases and insect-pests (Tefera et al. 2011). Great progress has been made by CIMMYT in the identification stable genotypic sources of resistance to main maize diseases using multi-location phenotyping (Prasanna 2015).

Egypt is one of the most vulnerable countries to the potential effects and risks of climate change. More studies were made to assess the potential impacts of climate change on crop productivity and reported that climate change impacts would adversely affect the yield of the crop due to plant exposure to pests and diseases (Fahim et al. 2013). Among various maize diseases, late wilt caused by *Cephalosporium maydis*, is one of the major diseases affecting maize yield productivity in Egypt. The best way of controlling this disease is through developing genetically resistance maize hybrids. However natural disease resistance mechanisms can be enhanced by plant nutrients. Nitrogen (N) fertilization is an important tool for increasing yield and resistance maize plant to diseases. Mosa et al. (2010) evaluated nine diverse yellow maize inbred lines, developed by maize research program, and all their possible combinations without reciprocal crosses (36 F1 hybrids) at Sakha Research Station. In their results they found some hybrids for grain yield gave 100% resistance to late wilt disease. El-Itriby et al. (1984) and Amer et al. (2002) reported that dominance and epistasis were the major contributors to the genetics of resistance to late wilt disease.

19.8   Milestones in Maize Breeding Advances

The traditional breeding applications and maize breeding methods development and the application techniques applications to maize improvement will be presented briefly under the following two subsections.

19.8.1   *Traditional Breeding Applications and Maize Breeding Methods Development*

19.8.1.1   The Initial Approaches for Maize Breeding

In the first half of the twentieth century, modernistic breeding started with the re-discovery of Mendel's laws. Fundamentally led to enhance of pure-lines, crosses, synthetic varieties and clones with more quality, stability, resistance to biotic and abiotic stresses and high yielding ability. The premature efforts were based on mass selection then later included ear to row selection (Hopkins 1896), high yielding ability hybrids from selected inbred lines (Shull 1908, 1909; East 1908) who suggested, the dominance theory of heterosis and production of single-cross hybrids followed by quite successful double cross hybrids by 4 inbred lines (Jones 1918, 1922).

Top cross (Line x Tester) test was proposed by Davis (1927) as a screening technique to overcome the problem of selecting some lines, out of an available huge number of inbred lines, which can be used to give superior hybrids. Top cross method is used to evaluate new inbred lines for general (GCA) and specific (SCA) combining ability using broad and narrow base testers for crossing with inbred lines producing high yielding ability hybrids. Companies such as Pioneer by the 1930s, had begun to influence long term development. These Companies committed to production of hybrid maize. Meanwhile, since the 1940s the best maize combinations have been first-generation hybrids made from inbred lines that have been optimized for specific characteristics, such as drought adaptability and pest and disease tolerance. Such conventional cross-breeding and genetic modification have succeeded in improving output and decreasing the need for cropland, pesticides, water, and fertilizer (Tina 2014).

In North America and Europe, modern crosses of maize achieve linear rise in grain yield/ unit area. Increasing plant density as a result of breeding programs for tolerance to high density played a main role up to more than 100,000 plants/ ha (Gozubenli et al. 2003). As indicated in Fig. 19.6 in the USA from 1865 till now much of the success of maize production is a result of the constant upward tendency in yields

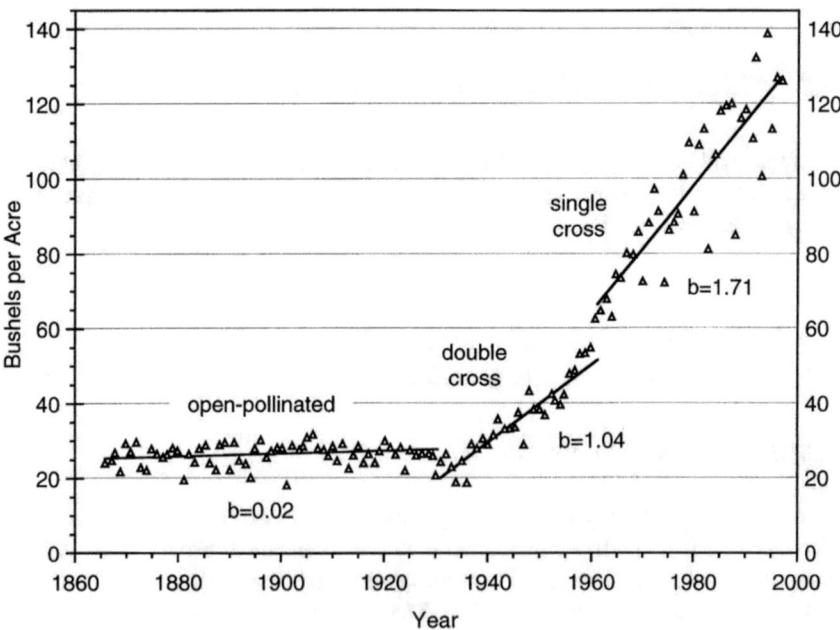

Fig. 19.6 Maize yield average trends in the United States from 1865 and regression (b) value for different types of grown cultivars (*Source* http://www.nass.usda.gov 2005 data prepared by Troyer and Wellin)

for different types of cultivated varieties started with open-pollinated varieties (b = 0.02) then double-crosses (b = 1.04) and finally single crosses (b = 1.71). In several producing areas of the world, similar trends can be seen.

The high yielding ability in maize-based on single crosses production due to using diallel cross mating proposed by Jensen (1970). This test gave an opportunity for additional recombination as a technique that helps select the best parents according to their performance for the selection of promising hybrids (Ramalho et al. 1993). Diallel crosses assess the heterotic potential, general, and specific combining ability of genotypes through evidence on the type of predominant gene action (Miranda and Chaves 1991).

In Egypt, at present, approximately all maize acreage (2.47 million feddan, ca. 1,039,241 ha) according to FAOSTAT (2017) is grown mostly through single and three-way cross hybrids and synthetics, developed mainly by the National Maize Breeding Program (NMBP), Agricultural Research Center (ARC). Under the Egyptian circumstances, the breeders succeeded in producing more than 50 single and three-way cross hybrids characterized by high productivity. The Maize Research Department at the Egyptian Agricultural Research Center provides all maize seeds for 3 million feddans (1,260,000 ha), of which 2 million white and 1 million feddans are yellow maize. The varieties allowed to be planted are white maize for single hybrid varieties 10, 128, 129, 130, 131 and 132, and three-way cross hybrids of 310, 314, 321, 324 and 329, and the yellow maize varieties, such as single hybrid 162,166,167,168,173,176,178 and 180 and three-way cross hybrids of 352, 353 and 368 (Anonymous 2020).

As a result of the using high yielding ability single and three-way crosses, the average yield increase of about 22.82 ard./fed. (7.6 ton ha^{-1}). According to Food and Agriculture Organization (FAO) reports, Egypt ranks between the fifth and the sixth place in the world with regard to average productivity after Germany, France, Canada, Italy and USA, as average yield for these developed countries reached 10–11 ton ha^{-1} (FAO statistics 2016). As shown in Fig 19.7, average maize grain yield per unit area in Egypt increased approximately linearly (R^2 0.861) during the period from 1960 to 2017, due to development in crop management practices and growing high yielding ability crosses.

19.8.1.2 Mutations in Maize

Many mutants of maize were specified in the early genetic studies (Neuffer et al. 1968). Suggestions for the use of dwarf (dw_i) genes, during the 1960s and 1970s of the twentieth century to reduce plant size, improve silage quality, change leaf orientation, and many other traits are examples of mutants that have been deliberated.

Also, many natural maize mutants were identified led to higher lysine and tryptophan in the 1960s and 1970s, viz., opaque-2 (o2), floury-2 (fl2), opaque-7 (o7), opaque-6 (o6), and floury-3 (fl3). Opaque-2 (o 2) mutation in the homozygous recessive o2 case identified to be the most suitable in breeding programs targeted to develop maize high in lysine and tryptophan (Vietmeyer 2000; Crow and Kermicle 2002).

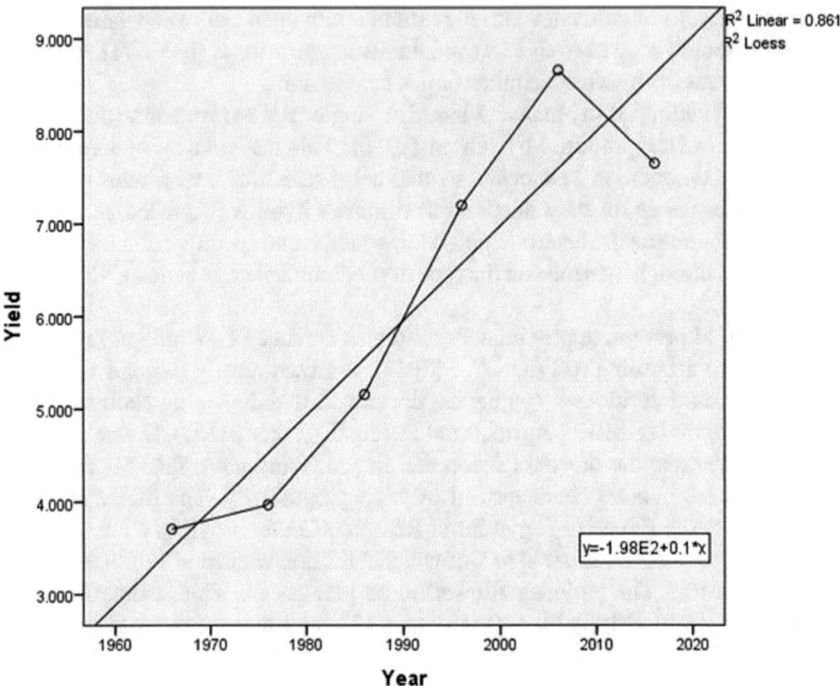

Fig. 19.7 Yield ton/ha of maize in Egypt, 1960–2016 (*Source* http://www.fao.org/faostat/en/#dat a/QC: 21 September 2017)

On the other hand, the opaque-2 maize grains were dull and chalky, had 15–20% less grain weight, and were more susceptible to numerous diseases and insects (Vivek et al. 2008).

To enhance the taste of sweet corn, breeders have depends on on mutant alleles (su 2 and sh 2, etc.). The recessive waxy (wx) allele was exploited to modify the starch to produce amylopectin, which has substantial properties for uses in foods and textiles (Hallauer and Carena 2009).

Currently, more than 23 countries in developing countries have produced quality protein maize QPM varieties for large-scale cultivation on an area over 2.5 million ha (Akande and Lamidi 2006).

19.8.1.3   Abiotic Stresses Detection Using Infrared Approach

Abiotic stress tolerance is required in plant breeding programs as a proper assessment method to overcome stress constraints. Such stresses stimulate various common physiological responses in plants and reduce CO_2 availability for photosynthesis via stomatal restriction and raise leaf temperature due to partly closed stomata. So, plants that can develop dynamic responses at the morphological, physiological, and biochemical levels to be adapted to unfavorable environmental conditions or enable them to escape. These physiological responses have an action to balance the temperature inside the plant. So, phenotyping through using infra-red thermography (heat-sensitive sensor), can be a beneficial tool to differentiate between the stress tolerant genotypes (Jones and Vaughan 2010; Taek-ryoun et al. 2015).

To study the variability of six inbred lines of maize selected by Aly 2004 and their F1's single crosses under four levels of soil salinity condition (T1) 1.5 ds/m (control), (T2) 2.5 ds/m, (T3) 3.5 ds/m and (T4) 5.5 ds/m in field trials were carried out by Emam et al. (2018) at east bitter lakes Experimental Farm Faculty of Agric. Suez Canal University (Sinai). The conducted field experiments presents visible and thermal maize images acquired by the fluke handheld camera for the studied genotypes under four salinity levels (Fig. 19.8, Photos 1–4). The results indicated that drought and salinity stress lead to an increase in leaf temperature by closed stomata. Temperature measurements using thermal infrared remote sensing were used to distinguish the change in canopy temperature over salinity levels and between

Photo: 1 (T1)   Photo: 2 (T2)

Photo:3 (T3)   Photo:4 (T4)

Fig. 19.8 T1–T4 Infrared thermal images for the different genotypes of maize under salinity levels (Emam et al. 2018)

different studied genotypes that influenced on change of maize behavior. All thermal infrared images in this study were acquired in visible and thermal infrared images in one shot between the hours of 1:00 and 3:30 PM to avoid the effect of shading. To compare among genotypes under salinity levels, the image analysis was used to determine the average temperature differences for each level. The results in Table 19.2 showed significant differences between salinity levels, genotypes, and interaction between them. Regarding salinity levels, the canopy temperature gradually increased from 33.2 in the control (T1) to 44.2°C in the highest level of salinity (T4). Where, canopy temperature was ranged from 31.1 (p_2 x p_3) to 35.3°C (p_4 x p_5) for control T1, 32.8 (p_4 x p_4) to 36.9°C (p_2 x p_5) for T2, 35.4 (p_2 x p_3) to 39.4°C (p_2 x p_6) for T3 and 41.3 (p_2 x p_3) to 45.8°C (p_3 x p_5) for T4. There is wide range in canopy temperature as a genotypic difference, from 31.1°C in cross p2 x p3 under control (T1) to 45.8°C in cross p3 x p5 in the highest salinity level (T4). Results also showed that cross p_2 x p_3 appears the lowest canopy temperature in average 35.3°C (Fig. 19.8,

Table 19.2 Average temperatures (C°) and stress indices for maize genotypes under salinity stress (Emam et al. 2018)

Genotypes	T1	T2	T3	T4	Mean
p_1 (Line 45)	32.7	35.3	37.5	43.4	37.2
p_2 (Line 7)	31.4	35.1	38.2	45.1	37.5
p_3 (Line 1)	33.3	36.4	38.6	45.7	38.5
p_4 (Line 8M)	34.3	35.4	39.4	45.1	38.6
p_5 (Line 5)	33.4	34.5	39.3	44.7	38.0
p_6 (Line 3)	34.5	35.4	38.2	45.6	38.4
$p_1 \times p_2$	31.4	32.8	37.5	42.3	36.0
$p_1 \times p_3$	32.3	34.4	36.6	42.1	36.4
$p_1 \times p_4$	31.6	35.8	38.4	43.5	37.3
$p_1 \times p_5$	32.3	33.5	37.5	44.4	36.9
$p_1 \times p_6$	35.2	36.1	38.4	43.1	38.2
$p_2 \times p_3$	31.1	33.5	35.4	41.3	35.3
$p_2 \times p_4$	34.4	35.8	37.7	42.4	37.6
$p_2 \times p_5$	33.4	36.9	38.1	44.2	38.2
$p_2 \times p_6$	32.1	34.4	39.4	45.6	37.9
$p_3 \times p_4$	32.3	35.2	38.4	44.8	37.7
$p_3 \times p_5$	34.5	36.5	39.1	45.8	39.0
$p_3 \times p_6$	34.3	34.6	37.4	44.3	37.7
$p_4 \times p_5$	35.3	35.8	37.2	44.1	38.1
$p_4 \times p_6$	32.7	34.4	39.3	45.6	38.0
$p_5 \times p_6$	35.2	36.5	38.5	45.2	38.9
Mean	33.2	35.2	38.1	44.2	
L.S.D 0.05	G = 0.28 & S = 0.2 & S*G = 2.5				

Photos 1–4). Also, Parent P1 appears the lowest canopy temperature (43.4°C) in the highest salinity level (T4). It is worth revealing that cross p_2 x p_3 gave the highest grain yield/plant in both seasons. Also, P1 produce the highest grain yield under the highest salinity level in both seasons. So, selecting tolerant genotypes to each of drought and salinity levels based on the absolute canopy temperatures may be an effective method. Many researchers recorded similar results on wheat, indicated that the physiological mechanism could be influenced on the behavior of genotypes due to water stress and subsequently leaf temperature (Bayoumi et al. 2015; Maes and Steppe 2012; Walter et al. 2012; Zarco-Tejada et al. 2012).

19.8.2   Technology Applications to Maize Improvement

Three main biotechnological methodologies for crop improvement include (a) molecular marker-assisted selection (MAS), (b) genetically modified food production (GM), and (c) tissue culture for micropropagation and other objectives.

New plant breeding methods that use high-density genotyping based on next-generation DNA sequencing technology, besides precision phenotyping, genomic selection (GS), and doubled haploid (DH) technology, can importantly increase speed the development of stress-tolerant and nutritionally enriched maize cultivars (Prasanna 2015).

19.8.2.1   Biotech Interventions in Maize

Technology enables gene transfer from one organism to another by passing biological barriers regardless of the species. There are numerous approaches for in vitro incorporation of an alien gene into maize: electroporation and direct DNA transfer, agrobacterium-mediated transformation, and biolistic method. The last one is the most effective method is employed by biotechnology corporations.

All techniques of inserting foreign DNA into maize plants rely on tissue culture as a methodology for producing whole plants from single cells, and an appropriate system to choice maize plants that had been successfully transformed via the incorporation of alien genetic material (Hoisington et al. 1998).

Transgenic approaches have provided high levels of pest resistance not achieved during several previous decades of breeding (Smith et al. 2005). *Bt* maize has revolutionized pest control, where it awarded considerable increases in the productivity of maize germplasm. *Bt* is a soil bacterium that produces crystalline inclusions during sporulation. *Bt* maize is maize that has been genetically modified to manufacture an insecticide to incorporate resistance to insect in transgenic plants. Although, there are many methods for incorporating insect resistance in transgenic plants, but the most successful method is using plants carrying the insecticidal protein gene transmitted from *Bacillus thuringiensis* (*Bt*). The resulted crystals dissolve in the alkaline medium of insect guts and releasing protoxin molecules that are processed by the gut proteases to produce active insecticidal proteins that interfere with the ion channel pumps and finally lead to insect larva death that ingest the crystal. Because these proteins are completely specific in their host, more than 50 diverse cry proteins have been identified that have various target insect specificity (Hilder and Boulter 1999).

Traditional breeding with up-to-date molecular techniques and biotechnology are helping to progress maize plant response to drought stress and introduce new drought tolerant genes into the plant. This could also ensure quick progress to improve maize for drought tolerance (Oikeh et al. 2014).

Maize hybrids carrying either insect tolerant or herbicide-resistant -traits have acquired constant acceptance by agricultural producers and consumers overall the world since its first introduction in the 1990s. The total maize area globally growing with biotech characters is 24% of the total acreage of maize and near 31% of globally planted biotech-characters crops in 2007 (James 2007).

19.8.2.2 Doubled Haploid (DH) Technology

A "doubled haploid" (DH) is a genotype formed after haploid (n) cells successfully subjected either naturally or artificially induced chromosome doubling. DH technology is a strong mean to produce improved products through short breeding cycle by speedy development of completely homozygous lines within 2–3 generations instead of 6–8 generations in the conventional inbred lines isolation (Chang and Coe 2009; Forster and Thomas 2005).

Using DH technology can potentially promote the efficiency of recurrent selection based program for characters with low heritability values (Bouchez and Gallais 2000). Also, it provides a chance to have an earlier idea about the potential of new lines, better knowledge about their ability to adapt to the environment before they are completely tested, and then used as parental lines for crosses development and commercial growing. Then finally, to reduce the time required producing new hybrids and coincidently reducing the time per breeding cycle (Li et al. 2018).

19.8.2.3 Molecular Marker-Assisted Breeding

Incorporating marker information into breeding programs has been widely studied to support identification and selection of superior genotypes (Ayoub et al. 2003; Jordan et al. 2003). Using marker assisted selection (MAS) would speed up improvement in selection programs and increase accuracy and differential of selection, particularly for quantitative traits (Van Arendonk et al. 1994). Preference to use MAS increases if selection pressure is increased or if the required populations' size in breeding programs, and the cost related with molecular marker assesses are decreased (Bernardo 2002).

Using microsatellite or simple sequence repeat (SSR) markers within the opaque-2 (O2) gene awarded the nutritional usefulness of quality protein maize (QPM) varieties contains nearly double as much lysine and tryptophan, amino acids. Such an approach provided a chance for accelerating the speed of QPM conversion programs by using marker-assisted selection (MAS) (Prasanna et al. 2010).

Molecular markers play a substantial role to improve the disease-resistance characteristics in maize follow polygenic inheritance, controlled by a little to numerous genes/QTL in each situation, with low G x E and rationally high heritability

Genomic Selection (GS), as a potential strategy based Breeding Complex characteristics, is controlled by numerous genes/QTL, with high G x E and low heritability for example drought stress tolerance. Given the great level of polymorphism establish even between highly related lines, maize was one of the first main crop species that a complete molecular marker map was developed (Helentjaris et al. 1986).

Challenges for future maize breeding in the current and future developments pertinent for maize breeding could concentrate on (1) development of inbred lines and hybrid seed production, (2) appreciative and prediction of hybrid performance, (3) exploiting genetic differences by transformation and genome editing, (4) germplasm diversity characterization and utilization and (5) genome analysis.

Finally, in light of the growing contribution of biotechnology to maize advances, about 75% of the maize in the United States enclosing biotech characters (James 2007). Moreover, continuing to develop new technologies will significantly contribute to food security and overcome poverty.

Programs of genetic engineering in Egypt began in 1990 on different crops included maize. These crops were improved with appropriate characters such as abiotic tolerance and biotic stress resistance. But, the developed genotypes did not reach the stage of commercial production due to the lack of national legislation regarding biotech plants, although, the fact that Egypt has recognized the Cartagena Protocol (Abdallah 2011). Through Agriculture Research Centre, Egypt successfully introduced the genetically engineered yellow corn variety Ajeeb-YG and had planted approximately 3800 acres of this variety, which awarded resistance to corn borers (USDA Foreign Agricultural Service 2015). Biotechnology has contributed enormous advances in maize production via various avenues, comprising the application of effective bio-fertilizers, plant growth promoter, and more significant development of transgenic characters resistant to herbicides and pests (Abdelhafez 2015).

19.9 Conclusions

From the above mentioned review, it is concluded that maize production will become more difficult with climate change, resource scarcity, and environmental degradation to follow the fastest and most practical path to improve yield. The close-term strategy is application and expansion using of agricultural technologies. In addition to the challenge of increasing productivity in the face of increasing demand for maize, both as food and feed, developing and deploying climate resistant germplasm adapted to the tropical/subtropical maize growing areas, has become one of the top priorities (Cairns et al. 2012).

The capability to quickly develop genotypes having tolerance to several complexes polygenic inherited abiotic and biotic pressures combined is critical to the resilience of cropping systems to cope with climate change. The proper challenge under climate change circumstances is to use adaptation planning, which is used to improve agricultural management procedures and the adoption of improved climate-resilient maize genotypes. Therefore, phenotyping via infra-red thermography (heat-sensitive sensor), can be an appropriate tool in selecting tolerant genotypes (Taek-ryoun et al. 2015).

19.10   Recommendations

The main restricted factors to maize production include drought during critical stages of crop growth, poor soil nutrient level (specially nitrogen and phosphorus), in addition to pest and diseases infestations. Other restrictions to maize production include poor management practices such as unsuitable planting time, low plant densities, inadequate control of weeds, inappropriate use of fertilizers and improved seeds. So, effective production of maize to a large extent depends on the timely commitment to all the recommended practices. Growing in deep, well-drained, fertile soils, enough irrigation or total seasonal rainfall exceeds 500 mm (in the rained regions), appropriate application of nitrogen and phosphorous, also potassium on some soils, Zinc in small amounts. Also, High plant densities are suitable for early-planted crops at high rainfall or irrigated environments where management is of a good standard.

Finally, molecular breeding and new technologies offers the tools to speed crop breeding process; however, appropriate phenotyping procedures are substantial to ensure that the long-awaited benefits of molecular breeding can be realized.

There is need for further research and breeding for short but high yielding cultivars of maize to accommodate high densities, facing climate change impacts using new tools and approaches of plant breeding.

References

Abdallah NA (2011) Biotech crops and the Egyptian revolution: where we stand. GM Crops 2(2):83–84 April/May/June 2011; © 2011 Landes Bioscience
Abdelhafez AA (2015) Biotechnology contribution to maize production. EC Agric 2(4):374–376
Abu GA, Demo-Choumbou RF, Okpacu SA (2011) Evaluating the constraints and opportunities of maize production in the west region of Cameroon for sustainable development. J Sustain Dev Afr 13 (4):189–197
Achiri D, Mbaatoh M, Njualem D (2017) Agronomic and yield parameters of CHC202 maize (*Zea mays* L) variety influenced by different doses of chemical fertilizer (NPK) in Bali Nyonga, North West Region Cameroon. Asian J Soil Sci Plant Nutr 2(4):1–9
Akande SR, Lamidi GO (2006) Performance of quality protein maize varieties and disease reaction in the derived-savanna agro-ecology of South-West Nigeria. Afr J Biotechnol 5:1744–1748

Al-Naggar A, Shabana R, Rabie AM (2011) Per se performance and combining ability of 55 new maize inbred lines developed for tolerance to high plant density. Egypt J Plant Breed 15(5):59–84

Aly AA (2004) Combining ability and gene action of new maize inbred lines (*Zea mays* l.) Using line x tester analysis. Egypt J Appl Sci 19(12B):492–518

Anonymous (2020) Recommendation techniques of field crops. ARC, Giza, Egypt

Amer EA, Mosa HE, Motawei AA (2002) Genetic analysis for grain yield, downy mildew, late wilt and kernel rot diseases on maize. J Agric Sci Mansoura Univ 27:1965–1974

Smith, AF (2013) Corn—Oxford reference. Publisher: Oxford University Press Print Publication Date: 2012 Print ISBN-13: 9780199734962

Araus JL, Slafer G, Royo C, Serret MD (2008) Breeding for yield potential and stress adaptation in cereals. Crit Rev Plant Sci 27:377–412

Aquino P, Carrion F, Calvo R (1999) Selected maize statistics. CIMMYT 1997/98 world maize facts and trends: Maize production in drought stressed environments. Technical options and research resource allocation. CIMMYT Mexico DF Mexico

Ayoub M, Armstrong E, Bridger G, Fortin MG, Mather DE (2003) Marker-based selection in barley for a QTL region affecting « -amylase activity of malt. Crop Sci 43:556–561

Bantte K, Prasanna BM (2003) Simple sequence repeat polymorphism in quality protein maize (QPM) lines. Euphytica 129(3):337–344

Bänziger M, Araus JL (2007) Recent advances in breeding maize for drought and salinity stress tolerance. In: Jenks MA, Hasegawa PM, Mohan S (eds) Advances in molecular breeding toward drought and salt tolerant crops. Springer, Dordrecht, The Netherlands, pp 587–601

Bayoumi TR, Mahmoud SA, Yousef MS, Emam MA (2015) Detecting drought tolerance in wheat genotypes using high throughput phenotyping techniques. In: The 3rd International Conference of Advanced Applied Sciences (ICAAS-III)

Barnabas BAT, Jager K, Feher A (2008) The effect of drought and heat stress on reproductive processes in cereals. Plant, Cell Environ 31:11–38

Bernardo R (2002) Breeding for quantitative trait in plants. Stemma Press Wood burry

Bouchez A, Gallais A (2000) Efficiency of the use of doubled haploids in recurrent selection for combining ability. Crop Sci 40:23–29

Bushamuka VN, Zobel RW (1998) Maize and soybean tap, basal, and lateral root responses to a stratified acid, aluminium-tox2015ic soil. Crop Sci 38:416–421

Cairns JE, Sonder K, Zaidi PH, Verhulst N, Mahuku G, Babu R, Nair SK, Das B, Govaerts B, Vinayan MT, Rashid Z, Noor JJ, Devi P, San Vicente F, Prasanna BM (2012) Maize production in a changing climate: impacts, adaptation, and mitigation strategies. Adv Agron 114:1–58

Chang MT, Coe EH (2009) Doubled haploids. In: Kriz AL, Larkins BA (eds) Biotechnology in agriculture and forestry. Molecular genetic approaches to maize improvement. Springer Verlag, Berlin, Heidelberg (63):127–142

Christou P, Twyman RM (2004) The potential of genetically enhanced plants to address food insecurity. Nut Res Rev 17:23–42

Crow JF, Kermicle J (2002) Oliver Nelson and quality protein maize. Genetics 160:819

Davis RL (1927) Report of the plant breeder. Rep Puerto Rico Agric Expt Stn 14–15

Dhillon BS, Prasanna BM (2001) "Maize. In: Chopra VL (ed) Breeding field crops: theory and practice. Oxford and IBH, New Delhi

Duvick DN, Cassman KG (1999) Post-green-revolution trends in yield potential of temperate maize in the north-central United States. Crop Sci 39(6):1622–1630

East EM (1908) Inbreeding in corn. Connecticut Agric Expt Stn 1907:419–428

Edmeades GO, Bolaños J, Lafitte HR (1992) Progress in breeding for drought tolerance in maize. In: Wilkinson (ed.) Proceedings of the 47th Annual Corn and Sorghum Ind Res Conf ASTA, Washington, pp 93–111

Edmeades GO, Bänziger M, Campos H, Schussler J (2006) Improving tolerance to abiotic stresses in staple crops: a random or planned process

EL-Itriby HA, Khamis MN, EL-Demerdash RM, EL-Shafey HA (1984) Inheritance of resistance to late wilt (*Cephalosporium maydis*) in maize. Proc 2nd Mediterranean Conf Genet Cairo March: 29–44

Emam MA, Ali AA, El-Ashry MA, Elshahat S, Ibrahem A (2018) Studies on breeding of some maize (*Zea mays*, L) inbred lines and their crosses under salinity conditions. Thesis Ph. D Suez Canal University Agric Faculty Agronomy Department Egypt

Energy Information Administration (EIA) (2018) U.S. fuel ethanol plant production capacity. Release Date July 30, 2018

Fahad S et al (2017) Crop production under drought and heat stress: plant responses and management options. Front Plant Sci 8:1147

Fahim MA, Hassanein MK, Khalil AA, Abou Hadid AF (2013) Climate change adaptation needs for food security in Egypt. Nat Sci 11(12):68–74

FAO (2005) Fertilizer use by crop in Egypt. First version, published by FAO, Rome

FAO statistics (2011) http://www.fao.org/faostat/en/#data/QC

FAO statistics (2016) http://www.fao.org/faostat/en/#data/QC

FAO statistics (2017) http://www.fao.org/faostat/en/#data/QC

Faske T Kirkpatrick T (2015) Corn Diseases and Nematodes Arkansas Corn Production Handbook. Chapter 7

Forster BP, Thomas WTB (2005) Doubled haploids in genetics and plant breeding. Plant Breed Rev 25:57–88

Granados G, Pandey S, Ceballos H (1993) Response to selection for tolerance to acid soils in a tropical maize population. Crop Sci 33:936–940

Gregory PJ, Johnson SN, Newton AC, Ingram JSI (2009) Integrating pests and pathogens into the climate change/food security debate. J Exp Bot 60:2827–2838

Hallauer, AR Carena MJ (2009) Maize breeding. Handbook of plant breeding. Cereals Springer, Cereals, pp 3–98

Hanway JJ (1966) How a corn plant develops. Iowa State University, Corn 17

Heba YA Enaam AM, Monia BH, Karima AM (2015) An economic analysis for maize market in Egypt. Middle East J. Agric. Res 4(4): 873 878. ISSN 2077 4605

Helentjaris T, Weber T, Wright S (1986) Use of monosomics to map cloned DNA fragments in maize. Proc Natl Acad Sci 83:6035–6039

Hilder VA, Boulter D (1999) Genetic engineering of crop plants for insect resistance: a critical review. Crop Prot 18:177–191

Hoisington D, listman GM, Morris ML (1998) Varietal development: applied biotechnology. In: Morries ML (ed) Maize seed industries in developing countries. Lynne Rienner Publishers/CIMMYT. Boulder, CO, USA, pp 77–102

Hopkins CG (1896) Improvement in the chemical composition of the corn Kernel. Ill Agric Exp Stn Bull 1896(55):205–240

Huang S, Frizzi A, Florida C, Kruger D, Luethy M (2006) High lysine and high tryptophan transgenic maize resulting from the reduction of both 19- and 22-kD alpha-zeins. Plant Mol Biol 2006 Jun 61(3):525–535

Gozubenli H, Konuskan O, Kilinc M (2003) Effect of hybrid and plant density on grain yield and yield components of maize (*Zea mays* L.). Indian J of Agron 48(3):203–205

ICAR (2006) Handbook of agriculture, New Delhi, pp 870–886

IITA (1975) Farming systems program. Annual Report 9 International Institute of Tropical Agriculture Ibadan Nigeria

James C (2007) Global status of commercialized biotech/GM crops: ISAAA Brief No. 37. ISAAA, Ithaca, NY. ASBN: 978-1-892456-42-7

Jenkins MT (1934) Methods of estimating the performance of double crosses in corn. J Amer Soc Agron 26:199–204

Jensen NF (1970) A diallel selective mating system for cereal breeding. Crop Sci 10:629 635

Jones DF (1918) The effects of inbreeding and crossbreeding upon development. Conn Agric Exp Stn Bull 107:100

Jones DF (1922) The productiveness of single and double first generation corn hybrids. J Am Soc Agron 14:242–252

Jones HG, Vaughan RA (2010) Remote sensing of vegetation: principles, techniques and applications. Oxford University Press, Oxford, UK. Int J Remote Sens ISSN 0143-1161 print/ISSN 1366-5901 online http://dx.doi.org/10.1080/01431161.2011.587097

Jordan DR, Tao Y, Godwin ID, Henzell RG, Cooper M, Mclntyre CL (2003) Prediction of hybrid performance in grain sorghum using RFLP markers. Theor Appl Genet 106:559–567

Lal R (1973) Effect of Seedbed preparation and time of planting on Maize (Zea mays) in western Nigeria. Exptl Aggrac 9:303–313

Li H, Rasheed A, Hickey LT, He Z (2018) Fast-forwarding genetic gain. Trends Plant Sci 23:184–186

Liu X, Zhang Y, Han W et al (2013) Enhanced nitrogen deposition over China. Nature 494:459–462

Louette D, Smale M (1998) Farmers' seed selection practices and maize variety characteristics in a traditionally-based Mexican community. CIMMYT Economics Working Paper No 98-04. CIMMYT, Mexico, DF

Maes WH, Steppe K (2012) Estimating evapotranspiration and drought stress with ground-based thermal remote sensing in agriculture: a review. J Exp Botany 63:4671–4712

Maize production manual (1982) International institute of Tropical Agriculture Oyo Road, PMB 5320. Ibadan, Nigeria. Man Ser 1(8):31

Mi G, Chen F, Zhang F (2005) Physiological and genetic mechanisms for nitrogen- use efficiency in maize. J Crop Sci Biotechnol 10(2):57–63

Miranda Filho JB, Chaves LJ (1991) Procedures for selecting composites based on prediction methods. Theor Appl Genet 81:265–271

Mosa HE, Motawei AA, Abd El-Aal AMM (2010) J Agric Res Kafer El-Sheikh Univ 36:278–291

Mouna L, Martin J, Raju D, Faouzi B (2018) Heat and drought stresses in crops and approaches for their mitigation. Published online: Agricultural Biological Chemistry 2018 Feb 19. https://doi.org/10.3389/fchem.2018.00026

Nafziger ED (2010) Growth and production of maize: Mechanized Cultivation. Crop Sci, University of Illinois, Goodwin, Urbana, IL, USA (438)

National Chamber of Milling, South Africa (2008) Cultivars: maize. Cited December 23, 2013. www.grainmilling.org.za

Neuffer MG, Jones L, Zuber MS (1968) Mutants of maize. CSSA Madison, WI

Nyamangara J, Nyaradzo Masvaya E, Tirivavi R, Nyengerai K (2013) Effect of hand-hoe based conservation agriculture on soil fertility and maize yield in selected smallholder areas in Zimbabwe. Soil & Tillage Res 126:19–25

Oikeh S et al. (2014) The water efficient maize for Africa project as an example of a public–private partnership. Springer Verlag, Berlin Heidelberg, pp 317–329

Piperno DR (2011) The origins of plant cultivation and domestication in the new world tropics: patterns, process, and new developments. Curr Anthropol 52(S4):453–470

Prasanna BM (2015) Maize in the developing world: globaltTrends, Challenges, and Opportunities for Maize. In: Proceeding 12th Asian Maize Conference and Expert Consultation on Maize for Food, Feed, Nutrition and Environmental Security. Bangkok, Thailand

Prasanna BM, Pixley KV, Warburton M, Xie C (2010) Molecular marker-assisted breeding for maize improvement in Asia. Mol Breed 26:339–356

Ramalho MAP, Santos JB and Zimmermann MJ (1993) Genética quantitativa em plantas autógamas. Editora UFG, Goiânia, p 271

Rosen and Scott (1992) Famine grips sub-Saharan Africa Outlook Agric, 191:20–24. View Record in Scopus

Roy AK, Sharma A, Taludker G (1988) Some aspects of aluminum toxicity in plants. Bot Rev 54:145–147

Salasya BS, Mwangi W, Verkuijl H (1998) An assessment of the adoption of seed and fertilizer package and role of credit in smallholder maize production in Kakamega and Vihiga districts, Kenya (http://www.cimmyt.org/) International Maize and Wheat Improvement Centre Library (http://ring.ciard.net/node/11021)

Samia ME, Hassanein MK, Ali AA (2012) Studies on Vulnerability of Egyptian Maize Varieties to Future Climatic Changes. American Eurasian J Agric & Environ Sci 12(9):1153–1161

Sanchez GJJ, Goodman MM (1992) Relationships among the Mexican races of maize. Econ Bot 46:72–85

Shiferaw B, Prasanna B, Hellin J, Banziger M (2011) Feeding a hungry world: past successes and future challenges to global food security in maize. Food Security 3:307–327

Shull GH (1908) The composition of a field of maize. Am Breeders' Assoc. Rep 4:296–301

Shull GH (1909) A pure line method of corn breeding. Am Breeders Assoc. Rep 5:51–59

Smith JSC, Smith OS, Lamkey KR (2005) Maize Breeding. Maydica 50:185–192

Snapp SS, Mafongoya PL, Waddington S (1998) Organic matter technologies for integrated nutrient management in smallholder cropping systems of southern Africa. Agric Ecosyst Environ 71:185–200

Sofi PA, Wani S, Rather AG, Wani SH (2009) Quality Protein Maize (QPM): genetic manipulation for nutritional fortification of maize. J Plant Breed Crop Sci 1:244–253

Stevenson JC, Goodman MM (1972) Ecology of exotic races of Maize I leaf number and tillering of 16 races under four temperatures and two photoperiods1. Crop Sci 12(6):864

Stockmeyer EW, Everett HL, Rehu RH (1978) Aluminum tolerance in maize seedlings as measured by primary root length in nutrient solution. Maize Genet Coop Newslett 52:15–16

Strange RN (2005) Plant disease: a threat to global food security. Annu Rev Phytopathol 43:83–116

Taek-ryoun K, Kyung-hwan K, Hae-Jin Y, Seung-kon L, Beom-ki K, Zamin S (2015) Phenotyping of plants for drought and salt tolerance using infra-red thermography. Plant Breed Biotech 3(4):299–307

http://dx.doi.org/10.9787/PBB.2015.3.4.299. Online ISSN: 2287-9366

Tefera T, Mugo S, Beyene Y, Karaya H, Tende (2011) Grain yield, stem borer and disease resistance of new maize hybrids in Kenya. African J Biotech (10):4777– 4783

Tina R (2014) A green revolution, this time for Africa. The opinion page: Opinionator: A gathering of opinion from around the WEB

Tollenaar M, Wu J (1999) Yield improvement in temperate Maize is attributable to greater stress Tolerance. Crop Sci 39(6):1597–1604

Turrent, A Serratos, JA (2004) Maize and biodiversity: the effects of Transgenic Maize in Mexico. Antonio. As part of the Article 13 initiative on Maize and Biodiversity. Commission for Environmental Cooperation of North America 8 November 2004

US Grains council (2017/2018) Corn Harvest Quality Report 3: https://grains.org/corn_report/corn-harvest-quality-report-2017-2018/

USDA Foreign Agricultural Service (2015) Agricultural Biotechnology Annual Report (2015) Global Agricultural Information Network 7/16/2015

Van Arendonk, JAM, TieBr, Kinghorn BP (1994) Use of multiple genetic markers in prediction of breeding values. Genetics (137):319–329

Vanlauwe B, Bationo A, Chianu J, Giller KE, Merckx R, Mokwunye U, Ohiokpehai O, Pypers P, Tabo R, Shepherd KD, Smaling EMA, Woomer PL, Sanginga N (2010) Integrated soil fertility management: Operational definition and consequences for implementation and dissemination. Outlook Agric 39(1):17–24

Verheye, W (2010) growth and production of maize: traditional low input cultivation. Soils, plant growth and crop production. In: Vol.II—Growth and Production of Maize: Traditional Low-Input Cultivation-January 2010. https://www.researchgate.net/publication/265978038_Growth_and_production_of_maize_traditional_low-input_cultivation

Vietmeyer ND (2000) A drama in three long acts: the story behind the story of the development of quality-protein maize. Diversity 16:29–32

Vivek BS, Krivanek AF, Palacios-Rojas N, Twumasi-Afriyie S, Diallo AO (2008) Breeding Quality Protein Maize (QPM): Protocols for Developing QPM Cultivars. CIMMYT 1, Mexico D.F

Walter A, Studer B, Kölliker R (2012) Advanced phenotyping offers opportunities for improved breeding of forage and turf species. Ann Bot 110:1271–1279

Zarco-Tejada PJ, González-Dugo V, Berni JAJ (2012) Fluorescence, temperature and narrow-band indices acquired from a UAV platform for water stress detection using a micro-hyper spectral imager and a thermal camera. Remote Sens Environment 117:322–337

Zhang X, Wang Q, Xu J et al (2015) In Situ Nitrogen Mineralization, Nitrification, and Ammonia Volatilization in Maize Field Fertilized with Urea in Huanghuaihai Region of Northern China. PLoS ONE 10(1):e0115649

Zohry A, Ouda S, Noreldin T (2016) Solutions for maize production-consumption gap in Egypt. In Conference: 4th African Regional Conferences on Irrigation and Drainage (ARCID) At Aswan Egypt

Chapter 20
Quinoa and Cassava Crops to Increase Food Security in Egypt

Abd El-Hafeez Zohry, Samiha Ouda, and Ahmed Sheha

Abstract A large gap between wheat production and consumption was estimated by 55% that negatively affects the production of breads in Egypt. Previous research by the Crop Intensification Research Department, Agricultural Research Center in Egypt indicated that 40% of quinoa flour could be mixed with the wheat flour for the bread making. Furthermore, investigations on cassava revealed that 30% of its flour could be mixed with wheat flour. Both crops can be cultivated in newly reclaimed sandy soils, where it can tolerate the low fertility of the soil. Cassava can grow in very low fertile soil. For quinoa plants with low soil nutrients content, grow well such as the sandy soil. Thus, both crops can contribute in the reduction of production-consumption gap in breads in Egypt. Thus, our objective in that chapter review and analysis the researches that were applied on quinoa and cassava internationally and nationally to help researchers in Egypt to expand and improve their work with these two important crops.

Keywords Quinoa · Cassava · Sandy soil · Flour · Bread making

20.1 Introduction

Egypt's growing population put a great pressure on the government to increase agricultural production. Intensive cultivation by using intercropping patterns and growing three crops annually is not enough to meet the growing population. Hence, it is necessary to horizontal expansion in the desert outside the cultivated area in the Nile Delta and the valley. The Nile Delta and valley represents about 4% of Egypt's total area, where agricultural lands are located. This region has one of the greatest population densities in the world. The new lands in Egypt are sited principally on both the

A. E.-H. Zohry (✉) · A. Sheha
Crop Intensification Research Department, Agricultural Research Center,
Field Crops Research Institute, Giza 12613, Egypt

S. Ouda
Water Requirements and Field Irrigation Research Department, Agricultural Research
Center, Soils, Water and Environment Research Institute, Giza 12613, Egypt

© Springer Nature Switzerland AG 2021
H. Awaad et al. (eds.), *Mitigating Environmental Stresses for Agricultural
Sustainability in Egypt*, Springer Water,
https://doi.org/10.1007/978-3-030-64323-2_20

east and west fringes of the Delta and valley. It scattered over different areas in the country, as it covers 1.01 million hectares. These lands are considered an opportunity to increase agricultural production and ensure food security in the country. Likewise, such regions provide a chance for absorbing the growing population and improving the demographic situation of the Egypt. The new lands are divided into sandy soils and salt-affected soils, where both types require reclamation before being suitable for agricultural production (FAO 2016).

Wheat flour is the main component in bread making in Egypt, also being very important for the Egyptians. Breads are considered the main constituent in the Egyptian diet, where the consumers have no other choice except consuming the breads, subsequently it is still the cheapest food. This situation puts wheat as a major and strategic crop. Wheat consumption is increasing due to the annual increase in the population which is close to 2.0 million/ year (Mansour 2012). The highest wheat per capita consumption levels in the world exists in Egypt (https://apps.fas.usda.gov/gai nfiles/200503/146119032.pdf). The large gap exists between wheat production and consumption, which estimated by 55%, negatively affects the production of breads in Egypt.

Integrating quinoa and cassava as new crops in the Egyptian cropping pattern has an important purpose. Both crops represent potential resources for new human foods. Additionally, both crops can contribute as livestock feeds, where both have high digestibility. Quinoa has the ability to grow as an alternative crop in areas suffering from drought, high temperature, salinity, and nutrient deficiencies that affect crop yields (Cocozza et al. 2013).

In developing countries, the import of wheat flour can be reduced with the use of composite flour, which encourages the use of flour from growing local crops (Hasmadi et al. 2014). The use of composite flour in various food products will result in an economic advantage, as the demand of bread and pastry products can be met by using locally grown products instead of wheat (Jisha et al. 2008). Composite flour is deliberated beneficial in developing countries as it reduces the importation of wheat flour and encourages the use of locally grown crops as flour (Hasmadi et al. 2014).

Although quinoa grains are gluten-free, it can be mixed with wheat flour to give bread with a high nutritional value (Morita et al. 2001). Previous research by the Crop Intensification Research Department, Agricultural Research Center in Egypt indicated that 40% of quinoa flour can be mixed with wheat flour to procedure composite flour to replace wheat flour in bread making (Shams 2011). Also, composite flour could also be prepared by addition 30% of cassava flour to wheat flour in bread making (Shams 2011).

Thus, the objective of this chapter was to focus on the research that was done on quinoa and cassava internationally and nationally to assistance researchers in Egypt to increase and improve their work on these two important crops.

20.2   Quinoa Crop

Quinoa (*Chenopodium quinoa* Willd.) is a unconventional Andean pseudo-cereal which is gaining interest quickly all over the world (Bhargava et al. 2006), due largely to its great nutritional value (Comai et al. 2007) and its adapted to less favorable soil and climatic conditions (Geerts 2008) (Fig. 20.1). Quinoa is considered one of the crops that have proven its capability to grow under marginal lands in Egypt, which are characterized by low fertility, salty soils, shortage water resources and severe weather conditions for use in food production (Zohry 2020). The food and agriculture organization (FAO) identified quinoa as one of the crops to deliver food security in the twenty-first century (Jacobsen 2003). Quinoa is categorized within the superfoods for its high content in protein. It also has lysine, which is expressed

Fig. 20.1 Quinoa plants in the field* (*From personal communications)

the first specific essential amino acid in cereal crops (Arendt and Zannini 2013). Quinoa seed, when treated and removed bitter saponins, has a mild flavor and could be consumed in the same way as cereal grains (Fuentes-Bazan et al. 2012).

Worldwide, there are in excess of 6000 varieties of quinoa grown by farmers (Rojas et al. 2015). Quinoa varieties can be categorized into five main classes or ecotypes, based on its adaptation to specific agro-ecological circumstances in major production regions (Bazile et al. 2015).The quinoa plants are described to be drought tolerant (Garcia et al. 2007) and cold before the flower-bud formation stage (Jacobsen et al. 2005). Moreover, quinoa can be cultivated in poor lands (Vilche et al. 2003). It adapt to less favorable soil and climatic conditions in marginal areas (Ruiz et al. 2016). Quinoa is considered a halophytic grain crop (Ruiz et al. 2016) grown in different climatic regions (Jacobsen et al. 2012). It can tolerate the salinity stress (Hariadi et al. 2011) and can tolerate drought and soil and water salinity (Wu et al. 2016). The cultivation of drought and salt-tolerant crops, such as quinoa has the possible to improve farm-level productivity and livelihoods in drought and salt prone areas (Kaya and Yazar 2016).

Chenopodium spp. has been cultivated for some centuries as a leafy vegetable (*Chenopodium album*) and an important secondary grain crop (*Chenopodium quinoa* and *C. album*) for human and animal foods due to the high content of protein and essential amino acids (Bhargava et al. 2003), where it contains a great range of vitamins (A, B2, E) and minerals (Ca, Fe, Cu, Mg, Zn) (Repo-Carrasco et al. 2003). Furthermore, it has very high lysine content. FAO (1990) showed that seeds of quinoa have high-quality proteins and greater levels of energy, calcium, phosphorus, iron, fiber and B-vitamins compared to rice, barley, oats, corn or wheat (Koziol 1992).

20.2.1   Effect of Salinity

Plants react to salinity stress in two stages. In the first stage, a rapid response to increased external osmotic pressure occurs as a result of the increased salt concentration round the roots to the threshold levels, which reduces the growth of the new shoot growth.

In the second phase, a slower response as a result of the accumulation of Na^+ in leaves (salt accumulation to toxic levels and increase the aging of the leaves (Munns and Tester 2008). Salinity tolerance in quinoa is inherited as a polygenic character (Flowers and Colmer 2008). Quinoa's tolerate high salinity at the primary stages of seed germination due to alterations in metabolites and enzyme activity (Adolf et al. 2013).

Algosaibi et al. (2015) experimented in a greenhouse and irrigated quinoa with four salinity levels (1.25, 4, 8 and 16 dS/m). They reported that the highest values of seed yield were recorded with 4 dS/m salinity level. Furthermore, 16 dS/m salinity level recorded the lowest seed yield of quinoa. Mahmoud (2017) conducted experiments in clay soil with salinity level equal to 3.37 dS/m. The plants were irrigated with three levels of saline water, namely 0.65, 10 and 20 dS/m. Her results indicated that quinoa seed yield was influenced by increasing water salinity, where about 12 to 21% reductions in seed yield were recorded in the plots irrigated with water having 10dS/m. Furthermore, 40 to 45% reductions in seed yield were recorded in the plots irrigated with water having 20 dS/m. Eisa et al. (2017) indicated that cultivation of quinoa under high saline soil conditions (ECe = 27 dS/m) resulted in 30% yield reduction compare to soil with low salinity level (ECe = 1.9 dS/m). On the other hand, Kaya and Yazar (2016) indicated that quinoa irrigated with salt water of 30 dS/m resulted in 18% decrease in the yield. The reduction was 7 and 14% with water salinity level ranges from 10 to 30 dS/m, respectively in comparison to fresh water. Furthermore, Eisa et al. (2017) reported that quinoa seed yield was reduced under EC value equal to 17.9 dS/m. However, seed quality was not highly affected. Similarly, Riccardi et al. (2014) reported that irrigation quinoa plants with saline water (ECe = 22 dS/m) did not affect seeds quality. Razzaghi et al. (2011) reported that the seed yield of quinoa cultivar Titicaca was reduced under salinity up to 20 dS/m of the irrigation water, however when exposed to salinity levels greater than 20 dS/m the plants adapted to it without yield component reduction. Hence, quinoa can be grown and produced successfully in soils affected by salt, where, most of traditional crops cannot grow. Although the yield was decresed, the seed quality was not greatly affected.

20.2.2 Effect of Water Stress

Water scarcity is one of the most serious problems affecting plant growth and yields in the world. Many species respond to drought conditions by increasing the ratio of assimilates transferred to root growth with the concomitant the ratio of root: shoot increase (Sharp and Davies 1989). A reduction in transpiration rates and CO_2 assimilation, were detected under water deficit as a result of low cell turgidity and close the stomata of the leaves (Waseem et al. 2011). Under decreased rate of CO_2 assimilation, leaf metabolism will be inhabited. Albert et al. (2011) stated that the balance between photochemical activity at photo system II and electron necessity for photosynthesis is influenced, causing an excitation on photosynthetic scheme and photoinhibitory injuries of photosystem II reaction centers". Also, water stress has been related with cell osmotic regulation, which is associated with accumulation of different organic compounds i.e. proline, glycine betaine, polyols, soluble sugars and others (Chai et al. 2001).

Soil moisture is an main factor in influential the time and rate seed germination and seedling growth of quinoa (Gonzalez et al. 2009). Quinoa develops unique tolerance devices to drought, allowing the plant to acclimatize with severe conditions in arid and semi-arid areas (Jacobsen et al. 2003). Additionally, the shortage of freshwater irrigation or the use of low-quality water for quinoa irrigation resulted in lower yields. Application of 50 and 75% of full irrigation resulted in 9 and 6% decrease, respectively in the yield of quinoa (Kaya and Yazar 2016). Algosaibi et al. (2017) indicated that decrease of irrigation water by 50% reduced the quinoa seeds by

14%. The level of the ionic, NH4-N and NO3-N were increased under water stress managements (Algosaibi et al. 2015). Geerts (2008) showed that water deficiency is a factor determining the stability of the yield of quinoa in different areas of Bolivian Altiplano, wherever intra-seasonal dry phases are of great importance.

Razzaghi et al. (2012) showed no statistical differences between Titicaca quinoa yield under full and shortage irrigated managements in a clay-loam soil. Pulvento, et al. (2012) revealed the ability of genotype to maintain yield under water stress and soil salinity at concentration of 16 dS/m in an open-field trial carried out under clay-loam soil in South Italy. Also, Cocozza et al. (2013) recorded a good tolerance to water and salt stresses through stomatal responses and severe dependencies between relative water content and potential water components that caused the osmotic adjustments, which is important to maintain leaf turgor satisfactory to plant growth.

20.2.3   Quinoa Insects and Diseases

Quinoa is infected by numerous pests in the Andean areas (Costa et al. 2009). The main pests exist in these regions are *Eurysacca melanocampta* Meyrick and *E. quinoae* Povolny (Lepidoptera Gelechiidae), which caused great losses to grain. Also, downy mildew caused by *Peronospora farinosa* is the greatest harmful disease of quinoa in Andean. Other quinoa diseases, viz. *Rhizoctonia* damping off, *Fusarium* wilt, leaf spot (*Ascochytahya lospora*), seed rot and damping off (*Sclerotiumrolfsii*, *Pythium zingiberum*), and brown stalk rot (*Phomaexigua* var. *foveata*) causing major loss in the yield of quinoa (Danielsen et al. 2003).

In Northern America, several plant bugs, leaf miners, aphids, and noctuids caused major injury in quinoa plants (Oelke et al. 1992). Some insects were found to exist in Northern Europe include *Cnephasia* sp. (Lepidoptera Tortricidae), *Aphis fabae* L. (Homoptera Aphididae) and *Lygus rugulipennis* Poppius Hemiptera Miridae), *Scrobipalpa atnplicella* (Röslerstamm) (Lepidoptera Gelechiidae) and *Cassidanebulosa* L. (Coleoptera Chrysomelidae) (Sigsgaard et al. 2008). In Southern Europe, further insects were observed include *Epitrix subcrinita* Le Conte (Coleoptera Chrysomelidae), and leafhoppers (Homoptera Cicadellidae) (El-Moity et al. 2015).

Assessments on diseases and pests attack quinoa in Egypt was performed by El-Moity et al. (2015) in five diverse Egyptian governorates viz. Giza, Faiyum, Ismailia, Beheira and Monufia to identify these predators. They indicated that, in all locations, *Rhizoctonia solani* and *Macrophomena phaseolina* seemed as aggressive causal organism for root rot and damping off disease for quinoa seedling. Also, *Peronospora farinosa* f. sp. *chenopodii* was discovered in all locations. This disease appeared through vegetative growth stage and before flowering or just after flowering. *Fusarium solani* was also isolated from samples collected from the Giza site. Also, two aphid species (*Myzus persicae* and *Aphis gossypii*, Hemiptera) were noticed in all locations. Two pests also belong to order Hemiptera were discovered in Ismailia governorate through vegetative and flowering stages. These two pests were recognized as (*Nysius cymoids*) while the other was (*Creontiades pallidus*). Cotton mealy bug (*Phenacoccus solenopsis*) was discovered in Giza on individual plants during seed formation stage. But, under Faiyum governorate, a weevil belongs to Coleoptera (*Sitophilus granaries*) was noticed during late flowering and grain formation stages (April–May), on little individual plants. A shoots feeder belongs to Diptera, *Atherigona theodori* was observed in Ismailia and Faiyum. This pest penetrates the stem of the plant and feeds on the content of the inner stem, which leads to wilting and death of the affected plants. *Tuta absoluta* (Lepidoptera) was also noticed in Giza, Faiyum and Ismailia governorates, it attacks fresh green leaves. Cotton leaf worm (*Spodoptera exigua*) was also identified in Faiyum on very few plants through early growth stages (October).

20.2.4 Use of Quinoa Internationally

In 2013, the United Nations blatant the International Year of Quinoa, also FAO organization identified quinoa to be the only crop food that provides all essential amino acids (González et al. 2012). Moreover, it is inexpensive in production due to their ability to adapt with different climatic conditions and soils, and their wide genetic variability (Romero and Shahriari 2011). Processing of quinoa gives pearled quinoa, particles, flakes, flour, nodules, oil, pasta and starch (Molina-Montenegro 2013).

Fresh leaves of quinoa and harvest husk are used to feed sheep, goats and fish. Quinoa leaves can also be used for silage (Bazile et al. 2015). The leftovers from milling have high nutritional value and are used for animal feed. Low quality broken grains are exploited for poultry feed, while stalks, pieces of leaf, remnants of the panicle, flowers and pedicels are utilized to feed sheep (Algosaibi et al. 2017).

20.2.5 Introduction of Quinoa in Egypt

Several studies were carried out on quinoa, hopefully, towards insertion it to innovate the crop composition of desert area of Egypt. Quinoa appeared to be very appropriate to grow under the Egyptian desert and new reclaimed sandy soil to combat degradation in these soils. In this connection, Shams (2011) indicated that thirteen genotypes of quinoa were planted in sandy soil of North Sinai in Egypt under two land preparation managements, with or without tillage. The results indicated that the European varieties out-yielded the Peru varieties. Moreover, cultivation quinoa plants under no tillage resulted in higher yield, in comparison of cultivation with tillage (1.59 ton/ha versus 1.49 ton/ha, respectively).

Moreover, Shams, (2011) tested the effected four dates of sowing quinoa seeds; 15th of November, 15th of December, 1st of February and 15th of February and two inter-row spacing, i.e. 15 and 20 cm apart in sandy soil. The results showed that the highest yield was obtained when 15 cm inter-row spacing was used, where yield reduction was reduced by 18%. Moreover, the yield of quinoa was reduced when planting date was delayed to February, where yield reduction was 63%.

Shams, (2012) conducted field trials in sandy soil in north Delta in Egypt to determine the adequate nitrogen fertilizer rate for higher yield in quinoa. The results showed that fertilizing quinoa with 360 kg N/ha increased plant hight to reach52.73 and 51.78 cm, and increase grain yield per plant to reach 10.07 and 8.17 g/plant, increased grain yield of 1.20 and 1.08 kg/ha and increased biological yield of 2.79 and 2.32 kg/ha.

According to Gomaa (2013), the highest values of quinoa yield in calcareous soil in Egypt were recorded with application of 240 kg/ha ammoniums nitrate in combination with bio-fertilizer (Nitrobin), in addition to application with 120 kg/ha calcium super phosphate in combination with bio-fertilizer (Phosphorin). She also indicated that the above treatment increased crude protein and mineral elements (phosphorus, potassium and calcium) in seeds.

Abdel-Rheem et al. (2014) indicated that the suitable sowing date for quinoa in Middle Egypt is 25th of November, where the highest yield was obtained. Furthermore, irrigation application when 60% of soil moisture was depleted (irrigation every 40 days) resulted in the highest yield. They also reported that water consumptive use for quinoa was between 215–226 mm.

Bazile et al. (2016) reported that "the FAO is actively involved in promoting and evaluating the cultivation of quinoa in 26 countries outside the Andean region with the aim to support food and nutrition, where Egypt was one of these countries". Field evaluations were conducted two seasons and in two sites using13 genotypes. A greater value than the weighted average yield for all countries was attained with 1.89 ton/ha at the first site and 2.35 ton/ha at the second site of Egypt.

Eisa et al. (2017) tested the response of quinoa plants to high salinity under the Egyptian conditions. They concluded that quinoa seed yield was decreased as a result of increasing soil salinity up to 17.9 dS/m. Their results also indicated that there was high ash content in seeds in saline conditions as a result of the increase of Na^+ and K^+, P_3^- and Fe^{++} concentrations. On the contrary, soil salinity led to significant reduction of Ca^{++} and Zn^{++} contents in seeds. Furthermore, Na^+ was principally accumulated in the pericarp followed by embryo tissues. They concluded that increasing most of essential minerals, particularly Fe, in quinoa seeds produced at high saline environments provides quinoa a unique value for human consumption.

According to Mahmoud (2017), sowing quinoa in North Egypt in the beginning of winter season (last quarter of November where maximum air temperature was 18.8°C and minimum temperature was 8.2°C) resulted in higher plant height, biomass and seed yields, compared to the sowing at second quarters of December and January. The optimum growth was associated with relative humidity about 68.8% and cumulative daylight hours about 1819 h and cumulative sunshine hours about 977 h. Moreover, symptoms of powdery mildew disease were observed on the quinoa leaves 53 days after sowing and progressed in appearance. The high percentages of relative humidity could be the major cause of spreading the disease in quinoa.

20.3   Cassava Crop

Cassava (*Manihot esculenta* Crantz) grows under wide range of soil and environmental circumstances. The crop is characterized by easy reproduction, tolerance to water deficiency and pests, also tuberous roots have a long viability up to 3 years when left in the soil (Nhassico et al. 2008) (Fig. 20.2). Cassava is broadly grown in tropical and sub-tropical Africa, Asia, and Latin America, between latitudes 0°N and 30°S, from sea level over 2000 m on marginal and highly-eroded low-fertility acidic lands, practically without the use of agrochemicals (El-Sharkaw 2004). Cassava is the third greatest vital source of calories under the tropics and the sixth most significant food crop after maize, rice, wheat, potato and sugar cane, in terms of worldwide annual production (FAOSTAT 2011). Ukwuru and Egbonu (2013) showed that cassava is a principal food offers carbohydrates and energy for over and above two billion people in the tropics.

Cassava is a product with higher carbohydrates per hectare than the main cereal crops, and it can be grown at a lower cost. This crop is mainly grown for its edible starchy tuberous roots (80% carbohydrates), and its leaves are rich in proteins, vitamins and minerals (UNCTAD 2012).

The growth cycle of cassava consist of two major alternating periods; the vegetative growth and carbohydrate storage in the roots. The main organ of photosynthetic production in cassava is the foliage. Early vegetative phase, flowering pattern and storage of assimilate in the vegetative parts are greatly affected by leaf loss (Alves 2002). Cassava has a high yield potentiality rather than sorghum, maize or rice under no production constraintsenvironments. Cassava can match or exceed these above crops in production of energy per hectare (De Vries et al. 1967), where it gives more energy per unit area as a result of its metabolic efficiency under marginal conditions,

Fig. 20.2 Cassava plants in the field* (*From personal communications)

compared to other crops under water stress conditions and in poor soils (El-Sharkawy et al. 1992). However, cassava stores nitrogen in the formula of cyanogenic glucosides and can cause cyanide poisoning if not properly treated (Nhassico et al. 2008). As a food crop, cassava can fit well into the farming scheme of the small holder farmers because it is available all year round, therefore providing household with food security (Ukwuru and Egbonu 2013). Although tubers of cassava are high in carbohydrate, it low in protein (1–3% dry matter) and micronutrients (Montagnac et al. 2009). Furthermore, is elastic to the time of harvest and it can be stored for a long time under the soil in the field. Cassava has a remarkable ability to tolerate and recuperate from biotic and abiotic pressures (van Oirschot et al. 2000).

20.3.1   Effect of Water Stress

Cassava has the ability to withstand long water stress, associated with decrease in growth and yields (Okogbenin et al. 2003). During prolonged water scarcities, cassava lessens its canopy by shedding older leaves and producing smaller new leaves leading to less light interception, which is another adaptive trait to water stress (El-Sharkawy 2007).The response of cassava plants to water pressure is affected by both the period and severity of stress and also depends on the genotype (Vandegeer et al. 2013). Also, water stress effects are most obvious when occurs in the first 1–5 months after planting, where plant canopy expansion occurs and the initiation and development of the tuberous roots (Bakayoko et al. 2009).

   Although the root yield decreases under conditions of water shortage, the crop can recover with water availability by rapidly creating new canopy leaves with greatly photosynthetic rates rather than unstressed crops (El-Sharkawy 2007). Cassava can get water slowly from deep soils, a trait of paramount importance in dry seasons and semiarid regions where stored water is obtained from the depth (El-Sharkawy 2010).

20.3.2   Effect of Growth Environment on Starch Accumulation

Starch production of cassava varieties are greatly affected by low subsoil density, great relative humidity and great pH in the upper layer of soil. Thus, it is important to recognize such conditions to enhance the quality and industrial yield of the cassava roots (Sholihin 2011).Water scarcity have a significantly effect on quality of cassava starch, particularly during early stage of development and before root harvest. In early plant development, water stress for the first six months caused a lower yield of starch in comparison with plants under normal conditions (Santisopasri et al. 2001). In the winter period, a substantial increase in the content of root starch was observed (De Proença et al. 2017). Furthermore, Chipeta et al. (2017) found great starch production in some cassava varieties during the 6–9 month period, while the best time for harvest cassava to attain acceptable starch levels depends on the variety and environment. However, when cassava plants are exposed to water stress, their hydrocyanic acid levels rise, and hence must be subjected to industrial methods to remove this toxin (Mtunguja et al. 2016).

20.3.3 Intercropping with Cassava

Intercropping is the cultivation of two or more crops in the same field, which allows water and nutrients to be used more efficiently (Kamel et al. 2010). Moreover, intercropping increase land productivity, increase water productivity and rise farmers income (Anderson 2007). Moreover, Pypers et al. (2011) recorded advantages varied from 60 to 90% in the productivity of cassava fields when intercropped with various crops. Furthermore, the architecture and growth habits are directly related to its ability of cassava to tolerate the competition with other plants in the early stage of growth (Silva et al. 2012).

Cassava is commonly intercropped with short duration crops due to its slow initial growth during the first 3–4 months of growth, which allows the farmer to utilize its place between the cassava stands and improving economic profits of the intercropping pattern (Salau et al. 2012). According to Adeniyan et al. (2014) in maize/cassava intercropping system, the highest cassava fresh tubers can be obtained when maize planting density was 20,000 plants per hectare (Fig. 20.3). Cowpea intercropping with cassava is another successful system. Intercropping cassava in double rows with two and three rows of cowpea permit attaining equivalent yield to the monoculture of cassava in single rows (Albuquerque et al. 2015). Osundare (2007) indicated that cassava tuber yield was increased by 25% when cowpea was intercropped with it.

Sorghum intercropping with cassava generally did not affect root yield compared to cassava mono cropping and gives an additional yield of the intercrop (Zinsoua et al. 2004).

Intercropping soybean with cassava resulted in yield increment for cassava ranged from 3-8% for fresh tuber yield per hectare (Fig. 20.4). The productivity of cassava/soybean mixture presented yield advantage of 59–84%, compared to mono cultivation (Mbah et al. 2003). Whereas, intercropping soybean with different plant density with cassava, namely 25, 50 and 100% from its recommended planting density resulted in increment in fresh tuber yield of cassava (Mbah and Ogidib 2012). Furthermore, Osundare (2007) reported that cassava tuber yield was increased by 24%, when soybean intercropped with it.

Adekunle et al. (2014) intercropped sesame or sunflower with cassava in different planting densities and reported that the biological efficiency was increased, compared to sole cassava planting. In this context, Langham and Schwarz (2008) reported that intercropping sesame on cassava loosens the soil by its tap root, forms a dense network of roots in the upper soil layer, helps keep the soil moisture well and its leaves works as a fertilizer when falls on the soil. Whereas, sunflower shows erect

Fig. 20.3 Maize intercropped with cassava* (*From personal communications)

Fig. 20.4 Soybean intercropped with cassava* (*From personal communications)

growth habit, comparable resistance to lodging, limited ground cover and has simply harvestable heads. These characteristics of the two crops make it good candidate to be intercropped with cassava (Adekunle et al. 2014).

Intercropping groundnut with cassava resulted in 24% increase in cassava yield, compared it cassava sole planting. This intercropping system resulted in a decrease in associated weeds by 27% (Osundare 2007).

20.3.4 Cassava Insects and Diseases

Cassava pests vary widely among the main areas of cassava cultivation in the Americas, Africa and Asia. The greatest variation of arthropod pests infects cassava is in the Neotropical Americas, the center of origin of this crop (Bellotti et al. 2011). Around 200 species of pests were recorded, some of which are specific to cassava (Bellotti et al. 2010) comprise laticifers and cyanogenic compounds (Bellotti and Riis 1994) and it is adapted in different degrees to the range of natural biochemical resistances in the host varieties. Bellotti (2008) stated that the arthropod pests that infect and injury cassava in the Americas include numerous species of white flies, mites, mealy-bugs, stem borers, thrips and the cassava hornworm. Whitefly *Bemisia tabaci* is considered the main pest of cassava in Africa, as it is the vector of Cassava Mosaic Disease (CMD) and Cassava Brown Streak Disease (Legg and Hillocks 2003).

Neosilbaperezi (Romero & Ruppell) is famous as the cassava shoot fly, and its larvae feed on cassava shoots. Moreover, its females lay eggs between the apical leaves of the shoots and in the larvae tunnel through the tissues of the plant. These injuries damage the terminal buds, which delay the normal growth of the plant and stimulate the lateral buds (Bellotti 2002). Liu et al. (2007) reported that great populations of *B. tabaci* lead to root yield decreases due to its feeding on the cassava crop.

Heat stress and changing rainfall levels can have significant effects on the cassava pest complex (Bellotti et al. 2011). Climate change predictions show that certain agricultural areas will receive fewer rainfalls in the future (IPCC 2013). Cassava crop can have an advantage because it can grow well under dry areas. Conversely, increased cassava yield in these dry areas of the Americas, Africa and Asia might result in hazardous pest outbreaks, decreasing yields and/or increasing pesticide use (Liu et al. 2007). Generally, Pillai et al. (1993) stated that arthropod pests viz. mites, mealybugs, thrips and lacebugs, are more harmful to cassava through the dry seasons. Warmer temperatures might also affect the developmental duration of the pest. It has been found that the intrinsic ratio of increase for mites, white-flies, hornworms and mealybugs are faster with increasing temperatures (Holguín et al. 2006).

20.3.5 Use of Cassava Internationally

Cassava ranks fourth among the staple crops with worldwide production of around 160 million tons annually. Cassava plants produce numerous products and there are still promising new products of the crop (Lawrence and Moore 2005). Normal annual cassava yield global reach 10 ton/ha of fresh root (nearly 65% moisture content), varied from 6 ton/ha in Mozambique to 26 ton/ha in India. In experimental trials in Colombia, high yields of some developed cassava cultivars grown under near-optimum environmental conditions, gave 90 ton/ha (El-Sharkawy 2005). Several studies have been implemented in many biotechnological methodologies to increase the safety and quality of cassava flour (Onitilo et al. 2007). Cyanide is a highly toxic substance for humans and animals and it is present in fresh, so cassava must be treated to decrease the cyanide content to a safe level (Eggleston et al. 1992). Ukwuru and Egbonu (2013) showed that traditional cassava processing technique involve some steps including peeling, soaking, grinding, steeping in water and left in air to permit fermentation to happen, drying, milling, roasting, steaming, pounding and mixing in cold or hot water.

Since cassava has high starch content, it is a good feed source to gave ethanol (Leng et al. 2008). Also, Cassava starch has many wonderful properties, including high paste clarity, high paste viscosity and high freeze-thaw stability that are useful to several industries (Zaidul et al. 2007). It can be used in several industries, viz. paper industry, adhesive industry and glucose production (Ukwuru and Egbonu 2013). Onuma et al. (1983) indicated that dried peels of cassava roots are exploited to feed sheep and goats. Also, boiled roots or raw in mash could be mixed with protein cotents such as corn, sorghum, peanuts, or palm oil meals as well as mineral salts to feed the livestock. Another important product of cassava is the high-quality cassava flour. Cassava flour is very white with low fat content. It can be mixed with wheat flour for use in making bread or cakes, and the product does not give a bad smell or taste to the food (IITA 2005).

20.3.6   Introduction of Cassava in Egypt

Several studies were done on cassava in Egypt indicated that the crop is very pertinent to grow in Egypt in areas under stress environmental condition in deserts and the new reclaimed sandy soil in rain fed and regions around Nasr Lake in south Egypt.

Sherif and Salem (2011) planted cassava in a newly reclaimed sandy soil in North Egypt to study the response of intercropping cowpea with cassava at different plant spacing X row width (1.00 m X 1.50 m; 1.00 m X 1.00 m and 1.00 m X 0.75 m) and three plant density of cowpea (50,400, 67,200 and 100,800 plant/ha) on cassava yield. The results point out that highest yield of cassava tubers was associated with lowest cassava spacing (1.00 m X 0.75 m, plant spacing X row width) and with highest cowpea plant density (100,800plant/ha). Furthermore, the highest value of land equivalent ratio was found under the above treatment.

Shams (2011) planted cassava in a newly reclaimed sandy soil in East Egypt to study the response of intercropping system of cowpea and cassava (1:2) to three rates of phosphorus and potassium fertilizers on cassava yield. The results showed that the highest cassava yield was obtained when 111.6 P_2O_5 kg/ha and 172.8 K_2O

kg/ha were applied. The results also indicated that intercropping cowpea with cassava attained yield advantage by 71%, compared to sole cultivation of cassava or cowpea. The lowest value of hydrocyanic acid concentration was found when the highest value of both phosphorus and potassium fertilizers was applied.

Shams (2011) studied the effect of four planting dates (first of October, first of November, mid of March and mid of April) on Brazilian and Indonesian cassava cultivars planted in Toshky region in south Egypt. He found that cassava planted in the third and fourth dates grow normally, whereas cassava plants grown in the first and the second dates failed. Additionally, cassava grown in the third date had better growth and yield. The results also indicated that Brazilian cultivar out-yielded the Indonesian cultivar.

Sherif and Nassar (2011) intercropped groundnut with cassava in two patterns (1:2 and 1:4) in sandy soil in north Egypt. The results indicated that there was an increase in branches number per plant, tubers number per plant, weight of fresh tubers per plant and tubers weight per hectare when groundnut was intercropped with cassava in 1:2 intercropping pattern.

Hasan (2012) studied the effect of different mineral fertilization doses and harvesting dates on the yield and quality of cassava tuber roots (Indonesian cultivar), grown under sandy soil conditions. She concluded that using 180 kg N/ha, 277.2 kg P_2O_5/ha and 345.6 kg K_2O/ha, in addition to implement harvest 10 month after planting gave marketable percentages of tuber roots. It also resulted inhigh weight and, diameter of tuber roots, high root/shoot ratio, high dry matter percentage and yield of tuber roots. It also produced high total carbohydrates, nitrogen, phosphorus and starch percentages in tuber roots, with decreased hydrocyanic acid concentration of tuber roots.

Helal et al. (2013) subjected cassava plants to water stress in pots experiment. The results indicated that application irrigation using 25% of field capacity achieved the highest cassava yield.

20.4   Conclusions

The productivity of the traditional crops in the cropping pattern in Egypt is reduced when cultivated in marginal soil as a result of low fertility and harsh weather conditions. Thus, both quinoa and cassava are good candidates for that. According to the implemented research in Egypt on quinoa and cassava, a summary about the two crops can be stated.

Concerning quinoa, it is a winter crop can be cultivated in both sandy and salt-affected soil. The suitable date for cultivation quinoa is between 15th and 21st of November. It can be grow with no tillage in the sandy lands of Sinai. The suitable inter-row spacing for quinoa is 15 cm. Quinoa could grow under rain-fed conditions and gave its highest yield with application of supplementary irrigation. Water consumptive use for quinoa is between 215–226 mm. It could be also cultivated successfully under irrigation, where it can produce its highest yield, when irrigated every

40 days in clay soil, with. In sandy soil, Nitrogen fertilizer requirement for quinoa is 360 kg N/ha depending on soil type. Furthermore, application of 240 kg N/ha to quinoa, in addition to biofertilizer (Nitrobin) is sufficient to fulfill its requirement. Phosphorus fertilizer requirement is 120 kg/P_2O_5, in addition to application to biofertilizer (phosphorin). Additionally, some quinoa cultivars can tolerate salinity levels up to 17.9 dS/m. Thus, quinoa can be cultivated in marginal lands in Egypt, namely sandy soil and salt-affected soils.

Regarding cassava, the suitable planting date that attained the highest yield is mid of April. The best planting method is 1.00 m X 0.75 m plant spacing X row width with planting density equal to 100,800 plants/ha. The crop can be successfully cultivated in sandy soil. Irrigation application when 25% of field capacity was depleted attained the highest yield. Nitrogen fertilizer requirement for cassava is 180 kg N/ha. Phosphorus and potassium fertilizer requirements are between 111.6–277.2 kg/P_2O_5 and 172.8–345.6 K_2O, respectively depending on soil type. The optimum date for harvest that attains the highest yield is 10 month after planting. Cowpea and ground nut can be intercropped with cassava with 1:2 intercropping pattern, which increased the yield and its components.

20.5 Recommendations

– The government should encourage the cultivation of quinoa, which can be successfully cultivated in both sandy and salt-affected soils. Whereas, cassava can be cultivated in sandy soil only and other crops can be intercropped with it.
– In these marginal soils, livestock production can be implemented as a result of using both crops in feed production.
– The government should build factories to process quinoa to produce flour.
– The government should build factories to process cassava and produce starch.
– Starch can be transform to ethanol and use as fuel to reduce air pollution.

References

Abdel-Rheem HA, Abdrabbo A, Aboukheira AA, Eisa NMA (2014) Response of quinoa crop for water stress and planting date in the middle Egypt. J Egypt Acad Soc Environ Develop 15(2):49–66
Adekunle YA, Olowe VI, Olasantan FO, Okeleye KA, Adetiloye PO, Odedina JN (2014) Mixture productivity of cassava-based cropping system and food security under humid tropical conditions. Food Energy Secur 3(1):46–60
Adeniyan ON, Ayoola OT, Ogunleti DO (2014) Evaluation of cowpea cultivars under maize and maize cassava based intercropping systems. Afr J Plant Sci 5(10):570–574
Adolf VI, Jacobsen SE, Shabala S (2013) Salt tolerance mechanisms in quinoa (*Chenopodium quinoa* Willd.) Environ Exp Bot 92: 43–54

Albert KR, Mikkelsen TN, Michelsen A, Ro-Poulsen H, van der Linden L (2011) Interactive effects of drought, elevated CO_2 and warming on photosynthetic capacity and photosystem performance in temperate heath plants. J Plant Physiol 168:1550–1561

Albuquerque A, Oliva LSdC, Alves JMA, Uchôa SCP, Melo DA (2015) Cultivation of cassava and cowpea in intercropping systems held in Roraima's savannah, Brazil. RevistaCiênciaAgronômica 46(2):388–395

Algosaibi AM, Badran AE, Almadini AM, El-Garawany MM (2017) The effect of irrigation intervals on the growth and yield of quinoa crop and its components. J Agric Sci 9(9):182–191

Algosaibi AM, El-Garawany MM, Badran AE, Almadini AM (2015) Effect of irrigation water salinity on the growth of quinoa plant seedlings. J Agric Sci (7)8:208–218

Alves AAC (2002) Cassava botany and physiology.In Hillocks RJ, Thresh JM and Bellotti AC (ed) Cassava: Biology, Production and Utilization. CABI, Wallingford, UK

Anderson RL (2007) Managing weeds with a dualistic approach of prevention and control: A review. Agron Sustain Dev 27:13–18

Arendt EK, Zannini E (2013) Cereal grains for the food and beverage industries. Woodhead Publ Ser Food Sci, Technol Nutr 248:409–438

Bakayoko S, Tschannen A, Nindjin C, Dao D, Girardin O, Assa A (2009) Impact of water stress on fresh tuber yield and dry matter content of cassava (*Manihot esculenta* Crantz) in Cote d'Ivoire. Afr J Agric Res 4:21–27

Bazile D, Bertero HD, Nieto C (2015) State of the Art Report on Quinoa Around the World in (2013) FAO; CIRAD, Rome. Available on line at http://www.fao.org/3/a-i4042e.pdf

Bazile D, Pulvento C, Verniau A, Al-Nusairi MS, Breidy J, Hassan L, Mohammed MI, Mambetov O, Otambekova M, Sepahv NA, Shams A, Souici D, Miri Kand Padulosi S (2016) Worldwide evaluations of quinoa: preliminary results from post international year of quinoa FAO Projects in Nine Countries. Front Plant Sci 7:850. https://doi.org/10.3389/fpls.2016.00850

Bellotti A, Campo BVH, Hyman G (2011) Cassava production and pest management: present and potential threats in a changing environment. Tropical Plant Biol. 10.1007/s12042-011-9091-4

Bellotti AC (2002) Arthropod pests, p 209-235. In: Hillocks RJ, Thresh JM, Bellotti AC (eds) Cassava biology, production and utilization. CAB International, Cali, p 480p

Bellotti AC (2008) Cassava pests and their management. In: Capinera JL (ed) Encyclopedia of entomology, 2nd edn. Springer, P O Box 17, 3300 AA Dordrecht, The Netherlands

Bellotti AC, Herrera CJ, Hernández M del P, Arias B, Guerrero JM, Melo EL (2010) Three major cassava pests in Latin America, 2417 Africa and Asia. In: Howeler RH (ed) A new future for cassava in Asia: its use as food, feed and fuel to benefit the poor. Proceedings of the Eighth Regional Workshop held in Vientiane, Lao PDR. Oct 20–24, 2008. NAFRI/CIAT. The Nippon Foundation, pp 544–577

Bellotti AC, Riis L (1994) Cassava cyanogenic potential and resistance to pests and diseases. Acta Hortic 375:141–151

Bhargava A, Shukla S, Ohri D (2003) Genetic variability and heritability of selected traits during different cuttings of vegetable Chenopodium. Ind J Genet Plant Breed 63(4):359–360

Bhargava A, Shukla S, Ohri D (2006) *Chenopodium quinoa*-an Indian perspective. Ind Crop Prod 23:73–87

Chai CL, Li SH, Xu YC (2001) Carbohydrate metabolism in peach leaves during water stress and after stress relief. Plant Physiol Commun 37:495–498

Chipeta MM, Melis R, Shanahan P, Sibiya J, Benesi IRM (2017) Genotype x environment interaction and stability analysis of cassava genotypes at different harvest times. J Anim & Plant Sci 27(3):901–919

Cocozza C, Pulvento C, Lavini A, Riccardi M, d'Andria R, Tognetti R (2013) Effects of increasing salinity stress and decreasing water availability on eco-physiological traits of quinoa grown in a Mediterranean-type agro-ecosystem. J Agron Crop Sci 199:229–240

Comai S, Bertazzo A, Bailoni L, Zancato M, Costa CVL, Allegri G (2007) The content of proteic and nonproteic (free and protein-bound) tryptophan inquinoa and cereal flours. Food Chem 100:1350–1355

CostaJF Cosio M, Cardenas E, Yábar E, Gianoli F (2009) Preference of Quinoa Moth: Eurysac-camelanocampta Meyrick (Lepidoptera: Gelechiidae) for two varieties of quinoa (*Chenopodium quinoa* Willd.) in olfactometry assays. Chil J Agric Res 69:71–78

Danielsen S, Bonifacio A, Ames T (2003) Diseases of Quinoa (*Chenopodium quinoa*). Food Rev Int 19(1–2):43–50

De Proença GG, Schmidt CAP, dos Santos JAA (2017) Seasonal effects on starch contents evaluated in cassava roots. J Agric Sci 9(10):523–601

De Vries CA, Ferwerda JD, Flach M (1967) Choice of food crops in relation to actual and potential production in the tropics. Netherlands J Agric Sci 15:241–248

Eggleston G, Onwaka PE, Ihedioha OD (1992) Development and evaluation of products from cassava flour as a new alternative to wheat bread. J Sci Food Agric 59:377–385

Eisa S, Abdel-Ati A, Ebrahim M, Eid M, Abd El-Samad E, Hussin S, El-Bordeny N, Ali S, El-Naggar A (2014) *Chenopodium quinoa* as a new non-traditional crop in Egypt. In: Proceeding of "Bridging the gap between increasing knowledge and decreasing resources" conference. Tropentag, September 17–19, Prague, Czech Republic

Eisa SE, Eid MA, Abd El-Samad EH, Hussin SA, Abdel-Ati AA, El-Bordeny NE, Ali SH, Hanan MAA, LotfyME, Masoud AM, El-Naggar1 AM, Ebrahim M, (2017) *Chenopodium quinoa* Willd. A new cash crop halophyte for saline regions of Egypt. Aust J Crop Sci 1(3):343–351

El-Moity THA, Badrawy, HBM, Ali AM (2015) Survey on diseases and pests attack quinoa in Egypt. In: Proceedings of the 6th International Scientific Agricultural Symposium, Bosnia, October 15–18

El-Sharkawy MA (2004) Cassava biology and physiology. Plant Mol Biol Faostat

El-Sharkawy MA (2005) How can calibrated research-based models be improved for use as a tool in identifying genes controlling crop tolerance to environmental stresses in the era of genomics-from an experimentalist's perspective. Photosynthetica 43:161–176

El-Sharkawy MA (2007) Physiological characteristics of cassava tolerance to prolonged drought in the tropics: Implications for breeding cultivars adapted to seasonally dry and semiarid environments. Braz J Plant Physiol 19(4):257–286

El-Sharkawy MA (2010) Cassava: physiological mechanisms and plant traits underlying tolerance to prolonged drought and their application for breeding cultivars in the seasonally dry and semiarid tropics. In: da Matta FM (ed) Ecophysiology of tropical tree crops. Nova Science Publishers, Hauppauge, NY, USA, pp 71–110

El-Sharkawy MA, Hernandez ADP, Hershey C (1992) Yield stability of cassava during prolonged mid-season water stress. Exp Agric 28:165–174

FAO (1990) Protein quality evaluation in Report of Joint FAO/WHO expert consultation. Food and Agricultural Organization of the United Nations, Rome, p 23

FAO (2016) Country profile—Egypt. Food and Agricultural Organization of the United Nations, Rome, Version 2016

FAOSTAT (2011) Food and agricultural commodities production. Food and Agriculture Organization of the United Nations Statistics Database. Available at http://faostat.fao.org/

Flowers TJ, Colmer TD (2008) Salinity tolerance in halophytes. New Phytol 179:945–963. https://doi.org/10.1111/j.1469-8137.2008.02531.x

Francis G, Kerem Z, Makkar HP, Becker K (2002) The biological action of saponins in animal systems: a review. Br J Nutr 88(6):587–605

Fuentes-Bazan S, Mansion G, Borsch T (2012) Towards a species level tree of the globally diverse genus Chenopodium (Chenopodiaceae). Mol Phylogenetics Evol 62:359–374

Garcia M, Condori B, Castillom CD (2015) Agro-ecological and Agronomic Cultural Practices of Quinoa in South America. In: Murphy K and Matanguihan J (eds) Quinoa: improvement and sustainable production, 1st edn. Wiley

Garcia M, Raes D, Jacobsen SE, Michel T (2007) Agro-climatic constraints for rain fed in the Bolivian Altiplano. J Arid Environ 71:109–121. https://doi.org/10.1016/j.jaridenv.2007.02.005

Geerts S (2008) Deficit irrigation strategies via crop water productivity modeling: field research of quinoa in the Bolivian Altiplano. Dissertationes de Agricultura. Faculty of Bio-Science Engineering, KU Leuven, Belgium

Gomaa EF (2013) Effect of nitrogen, phosphorus and biofertilizers on quinoa plant (*Chenopodium quinoa*). J Appl Sci Res 9(8):5210–5222

Gonzalez JA, Gallardo M, Mirna Hilal M, Rosa M, Prado FE (2009) Physiological responses of quinoa (*Chenopodium quinoa* Willd.) to drought and waterlogging stresses: dry matter partitioning. Bot Stud 50:35–42

González JA, Konishi Y, Bruno M, Valoy M, Prado FE (2012) Interrelationships among seed yield, total protein and amino acid composition of ten quinoa (*Chenopodium quinoa*) cultivars from two different agroecological regions. J Sci Food Agric 92:1222–1229

Hariadi Y, Marandon K, Tian Y, Jacobsen SE, Shabala S (2011) Ionic and osmotic relations in quinoa (*Chenopodium quinoa* Willd.) plants grown at various salinity levels. J Exp Bot 62:185–193

Hasan SKH (2012) Effect of some mineral fertilization rates and harvesting dates on productivity and quality of cassava yield under sandy soil conditions. Ph D thesis. Faculty of Agriculture, Cairo University, Egypt

Hasmadi M, Faridah S, Salwa A, Matanjun IP, Abdul Hamid M, Rameli AS (2014) The effect of seaweed composite flour on the textural properties of dough and bread. J Appl Phycol 26:1057–1062

Helal NAS, Eisa SS, Attia A (2013) Morphological and chemical studies on influence of water deficit on cassava. World J Agric Sci 9(5):369–376

Holguín CA, Carabalí A, Bellotti AC (2006) Tasaintrínseca de crecimiento de Aleurotra-chelussociales (Hemiptera: Aleyrodidade) enyuca, *Manihot esculenta*. Rev Colomb Entomol 32(2):140–144

IITA (2005) Pre-emptive management of the virulent cassava mosaic disease in Nigeria. Annual report, July 2003–June, 2004 International Institute of Tropical Agriculture, Ibadan, Nigeria

IPCC: Intergovernmental Panel on Climate Change (2013) Summary for Policymakers. In: Stocker, T F, Qin D, Plattner GK, Tignor M, Allen SK, Boschung J, Nauels A, Xia Y, Bex V, Midgley PM (eds) Climate change. The Physical Science Basis. Contribution of Working Group I to the Fifth Assessment Report of the Intergovernmental Panel on Climate Change, Cambridge University Press, Cambridge, United Kingdom and New York, NY, USA

Jacobsen SE (2003) The worldwide potential for quinoa (*Chenopodium quinoa* Willd.). Food Rev Int 19(1–2):167–177

Jacobsen SE, Jensen CR, Liu F (2012) Improving crop production in the arid Mediterranean climate. Field Crops Res 128:34–47

Jacobsen SE, Monteros C, Christiansen JL, Bravo, LA, Corcuera LJ, Mujica A (2005) Plantresponses of quinoa (*Chenopodium quinoa* Willd.) to frost at various phonological stages. Eur J Agro 22(2):131–139 http://dx.doi.org/10.1016/j.eja.2004.01.003

Jacobsen SE, Mujica A, Jensen CR (2003) The resistance of quinoa (*Chenopodium quinoa* Willd) to adverse abiotic factors. Food Rev Int 19:99–109

Jisha S, Padmaja G, Moorthy SN, Rajeshkumar K (2008) Pre-treatment effect on the nutritional and functional properties of selected cassava-based composite flours. Innov Food Sci Emerg Technol 9:587–592

Kamel AS, El-Masry ME, Khalil HE (2010) Productive sustainable rice based rotations in saline-sodic soils in Egypt. Egypt J Agron 32(1):73–88

Kaya CI, Yazar A (2016) Saltmed model performance for quinoa irrigated with fresh and saline water in a Mediterranean environment. Irrig Drain 65:29–37

Koziol MJ (1992) Chemical composition and nutritional evaluation of quinoa (*Chenopodium quinoa* Willd). J Food Compos Anal 5:36–68

Langham R, Schwarz F (2008) Quick facts about sesame. American sesame growers association. Available at www: file://C:\Documents and settings\user1\Desktop\Quick facts on sesame from ASGA.mht

Lawrence JH, Moore LM (2005) United States department of agriculture plant guide. Cassava, *Manihot esculenta* Crantz. USDA, Washington, DC. http://plants.usda.gov/plantguide/doc/pg_maes.doc

Legg JP, Hillocks RJ (2003) Cassava brown streak virus disease: past, present and future. In: Aylesfor, UK: Natural Resources International Limited. Proceeding of an International Workshop, Mombasa, Kenya, 27–30 October 2002, p 100

Leng R, Wang C, Zhang C, Dai D, Pu G (2008) Life cycle inventory and energy analysis of cassava-based Fuel ethanol in China. J Clean Prod 16:374–384

Liu SS, De Barro PJ, Xu J, Luan JB, Zang LS, Ruan YM (2007) Asymmetric mating interactions drive widespread invasion and displacement in a whitefly. Science 318(5857):1769–1772

Mahmoud AH (2017) Production of Quinoa (*Chenopodium quinoa*) in the marginal environments of South Mediterranean Region: Nile Delta, Egypt. Egypt J Soil Sci 57(3):329–337

Mansour S (2012) Global agriculture information system; Egypt: wheat and corn production on the rise: grain and feed annual. Annual report

Mbah EU, Muoneke CO, Okpara DA (2003) Evaluation of cassava/soybean intercropping system as influenced by cassava genotypes. Niger Agric J Niger Agric J 34:11–18

Mbah EU, Ogidib E (2012) Effect of soybean plant populations on yield and productivity of cassava and soybean grown in a cassava based intercropping system. Trop Subtrop Agroecosystems 15:241–248

Molina-Montenegro MA (2013) Quinoa biodiversity and sustainability for food security under climate change. A Rev Agron Sustain Dev 34:349–359

Montagnac JA, Davis CR, Tanumihardjo SA (2009) Nutritional value of cassava for use as a staple food and recent advances for improvement. Compr Rev Food Sci Food Saf 8:181–194

Morita N, Hirata C, Park SH, Mitsunaga T (2001) Quinoa flour as new food stuff for improving dough and bread. J Appl Glyco Sci 48(3):263–270

Mtunguja MK, Henry SL, Edward K, Joseph N, Yasinta CM (2016) Effect of genotype and genotype by environment interaction on total cyanide content, fresh root, and starch yield in farmer-preferred cassava landraces in Tanzania. Food Sci & Nutr 4(6):791–801. https://doi.org/10.1002/fsn3.345

Munns R, Tester M (2008) Mechanisms of salinity tolerance. Annu Rev Plant Biol 59:651–681

Nhassico D, Muquingue H, Cliff J, Cumbana A, Bradbury JH (2008) Rising African cassava production, diseases due to high cyanide intake and control measures. J Sci Food Agric 88:2043–2049

Oelke EA, Putnam DH, Teynor TM, Oplinger ES (1992) Quinoa. Available from https://www.hort.purdue.edu/newcrop/afcm/quinoa.html

Okogbenin E, Ekanayake IJ, Porto MCM (2003) Genotypic variability in adaptation responses of selected clones of cassava to drought stress in the Sudan savanna zone of Nigeria. J Agron Crop Sci 189:376–389

Onitilo MO, Sanni LO, Daniel I, Maziya Dixon B, Dixon A (2007) Physicochemical and functional properties of native starches from cassava varieties in Southwest Nigeria. J Food Agric Environ 5(3&4):108–114

Onuma O, Kosikowski FV, Markakis P (1983) Cassava as a food. Critical Reviews in Food Sci Nutr 17(3):259–275

Osundare B (2007) Effects of different interplanted legumes with cassava on major soil nutrients, weed biomass, and performance of cassava (*Manihot esculenta* Crantz) in the southwestern Nigeria. Inter J Agric Sci (ASSET Series A). 7(1):216–227

Pillai KS, Palaniswami MS, Rojamma P, Mahandas C, Jayaprakas CA (1993) Pest management in tuber crops. Indian Hortic 38(3):20–23

Pulvento C, Riccardi M, Lavini A, Iafelice G, Marconi E, d'Andria R (2012) Yield and quality characteristics of quinoa grown in open field under different saline and non-saline irrigation regimes. J Agron Crop Sci 198:254–263

Pypers P, Sanginga JM, Kasereka B, Walangululu M, Vanlauwe B (2011) Increased productivity through integrated soil fertility management in cassava–legume intercropping systems in the highlands of Sud-Kivu, DR Congo. Field Crop Res 120:76–85

Rasmussen C, Lagnaoui A, Esbjerg P (2003) Advances in the knowledge of Quinoa pests. Food Rev Int 19:61–75

Razzaghi F, Ahmadi SH, Adolf SH, Jensen CR, Jacobsen SE, Andersen MN (2011) Water relations and transpiration of quinoa (*Chenopodium quinoa* Willd.) under salinity and soil drying. J Agr Crop Sci 197:348–360. https://doi.org/10.1111/j.1439-037X.2011.00473.x

Razzaghi F, Plauborg F, Jacobsen SE, Jensen CR, Andersen MN (2012) Effect of nitrogen and water availability of three soil types on yield, radiation use efficiency and evapotranspiration in field-grown quinoa. Agric Water Manag 109:20–29

Repo-Carrasco R, Espinoza C, Jacobsen SE (2003) Nutritional value and use of the Andean Crops Quinoa (*Chenopodium quinoa*) and kaniwa (*Chenopodium pallidicaule*). Food Rev Int 19:179–189. https://doi.org/10.1081/FRI-120018884

Riccardi M, Pulvento C, Lavini A, d'Andria R, Jacobsen SE (2014) Growth and ionic content of quinoa under saline irrigation. J Agro Crop Sci 200:246–260

Rojas W, Pinto M, Alanoca C, Gomez Pando L, Leon-Lobos P, Alercia A (2015) Quinoa genetic resources and ex situ conservation: In: Bazile D, Bertero HD, Nieto C (eds) State of the art report on Quinoa Around the world in 2013. FAO & CIRAD, Roma, pp 56–82

Romero S, Shahriari S (2011) Quinoa's global success creates quandary at home. *The New York Times*, 19 March 2011. http://www.nytimes. com/2011/03/20/world/americas/20bolivia.html?_r = 2&hp = &pag&

Ruiz KB, Aloisi I, Duca SD, Canelo V, Torrigiani P, Silva H, Biondi S (2016) Salares versus coastal ecotypes of quinoa: salinity responses in Chilean landraces from contrasting habitats. Plant Physiol and Biochem 101:1–13

Salau AW, Olasantan FO, Bodunde JG (2012) Effects of time of introducing okra on crop growth and yield in a cassava-okra intercrop. Nigeria J Hortic Soc 17:57–67

Santisopasri V, Kurotjanawong K, Chotineeranat S, Piyachomkwan K Sriroth K, Oates CG (2001). Impact of water stress on yield and quality of cassava starch. Ind Crop Prod 13(2):115–129

Shams A (2011) Combat degradation in rain fed areas by introducing new drought tolerant crops in Egypt. Int J Water Resour Arid Environ 1(5):318–325

Shams AS (2012) Response of quinoa to nitrogen fertilizer rates under sandy soil conditions. In: Proceeding of the 13th International Conference in Agronomy, Benha, Egypt, 9–10 September, pp 195–205

Sharp RE, Davies WJ (1989) Regulation of growth and development of plants growing with a restricted supply of water. In: Jones HG, Flowers TJ, Jones MB (eds) Plants under stress. Cambridge University Press, Cambridge, UK, pp 71–93

Sherif SA, Nassar NMA (2010) Introducing cassava into Egypt. Gene Conserv 9(35):118–123

Sherif SA, Salem AK (2011) Studies on cassava (*Manihot esculenta* Crantz) intercropped with fodder cowpea (*Vigna sinensis* L.) in sandy soil. Egypt J Agron 33(1):95–111

Sholihin (2011) AMMI model for interpreting clone-environment interaction in starch yield of cassava. HAYATI J Biosci 18(1):21–26. https://doi.org/10.4308/hjb.18.1.21

Sigsgaard L, Jacobsen, S-E, Christiansen, JL (2008) Quinoa, Chenopodium quinoa, provides a new host for native herbivores in northern Europe: Case studies of the moth, Scrobipalpa atriplicella, and the tortoise beetle, Cassida nebulosa. J Insect Sci 8:50–54

Silva DV, Santos JB, Ferreira EA, Silva AA, França AC, Sediyama T (2012) Manejo de plantas daninhasna cultura da mandioca. Planta Daninha 30(4):901–910

Ukwuru MU, Egbonu SE (2013) Recent development in cassava-based products research. Acad J Food Res 1(1):001–013

UNCTAD, United Nations Conference on Trade and Development (2012) Infocomm commodity profile Cassava. Available at http://www.UNCTAD.info/en/infocomm/AACP=products/Commodity+profile.cassava/

Vandegeer R, Miller RE, Bain M, Gleadow RM, Cavagnaro TR (2013) Drought adversely affects tuber development and nutritional quality of the staple crop cassava (*Manihot esculenta* Crantz). Funct Plant Biol 40:195–200

Van Oirschot QE, O'Brien GM, Dufour D, El-Sharkawy MA, Mesa E (2000) The effect of pre-harvest pruning of cassava upon root deterioration and quality characteristics. J Sci Food Agric 80:1866–1873

Vilche C, Gely M, Santalla E (2003) Physical properties of quinoa seeds. Biosys Eng 86(1):59–65

Waseem M, Ali A, Tahir M, Nadeem M, Ayub M, Tanveer A (2011) Mechanisms of drought tolerance in plant and its management through different methods. Cont J Agric Sci 5:10–25

Wu G, Peterson AJ, Morris CF, Murphy KM (2016) Quinoa seed quality response to sodium chloride and sodium sulfate salinity. Front Plant Sci 7(790):1–8

Zaidul ISM, Yamauchi H, Kim SJ Hashimoto N, Noda T (2007) RVA study of mixtures of wheat flour and potato starches with different phosphorus content. Food Chem 102:1105–1111

Zinsoua V, Wydra K, Ahohuendoa B, Haub B (2004) Effect of soil amendments, intercropping and planting time in combination on the severity of cassava bacterial blight and yield in two ecozones of West Africa. Plant Pathol 53:585–595

Zohry A (2020) Prospects of quinoa cultivation in marginal lands of Egypt. Mor J Agri Sci 1(3):132–137

Part VI
Conclusions

Chapter 21
Update, Conclusions, and Recommendations of "Mitigating Environmental Stresses for Agricultural Sustainability in Egypt"

Hassan Auda Awaad, Abdelazim M. Negm, and Mohamed Abu-hashim

Abstract This chapter focused on the main conclusions and recommendations that prepared in the chapters existing in the book. In addition, some results from recently scientific published research works that considered the practical approaches for agricultural sustainability in Egypt and alleviating environmental stresses for enhancing sustainable crop production. Therefore, the chapter in Part I includes an introduction about mitigating environmental stresses for agricultural sustainability in Egypt and Part II contains information about improve the crop tolerance to abiotic stresses such as drought, high temperature, environmental pollution and salinity in crop plants. Whereas, Part III explained the recent approaches for biotic stresses, and contain varietal differences and their relation to brown rot disease resistance and effect of soil type and crop rotation on the disease in potato, advanced methods in controlling late blight in potato and importance of faba bean diseases besides developing rust resistance in wheat. Moreover, Part IV is about advanced procedures in improving crop productivity. It includess the role of He–Ne LASER, seed technology, biochemical and molecular tools and importance of mycorrhizae in improving crop productivity. Part V is devoted to sustainability of environmental resources from a crop production perspective. This part includes sustainable use and optimizing inputs management for natural resources by crop rotations and bio-fertilizer in different climatic zones that relevant for agriculture in Egypt, and concentrate on different field crops, as well as highlights on Quinoa and Cassava as promising crops to increase food security in Egypt. Finally, set of recommendations were performed in this chapter for future research work to motivate the future researches towards sustainability approach which is a main strategic concept of the Egyptian government.

H. A. Awaad (✉)
Crop Science Department, Faculty of Agriculture, Zagazig University, 44511, Zagazig, Egypt

A. M. Negm
Water and Water Structures Engineering Department, Faculty of Engineering,
Zagazig University, 44519, Zagazig, Egypt
e-mail: amnegm85@yahoo.com; amnegm@zu.edu.eg

M. Abu-hashim
Soil Science Department, Faculty of Agriculture,
Zagazig University, 44511, Zagazig, Egypt

© Springer Nature Switzerland AG 2021
H. Awaad et al. (eds.), *Mitigating Environmental Stresses for Agricultural Sustainability in Egypt*, Springer Water,
https://doi.org/10.1007/978-3-030-64323-2_21

561

Keywords Environmental stresses · Field crops · Drought tolerance · Heat stress · Pollution tolerance · Salt stress · Natural resources · Biotic stresses · He-Ne LASER · Seed technology · Molecular tools-Mycorrhizae

21.1 Introduction

The world's population has been estimated to be around 9.8 billion by 2050, which will result in severe food shortages and quality under agricultural adjustments. Egyptian population is assessed to be nearly 100 million. It is alarming that Egypt has suffered a severe water shortage in recent years, and this shortage could be expected to increase within the near future due to the negative effects of the Ethiopian renaissance. Besides water, heat, pollution, and salinity are major environmental factors affecting crop productivity. Hereby, integrating quinoa and cassava as new crops tolerant to stresses in the Egyptian cropping pattern has an important purpose of coping with the increase in population and dealing with the adverse environmental factors. Furthermore, recent approaches using the arbuscular mycorrhizal fungus, seed quality, biochemical and molecular tools, and helium-neon laser are important technologies in improving crop productivity. The next section will display a transitory of the main results of the recently published research on the mitigating environmental stresses and improving sustainable crop production. Finally, Agro-Environmental policies and comparison between Egyptian and EU structure considering the recommendations for researchers and decision makers.

21.2 Update

The current world population is approximately 7.6 billion and predicted to be 8.6 billion in 2030, 9.8 billion in 2050, and to be 11.2 billion in 2100, in the light of a new UN report. With an 83 million inhabitant being added yearly to the world's population, the exponential increase in population rate is expected (https://www.un.org/development/desa/en/news/population/world-popula tion-prospects-2017.html). This will result in severe food shortages and quality under agricultural adjustments. Herby, crops adapted to the adverse environmental factors should be developed to cope with the increase in the world population.

One of the main challenges facing the agriculture sector in Egypt is environmental changes and its impression on the cultivation and production of crop plants and food security. Six key environmental induced hazards and their significances in Egypt were recognized. These threats include: (1) Drought (2) High temperature, (3) Pollution, (4) Salinity, (5) Gap between production and consumption and (6) biotic stresses of diseases attack the field crops.

For further evidence about mitigating environmental stresses and improving sustainable crop production, a recent study on challenges facing agriculture in Egypt,

Osman et al. (2016) stated that Egypt has been suffering from a severe water shortage in recent years, facing water shortage about 7 billion cubic meters annually. This shortage will increase in the future because of the Ethiopian Renaissance Dam effects on Egyptian water resources. Beside this, a gradual increase in global temperature and environmental pollutants are observed as a result of human activities and increased levels of atmospheric CO_2 which absorb and capture a large amount of infrared and geothermal radiation that emitted from the soil surface. Subsequently, evaluating the impact of pollutants on crop plants is an essential objective to improve the quality of food. So, on the light of climatic changes, release high resilient cultivars and follow crop relevant rotation that could reduce the spread of pests and provide efficient week control and interrupts pest cycles.

An alternative solution to the potential for cultivating drought-affected land is to cultivate varieties that use water more efficiently to reduce the water consumption of the crop, and capable of growing, developing and producing accepted yield levels under water shortages, this is beside proper agricultural management. This goal is one of the efficient strategies for plant breeders and crop scientists to meet the growing human needs of food and clothing and reducing the problem of food shortages suffered by humans in many parts of the world and in Egypt (Awaad 2009).

The performance of crop varieties suffers from drought stress and its effects on various plant traits. Bread and durum wheat genotypes varied in their reaction to water stress conditions and showed various degrees of tolerance. Also, stability analysis revealed several responses of wheat genotypes to different drought stress environments (Ali and Abdul-Hamid 2017; Abdel-Motagally and Manal El-Zohri 2018). Besides this irrigation systems play an essential role in improving the water use efficiency under limited water supply. For example, a subsurface drip irrigation system enhance the water use efficiency which resulted in low evaporation from the soil surface, that the root and/or partial root areas are irrigated. In contrast, in a sprinkler irrigation system, moisturizing the entire field area (Mansour et al. 2014).

On the other hand, surface irrigation causes significant loss of water. Under sandy loam soil, Nubaria, Behaira Governorate, Egypt, Mansour and Abd El-Hady (2014) mentioned that subsurface drip irrigation at 20 cm has a promoting impact on wheat grain and straw yield with increasing value of 6.9 and 5.7%, respectively of wheat cultivar Gemmeiza 9, compared to surface drip system. Nevertheless, the percentage of the increase was 1.7 and 1.8% for grain and straw yield, respectively when comparing subsurface drip irrigation at 10 cm with surface drip irrigation system. Furthermore, the comparison among 29 bread wheat lines (*Triticum aestivum* L.), 5 durum wheat lines (*Triticum durum* L.) and 4 commercial check varieties under different irrigation systems for drought stress. Under El-Kattara and Ghazalla regions of Egypt, Ali and Abdul-Hamid (2017) showed that wheat genotypes exhibited higher grain yield under drip irrigation than sprinkler and surface flood irrigation systems.

Furthermore, a gradual increase in global temperature is observed as a result of human activities and increased levels of atmospheric CO_2 which absorb and capture a large amount of infrared and geothermal radiation emitted from the soil surface. This would be resulted in occurrence of global warming and the impact of Greenhouse effect. Within the target set by the 2016 Paris Agreement, Hickey

(2017) mentioned that a mere 1 percent possibility that cause warming could be at or below 1.5 degrees. The statistically-based projections that published at July 31 in Nature Climate Change, showed that 90 percent possibilities the temperatures will be increased this century by 2.0–4.9°C (http://www.washington.edu/news/2017/07/31/earth-likely-to-warm-more-than-2-degre). According to data of FAO (2015), Fig. 21.1 shows that the current world emission worst scenario that projected the global warming.

Heat stress affects crop productivity in several regions, especially arid, semi-arid, and hot regions such as the Arab region. The areas of horizontal expansion in the Egyptian deserts are typical examples. In the light of the increase in Egyptian population and extension of agriculture to marginal lands, various environmental stresses face these hot spots in Egypt and in the globe. Water and heat are major environmental factors affecting crop productivity. Therefore, breeding tolerant varieties to high temperature and follow suitable agricultural practices are considered as important procedures to cope with heat stress. The progress achieved in this area is still limited because heat stress tolerance is a complex function controlled by polygenes (Irmak 2016). The amount of yield obtained under heat stress depends on several major determinants. Crop plants have many morpho-physiological and biochemical adaptations which help them to adapt with high temperature. Earliness, cooler canopies, stay green, transpiration rate, water use efficiency, photosynthetic rate, and the contribution of stored products in stems, bases and sheaths of leaves are an important abilities to support plants under stress conditions to improve grain filling

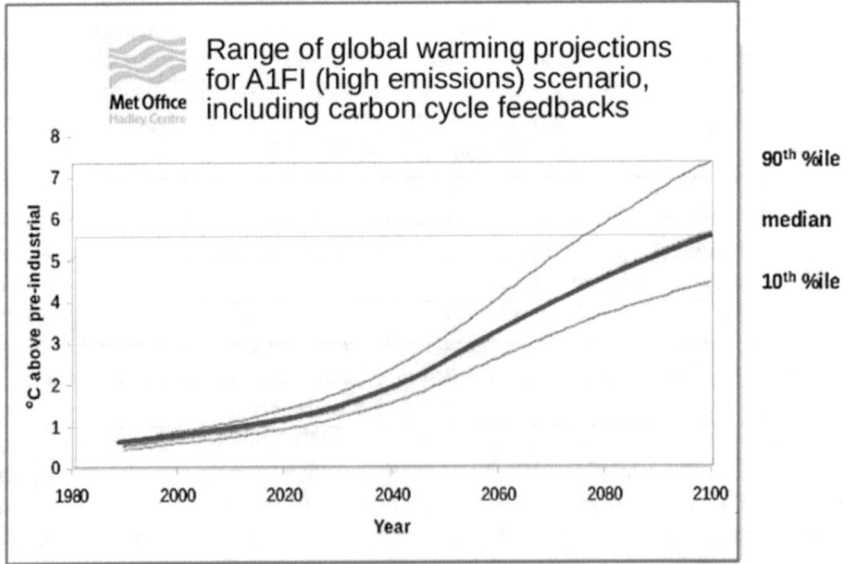

Fig. 21.1 The current world emission worst scenario to project the global warming (FAO 2015)

under heat stress conditions (Reynolds et al. 1998; Moursi 2003). In addition, maintaining high leaf chlorophyll content is a desirable trait as it reflects the low degree of photo inhibition that results in the photosynthetic apparatus at high temperature (Ristic et al. 2007; Talebi 2011). Whereas, Ali et al. (2010) mentioned that the plant ability to keep the content of the leaf chlorophyll under high-temperature stress is correlated with grain yield and its components.

Another challenge is pollution. With rapid urbanization and the development of modern industry, many soil pollution cases have been monitored with heavy metals and other pollutants in several regions of the world. Assessing the impact of pollutants on crop plants is an essential objective to improve the quality of food. FAO (2014) revealed that the actual global greenhouse gas that emitted "from agriculture, forestry, and other land use" were over 8 billion metric tons of CO_2 (https://19january201 7snapshot.epa.gov/ghgemissions/global-greenhouse-gas-emissio.). That a significant share of the emitted global greenhouse gas would be produced and destroys the habitats. Thus, these increasing trends could be threaten the sustainability of the world food systems and undermine its capacity to meet the food needs. In another report of FAO (2018) stated that increasing use of nitrogen fertilizers and spreading the environmental pollutants from heavy metals such as lead, arsenic, cadmium, and mercury lead to reduce in crop productivity and quality, and to harmful effects on plant metabolism. As these pollutants get into the food chain, they cause risks to food security, rural livelihoods, water resources, and human health. Assessing the impact of pollutants is important in examining the quality of the environment and how to deal with sources of pollution (Pal 2016). The high concentrations of ozone, sulfur and nitrogen oxides, and the diffusion of heavy element particles from environmental contaminants affect significantly on the physical and chemical properties of the air. Pollutants also affect plants, human and animal health (Awaad et al. 2013). In different literatures, the decrease in wheat yield due to heavy metals pollution is more than 20% besides the effect on quality (EL-Gharbawy 2015). Nevertheless, certain heavy metals such as Zn, Fe, Ni, and Cu are essential micronutrients for certain physiological processes in the plant as enzyme metabolism and photosynthesis. However, at high levels, they become toxic pollutants and effect negatively on plant growth (Ali et al. 2018). Heavy metals phytotoxicity inhibits transpiration, chlorophyll synthesis, photosynthesis, disturbing carbohydrate metabolism, damage of lipids, proteins and DNA and hence adverse effects on crop productivity (Lequeux et al. 2010). Hence it is of the importance of reducing the factors that lead to environmental pollution and reduce their risks through plant breeding and biotechnology as well as procedures to mitigate the impact of the pollutants.

Salinity is one of the main abiotic stresses that affect crop productivity. Salinity with high level in the crop environment are well known on several physiological and metabolic processes which will lead to reduction in growth and yield. Maggio et al. (2000) mentioned that high concentration of NaCl would leads to osmotic stress and ion imbalance in numerous crop species. Beside to these primary effects, secondary stresses happened such as oxidative injury (Zhu 2001). Genotypes that have high levels of antioxidant activity have better resistance to the oxidative injury (El-kahoui et al. 2005). In addition, plant cells which contain protective and repair schemes controlled the metabolism under stress environments. Stimulation of secondary metabolism is an adjusting system for the crop genotypes that involved in the defense mechanism (Singh et al. 2015). Proline accumulation is one of the most important amino acids whose production increases repeatedly due to salinity and water shortage in crop plants. (Ramanjulu and Sudhakar 2001), and it is frequently expressed to be involved in stress resistance. Environmental stress tolerant genotypes have the enzymatic and non-enzymatic anti-oxidant systems to protect cell membranes and organelles from the harmful effects of ROS. A significant correlation was recorded between the antioxidative enzymes activities and the salt tolerance of crop genotypes (Nader et al. 2005). Therefore, it is essential to develop, modify, and release new cultivars and determine their adaptability and stability to cope with various environmental conditions (Aly and Awaad 2002; Abd El-Shafi et al. 2014; Ali 2017; Ali and Abdul-Hamid 2017; Siddhi et al. 2018).

The great gap occurs between the production and consumption of some strategic crops in Egypt are topped by wheat, which estimated by 55%, negatively affects the production of bread in Egypt. FAOSTAT (2017) showed that the area of wheat amounted to approximately 3.39 million feddan (Feddan $= 4200$ m^2) and total production was 9.28 million tons of grains.

On the other hand, domestic consumption is referring to 16 million tons, which pointing to a gap between yield production and inhabitant consumption. Wheat consumption has been increased as a result of increasing the annual population to 2.0 million/year. That Egypt has one of the highest wheat consumption levels per capita in the world. Also, the amount of imported maize was raised from about 4.96 million tons in 2000 to about 10.81 in 2014 (Heba et al. 2015). The estimated gap between the production and consumption of maize was 45%. This gap is compensated by import, which consume the state budget (Zohry et al. 2016). Generally, in order to balance the global population increase with the global food demand, it is a must to use improved agricultural technologies that help in boosting crop production and productivity (FAO 2011). Also, a better identifying of plant responses to the stresses pests, drought, heat, poor soil fertility, and low soil nitrogen tolerance has pragmatic implication for remedies and management. It seems important to increase productivity either through vertical expansion by enhancing the productivity of the unit area and/or by horizontal expansion by increasing the area cultivated in the new lands. Also, increase production by developing high yielding varieties tolerant to environmental stresses and simultaneously applying good agricultural management to ensure high production at different locations. Besides increasing the irrigated zone and intensive agriculture, merging large areas to grow one crop, crop cultivation in non-traditional spaces and changing the agricultural cycle (Magdoff and van 2000). Therefore, the authors suggested optimizing inputs management for sustainable agricultural development. That the rate of population are increasing rapidly in the developing countries, leading to starvation if the food output does not grow by 50–60%. A fundamental increase in crop production was achieved through new cultivars and application of high fertilization rates (Francis and Daniel 2004).

Biotic stresses are another challenge on the light of climatic change. Potato plants in fields are attacked by many pathogens such as fungi, bacteria, viruses, phytoplasmas, viroids, and nematodes, which reduce and affect the yield quality and quantity. Another phenomena that has been reported in Egypt many years ago was Potato brown rot, caused by *R. solanacearum* (Briton-Jones 1925). That disease has created one of the main quarantine problems through the course of exportation of table potatoes to Europe (Farag 2000). Bacterial wilt is an important soil-borne disease that affects potato production with a significant loss of about 15% (Zehr 1969). Yield losses are mostly affected by tuber rotting. If weather environments are relevant for the development of the disease, the yield might decline by 50% (Anonymous 2015). In addition, this disease has been recognized to effect on 1.7 million hectares in 80 countries worldwide, with global injury that estimated annually by USD 950 million (Champoiseau et al. 2009). That direct yield losses triggered by *R. solanacearum* could be varied from 33 to 90% dependent on the diverse reasons such as climate, soil type, cropping pattern, cultivar and the strain of the bacteria (Karim et al. 2018). Identification of pathogens is important to control the disease. Isolates of *R. solanacearum* in Egypt were identified as race 3 biovar 2 using morphological, physiological and biochemical tests. Also, advanced Immunofluorescence Antibody Stains (IFAs), Polymerase Chain Reaction (PCR) as well as Real-time PCR (Taq-Man) techniques were applied and proved that the isolates are related to race 3 biovar 2 of R. *solanacearum* (Hamad 2016). High levels of varietal differences to brown rot disease resistance in potato were registered by many investigators of them Fahmy and Mohamed (1990), El-Didamony et al, (2003) and El-Halag (2008) and Muthoni et al., (2014). Several investigations clearly mentioned the antibacterial activity of plant extracts against pathogens (Gottlieb et al. 2002; Slusarenko et al. 2008 and Hamad 2016). Furthermore, Malafaia et al. (2018) reported that the usage of antimicrobial agents techniques that could effect on the planktonic bacteria to be as an alternate method to control the plant diseases. Plant extracts used for medicinal purposes as *Croton heliotropiifolius*, *Eugenia brejoensis*, and *Libidibia ferrea* could be considered to be favorable alternatives to control of *R. solanacearum*.

The effect of soil type on disease severity has been repeatedly determined (Moffett et al. 1983; van Elsas et al. 2000; El-Halag et al. 2015; Messiha et al. 2019). However, the actual effects are determined by geographical place, race, biovar of the pathogen, and the crops. In some cases, bacterial wilt was more severe in well-drained sandy loams, but the reduction rate of the pathogen was higher in clay loam compared to sandy loam (Moffett et al. 1983). Moreover (van Elsas et al. 2005) mentioned that there are several parameters could effect on the suppressiveness of bacterial wilt in the soil such as soil texture, pH, organic matter content and microbial populations. Disease dominance was decreased after soil treating with methyl bromide or heat, demonstrating that a biological source could be the main factor. Resistance cultivars and diverse cultural measures including the use of certified seed, soil type, crop rotations with non-host crops, and careful water management has been practically successful against racing the pathogen (Hartman and Elphinstone 1994; French 1994; El-Halag et al. 2015). Potato – Rice crop rotation was more operative in decreasing the population density of *R. solanacearum* compared to Potato—Maize and Potato—Peanut crop rotations (Hamad 2008).

Furthermore, potato late blight disease caused by the *Phytophthora infestans* is deliberated the highest known and remains among the most damaging plant diseases. Late blight affects humans because foliar phase limits production of tubers that are a food source. Human suffering caused by dramatic epidemics was aggravated in the 1840s by a lack of potatoes (Mizubtui and Fry 2006). Recently, new strains with the ability to reproduce sexually are spreading that are related with increased genetic variation and survival in some parts of the world (Fry 2008). Control of this disease is remains dependent on multiple applications of chemical fungicides through flowering and fruiting (Shattock 2002). A combination of high inoculum pressure, humid conditions that favor pathogen growth and repeated pesticides applications have resulted in the emergence of resistant pathogen strains in several countries (Gisi and Cohen 1996). Further, the use of some synthetic chemicals to control fungal disease of food commodities is restricted due to their possible carcinogenicity, high and acute toxicity, long degradation periods and environmental pollution. Hence, cultivating high-tolerant varieties is essential in this respect.

To complement biotic stresses, the wheat crop is exposed to the injury by three rust diseases attack the crop, *i.e.* stripe, leaf and stem rusts, caused by *Puccinia striiformis* Westend. f. sp. *tritici* Erikss., *Puccinia triticina* Erikss. and *Puccinia graminis* f. sp. *tritici*, respectively, are severe foliar diseases in wheat growing regions worldwide cause vital losses of yield. Under Egyptian conditions, stripe rust is considered the most common disease is producing significant injuries in wheat production (Abu El-Naga et al. 2001). Stripe rust can cause 100% yield loss but often different from 10% to 70% (Chen 2005). Early infection of leaf rust usually causes higher yield losses varied from 60 to 70% based on the sensitivity of the wheat genotypes and the severity of epidemics (Appel et al. 2009). Bajwa et al. (1986) stated that losses in kernel weight of wheat varieties due to leaf rust infection varied from 2.0% and 41% according to the level of susceptibility. Leaf rust can cause yield losses of at least 25% (Kolmer et al. 2009). Whereas, yield loss in wheat from natural infections with *Puccinia graminis* f. sp. *tritici* fluctuated from 10 to 45% in three experiments across two years (Loughman et al. 2005). Plant breeding for disease resistance is one of the most important strategies for reducing the yield loss of wheat. The importance of disease depends principally on the dominance of virulent races and their compatibility with the genetic makeup of the host variety under specific condition. Therefore, the Egyptian wheat varieties have suffered from sudden epidemics through the last decades as a result of environmental changes in relation to the genetic makeup of both host and pathogen (El-Daoudi et al. 1987). New sources of resistance could be combined into wheat to different the present gene pools for rusts resistance. The active resistant genes against rust can be used singly or in combination with high yielding genes to increase high-yielding resistant cultivars in wheat-growing areas. Genetic resistance is the most economic and effective approaches for decreasing yield losses caused by rust diseases (Singh et al. 1991; Liu and Kolmer 1997). However, breeding genotypes more resistant to rust diseases is a continuous process, and plant breeders need to increase new active sources to their breeding materials (Draz et al. 2015).

Faba bean cultivated in 56 countries, and the total production was estimated as 4.46 million tons of dry grains (FAOSTAT 2018). In Egypt, the grown areas of faba bean were decreased progressively during the last two decades from around 220.00 to 80.00 feddan (Anonymous 2017). The shortage of the grown areas is due to several factors i.e. plant diseases, pests, weeds, parasitic plants, nutrition's and cost of production materials, etc. These constrains led to import a considerable amount of faba bean seeds to fulfill the requirements and demands of the consumers. In addition, Chocolate spot *Botrytis fabae* Sard and rust *Uromyces vicia fabae* (Pers.), Root rot *Rhizoctonia solani, Microphomina phasolina, Fusarium solani, Fusarium avenacerum, Fusarium oxysporum, pythium sp, Sclerotium rolfsii, and Sclerotinia sclerotiorum*, wilt *Fusarium* oxysporium *f. sp. fabae* Yu and Fang are considered the most important diseases in the Mediterranean region. Breeding for disease resistance with high yielding ability using traditional breeding programs (Khalil et al. 1993) as well as advanced molecular studies (EL-Badawy 2008) are of great importance. Results of detached leave technique, greenhouse and field experiments revealed that faba bean cultivars Sakha 1, Sakha 3, Sakha 4, Giza 3 and Giza 716 proved to be resistant to the diseases. Fungicides treatment, application of other treatments, *i.e.* growth regulators, biological control, plant extracts and some agricultural practices offered a degree of tolerance to the disease and increased yield (Eisa et al. 2006; Baraka et al. 2008; Salem et al. 2012; EL-Wakil et al. 2016).

A Helium-Neon Laser was discovered by Maiman since 50 years ago using a flash lamp pumped ruby crystal as the medium as advanced procedures in improving crop productivity (Maiman 1960). That the integration of existing and recent technologies to induction and exploit genetic variability is an important objective in the development of healthy, nutritious and productive crops. Laser is a proven technology for improving crop genotypes with desirable qualities (Suprasanna and Jain 2017). Since the discovery of the laser in the last century, this has made great advances in biological science. Agriculture has had a good deal of laser contribution in promoting and improving productivity and quality. Thus, apply the Laser technology in the field of agriculture have contributed to increased tolerance to a biotic and biotic stresses and enhancing yield of crop plants. Joanna et al. (2016) showed that genetic variation is considered as the main parameter of phenotypic diversity and is the main engine of evolutionary diversification. The laser-induced differences were widely applied to investigate the genome of different plant species and to improve many crop geniuses like Arabidopsis, tomato, barley, soybean, and wheat.

The technology of seed quality of field crops plays essential effects in the speed of germination, field emergence capacity, seedling vigor and subsequent growth of crop plants under stressed environment. Weather, fertilization, cross-pollination and agronomic practices all influence final seed quality of field crops. On other context, the study of seed is a study of life itself. Studying the seed and/or the germinating seedling, more knowledge could be obtained about growth regulators, cell division, respiration, morphogenesis, photosynthesis and further metabolic processes (Copeland and McDonald 1995). Good seed or high-quality seed is an essential input in agriculture production that the successful agriculture production based on the seeds quality that used for sowing. So, the seed producers perform greater responsibility in maintaining genetically pure seed and preserving the quality of seeds through different stages and periods to be free from various contaminants and in healthy condition. Planting a superior seed lot generally results in a more uniform stand that allows better secondary tillage and weed/pest control (http://www.sciencepub.net/nat ure/0705/11_0768_Soybean_seed_quality_ns0705.pdfThese factors promote earlier and more uniform emergence, which lead to increase yields and more economic return to the seed consumers and national production income (Eraky Hania 2015).

Legume crop is grown worldwide, and peanut (*Arachis hypogaea* L.) is important as it contains high oil content and vegetable protein (Reddy et al. 2003). Salinity is a restricted production of legumes and determines its distribution through environments (Giannakoula et al. 2012). Plant organs are greatly affected by salt pressure and roots are the first organ that interacts with salt stress (Delgado and Sanchez-Raya 2007). Negrao et al. (2017) discussed the relevant measurements for the effect of salinity on different plant traits; relative growth rate, water relations, transpiration use efficiency, transpiration, ionic relations, senescence, photosynthesis, yield and yield components. Davenport et al. (2005) described the plant's mechanics to tolerate salinity stress excluding Na^+ from leaves. In addition, under different salinity levels on peanut seedlings, Hossain et al. (2011) investigated the magnitude of proline accumulation. They showed a significant increase in proline after 7 days of exposure to NaCl stress. Proline acting as an significant osmolyte, and plays three essential roles during stress; metal chelator, an antioxidative protection molecule, and a signaling molecule (Hayat et al. 2012). Moreover, Musa et al. (2015) showed that proline level and the activity of glutathione reductase considered the main parameters that play a main role in salt stress protection in peanut.

Supporting the crop plants and increasing their ability to tolerate severe stress conditions and improve plant growth and soil fertility, mycorrhizal fungi played an importance role in this respect (Gosling et al. 2006; Smith et al. 2011). Arbuscular mycorrhizal fungi can enhance the host plant resistance to drought or diseases (Smith and Read 2008). Plant roots are the location where the carbohydrate and mineral nutrient exchange between the host plant cells and arbuscular mycorrhizal fungi can occur. Arbuscular mycorrhizal fungi get the photosynthetic carbon from the host plant species, then they gave the host plants with the acquisition and transport of nitrogen (Fellbaum et al. 2012; Liu et al. 2013). In the rizhosphere, there are six types of mycorrhiza fungi, viz. arbuscular, arbutoid, ecto, orchid, ericoid and monoptropoid, which are classified based on their morphological characteristics (Garg et al. 2006; Wang and Qiu 2006). Majority of the terrestrial plant species (250 000) have almost symbiosis process with mycorrhizal fungi (Harley and Harley 1987; Peterson et al. 2004; Wang and Qiu 2006; Smith and Read 2008; Helgason and Fitter 2009). Plant species belong to *Brassicaceae* and *Chenopodiaceae* families do not have a symbiotic association with mycorrhizal fungi (Newman and Reddell 1987; Peterson et al. 2004). Following root colonization of host plants, the mycelium grows out of the root exploring the soil, up-taking different nutrients and water, and then translocating them to the roots, and it can colonize other susceptible roots (Smith and Read

2008, Smith et al. 2010 and Smith and Smith 2011). The arbuscules of mycorrhizal fungi are considered the location of the nutrient transfer between the fungus and the host plant (Taiz and Zeiger 2006). In addition, arbuscular mycorrhizal fungi colonization has several benefits *i.e.* plant uptake for nitrogen, phosphorus, potassium and manganese (Smith and Read 2008; Smith et al. 2011; Seleiman et al. 2013), and copper and zinc (Li et al. 1991; Azaizeh et al. 1995; Taiz and Zeiger 2006). The host plant provides up to 20% of the organic carbon for the function of arbuscular mycorrhizal fungi (Smith and Read 2008; Smith and Smith 2011).

Sustainability of environmental resources from a crop production perspective represents great importance to overcome the food gap in Egypt. Proceeding from this, we must focus on the main principles of sustainability in agriculture which include the following:

a. Integration biological and ecological processes for instance N fixation, different nutrient cycling and soil regeneration,
b. Reducing the usage of pollutants inputs that can pose harm effect to the environment, animals and humans,
c. Solving common agricultural issues as well as natural resource problems for instance pest, irrigation and forest management.

So, sustainable agriculture has become very popular recently. It means the production of crops without depending on the toxic chemical pesticides or herbicides, synthetic fertilizers, genetically modified grains or agricultural practices that pollute water, soil, or other natural resources. Sustainable of agriculture can be achieved by cultivating high-yielding cultivars and using sustainable techniques such as crop rotation and tillage (DeLonge et al. 2016). Using the agriculture technologies that can protect the environment and public health from hazardous pesticides and herbicides are very important to grow crops that are safe for consumers and farmers (Gallandt 2014). In addition, sustainable farming also able to protect humans from exposure to pathogens, toxins, and other hazardous pollutants. This concept of agriculture could enables us to produce healthy food without compromising future generations (Ryals et al. 2014). It is important to focus on the key principles of sustainability in agriculture, and the integration of biological and ecological processes, for example, N fixation, various nutrient cycles, and soils. Also, the availability of new tools to enhance genetic gains, and the continued and intensive application of crop productivity. For example, in maize, there is a shortage between the production and consumption of maize estimated at about 45%. This gap is compensated by imports, which consumed the country's budget (Zohry et al. 2016). In order to balance global population growth with global food demand, it is necessary to use improved agricultural technologies that help promote crop production and productivity (FAO 2011). The new land in Egypt is mainly on the eastern and western sides of Delta and Wadi that covering 1.01 million hectares. These lands considered the main opportunity to increase the agricultural yield and ensure food security in the country. Similarly, these regions provide an opportunity to accommodate population growth and to improve the demographic situation of the country. Therefore, the integration of new crops into agriculture such as quinoa and cassava in the Egyptian cropping pattern is important.

Both crops represent potential resources for new human food. In addition, both can contribute as food for livestock, both of which have high digestion. Quinoa is considered one of the crops that have proven its capability to grow under marginal lands in Egypt, which are characterized by low fertility, salty soils, shortage water resources and severe weather conditions for use in food production (Zohry 2020). Compound flour is useful in developing countries that it could reduce the importation of wheat flour and encourages the use of locally grown crops as flour (Hasmadi et al. 2014). The application of compound flour in several food products will be economically beneficial, as demand for bread and pastry products can be met using locally grown products instead of wheat (Jisha et al. 2008). Even the quinoa granules do not contain gluten, they can be combined with wheat flour in preparing high-value bread (Morita et al. 2001). A previous search for Crop intensification research, ARC pointed out that 40% of quinoa flour can be mixed with wheat flour to form a compound flour to replace the wheat bottom in the baking industry. Similarly, compound flour can also be prepared by adding 30% cassava flour to wheat flour in the baking industry (Shams 2011).

21.3 Conclusions

The main conclusions are extracted from the presented chapters:

1. In the introduction, we presented the main parts and the main technical approaces in this book. The chapters address the research areas and findings that represent different challenges facing Egyptian agriculture under environmental changes. The book also deals with how to face changes in the environmental stresses, and present agricultural procedures and techniques to mitigate the impact of environmental changes on agricultural crops to serve agricultural sustainability in Egypt. The main themes included in the book are (a) improve crop tolerance for abiotic stresses, (b) recent approaches for biotic stress tolerance, (c) advanced procedures in improving crop productivity and (d) sustainability of environmental resources from a crop production perspective. The extracted conclusions from the chapters under each theme will be presented in the next subsections.

21.3.1 Improve Crop Tolerance for Environmental Abiotic Stresses

2. Egypt suffers from severe water deficit in recent years, facing water shortage amounted about 7 billion cubic meters annually, and this may be enhanced in the near future due to the effect of Ethiopian Renaissance Dam. Drought stress has a significant negative effect on plant growth, yield and its components as well as grain quality, particularly under current climate change. This part focuses on rice,

maize, barley, and sunflower crops. Drought-tolerant genotypes produce considerably greater yield compared with drought-sensitive genotypes, especially under water deficit conditions. Therefore, utilizing drought-tolerant genotypes is essential to enhance productivity in arid and semi-arid regions of the Mediterranean regions. There are various morphological and physiological traits associated with drought tolerance could be exploited in selecting drought tolerant-genotypes; as the deep and prolific root system, leaf rolling, erect leaves, peduncle length, leaf cuticular wax, relative water content, proline accumulation, osmotic adjustment, and antioxidants. Integration of recent advances breeding methods (as quantitative trait loci, Marker-assisted selection and genetic engineering technique with classic plant breeding help significantly in developing drought tolerant genotypes accurately and rapidly.

3. Water stress affects wheat productivity in many regions of the world and caused yield reduction by about 25%. In the light of the increase in Egyptian population and extension of agriculture to marginal lands where wheat plants face the effects of water shortage. Hence, approaches should be advanced to manage with the climate change for alleviating the harmful effects of water stress on wheat production. There are several practical options for adapting to water stress. The first option is exploited genetic diversity and sources of drought tolerance in releasing new cultivars. The second option is focused on biotechnology in improving water stress tolerance by using molecular markers and gene transfer technology. The third option is adjusted agricultural techniques with meteorological data and follows appropriate fertigation programs to avoid the harmful effects of water stress.

4. High temperatures affect crop productivity in many regions of the world, especially arid, semi-arid and hot regions such as the Arab region. The areas of horizontal expansion in the Egyptian deserts are one of these areas. In the light of the extension of agriculture to marginal lands where crop plants face various adverse effects. Water and heat are major environmental factors affecting crop productivity. Therefore, approaches should be developed to cope with climate change for mitigating the negative impacts of heat stress on crop productivity. There are some important strategies for adapting to increased temperatures. The first strategy is growing early maturing and tolerant crop cultivars to heat stress. The second strategy is adjusted crop irrigation dates with meteorological data and follows proper fertilization and irrigation programs to evade the harmful effects of high heat waves. The third strategy is to exploit the technology of DNA markers to diagnose and isolate the genes associated with adaptation to high temperature and to confirm their transfer to sensitive cultivars

5. It is known that the environment suffers from several kinds of environmental pollutants which cause different types of pollution for the air, soil, and water. Despite the great importance of reducing the sources of pollutants and their damage on crop plants. Identify mechanisms responsible for crop tolerance to environmental pollutants is of particular importance. Understanding morpho-physiological and biochemical bases help to improve tolerance of crop plants to

environmental pollutants. The genetic system and nature of gene action controlling the inheritance of environmental pollutants tolerance give plant breeders the
ability to choose the appropriate breeding program to release new cultivars more
tolerant to environmental pollutants. DNA- markers and gene transfer in addition
to bioremediation manners help to reduce the harmful effects of environmental
pollutants on plants that reach the animals and humans through the food chain.

6. The heavy metals; zinc (Zn), lead (Pb), cadmium (Cd) and their mixture had a
    significant drastic effect on seed germination, root and shoot growth. The mixture
    of the three elements has the most diminishing effects followed by cadmium (Cd)
    and then lead (Pb), while zinc (Zn) has the lowest negative effect. Wheat geno-
    types differ in their response to heavy metal stress. Gemmiza-11 and Giza-168
    exhibited the maximum tolerance against the three elements and their mixture,
    while Misr-1 and Sids-13 displayed the lowest tolerance.

7. In light of climate change and its negative impacts on the agricultural sector. It
    is important to develop and release new genotypes and cultivars of wheat and
    determine their adaptability and stability to cope with environmental changes.
    Highly significant G x E "linear" was registered for days to 50% heading, grain
    protein content and grain yield. Phenotypic stability parameters models of Eber-
    hart and Russell (1966) showed that wheat genotypes Sakha 94 and Giza 168
    were highly adapted to stress environments for days to 50% heading; Line 7 and
    Misr 1 for grain protein content and Line 2, Line 4, Line 6, Sakha 94 and Misr 1
    for grain yield. The most desired and stable genotypes were Line 1, Line 4 and
    L8 for earliness, Line 8 for grain protein content and Line 1 and Line 5 for grain
    yield. According to GE biplot and ASV of AMMI (Gauch, 1992), the most stable
    genotypes were Line 2, Line 3, Line 5 and Line 4 for days to 50% heading; Line
    1, Line 4, Line 8 and Line 2 for grain protein content as well as Line 4, Line
    1, Line 2, Line 5 and Misr 1 for grain yield. Tolerance index revealed that the
    most tolerant wheat genotypes to environmental stress were Sakha 94 followed
    by Line 6, Misr 1, Line 2 and Line 4, while the other wheat genotypes exhibited
    various degrees of sensitivity.

8. Salinity is one of the major and increasing abiotic stress. The rate of proline
    accumulation in *Parkia biglobosa* suspension cells was higher at moderate NaCl
    stress. The electrophoretic analysis is considered one of the most important
    biochemical and molecular markers to the variations between the callus and
    NaCl treated cell suspension cultures of *Parkia biglobosa* produced in vitro.
    Also, phenolic constituents may have an antioxidant role during salt stress and
    could play a role in adaptation processes. By using HPLC could separate the
    phenolic constituents that responsible for *Parkia biglobosa* to tolerate salinity
    such as gallic, caffeic, vanillic, ferulic, p-coumaric and salicylic acids, it is used
    as a marker for salt tolerance.

21.3.2 Recent Approaches for Biotic Stresses

9. Potato (*Solanum tuberosum* L.) is an important vegetable cash crop in Egypt. It is considered an important crop for exportation. Bacterial wilt caused by *Ralstonia solanacearum* is an important soil-borne disease that affects potato production with a significant loss of about 15% reached to 50% if weather environments are favorable for the development of the disease. Identification and race determination based on the physiological and bacteriological characteristics and PCR technique play an important role in the diagnosis of pathogens. These techniques proved that the pathogen cause of brown rot disease in potatoes under Egyptian conditions belongs to race 3 biovar 2 of *R. solanacearum*. High degree of genetic variability has been recorded between potato cultivars in their reaction to *R. solanacearum*. Crud of plant extracts from *Corchorus olitorius, Solanum nigrum, Portulica oleracea,* and *Ricinus communis* were found to be more effective against the pathogen.

10. Many factors play an important role in potato brown rot disease and then affect quality and export of the potato. Soil type, soil pH and crop rotation represent a significance role in the persistence of the causal agent of brown rot *R. solanacearum* in potato. The decline rate of the pathogen was greater in clay loam compared to sandy loam. Soil pH value lower than 7.2 or more than 7.7 and potato – rice crop rotation were more effective in reducing the pathogen.

11. Also, late blight is considered the most serious fungal disease in potatoes. The disease spreads almost everywhere grown in potatoes. The disease is of importance worldwide, particularly in traditional potato growing areas due to the disruption of potato production. Losses may be as high as 100 percent if the disease is not controlled, and low levels of infection affect the crop validity for storage. The disease can relatively easily be controlled by the use of fungicides. In many developing areas, however, chemical control is hardly feasible for the subsistence farmer due to the high cost of fungicide applications. In addition, in nearly all developing countries, the fungicides or their active ingredients are imported, making the potato an expensive vegetable and at the same time costing the country valuable foreign exchange. The present finding revealed management strategies for its effective control to late blight include the use of alternatives of fungicides; resistance cultivars, biological control, plant oils, plant extracts, and nanotechnology.

12. Wheat productivity and disease factors are influenced by environmental elements, among which temperature and moisture are the utmost vital in determining disease severity and yield loss. Yield loss affected by rusts is determined by numerous aspects, including the degree of sensitivity, infection time, rate of disease development and duration of disease, causing severe infection and consequently high losses. The effective resistant genes against rust may be deployed singly or in combination with high yielding genes to develop high-yielding resistant wheat cultivars in wheat-growing areas. Genetic resistance is one of the most economic and effective means of reducing yield losses

caused by rust diseases. Breeding strategy and growing resistant cultivars to rusts are considered one of the most sustainable, cost-effective and environmentally friendly approaches for controlling rust diseases. To date, more than 187 rust resistance genes (80 leaf rust, 58 stem rust, and 49 stripe rust) have been derived from diverse bread or durum wheat cultivars and the related wild species using different molecular methods. Classical breeding approaches like the introduction of exotic germplasm, hybridization, composite crossing, multiline, and backcross breeding were exploited for this purpose. Speed breeding and recent advances like molecular marker and gene transfer have created operative tools for resolving many difficulties and improving the resistance. The National Wheat Improvement Program in Egypt over the last 25 years has succeeded in developing a collection of bread wheat cultivars that resistant to rusts such as Giza 168, Giza 171, Gemmeiza 7, Gemmeiza 9, Gemmeiza10, Gemmeiza 11, Gemmeiza12, Sids 12, Sids 13, Sids 14, Sakha 94, Sakha 95, Misr 1, Misr 2, Shandweel 1, and durum wheat cultivars *i.e.* Sohag 3, Sohag 4, Sohag 5, Beni Suef 1, Beni Suef 4, Beni Suef 5 and Beni Suef 6.

13. Faba bean (*Vicia faba* L) is a major food crop in Egypt. The crop grown in the field came under the bombardment of wild arrays of the pathogen(s) causing various diseases to the plants. Faba bean plants attacks by fungal diseases (chocolate spot, rust, root rot, and wilt) and viral diseases (Faba Bean Necrotic Yellow Virus, Bean Yellow Mosaic Virus and Bean Leaf Roll Virus). These pathogens considered as a main constrains affected growth of the plant and contributed significantly to causing great yield loss both in quantity and quality. Yield loss caused by diseases estimated by about (25% - 80%) and reduced protein content by about 4%. It is worth mention that combined infections with more than one pathogens on faba bean plant caused substantial damage and significant yield loss than infection by a single pathogen by 85%. Such effect depends mainly on the aggressiveness of the pathogens, the susceptibility of the host associated with favorable environmental conditions. However, some control measures for minimizing these diseases are available. These involve breeding for disease resistance, fungicides treatment, induce disease resistance, biological control, plant extracts, growth regulators, and agricultural practices.

21.3.3 Advanced Procedures in Improving Crop Productivity

14. Application of laser irradiation is considered as a relatively new technique in agriculture. The effect of helium-neon laser on the seed germination and acceleration of the growth rate of crop plants has been observed. Also, the laser effect was noticed to improve early maturity and increase leaf area, plant height, yield, yield components and quality characters of crop plants. The advantages of the laser in agriculture include many aspects which include encouraging plants to tolerate environmental effects and safely mitigate the effects of environmental changes, such as drought, high temperature, salinity, and pollution. Also, it

biostimulates plant resistance to biotic stresses of disease injuries. Moreover, the application of laser technology is considered a sustainable, secure and clean means to improve growth, yield, and quality of crop plants. In wheat, the laser also has a positive physiological effect on morph physiological, biochemical, enzymatic activity, yield, and grain quality, also enhances its tolerance to stress environmental conditions.

15. The availability of high-quality seed of improved crop varieties, along with modern power equipment, improved fertilizers and better method of weed and pests control, has revolutionized farming resulted in maximizing the productivity and quality of field crops. The availability of new varieties demanded a strong seed industry, with the ability for rapid seed increase and distribution with adequate safeguards for varietal purity. Seed vigor and longevity would be able to establishment good seedlings and stand per field under a wide range of environmental conditions. The improved varieties and local ones might be deteriorated when grown for generation after generation in general cultivation due to several factors during production cycles. Therefore, the breeding programmer must be continuing in order to produce new improved varieties to be replaced by the deteriorated varieties. It is of great important to continue to produce high-quality seeds with safety storage; protected from mixture and deterioration; as well as better agronomic practices and protection from pests, with much care at harvesting, handling, and processing to reaching the maximum productivity and quality from growing field crops.

16. Using biochemical and molecular tools is considered a recent trend in the identification of salt tolerant genotypes. High significance increase in proline accumulation and Na^+ concentration in Egyptian and Nigerian peanut genotypes treated with salinity compared with control. KAT1 gene was detected by specific primer and amplified using PCR at 554pb. BLAST data showed identical 100% between the sequence of the Egyptian peanut genotype (Giza 6) and the sequence of KAT1 gene of *Arapidopsis thaliana*, and the similarity was 99% between Ismailia1 and Samnut 22. Giza 6 and Ismailia1 share in one SNP (A) in codon 195, while Ismailia1 and Samnut 22 share in four SNPs (T) in codon 258, 259, 263 and 406. The phylogenetic tree shows that Giza 6 is more close to Ismailia 1 than Samnut 22, while Ismailia 1 and Samnut 22 very close, although Ismailia 1 and Giza 6 are Egyptian peanut genotypes and Samnut 22 is Nigerian one. Evolutionary tree emphasized the data of screening of all investigated genotypes for their salt tolerant, which reported that Ismailia 1 and Samnut 22 were tolerance for salinity while Giza 6 was a sensitive one. Ismailia1 (for Egyptian genotypes) and Samnut 22 (for Nigerian genotypes) are more tolerance for salinity than other varieties.

17. Arbuscular mycorrhizal fungi are considered the most vital type among different types of mycorrhizae fungi that can enhance and improve plant productivity and soil fertility. They uptake the nitrogen in the form of nitrate and ammonium as well as they can uptake the nitrogen from different organic sources and change it into mineral nitrogen for plant uptake. Arbuscular mycorrhizal fungi enhance plant uptake and use efficiency of phosphorus and some micronutrients, and

consequently can result in an improvement in growth and yield of plants. Moreover, they can improve protein and lipid contents of plants through prompting phosphorus nutrition or prompting other metabolic reactions in the host plant species. Organic sources of different nutrients and slow-release chemical fertilizers can enhance arbuscular mycorrhizal fungi activity, while synthetic fertilizers can suppress mycorrhizal colonization activity with the roots of host plants. A reduction of more than 50% of the recommended synthetic phosphorus fertilizer can be replaced through the inoculation of plant seeds with arbuscular mycorrhizal fungi, and this is an important outcome for the economic and environmental impacts. Also, Arbuscular mycorrhizal fungi can increase the host plant resistance to drought stress through enhancing the water relations. Also, they are important bio-ameliorators in saline soils. The mechanism of arbuscular mycorrhizal fungi on the improvement of plant growth in saline soils is associated with the physiological processes such as increasing exchange rate of carbon dioxide, stomatal conductance, transpiration rate, and water use efficiency instead of the nutrient uptake particularly nitrogen or phosphorus. Therefore, inoculation of plant seeds with arbuscular mycorrhizal fungi and reducing synthetic fertilizer is highly recommended for suitability of agriculture.

21.3.4 Sustainability of Environmental Resources from a Crop Production Perspective

18. Principles of sustainability in agriculture include the integrated biological and ecological processes, for instance, N fixation, different nutrient cycling, crop rotation, high-productivity cultivars as well as reducing the use of pollutants inputs that can pose harm effect to the environment, animals, and humans. Agriculture system is one of the most critical human activities that can increase the chemical pollutants through the excessive use of different synthetic chemical fertilizers and pesticides. Excessive use of chemical fertilizer can reduce the biological nitrogen fixation in the soil and increase the chemical pollutants of nitrous oxide. Thus, organic farming including organic and bio-fertilizers, crop rotations, high yield cultivars, and nano-fertilizer are highly recommended for sustainability in agriculture.

19. With climate change, resource scarcity like land, water, energy and nutrients and environmental degradation, *i.e.* a decline in soil quality, an increase in greenhouse gas emissions, and surface water eutrophication crop production will become more difficult. So the near-term strategy is the application and extension of existing agricultural technologies. Further research efforts are needed to focus on yield improvement and further shrinking of environments via improved hybrid and crop management technologies. The need to improve productivity in the face of the increasing demand for maize, whether as food or feed, through the development and dissemination of climate-resistant and adaptive hybrids of tropical/ subtropical maize farming areas. Increased stress tolerance, combined

with increased stand uniformity under stress conditions, will probably continue to provide the highest potential for yield improvement in maize in the next decade. A large proportion of the genetic improvement of maize in the past is attributable to increase stress tolerance, pest resistance. Molecular breeding offers the tools to accelerate cereal breeding. However, suitable phenotyping protocols are essential to ensure that the much-anticipated benefits of molecular breeding can be realized. The real challenge under climate change is to improve agricultural management practices and adopt improved maize varieties that are tolerant to climate change. So the phenotype through infrared (heat sensitive) can be a beneficial tool in selecting tolerant genotypes.

20. The productivity of the traditional crops in the cropping pattern in Egypt is reduced when cultivated in marginal soil as a result of low fertility and harsh weather conditions. Thus, both quinoa and cassava are good candidates for that. With respect to quinoa, it is a winter crop that can be cultivated in both sandy and salt-affected soil. The suitable date for cultivation quinoa is between 15th and 21st of November. It can grow with no-tillage in the sandy lands of Sinai. The suitable inter-row spacing for quinoa is 15 cm. Quinoa can grow under rain-fed conditions and gave its highest yield with application of supplementary irrigation. Water consumptive use for quinoa is between 215–226 mm. It can also be cultivated successfully under irrigation, where it can produce its highest yield when irrigated every 40 days in clay soil. In sandy soil, Nitrogen fertilizer requirement for quinoa is 360 kg N/ha depending on soil type. Furthermore, application of 240 kg N/ha to quinoa, in addition to biofertilizer (Nitrobin) is sufficient to fulfill its requirement. Phosphorus fertilizer requirement is 120 kg/P_2O_5, in addition to the application to biofertilizer (phosphorin). Additionally, some quinoa cultivars can tolerate salinity levels up to 17.9 dS/m. Thus, quinoa can be cultivated in marginal lands in Egypt, namely sandy soil and salt-affected soils.

Regarding cassava, the suitable planting date that produced the highest yield is mid of April. The results indicated that highest yield of cassava tubers was associated with lowest cassava spacing (1.00 m × 0.75 m, plant spacing X row width) and with highest cowpea plant density (100, 800 plants/ha). The crop can be successfully cultivated in sandy soil. Irrigation application when 25% of field capacity was depleted attained the highest yield. Nitrogen fertilizer requirement for cassava is 180 kg N/ha. Phosphorus and potassium fertilizer requirements are between 111.6–277.2 kg/P_2O_5 and 172.8–345.6 K_2O, respectively depending on soil type. The optimum date for the harvest that gave the highest yield is 10 month after planting. Cowpea and groundnut can be intercropped with cassava with 1:2 intercropping pattern, which resulted in an increase in the yield and its components.

21.4 Recommendations

The following recommendations are mainly extracted from the chapters:

1. In general, in the light of the research results included in the book, that it is recommended to cultivate the most tolerate genotypes to abiotic and biotic stresses, application of the appropriate agricultural procedures and utilize of environmental-friendly technologies are essential. Also, crop rotation is highly recommended in Egyptian agriculture because it results in a reduction in the application of herbicide, fertilizers, the growth of weeds and water pollution as well as improves soil fertility, micro-organisms activity and productivity of field crops. And encourage the cultivation of quinoa and cassava, which successfully cultivated in both sandy and salt-affected soils. Application of DNA-Markers, gene transfer and laser technologies and use of high-quality seeds are important in improving plant productivity. It is also necessary to work more books in these areas of research to provide more ideas to deal with environmental changes in light of the rapid scientific developments in the world.

21.4.1 Improve Crop Tolerance for Abiotic Stresses

2. To tolerate drought stress, combining morpho-physiological and biochemical traits can provide a complete model of gene-to-phenotype relationships and genotype-by-environment interactions to tolerate environmental stress. Integration of recent improvement procedures as DNA-marker assisted selection and gene manipulation with classic plant breeding aids in improving drought tolerant genotypes accurately and rapidly.
3. The effects of water stress can be reduced through exploited genetic diversity and sources of drought tolerance in releasing new cultivars, cultivate the most tolerate genotypes to water stress, application the appropriate agricultural procedures. Also taking into account the critical periods of wheat life and avoid stress through irrigation. Moreover, applying for the appropriate fertilization programs and modern irrigation systems are of great importance.
4. The effects of heat stress can be addressed through the following procedures, i.e. cultivate the most tolerate genotypes to heat stress at the appropriate time, application the appropriate agricultural procedures and follow appropriate fertilization and irrigation programs. Besides, consider the most sensitive periods in the life of the plant and evade the cultivated varieties the negative effects of high temperature on the yield and its components through appropriate agricultural procedures, including irrigation. Furthermore, link crop irrigation dates with meteorological data to avoid the harmful effects of high heat waves on the crop.
5. To reduce the impact of environmental contaminants should develop utilize of environmental-friendly technologies for cleaning up polluted soils. Also, the

development of crop cultivars tolerant to the environmental pollutants. Moreover, it is of interest to exploit of genetically engineered plants that have enzymatic systems to detoxification of pollutants. Furthermore, the application of environmental safety measures in industrial facilities. Besides spreading environmental awareness among community groups and develops regulations to reduce greenhouse gas emissions

6. The tolerant genotypes, as Gemmiza-11 and Giza-168, which exhibited maximum tolerance against heavy metals could be recommended to be cultivated commercially under heavy metals stress. While it should avoid cultivation of the sensitive cultivars under heavy metals stress as Misr-1 and Sids-13. The tolerant genotypes to heavy metal stress could be exploited as donors in a wheat breeding program for developing promise cultivars destined for agricultural production under heavy metals stress.

7. In the light of climate change and its negative impacts on the agricultural sector, it is important to develop and release new genotypes and cultivars of wheat and determine their adaptability and stability to cope with environmental changes. Also, understanding the relationship between climate and wheat crop yield is fundamental to identify possible impacts of future climate and to develop adaptation measures. Increasing production per unit area and grain protein content with a good level of earliness is very important to avoid terminal drought, and heat stresses and achieves global food security. Therefore, the selection of various wheat genotypes under stress environments is one of the main tasks of plant breeders for utilizing the genetic variation to produce stress tolerant wheat cultivars

8. It can introduce African locust bean plant (*Parkia biglobosa*) as a promising crop in Egyptian agriculture and expanding its cultivation and recognition of Egyptian farmers by this crop for its important uses in many industries especially pharmaceuticals, food, textile, and other industries. Also, it exploits as a windbreak and to shade field crops from high temperature, and also its ability to salinity. From this plant, it can get many secondary metabolites used in medicinal applications and make use of increasing its production by using tissue culture techniques as callus and cell suspension cultures. By using genetic engineering can reveal the gene responsible for the ability of this legume plant to tolerate the salinity stress conditions and hence transforming it to another legume plants to resist such conditions.

21.4.2 Recent Approaches to Biotic Stress Tolerance

9. It is important to reduce the incidence of brown rot under Egyptian conditions, taking into account the following recommendations:

   1. Detect the causal organism of potato brown rot bacterium in different hosts,
   2. Study the cultivar reaction of potato brown rot,
   3. Determine the effect of plant extracts on *R. solanacearum*,

    4. Cultivate high-resistance varieties of the disease to reduce the causative damage and

    5. Applied integrated pest management and applied the rules of plant quarantine.

10. To diminish the spread of brown rot disease caused by *R. solanacearum* in potatoes, the subsequent recommendations can be followed *i.e.* follow the integrated pest management, potato cultivation in pest-free area of pathogen, cultivation in clay loam soil with pH value ranged from 7.2 to 7.7, and application of the suitable crop rotation such as potato – rice.

11. In some countries such as Egypt, the use of fungicides excessively leading to the impact on human health and harm to the environment. This uses lead to the emergence of new strains of pathogens of the plant. All this makes us underestimate and guide the fungicides and use them only when necessary and then follow other ways to resist or combat pathogens. Uses of resistant cultivars, biological control, plant oils, plant extracts, and nanotechnology are alternatives to synthetic chemical pesticides in controlling the late blight disease. However, pesticides will continue to be required in many production systems if quality and competitiveness are to be maintained by producers. Therefore, within the foreseeable future resistance cultivars, biological control, plant oils, and extracts and nanotechnology will be utilized in management system as alternatives and supplement to pesticides.

12. Applied integrated pest management and applied the rules of plant quarantine.

13. The safe, economical and stable mean of controlling faba bean pathogens is the breeding for disease resistance. The breeding program showed an involvement of single and multiple disease resistance (since the plant can infect by more than one pathogen). This achieved by corroboration between plant breeders and pathologists of faba bean plants. Other control measures, i.e., fungicides, biological control, induce disease resistance, growth regulators, plant extracts and agricultural practices also to be considered.

21.4.3 Advanced Procedures in Improving Crop Productivity

14. The safety of helium-neon laser in improving productivity is recommended as new technology in agriculture. Application of laser technology is considered a sustainable, secure and clean means to improve growth, yield, and quality of crop plants. Application of helium-neon laser has shown to improve morph physiological, biochemical, enzymatic activity, yield and grain quality in wheat. It is recommended to use the helium-neon laser in the field of agriculture, and investigate the criteria to produce desirable effects of biostimulation conditions besides setting up specialized research centers more in Egypt.

15. The use of high-quality seeds increases the productivity and quality of field crops. Thus, the development of new varieties characterized by seed vigor and viability with high genetic purity will contribute to increasing production. The

storage of high-quality seed under controlled suitable conditions should be applied in order to keep such seeds in the best case until next planting. Also, application of seed pre-sowing treatments will improve field germination and obtained good healthy stand which led to faster plant growth, flowering and maturity and an increase in productivity. It is important to treat seeds whichever before or after harvesting to obtain clean, up-graduated, safe moisture content and treated seed before storage.

16. Ismailia 1 and Samnut 22 genotypes could be involved in peanut breeding programs. Both peanut genotypes could also be cultivated in salt-affected soils and regions that depend on irrigation with salty water up to 3000 ppm in Egypt and Nigeria.

17 Arbuscular mycorrhizal fungi application should be encouraged in Egypt since they can enhance and improve plant productivity, quality, and soil fertility, plant nutrients uptake. Arbuscular mycorrhizal fungi application should be used with less synthetic fertilizers because synthetic fertilizers can cause an adverse effect on the roots colonization with mycorrhiza fungi. Also, application of arbuscular mycorrhizal fungi enhances plant tolerance against abiotic stress such as drought and salt stress, so it is highly recommended in arid and semi-arid regions.

21.4.4 Sustainability of Environmental Resources from a Crop Production Perspective

18 Excessive use of chemical fertilizer should not be used, because it reduces the biological nitrogen fixation in the soil and increases the chemical pollutants of nitrous oxide. Crop rotation is highly recommended in Egyptian agriculture because it results in a reduction in the application of herbicide, fertilizers, the growth of weeds and water pollution as well as improves soil fertility, micro-organisms activity and productivity of field crops. Bio-fertilizers are highly recommended as single or as a complementary with synthetic fertilizer because they are energy-efficient, free-pollution, low-cost plant nutrient source and eco-friendly. Biochar and compost are also highly recommended particularly in poor soils with 50% of the recommended NPK because their application improves the soil properties and fertility as well as enhance the yield and quality of field crops. It is recommended to use Nano-fertilizers because of their new mechanisms of actions, lessen nutrient loss, enhance use efficiency, and reduce worsening of the environment. Also, nano-fertilizers have the capability of entering the plant cells straight due to their small size, and this decreases the required energy of their uptake and transportation into the plant cells.

19. The yield of a corn plant depends on the genetic potential of a given hybrid, in addition to a number of environmental conditions and in-season management" (http://www.uaex.edu/publications/pdf/mp437/mp437.pdf). So it is important to address the factors that limit the maize production which include drought during critical early stages of crop growth, low soil nutrient level (particularly

nitrogen and phosphorus), and pest and diseases infestations. Other limitations to maize production include poor management practices such as low plant populations, inappropriate planting time, inadequate control of weeds, limited use of inputs (especially fertilizer and improved seeds) as well as the untimely application of adequate quantities of fertilizers. Successful production of maize to a large extent depends on the timely adherence to all the recommended steps. Growing in fertile soil with appropriate water irrigation requirements and add of the most important nutrients for the maize, i.e. nitrogen, phosphorus as well as potassium in some soils, zinc in small quantities. Also, high plant densities are appropriate for early-planted crops under high rainfall or irrigated conditions where management is of a good standard. There is a need for further research and breeding for short but high yielding varieties of maize to accommodate high densities, facing climate change impacts using new tools and approaches of plant breeding.

20 The government should encourage the cultivation of quinoa, which can be successfully cultivated in both sandy and salt-affected soils. Whereas, cassava can be cultivated in sandy soil only and other crops can be intercropped with it. In these marginal soils, livestock production can be implemented as a result of using both crops in feed production. The government should build factories to process quinoa to produce flour. The government should build factories to process cassava and produce flour and starch. Starch can be transformed into ethanol and use as fuel to reduce air pollution.

Acknowledgements Hassan Awaad, Mohamed Abu-hashim and Abdelazim Negm acknowledge the partial support of the Science and Technology Development Fund (STDF) of Egypt in the framework of the grant no. 30771 for the project titled "A Novel Standalone Solar-Driven Agriculture Greenhouse - Desalination System: That Grows Its Energy And Irrigation Water" via the Newton-Musharafa funding scheme.

References

Abd El-Shafi MA, Gheilth EMS, Abd El-Mohsen AA, Suleiman HS (2014) Stability analysis and correlations among different stability parameters for grain yield in bread what. Sci Agric 2(3):135–140

Abdel-Motagally FMF, El-Zohri Manal (2018) Improvement of wheat yield grown under drought stress by boron foliar application at different growth stages. J Saudi Soc Agric Sci 17(2):178–185

Abu El-Naga SA, Khalifa MM, Sherif S, Youssef WA, El-Daoudi YH, Shafik I (2001) Virulence of wheat stripe rust pathotypes identified in Egypt during 1999/2000 and sources of resistance. First Regional Yellow Rust Conference for Central & West Asia and North Africa, 8–14 May, SPH, Karj, Iran

Ali MB, Ibrahim AMH, Hays DB, Ristic Z, Jianming F (2010) Wild tetraploid wheat (*Triticum turgidum* L) response to heat stress. J Crop Imp 24:228–243

Ali MMA (2017) Stability analysis of bread wheat genotypes under different nitrogen fertilizer levels. J Plant Production, Mansoura Univ 8(2):261–275

Ali MMA, Abdul-Hamid MIE (2017) Yield stability of wheat under some drought and sowing dates environments in different irrigation systems. Zagazig J Agri Res 44(3):865–886

Ali Z, Mujeeb-Kazi A, Quraishi UM, Malik RN (2018) Deciphering adverse effects of heavy metals on diverse wheat germplasm on irrigation with urban wastewater of mixed municipal-industrial origin. Environ Sci Pollut Res 1–14

Aly AA, Awaad HA (2002) Partitoning of genotype x environment interaction and stability for grain yield and protein content in bread wheat. Zagazig J Agric Res 29(3):999–1015

Anonymous (2015) Brown rot disease mars export prospects of Indian potatoes. https://www.the hindubusinessline.com/economy/agri-business/brown-rot-disease-mars-export-prospects-of-ind ian-potatoes/article7185569.ece

Anonymous (2017) Field Crops. Statistics Year Book. Ministry of Agriculture and Land Reclamation, Egypt

Appel JA, DeWolf E, Bockus WW; Odd TT (2009) Preliminary 2009 Kansas wheat disease loss estimates. Kansas Cooperative Plant Disease Survey Report

Awaad HA (2009) Genetics and breeding crops for environmental stress tolerance, I: drought, heat stress and environmental pollutants. Egyptian Public library

Awaad HA, Morsy AM, Moustafa ESA (2013) Genetic system controlling cadmium stress tolerance and some related characters in bread wheat. Zagazig J of Agric Res 40(4):647–660

Azaizeh HA, Marschner H, Römheld V, Wittenmayer L (1995) Effects of a vesicular-arbuscular mycorrhizal fungus and other soil microorganisms on growth, mineral nutrient acquisition and root exudation of soil grown maize plants. Mycorrhiza 5:321–327

Bajwa MA, Aqil KA, Khan NI (1986) Effect of leaf rust on yield and kernel weight of spring wheat. RACHIS 5:25–28

Baraka MA, Omar SA, Mazen MM, Soltan HH (2008) Enzymatic activities in faba bean plants induced by some biotic and abiotic agents. Egypt J Basic Appl Sci 23(2):59–68

Briton-Jones HR (1925) Mycological work in Egypt during the period 1920–1922. Egypt Min Agric Tech And Sci Serv Bul 49:129

Champoiseau PG, Jones JP, Allen C (2009) *Ralstonia solanacearum race 3 biovar 2 causes tropical losses and temperate anxieties [Online]*. Madison: American Phytopathological Society. Available at http://www.apsnet.org/online/feature/ralstonia/. Accessed 25 June 2010

Chen XM (2005) Epidemiology and control of stripe rust (*Puccinia striiformis* f. sp. *tritici*) on wheat. Can J Plant Pathol 27(3):314–337

Cocozza C, Pulvento C, Lavini A, Riccardi M, d'Andria R, Tognetti R (2013) Effects of increasing salinity stress and decreasing water availability on eco-physiological traits of quinoa grown in a Mediterranean-type agro-ecosystem. J Agron Crop Sci 199:229–240

Copeland LO, McDonald MB (1995) Principles of seed science and technology: Burgess Publishing co. Minneapolis, Minneapolis, MI, USA

Davenport R, James R, Zakrisson-Plogander A, Tester M, Munns R (2005) Control of sodium transport in durum wheat. Plant Physiol 137:807–818

Delgado IC, Sanchez-Raya AJ (2007) Effects of sodium chloride and mineral nutrients on initial stages of development of sunflower life. Commun Soil Sci Plant Anal 38:2013–2027

DeLonge MS, Miles A, Carlisle L (2016) Investing in the transi-tion to sustainable agriculture. Environ Sci Pollut 1:266–273

Draz IS, Abou-Elseoud MS, Kamara AM, Alaa-Eldein OA, El-Bebany AF (2015) Screening of wheat genotypes for leaf rust resistance along with grain yield. Ann Agric Sci 60:29–39

Eberhart SA, Russell WA (1966) Stability parameters for comparing varieties. Crop Science 6:36–40

Eisa NA, EL-Habbaa GM, Omar SA, EL-Sayed SA (2006) Efficacy of antagonists, natural plant extract and fungicides in controlling wilt,root rot and chocolate spot pathogens of faba bean *in vitro*. Egyptian J of Applied science 44(4):1547–1570

El-Halag KM (2008) Studies on the interaction between potato brown rot bacterium and root exudates in certain crops. MSc Thesis. Department of Botany, Faculty of Science, Benha University, pp 96–98

El-Halag KM, Hassan ME, Messiha Nevein AS, Elhadad SA, Abdallah SA (2015) The relation of different crop roots exudates to the survival and suppressive effect of *Stenotrophomonas maltophilia* (PD4560), Biocontrol Agent of Bacterial Wilt of Potato. J Phytopathlogy 163(10):829–840

EL-Badawy NFA (2008) Biochemical and molecular studies on some genes that induce after infection of faba bean plants by chocolate spot disease. Ph D Thesis, Faculty of Agriculture, Cairo University, Egypt

El-Daoudi YH, Shenoda Ikhals S, Bassiouni AA, Sherif SE, Khalifa MM, (1987). Genes conditioning resistance to wheat leaf and stem rust in Egypt. In: Proceedings 5th Egypt Phytopathol Soc, Giza, pp 387–404

El-Didamony G, Ismail AEA, Sarhan M, Abdel-Azez SS (2003) Idenification and pathogenicity of bacterial wilt of potatoes in some types of Egyptian soil. Egypt J Micropiology 38(1):89–103

EL-Gharbawy SS (2015) Wheat breeding for tolerance to heavy metals pollution. Agron Department, Faculty of Agriculture, Zagazig University, Egypt

El-kahoui S, Hernandez JA, Abdelly C, Ghrir R, Limam F (2005) Effects of salt on lipid peroxidation and antioxidant enzyme activities of *Catharanthus roseus* suspension cells. Plant Sci 168:607–613

EL-Wakil MA, Abass MA, EL-Metwally MA, Mahmoud MS (2016) Green chemistry for inducing resistance against chocolate spot disease of faba bean. J Environ Sci Technol 9(1):170–187

Eraky Hania AME (2015) Influence of specific seed treatments on field emergence and seedling growth of maize (Zea mays, L.) varieties. Ph D Thesis, Faculty of Agriculture, Zagazig University, Egypt

Fahmy FG, Mohamed MS (1990) Some factors affecting the incidence of potato brown rot. Assuit J Agric Sci 2:221–230

FAO (2014) Agriculture, Forestry and Other Land Use Emissions by Sources and Removals by Sinks (89 pp, 3.5 M, About PDF) Exit Climate, Energy and Tenure Division, FAO

FAO (2015) Climate Change and Food Security, UN Food & Agricultural Organization FAO. http://www.climatechange-foodsecurity.org/fao.html

FAO (2018) Soil pollution comes under scrutiny: Global Soil Partnership annual meeting focuses on "black soils" and data-sharing initiatives

FAO statistics (2011) http://www.fao.org/faostat/en/#data/QC

FAOSTAT (2017) Food and Agriculture Organization of the United Nations (FAO), FAO Statistical Database. http://faostat3.fao.org/dowload/Q/QC/E

FAOSTAT (2018) Statistical Database of The United Nation Food and Agriculture Organization (FAO), Rome. http:www.Fao. org/faostat / en/ # data/QC

Farag NS (2000) Spotlights on potato brown rot in Egypt. In: Proceedings 9th Congress of the Egypt. Phytopathology Soc, May, 405–408

Fellbaum CR, Gachomo EW, Beesetty Y, Choudhari S, Strahan GD, Pfeffer PE, Kiers ET, Bucking H (2012) Carbon availability triggers fungal nitrogen uptake and transport in arbuscular mycorrhizal symbiosis. Saudi J Biol Sci 109:2666–2671

Francis CA, Daniel H (2004) Organic farming. Encyclopedia of soils in the environment. Elsevier, Oxford, UK, pp 77–84

French ER (1994) Strategies for Integrated Control of Bacterial Wilt of Potatoes. In: Hayward AC, Hartman GL (eds) Bacterial wilt: The Disease and its Causative Agent, *Pseudomonas solanacearum*. CAB International, Wallingford, UK, pp 98–113

Fry W (2008) *Phytophthora infestans*: the plant (and R gene) destroyer. Mol Plant Pathol 9:385–402

Gallandt E (2014) Weed Management in Organic Farming. In: Chauhan B, Mahajan G (eds) Recent advances in weed management. Springer, New York, NY

Garg N, Geetanjali Kaur A (2006) Arbuscular mycorrhiza: nutritional aspects. Arch Agron Soil Sci 52:593–606

Giannakoula A, Ilias IF, Maksimović JJD, Maksimović VM, Živanović BD (2012) Does overhead irrigation with salt affect growth, yield, and phenolic content of lentil plants. Arch Biol Sci 64(2):539–547

Gisi U, Cohen Y (1996) Resistance to phenylamide fungicides: a case study with *Phytophthora infestans* involving mating type and race structure. Annu Rev Phytopathol 34:549–572

Gosling P, Hodge A, Goodlass G, Bending GD (2006) Arbuscular mycorrhizal fungi and organic farming. Agric, Ecosyst & Environ 113:17–35

Gottlieb OR, Borin MR, Brito NR (2002) Integration of ethnobotany and phytochemistry: dream or reality? Phytochemistry 60:145–152

Hamad YI (2008) Studies on the transmission of potato brown rot causal organism through weeds in the Egyptian fields M.Sc. Thesis, Plant Pathology Department, Faculty Agriculture, Zagazig University, Egypt

Hamad YI (2016) Pathological studies on Potato Brown Rot under Egyptian Conditions. PhD Plant Pathology, Agric Botany Department, Faculty Agriculture, Suez Canal University, Egypt

Hamad YI, Tohamy MRA, El-Morsy GA (2008) Detection of *Ralstonia solanacearum* on some crops and weeds under Egyptian conditions. Zagazig J Agric Res 35(4):769–788

Harley JL, Harley EL (1987) A check-list of mycorrhiza in the British Flora. New Phytol 105:1–102

Hartman GL, Elphinstone JG (1994) Advances in the control of Pseudomonas solanacearum Race 1 in major food crops. In: Hayward AC, Hatman GL (eds) Bacterial Wilt: the disease and its causative agent, *Pseudomonas solanacearum*. CAB International, Wallingford, UK, pp 157–177

Hasmadi M, Faridah S, Salwa A, Matanjun IP, Abdul Hamid M, Rameli AS (2014) The effect of seaweed composite flour on the textural properties of dough and bread. J Appl Phycol 26:1057–1062

Hayat Sh, Hayat Q, Alyemeni MN, Wani AS, Pichtel J, Ahmad A (2012) Role of proline under changing environments. Plant Signal Behav 7(11):1456–1466

Hayward AC (1991) Biology and epidemiology of bacterial wilt caused by *Pseudomonas solanacearum*. Annu Rev Phytopathol 29:65–87

Heba YA, Enaam AM, Monia BH, Karima AM (2015) An Economic Analysis for Maize Market in Egypt. Middle East J Agric Res 4(4):873–878

Helgason T, Fitter A (2009) Natural selection and the evolutionary ecology of the arbuscular mycorrhizal fungi (*Phylum Glomeromycota*). J Exp Bot 60:2465–2480

Hickey H (2017) Earth likely to warm more than 2 degrees this century. http://www.washington.edu/news/2017/07/31/earth-likely-to-warm-more-than-2-degrees-this-century/ Search. UW News, NIH grants: HD054511, HD070936

Hossain MA, Ashrafuzzaman M, Ismail MR (2011) Salinity triggers proline synthesis in peanut leaves. Maejo Int J Sci Technol 5(1):159–168

Irmak S (2016) Impacts of extreme heat stress and increased soil temperature on plant growth and development. Published by IANR Media CropWatch Privacy Policy. https://cropwatch.unl.edu/2016/impacts-extreme-heat-stress-and-increased-soil-temperature-plant-growth-and-development

Jisha S, Padmaja G, Moorthy SN, Rajesh K (2008) Pre-treatment effect on the nutritional and functional properties of selected cassava-based composite flours. Innov Food Sci Emerg Technol 9:587–592

Joanna JC, Mba C, Bradley JT (2016) Mutagenesis for crop breeding and functional genomics. Biotechnologies for Plant Mutation Breeding, pp 3–18

Karim Z, Hossain MS, Begum MM (2018) *Ralstonia solanacearum*: A threat to potato production in Bangladesh. Fundam Appl Agric 3(1):407–421

Khalil SA, EL-Hady MM, Dissouky RF, Amer MI, omar SA (1993) Breeding for high yielding ability with improved level of resistance to chocolate spot (*Botrytis fabae*) disease in faba bean (*Vicia faba* L). The Journal of Agricultural Science, Mansoura University 18(5):1315–1328

Kolmer J, Chen XM, Jin Y (2009) Diseases which challenge global wheat production. The wheat rusts. pp 89–124. In: Carver BF (ed) Wheat: science and trade. Wiley-Blackwell, Wiley, Ames, Iowa p 569

Lequeux H, Hermans C, Lutts S, Verbruggen N (2010) Response to copper excess in Arabidopsis thaliana: Impact on the root system architecture, hormone distribution, lignin accumulation and mineral profile. Plant Physiol Biochem 48(8):673–682

Li XL, Marschner H, George E (1991) Acquisition of phosphorus and copper by VA-mycorrhizal hyphae and root-to-shoot transport in white clover. Plant Soil 136:49–57

Liu JQ, Kolmer JA (1997) Genetics of leaf rust resistance in Canadian spring wheats AC Domain and AC Taber. Plant Dis 81:757–760

Liu ZL, Li YJ, Hou HY, Zhu XC, Rai V, He XY, Tian CJ (2013) Differences in the arbuscular mycorrhizal fungi-improved rice resistance to low temperature at two N levels: aspects of N and C metabolism on the plant side. Plant Physiol Biochem 71:87–95

Loughman R, Jayasena K, Majewski J (2005) Yield loss and fungicide control of stem rust of wheat. Aust J Agric Res 56:91–96

Magdoff F, van Es H (2000) Building soils for better crops. Sustainable Agriculture Network, National Agriculture Library, Beltsville, MD

Maggio A, Reddy MP, Joly RJ (2000) Leaf gas exchange and solute accumulation in the halophyte *Salvadora persica* grown at moderate salinity. Environ Exp Bot 44:31–38

Maiman T (1960) Stimulated optical radiation in ruby. Nature 187:493–494

Malafaia CB, Ana Cláudia SJ, Alexandre GS, Elineide B de Souza, Alexandre JM, Maria T dos Santos C, Márcia VS (2018) Effects of Caatinga Plant Extracts in Planktonic Growth and Biofilm Formation in *Ralstonia solanacearum*. Microb Ecol 75(3):555–561

Mansour HA, Abd El-Hady M (2014) Performance of irrigation systems under water salinity in wheat production. J Agric VetY Sci 7(7):19–24

Mansour HA, Gaballah MS, Abd El-Hady N, Eldardiry Ebtisam I (2014) Influence of different localized irrigation systems and treated agricultural wastewater on distribution uniformities, potato growth, tuber yield and water use efficiency. Int J Adv Res 2(2):143–150

Messiha NAS, Elhalag KMA, Balabel NM, Farag SMA, Matar HA, Hagag MH, Khairy AM, Abd El-Aliem MM, Eleiwa E, Saleh OME, Farag NS (2019) Microbial biodiversity as related to crop succession and potato intercropping for management of brown rot disease. Egypt J Biol Pest Control 29(84):1–16

Mizubtui ESG, Fry WE (2006) Potato Late Blight. The Epidemiology of Plant Diseases, 2nd edn, 445–471

Moffett ML, Giles JE, Wood BA (1983) Survival of *Pseudomonas solanacearum* biovars 2 and 3 in soil: Effect of moisture and soil type. Soil Biol Biochem 15:587–591

Morita N, Hirata C, Park SH, Mitsunaga T (2001) Quinoa flour as a new food stuff for improving dough and bread. J Appl Glyco Sci 48(3):263–270

Moursi AM (2003) Performance of grain yield for some wheat genotypes under stress by chemical desiccation. Ph D Thesis, Agron Department, Faculty of Agric, Zagazig University, Egypt

Musa K; Oya EA, Ufuk CA, Begüm P, Seçkin E, Hüseyin AÖ, Meral Y (2015) Antioxidant responses of peanut (*Arachis hypogaea* L.) seedlings to prolonged salt-induced stress. Arch Biol Sci Belgrade 67(4):1303–1312

Muthoni J, Shimelis H, Melis R, Kinyua ZM (2014) Response of potato genotypes to bacterial wilt caused by *Ralstonia solanacearum* Smith (Yabuuchi *et al.*) in the tropical highlands. Am J Potato Res 91:215–232

Nader BA, Karim BH, Ahmed D, Claude G, Chedly A (2005) Physiological and antioxidant responses of the perennial halophyte *Crithmum maritimum* to Salinity. Plant Sci 168:889–899

Negrao S, Schmockel SM, Tester M (2017) Evaluating physiological responses of plants to salinity stress. Ann Bot 119:1–11

Newman EI, Reddell P (1987) The distribution of mycorrhizas among the families of vascular plants. New Phytol 106:745–751

Osman R, Ferrari F, McDonald SM (2016) Water scarcity and irrigation efficiency in Egypt—global.Water Econ Pol 2(4):165009, 1–28

Pal P (2016) Detection of environmental contaminants by RAPD method. Int J Curr Microbiol App Sci 5(8):553–557

Peterson RL, Massicotte HB, Melville LH (2004) Mycorrhizas: Anatomy and cell biology. NRC Research Press, Ottawa/CABI Publishing, Wallingford, UK 173 p

Prior P, Fegan M (2005) Recent developments in the phylogeny and classification of Ralstonia solanacearum. Acta Hortic 695:127–136

Ramanjulu S, Sudhakar C (2001) Alleviation of NaCl salinity stress by calcium is partly related to the increased proline accumulation in mulberry (*Morus alba* L.) callus. J Plant Biol 28:203–206

Reddy TY, Reddy VR, Anbumozhi V (2003) Physiological responses of groundnut (*Arachis hypogea* L.) to drought stress and its amelioration: a critical review. Plant Growth Regul 41:75–88

Reynolds MP, Singh RP, Ibrahim A, Ageeb OAA, Larque-Saavedra A, Quick JS (1998) Evaluating physiological traits to complement empirical selection for wheat in warm environments. Euphytica 100:84–95

Ristic Z, Bukovnik U, Vara Prasad PV (2007) Correlation between heat stability of thylakoid membranes and loss of chlorophyll in winter wheat under heat stress. Crop Sci 47:2067–2073

Ryals R, Kaiser M, Torn S, Berhe AA, Silver WL (2014) Impacts of organic matter amendments on carbon and nitrogen dynamics in grassland soils. Soil Biol & Biochem 68:52–61

Salem MF, Rizk NM, Omar SA, Abdel Hamed AS (2012) The Potential utilization of alfalfa root saponins in controlling chocolate spot disease in faba bean. J Biol Chem Environ Sci 7(3):157–171

Seleiman MF, Santanen A, Kleemola J, Stoddard FL (2013) Improved sustainability of feedstock production with sludge and interacting mycorrhiza. Chemosphere 91:1236–1242

Shams A (2011) Combat degradation in rain fed areas by introducing new drought tolerant crops in Egypt. Int J Water Resour Arid Environ 1(5):318–325

Shattock R (2002) *Phytophthora infestans*: populations, pathogenicity and phenylamides. Pest Manag Sci 58:944–950

Siddhi S, Patel JB, Patel N (2018) Stability analysis in bread wheat (*Triticum aestivum* L.). J Pharmacogn Phytochem 7(4):290–297

Singh PK, Shahi SK, Singh AP (2015) Effects of salt stress on physicochemical changes in maize (*Zea mays* L.) plants in response to salicylic acid. Indian J of Plant Sci 4(1):69–77

Singh RP, Payne TS, Figueroa P, Valenzuela S (1991) Comparison of the effect of leaf rust on the grain yield of resistant, partially resistant, and susceptible spring wheat cultivars. Am J Altern Agric 6(3):115–121

Slusarenko AJ, Patel A, Portz D (2008) Control of plant diseases by natural products: allicin from garlic as a case study. Eur J Plant Pathol 121:313–322

Smith SE, Christophersen HM, Pope S, Smith FA (2010) Arsenic uptake and toxicity in plants, integrating mycorrhizal influences. Plant Soil 327:1–21

Smith SE, Jakobsen I, Grønlund M, Smith FA (2011) Roles of arbuscular mycorrhizas in plant phosphorus nutrition: Interactions between pathways of phosphorus uptake in arbuscular mycorrhizal roots have important implications for understanding and manipulating plant phosphorus acquisition. Plant Physiol 156:1050–1057

Smith SE, Read DJ (2008) Mycorrhizal Symbiosis. Elsevier, London, UK, p 787

Smith SE, Smith FA (2011) Roles of arbuscular mycorrhizas in plant nutrition and growth: new paradigms from cellular to ecosystem scales. Annu Rev Plant Biol 62:227–250

Suprasanna P, Jain SM (2017) Mutant resources and Mutagenomics in crop plants. Emir J Food Agric 29(9) https://doi.org/10.9755/ejfa.2017.v29.i9.86

Taiz L, Zeiger E (2006) Plant Physiology. Sinauer Associates, Cambridge, MA, USA, p 764

Talebi R (2011) Evaluation of chlorophyll content and canopy temperature as indicators for drought tolerance in durum wheat (*Triticum durum* desf.). Austr Basic Applied Sci 5:1457–1462

Van Elsas JD, Kastelein P, Van Bekkum P, Van Der Wolf JM, De Vries PM, Van Overbeek LS (2000) Survival of *Ralstonia solanacearum* biovar 2, the causative agent of potato brown rot, in field and microcosm soil in temperate climates. Phytopathology 90(12):1358–1366

van Elsas JD, van Overbeek LS, Bailey MJ, Schönfeld J, Smalla K (2005) Fate of *Ralstonia solanacearum* biovar 2 as affected by conditions and soil treatments in temperate climate zones. In: Allen C, Prior P, Hayward AC (eds) Bacterial wilt disease and the *Ralstonia solanacearum* species complex. American Phytopathological Society, Saint Paul, MI, USA, pp 39–49

Wang B, Qiu YL (2006) Phylogenetic distribution and evolution of mycorrhizas in land plants. Mycorrhiza 16:299–363

Zehr EI (1969) Bacterial wilt of ginger in the Philippines. Philippines Agriculturist 53(3&4):224–227

Zhu JK (2001) Plant salt tolerance. Trends Plant Sci 6:66–71

Zohry A (2020) Prospects of quinoa cultivation in marginal lands of Egypt. Mor J Agri Sci 1(3):132–137

Zohry A, Ouda S, Noreldin T (2016) Solutions for maize production-consumption gap in Egypt. In: Conference: 4th African Regional Conferences on Irrigation and Drainage (ARCID), At Aswan, Egypt https://www.researchgate.net/publication/301779286

Printed in the United States
by Baker & Taylor Publisher Services